An Ecosystem Approach to Aquatic Ecology

AN ECOSYSTEM APPROACH TO AQUATIC ECOLOGY

Mirror Lake and Its Environment

Edited by Gene E. Likens

With 197 Figures

Springer-Verlag New York Berlin Heidelberg Tokyo

Gene E. Likens
Director, Institute of Ecosystem Studies
The New York Botanical Garden
Mary Flagler Cary Arboretum
Millbrook, New York 12545/U.S.A.

Cover photo by Ashton R. Hallett.

Library of Congress Cataloging in Publication Data
Main entry under title:
An Ecosystem approach to aquatic ecology.
    Bibliography: p.
    Includes index.
    1. Lake ecology—New Hampshire—Mirror Lake (Grafton
County) 2. Limnology—New Hampshire—Mirror Lake
(Grafton County) 3. Freshwater ecology—New Hampshire—
Mirror Lake Watershed (Grafton County) 4. Mirror Lake
(Grafton County, N.H.) I. Likens, Gene E., 1935–
QH105.N4E3   1985   574.5′26322′097423   84-26686

© 1985 by Springer-Verlag New York, Inc.

All rights reserved. No part of this book may be translated or reproduced in any form without written permission from Springer-Verlag, 175 Fifth Avenue, New York, New York 10010, U.S.A. The use of general descriptive names, trade names, trademarks, etc., in this publication, even if the former are not especially identified, is not to be taken as a sign that such names, as understood by the Trade Marks and Merchandise Marks Act, may accordingly be used freely by anyone.

Typeset by Bi-Comp, Inc., York, Pennsylvania.
Printed and bound by Halliday Lithograph, West Hanover, Massachusetts.
Printed in the United States of America.

9 8 7 6 5 4 3 2 1

ISBN 0-387-96106-2 Springer-Verlag New York Berlin Heidelberg Tokyo
ISBN 3-540-96106-2 Springer-Verlag Berlin Heidelberg New York Tokyo

Preface

Why study a small lake in the White Mountains of north-central New Hampshire? Better yet, why write a book about those studies? We wrestled with such questions for years, and the answers lie in the overall ecosystem focus and approach we have taken. At the same time that we studied Mirror Lake, numerous and comprehensive studies were done of the surrounding terrestrial and stream ecosystems of the Hubbard Brook Valley. These associated studies complemented those done in the lake and provided unique information about air-land-water linkages in the Valley. Many of the studies conducted in Mirror Lake were about organisms or about the processes carried out by these individual organisms or communities. Nevertheless, an abiding objective always was to determine the significance that these individuals and individual processes had to the overall structure, metabolism and biogeochemistry of the Mirror Lake ecosystem. To wit, what is the significance for the ecosystem? In some cases the ecosystem role was clear, for many it was not. But the hope is that our attempts to unravel and understand the whole will be informative and, more importantly, that it will stimulate others to study the holistic and functional relationships in entire landscapes.

The book is more than a case history of a single lake. We have written it not only for the serious student of lakes, but also for those interested in ecosystems and their interactions. Indeed, the casual observer of nature should find much of interest in this book.

Many comparisons have been made in this book to studies done at the Experimental Lakes Area (ELA) in Ontario, Canada. Not only are the ELA studies exceptionally detailed and limnologically important in their own right, but there are many biogeochemical and biological similarities between Mirror Lake and the lakes in the ELA. Thus, we have endeavored to expand the relevance of our results by these comparisons.

Our studies of Mirror Lake began very slowly, with almost casual limnological observations and measurements during the early 1960s; these early observations were followed by more detailed and comprehensive studies. In fact, studies of Mirror Lake are still ongoing and are much more active currently than in those early years. For example, sophisticated studies of the hydrological regime for Mirror Lake were initiated five years ago in cooperation with the U.S. Geological Survey and will not reach full fruition for several years.

All of the scientists, assistants and families associated with the Hubbard Brook Ecosystem Study have enjoyed the aesthetic and recreational beauty of Mirror Lake. It has been our centerpiece for summer living and winter visits to New Hampshire for more than 20 years! It is unexcelled for swimming during July and August. Its popularity is attested to by the crowds of swimmers on the public beach during the summer. It is only fitting that we should have attempted to understand this lake and its environment, and to provide some ecological guidelines for protection of its quality into the future.

F. Herbert Bormann and I began the Hubbard Brook Ecosystem Study in 1963 when we were both faculty members at Dartmouth College. The Study was developed in cooperation with Robert S. Pierce of the U.S. Forest Service. Major acknowledgment and credit for helping to initiate and to sustain the study also is due John S. Eaton and Noye M. Johnson. Numerous other colleagues, postdoctoral associates and graduate students have contributed ideas, talent and time to our study of the Hubbard Brook ecosystem over the years. Herb moved to Yale University in 1966 and I moved to Cornell University in 1969, but we continued our joint studies at Hubbard Brook much as we had before. The long hours of travel between Ithaca (now Millbrook) or New Haven and Hubbard Brook provided a valuable opportunity to think and to plan.

Until last year, our living and laboratory conditions at Pleasant View Farm in West Thornton, New Hampshire were primitive and crowded, but a deep bond developed among most of the Hubbard Brook "alumni" because of the close scientific and personal exchanges that were possible from living and working in a spartan, but friendly old farmhouse during the year. Last year, with the help of Cornell University, Yale University and the National Science Foundation, we renovated our living facilities and built a small, but modern, laboratory at Pleasant View Farm.

Acknowledgments

In the first two volumes (*Biogeochemistry of a Forested Ecosystem* and *Pattern and Process in a Forested Ecosystem* [Springer-Verlag, New York]) describing the Hubbard Brook Ecosystem Study, F. Herbert Bormann and I acknowledged approximately 200 persons that had contributed to our studies and to our development of these volumes. To that list we now add: C. Asbury, E. Beecher, A. Blair, N. Bloch, R. Bowden, N. Caraco, J. Confer, R. Edwards, C. Faust, M. Havas, L. Hedin, G. Helfman, J. Lowe, M. Mattson, M. Mayer, S. Nolan, J. O'Brien, C. Ochs, F. Oldfield, R. Santore, N. Scott, J. Segal, W. Shoales, J. Sobel, D. Strayer, G. Vostreys, F. Vertucci, C. Weiler, G. Wollney and T. Winter for their contributions to the studies of Mirror Lake.

I especially want to thank J. Cole, R. Moeller and D. Strayer for reviewing and commenting on large sections of the manuscript for this book. Their thoughtful and dedicated effort is very much appreciated.

I thank the members of my family for their patience as long hours were taken away from family activities. I also acknowledge the patience and cooperation of the various authors contributing to this volume as we struggled to make the manuscript into a book rather than just a collection of articles. We were only partly successful in achieving this difficult goal.

A special acknowledgment is given to my wife, Phyllis, who worked diligently with me throughout the preparation of this book. The manuscript was typed, retyped, prepared and cared for by her. In addition, she took full responsibility for assembling an accurate and consistent *References* section for the entire book, using inputs from 23 authors. A formidable task! I am extremely grateful for these efforts.

Another special note of thanks is owed to John S. Eaton, who carefully and painstakingly has maintained a high-quality laboratory over the years where baseline data could be believed. He provided counsel to numerous students and colleagues and engendered loyalty and stability within the group.

I thank the numerous postdocs, graduate students and research assistants for their good humor and dedication to excellence over the years.

Overall financial support for our studies of Mirror Lake was provided from the National Science Foundation. Assistance in gauging the amounts of precipitation, runoff and loss at the outlet was provided by personnel of the U.S. Forest Service, primarily C. W. Martin and R. S. Pierce. R. E. Munn provided an analysis of local wind patterns for the Hubbard Brook Valley. D. Buso collected numerous data and samples, often under very arduous conditions. Data on the chemistry of particulate matter in Mirror Lake were provided by B. Peterson.

We are aware of the new terminology for describing soil profiles. Nevertheless, the old terminology is used in this book. The equivalent designations are A_o and A_{oo} horizons = O_i, O_e and O_a horizons, A_1 = A, A_2 = E, B_1 = BA and C = C.

The analyses for *Chapter VII.B* and *C* were funded by the Supporting Research Fund of the Division of Limnology and Ecology of the Academy of Natural Sciences of Philadelphia. Dr. Sherman acknowledges Dr. Ruth Patrick and Dr. Charles Reimer of the Academy of Natural Sciences and Dr. Frederick Swain of the University of Delaware for their help and encouragement throughout his study. Dr. Jordan acknowledges Dr. H. A. Lechevalier and Dr. J. Hobbie for reviewing and providing advice on the bacterial aspects of *Chapter V.A.1*. Dr. Winter thanks Dr. N. Johnson and Dr. deNoyelles thanks Dr. D. Gerhart for numerous contributions and long hours they put into *Chapters III.A; V.A.2* and *V.B.2*.

The staff of Springer-Verlag New York Inc. provided excellent advice in helping to edit and prepare the manuscript for typesetting.

Gene E. Likens
August 1985

Contents

PREFACE v

CONTRIBUTORS xiii

CHAPTER I: AN ECOSYSTEM APPROACH

A. Ecosystem Analysis... 1
   1. The Hubbard Brook Ecosystem Study..................... 3
   2. Other Interactions Between Ecosystems 7

CHAPTER II: THE HUBBARD BROOK VALLEY

A. Environmental Parameters 9
B. Biogeochemistry... 15

CHAPTER III: MIRROR LAKE AND ITS WATERSHED

A. Physiographic Setting and Geologic Origin of Mirror Lake 40
   Location and General Physiographic Setting 40
   Geology 40
   Glacial History and Origin of Lake Basin..................... 48
   Topography and Drainage Basin Characteristics 49
   Comparison of Mirror Lake and Hubbard Brook Drainage
   Basins 52
B. Historical Considerations 53
   1. History of the Vegetation on the Mirror Lake Watershed..... 53
   2. Catastrophic Disturbance and Regional Land Use............ 65
   3. Mirror Lake: Cultural History 72

CHAPTER IV: MIRROR LAKE—PHYSICAL AND CHEMICAL CHARACTERISTICS

A. Importance of Perspective in Limnology 84
B. Physical and Chemical Environment 89
C. Stability, Circulation and Energy Flux in Mirror Lake 108
D. Approaches to the Study of Lake Hydrology 128
E. Flux and Balance of Water and Chemicals.................... 135

CHAPTER V: MIRROR LAKE—BIOLOGIC CONSIDERATIONS

A. Species Composition, Distribution, Population, Biomass and
   Behavior 156
   1. Bacteria 156
   2. Phytoplankton 161
   3. Periphyton 175
   4. Macrophytes 177
   5. Zooplankton 192
   6. Benthic Macroinvertebrates 204
   7. Benthic Microinvertebrates 228
   8. Salamanders 234
   9. Fishes 236
B. Production and Limiting Factors 246
   1. Bacteria 246
   2. Phytoplankton 250
   3. Macrophytes 257
   4. Periphyton Production 265
   5. Zooplankton 268
   6. Benthic Macroinvertebrates 280
   7. Vertebrates 288
C. Organic Carbon Budget 292
D. Decomposition 302

CHAPTER VI: MIRROR LAKE—ECOLOGIC INTERACTIONS

A. The Littoral Region 311
B. The Profundal Region 317
C. The Pelagic Region 322
D. The Lake-Ecosystem 337

CHAPTER VII: PALEOLIMNOLOGY

A. Sedimentation 345
   1. Late-Glacial and Holocene Sedimentation 346
   2. Contemporary Sedimentation 355
B. Diatoms 366
C. Animal Microfossils 382
D. Fossil Pigments 387
E. Chemistry 392
F. Paleoecology of Mirror Lake and Its Watershed 410

CHAPTER VIII: THE AQUATIC ECOSYSTEM AND AIR-LAND-WATER INTERACTIONS

A. Direct Atmospheric Input to Lakes 431
B. Fluvial Input to Lakes 432
C. Effects of Disturbance on the Land-Water Linkage 433
D. Atmospheric Inputs to Watersheds Affecting Inputs to Lakes 434

CHAPTER IX: AIR AND WATERSHED MANAGEMENT AND THE AQUATIC ECOSYSTEM

A. Developmental State of the Forested Watershed-Ecosystem 436
B. Forest Harvesting.. 436
C. Highway Construction and Maintenance...................... 440
D. Air and Water Pollution 440
E. Eutrophication Trends in Mirror Lake 441

REFERENCES 445

INDEX OF GENERA AND SPECIES 497

INDEX OF LAKES AND STREAMS 504

GENERAL INDEX 506

Contributors

F. HERBERT BORMANN
School of Forestry and Environmental Studies, Yale University, New Haven, Connecticut 06511

REBECCA E. BORMANN
1031 Fifth Street, S.E., Minneapolis, Minnesota 55414

THOMAS M. BURTON
Department of Fisheries and Wildlife, Natural Science Building, Michigan State University, East Lansing, Michigan 48824

JONATHAN J. COLE
Institute of Ecosystem Studies, The New York Botanical Garden, Mary Flagler Cary Arboretum, Millbrook, New York 12545

MARGARET B. DAVIS
Department of Ecology and Behavioral Biology, 108 Zoology Building, University of Minnesota, Minneapolis, Minnesota 55455

JOHN S. EATON
Institute of Ecosystem Studies, The New York Botanical Garden, Mary Flagler Cary Arboretum, Millbrook, New York 12545

M. S. (JESSE) FORD
Ecosystems Research Center, Dale Corson Hall, Cornell University, Ithaca, New York 14853

CLYDE E. GOULDEN
Division of Limnology and Ecology, The Academy of Natural Sciences, Philadelphia, Pennsylvania 19103

GENE S. HELFMAN
Department of Zoology, University of Georgia, Athens, Georgia 30602

NOYE M. JOHNSON
Department of Earth Sciences, Dartmouth College, Hanover, New Hampshire 03755

MARILYN J. JORDAN
246 Woods Hole Road, Woods Hole, Massachusetts 02543

GENE E. LIKENS
Institute of Ecosystem Studies, The New York Botanical Garden, Mary Flagler Cary Arboretum, Millbrook, New York 12545

JOSEPH C. MAKAREWICZ
Department of Biology, State University College at Brockport, Brockport, New York 14420

ROBERT E. MOELLER
Division of Limnology and Ecology, The Academy of Natural Sciences, Philadelphia, Pennsylvania 19103

F. DENOYELLES
Department of Systematics and Ecology, Snow Hall, University of Kansas, Lawrence, Kansas 66045

W. JOHN O'BRIEN
Department of Systematics and Ecology, Snow Hall, University of Kansas, Lawrence, Kansas 66045

BRUCE J. PETERSON
The Ecosystems Center, Marine Biological Laboratory, Woods Hole, Massachusetts 02543

ROBERT S. PIERCE
Northeastern Forest Experiment Station, Forestry Sciences Laboratory, Durham, New Hampshire 03824

JOHN W. SHERMAN
Division of Limnology and Ecology, The Academy of Natural Sciences, Philadelphia, Pennsylvania 19103

DAVID L. STRAYER
Institute of Ecosystem Studies, The New York Botanical Garden, Mary Flagler Cary Arboretum, Millbrook, New York 12545

GAY VOSTREYS
Department of Geology, The Academy of Natural Sciences, Philadelphia, Pennsylvania 19103

RHODA A. WALTER
1650 Lakeview Avenue, North East, Pennsylvania 16428

THOMAS C. WINTER
U.S.D.I. Geological Survey, Box 25046, Denver Federal Center, Mail Stop #413, Denver Colorado 80225

Chapter 1

An Ecosystem Approach

GENE E. LIKENS and F. HERBERT BORMANN

A. Ecosystem Analysis

Ecosystems have been defined as the *functional units of the landscape* (Odum 1971). As such, they provide both conceptual and practical units of organization and of study.

The popular, but often misused term—ecosystem—apparently was first introduced by the British ecologist, A. G. Tansley (1935). However, the fundamental aspects of the concept had been discussed long before. For example, the *idea* of a biosphere (the largest possible ecosystem on the planet Earth) may be credited to the French naturalist, Jean Lamarck (e.g. 1802); but the *term*—biosphere—was introduced by Edvard Suess, an Austrian geologist (Suess 1875). About 50 yr later the Russian mineralogist, W. I. Vernadsky developed the biospheric *concept* in his book entitled, *La Biosphere* (Vernadsky 1929).

Tansley (1935) stressed the interaction between living and nonliving components of whole systems. He stated, ". . . it is (these) systems . . . which . . . are the basic units of nature. . . ." In an important paper, Francis Evans (1956) extended and refined the ecosystem concept. He identified the major components of an ecosystem as (1) the living parts where energy and matter are fixed, circulated, transformed and accumulated by photosynthesis, decomposition, herbivory, predation, parasitism and other activities forming complex food webs, and (2) nonliving parts where energy and matter are fluxed and cycled by evaporation, precipitation, erosion and deposition. He also stressed that ecosystems are open relative to these fluxes and that the inputs and outputs of energy and matter link individual ecosystems together within the biosphere of our planet.

However, it was Odum's (1953) textbook, *Fundamentals of Ecology* (particularly the second edition), which really laid the foundation and promoted an enthusiasm for an ecosystem approach to ecology in the United States and elsewhere.

It is difficult to identify naturally-occurring ecosystem units that can be neatly classified in a taxonomic sense. This difficulty arises because populations or communities of organisms are never coincidental in the sense of Clements (1916), but vary somewhat independently in the sense of Whittaker (1951). Furthermore, environmental factors governing ecosystem structure and function, such as slope, aspect, drainage and elevation, vary continuously, making it exceedingly difficult to define sharp boundaries on the basis of physiographic factors. Nevertheless, the ecosystem concept provides a theoretical framework for the study of the interactions between individuals, populations, communities and their abiotic environments, and for the study of the change in these relationships with time.

Ecosystems are considered to have structure and function (e.g. Odum 1971). The structural components of an aquatic ecosystem include the physical features such as the water and sediments of a lake, the gradient of conditions such as found in the epilimnion, metalimnion and hypolimnion of a thermally-stratified lake, or the biotic composition such as species and numbers of individuals and their biomass. Metabolic and biogeochemical activities characterize the functional aspects of ecosystems, such as photosynthesis, decomposition, flux and cycling of water and chemicals.

At this point it is useful to distinguish between flux and cycling in ecosystems as these two concepts are often confused. *Flux* refers to movement, input and output, across the boundaries of an ecosystem. Quantification of flux is the basis for constructing budgets (input, output, net change) and for identification of flux patterns, which give rise to hypotheses that may be tested by detailed studies within the boundaries of the ecosystem. The biota and nonliving components within the boundaries of the ecosystem respond to, as well as alter, these external fluxes. *Cycling* refers to the exchanges of materials among various living and nonliving components within the boundaries of the ecosystem (see Likens 1975a). An understanding of cycling provides an explanation for the factors that give rise to particular flux patterns.

The terms allochthonous and autochthonous are commonly applied to the origin of substances relative to ecosystem boundaries. *Allochthonous* refers to sources of materials that originated outside of the boundaries of an ecosystem, such as dissolved organic matter added to a lake in precipitation. In contrast, *autochthonous* refers to sources within the boundaries of the ecosystem, e.g. production of dissolved organic matter by plants within the ecosystem.

In general, the boundaries of an ecosystem must be carefully and explicitly defined before quantitative determinations of flux can be made. Ecosystem boundaries are usually determined for the convenience of the investigator rather than on the basis of some obvious, generally agreed-upon, naturally-occurring unit or some functional discontinuity with an adjacent ecosystem. This artificiality occurs because the actual functional relationships between ecosystems are usually unknown, but it may or may not represent a serious analytical problem depending on the system and on the objectives of the study (cf. Bormann and Likens 1967, 1979). The problem of determining boundaries for ecosystems selected for intensive study is more obvious in terrestrial ecosystems than it is in lake ecosystems where the lateral boundaries are ostensibly obvious—that is, the shoreline. The upper boundary for the lake ecosystem can be taken as the air-water interface and the lower boundary as the maximum sediment depth, which is normally inhabited or utilized by benthic organisms. But are these practical boundaries the usual functional boundaries for the ecosystem? More on this matter later. Ideally, all boundaries should represent the plane (boundary) at which short-term exchanges of chemicals or other matter are irreversible relative to the functional ecosystem, i.e. cycling exchanges become fluxes (see Likens 1975a).

Because ecosystem structure includes abiotic and biotic components, ecosystems have volume. Volumetric relationships may change with time. Likewise, the structural and functional components may be distributed throughout the volume in different ways and/or at different times (cf. Likens and Bormann 1972).

Ecosystems also have a history. An understanding of history is vital for evaluating present-day relationships or for predicting the future state of an ecosystem. For example, long-term historical considerations such as changes in climate, soil development, successional patterns of vegetation, eutrophication and pond filling, or short-term factors such as changes in water quality and thermal structure, all affect the current structure and function of the ecosystem, as well as the potential for future growth and development of an ecosystem. To date, most scientists have been overwhelmed by the complexity of the extant natural ecosystem and thus have given little attention to the developmental aspects. However, it is of utmost importance for quantitative or particularly for comparative studies to identify what is the current developmental state of an ecosystem and how it has changed with time.

Currently there are several approaches to the study of ecosystems. They may be characterized as follows: (1) empirical studies where bits of information are collected and an attempt is made to integrate and assemble these into a whole; (2) comparative studies where one (or possibly a few) structural or one functional component is compared for a range of ecosystem types (a major pitfall is the mixing of different developmental stages); (3) experimental studies where manipulation of whole ecosystems is done to identify and elucidate mechanisms; and (4) modeling or computer simulation studies. The motivation in all of these approaches is to learn how an understanding of the entire ecosystem gives more insight than the sum of knowledge about its parts relative to the structure, metabolism and biogeochemistry of the landscape.

Many years ago Forbes ([1887] 1925, 537) discussed interrelationships between components of a lake-ecosystem and thus made a historic contribution to our understanding of lake ecology. However, his concept of the lake as a microcosm (the lake "forms a little world within itself—a microcosm") for many years focused attention on the lake as a self-sustaining unit without adequate consideration of the vital interactions with the surrounding watershed and airshed. Now we know that a lake's structure, metabolism and biogeochemistry reflect the inputs from its watershed and airshed, often in very complex ways. The inputs and outputs of lakes are the links with the surrounding biosphere. Moreover, a lake reflects the developmental state and condition (including disturbance) of its watershed, drainage streams and airshed (cf. Likens and Bormann 1974b, 1979). Careful and detailed studies of individual aquatic ecosystems are important and necessary, but researchers need to take their blinders off! To understand a lake-ecosystem, the view must be as large as the watershed, the airshed, the landscape and eventually as large as the biome or planet. Hence, the central theme of this book is to consider the details of Mirror Lake, a small lake in New Hampshire, in relation to its surroundings and history. Attempts to understand or manage environmental problems are usually based on information pieced together from studies isolated in time, space and/or habitat. This approach ignores important functional linkages that are vital to natural ecosystems. Our purpose is to consider and evaluate historical and present-day linkages between the atmosphere and terrestrial and aquatic components comprising the Mirror Lake watershed-ecosystem.

1. The Hubbard Brook Ecosystem Study

For two decades ecologic systems within the Hubbard Brook Valley of New Hampshire have been the subject of continuous and detailed scientific study. In 1955 the U.S. Forest Service established the Hubbard Brook Experimental Forest (HBEF) as a major hydrologic facility in northern New England. The HBEF comprises the majority of the Hubbard Brook Valley (see *Chapter II*). In 1963, in close cooperation with scientists of the U.S. Forest Service, we began the Hubbard Brook Ecosystem Study (HBES), which is a detailed study of the biogeochemistry and ecology of northern hardwood forest-ecosystems and associated aquatic ecosystems of the Valley.

Thousands of water samples—including rain, snow, stream water, ground water, soil solution, stemflow, throughfall and lake water—have been collected and chemically analyzed. Thousands of biologic samples have been collected and evaluated. Likewise, numerous samples of the geologic substrate, soil, lake sediments and air have been analyzed. These extensive data provide the means and perspective to identify and quantify some of the complex ecologic and biogeochemical relationships within and between air, land and water systems of the Valley. Thus the HBES has been a long-term, multidisciplinary investigation of the structure, metabolism, biogeochemistry and ecologic interactions between atmospheric, terrestrial and aquatic ecosystems.

At the beginning of our studies we developed a conceptual model to facilitate the quantitative biogeochemical study of terrestrial ecosystems (Bormann and Likens 1967) and associated aquatic systems (Likens and Bormann 1972). Our ecosystem unit is a watershed or drainage area, with vertical and horizontal boundaries defined functionally by biologic activity and the drainage of water. Various adjacent, small watersheds within the HBEF with similar climate, vegetation and geology have been used in these studies. The continuous flux of energy, water, nutrients and other materials across the boundaries of the ecosystem are considered to be inputs and outputs, which are transported by meteorologic, geologic and biologic vectors (*Figure I.A.1*–1). Meteorologic inputs and outputs consist of wind-borne particulate matter, dissolved substances in rain and snow, aerosols and gases (e.g. CO_2). Geologic flux includes dissolved and particulate matter transported by surface and subsurface drainage, and the mass movement of colluvial materials. Biologic flux results when chemicals or energy gathered by

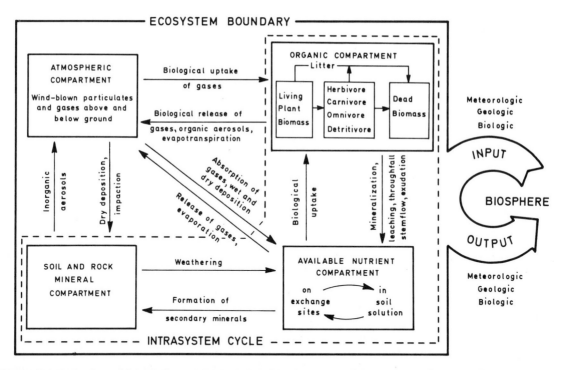

Figure I.A.1–1. A model depicting nutrient relationships in a terrestrial ecosystem. Inputs and outputs to the ecosystem are moved by meteorologic, geologic and biologic vectors (Bormann and Likens 1967; Likens and Bormann 1972). Major sites of accumulation and major exchange pathways within the ecosystem are shown. Nutrients that, because they have no prominent gaseous phase, continually cycle within the boundaries of the ecosystem between the available nutrient, organic matter and primary and secondary mineral components tend to form an intrasystem cycle. Fluxes across the boundaries of an ecosystem link individual ecosystems with the remainder of the biosphere. (Modified from Likens et al. 1977. Used with the permission of the publisher and the authors.)

animals in one ecosystem are deposited in another (e.g. local exchange of fecal matter or mass migrations). These input-output categories are therefore defined as vectors or "vehicles" for transport of nutrients, matter or energy, rather than sources; i.e. a leaf blown into an ecosystem represents meteorologic input rather than biologic input.

Within the ecosystem, the nutrients may be thought of as occurring in any one of four basic compartments: (1) atmosphere, (2) living and dead organic matter, (3) available nutrients, and (4) primary and secondary minerals (soil and rock). The atmospheric compartment includes all elements in the form of gases or aerosols both above and below ground. Available nutrients are ions that are absorbed on or in the humus or clay-humus complex or mineral surfaces, or dissolved in the soil solution. The organic compartment includes all nutrients incorporated in living and dead biomass. The primary and secondary minerals contain nutrients that comprise the inorganic soil and rock portions of the ecosystem.

The biogeochemical cycling of elements involves an exchange between the various compartments within the ecosystem. Available nutrients and gaseous nutrients may be taken up and assimilated by the vegetation and microorganisms; some nutrients incorporated in organic compounds may be passed on to heterotrophs and then made available again through respiration, biologic decomposition and/or leaching from living and dead organic matter. Insoluble primary and secondary minerals may be converted to soluble available nutrients through the general process of chemical weathering; soluble nutrients may be redeposited as secondary minerals.

Knowledge of hydrologic inputs and outputs

A. Ecosystem Analysis

for a lake-ecosystem is basic to an evaluation of its biogeochemistry. Inputs of water occur as: (1) precipitation falling directly on the lake's surface, (2) drainage of surface water into the lake, (3) seepage of ground water through the sediments of the basin, and (4) discharge of lacustrine springs. These sources of inflowing water may add nutrients as well as dissolved or suspended impurities. Water losses occur through: (1) evaporation and transpiration by emergent or shoreline vegetation, (2) surface discharge, (3) seepage through the sediments, and (4) discrete subsurface flows (see *Chapter IV.D and E* for more details).

Nutrient inputs and outputs for the lake-ecosystem as a whole can be categorized as was done in the terrestrial ecosystem into geologic, meteorologic and biologic vectors (*Figure I.A.1–2*). Nutrients without a prominent gaseous phase at normal biologic temperatures (e.g. Ca, K) are input to freshwater ecosystems primarily in precipitation (meteorologic vector) and in drainage waters (geologic vector); output losses are primarily as accumulations in deep sediments below the activity zone of organisms and in outflows (geologic vectors). Nutrients with a normal gaseous phase (e.g. N, C, S, H, O) move across the boundary of the ecosystem as gases (meteorologic vector) in addition to the other input and output fluxes.

The geologic input to a lake-ecosystem is derived primarily as the geologic output from the surrounding terrestrial (forest-stream) ecosystem and, therefore, represents one of the important linkages between ecosystems in the biosphere. If the drainage area for the lake is disturbed for agricultural or urban purposes, then the linkage may become sociologically and politically more critical. From a pragmatic viewpoint, it is necessary to study the terrestrial watershed to obtain an accurate prediction of geologic inputs into the lake. Meteorologic inputs are similar to those for terrestrial ecosystems. Biologic inputs and outputs may be larger and more directional, relative to the total budgets, than they are for many terrestrial ecosystems. For example, terrestrially-based predators (e.g. kingfishers) feeding on aquatic fauna, or spawning migrations of salmon, or emerging insects (e.g. mayflies) may contribute to significant removal of matter from or addition of matter to a lake-ecosystem.

Meteorologic output of dissolved or particulate matter from a lake-ecosystem is usually relatively small. Spray or aerosols may be generated from large lakes on windy days, but the chemical flux is relatively low. There may be appreciable gaseous flux such as carbon dioxide, methane, ammonia, hydrogen sulfide, etc., particularly from shallow anaerobic lake-ecosystems.

Geologic outputs occur through losses of dissolved or particulate matter in any drainage water from lakes, although the particulate matter losses would usually be relatively small since the lake acts as a settling basin for suspended matter. In this regard, geologic outputs from a lake-ecosystem represent a special problem in relation to the concept of ecosystem boundaries. Sediments are constantly accumulating in the lake basin. Since the living biota normally do not penetrate these sediments to depths greater than a few tens of centimeters, sediments below a relatively shallow depth of the lake bottom are essentially removed from fur-

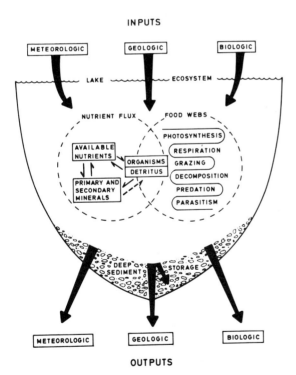

Figure I.A.1–2. A diagrammatic model for nutrient cycling in a freshwater ecosystem. (From Likens and Bormann 1975. Used with the permission of the publisher and the authors.)

ther cycling within the lake and must be considered as geologic output in a functional sense. However, these deeper sediments represent long-term storage in terms of the original ecosystem boundaries because conditions governing the ecology of the lake can change and nutrients in deep sediments could be brought back into circulation. They then would be considered as geologic input. Such "stored" nutrients in deep sediments could be made available by human activities such as artificial stirring of the sediments or by natural events such as upheaval.

Under natural conditions, long-term sediment storage may not be quantitatively important to considerations of nutrient cycling during an annual cycle, but such storage in the bottom sediments progressively decreases the volume of the lake-ecosystem. Accumulation of such sediments could eventually obliterate the lake-ecosystem and provide a substrate for the development of a terrestrial ecosystem.

Within the ecosystem's boundaries, water provides the medium whereby nutrients may be dissolved, adsorbed, absorbed, complexed, ingested, fixed, exchanged, excreted, exuded, leached, precipitated, oxidized, reduced, etc.— i.e. cycled. Nutrients tend to occur or exchange between three basic ecosystem compartments: available inorganic nutrients, organic matter and primary and secondary minerals (*Figure I.A.1–2*). Note that organic matter (living and dead) links two major processes of aquatic ecosystems, energy flow and nutrient cycling.

Available nutrients are those dissolved in water or on exchange surfaces of pelagic particulate matter or bottom sediments. Nutrients incorporated in living or dead organic matter, both in the pelagic region and in the functional volume of the sediments, comprise the organic matter compartment. Nutrients incorporated in rocks, as primary and secondary minerals in the sediments or suspended in the water, constitute the primary and secondary mineral compartment.

Aquatic organisms absorb and assimilate available nutrients from the pelagic region as well as from the sediments. Available nutrients are released from organic matter by excretion, exudation, leaching, respiration and decomposition (*Figure I.A.1–2*). Primary and secondary minerals may chemically decompose to form available nutrients, or secondary minerals may be reformed from available nutrients depending on environmental conditions.

Secondary minerals may be formed from available nutrients by the activity of organisms. The very large deposits of marl in some lakes attest to the importance of such mechanisms in lakes. Conversely, some organisms may incorporate certain primary or secondary minerals directly (Likens and Bormann 1972).

Lakes within a terrestrial landscape have a variety of linkages for energy and nutrient exchange with surrounding watersheds and airsheds (*Figure I.A.1–3*). Overall, relatively more effort is required to measure quantitatively the nutrient budget for a lake-ecosystem

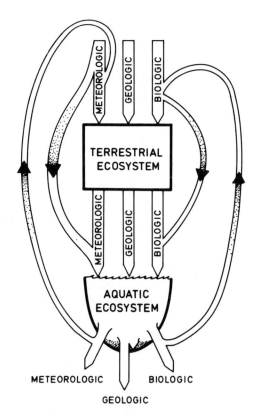

Figure I.A.1-3. Diagrammatic model of the functional linkages between terrestrial and aquatic ecosystems. Vectors may be meteorologic, geologic or biologic components moving nutrients or energy along the pathway shown. (Modified from Likens and Bormann 1974b, *Bioscience* 24(8):447, Fig. 1. Copyright © 1974 by The American Institute of Biological Sciences. Used with permission.)

A. Ecosystem Analysis

than for a terrestrial ecosystem because there are a larger number of parameters to be considered. For example, inputs and outputs for the terrestrial drainage area must be evaluated to determine the geologic input for the lake, and the difficult-to-measure geologic outputs for the lake, including seepage losses and deposition of functionally inactive deep sediments, must be determined (see *Chapter VII*).

Our ecosystem model for flux and cycling allows for development or degradation of the ecosystem. Such changes with time are critically important to comparisons of ecosystems of different structure and history. These changes may be expressed in several ways: (1) by an increase or decrease in the functional volume of the ecosystem, (2) by changes in the rate of cycling between compartments, (3) by changes in the size of the compartments, and (4) by a change in output relationships (Likens and Bormann 1972; Bormann and Likens 1979).

Very early in our studies we recognized that the watershed-ecosystems in the Hubbard Brook Valley had a great potential for multifaceted ecosystem analysis. Major virtues were: (1) the possibility to construct fairly complete nutrient budgets, considering not only meteorologic input and geologic output, but also internal release from weathering and gaseous exchange, (2) an accumulating wealth of background information in precipitation and stream-water chemistry, meteorologic and hydrologic data and biologic information, (3) several adjacent, relatively similar forested, watershed-ecosystems that provide statistically valid reference data for evaluation of experimental manipulations of natural ecosystems, (4) a variety of undisturbed stream-ecosystems that connect the forest with various aquatic systems, and (5) the inclusion of a small lake in the matrix of the forest that permits study of air-land-water interactions. Finally, there was the unusual opportunity to conduct experiments at the ecosystem level and to test these results against reference data collected from adjacent watershed-ecosystems.

For more information about the use of the small watershed approach in quantitative studies of ecosystems, see Bormann and Likens (1967), Likens and Bormann (1972), Likens et al. (1977) and Bormann and Likens (1979).

2. Other Interactions Between Ecosystems

Riverine ecosystems may also play an important role in the interactions between terrestrial and aquatic ecosystems. The geologic output of water, dissolved chemicals and particulate matter from the drainage basin is the main geologic input for most aquatic ecosystems, and one of the most important land-water linkages in the biosphere. Thus, rivers are the primary links between terrestrial and standing-water systems. However, since rivers contain a diverse assemblage of interacting biotic and abiotic components, they may be considered as ecosystems in their own right, and should not be thought of as conduits simply for conveyance of fluids and matter. Rivers may transform, store or release materials and thus alter the output-input relations between lakes and their drainage basins (e.g. Meyer and Likens 1979; Fisher and Likens 1972; Meyer et al. 1981; Likens 1984).

Aldo Leopold (1941) describes these linkages with characteristic eloquence:

Soil and water are not two organic systems, but one. Both are organs of a single landscape; a derangement in either affects the health of both. . . . All land represents a downhill flow of nutrients from the hills to the sea. This flow has a rolling motion. Plants and animals suck nutrients out of the soil and air and pump them upward through the food chains; the gravity of death spills them back into the soil and air. Mineral nutrients, between their successive trips through this circuit, tend to be washed downhill. Lakes retard this downhill wash, and so do soils. . . . The continuity and stability of inland communities probably depend on this retardation and storage. . . . The downhill flow is carried by gravity, the uphill flow by animals. There is a deficit in uphill transport, which is met by the decomposition of rocks.

In the short term, the only means whereby material is continually moved "uphill" (Leopold 1941, 1949) to terrestrial ecosystems are biologic or meteorologic vectors (*Figure I.A.1–3*). Meteorologic return of materials, other than gases, from aquatic to terrestrial systems is normally very small. However, the flux of nutrients and organic matter by animals may be large, although data are limited (cf. *Chapter V.B.6*; Leopold 1941; Vallentyne 1952; Dugdale 1956; Hasler and Likens 1963; Donaldson 1967;

Krokhin 1967; Kormondy 1969; Hall 1972). In the long term, catastrophic events such as geologic uplift can transfer massive amounts of material from aquatic to surrounding terrestrial systems and are major, although periodic, means of "uphill" movement.

Summary

Ecosystems are comprised of living and nonliving components and their functional interactions. Materials can flux across the boundaries or cycle within the boundaries. Ecosystems have volume and change with time.

A lake and its drainage area (watershed) represents a functional ecosystem unit of the landscape. However, the structure, metabolism, biogeochemistry and particularly the management of a lake, cannot be understood without adequate consideration of both its watershed and its airshed. Conceptual models are presented for the various components of flux and cycling for a terrestrial and a lake-ecosystem as well as the air-land-water linkages.

This book is one of the first attempts to evaluate the historical and current limnology of a lake relative to the influences of its watershed and airshed. In this sense, the Mirror Lake ecosystem extends to the watersheds of the drainage area and to the airsheds for the meteorologic inputs. Thus, wherever and to whatever extent possible, we have made an effort to place the individual, detailed studies (found in the ensuing chapters) within this context. This new approach promises to be a powerful tool for integrative studies of aquatic ecology and management of aquatic ecosystems.

Chapter II

The Hubbard Brook Valley

GENE E. LIKENS, F. HERBERT BORMANN, ROBERT S. PIERCE, and JOHN S. EATON

A. Environmental Parameters

Mirror Lake and its drainage basin (43° 56.5′N, 71° 41.5′W) are located in the town of Woodstock (Grafton County), within the Hubbard Brook Valley of a mountainous region in northcentral New Hampshire. The Hubbard Brook Valley lies mostly within the White Mountain National Forest near the small village of West Thornton. The Valley is drained by Hubbard Brook, a fourth-order (Strahler 1957) stream that generally flows from west to east for about 10 km before bending sharply to the south for another 3 km where it joins the Pemigewasset River (*Figure II.A–1*). More than 20 tributaries flow into Hubbard Brook along the west to east orientation; Mirror Lake, which receives three tributaries, drains into Hubbard Brook at the beginning of the north-south orientation. The Pemigewasset River eventually becomes the Merrimack River, which discharges into the Atlantic Ocean at Newburyport, Massachusetts. The Hubbard Brook Valley covers an area of more than 3,300 ha and ranges in altitude from 173 m MSL at the junction with the Pemigewasset River to 1,015 m at the summit of Mt. Kineo on the southwestern border. The upper 3,076 ha of the Valley comprise the Hubbard Brook Experimental Forest (HBEF) established in 1955 and operated by the U.S. Forest Service as a principal research area for watershed management in New England and as the site of the Hubbard Brook Ecosystem Study (Bormann and Likens 1967; Likens et al. 1977; Bormann and Likens 1979). The Ecosystem Study has been underway since 1963.

Climate

Measurements of climatic variables have been made at the HBEF since 1956. The climate varies with altitude, but is classified as humid continental with short, cool summers and long, cold winters (Trewartha 1954). Characteristics of the climate at the HBEF are (1) changeability of the weather, (2) a large range in both daily and annual air temperatures, and (3) equable monthly precipitation.

Solar Radiation

The amount of solar radiation in the Valley is appreciably less than the potential for a flat surface at 44°N latitude, although the seasonal pattern is similar (*Figure II.A–2*). Measured average monthly solar radiation during the summer months (June through September) varied from about 325 to 475 ly/day, whereas average wintertime (December through February) values varied from about 100 to 225 ly/day during 1960 through 1978. Solar radiation on specific days, however, is highly variable. For example, inputs of solar radiation for cloudy summer days and clear winter days are approximately the same. Moreover, some seasons and even years are exceptionally sunny or cloudy. It also should be noted that on sunny days, the solar radiation received in the morning by the east-facing slopes exceeds the amount on a horizontal surface. The net loss is due to shadows that arrive early in the afternoon.

Annual incident solar radiation was about 11.1×10^9 kcal/ha-yr, and 5.1×10^9 kcal/ha

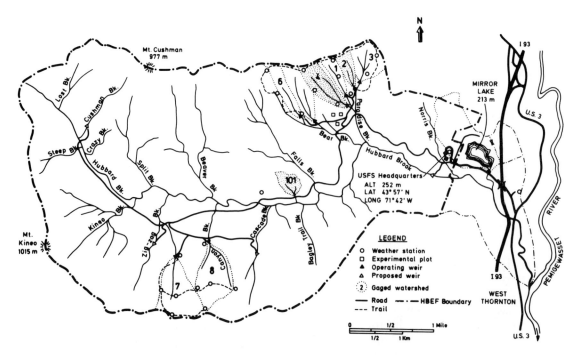

Figure II.A–1. Map of the Hubbard Brook Valley.

during June through September, for the Hubbard Brook Valley during 1960 through 1978. (Earlier estimates Gosz et al. [1978] and Bormann and Likens [1979] were based on one year, 1969–70.) During the summer, some 15% of this radiational input is reflected, 41% is lost as heat, 42% is used in evaporation and transpiration of water, and about 2% is fixed by photosynthesis (Gosz et al. 1978). These proportions are very different for Mirror Lake (e.g. percentage reflected by lake is greater) but the lake surface area represents less than 0.5% of the

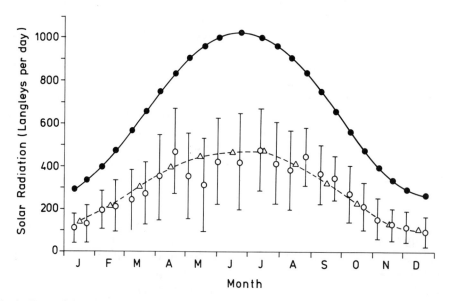

Figure II.A–2. Potential and measured (1960 to 1978) solar radiation at the U.S. Forest Service Headquarters site within the Hubbard Brook Valley. ●——●, potential solar beam radiation on a flat surface at 44°N for the same days as the potential insolation days, from Frank and Lee (1966); △---△, measured average monthly values; ⌀, mean values ± one standard deviation for selected days.

A. Environmental Parameters

total area of the Valley. The flux of energy for Mirror Lake will be considered in more detail in *Chapter IV.C*.

Air Temperature

Mean air temperature within the forest and above Mirror Lake in the Hubbard Brook Valley strongly reflects the seasonal pattern of solar radiation (*Figure II.A–3*). Mean daily air temperature (1.5 m above the ground in the forest) ranged from about −12°C in January to about +18°C in July during 1960 through 1970 on a south-facing slope at 450-m elevation (Federer 1973). In contrast, air temperature ranged from −27°C in December to about +24°C in July at the shoreline of Mirror Lake during 1980 (*Figure II.A–3*).

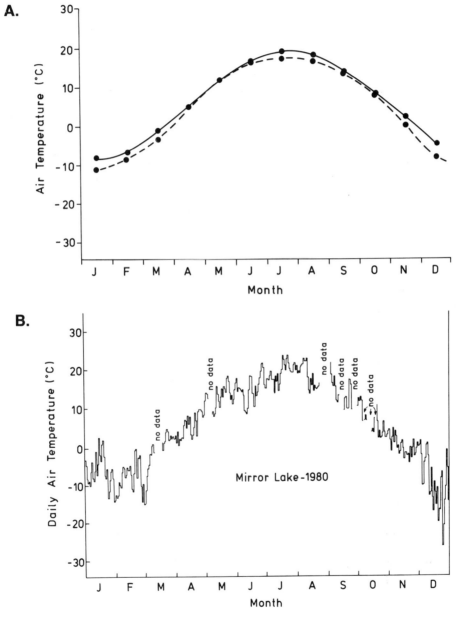

Figure II.A–3. A. (---) Annual cycle of mean daily air temperatures during 1960 to 1970 at 1.5 m above the forest floor on a gentle, south-facing slope at 450-m elevation (from Federer 1973); (——) annual cycle of mean daily air temperatures during 1964 to 1978 at 1.5 m above the ground at the U.S. Forest Service Headquarters at 252 m. **B.** Daily air temperatures at the shoreline of Mirror Lake during 1980.

Wind

Air masses generally move from west to east in eastern North America, but cyclonic storm tracks tend to converge in New England and move toward the northeast. The Hubbard Brook Valley lies within the zone of convergence. Occasionally, intense tropical or extratropical cyclones may affect the area with exceptionally high wind velocities. A hurricane in 1938 uprooted a large number of trees in New England, but in general, hurricanes are not considered to be a major ecologic factor in the Hubbard Brook Valley (Bormann and Likens 1979).

The surface wind in the Hubbard Brook Valley is determined not only by the large-scale, synoptic pressure field, but also by mesoscale features of the terrain. The resulting wind patterns are of two types.

1. *Slope winds occurring during clear skies and weak, large-scale pressure fields.* At such times the air drains downhill at night and moves up heated sunlit slopes in the daytime. The mesoscale wind field will be most pronounced in the main Pemigewasset River Valley (north winds at night, south winds during the day) but there will be subsidiary flows in tributary valleys (for the U.S. Forest Service Headquarters anemometer site, this means nighttime northwesterly components and daytime southeasterly winds). Depending on location in the Valley, time of day, season and pressure gradient, many modifications of the general picture are possible.

2. *Turbulent vortices occurring during strong large-scale pressure fields.* Particularly when upper winds are blowing at right angles to the main topographic features, downwind vortices are shed, creating complex wind patterns in the Valley. The net result will be a very gusty surface wind, much more turbulent than over a flat plain. On such occasions, lake evaporation and terrestrial evapotranspiration will be considerably increased.

As might be expected, these several influences are masked in annual and monthly wind roses (see *Figure II.A–4* and *Table II.A–1*). The Hubbard Brook surface wind (as measured at a height of 2 m at the U.S. Forest Service Headquarters) (*Figure II.A–1*) comes most frequently from the north-northwest and only infrequently comes from the northeastern through the southwestern quadrants (*Figure II.A–4*). Northerly flows predominate from November through April and northwesterly flow predominates from May through October (*Figure II.A–4*). Flow from the southerly directions is more common during the summer, but

Table II.A–1. Average Monthly Surface Wind Direction and Speed in the Hubbard Brook Valley (U.S. Forest Service Headquarters Station)

Rank	Month	Average km/month	Predominant Direction	% of April
1	April	3,644	NW (34%); N (32%)	100
2	March	3,564	NW (35%); N (26%)	98
3	January	3,324	N (45%); NW (27%)	91
4	May	3,101	NW (28%)	85
5	February	2,774	N (35%); NW (30%)	76
6	December	2,673	N (41%); NW (33%)	73
7	October	2,521	N (33%); NW (27%)	69
8	November	2,315	NW (29%); N (29%)	64
9	June	2,183	NW (30%); N (23%)	60
10	July	2,055	NW (34%); N (24%)	56
11	September	1,842	NW (30%); N (28%)	51
12	August	1,569	NW (39%)	43
	Σ	31,565		
	$\bar{x}$	2,630		
	$s_{\bar{x}}$	195		

A. Environmental Parameters

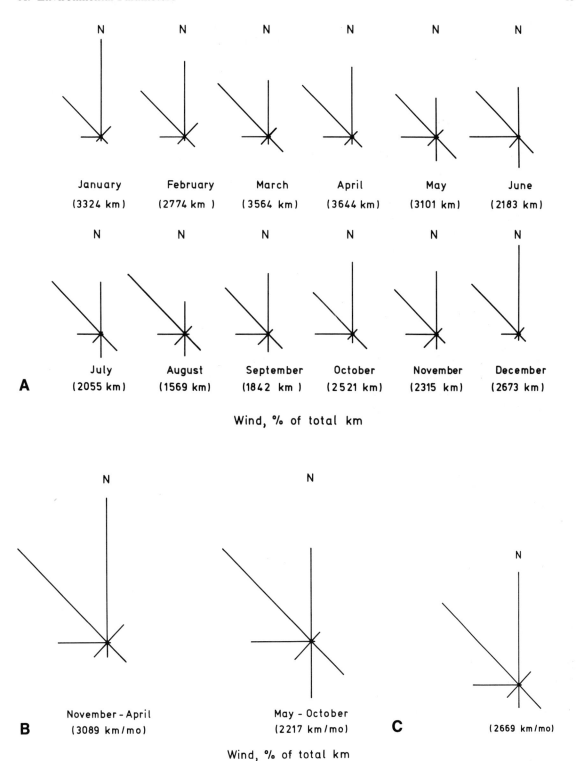

Figure II.A–4. Average monthly (**A**), seasonal (**B**), and annual (**C**) surface wind roses for the U.S. Forest Service Headquarters site during November 1965–August 1967, plus January 1975–August 1977.

never dominates the monthly average (*Figure II.A–4*). April is the windiest month and August has the lightest winds on average (*Table II.A–1*).

Vegetation and Animals

The present vegetation in the Hubbard Brook Valley is fairly typical of similarly aged, second-growth hardwood forests found throughout northern New England (Bormann et al. 1970; Siccama et al. 1970; Bormann and Likens 1979). The principal deciduous species for the Valley include American beech (*Fagus grandifolia*), sugar maple (*Acer saccharum*), yellow birch (*Betula alleghaniensis*), white ash (*Fraxinus americana*), basswood (*Tilia americana*), red maple (*Acer rubrum*), red oak (*Quercus borealis*) and American elm (*Ulmus americana*); the principal coniferous species are hemlock (*Tsuga canadensis*), red spruce (*Picea rubens*), balsam fir (*Abies balsamea*) and white pine (*Pinus strobus*). Herbaceous species in W6 are dominated by the evergreen fern *Dryopteris spinulosa*, the shrub *Viburnum alnifolium* and seedlings of sugar maple and beech (Bormann et al. 1970; Siccama et al. 1970; Forcier 1973, 1975) and includes *Erythronium americanum*, a vernal photosynthetic (Muller 1975), *Oxalis montana, Maianthemum canadense, Dennstaedtia punctilobula, Clintonia borealis, Lycopodium lucidulum, Aster accuminatus* and *Smilicina racemosa* (Siccama et al. 1970). The vascular flora of the northern hardwood forest in the Valley contains about 96 species: 14 trees, 11 shrubs and 71 herbs (Likens 1973).

At Hubbard Brook, the vegetation is characteristic of an aggrading northern hardwood forest-ecosystem (Bormann and Likens 1979). The forest was logged in 1910–1919 but has not been disturbed by logging since (Bormann et al. 1970; *Chapter III.B.2 and 3*) and there is no evidence of fire in the history of these stands (*Chapter III.B.2*). Currently, the forest is accumulating both living and dead biomass. The forest is unevenly aged and well stocked, primarily with sugar maple, American beech and yellow birch. Red spruce, balsam fir and hemlock are prominent on north-facing slopes, on ridge tops, on rock outcrops and along the main channel of Hubbard Brook. The forest has a basal area of about 24 m^2/ha (Bormann et al. 1970).

Net primary productivity and living biomass accumulation for the forest averaged about 1,040 g/m^2-yr and 360 g/m^2-yr dry weight, respectively, during 1956–1965 (Whittaker et al. 1974). Forest floor accumulation averaged about 33 g dry weight/m^2-yr, some 55 years after the forest was logged (Covington 1976).

Fauna common to northern hardwood forests occur within the Hubbard Brook Valley. More than 90 species of birds have been observed (Likens 1973) and snowshoe hare (*Lepus americanus*), beaver (*Castor canadensis*), red fox (*Vulpes fulva*), black bear (*Ursus americanus*) and white-tail deer (*Odocoileus virginianus*) are present.

The maximum number of deer in 1976 was estimated to be about 200 or five deer per km^2. However, the size of the herd in the HBEF has declined steadily since 1976, with the population size in 1980 only about one-tenth of that in 1976. Apparently, the decline was due to high mortality during winter (D. Pletscher, personal communication).

Geology

All of the eastern half of the Hubbard Brook Valley (including south-facing Watersheds 1–6) is underlain by bedrock of the Littleton Formation. This bedrock is derived from highly metamorphosed and deformed sedimentary sandstone and mudstone (a mixture of gneiss and schist) with coarse-grained minerals of quartz, plagioclase, biotite and sillimanite. Except for the southwest rim of the Valley, the bedrock in the western portion of the Valley (east of Watersheds 7 and 8) is largely Kinsman quartz monzonite, a granitic material that intruded into the metamorphic Littleton Formation (Billings 1956). The Kinsman Formation is composed of a metamorphosed granitic material containing medium- to coarse-grained minerals such as oligoclase-andesine, potash feldspar, biotite and muscovite. Glacial till overlaps most bedrock throughout the Valley. Materials deposited by the glacier as it receded were highly variable in size, from particles of clay to large boulders. The depths of deposits now vary from complete absence of material on the ridgetops and rock outcrops to 9 m or more in the valleys, with an average of about 2 m.

The bedrock has few fissures, and these diminish rapidly with depth. Major faults have not been observed. It is believed that deep seepage is negligible over the course of one year, compared with precipitation, streamflow and evapotranspiration (Likens et al. 1977).

Although the general slope of the entire tributary basin may not appear steep, i.e. the slope of the lid for Watershed 3 is about 21%, internal slopes within the watershed (particularly at ridgetop edges and along stream channels) may have slopes of 70% or greater. Alluvial fans along Hubbard Brook and other terrace deposits in some tributary streams tend to have lower gradients.

Details about the geology as it specifically applies to the Mirror Lake watershed are discussed in *Chapter III.A*.

Soils

Soils in the Valley are mostly well-drained spodosols (haplorthods) derived from glacial till, which overlies most of the bedrock. The soils are characterized by sparse clay, a sandy-loam texture and with a thick forest floor, a 3 to 15 cm layer of organic matter at the surface. The soil is very heterogeneous, but generally is acid, pH $\leq$ 4.5, and infertile (cf. Pilgrim and Harter 1977) and has a very high infiltration capacity (Pierce 1967). The major soil series represented are Hermon (most extensive), Becket, Waumbek, Canaan, Berkshire, Peru, Leicester and Colton. The separation between the soil mantle and the virtually unweathered bedrock below is distinct. Rocks of all sizes abound throughout the soil profile and, in many places, boulder fields are prominent. Soil depths are highly variable ranging from zero in some areas to several meters in the valley bottoms. In some places, impermeable pan layers, fragipan, at depths of about 0.6 m, restrict vertical water movement and root development. Soil on the ridgetops may consist of a thin accumulation of organic matter, resting directly on bedrock.

B. Biogeochemistry

A detailed description of the biogeochemistry of the HBEF through 1974 was given in our volume, *Biogeochemistry of a Forested Ecosystem* (Likens et al. 1977). These data have been updated and elaborated and are summarized in the following pages.

Hydrologic Flux

Precipitation is measured in the Valley by a fixed network of standard U.S. Weather Bureau precipitation gauges, approximately one for every 13 ha of area within the Experimental Watersheds, and two near the lake (*Figure II.A*–1). The Thiessen polygon method (Thiessen 1911) is used to determine amounts of precipitation over specified areas, such as the Experimental Watersheds.

The concept of *normal* or *typical* is usually difficult to define relative to long-term data. Ostensible *trends* from short-term records may be only brief fluctuations in some long-term pattern. Thus, analyses or interpretations of typical or mean conditions based on short-term records may be quite misleading (see Likens 1983). Hydrologic measurements for the Hubbard Brook Valley began in 1955. This continuous record, although not exceptionally long, provides a view that is longer than for most areas of the pattern and variability associated with water flux for a terrestrial ecosystem.

On the average, 131.1 cm of water falls annually on the Experimental Watersheds (*Table II.B*–2). Somewhat less, 121.5 cm (1970–71 to 1979–80) falls on the Mirror Lake drainage basin each year (see *Chapter IV.E*). Of the amount falling on the forested areas, some 82.1 cm or 63% becomes streamflow and the remaining 49.0 cm or 37% is lost through evapotranspiration.

Because loss to deep seepage is negligible (Likens et al. 1977), evapotranspiration can be estimated as the difference between precipitation and streamflow for the water-year (1 June through 31 May). Annually, the evapotranspiration value is much less variable than the annual amount of precipitation and streamflow (*Table II.B*–2). The amount of precipitation per year has varied by 2.0-fold, streamflow by 2.7-fold and evapotranspiration by 1.3-fold from 1956 through 1980. Because annual evapotranspiration is so constant, amount of annual streamflow is highly and directly correlated with the amount of annual precipitation (Likens et al. 1977).

Table II.B–2. Average Annual Hydrologic Budget for Watersheds 1–6 of the Hubbard Brook Experimental Forest during 1956 to 1980

	Precipitation (cm)	Streamflow[a] (cm)	Evapotranspiration (cm)
$\bar{x}$	131.1	82.1	49.0
$s_{\bar{x}}$	4.25	4.14	0.65
% of total	100	63	37
range	95.1–185.7	50.1–136.1	41.4–53.7

[a] Watersheds subjected to experimental treatment (e.g. W2 in 1965–66 and W4 in 1969–70) are not included after disturbance.

There is variation in amount of precipitation at various locations in the Hubbard Brook Valley (e.g. *Table II.B*–3). Of the watersheds located on the same south-facing slope of the Valley, the more westerly ones (W6 and W5) receive, in general, slightly more precipitation than the easterly ones (W1–W4). Watersheds 7 and 8 on the north-facing slopes at slightly higher elevations receive somewhat more precipitation than the south-facing watersheds. The drainage area for Mirror Lake receives less precipitation than any of the watersheds at higher elevations (see *Chapter IV.E*). This variation illustrates the need for rigorous hydrologic studies that directly measure the precipitation impinging on the area of study.

Not only is there spatial variation, but temporal variation as well. For example, the average annual precipitation for the decade of 1966–67 to 1975–76 was 1.2-fold greater than for the previous decade; streamflow was 1.3-fold greater (*Table II.B*–4). The water-year 1964–65 was the driest (95.1 cm) on record. The mean for the period of biogeochemical study (1963 through 1980), however, agrees well with the mean for the full period, 1956 through 1980 (*Tables II.B*–2 and 4).

On the average, precipitation (i.e. > a trace)

Table II.B–3. Annual Hydrologic Budget for Four Experimental Watershed-ecosystems of the Hubbard Brook Experimental Forest during 1979–80[a]

Experimental Watershed-ecosystem	Precipitation (cm)	Streamflow (cm)	Evapotranspiration[b] (cm)
W1	113.7	71.4	42.4
W3	113.9	69.3	44.6
W5	115.5	70.2	45.3
W6	119.3	70.8	48.5
$\bar{x}$	115.63	70.41	45.21
$s_{\bar{x}}$	1.30	0.45	1.27
% of total	100	61	39

[a] The estimates for annual evapotranspiration do not consider differences in soil water storage from year to year. Estimates of such storage differences between years, based on the BROOK model (Federer and Lash 1978), rarely exceed 4 cm and commonly are less than 1 cm. Neglecting changes in soil water storage could lead to errors of up to 7% in estimating annual evapotranspiration. Further, corrections for soil water storage should lessen the variability of annual evapotranspiration values as shown in *Table II*–2.
[b] Precipitation minus streamflow.

B. Biogeochemistry

Table II.B-4. Annual Hydrologic Budgets for the Hubbard Brook Experimental Forest during Different Periods from 1956 to 1980

		Precipitation (cm/yr)	Streamflow (cm/yr)	Evapotranspiration (cm/yr)
Mean for HBES Period 1963 to 1980				
n = 17	$\bar{x}$	133.6	85.0	48.6
	$s_{\bar{x}}$	5.03	4.81	0.75
	% of total	100	64	36
Mean for 1956–57 to 1965–66				
n = 10	$\bar{x}$	121.1	71.6	49.5
	$s_{\bar{x}}$	6.29	6.00	1.02
	% of total	100	59	41
Mean for 1966–67 to 1975–76				
n = 10	$\bar{x}$	141.0	92.3	48.6
	$s_{\bar{x}}$	6.65	6.49	1.09
	% of total	100	65	34

occurs on 111 days each year. This high frequency of rain or snow events and associated cloudiness account for the relatively small proportion of potential solar radiation that actually falls on the Hubbard Brook Valley (*Figure II.A–2*). Although individual months may vary drastically in amount of precipitation, on average, over the long term, about the same amount of precipitation falls in each month (*Figure II.B–5*). However, within this pattern, the amount of water delivered in individual precipitation events may vary by several orders of magnitude. Some 30% of the annual precipitation falls as snow and accumulates as a snowpack that sometimes reaches a depth of 2 m, but averages about 1.2 m in most winters (Likens et al. 1977).

In contrast to the uniformity of monthly precipitation, most of the streamflow occurs during the spring when the snowpack melts (*Figure II.B–5*). Some 53% of the annual streamflow occurs during March, April and May. In sharp contrast, streamflow during the summer months is very low, with water loss occurring primarily

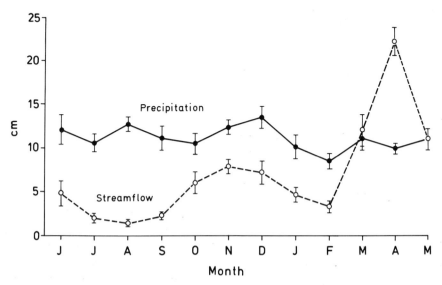

Figure II.B–5. Average monthly precipitation (●——●) and streamflow (○---○) for the Hubbard Brook Experimental Forest during 1956 to 1980. The vertical bars are ± one standard deviation of the mean.

through evapotranspiration. Although annual streamflow has varied by as much as 2.7-fold (*Table II.B–2*) during 1956 through 1980, stormflow may exceed average flows by several orders of magnitude. Stormflow is the major vehicle for the erosion and transport of particulate matter (Bormann et al. 1974; Likens et al. 1977).

Soils in the Hubbard Brook Valley have a high infiltration capacity for water (76 cm/ha, Trimble et al. 1958). Thus during the most intense rainstorms, overland runoff has not been observed, except for a few boggy areas and in stream channels (see Pierce 1967). Soil frost is an uncommon feature even during the coldest months because of a thick humus layer and deep snow cover (Hart et al. 1962; Federer 1973). However, the surface soil may freeze during winters when the snowpack develops late (e.g. 1969–70) or is essentially nonexistent (e.g. 1979–80). Infrequently, mid-winter thaws may reduce the snowpack sufficiently to allow some soil frost to occur. There is appreciable micro-relief of the forest floor produced by large boulders and by throw mounds that occur when trees are uprooted (pit and mound topography). This surface topography favors infiltration rather than overland flow of water.

Streamflow is measured continuously throughout the year in the Experimental Watersheds by automatic recorders at stream-gauging stations, which include either a V-notch weir or a combination of V-notch and modified San Dimas flume attached to the bedrock at the base of each watershed (Likens et al. 1977). Some of the ponding basins upstream of the weirs are heated during winter to facilitate continuous measurement of streamflow during the coldest weather.

Streamflow and direct drainage into Mirror Lake were calculated as the yield of water from the watersheds draining into the lake by multiplying the unit area, precipitation-corrected yield of water from W3 by the area of the Mirror Lake watersheds (see *Chapter IV.E*).

Chemical Flux—Chemistry of Precipitation and Stream Water

Measurement of the concentration of dissolved substances in precipitation and stream water has been done at the HBEF since 1963. Detailed chemical analyses have been done on weekly samples of precipitation and water from streams draining the Experimental Watersheds. and on frequent samples from Mirror Lake, its tributaries and its outlet.

Rain and snow are collected as bulk precipitation with continuously open plastic collectors (Likens et al. 1967). Samples of stream water are collected approximately 10 m upstream from each weir or flume in clean polyethylene bottles. The concentrations of dissolved chemicals characterizing a period of time are reported as volume-weighted averages. Nutrient flux is determined by multiplying appropriate concentrations of dissolved chemicals by the amounts of water moving across the ecosystem boundary during the period. Details concerning these calculations and the routine methods used in collecting samples of precipitation and stream water, as well as analytical procedures are given by Likens et al. (1977), Bormann and Likens (1967), Likens et al. (1967), Fisher et al. (1968), Bormann et al. (1969) and Hobbie and Likens (1973).

Only the long-term data from the southeast-facing watersheds of the HBEF are summarized here. Data relative to watersheds draining into Mirror Lake are treated in detail in *Chapter IV.E*.

Chemical Concentrations in Precipitation

The rain and snow at Hubbard Brook are dominated by hydrogen and sulfate ions (*Table II.B–5*). Nitrate, chloride and ammonium ions are the next most abundant on an equivalent basis, and the remaining ions are present in low concentrations. Thus, the long-term, average bulk precipitation at Hubbard Brook can be characterized as a dilute solution of sulfuric and nitric acids at a pH of about 4.15.

We have found no significant correlation for concentraations of Ca^{2+}, Mg^{2+}, Na^+, K^+, NH_4^+ or Cl^- in precipitation with elevation in the Hubbard Brook Valley (Likens et al. 1977). However, concentrations of SO_4^{2-} and NO_3^- were significantly higher at lower elevations (i.e. at the U.S. Forest Service Headquarters site at 252 m MSL) and these data from the

B. Biogeochemistry

Table II.B–5. Volume-weighted Annual Mean Concentration of Dissolved Substances in Bulk Precipitation and Stream Water for Watershed 6 of the Hubbard Brook Experimental Forest

Substance	Precipitation 1963 to 1980		Stream Water 1963 to 1980	
	mg/l	μEq/l	mg/l	μEq/l
H^{+a}	0.070^b	70	0.014^c	14
$NH_4^{+\ b}$	0.19	11	0.027	1.5
Ca^{2+}	0.13	6.5	1.31	65
Na^+	0.11	4.8	0.76	33
Mg^{2+}	0.04	3.3	0.32	26
K^+	0.06	1.5	0.22	5.6
$Al^{n+\ d}$	0.02	2.2^j	0.38	20^k
$SO_4^{2-\ b}$	2.6	54	5.9	123
$NO_3^{-\ b}$	1.47	24	1.73	30
Cl^-	0.40^i	11	0.50	14
$PO_4^{3-\ e}$	0.0097	0.31	0.0029	0.092
HCO_3^-	$\sim 0.006^f$	0.098	—	—
Dissolved Si^e	0.02	—	4.01	—
DOC	1.08^g	—	2.0^h	—
pH	4.15		4.92	
Total	(+)	99.3	(+)	165
	(−)	89.4	(−)	167

[a] Calculated from weekly measurements of pH.
[b] 1964 to 1980.
[c] 1965 to 1980.
[d] 1977 to 1980.
[e] 1972 to 1980.
[f] Calculated from H^+–HCO_3^- equilibrium.
[g] 1976–77 (Likens et al. 1983).
[h] 1976–77, 1979 to 1981.
[i] 1967 to 1980.
[j] Based on 3.0 charges/mole.
[k] Based on 1.4 charges/mole.

lower elevation are used as inputs to Mirror Lake (*Chapter IV.E*).

Various seasonal patterns are apparent from the long-term record of bulk precipitation chemistry (*Figures II.B–6 and 7*). Monthly volume-weighted concentrations of hydrogen ion, sulfate, ammonium, DOC, phosphorus and to lesser extent K^+, have maximum values in the summer and minimum values in the winter; Mg^{2+}, Ca^{2+}, Na^+, K^+ and Cl^- have peaks in concentration during the autumn, possibly in relation to the occurrence of coastal storms; Ca^{2+} and NO_3^- concentrations are relatively low from June to February and then reach maximum concentrations during the spring. These concentration patterns are largely independent of amount of precipitation, which varies relatively little on a monthly basis throughout the year. Apparently, the chemical composition of precipitation is influenced by a variety of sources for atmospheric contaminants, including gaseous emissions from natural ecosystems, terrestrial dust, oceanic spray, volcanic emissions and anthropogenic emissions (see Likens et al. 1977, 1979), as well as by the trajectory of the air mass preceding the precipitation event.

Rain and snow at Hubbard Brook contain surprisingly large amounts of organic matter (Likens et al. 1983). These inputs are not large in relation to organic carbon fixed photosynthetically by the terrestrial ecosystem, but may represent significant allochthonous inputs to the

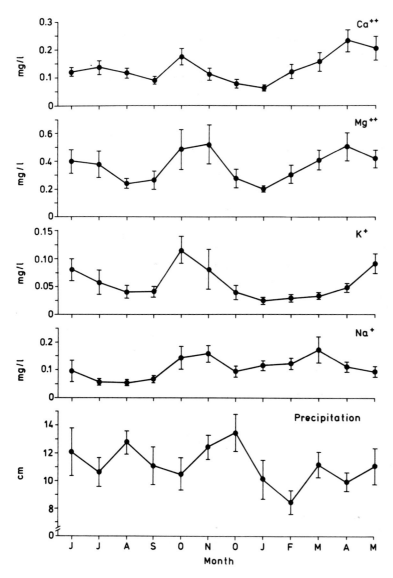

Figure II.B–6. Average monthly volume-weighted concentrations in bulk precipitation and amount of precipitation during 1963 to 1980. Samples were collected in the south-facing Experimental Watersheds. The vertical bars are ± one standard deviation of the mean.

lake ecosystem (see *Chapter V.C*). Based on wet-only samples of precipitation, the annual volume-weighted concentration of total organic carbon during 1976–77 was 1.28 mg C/liter. Some 85% of this carbon was in dissolved form. Particulate plus dissolved macromolecular (>1,000 MW) organic material accounted for 51% of the total organic carbon. Most of the remainder (<1,000 MW) was composed of carboxylic acids, aldehydes (mostly formaldehyde), carbohydrates (mostly polysaccharides) and tannin/lignin. Small amounts of primary amines and phenols also were found. Based on only a few samples, carbohydrates and tannin/lignin predominated (up to 50% of total) in the macromolecular fractions, and only a small portion (<10%) of this fraction was proteinaceous. The majority of these organic compounds found in precipitation are biologically active and might have some metabolic effect on receiving systems. Carbon of high molecular weight may be less available as an energy source for organisms, but may provide sites for metal-complexation and pH-buffering capacity for natural ecosystems. Although a diverse number of carboxylic acids were present in almost every

B. Biogeochemistry

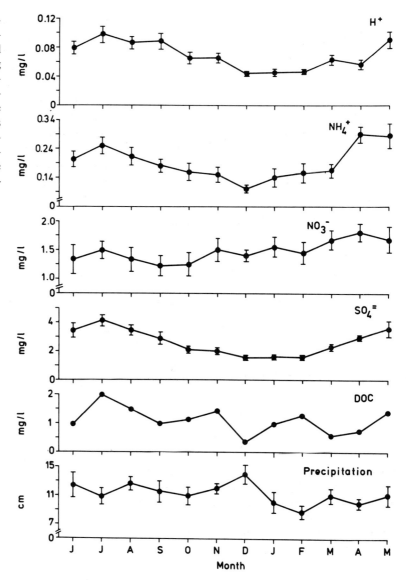

Figure II.B–7. Average monthly volume-weighted concentrations in bulk precipitation and amount of precipitation during 1964 to 1980. Samples were collected in the south-facing Experimental Watersheds. The average DOC concentrations are for 1976–77 in samples collected near the U.S. Forest Service Headquarters station. The vertical bars are ± one standard deviation of the mean.

sample, they did not contribute significantly (<5%) to the free proton concentration of the precipitation at Hubbard Brook.

The volume-weighted annual concentrations for some ions in precipitation have changed significantly since the beginning of our study in 1963 (*Figures II.B–8* and 9). Most notably, based on regression analysis, Ca^{2+} has decreased by 87%, Mg^{2+} by 81%, NH_4^+ by 42% and SO_4^{2-} by 35%. Nitrate concentrations increased from 1964 to 1971 and H^+ concentrations decreased after 1970.

The major source of Ca^{2+} and Mg^{2+} in precipitation is thought to be terrestrial dust, and local sources of dust are negligible and seemingly unchanged during the period. Hence the explanation for the decline in these two basic cations is unknown. The pattern for K^+ is largely determined by the values in 1963–65, a drought period.

Occurrence frequency distributions of the concentrations of the chemical components in precipitation are highly skewed; small concentrations occur most frequently. Highest concentrations generally occur with lowest amounts of precipitation for most ions, but low concentrations can occur with either low or high amounts of precipitation (Likens et al. 1984).

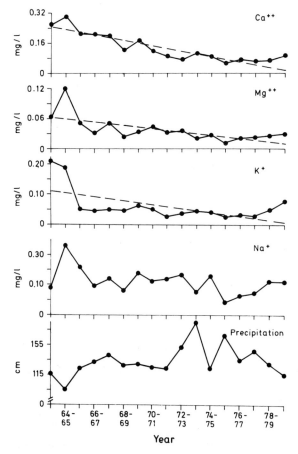

Figure II.B–8. Average annual volume-weighted concentrations in bulk precipitation and amount of precipitation during 1963 to 1980. Samples were collected in the south-facing Experimental Watersheds. The dashed regression lines have a probability for a larger F-value of <0.05.

our study in *Figure II.B–10* (see also *Table II.B–6*). As was the case with annual concentrations, annual inputs of Ca^{2+}, Mg^{2+} and K^+ decreased from 1963 to 1980. Inputs of hydrogen ion, NH_4^+ and NO_3^- generally increased from 1964–65 to 1973–74, but have decreased or remained about the same since. Although volume-weighted concentrations of NH_4^+ and SO_4^{2-} decreased on a statistical basis during the period of study (*Figure II.B–9*), annual inputs did not (*Figure II.B–10*), showing the influence of the amount of precipitation on long-term trends. Chemical deposition generally increases

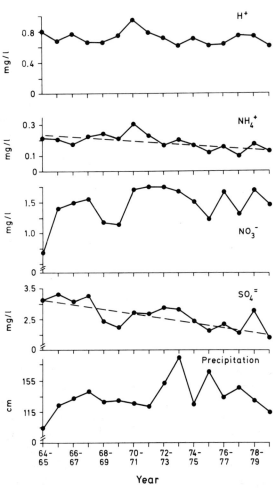

Figure II.B–9. Average annual volume-weighted concentrations in bulk precipitation and amount of precipitation during 1964 to 1980. Samples were collected in the south-facing Experimental Watersheds. The dashed regression lines have a probability for a larger F-value of <0.05.

Mass Input of Chemicals

The meteorologic input (*Chapter I*) of chemicals for the HBEF is calculated by multiplying the measured concentrations of each substance in the weekly precipitation sample (mg/liter or mEq/liter) by the volume of precipitation that fell on each unit area of the watersheds (e.g. liter/ha). The meteorologic input can then be given in units of mass (e.g. kg) or equivalency (e.g. Eq) per unit area (kg/ha or Eq/ha) per week. The weekly values can be summed to give monthly, seasonal or annual values.

Annual inputs of chemicals and water for Watershed 6 are shown throughout the period of with an increasing amount of precipitation in weekly samples.

In addition to materials added to an ecosystem in rain, snow, sleet, etc. (wet deposition), chemicals may be added directly by dry deposition. Dry deposition may occur through gravity (e.g. settling of dust particles), impingement of small particles (aerosols) on surfaces, or by gaseous adsorption on surfaces or through uptake by organisms. Dry deposition of chemical compounds is exceedingly difficult to quantify for natural ecosystems, but cannot be ignored. Bulk deposition measurements may include some dry deposition, but bulk collections are generally thought to provide gross underestimates of dry deposition (e.g. Eaton et al. 1980). At Hubbard Brook, at least one-third of the total meteorologic input of sulfur (Eaton et al. 1978; Likens et al. 1977) and two-thirds of the total input of nitrogen (Bormann, et al. 1977) occur by dry deposition or biologic fixation (i.e. in excess of bulk deposition).

Based on preliminary calculations using atmospheric concentrations and deposition velocities for both SO_2 and sulfur aerosols, SO_2 should account for 96% of the gravitational dry deposition of sulfur at Hubbard Brook and sulfur aerosols should contribute the remaining 4% (Eaton et al. 1980). Since SO_2 concentrations are maximal in winter at Hubbard Brook (Eaton et al. 1978), these calculations suggest major dry deposition during winter months.

In addition, studies of the dry deposition of sulfur in comparative studies between bulk and wet-only precipitation collectors at Hubbard Brook show that there is a strong seasonal aspect in the dry deposition of sulfur. Calculations based on these two approaches suggest that 60% of the annual dry deposition of sulfur occurs during the winter months of December through March. We emphasize that these calculations are preliminary and are based on relatively sparse data on atmospheric concentrations and on the assumption of uniform deposition velocities throughout the seasons. Moreover, large amounts of SO_4^{2-} in throughfall during the summer months (Eaton et al. 1973) suggest that dry deposition to the summertime forest canopy is significant. In contrast to the winter peak in dry deposition of sulfur, maximum wet (bulk) deposition occurs during the summer (Hornbeck et al. 1976).

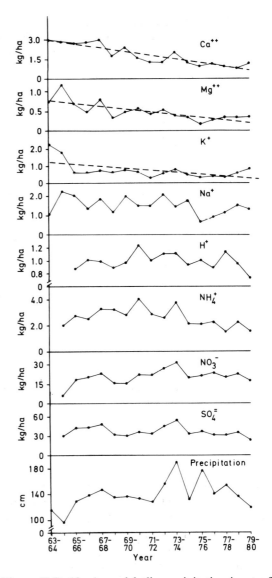

Figure II.B–10. Annual bulk precipitation input of chemicals (kg/ha) and amount of precipitation for Watershed 6 during 1963 to 1980. The dashed regression lines have a probability for a larger F-value of <0.05.

We estimate that in addition to sulfur, the dry deposition of nitrogen, calcium, magnesium and potassium may be significant. In contrast, the dry deposition of sodium, chloride and silica does not appear to be a significant input to the ecosystem. Investigations of the dry deposition of nitric acid vapor by B. Huebert and J. M. Roberts (personal communication) near Champaign, Illinois indicated an average deposition velocity of approximately 2.5 cm/s. This veloc-

Table II.B–6. Annual Mean Budgets of Bulk Precipitation Inputs and Stream-water Outputs of Dissolved Substances for Forested Watershed 6 of the Hubbard Brook Experimental Forest (kg/ha)

Substance	Precipitation Input	Stream-water Output	Net Gain (+) or Loss (−)
Dissolved Silica	0.37^e	35.6^b	−35.2
Sulfate^{b,f}	36.6	52.2	−15.6
Calcium^a	1.8	11.5	− 9.7
Sodium^a	1.5	6.6	− 5.1
Aluminum^h	0.23	3.2	− 3.0
Magnesium^a	0.5	2.8	− 2.3
Dissolved Organic Carbon^g	14.2	16.4	− 2.2
Potassium^a	0.8	1.9	− 1.1
Nitrate^{b,f}	20.6	15.3	+ 5.3
Ammonium^{b,f}	2.6	0.24	+ 2.36
Chloride	5.8^d	4.4^b	+ 1.4
Hydrogen ion	1.000^b	0.127^c	+ 0.87
Phosphate^e	0.145	0.028	+ 0.12

^{a} 1963 to 1980.
^{b} 1964 to 1980.
^{c} 1965 to 1980.
^{d} 1967 to 1980.
^{e} 1972 to 1980.
^{f} Dry deposition could add an additional input of 6.1 S kg/ha and 14.2 N kg/ha-yr.
^{g} Based on 131.1 cm of precipitation and 82.1 cm of streamflow.
^{h} 1977 to 1980.

ity translated into a dry deposition of nitrogen that was equivalent to the wet deposition for the month of June at their site. If our bulk precipitation collector is as poor at collecting nitrogen aerosols as it is at collecting the dry deposition of sulfur, we may be able to account for much of the discrepancy in the nitrogen budget for the watersheds (Bormann et al. 1977) by dry deposition of various nitrogenous compounds.

Chemical Concentrations in Stream Water

Some watersheds (W2, W4, and W101) in the HBEF were disturbed by experimental study after 1963. Because the watersheds have somewhat different chemistry, e.g. presence of bicarbonate in W4 (Likens et al. 1967), it was not possible to include data from these disturbed watersheds for only a part of the period for which volume-weighted averages were calculated. By so doing, the partial data could have produced artificial temporal trends in *average* stream-water chemistry. The only undisturbed watershed with data for the entire period, 1963 to 1980, is W6. Thus, data from W6 have been used here to characterize the chemistry of drainage waters from south-facing Experimental Watersheds of the Hubbard Brook Valley.

As water from precipitation passes through the ecosystem (vegetation canopy, forest floor, soil) its chemistry is altered appreciably (*Table II.B–5*). Whereas hydrogen ion and sulfate dominate the chemistry of precipitation, calcium and sulfate dominate the chemistry of stream water issuing from these aggrading forested watersheds. Sodium, magnesium, nitrate and aluminum are also important ions in stream water on an equivalent basis. The ionic strength of stream water (166 μEq/liter) is 1.8 times more than that of precipitation (94 μEq/liter). Part of this concentration effect results because 37% of the incoming water is lost as water vapor through evaporation and transpiration (*Table II.B–2*). However, additional amounts of chemicals are acquired by drainage water as it passes through the canopy of vegetation and soils of the ecosystem (see Likens et al. 1977).

In general, the measured concentrations in stream water change very little (less than a factor of 2 or 3) even though streamflow may fluctuate over 4 to 5 orders of magnitude during an annual cycle at Hubbard Brook (Fisher and Likens 1973; Johnson et al. 1969; Likens et al. 1967, 1977). This relation is particularly true for Mg^{2+}, SO_4^{2-}, Cl^- and Ca^{2+}. Concentrations of Na^+ and dissolved Si may be diluted up to threefold during periods of high flow, whereas Al^{n+}, H^+, DOC, NO_3^- and K^+ concentrations may be increased slightly with increased flow.

Monthly patterns of stream-water concentrations are variable for a few ions (*Figures II.B–11, 12 and 15*). The most striking patterns are shown by NO_3^- and K^+, which normally have very low values during the growing season (~15 May to ~15 October) and then increase to much higher concentrations during the nongrowing season. The pattern for sulfate is generally the converse for nitrate although the relative amount of change during the year is much less (see also Likens et al. 1977). Hydrogen-ion concentrations are generally higher in spring streamflow during snowmelt (*Figure II.B–12*).

Based on regression analysis, volume-weighted annual concentrations of Mg^{2+}, Na^+, NH_4^+ and SO_4^{-2} have decreased in stream water since 1963 or 1964 (*Figures II.B–13 and 14*). This general pattern compares with the decreased inputs of these ions in precipitation during the same period (*Figures II.B–8 and 9*).

In spite of the peak concentrations during 1968–69 through 1973–74, the overall annual volume-weighted concentrations of hydrogen ion in stream water have been relatively constant since 1965–66. We do not understand the

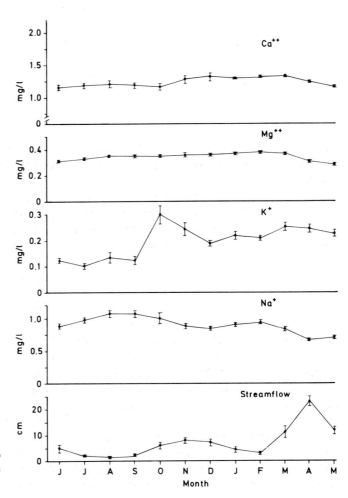

Figure II.B–11. Average monthly volume-weighted concentrations (mg/liter) and amount of streamflow in Watershed 6 during 1963 to 1980. The vertical bars are ± one standard deviation of the mean.

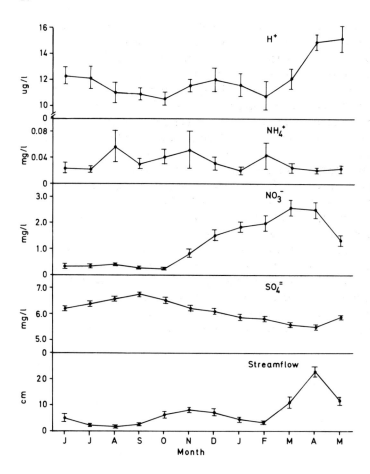

Figure II.B–12. Average monthly volume-weighted concentrations (mg/liter) and amount of streamflow in Watershed 6 during 1964 to 1980. The average H-ion concentrations are for 1965 to 1980. The vertical bars are ± one standard deviation of the mean.

cause for the hump (1968–69 to 1973–74) in the otherwise rather consistent pattern during the period of our measurements.

Although there is a consistent seasonal pattern (*Figure II.B*–15), annual nitrate concentrations in stream water have been highly variable during the study (*Figure II.B*–14).

Mass Output of Chemicals

The geologic output (*Chapter I*) of dissolved chemicals for the HBEF is calculated by multiplying measured concentrations of each substance in stream water by the volume of streamflow. Outputs are expressed as kilograms or equivalents per hectare of watershed-ecosystems.

Annual outputs of dissolved chemicals in streamflow are given in *Figure II.B*–16 and *Table II.B*–6). Because the chemical concentrations in stream water are relatively constant during the course of a year, the gross annual export of most ions is directly related to the annual streamflow and is highly predictable (Likens et al. 1977). The controlling influence of amount of streamflow on mass output of NH_4^+, NO_3^- and H^+ is less clear (*Figure II.B*–16).

In addition to dissolved chemical substances, both organic and inorganic particulate matter may be exported from ecosystems in stream water. Particulate matter is removed from the watershed as suspended load carried by turbulent water and as bedload rolled, slid or bounced along the streambed. Heavier particulate matter collects in the ponding or stilling basin behind the weir where it can be periodically dug out, weighted, proportionally sampled, oven-dried and analyzed for dry weight and organic and inorganic content (Bormann et al. 1969, 1974). Suspended particulate matter that passed over the weir was periodically sampled by using a 1-mm mesh net and passing a sample of water through the net and then filter-

B. Biogeochemistry

Figure II.B–13. Average annual volume-weighted concentrations (mg/liter) and amount of stream water in Watershed 6 during 1964 to 1980. The dashed regression lines have a probability for a larger F-value of <0.05.

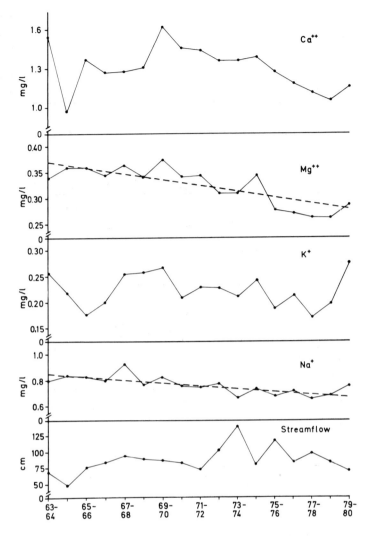

ing it through a 0.45-μm Millipore filter at 40 psi. Samples were then analyzed for dry weight and organic and inorganic content, expressed as concentrations per unit volume of water. These concentrations were used in combination with the record of hydrologic discharge to determine the export of materials over the weir (see Likens et al. 1977). These three components—basin, netted and filtered—are combined to give total particulate export from the ecosystem (*Table II.B–7*). Erosion and transport of particulate matter from our forested watersheds is relatively low (Bormann et al. 1974).

Removal of both inorganic and organic particulate matter from second-growth aggrading forests at Hubbard Brook is primarily an exponential function of flow rate (*Figure II.B–17*).

The shape of the concentration-flow rate curve is determined by the erodibility of the aggrading ecosystem and the fact that the capacity of running water to do work increases roughly as the square of its velocity (Bormann et al. 1969).

Because of the close relationship of particulate matter removal to flow rate, we find that occasional storms with high storm peaks do most of the work of removal. During a 50-month period, 77% of the total stream water was discharged at rates between 0 and 30 liters/s, but only 14% of the particulate matter was removed. At discharge rates of >85 liters/s, 3.7% of the water was discharged, while 52% of the total particulate matter was lost. Sixteen percent of the particulate matter was eroded by one flow ranging from 310 to 340 liters/s. This

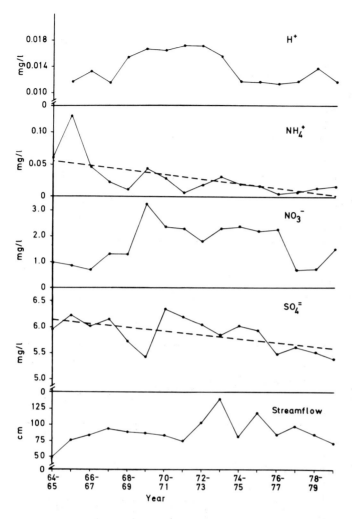

Figure II.B–14. Average annual volume-weighted concentrations (mg/liter) and amount of stream water in Watershed 6 during 1964 to 1980. The dashed regression lines have a probability for a larger F-value of <0.05.

Table II.B–7. Average Annual Particulate Matter Output in Kilograms of Oven-Dry Weight Organic and Inorganic Materials per Hectare for Watershed 6 from 1965 to 1973[a,b]

	Source of Output	Organic	Inorganic	Total
8-yr Average	Ponding Basin	7.44	18.30	25.74
	Net	0.42	0.01	0.43
	Filter	3.27	3.84	7.11
	Total	11.13	22.15	33.28

[a] Ponding basin output = particulate matter collected in a ponding basin of the weir upstream from the V-notch; Net output = suspended particulate, >1 mm, that passes over the V-notch of the weir; Filter output = suspended particulate matter, >0.45 μm <1 mm, that passes over the V-notch of the weir.
[b] From Likens et al. 1977.

flow carried only 0.2% of the total streamflow and comprised less than one hour's time during the entire 50-month period. Clearly, the annual variability in particulate matter output from the aggrading forest is primarily due to the occurrence of large storms (Bormann and Likens 1979).

Input-Output Budgets

During 1963 to 1980 there was an average annual net loss of 64 kg/ha of dissolved substances from the forested watersheds at Hubbard Brook (*Table II.B–6*). Of this, some 55% was as dissolved silica. The net loss averaged 85 Eq/ha-yr for cations and 197 Eq/ha-yr for anions. On a mass basis, dissolved silica and sulfate dominated the net change values, whereas on an

B. Biogeochemistry

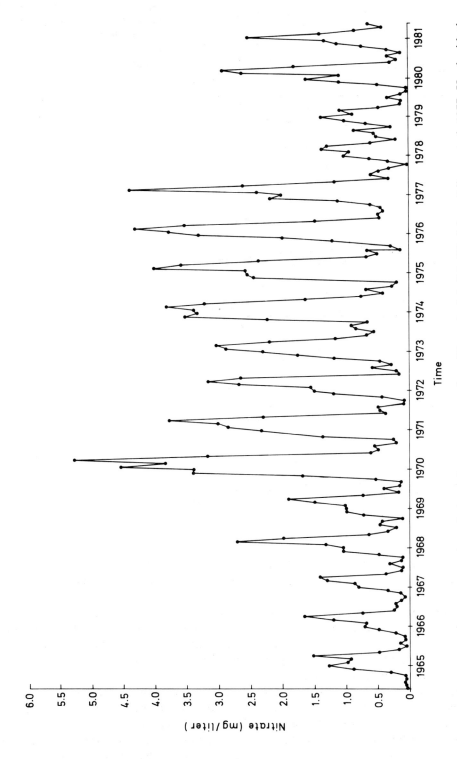

Figure II.B–15. Monthly concentration of nitrate in stream water from Watershed 6. (Modified from Likens et al. 1977. Used with the permission of the publisher and the authors.)

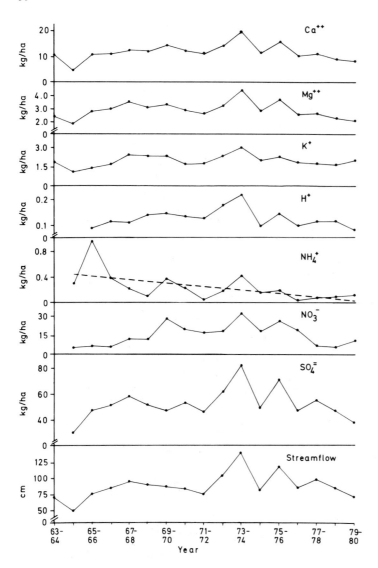

Figure II.B–16. Annual output in kg/ha and amount of stream water in Watershed 6 during 1963 to 1980. The dashed regression line has a probability for a larger F-value of <0.05.

ionic basis, hydrogen ion (+863 Eq/ha-yr) and calcium (−484 Eq/ha-yr) and sulfate (−325 Eq/ha-yr) dominated.

The annual net loss or gain values for some of the dissolved substances were quite variable, e.g. NO_3^-, Cl^-, K^+ and SO_4^{2-} (*Figure II.B–18*). During the period of study there was a net gain in nitrate in 12 years, a net loss in three years and in one year, the budget was approximately balanced (*Figure II.B–18*). Overall, there were 330 kg NO_3^-/ha input and 245 kg NO_3^-/ha output from Experimental Watershed 6 during 1964–65 through 1979–80.

Trace elements may serve as essential nutrients for plants and animals or as anthropogenic environmental pollutants. We measured the flux of Fe, Mn, Zn, Cu, Ni and Pb in the Experimental Watersheds at Hubbard Brook during 1975–79 (Siccama and Smith 1978; Smith and Siccama 1981). Two major ecologic findings were that during 1975 to 1982 the concentration of lead in bulk precipitation declined by more than 70% (*Figure II.B–19*) and that the forest floor layer of the soil acted as an exceedingly efficient sink for lead (*Table II.B–8*). The input of lead in bulk precipitation was estimated at 266 g/ha-yr during 1975–79; these annual average inputs exceeded outputs in stream water by 44:1. Thus, lead has been accumulating in this terrestrial ecosystem at a rate of about 27 mg/m²-yr. In 1979, the total lead content in the forest floor was approximately 8.6 kg/ha. These

B. Biogeochemistry

data clearly illustrate the effective filtering capacity by the terrestrial ecosystem and its role in regulating drainage inputs to associated aquatic ecosystems.

Regulation of Flux Rates

To this point, we have discussed the input-output relationships of our monitored watershed-ecosystems, which we use heavily to compute inputs into the Mirror Lake ecosystem. Outputs have been discussed in terms of amounts of water, particulate matter and dissolved substances transported out of the watersheds, with little mention of how concentrations and amounts are regulated by the terrestrial ecosystem. In fact, we may consider two steps in regulation that ultimately determine the timing, quality and quantity of flux to the lake-ecosystem. These are (1) regulation (biotic and abiotic) occurring in the forest-ecosystem that determines the nature of drainage that reaches stream-ecosystems or ground water that may reach the lake directly, and (2) regulation by the stream-ecosystem, which further modifies outputs of the forest-ecosystem before discharge into the lake.

Regulation is a product of biotic and abiotic features of the ecosystem and is a function of the developmental state of the system. Regulatory powers of the ecosystem can be greatly changed by disturbance of either the forest- or stream-ecosystem, and as a consequence disturbance of the watershed can have a major effect on input into and activities within the lake-ecosystem. We discuss some of these effects in *Chapter IX*.

Forest-Ecosystems—Biotic Regulation

We have discussed the topic of biotic regulation of outputs by the forested ecosystem in our volume, *Pattern and Process in a Forested Ecosystem* (Bormann and Likens 1979). In summary, we proposed a model of forest development (to be discussed more fully in *Chapter IX*) that envisages four phases of development: Reorganization, Aggradation, Transition and Steady State following the disturbance of the forest-ecosystem by, for example, clear cutting or hurricanes.

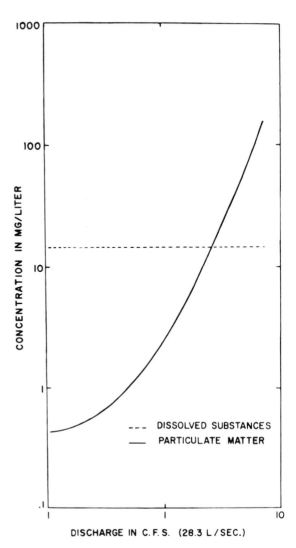

Figure II.B–17. General relationship between the concentration of dissolved substances and particulate matter and streamflow in the Hubbard Brook Experimental Forest. (After Bormann et al. 1969, *Bioscience* 19(7):608, Fig. 8. Copyright © 1969 by The American Institute of Biological Sciences. Used with permission.)

The greatest regulation over ecosystem outputs occurs when the forest is in the Aggradation Phase. The aggrading ecosystem has a remarkable effect on the chemistry of water moving through it, and stream water leaving the terrestrial ecosystem is chemically very different from precipitation entering it (*Table II.B–6*). The output data in the preceding pages are from aggrading second-growth forested eco-

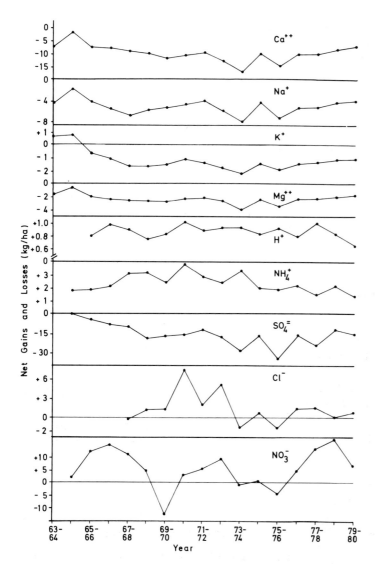

Figure II.B–18. Annual net gains and losses of dissolved substances (bulk precipitation inputs minus stream-water outputs) for Watershed 6 of the Hubbard Brook Experimental Forest during 1963 to 1980.

Table II.B–8. Estimated Input and Output Budgets for Lead in Forested Experimental Watershed 6 of the Hubbard Brook Experimental Forest (Values in g/ha-yr)[a]

| Watershed | Year | Bulk Precipitation Input | Output | | | Net Gain in Soil |
			Unfiltered Stream Water	Particulate Matter	Total	
6	1975	318	6.2	0.43	6.6	+311
	1976	373	1.6	0.72	2.3	+371
	1977	219	5.3	1.80	7.1	+212
	1978	156	6.7	1.68	8.4	+148
mean ± $s_{\bar{x}}$	1975 to 1978	266 ± 48.7	4.9 ± 1.15	1.2 ± 0.34	6.1 ± 1.32	260 ± 49.8

[a] Modified from Smith and Siccama 1981.

B. Biogeochemistry

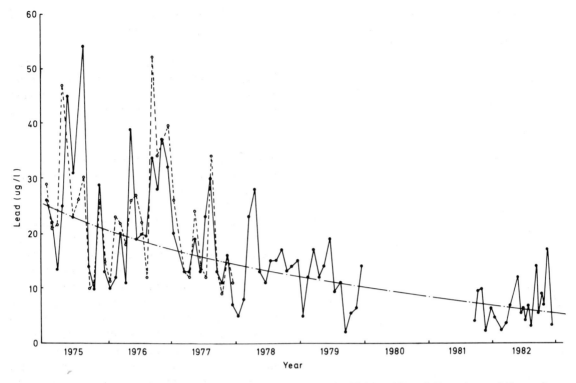

Figure II.B–19. Concentration of lead in bulk precipitation at the Hubbard Brook Experimental Forest from 1975 through 1982. The regression line, $Y = ae^{-bx}$, has coefficients for a of 25.53 and for b of -0.01589, where $Y = \mu$g Pb/liter and $x =$ months. The dashed line is for a collector near the ridge top (750 m), and the solid line is for a collector at 490-m elevation in the south-facing watersheds. The break in the data occurred because of an interruption in funding for the study. (From Likens 1983; data provided by T. Siccama.)

systems at the HBEF and reflect the extensive regulation these systems have over the outputs of water, particulate matter and dissolved substances. As a consequence, the amount of water draining from those ecosystems is minimal, clear, low in nutrients and generally considered of *high quality*. Because of regulation, dissolved substance concentrations are relatively constant (*Figure II.B*–17) despite large variations in volume of streamflow. This relationship makes output of dissolved substances highly predictable based on annual streamflow alone and represents one of the major findings of the Hubbard Brook Ecosystem Study.

Living biomass provides an important regulating component. For example, the ability of the dominant plants of the ecosystem to utilize solar energy to convert liquid water to vapor through transpiration results in the conservation of nutrients because water lost through evapotranspiration cannot transport dissolved or particulate matter out of the terrestrial ecosystem. Also, the reduction in streamflow has the direct effect of regulating the timing and amount of nutrient output in the aggrading, forested ecosystem as concentrations of dissolved substances in drainage water are highly predictable and strongly related to total flow. As water percolates through the canopy and soil zone, its chemistry is altered. Phosphorus and DOC provide an example (*Figure II.B*–20). Both phosphorus and DOC concentrations in soil solution increase dramatically in the forest floor, but rapidly decrease to concentrations (~3 mg C/ liter) in the mineral soil, which are similar to stream-water concentrations (Wood 1980; McDowell 1982). The P and DOC from forest floor soil solution are rapidly adsorbed by the upper B-horizon of the mineral soil, which thereby provides a powerful regulating mechanism for stream or drainage water chemistry and watershed-ecosystem outputs. Subsurface drainage

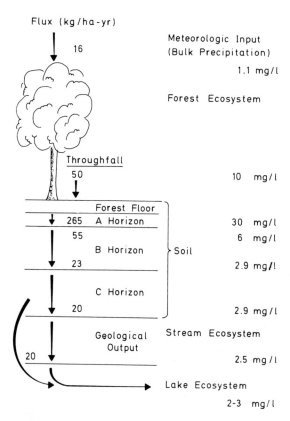

Figure II.B–20. Flux of dissolved organic carbon through the forest-ecosystem. (Modified from McDowell 1982.)

water contains negligible amounts of particulate matter because of the filtering effect of the soil.

Factors regulating concentration in drainage water are only partially understood, but both biotic and abiotic functions, in the canopy and in the soil, play a role. The net result is a highly predictable chemistry of drainage water. That these relationships are regulated (biotic and abiotic) by the aggrading forested ecosystem is amply shown by experimental work where these predictable relationships are completely altered by severe disturbance (e.g. clear cutting) of the aggrading forest (Bormann and Likens 1979).

Particulate matter losses are a nonlinear, direct function of discharge rates (*Figure II.B*–17). In the aggrading ecosystem, transpiration and interception can damp discharge rates by lowering the quantity of water stored within the system. Thus, a heavy rain when the storage volume of the system is relatively empty will produce relatively little flow and modest discharge rates so long as storage capacity is unsatisfied. Transpiration tends to keep hydrologic storage at a minimum during the growing season, hence streamflow during summer months is typically low even though precipitation is evenly distributed throughout the year.

Particulate matter losses are a function not only of discharge rate but also of erodibility or the capacity of the ecosystem to resist erosion. Although geologic and other abiotic factors are important in determining the erodibility of an ecosystem, living biota and organic debris exert considerable control over this phenomenon. Biotic control of erodibility in the aggrading forest is achieved by: (1) organic matter enhancement of infiltration and percolation, which converts surface flow to subsurface flow (surface flow does the work of particulate matter erosion); (2) the binding together of soil and organic debris in drainage channels and along stream banks by roots; (3) protection (interception) of mineral soil from direct action of falling raindrops by canopy and forest floor layers; (4) organic-debris dams in stream channels, which tend to regulate streamflow and retain particulate matter within the ecosystem; and (5) hardwood leaves, which form a protective shield against erosion of exposed stream banks (Bormann et al. 1969; Fisher 1970; Bormann and Likens 1979).

As a consequence of biotic and abiotic controls, aggrading watershed-ecosystems at Hubbard Brook have extremely low rates of erosion, 3.3 ± 1.3 t/km^2-yr, despite the fact that these forests occur on rather steep slopes, and are subject to more than 130 cm of precipitation per year.

Stream-Ecosystems— The Land-Water Link

More than 20 streams of varying size drain into Hubbard Brook by a dendritic drainage pattern (*Figure II.A*–1). The total length of stream channels in the Valley is approximately 75 km or 2.4 km/km^2. Because of the shallow soil and high infiltration capacity, large storms, e.g. 100 mm or more, quickly fill the stream channels and 60 to 80% of the precipitation may pass through the stream as storm flow. In a record storm the streamflow from an 11.8-ha watershed yielded maximum flows of 51 liters/s-ha, and the main Hubbard Brook draining 3,076 ha yielded 45 liters/s-ha.

Although stream channels occupy only 1 to 2% of the land area in the HBEF, they (particularly headwater streams) play a crucial role in the linkages between terrestrial and lake-ecosystems.

Headwater streams (first- to third-order streams) constitute 85% of the total stream length in the U.S. (Leopold et al. 1964), and biogeochemical events occurring in these headwaters often determine the chemical milieu of large river systems (e.g. Gibbs 1967), and as a result affect the input into lakes. Outputs of water, particulate matter and dissolved substances from the forest-ecosystem are inputs to stream-ecosystems, where they are transported downstream (to the lake in this discussion). Streams are not merely passive conduits, but are active ecosystems exercising some degree of biotic and abiotic regulation over the timing, quantity and quality of water, particulate matter and dissolved substances moving downstream (Likens 1984). We have extensively studied these important land-water links (Bormann et al. 1969, 1974; Bilby and Likens 1979, 1980; Meyer and Likens 1979; Hall et al. 1980; Johnson et al. 1981; Meyer et al. 1981).

Headwater streams draining in the Valley are small but mostly perennial. Flow ranges from zero for some streams during occasional summer droughts to hundreds of m^3/ha-day during snowmelt and storm events.

From June to October the streams are heavily shaded by the forest canopy. As a result, daily water temperatures do not vary more than a degree or two Celsius. In winter the streams are covered by ice and snow. The annual temperature range in these streams is from 0 to approximately 18°C. The stream water is normally saturated or supersaturated with dissolved oxygen.

The streambed is covered by organic debris, fine sand, gravel, cobbles and small boulders; there are occasional bedrock exposures in the stream channel. Numerous dams of organic debris occur throughout the length of the stream (Bilby and Likens 1980). Therefore, the stream has a stair-step appearance of alternating "waterfalls" and pools.

Organic matter may be stored in streams, primarily in organic debris dams (*Table II.B–9*), but at Hubbard Brook the streams have a relatively small capacity, for storage of organic matter, and the amount is thought to be in Steady State from year to year (Bormann et al. 1969, 1974; Fisher and Likens 1973).

An organic debris dam is an accumulation of organic matter in a stream that obstructs the flow of water. Sediments are trapped both upstream and downstream of dams. Dams also reduce the potential for stream water to transport sediments by causing a rapid dissipation of kinetic energy as water flows over the dam (Heede 1972a,b; Bormann et al. 1969, 1974; Fisher and Likens 1973). Thus, any change in the density or efficiency of organic debris dams can greatly alter export relationships for stream-ecosystems (Bilby and Likens 1980). Likens and Bilby (1982) have proposed that such changes in export can occur over time as vegetational and hydrologic conditions change following disturbance.

Owing to organic debris dams and other features of stream-ecosystems, streams can significantly alter or process inputs from the forest-ecosystem. Headwater stream-ecosystems draining forests are primarily heterotrophic, deriving a very large fraction of their total energy input from allochthonous (terrestrial) sources. Organic carbon budgets calculated for Bear

Table II.B–9. Distribution of Organic Debris Dams, Organic Matter and the Proportion of the Total Standing Stock of Organic Matter Contained in Organic Dams in First-, Second- and Third-order Streams in the Hubbard Brook Watershed[a]

	Stream Order		
	1st	2nd	3rd
Number of dams per 100 m of streambed	33.5	13.7	2.5
Organic matter per m^2 of streambed (kg)	3.52	2.58	1.64[a]
% of total organic matter in dams	74.5%	58.4%	20.0%
% of streambed area covered by dams	8.1%	4.9%	1.2%

[a] From Bilby and Likens 1980.

Brook, a tributary to Hubbard Brook (Fisher and Likens 1973) and a stream in a coniferous ecosystem (Sedell et al. 1973) have shown that more than 30% of the organic matter (energy) entering the stream was respired to CO_2 and the remainder was transported downstream.

Phosphorus and nitrogen, which are chemically reactive as well as limiting nutrients for plant growth in Mirror Lake (*Chapter V.B.2*), may be retained or exported from stream-ecosystems in varying amounts depending upon environmental conditions (Meyer et al. 1981). Meyer and Likens (1979) showed that the export of phosphorus from Bear Brook varied from year to year and was related to amount and timing of peak streamflow, because the majority of the phosphorus was transported as particulate P. During years with excessive stormpeaks, exports of P tend to exceed inputs to the stream-ecosystem, whereas in average years with few and small peaks, inputs may be stored or balanced by outputs. Our data for 1968–69 suggest that inputs of nitrogen may exceed outputs from the stream-ecosystem by small amounts <10% (Meyer et al. 1981). These net gains have been interpreted as evidence for temporary storage of N in the stream-ecosystem. However, denitrification losses were not evaluated carefully and relationships may change from year to year depending on hydrologic patterns.

Matter entering stream-ecosystems may be transformed by physical, chemical and biologic processes (Fisher and Likens 1973; Sedell et al. 1973; Boling et al. 1975; Meyer et al. 1981). For example, nutrients that enter the stream-ecosystem incorporated in particulate matter may leave the stream as dissolved substances or as a gas. Transformations that affect the rate of storage and utilization within the stream-ecosystem as well as the rate of transport can affect quality and quantity of inputs to downstream ecosystems.

Finally, water chemistry varies with stream length or stream order in the Hubbard Brook Valley. For example, the pH of headwater tributaries may be 4.0 to 4.5 (especially during spring) and then increase to pH 6.0 and 6.5 on the main Hubbard Brook (Johnson et al. 1981). pH changes are accompanied by changes in concentrations of other ions such as aluminum and calcium. Some of the chemical changes are thought to be due to additions of ground water, which has had variable residence times in the soil, but other changes are due to biotic and abiotic activities within the stream itself.

Acid Precipitation

We must consider a final point about regulation of inputs to the lake. The rain and snow falling on the Hubbard Brook Valley are highly acidic. The volume-weighted annual average pH is 4.15 (*Table II.B–5*) and the lowest value recorded for a single rain event is 2.85. The discovery that precipitation was so acid and was distributed throughout the northeastern U.S. has been one of the more interesting and ecologically significant findings of the Hubbard Brook Ecosystem Study (Likens et al. 1972; Likens and Bormann 1974a; Cogbill and Likens 1974).

Sulfur and nitrogen emissions to the atmosphere, primarily from the combustion of fossil fuels, are converted to sulfuric and nitric acids, which lower the pH of rain and snow well below the ambient levels that would be expected without these pollution inputs (see Likens and Bormann 1974a; Likens et al. 1979). Currently sulfuric acid contributes some 55 to 60% and nitric acid some 30 to 35% of the total acid equivalents in precipitation at Hubbard Brook. The contribution of nitric acid has doubled since our initial measurements in 1964–65 (see Likens et al. 1977). The increased proportion of nitric acid represents an ecologic dilemma because nitrogen may have a fertilizing effect on terrestrial and aquatic ecosystems at Hubbard Brook, but the accompanying hydrogen ion (HNO_3) may have direct and indirect (e.g. mobilizing Al^{n+}) toxic effects.

There is considerable controversy about historical trends in acid precipitation because data are few. The record of measurements at Hubbard Brook is the longest in North America. Whereas at some other sites there has been some increase or decrease in pH values with time, only small changes have been observed at Hubbard Brook (see Likens et al. 1979, 1984). Although the average annual pH of precipitation at Hubbard Brook has remained at approximately 4.1 since 1964 (*Figure II.B–9*), there was a significant ($p < 0.10$) decline in annual volume-weighted H^+ concentration after 1970–

71. Also the range of concentration in weekly samples has narrowed, and the distribution has shifted towaard higher concentrations (lower pH) during the past 16 yr; pH values for weekly bulk samples have not been above 4.8 since 1968. Because of the logarithmic nature of the pH scale, it is difficult to identify a ready source for a large increase in hydrogen ions that would be necessary to lower the *average* pH of rain and snow much below the current value of 4.1. For example, a doubling of hydrogen ions would be required to lower the pH from 4.1 to 3.8. Based on regression analysis, there has been a significant decrease in concentrations of Ca^{2+}, Mg^{2+}, K^+, NH_4^+ and SO_4^{2-} during this period (*Figures II.B–8* and 9).

Precipitation in summer is much more acid than in winter (*Figure II.B–7*). Ecologically, this finding is significant because summer is the more active growing season for organisms in both aquatic and terrestrial ecosystems. Also, cloud water may be three to five times more acid than rain water.

Snow accumulates throughout the winter and on melting in the spring yields a flush of acid water to headwater streams (*Figure II.B–12*; Johnson et al. 1981) and to lakes (Hultberg 1976). Because of differential melting of snow and ice, the first portions of melt water may be appreciably more acid than the remaining portions (Hornbeck et al. 1976; Johannesen et al. 1976).

The terrestrial ecosystem has a large capacity to regulate stream-water acidity. In contrast to the summer-winter seasonal pattern of acidity in precipitation (*Figure II.B–9*), stream-water acidity is highest during the spring snowmelt period (*Figure II.B–12*). Not only are the fluctuations damped as the water flows out of the terrestrial ecosystem, but the concentrations are much lower than those in precipitation (*Figure II.B–21*). This regulation of terrestrial outputs for aquatic ecosystems is obviously an important one.

Nevertheless, our studies of cationic denudation within the terrestrial ecosystem suggest that acids in precipitation provide 50 to 52% of the protons for the weathering reactions in the soil zone (Likens et al. 1977; Driscoll and Likens 1982). These acids in rain and snow are over and above those generated within the terrestrial ecosystem and, thus, are potentially available for acidification of drainage waters. In fact, we calculate that headwater streams at Hubbard Brook would have a pH of approximately 6.5 if it were not for acid deposition from the atmosphere (Driscoll and Likens 1982).

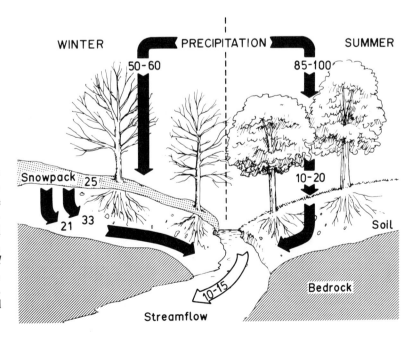

Figure II.B–21. General relationships for two seasonal periods at Hubbard Brook. The values are H-ion concentrations in μEq/liter. (Modified from Hornbeck et al. 1976, *Journal of Water, Air, and Soil Pollution* 7:364, Fig. 5. Copyright © 1977 by D. Reidel Publishing Co. Used with permission.)

Long-Term Biogeochemical Data

The biogeochemical data at Hubbard Brook represent the longest continuous record of precipitation chemistry and stream-water chemistry in North America. The immense value of such long-term data has only gradually revealed itself.

Understanding of the basic functions of both terrestrial and aquatic ecosystems requires adequate knowledge of their connection with global biogeochemical cycles. These connections are the input, output and net change relations portrayed earlier in this chapter. Our long-term record clearly illustrates the danger to investigators of assuming that two or three years of data or sporadic measurements are adequate to characterize the important links of local ecosystems to global cycles. For example, the annual volume-weighted concentrations of total inorganic nitrogen in precipitation and stream water have varied dramatically during the period of our study (*Figure II.B*–22). From 1964–65 to 1968–69, annual volume-weighted stream-water concentrations were consistently less than precipitation, but then concentration in stream water increased significantly during 1969–70 to 1976–77 and was similar to that in precipitation. After 1976–77, the pattern reverted to what it had been originally. Such changes have significant effects on input-output budgets, interpretation of ecologic concepts (e.g. Vitousek and Reiners 1975) and predictions of inputs for aquatic ecosystems.

Although at the start of the HBES air pollution was of minor concern, the chemical record has become an important national asset in our effort to understand the impact of air pollution on the structure and function of natural ecosystems (see Calvert 1983). The record of precipitation chemistry provides one measure of the effect of air pollution on input chemistry and an evaluation of the effectiveness of the national effort to reduce air pollution. The stream-water outputs and net change data provide some measure of the effect of pollution on ecosystem function. For example, our data show the important filtration effect of the forest-ecosystem on H^+ and Pb cycles, while our long-term data on cation output in stream water may indicate some diminishment of the neutralizing capacity of the terrestrial system. On a short-term basis, these trends are ambiguous.

It seems clear that these aspects of long-term

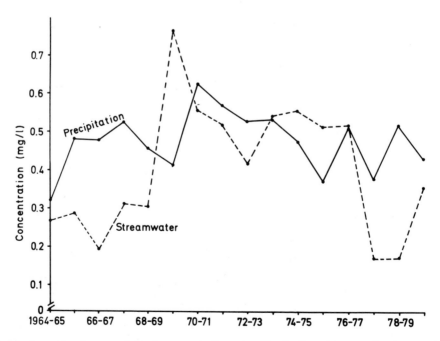

Figure II.B–22. Annual volume-weighted concentrations (mg/liter) of total inorganic nitrogen in precipitation and stream water in Watershed 6 during 1964–65 to 1979–80.

ecologic research are not only of major importance in developing questions or models of basic functions of ecosystems (see Likens 1983), but are even more important in helping to evaluate the consequences of economic growth on our basic life support systems.

Summary

The Mirror Lake ecosystem is located within the Hubbard Brook Valley. Data characterizing the continental climate, solar flux, air temperature, and wind speed and direction in the Valley are given. The structure and composition of the second-growth northern hardwood forest and the well-drained spodozols covering most of the Valley are described.

Hydrology. Annual precipitation ranges from 131 cm/yr at higher elevations to 122 cm/yr at Mirror Lake. Monthly precipitation averages approximately 11 cm. Evapotranspiration in the Experimental Watersheds is relatively constant at 49 cm/yr, while the remainder constitutes streamflow, almost all of which moves through the soil rather than over it; 30% of the annual precipitation falls as snow.

Precipitation chemistry is dominated by H^+ and SO_4^{2-} with an average pH of 4.1 and is under the strong influence of air pollution. Seasonal and annual patterns of dissolved chemicals covering 1963 to 1981 are presented. H^+ and SO_4^{2-} concentrations peak in the summer during maximum biologic activity. Ca^{2+}, Mg^{2+}, Cl^-, NH_4^+ and SO_4^{2-} show statistically significant declines in concentrations and input since 1963.

Stream Water–Dissolved Substances. Streamwater chemistry differs from precipitation chemistry. This difference is due to a variety of processes such as weathering, evapotranspiration, biomass accumulation and exchange processes within the ecosystem. K^+ and NO_3^- have higher concentrations during the dormant season.

Particulate Matter. Loss of particulate matter is strongly related to storm peaks. The export of phosphorus is highly discharge dependent because it is moved primarily in particulate form.

Budgets. Exports of Ca^{2+}, Na^+, K^+, Mg^{2+} and SO_4^{2-} consistently exceed meteorologic inputs in bulk precipitation. Exports of H^+, NH_4^+, P and Pb are less than inputs in bulk precipitation. NO_3^- and Cl^- fluctuate between net accumulation and loss, on an annual basis.

Dissolved substance concentrations in stream water draining aggrading forests remain relatively constant despite extremes in hydrologic output and extreme climatic variations. This relationship allows for general predictions of dissolved substance output based on the hydrologic record. Particulate output, on the other hand, is dependent on storm peaks and is relatively independent of quantity of hydrologic output.

Regulation of Flux. Output characteristics of water and chemicals are under strong regulation by both forest- and interconnected stream-ecosystems. Regulation is a function of the developmental condition of the ecosystem and can be appreciably changed by disturbance of the ecosystem.

Long-term Data. Short-term biogeochemical data may provide ambiguous or misleading information about the function of natural ecosystems or about temporal trends in measures of pollution.

CHAPTER III

Mirror Lake and Its Watershed

A. Physiographic Setting and Geologic Origin of Mirror Lake

THOMAS C. WINTER

Understanding the physiographic setting of a lake is basic to understanding the physical, chemical and biologic processes that act within that setting. For example, knowledge of the movement of water overland, in streams, and through the subsurface near a lake is of primary importance in understanding processes acting upon and within the lake. The two physiographic characteristics of the land that have a major part in controlling the movement of water are geology and topography.

Surface runoff is dependent on infiltration, which is related to soil characteristics and parent rock, as well as on the slope of the land, development of drainage patterns and channel characteristics of streams. Movement of ground water is controlled to a large extent by hydraulic conductivity of the geologic units through which it flows, as well as by hydraulic gradients within the ground-water system. Interchanges of precipitation and evaporation with the lake are affected by the wind regime in the area, which is also affected by topography (see *Chapter II*).

Location and General Physiographic Setting

Mirror Lake is located near the mouth of the Hubbard Brook Valley (*Figure III.A–1*). The lake lies on the north side of Hubbard Brook, which trends east-west; therefore, most of the drainage basin of the lake faces south-southeast (*Figure III.A–2*). The lake is 15 ha in area, and it has a maximum depth of 11 m and an average depth of 5.75 m. Outflow from the lake drains into Hubbard Brook, which is a tributary of the Pemigewasset River.

The drainage basin of Mirror Lake is characterized by high knobs and ridges, and by steep land slopes. The lake lies at an altitude of about 213 m. The highest elevation on the watershed of Mirror Lake is approximately 469 m. In contrast, the watershed of Hubbard Brook, west of the Mirror Lake area, reaches altitudes greater than 1,006 m (see *Chapter II*).

Bedrock underlying Mirror Lake consists of crystalline rocks of Early Devonian age, which date to about 415 million yr B.P. Unconsolidated glacial deposits (drift) overlie the bedrock in much of the topographically lower parts of the area. The drift was deposited within the last 100,000 yr. Mirror Lake was formed approximately 14,000 yr B.P. (see *Chapter VII.A*) when an ice block buried by glacial drift melted, leaving a depression in the land surface, which then filled with water.

Geology

Bedrock

Bedrock in the Hubbard Brook area consists of igneous and metamorphic crystalline rocks of earliest Devonian age (Billings 1956). The rocks consist principally of the Littleton Formation and Kinsman quartz monzonite (see *Chapter II*).

A. Physiographic Setting and Geologic Origin of Mirror Lake

Figure III.A-1. Location of Mirror Lake and the Hubbard Brook Valley.

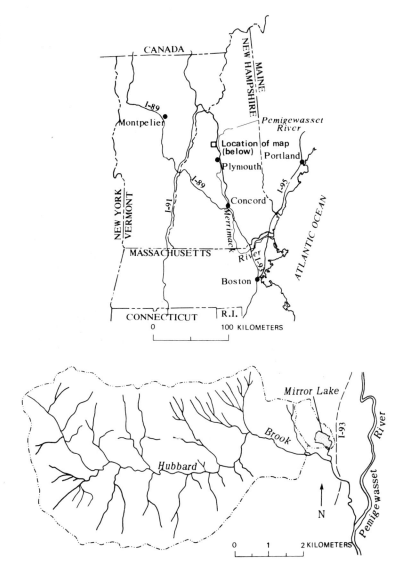

The rocks originated about 415 million yr B.P. when sediments consisting of clay, silt and lesser amounts of sand were deposited in a shallow marine environment that was probably intracontinental. By 410 million yr B.P., mountain-building processes had begun that caused the sediments to be metamorphosed (Littleton Formation) and intruded by the igneous Kinsman quartz monzonite. Deformation of the rocks, principally folding, also took place at this time. During the Carboniferous Period (approximately 330 million yr B.P.) further mountain-building stresses resulted in considerable fracturing and faulting, and possibly additional metamorphism of the rocks. Still later, during the early Mesozoic Era, approximately 180 to 190 million yr B.P., the rocks were again subjected to fracturing and faulting associated with further uplift of the area.

During and after the processes described above, the area was subjected to long periods of erosion. The only significant evidence of widespread depositional processes in the Mirror Lake area since the Mesozoic Era is the presence of glacial drift.

Although the bedrock geology of the Hubbard Brook Valley has been determined in general (*Chapter II*), more detailed geologic studies specific to Mirror Lake were needed to determine the geology as well as the hydrology

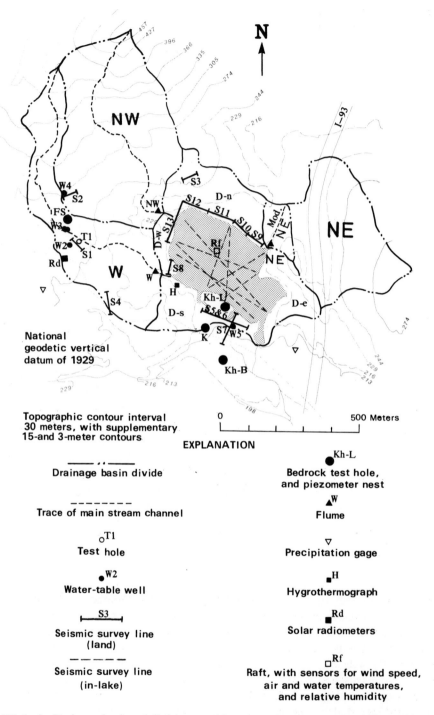

Figure III.A–2. Drainage-basin subdivisions and location of test and instrumentation sites.

of the lake and its drainage basin. The type of information needed was the configuration of the bedrock surface, and mineralogic and hydraulic properties of the bedrock, including distribution and hydraulic properties of the fractures. Because drift overlies the bedrock, these data were obtained by test drilling and geophysical methods.

As part of the Mirror Lake ecosystem studies, three holes were drilled through the drift and at least 30 m into the bedrock; a fourth hole was drilled 61 m into the bedrock. Geologic descriptions of the samples of drilled rock and geophysical data were obtained for each of the holes. Geophysical surveys using seismic refraction were done at the land surface around most of the perimeter of Mirror Lake and at a few locations higher in the drainage basin. A geophysical survey using seismic reflection was made in Mirror Lake (sites are shown in *Figure III.A–2*).

Data on the configuration of the bedrock surface obtained from these studies show that the south-central part of Mirror Lake overlies a saddle in the bedrock surface (*Figure III.A–3*). A south arm of a bedrock valley begins at this saddle and descends to the north. A west arm of this same valley begins beneath the lower part of subbasin NW and joins the south arm, north of Mirror Lake. A south-trending bedrock valley also begins at the saddle underlying the south-central part of Mirror Lake, and descends toward Hubbard Brook. The southern extension of this valley is not defined, and it is possible that Hubbard Brook does not flow directly on bedrock in the reach where the bedrock valley crosses the brook.

The bedrock surface rises in altitude to the east and to the west from the bedrock saddle. Bedrock outcrops occur on the east shore of Mirror Lake; highway I-93 cuts through bedrock east of the lake.

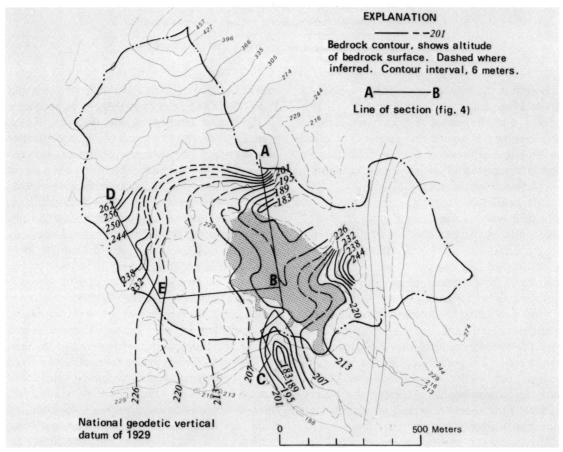

Figure III.A–3. Bedrock topography.

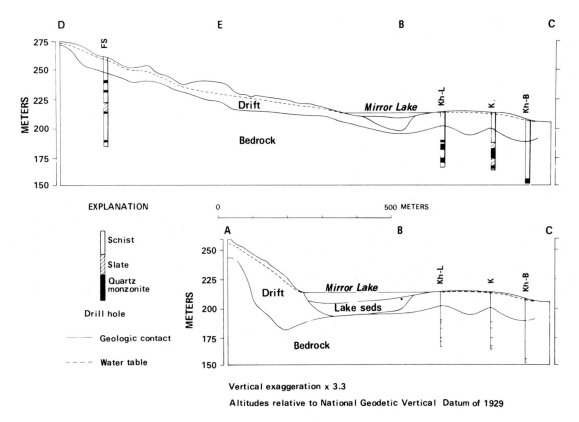

Figure III.A–4. Geologic sections (see *Figure III.A–3* for locations).

Based on the four holes drilled into bedrock, schist of the Littleton Formation consists principally of biotite, but it also includes considerable amounts of quartz and garnet. The quartz monzonite is white to slightly greenish in color, and is composed of nearly equal amounts of microcline, plagioclase and quartz.

Distribution of rock types penetrated in the four drill holes is shown in *Figure III.A–4*. In hole Kh-B, approximately 27 m of schist overlie quartz monzonite. In hole K, only the upper 12 m of the bedrock consist of schist. In the remaining 18 m of the hole, black, gray and greenish-black slate that is part of the Littleton Formation is intruded by quartz-monzonite sheets. Bedrock in hole Kh-L consists mostly of schist that is intruded by several sheets of quartz monzonite. Hole FS penetrates mostly schist, but an 8-m thick section of greenish-gray slate occurs approximately 26 m below the bedrock surface. Four sheets of quartz monzonite, each less than 2 m thick, occur in the hole.

Movement of ground water in crystalline rocks commonly is dependent on fractures in the rocks. Moke (1946) and Fowler-Billings and Page (1942) indicated that there is no evidence of significant faulting in the Hubbard Brook area. Bradley and Cushman (1956) state further that openings of the fractures present are not large, that the fractures extend to only a few meters below the bedrock surface, and that their size and numbers decrease with depth.

Because regional studies commonly do not provide information adequate for local site-specific purposes, the four holes drilled into bedrock were intended primarily to assess the significance of ground-water movement through fractures near Mirror Lake. The four test holes were constructed into bedrock wells by casing and cementing the part that penetrated drift and leaving the hole open as it penetrated through the bedrock. Geophysical data, obtained from the four bedrock holes, including televiewer logs of holes Kh-L and K, clearly show fractures at several depths throughout the intervals drilled (as much as 61 m in hole FS), and large numbers of fractures, particularly in hole Kh-B. Further, the water level in hole Kh-B is consis-

A. Physiographic Setting and Geologic Origin of Mirror Lake

tently above land surface, and water levels in all the bedrock wells fluctuate seasonally indicating dynamic ground-water movement through the bedrock at Mirror Lake.

Glacial Drift

Because Mirror Lake is located principally in glacial drift, an understanding of ground-water movement near the lake requires knowledge of hydrogeologic characteristics of drift. The geometry of the drift needs to be known so boundaries of ground-water flow systems can be ascertained. Type and texture of the drift, including distribution of the different units, need to be known, so hydraulic properties, such as hydraulic conductivity and storage coefficient, can be determined.

Prior to 1978, little was known of glacial drift characteristics near Mirror Lake. Goldthwait et al. (1951) indicated that a gravel deposit on the southeast side of Mirror Lake is related to delta or terrace deposits associated with the Pemigewasset River. However, nothing was known of the type, texture or thickness of drift deposits in the area. Beginning in 1978, drilling and geophysical data as part of the Mirror Lake ecosystem studies resulted in considerable information on hydrogeologic characteristics of drift near the lake.

Geometry

The data indicate that drift is thickest (>30 m) along the base of the bedrock knob on the northwest side of Mirror Lake (*Figures III.A–4 and 5*). The data also indicate that the topographic ridge on the north side of area D-n (*Figure III.A–2*) may consist entirely of drift; however, further verification is necessary. The ridge probably is a moraine that filled the north-trending bedrock valley previously described. Drift as much as 24 m thick also fills the bedrock valley on the south side of Mirror Lake. This drift probably is a combination of moraine, valley-train and delta deposits associated with glaciation, and with subsequent fluvial pro-

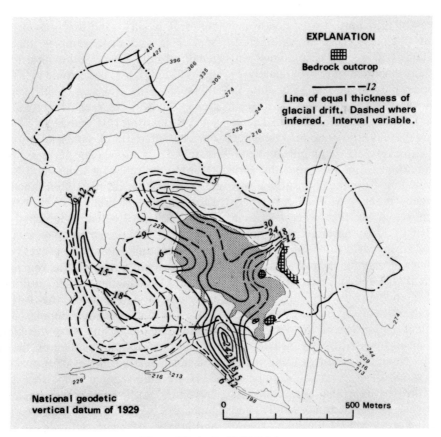

Figure III.A–5. Drift thickness.

cesses in the Hubbard Brook and Pemigewasset Valleys.

These data indicate that geologic materials along most of the south side of Mirror Lake, except near the outlet dam, consist of glacial drift. The thick (>18 m) section of drift southwest of Mirror Lake also probably is a morainal (including ice-contact) deposit. In much of the remainder of the Mirror Lake drainage basin, drift is generally between 6 and 12 m thick.

An important question concerning the relationship of Mirror Lake to the ground-water system is, What is the thickness and texture of drift underlying the lake sediments? For example, if there is little or no drift between the lake sediments and bedrock, the only possible route for ground water to pass beneath the lake would be through fractures in bedrock. On the other hand, if drift is thick beneath the lake sediments, it is possible that considerably more ground water could pass beneath the lake and discharge into Hubbard Brook.

Thickness of all materials beneath the water surface, including lake water, lake sediments and drift is shown in *Figure III.A–5*. To determine thickness of drift only, water depth and lake sediment thickness must be subtracted. Likens and Davis (1975) show a combined water and sediment thickness of approximately 23 m at the deepest part of the lake. The data in *Figure III.A–5* indicate a thickness of drift at this same point of between 18 and 24 m. Therefore, it appears there is little or no drift between the base of the thickest lake sediments and bedrock. It is important to mention that the geophysical survey done from the lake surface was successful in penetrating to bedrock only where organic sediments in the lake were less than approximately 6 m thick; therefore, the altitude of the bedrock surface had to be inferred from nearby geophysical data for deeper areas. (Areas of uncertainty are shown by dashed lines in *Figures III.A–3 and 5*).

Comparing the data of Davis and Ford (1982) on lake-sediment thickness with the data on total thickness of all material overlying bedrock throughout the remainder of Mirror Lake as obtained from geophysical surveys, it appears there is little drift between the lake sediments and bedrock east of the bedrock saddle, and also toward the west end of the lake, where stream W enters Mirror Lake. At the latter locality, the inlake geophysical survey shows a bedrock knob underlying the lake, which is marked by boulders on the map of lake sediments (*Figure III.A–6*). Probably a considerable thickness (perhaps as much as 10 m) of drift underlies the sediments on the northwest part of Mirror Lake and near the bedrock saddle on the south side.

Type and Texture

Throughout much of the drainage basin of Mirror Lake, the drift consists of silty, sandy till, and it contains numerous cobbles and boulders derived mostly from local crystalline bedrock. Deposits of ice-contact, stratified drift are scattered throughout the area. Between Mirror Lake and Hubbard Brook, the drift is a complex mixture of till, sand and gravel.

Based on drilling data, the drift at hole Kh-B is predominantly medium sand, but it also contains much coarse sand to coarse gravel. At hole K, the drift is largely silty, fine sand to medium gravel (probably till), and it contains scattered cobbles and boulders. At hole Kh-L, the drift is silty sand and gravel; silt was observed in the drift samples throughout the length of hole Kh-L. Although drift samples in holes K and Kh-L have some characteristics of till (such as the presence of silt), seismic velocities in the drift between Mirror Lake and Hubbard Brook (lines S5 and S6 in *Figure III.A–2*) are characteristic of sand and gravel (approximately 1,524 m/s). (The term *seismic* velocity refers to the rate at which sound is transmitted through the rocks; it is determined by seismic geophysical measurements.) However, the seismic velocity of 1,829 m/s along line S7 indicates that the drift here is a complex mixture of both till and stratified sand and gravel.

Throughout much of the remainder of the Mirror Lake drainage basin, drilling and seismic-velocity data indicate the drift is silty till; however, sand and gravel deposits are scattered throughout the basin. Near the lake, seismic-velocity information indicates that the north and northwest perimeter of the lake (lines S9–S13) are till; seismic velocities are 2,134 to 2,256 m/s. At line S8, however, the velocity is 1,524 m/s, which is more characteristic of sand and gravel. At higher altitudes in the drainage basin, drilling data at sites FS and W4 and seis-

Figure III.A-6. Bathymetry and surficial sediments in Mirror Lake. (From Moeller 1978b.)

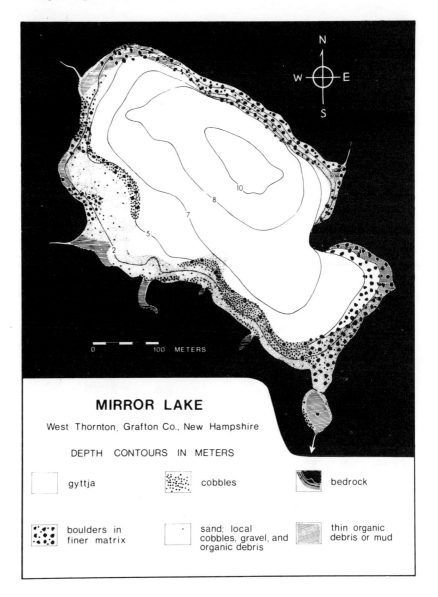

mic-velocity data along line S4 indicate the drift is till; but seismic-velocity data along line S2 indicate the drift is sandy; velocity is 1,524 m/s. At seismic line S3, the drift also has seismic velocities characteristic of sand and gravel (1,463 m/s). At test hole W2 and seismic line S1, the drift is clearly sand and gravel. It is interesting to note that many of the sandy deposits occur at an altitude of approximately 253 m.

As indicated by the above description of glacial deposits, at a few localities the drilling data and the geophysical data do not appear to be consistent. For example, the samples from test holes K and Kh-L indicate mostly till, but the seismic data from lines S5 and S6 indicate sand and gravel. The same is true of the area by test holes FS and W4 and seismic line S2. The apparent disagreement in texture is probably related to the indirect nature of geophysical data. For example, because the till in the Mirror Lake area is sandy, it is difficult to distinguish from seismic data alone between sandy till and silty sand and gravel. In addition, geologic samples from drilling are direct evidence from a very small part of an area, and seismic data are indirect evidence from a larger area that integrates all deposits in that area. Both drill samples and geophysical data are useful, but it will be necessary to do more intensive study at selected localities to fully understand the geologic deposits in the Mirror Lake drainage basin.

Glacial History and Origin of Lake Basin

During the ice age, glaciers moved south-southeast across New Hampshire. The ice-cover was continuous at the maximum glacial advance (Goldthwait et al. 1951) and is estimated to have covered nearby Mt. Washington (the highest point in New England at an altitude of 1,917 m) to a depth of at least 305 m. Although the general direction of ice movement was south, in the early stages of glaciation local topography probably caused ice to move down the Hubbard Brook Valley and coalesce with ice moving down the Pemigewasset Valley. Till that blankets much of New Hampshire was deposited by the advancing glacier, and the ridge on the north side of subbasin D-n probably was formed as a medial moraine between the two tongues of ice (*Figure III.A–7A*).

Ice-contact geomorphic features consisting largely of sand and gravel are characteristic deposits of melting ice sheets. As ice in mountain valleys melts, materials in the ice wash off and are deposited as stratified sand and gravel, in places intermixed with till, directly against the ice. This process probably accounts for the scattered sand and gravel deposits at high altitudes in the Mirror Lake drainage basin. The presence of sand and gravel in several places at an altitude of approximately 253 m (test sites W2, S1 and S3 in *Figure III.A–2*) indicates that the ice in the Hubbard Brook Valley might have persisted at about this level for a short time; the

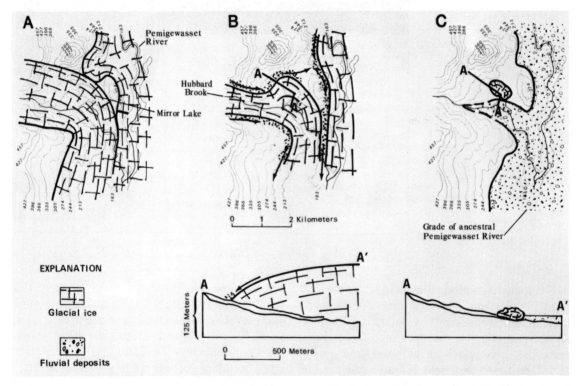

Figure III.A–7. Glacial history of the Mirror Lake area. **A.** Glaciers moving down the Hubbard Brook and Pemigewasset Valleys coalesced on the north side of Mirror Lake to form a medial moraine. **B.** As the ice lobes melted, material washed off the ice and the melt water flowing between the ice and valley walls deposited the sand and gravel at distinct levels on the valley side (valley-train deposits). **C.** As the glaciers receded, melt water carried and deposited sediments in the Pemigewasset Valley at several prominent levels. Side valleys like Hubbard Brook adjusted their grade to the mainstream valley. An ice block, which later melted to form Mirror Lake, existed in the Mirror Lake location at this time.

deposits may represent valley-train deposition, as depicted in *Figure III.A-7B*.

The low ridge between Mirror Lake and Hubbard Brook that contains some till and a large amount of sand and gravel may in part be a small lateral moraine; however, some of the geologic material is also a result of ice-contact deposition, as well as terrace and delta deposits associated with fluvial conditions in the Pemigewasset Valley.

Well-defined terraces in the Pemigewasset Valley indicate that one grade (flood plain) of the ancestral Pemigewasset River had an altitude of approximately 213 m in the Mirror Lake area. At that time, probably about 15,000 yr B.P., melt water from the retreating ice sheet discharged into glacial Lake Merrimack, the north end of which was approximately 20 km downstream, near present-day Plymouth, New Hampshire. Glacial Lake Merrimack extended another 100 km to the south, filling the valley from Plymouth to Manchester, New Hampshire. (Large glacial lakes similar to glacial Lake Merrimack occupied major river valleys throughout much of New England at this time.) Hubbard Brook, which was tributary to the glacial Pemigewasset River, adjusted its grade to that of the main stream by depositing deltaic and outwash materials (*Figure III.A-7C*). These deposits are mostly sand and gravel, but also include silt- and clay-sized material. An example of the texture of these terrace-type deposits is the well-sorted sand and gravel at hole Kh-B, which is located on a small terrace of Hubbard Brook.

At the time Hubbard Brook was adjusting to the grade of the glacial Pemigewasset River, an ice block must have existed at the present location of Mirror Lake (*Figure III.A-7C*), as evidenced by the small amount of drift underlying Mirror Lake sediments. Without the ice block, the entire area would have been filled with sediment to an altitude of approximately 213 m. The kettle left after the ice block melted became the basin of Mirror Lake.

Topography and Drainage Basin Characteristics

Drainage basin characteristics of the Mirror Lake area result from the geologic processes discussed previously, as well as from subsequent modification of the landscape by erosion and development of vegetation and soils. As mentioned earlier, knowledge of physiographic characteristics of a drainage basin is essential to understanding both surface-water and groundwater movement through the basin. The following discussion provides some quantitative measures of the drainage basin of Mirror Lake that are used in studies of the hydrology of the lake (see *Chapter IV.D*).

Mirror Lake is 15 ha in area and the area of its drainage basin (excluding the lake) was approximately 103 ha prior to the construction of highway I-93; thus, the ratio of drainage area to lake area was 6.9. As part of the construction of highway I-93, much of the surface drainage east of Mirror Lake was isolated from the lake by a small earthen dam (Likens 1972c). As a result, the area of subbasin NE was decreased from 20 ha to 2.5 ha. The present total area of drainage to Mirror Lake is 85 ha, and the modified ratio of drainage area to lake area is 5.7. Subdivisions and designations of the subbasins associated with Mirror Lake are shown in *Figure III.A-2*.

Morphometric characteristics of Mirror Lake itself are given in *Table III.A-1* and *Figure III.A-6*. Lake-depth contours show that the lakebed is asymmetric in shape; its deepest part (11 m) is much closer to the north shore than to the south shore. The north side of the lakebed also contains numerous boulders (*Figure III.A-6*). Most of the more gently sloping lakebed on the south side of the lake is sandy.

Three small streams flow into Mirror Lake (*Figure III.A-2*). The two streams, NW and W, that enter the west side of the lake drain the south slopes of a bedrock knob that reaches an altitude of 469 m north of Mirror Lake. The third stream enters the northeast side of the lake and drains modified subbasin NE. Quantitative descriptors of subbasins NW and W, as well as of subbasin NE, for before and after construction of the earthen dam along highway I-93, are given in *Table III.A-2*.

Subbasins W and NW have considerably different quantitative topographic characteristics. These characteristics are useful for quantitative comparisons of basins, and they are commonly used in hydrologic studies, such as rainfall-runoff analysis (Betson et al. 1978). Subbasin W is approximately 11% longer and about 37% nar-

Table III.A-1. Morphometric Characteristics of Mirror Lake, New Hampshire

43° 56.5′ N, 71° 41.5′ W

Maximum Effective Length	610 m	Average Depth	5.75 m
Maximum Effective Width	370 m	Length of Shoreline	2,247 m
Area	15.0 ha	Shore Development	1.64
Maximum Depth	11.0 m	Volume Development	1.57
		Relative Depth	2.5%

Depth (m)	Area (m² × 10⁴)	% of Total	Stratum (m)	Volume (m³ × 10³)	Volume (% of Total)
0	15.0	100.0	0–1	142.9	16.6
1	13.6	90.5	1–2	130.0	15.1
2	12.4	82.9	2–3	119.5	13.9
3	11.5	76.5	3–4	110.0	12.8
4	10.5	70.1	4–5	101.8	11.8
5	9.86	65.7	5–6	94.1	10.9
6	8.96	59.7	6–7	78.5	9.1
7	6.79	45.2	7–8	48.9	5.7
8	3.21	21.4	8–9	23.6	2.7
9	1.61	10.7	9–10	10.7	1.2
10	0.609	4.06	10–11	2.0	0.2
11	0	0			
			Total	862.0	100.0

Table III.A-2. Topographic Characteristics of the Mirror Lake Drainage Basin

Characteristic	Total	NW	W	NE Total	NE Modified	D-n	D-w	D-s	D-e
BASIN									
Area (A), ha	103	34.6	24.0	20.0	2.6	9.3	1.9	6.1	4.6
Basin Length (L_B), m		911	1,012	649	162	198	61	122	110
Basin Width (W_B), m[a]		378	238	308	159	442	259	488	424
Basin Perimeter (P_B), m		2,393	3,024	2,103	689	1,542	668	1,768	1,308
Basin Land Slope (S_{B1})		0.2114	0.1241	0.1202	0.0733	0.1300	0.1465	0.0533	0.0934
Basin Diameter (B_D), m		457	594	354	149	293	137	265	216
Basin Shape (SH_{B1})		2.40	4.27	2.11	1.02	2.21	4.25	4.03	3.89
Compactness Ratio (SH_{B4})		1.15	1.74	1.33	1.22	1.46	1.50	2.05	1.72
STREAM									
Main Channel Length (L_{C2}), m		1,070	1,314	741	171				
Main Channel Slope (S_{C1})		0.2245	0.1290	0.1213	0.0119				
Sinuosity Ratio (P)		1.17	1.30	1.14	1.06				
LAKE									
Shoreline Length (L_S), m						488	287	896	576
Shoreline Sinuosity (P_S)						1.10	1.11	1.84	1.36

[a] For direct runoff areas: W_D = Straight line distance from stream mouth to stream mouth
(D-n, D-w, D-s, D-e) $L_B = A/W_D$
                       $SH_{B1} = W_D^2/A$

Shoreline Length (L_S) = Actual length of shoreline from stream mouth to stream mouth
Shoreline Sinuosity (P_S) = L_S/W_D

A. Physiographic Setting and Geologic Origin of Mirror Lake

Table III.A–2. *Continued*

	Definitions of Topographic Characteristics
Basin Length (L_B)	Straight-line distance from outlet to the point on the basin divide used to determine main channel length, L_{C2}.
Basin Width (W_B)	Average width of the basin determined by dividing the area, A, by the basin length, L_B: $$W_B = A/L_B$$
Basin Perimeter (P_B)	The length of the curve that defines the surface divide of the basin.
Basin Land Slope (S_{B1})	Average land slope calculated at points uniformly distributed throughout the basin. Slopes normal to topographic contours at each of 50 and preferably 100 grid intersections are averaged to obtain S_{B1}. The difference in altitude for the two topographic contours nearest a grid intersection is determined and the normal distance between these contours is measured.
Basin Diameter (B_D)	Diameter of the smallest circle that will encompass the entire basin.
Basin Shape (SH_{B1})	A measure of the shape of the basin computed as the ratio of the length of the basin to its average width: $$SH_{B1} = \frac{(L_B)^2}{A}$$
Compactness Ratio (SH_{B4})	The ratio of the perimeter of the basin to the circumference of a circle of equal area. Computed from A and P_B as follows: $$SH_{B4} = \frac{P_B}{2(\pi A)^{1/2}}$$
Main Channel Length (L_{C2})	Length of main channel from mouth to basin divide.
Main Channel Slope (S_{C1})	An index of the slope of the main channel computed from the difference in streambed altitude points 10% and 85% of the distance along the main channel from the outlet to the basin divide. Computed by the equation: $$S_{C1} = \frac{(E_{85} - E_{10})}{0.75 \, L_{C2}}$$
Sinuosity Ratio (P)	Ratio of main channel length to the basin length: $$P = \frac{L_{C2}}{L_B}$$

rower; its perimeter (P_B) is about 26% greater than subbasin NW. Basin length (L_B) is somewhat misleading in the case of subbasin W, because it is defined as a straight line from the stream outlet to the basin divide; therefore, it cuts across subbasin NW and actually is not much longer than subbasin NW. Main channel length (L_{C2}) gives a more accurate description of the actual length of the subbasin.

Quantitative values related to the shapes of subbasins also show striking differences between subbasins W and NW (*Figure III.A–2*). To describe basin shape, many studies use basin length (L_B) in the calculation. Even though basin length is not a good descriptor for subbasin W, a shape factor using basin length, such as basin shape (SH_{B1}), clearly shows the generally elongate shape of subbasin W relative to subbasin NW; values differ by approximately 80%. Compactness ratio (SH_{B4}) is the easiest shape factor to visualize because it compares the shape of a basin with a circle. A perfectly round basin, for example, has a compactness ratio of 1.0; in the case of Mirror Lake subbasins, the compactness ratio clearly shows the more round shape of subbasin NW compared with subbasin W.

Main-channel slope (S_{C1}) as well as compactness ratio are drainage-basin characteristics that are potentially useful in precipitation-runoff analyses. The relatively large difference in these characteristics for subbasins NW and W may be expected to be reflected in runoff characteristics of the subbasins. For example, timing of peak discharge may be expected to be faster in a stream that has a steep slope and short channel lengths compared with one that has a more gentle slope and longer channel lengths.

Part of the drainage basin of any lake cannot be included as part of the basins of inflowing streams. Runoff from these areas does not collect in channels before entering the lake; water flows directly to the lake either as overland flow or, if the water infiltrates, as subsurface flow in the unsaturated and ground-water zones. Areas of direct runoff are particularly important to lakes because they are always directly adjacent to the lake. Because of this proximity and because human development is commonly most intense in these areas, they can be the most critical parts of the drainage basin to manage.

Water from the area of surface drainage into Mirror Lake that does not become channelized, but flows directly into the lake, encompasses 22 ha or approximately 26% of the modified total drainage area. This area consists of four separate tracts, separated by the stream inlets and outlet, identified as D-n (north), D-w (west), D-s (south) and D-e (east) (*Figure III.A–2*). Quantitative measures of topographic characteristics of these areas of direct drainage are given in *Table III.A–2*.

Comparison of Mirror Lake and Hubbard Brook Drainage Basins

Because of the long-term studies on the hydrology and biogeochemistry of the nearby Hubbard Brook Experimental Watersheds, it is interesting to compare the Hubbard Brook and the Mirror Lake drainage basins.

A basic requirement of the Hubbard Brook Ecosystem Study was to isolate Experimental Watersheds so all chemical inputs and outputs could be accurately measured. It was most convenient to isolate the higher parts of drainage basins where it would not be necessary to gauge a surface inflow, where surface-water divides could be clearly defined, and where geologic materials overlying the bedrock would be thin. It was assumed that if weirs could be placed directly on bedrock, ground water could be ignored, because any subsurface water moving through unconsolidated deposits would be forced to the surface and could then be measured as streamflow. This assumes that the bedrock is watertight or that water movement through fractures is minimal (Bormann and Likens 1967; Likens et al. 1977). Although test drilling was not done, various geologic, hydrologic and chemical data were used to support this assumption (Bradley and Cushman 1956; Likens et al. 1977).

Mirror Lake has a considerably different physiographic setting than the Hubbard Brook Experimental Watersheds. Mirror Lake is in the lower part not only of its own watershed, but of the entire Hubbard Brook Valley. Because of this, land slopes are gentler and the unconsolidated deposits (glacial drift) are much thicker than in the Hubbard Brook Experimental Watersheds. Therefore, surface runoff, unsaturated-zone flow and ground-water flow may be expected to be markedly different in the two areas. Because a principal focus of the Mirror Lake studies is interaction of the lake with ground water, entirely different approaches to gauging streams are needed. Unlike the Hubbard Brook studies, where weirs were used partly to intercept subsurface flow (*Chapter II*), flumes are used in the Mirror Lake studies (*Chapter IV.C*), because it was believed necessary to measure open-channel flow separately from ground-water flow. In addition, because of particular interest in ground-water flow, it was necessary to assess the hydraulic properties of the bedrock. Bedrock drilling indicated the existence of fractures throughout the intervals drilled and dynamic ground-water movement through the fractures.

Summary

Mirror Lake is located at the lower end of the Hubbard Brook Valley, in the White Mountains of central New Hampshire. The lake is situated largely within glacial drift, which is as much as 30 m thick in parts of the Mirror Lake drainage basin. The basin of Mirror Lake was formed in a kettle that resulted from the melting of an ice block about 14,000 yr B.P.

Drilling and seismic-velocity data indicate the drift is till on the north and northwest sides of Mirror Lake, and principally sand and gravel on the south side, between the lake and Hubbard Brook. However, till is also present on the south side, indicating complex glacial deposits in this area. Elsewhere in the drainage basin, the drift is principally till, but in places it is sandy; scattered areas of sand and gravel also occur, possibly from local ice-contact deposition. There is little or no drift between lake sediments and bedrock at the point of maximum sediment thickness.

Crystalline bedrock underlying the drift is composed of schist, slate and quartz monzonite of earliest Devonian age (about 415 million yr B.P.). These rocks are folded and fractured, and ground water moves through them.

Because of the erosion-resistant properties of the bedrock, the drainage basin of Mirror Lake is characterized by high knobs and ridges, and steep land slopes. In the lower parts of the basin, the steepness of the slopes is modified by glacial deposits. The total area of the drainage basin, excluding the lake, is approximately 103 ha.

Three small streams flow into Mirror Lake; the only stream flowing out of the lake joins Hubbard Brook approximately 0.4 km from the outlet of the lake. Although the drainage basins of the inlet streams mostly have a south aspect, quantitative measures of the various basins are considerably different.

Mirror Lake, which lies at an altitude of approximately 213 m, is 15 ha in area, has a maximum depth of 11 m and an average depth of 5.75 m. Total water volume of the lake is approximately 860,000 m^3 and the volume development is 1.6.

B. Historical Considerations

1. History of the Vegetation on the Mirror Lake Watershed

MARGARET B. DAVIS

Mirror Lake came into existence 14,000 yr ago when the melting of a buried ice block formed the lake basin and sediment deposition began (see *Chapter III.A*). These early sediments, and all sediments deposited subsequently, contain fragments of terrestrial plants that were blown or washed into the lake from the surrounding watershed. They provide a continuous record of vegetation on the watershed that can be related to the development of the lake. The most numerous terrestrial fossils are pollen grains, which are produced abundantly and dispersed by the wind. These have been the most valuable fossils for reconstructing forest history, because the quantities of pollen grains in sediments depend directly on the quantities of source plants in the surrounding landscape (Davis et al. 1973). Of course, sedimentary processes distort the input of pollen from the air, concentrating pollen grains and sediment in the deeper part of the lake basin. For this reason we have supplemented the pollen record from the Central core in Mirror Lake (Likens and Davis 1975; *Figure III.B.1*–1) with analyses of pollen in several additional cores, and with counts of macrofossils and charcoal in one of the cores from 6.8-m water depth near shore (core #783). Two deep-water cores (#782, P), which included older sediment than the deepest levels in the Central core, were analyzed for the early pollen record (*Figure III.B.1*–1). Supplementary information on regional vegetation has been obtained from pollen and macrofossils from eight ponds and bogs at higher elevations in the White Mountains (Davis et al. 1980; Spear 1985). To calibrate the ancient pollen record, modern pollen has been sampled from the surface sediment in 34 lakes at various elevations in New Hampshire and Massachusetts. Additional samples from the eastern United States and Canada are available in the literature (Davis and Webb 1975).

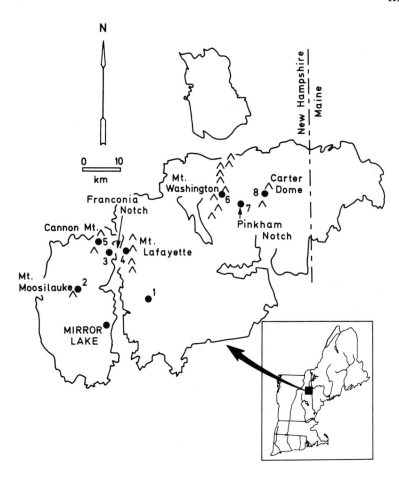

Figure III.B.1–1. Outline map of the White Mountain National Forest, showing the location of the major mountain peaks, Franconia Notch, Pinkham Notch, and Mirror Lake. Black dots show the location of lakes or bogs from which sediment cores have been analyzed for pollen and macrofossils: 1. Little East Pond; 2. Deer Lake Bog; 3. Lonesome Lake; 4. Eagle Lake Bog; 5. Kinsman Pond; 6. Lakes-of-the-Clouds; 7. Lost Pond; 8. Carter Pond. Reprinted from Spear (1981).

Tundra Period

Sedimentation began in Mirror Lake 13,870 ± 560 yr ago, the radiocarbon age of silty sediment near the bottom of core #782 (Davis and Ford 1982). Coarse inorganic sands that underlay the silt were presumably deposited before the lake came into existence, while rapidly moving water still flowed through the area (*Chapter III.A*). Sedimentation of fine-grained silts began at other lakes and ponds in the region within a few hundred years of the origin of Mirror Lake. For example, basal sediments from Deer Lake Bog at 1,325-m elevation on Mt. Moosilauke, and from Lost Pond at 650-m elevation in Pinkham Notch near Mt. Washington (*Figure III.B.1–1*), have been carbon dated at about 13,000 yr B.P. The similarity in age means that the ice sheet covering the mountains in New Hampshire downwasted within a few hundred years and melted away rapidly from the entire region (Davis et al. 1980). There is evidence for a residual ice mass only in Franconia Notch (20 km north of Mirror Lake; *Figure III.B.1–1*), where basal dates from three lakes and bogs (Kinsman Pond, Lonesome Lake and Eagle Lake Bog) are approximately 2,000 yr younger than at other sites in the region (Spear 1985).

As soon as the ice melted from the landscape, tundra plants became established. The tundra vegetation was sparse at first, dominated by sedges, with grass, abundant arctic sage and a variety of herbaceous weeds, such as caryophylls and various species of composites, legumes and chenopods. *Thalictrum* (foamflower), buttercups and sorrel were also present. After about a thousand years, a shrub-tundra developed, with willow, heath, juniper and *Shepherdia canadensis*. A number of species characteristic of arctic regions are known to have been present in the tundra vegetation. Macrofossils in a Mirror Lake core (#783) demonstrate the presence locally of *Dryas integrifolia*. *Dryas* is

B. Historical Considerations

a pioneer plant of subarctic and arctic regions that grows in open, treeless habitats on bare soils. It is capable of fixing nitrogen and characteristically occurs on neutral or basic soil (Miller and Thompson 1979). *Dryas* no longer grows in New Hampshire; its nearest station is on the Gaspé Peninsula of Quebec. Arctic-alpine species that grow today above treeline on Mt. Washington were probably present at low elevations just after glacial retreat. These include bog bilberry and dwarf willow, both of which, together with *Dryas integrifolia,* are known from late-glacial sediments in Massachusetts (Argus and Davis 1962). Spores from *Lycopodium selago* and seeds from *Potentilla tridentata* have been identified in late-glacial sediment from Lakes-of-the-Clouds on Mt. Washington (Spear 1985). Many more plant species present in the New England late-glacial flora have been identified from a rich macrofossil deposit near the Connecticut River in northern Vermont, 50 km north of Mirror Lake (Miller and Thompson 1979). The absence of trees was important to the pioneer herbaceous vegetation, as most of the species that were present cannot tolerate shade. However, *Dryas,* bog bilberry, dwarf willow, *Lycopodium selago* and *Potentilla tridentata* no longer grow at low elevations in New Hampshire, even though open places are present where they would meet little competition. Climatic conditions also must be critical for the growth of these species; temperatures during the tundra period were probably several degrees Celsius lower than today. The vegetation was affected also by the condition of the soil. Recently exposed glacial till and outwash sand were unleached, and devoid of any humus or organic material. Furthermore, the combination of cold climate and high soil moisture (because there were no trees to transpire water) must have resulted in soil instability, with intense frost action, solifluction and overland flow of water.

Supplementing the floristic evidence for cool climate is geologic evidence in the form of "fossil" frost features. Massive stone stripes and polygons occur in the alpine zone of Mt. Washington. Frost features of such magnitude are not forming in the White Mountains at the present time; they must have formed while temperatures were still well below modern levels, probably within the first 2,000 yr after glacial ice left the area (Goldthwait 1976; Davis et al. 1980; Spear 1985).

Woodland and Forest Development

The first trees to grow in New Hampshire invaded the area 2,000 yr after glacial retreat. Spruce needles in the sediments of Mirror Lake and Lost Pond in Pinkham Notch demonstrate that spruce grew in the area as early as 11,300 yr ago (Davis et al. 1980). At this time, the accumulation rate for spruce pollen grains rose steeply at all the low and mid-elevation lakes that we have studied, indicating that spruce trees were widespread throughout the region (*Figure III.B.1–2*). The earliest spruce trees were probably white spruce (Davis 1958), the dominant species today in many parts of the boreal forest. White spruce no longer occurs in the White Mountains, although it grows at the same latitude on the calcareous soils of Vermont. Species identifications of pollen grains from the Mirror Lake core have been made by William A. Watts (personal communication), using identification criteria established by Richard (1970) and Birks and Peglar (1980). Watts believes that the 11,000-yr-old sediment contains pollen of white spruce and the 10,000-yr-old sediment contains pollen of red spruce, one of the two species that still grow in the region today. But unlike spruce forests in New England today, the late-glacial spruce woodland was semi-open and park-like with scattered trees, or it consisted of groves of trees interspersed with tundra openings. Spruce pollen influx was only about 9,000 grains/cm^2-yr, less than recorded in areas of boreal forest, and the high percentages of herbaceous pollen throughout the spruce zone sediments argue for the continued presence of large numbers of herbaceous plants. By 10,000 yr ago, spruce was widespread in the White Mountain region, growing at low elevations and up to the highest elevations that occur within the Mirror Lake watershed. Black spruce needles have been found in sediment of this age at Lakes-of-the-Clouds at 1,542-m elevation. Black spruce Krummholz grew near this lake, which is above the modern limit for trees on Mt. Washington, and near Deer Lake Bog, which is just below the modern tree line, indicating that 10,000 yr ago the tree line had already risen to elevations

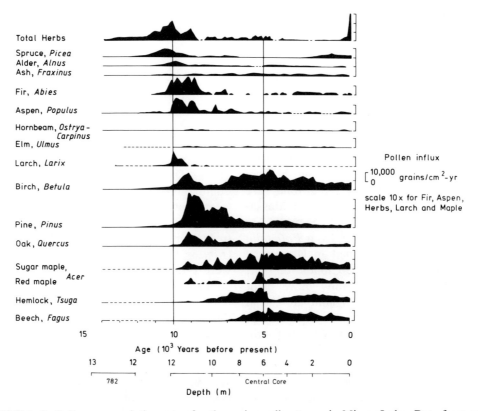

Figure III.B.1–2. Pollen accumulation rates for the major pollen types in Mirror Lake. Data from sediment younger than 11,000 yr are from the Central core; the older analyses are from core #782, which was collected in 8.9-m water depth. Note that two different scales have been used: pollen accumulation rates for herb pollen, fir, aspen, larch and maple were magnified ten times in order to make the changes through time more visible. These data are shown as percentages vs. depth in Likens and Davis (1975). Percentages are plotted against age in *Figure VII.F*–1.

close to its modern position (Spear 1985). This pattern implies that by 10,000 yr ago the climate was not very different from today.

The arrival of trees had an important effect on the ecosystem. Soil stabilization occurred rapidly between 12,000 and 11,000 yr ago, as indicated by the sharp decline in export of particulate inorganic sediments from the watershed to the lake (*Chapter III.A*). Exports started to decrease around 12,000 yr ago and matched modern levels by 10,000 yr ago (Likens and Davis 1975; Davis and Ford 1982). From the standpoint of vegetation, the instability of the landscape prior to 10,000 yr ago guaranteed open sites for the establishment of invading plants, whereas later arrivals had to compete with existing herbaceous or arboreal vegetation. Miller and Thompson (1979) considered late-glacial soil instability an explanation for the occurrence in the same fossil site of tundra plants like *Dryas* and *Vaccinium uliginosum* that now (at least at their southernmost stations) exhibit different edaphic specializations.

Climatic changes 10,000 yr ago are probably responsible for the decline of spruce, because spruce declined more or less simultaneously over a wide region, and at all elevations below 1,300 m in the White Mountains (Davis 1983). At this time, spruce needles appeared at higher elevations, close to modern tree line (Spear 1985). Computer simulations of forest history in Tennessee, where the decline of spruce occurred somewhat earlier, suggest that minor climatic change could have given competing species a strong competitive advantage over spruce (Solomon et al. 1981).

A number of additional tree species had arrived by 10,000 yr ago (*Figure III.B.1*–2). All

B. Historical Considerations

of these early migrants were pioneers, able to germinate under conditions of high light intensity. The first species was alder (mostly *Alnus crispa*), which began to increase in abundance when spruce woodland was still present but after spruce had begun to decline. The increase of alder *after* spruce is the reverse of the famous Glacier Bay, Alaska, successional sequence, where the growth of alder, a nitrogen-fixer, and the resulting increase of nitrogen in the soil is a necessary precedent to the growth of spruce (Crocker and Major 1955). This pattern suggests that other mechanisms of nitrogen fixation may have been at work in these early forests. *Dryas* may have played a critical role in the tundra. *Shepherdia canadensis,* another nitrogen-fixer (Miller and Thompson 1979), occurred as pollen throughout southern New England in the late tundra period, before spruce pollen became abundant (Davis 1958, 1969); it may have been an important successional species in late-glacial time.

In the vicinity of Mirror Lake, alder and birch increased, followed by aspen, balsam fir and cedar/juniper (*Figure III.B.1*–2). Between 10,000 and 9,000 yr ago, a mixed conifer-aspen forest developed at low elevations. The percentage of herbaceous plant pollen declined, but continued high deposition rates suggest that open areas were present. Farther south in New England, there were appreciable quantities of jack pine at an equivalent stage (Davis 1958). Jack pine pollen is infrequent in sediments from Mirror Lake. Presumably, jack pine was present, but not at all abundant. An isolated population of this species grows today only a few kilometers away on the south-facing granitic slopes of Welch Mountain, near Campton, New Hampshire. This population may be a relic from the ancient movement of jack pine into this region.

The major factor affecting the development and composition of the forest over the next few millenia was the order of arrival of tree species migrating northward. Both oak (presumably northern red oak) and white pine arrived in the Mirror Lake region 9,000 yr ago; sugar maple, 8,000 yr ago; hemlock, 7,000 yr ago; and beech, 6,000 yr ago. Comparisons of the Mirror Lake diagram with pollen diagrams from other sites shows earlier arrival times in the south and later arrival times in the north, as these species moved from south to north during the early post-glacial period (Davis 1976b, 1981b).

Decline of Hemlock

A major event in the history of the forest surrounding Mirror Lake was the decline of hemlock 4,800 yr ago (Likens and Davis 1975; Davis 1981a). Within a 6-cm-thick layer of Mirror Lake sediment, estimated to have been deposited within 30 yr, hemlock pollen influx declined to one-tenth its previous value. A similar sharp drop can be seen at all sites throughout eastern U.S. and Canada (Davis 1981a). The numbers of hemlock needles deposited in Mirror Lake also show a sharp decline (Davis et al. 1980). The sudden decline of hemlock was unique; other species were not affected, making it unlikely that the decline was the result of climatic change. In recent years, outbreaks of insect-vectored fungal pathogens have caused a similar decline of chestnut, and more recently, elm. Chestnut provides an appropriate comparison, because the species persists as a living rootstock, while above ground sprouts, which now occur in low frequency in the forest, are attacked by the fungus. Sprouts occasionally grow large enough to flower before they are attacked by the fungal pathogen, which is still endemic 75 yr after the initial outbreak. The fossil record in Mirror Lake shows that hemlock likewise persisted in low numbers after the decline. For approximately 2,000 yr, pollen was continuously produced in small amounts, and small numbers of needles were continuously deposited in Mirror Lake sediment. During this long interval, populations were kept in check, perhaps by the continued presence of a pathogen or insect. Competing species, first birch and later sugar maple, beech, pine and oak, increased in frequency during this period. As soon as hemlock began to recover in abundance (because it developed resistance, or because the virulence of the pathogen was attenuated?), pollen from competing species began to decline again in both absolute and relative abundance (Davis 1981a). By 3,000 yr ago, the pollen percentages of hemlock and its co-dominants were almost exactly as they had been 5,000 yr ago, before the hemlock decline. The hemlock decline and the sequence following it is a rare example in the fossil record of a biotic community changed by disturbance, and subsequently returning through biologic processes to its original state.

In areas near Mirror Lake where hemlock was abundant 4,800 yr ago, many of the canopy trees must have died over a very short time interval. A short-lived peak in birch pollen, occurring at the same level in the sediments where hemlock pollen declines, indicates that the increase in light favored birch establishment. But after a century, birch was itself replaced by white pine and by shade-tolerant trees like beech and maple that competed successfully with birch for the space given up by hemlock. A similar succession caused an increase in birch following cutting of hemlock for tanbark in the nineteenth century (*Chapter III.B.3*); the second phase of succession, involving the replacement of birch, is just beginning to occur in the modern forest (Bormann and Likens 1979). Several lines of evidence (see below) suggest that hemlock was much more abundant and widespread in the forest 5,000 yr ago than it was in presettlement forests. Its pollen percentage fell from 28 to 5% of the total 4,800 yr ago, whereas the nineteenth-century harvest caused a decline from 15 to 8%. The hemlock decline and the sequence of changes in abundances of forest species that followed it, including the eventual return of hemlock, were, therefore, major events in the forest history of the Mirror Lake watershed. These events were comparable in magnitude on the population and community level (although not on the ecosystem level) to changes that occurred as the result of human disturbance.

The Climatic Optimum

A climate change in many regions of the world between 8,000 and 5,000 yr ago caused temperature to rise several degrees Celsius higher than today. This climatic episode, variously called the "climatic optimum," "hypsithermal interval," "xerothermic interval" or "Prairie Period" is documented by a variety of paleontological and geologic evidence (Deevey and Flint 1957; Wright 1966). Changes in vegetation document a similar climatic change in the vicinity of Mirror Lake (Davis et al. 1980). There are two lines of evidence. The first involves spruce, which was abundant at many elevations 10,000 yr ago. At that time, spruce pollen declined in frequency at all sites. For the next 8,000 yr, spruce pollen at Mirror Lake and other White Mountain sites, even those at high elevations, was quite rare. No surface sediments have been found anywhere in New Hampshire that have so little spruce pollen as Mirror Lake during this interval (M. Davis, unpublished data; Spear 1985), suggesting that spruce was very much less abundant then than it is today. Two thousand yr ago, there was a sudden resurgence of spruce pollen and macrofossils especially at sites at 650- and 1,000-m elevation. Pollen and fossil needles increased to modern levels of abundance. Because spruce grows today most abundantly at relatively high elevations on the mountain slopes (750 to 1,000 m), where the microclimate is cool and relatively moist, the minimum in its abundance implies a shrinking of the area where climate favored spruce over competing species. The shrinking in area could have been caused by a steepening of the climatic gradient on the mountain slopes, with a rise in temperature and a decrease in moisture at low and mid-elevations (Davis et al. 1980).

A second line of evidence involves tree species that are presently confined to low elevations where the climate is warm and relatively less moist. Several of those species became more abundant, and ranged more widely to higher elevations than they do today. Pollen from white pine and oak, for example, reached maximum abundance 8,000 to 5,000 yr ago. Eight thousand years ago, they contributed 70% of the pollen in Mirror Lake sediments, whereas today they make up 10%. Today they occur in abundance (except on ledges and other unusual habitats) only at elevations below 450 m (Bormann et al. 1970), apparently because they grow poorly under the conditions of low temperature, fogginess and deep snow cover that now occur at high elevations on the mountain slopes. We have not yet found any macrofossils to show exactly where oak trees were growing 8,000 yr ago; oak pollen is carried many kilometers from source plants, and consequently cannot be used as definitive evidence of the presence of oak trees. White pine, however, leaves a clear fossil record in the form of needles and needle fragments preserved in lake sediments. Seeds also are found occasionally, as are bracts from the cone scales. White pine needles occur abundantly in sediments from Mirror Lake, where white pine grows today, but we have also found significant numbers of fossil white pine needles at lakes as high as 800 m, 350 m above the elevation limit for forest where white pine is now abundant. The lakes are in areas of glacial till, not extensive granite outcrops where white pine might be expected to occur above its usual limit. The needles first occurred at high elevation 9,000 yr ago when white pine arrived in the region, and they were deposited repeatedly in the sediment at high-elevation lakes throughout the Holocene until just a few hundred yr before settlement. Between 1450 and 1850 A.D., temperatures decreased 1–2°C, causing the advance of glaciers in many alpine regions. This climatic episode is known as the "Little Ice Age" (Mitchell 1977; Porter and Denton 1967). Apparently, this climatic change occurred in New England as well, causing white pine to contract its range downward to lower elevations, where it grows today.

Hemlock is a significant component of forests in northern New Hampshire at sites below 650-m elevation (Bormann et al. 1970). Fossil hemlock needles show, however, that this species occurred over a much wider range of elevations 7,000 yr ago, when it first appeared in the region, until 5,000 yr ago. Throughout this interval its needles were deposited in sediments at several lakes as high as 1,000 m, 350 m higher than the elevations where it is presently abundant. Needles disappeared from the fossil record at high elevation sites 5,000 yr ago, perhaps as a result of the same factor responsible for the hemlock decline. When regional populations recovered abundance at low elevations 3,000 yr ago, hemlock failed to recolonize the marginal sites at high elevation. Hemlock needles did occur continuously at Lost Pond at 650-m elevation, a few hundred meters above the elevations where it grows in Pinkham Notch today, until the time of the Little Ice Age, when hemlock contracted to its present range (Davis et al. 1980).

The occurrence of pine and hemlock 350 m above the elevations where they commonly occur today, suggests a change in climate 9,000 to 5,000 yr ago. A possible change was a reduction in moisture and a rise in temperature, probably a 2°C rise in mean annual temperature. This estimate is based on the difference in mean annual temperature between sites 350 m different in elevation, calculated from the lapse rate that prevails on the mountain slope today. The use of lapse rates assumes that the prevailing climate can be displaced upward on the mountain slopes. The tree line does not seem to have changed position very much during the climatic optimum (Spear 1985), suggesting that the climate gradient was steepened, rather than displaced. A steepened gradient on the mountain slopes could have been caused by a change in the source of the prevailing air masses reaching New Hampshire, because air with different moisture content and stability characteristics would cool at a different rate as it rose over the mountain massif, producing a different lapse rate than occurs today (Davis et al. 1980). Spear (1985) argues that the position of tree line is determined by exposure, rather than moisture and temperature. He suggests that the climatic optimum, regardless of lapse rate, would have had little effect on the position of the tree line.

History of Disturbance

The important role that perturbations play in forest ecosystems has been emphasized by Bormann and Likens (1979). The record of disturbance in the White Mountains is reviewed in *Chapter III.B.2*, concluding that large-scale disturbance by fire and wind rarely occurred in the presettlement hardwood forest, assuming a climate similar to today's. Obviously, this generalization cannot be applied to the vegetation surrounding Mirror Lake during the entire 14,000 yr of its history. The fossil record has been inspected for changes in the disturbance regime in the past.

Large-scale disturbances that initiate biologic succession can be inferred in sedimentary deposits from increased pollen from successional species, provided the succession was synchronized over a large enough area to affect regional pollen deposition. The increase in birch pollen 4,800 yr ago at the time of the hemlock decline is an example. Another example is the increase in birch following widespread forest clearance and logging in the nineteenth century (see below). Small-scale disturbances, however, caused by an individual treefall or storm damage to a small stand of trees, can be detected only under special circumstances where local pollen alone is collected, e.g. mor layers in forest soils (Iversen 1973) or small forest bogs less than 15 m in diameter (Andersen 1973). Lakes as large as Mirror Lake receive pollen from an area 5 to 15 km in diameter, and although trees growing along the lake shore or near inflowing streams are best represented, a considerable (although unmeasured) proportion of the pollen comes from trees growing more than 5 km away (Jacobson and Bradshaw 1981; Tauber 1967; Berglund 1973; Andersen 1970; Webb et al. 1981). As a result, the pollen represents an average of plant abundances over a large region, and changes within small areas are averaged out. Windthrow from hurricanes, for example, is patchy, depending on slope orientation and the sizes of trees (*Chapter III.B.2*). A series of storms creates a series of age classes within the forest, or a mosaic of successional forest stands of various ages (Bormann and Likens 1979). Although these would not be detectable individually in a pollen diagram from a lake the size of Mirror Lake, a change through time in the frequency of storms would shift the mean age of successional stands. This shift could be recorded by a change in the relative abundance of pollen from successional species (Swain 1973). A change of this kind does appear to be recorded at Mirror Lake because the frequency of successional species is highest in the early Holocene. The abundance of successional species may suggest increased disturbance at that time period, although alternative explanations must be considered as well. The immature soils at that time might have given an advantage to species that are characteristic of early stages in succession. These species also may be able to extend their ranges more rapidly than others, and they may have arrived sooner, thus making up a higher proportion of the total flora early in the Holocene. The possibility remains, however, that tropical storms that result in large-scale damage to forests (*Chapter III.B.2*) were more frequent in New Hampshire in the early Holocene than in recent centuries. A climatic regime that was warmer than today's climate could have been coupled with increased incursions of tropical air to New England.

Fire is another kind of perturbation that is easier to evaluate because it leaves a fossil record in the form of fragments of charcoal blown or washed into lakes and incorporated into the sediment (Swain 1973; R. B. Davis 1967). Because Mirror Lake sediments are well mixed by chironimids (*Chapter VII.A*), there is little hope of detecting individual fire events, but the overall abundance of charcoal can be compared with charcoal counts from regions where fire is an important ecosystem component. We were particularly interested in comparing Mirror Lake to a lake in the Boundary Waters Area in northern Minnesota, because the fire frequency there is well known (Heinselman 1973). Swain (1973) counted charcoal in recent sediments of the Minnesota lake; he found an accumulation rate of 5 to 15 mm^2 charcoal fragments/cm^2-yr. These counts were from a core collected below 30 m of water near the center of the 12.5-ha lake. Major fires have occurred in the region at intervals ranging from 10 to 50 yr. Because some areas burn more frequently than others, on the average an area equivalent to the entire area would burn about every 100 yr (Heinselman 1973).

By comparison, the Central core from Mirror Lake (11-m depth, 15-ha area) contains very little charcoal—only about 2% as much as the Minnesota lake, suggesting much lower fire frequency. This finding corroborates Bormann and Likens' (1979) suggestion that fire was unimportant as a disturbance factor in the presettlement hardwood forest-ecosystem in the White Mountains.

To look for changes in the abundance of charcoal through time, we counted charcoal using a standard method (Swain 1973) at 15 different levels in core #783, a core from 6-m water depth near the southwest inlet (see map in *Chapter VII.A*). These sediments were deposited in relatively shallow water and contained coarse-grained material, such as plant macrofossils. The record was, therefore, biased toward coarse debris like charcoal fragments, which were much more abundant than in the Central core. Even so, the accumulation rate for charcoal for the last 7,000 yr was only 10% as high as in the core from the Minnesota lake. Of major interest was the higher rate of accumulation of charcoal in sediments deposited prior to 7,000 yr ago (*Figure III.B.1–3*). At this time, conifers were abundant on the landscape, primarily spruce and fir 11,000 to 9,000 yr ago and white pine 9,000 to 7,000 yr ago. Conifers are often dominant in fire-prone vegetation.

B. Historical Considerations

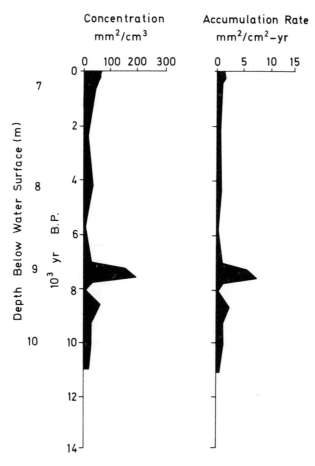

Figure III.B.1–3. Charcoal concentration and accumulation rate in core #783, Mirror Lake, plotted against the age of the sediment. Core #783 was collected in 6-m water depth.

Oak, aspen and birch, which occur as successional trees in burned-over areas, were also abundant at that time (*Figure III.B.1–2*). Charcoal was especially abundant 8,000 to 7,000 yr ago. Additional data are needed to test for regional significance, but the peak does occur at a time when range extensions of white pine and hemlock, and abundance maxima for white pine and oak pollen, suggest a climate that was warmer and drier than today.

A special search was made for charcoal in the Central core just at the levels where hemlock declined 4,800 yr ago. There is no evidence from charcoal that this event was accompanied by fire. The dead and dying hemlock trees must have provided a fuel source, but a source lasting only a short time, a few decades at most.

Individual sediment samples in core #783 contained large fragments of charcoal, apparently derived from local fires. We have also found charred conifer needles in sediment of various ages at other sites in the White Mountains (Davis et al. 1980; Spear 1985). These finds mean that forest fires have occurred throughout the history of this region. But the overall paucity of charcoal suggests fire frequencies far below the primeval frequency in northern Minnesota. Even 7,500 yr ago, when fires were apparently more frequent (perhaps similar to primeval fire frequencies in southern New Hampshire?), the influx of charcoal is significantly lower than in Minnesota. The evidence from the sediments suggests that fires were two to five times more frequent 11,000 to

7,000 yr ago than they have been since, but that fires have never been a dominant factor controlling the Mirror Lake ecosystem.

Impact of Human Disturbance

Low-elevation forests in the Mirror Lake region were cleared for farming in the early and mid-nineteenth century, and hemlock trees were harvested for tanbark. The upper slopes were logged in the late nineteenth and early twentieth centuries (Likens 1972c; *Chapter III.B.3*). The long-lasting effects of these events on the average composition of the second-growth forest are reflected by changes in the regional deposition of pollen. We have compared the pollen produced by modern forest and preserved at the sediment surface with pollen produced by the primeval forests and preserved in sediment underlying the culture horizon. The latter is the sediment layer containing elevated percentages of grass and weed pollen that reflects nineteenth-century farming (*Figure III.B.1–4*; *Chapter VII.A*). A comparison of tree pollen percentages at these two stratigraphic levels at 13 sites at various elevations in the White Mountains is shown in *Figure III.B.1–5*. The silhouettes show differences in percentages plotted against elevation: a decrease in pollen percentage for those genera that have declined in relative abundance, and an increase in percentage for those genera that have become more abundant. The response is different at different elevations.

Lakes above 1,000-m elevation show smaller differences in birch pollen percentages than lakes at lower elevations (*Figure III.B.1–5*). Logging at these elevations (mainly for spruce in the early decades of this century) was followed by biologic succession, first paper birch, then balsam fir, then spruce (Botkin et al. 1972). Sediment cores at individual sites show a series of stratigraphic changes in the uppermost centimeters of sediment that reflect these changes. However, both paper birch and balsam fir are short-lived species, and the succession has progressed to spruce at some sites. Percentages of fir pollen are higher than in presettlement sediment at other sites. Pollen from balsam fir is not produced in great quantity even when fir trees are abundant, so the difference in pollen tends to underestimate the difference in abundance of trees. Paper birch is no longer as abundant as it was immediately following logging, and succession has proceeded to primarily coniferous forest at high elevations.

At mid- and low elevations, striking differences in pollen percentages show that the modern forest is quite changed from its primeval counterpart. Sites at mid-elevations, where spruce grows mixed with hardwoods, display much lower percentages of spruce pollen. Spruce trees grow slowly relative to deciduous trees at these elevations; spruce, although it has regenerated, has not yet gained its former prominence in the canopy and therefore is not yet producing pollen. From 750-m elevation down to the lowest elevations, pollen production from birch is much higher than in presettlement time, and sugar maple, beech and hemlock are much reduced. The latter species have been slow to recover their former abundance in the canopy following logging in the forest.

Bormann and Likens (1979) show that after

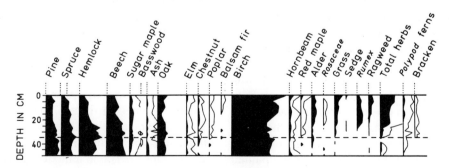

Figure III.B.1–4. Percentages of pollen in closely spaced samples from a short core. Percentages are plotted against depth. The dashed line at 35-cm depth represents the base of the settlement horizon, which is characterized by increased percentages of pollen from weedy plants.

Figure III.B.1–5. Differences in the percentages of pollen in presettlement sediment, deposited more than 200 yr ago, and at the sediment surface. The differences are plotted against the elevation of the site from which the data were obtained.

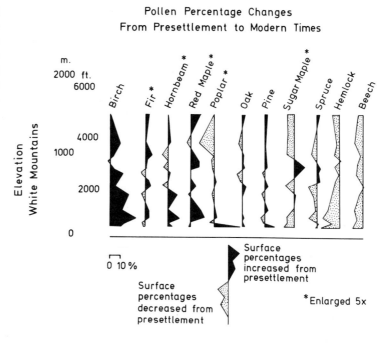

clear cutting, all species of the forest regenerate rapidly, but the forest undergoes differentiation marked by a sequence of tree dominants. The first dominants are short-lived, rapidly growing exploitative species. Some of these, like cherry, produce few pollen grains and are poorly represented in the pollen record. They are followed by dominants of intermediate light tolerance (e.g. birch) and finally by tolerant, long-lived species (e.g. beech and sugar maple) mixed with intermediate long-lived species (yellow birch). The authors project that within approximately 120 yr the forest will be dominated by a mixture of tolerant and intermediate species. A similar, but somewhat slower differentiation of the forest is predicted by computer simulation: Botkin et al. (1972) predict from that simulation that yellow birch will dominate the canopy 100 yr after clear cutting. Because yellow birch is a long-lived species, once it has attained importance in the canopy, it retains it for several centuries while shade-tolerant sugar maple and beech grow in the understory beneath it. The simulation predicts that 600 or 700 yr will elapse following clear cutting before beech and sugar maple increase to the *equilibrium* abundances predicted in the absence of disturbance. The pollen in sediments corroborates the greater importance of beech, maple, hemlock and spruce in the presettlement forest than in the modern forest. It shows little detail concerning the short-lived exploitative species in the early stages of forest differentiation, but it does record the predicted dominance of yellow birch, which has increased in pollen abundance at the sediment surface 75 to 100 yr after the upland forest was cut.

Discussion

The plant communities surrounding Mirror Lake have changed dramatically many times during the last 14,000 yrs. Pollen deposition (*Figure III.B.1–2*) records the fluctuations in population size for the various species as forest communities adjusted to changes in climate, to the arrival of invading species, to changes in soil, to changes in the frequency of disturbance, or to the apparent outbreak and attenuation of a pathogen. Periods of stability have been rare, and when they did occur, they lasted 1,000 to 2,000 yr—only long enough for the growth of four to six generations of canopy trees. Because the forest community adjusts so slowly to change, taking as long as 700 yr to reach equilibrium following a major disturbance like clear cutting (Botkin et al. 1972), and because changes in climate or biotic disturbances have occurred frequently (once per millennium on the average), the forest community has been undergoing an almost continuous series of changes and adjustments during the Holocene.

Bormann and Likens (1979) discuss the concept of Steady State as it applies to the forest-ecosystem. They describe the Steady-State forest as a mosaic of stands, some of which include intolerant, early successional species, others of which include species of intermediate- or high-shade tolerance, and they emphasize that the Steady State is not necessarily stable, but may encompass slow unidirectional change. They do not discuss the possibility of climatic change. Their concept of Steady State is very different and much more sophisticated than older ideas of *mature, equilibrium* or *climax* ecosystems. Bormann and Likens conclude that major disturbances such as windstorms and fires are rare in New England; their discussion implies, therefore, that given no change in climate the Steady State is the normal condition for the forest-ecosystem at Hubbard Brook.

The fossil record offers only partial support for this idea. Pollen frequencies suggest that forests prior to nineteenth-century cutting were similar to the predicted Steady-State community. Spruce, hemlock, beech and sugar maple, all slow-growing tolerant species, were more abundant than they are in modern second-growth forests. In addition, the continuous occurrence throughout the Holocene of very small amounts of pollen from intolerant trees like *Populus* (*Figure III.B.1*–2) supports the idea that a mosaic of communities has been present through time. But the overall record shows such frequent, rapid changes in the forest community that the concept of Steady State seems an abstraction rather than a useful descriptive term. True, the forest community may have been tending toward a Steady State, but if the condition was actually attained, it persisted for only a few generations of trees before the onset of further change. In terms of the years spent in one condition or another, the Reorganization, Aggradation and Transition states (Bormann and Likens 1979) are much more important descriptions of the majority of sites within the forest-ecosystem than the Steady State.

Mirror Lake has provided an opportunity to study the development of a pristine ecosystem, unaffected by human influences during most of its history. It contrasts with European lakes, where the disturbance of vegetation and soil by agriculture has been such an important factor over the last 5,000 yr (Mackereth 1966; Berglund 1969; Tutin 1969; Pennington 1981a). Even in the absence of human influence, however, the history of the forest surrounding Mirror Lake has been eventful, with a series of changes in community composition caused by frequent changes in the physical and biologic environment. Within the last two centuries, the vegetation has been altered by human activity as well. Paleolimnology provides an opportunity to observe which of these changes affected Mirror Lake, allowing unique insight into the important processes in the linkage of terrestrial and aquatic systems. The paleolimnology section of this book (*Chapter VII.F*) discusses in detail the historical changes in the lake. The major changes that can be related to events on land occurred when trees were first established on the landscape 12,000 to 11,000 yr ago, and when trees were removed from the landscape 100 to 200 yr ago. Both of these changes, which had far-reaching effects, involved the continuity of the soil cover. In contrast, changes in population sizes of tree species or in community composition within the forest either had no effect, or only subtle effects on the chemistry and biology of the lake.

Summary

The vegetation surrounding Mirror Lake has undergone a series of major changes since glacial ice melted from the region and the lake came into existence. From 14,000 to 12,000 yr ago, tundra surrounded Mirror Lake, including many herbaceous species that presently grow in arctic or alpine regions. Soils were unstable, with neutral or basic pH indicated by calcicolous plants, and the climate was cold. Willows, heaths, juniper and *Shepherdia canadensis* became more common approximately 12,000 yr ago. By 11,300 yr ago, spruce had arrived in the region. It grew interspersed with herbaceous plants and shrubs in the valleys and on the lower slopes. By 10,000 yr ago, spruce woodland grew at all elevations at least up to 1,000 m, just 500 m below the modern tree line. At this time, the climate must have been only slightly cooler than today. Soils were becoming stabilized and exports of particulate matter from the watershed decreased to levels that characterize forested watersheds in the present-day Hubbard Brook Valley.

A series of rapid changes in forest composition occurred 10,000 to 9,000 yr ago as spruce declined in abundance and additional tree species arrived from the south. The prevalence of early successional species like aspen and the continuing high deposition rates of pollen from herbs suggest a mosaic of open and forest habitats, at least at first. Alder, fir, larch and aspen were soon succeeded by birch, pine, oak and maple. The rate of charcoal accumulation in the sediment at this time suggests that fires were two to three times more common than in recent centuries, although still very much less important than in fire-prone regions like northern Minnesota. Fires had become less frequent by the time of arrival of hemlock, 7,000 yr ago, and beech, 6,000 yr ago. At this time, mean annual temperatures may have been approximately 2°C warmer, and precipitation 125 mm less than today. White pine, oak and hemlock were all much more abundant and more widely distributed than they were at the time of settlement.

The forest community was strongly influenced by biotic events, as well as changes in climate. For example, a series of changes was initiated by the decline of hemlock 4,800 yr ago. The rapid regional decrease in abundance of pollen of this species, which occurs throughout the eastern U.S. and Canada, is similar to a decline caused by the outbreak of a pathogen. Hemlock was rare for a millenium, then began to increase slowly, returning to its former abundance 2,000 yr after its original decline. At the time of the hemlock decline, birch increased sharply, followed by increased percentages of beech, sugar maple, pine and oak, which expanded to occupy the space vacated by hemlock. All of these species decreased again as hemlock returned to its position of dominance in the forest.

In the late Holocene, as temperatures fell to modern levels, climatic changes have again had important effects on forest composition. Low-elevation tree species have contracted to the restricted ranges they occupy today, and spruce has become more abundant, increasing at the expense of fir and hardwoods at 750- to 1,200-m elevation on the mountain slopes.

Within the last two centuries, the forest has been changed by human disturbance. The forests were cut for timber everywhere in the White Mountains, and at low elevations the land was cleared for farming. The percentages of tree species in the modern forest have been changed as the result of succession following logging. At low and intermediate elevations, birch is more abundant, and hemlock, beech and spruce, all slow-growing trees, are less abundant. At high elevations, paper birch increased following logging, but because it is short-lived, birch has been succeeded by fir. On some sites, fir has in turn been succeeded by spruce, and species frequencies have nearly returned to primeval abundances.

During most of the last 14,000 yr, the vegetation surrounding Mirror Lake has been changing and adjusting to changes in the physical and biotic environment, as the climate has changed, or as new species have arrived and have invaded the forest communities. Species abundances have seldom been stable for more than 1,000 yr at a time. A goal of the present multidisciplinary study of Mirror Lake has been to assess the effects that these changes in terrestrial vegetation may have had on the aquatic ecosystem.

2. Catastrophic Disturbance and Regional Land Use

REBECCA E. BORMANN, F. HERBERT BORMANN, and GENE E. LIKENS

The historical importance of wind and fire in shaping the structure and function of North American forests is increasingly recognized. A number of researchers have suggested that catastrophic exogenous disturbances that occur irregularly but relatively frequently, together with endogenous disturbance, restrict the development of a Steady State in most forest-ecosystems (see Bormann and Likens 1979). These disturbances affect lake inputs. This concept has primary importance for paleolimnologists because sediment cores from lake bottoms represent not only the history of biologic and physical interaction within the lake-ecosystem but also the history of inputs into the system. The most notable of these inputs are geologic and are generated on the watershed of the lake (Likens and Bormann 1974b, 1979).

Fires can have dramatic impacts on aquatic habitats through their effect on the terrestrial watershed. Vegetative cover is destroyed, resulting in increased runoff and potential flooding. Infiltration may be reduced and erosion increased, causing higher sediment loads and turbidity in recipient streams and lakes. Ash, which is either washed or blown into the lake, can bring about drastic, though usually short-lived, chemical changes (Bullard 1965).

Strong winds, particularly when they are associated with heavy rains, can also wreak havoc with the terrestrial watershed and have a substantial impact on the lake. During the actual storm, trees may be blown down and extensive areas of soil exposed, while torrential rains associated with storms may carry large amounts of sediment into nearby lakes (Bloomfield 1978). Following extensive blowdown, the damaged watershed may export increased amounts of sediments and nutrients until an equilibrium can be reestablished. On the other hand, trees blown down across stream channels may enhance the development of organic debris dams, which retard the export of particulate matter from the watershed to the lake (Bilby and Likens 1980; Likens and Bilby 1982).

The results of human activity on the watershed may parallel or exceed the disturbance engendered by natural catastrophes. Wright (1974) suggested that the disturbance effects of clear cutting are quite similar to the effects of fire. Agricultural activities, with their concomitant prolonged deforestation of the landscape and associated accelerated erosion, would be expected to have an even greater effect.

The goal of this chapter is to explore, as far as the regional land history permits, the potential impact of catastrophic disturbance resulting from fire, wind or aboriginal land use on the paleolimnological history of Mirror Lake. Such information may improve the interpretation of data from lake sediments or even suggest fruitful lines of future paleoecologic analyses.

Fire

Catastrophic fires, both natural and man-made, have been common for thousands of years throughout much of the northern hardwood forest (MacLean 1960; Heinselman 1973; Rowe and Scotter 1973). A detailed fire history of the Boundary Waters Canoe Area in northwestern Minnesota, reconstructed by Heinselman, indicated that on the average an area equivalent to the whole area was burned every 100 yr prior to European disturbance. Between 1681 and 1894, 83% of the area burned during nine fire periods. Collectively, these data suggest that only a small proportion of the transitional northern hardwood forest is free from fire for periods exceeding several hundred years.

Even-aged stands of white pine occurring throughout the Great Lakes region are generally thought to have originated after intense fires (Graham 1941; Maissurow 1941; Hough and Forbes 1943; Chapman 1947). Frissell's (1973) detailed study of Itasca Park in central Minnesota reported that, on the average, fires occurred every 10 yr between 1650 and 1922. Like Heinselman, Frissell found that fires often resulted in subsequent establishment of a pre-settlement mosaic of largely even-aged pine stands dating from different fires. Lutz (1930) identified 40 fires between 1687 and 1927 at Heart's Content in northwestern Pennsylvania following a study of fire scars. Five of these fires—1749, 1757, 1872, 1903 and 1911—were considered to be severe and two were thought to be associated with drought years.

Swain (1973) analyzed the sediment core from Lake of the Clouds in the Boundary Water Canoe Area for evidence of fires and subsequent vegetation changes using paleoecologic techniques. Local fires resulted in prominent charcoal peaks associated with an increase in varve thickness in the lake core. Regional fires also resulted in a high influx of charcoal, distinguished from local fires by the smaller fragment size. The ratio of charcoal influx to pollen influx rose following fire events. Changes in the species composition of pollen following fires were not obvious but were detectable when groups of species, such as conifers, hardwoods that sprout after fires and other groups, were compared.

B. Historical Considerations

The record of fires in forests of the northeastern U.S. based on even-aged stands, fire scars or charcoal is far less obvious than that for the Great Lakes area. Cline and Spurr (1942) suggest that more or less even-aged stands of old-growth white pine in southwestern New Hampshire originated from fires. Henry and Swan (1974), working in the same area, were able to assign one of these fires to the year 1665.

In evaluating the frequency of fires in presettlement times, it is important to bear in mind that Indians often deliberately set fires to clear land for agriculture, drive game and keep trails passable. Based on early historical accounts, such fires were common in southern and central New England and westward to Pennsylvania (Lutz 1930; Day 1953; Spurr 1956a; Thompson and Smith 1970). However, in northern New England, the locale for Mirror Lake, historical evidence for the widespread deliberate use of fire by Indians is lacking (Day 1953; J. W. Brown 1958), perhaps because the usual incentives for burning were absent. Indian populations in the mountainous regions were small and migratory, and agriculture was largely confined to the cultivation of maize on a limited scale (Kilbourne 1916). Travel, by canoe in summer and by snowshoe in winter, was not hindered by underbrush, nor was deer hunting, which took the form of stalking or still hunting rather than driving (Day 1953).

In river valleys near the Mirror Lake area, early explorers noted small stands of white pine (J. W. Brown 1958), which may have originated after abandonment of Indian garden sites or after fire had been used to clear the land. An early traveler, Peter Kalm, reported extensive Indian fires along the southern Champlain Valley of New York and Vermont in 1750 (Benson 1937). However, comments by land surveyors from 1783 to 1787 on the nature of the land along survey lines of township borders in the presettlement forest of northern Vermont include no mention of either fires or fire scars (Siccama 1971).

While the available literature suggests that fires were common in presettlement times throughout much of the northern hardwood and boreal region surrounding the area of the White Mountains, New Hampshire and northern Vermont, there is very little evidence, vegetational or historical, for the common occurrence of widespread presettlement fires within the White Mountain area. Certainly, there are no fire histories comparable to those found by Heinselman (1973) and Frissell (1973) in Minnesota (see *Chapter III.B.1*).

Following settlement, and more precisely the advent of widespread commercial logging, the incidence and extent of forest fires in the White Mountain region greatly increased. The New Hampshire Forestry Commission commented in 1885 that forest fires had destroyed as much timber as had been harvested in the previous 25 yr (Hale et al. 1885). In 1891 the Forestry Commission reported that extensive tracts within the White Mountains had been burned (Walker et al. 1891). These fires included one of the largest fires recorded for the region, the Zealand fire of 1888, which covered 5,000 ha (Chittendon 1905; Kilbourne 1916). This fire occurred in 1886, according to Belcher (1961).

Chittendon (1905), in what was probably the most reliable survey of the time, mapped burned areas throughout the northern third of New Hampshire. About 6% (34,000 ha) of the total area was burned in scattered fires in 1903, an exceptional fire year in the northeastern U.S. Kilbourne (1916) suggests that logging railroads may have been a primary agent for starting these fires. Over 550 fires were reported along railroad tracks in the White Mountain area in 1903 prior to October. Chittendon also reported that the *virgin* forest was strikingly free of fire, despite the exceptional fire year. He offered the opinion that the mountainous topography precludes fires from sweeping unhindered over vast areas.

Since the advent of fire protection, fire is a relatively rare occurrence in the White Mountain region, despite the fact that the region is heavily used by tourists, campers and hikers, as well as by wood-using industries. In the period 1946 to 1976, only 0.025% of the 275,000 ha in the White Mountain National Forest was subject to forest fire damage.

Wind

Wind, unlike fire, is relatively independent of control by humans. Based on a study of land survey records, Stearns (1949) reported considerable wind damage in presettlement northern hardwood forests in northeastern Wisconsin. Surveyors' maps made in about 1859 show areas of blowdown from one to several kilometers in size, and the surveyors recorded many instances of blowdown in small areas. In nearby Menominee County, Wisconsin, 80 million board feet (1 board foot = 2,359 cm^3) of drought-weakened timber blew down between 1930 and 1937 (Secrest et al. 1941). Extensive windstorms on 4 July 1977 (Fujita 1977) blew down more than 100 mi^2 of northern Wisconsin forest (P. Michalls, personal communication). While these data emphasize the importance of wind in the western extension of the northern hardwood forest, differences in topography and wind patterns make it difficult to relate them directly to forests in New England.

New England forests are subject to three major classes of destructive storms—tropical cyclones (hurricanes), extra-tropical cyclones and localized intense winds associated with storm fronts and thunderstorms (Smith 1946). Since 1492, severe hurricane-force winds have struck northern New Hampshire and Vermont only twice—in 1815 and again in 1938. A hurricane in 1788 seems to have been confined to southeastern Vermont and central New Hampshire (Ludlum 1963).

The effects of the 1938 hurricane are reasonably well known from survey records of the U.S. Forest Service (Spurr 1956b; Smith 1946). About 4.5 million ha out of a total 5.3 million ha of forestland in Vermont, New Hampshire, Massachusetts and Connecticut were in the zone subjected to the hurricane (Bormann and Likens 1979). Nearly 3 billion board feet of timber were blown down, and 243,000 ha of timberland were severely damaged, but this represents only about 5% of the affected zone. Although many stands were completely destroyed or lost patches of trees, most stands within the 4.5 million ha region escaped with injury only to isolated trees. Most observers agree that topography was a major factor in limiting damage from this hurricane (Smith 1946). In the White Mountains, damage was principally confined to defective trees or sites exposed to wind. Most old-growth and even-aged northern hardwoods were relatively unaffected (Jensen 1939).

Since there have been only two severe hurricanes during 487 yr and since the greatest devastation was restricted to particular sites close to the paths of these storms, hurricanes probably are not a major disturbance factor for most of the hardwood forest in northern Vermont and New Hampshire. This area also seems to be at low risk from large-scale storms, including extra-tropical storms (Smith 1946).

When hurricanes or extra-tropical storms do occur they may cause a number of changes in lakes within the area. Tropical storm Agnes in June of 1972 resulted in increased inputs of allochthonous materials in several lakes in New York State (Oglesby 1978; Mills et al. 1978). Increases in sediment deposition, turbidity and water levels were also noted (see Bloomfield 1978).

Intense, localized winds associated with frontal showers and thunderstorms may be a more important factor than large-scale storms since they occur more frequently and probably affect a larger total area. There are few data on these local winds, but observations over the last 20 yr of forests in the White Mountains suggest that single trees and some small- to moderate-sized patches of trees are commonly lost to windthrow (see Chapter 7 in Bormann and Likens 1979).

Apparently there was relatively little loss of trees to wind in the presettlement forest of northern Vermont. Only 5 of 163 comments on vegetation along land survey lines mention windfalls (Siccama 1971). Surveys of northeastern Maine over the period from 1793 to 1827 reported that windfalls covered only 2.2% of the total surveyed distance. Most of these windfalls were confined to conifer forests, with only a few occurring in mixed or hardwood stands (Lorimer 1977).

In Massachusetts, a 500-yr history of a 0.4-ha plot was reconstructed through a careful analysis of throw mounds (small hills of earth thrown up by roots of falling trees, usually associated with pits), recent and remnant dead wood and living trees (Stephens 1955, 1956). About 14% of the area of the plot was covered with mounds and pits. Stephens concludes that during the 500-yr period there were six years during which many throw mounds were formed. Mound formation was major in four of the six years. One

B. Historical Considerations

was in the fifteenth century and the three others—1815, 1936 and 1938—corresponded to dates of major hurricanes.

Henry and Swan (1974) applied Stephens' technique to a 0.04-ha plot in Pisgah Forest in southwestern New Hampshire. The plot was chosen in an area where rotting material was abundant to optimize the conditions for reconstructing the forest history. Henry and Swan concluded that a severe fire in about 1665 resulted in a more or less even-aged conifer forest, which grew without major disturbance for 262 yr. Between 1897 and 1938, four wind storms, culminating in the 1938 hurricane, completely destroyed the larger trees.

Both of these studies emphasized the frequency and magnitude of wind as a source of disturbance in forests of central New England, but it should be noted that the degree of disturbance per unit of land reported is very much a function of the study site selected by the investigator. Nevertheless, throw mounds are a widespread phenomenon. Lutz (1940) determined that over long periods, the soil under forest stands is repeatedly subjected to disturbance when trees are uprooted. The prevalence of mounds in a northern hardwood forest in Pennsylvania led Goodlett (1954) to single out windthrow of individual trees or small groups of trees as an important factor in maintaining shade-tolerant species in the presettlement northern hardwood forest. Stephens (1956), in an 8,000-km journey to the Cumberland and Smoky Mountains, south through the southern Piedmont, west to the Ouachita and Boston Mountains and back to Massachusetts, found mounds and pits of uprooted trees nearly everywhere. Pit and mound topography is also prevalent at Hubbard Brook (Bormann et al. 1970; *Chapter II*). The occurrence of pits and mounds, however, is not indisputable evidence of the occurrence of hurricanes or broad-scale winds. In fact, most mounds probably result from the sporadic fall of individual trees attributable to local causes such as age, disease, ice, snow or modest winds acting on weak trees (Bormann and Likens 1979).

Scattered tree throws of local origin probably would not have a major impact on nutrient inputs or lake sediments. While some new soil might be exposed and nutrients concentrated in the fallen tree trunk would eventually be released, the small area of the disturbance, particularly if it were surrounded by undisturbed forest, would limit any additional inputs to the lake-ecosystem.

It seems reasonable to conclude that presettlement northern hardwood forests in the White Mountains were subjected to frequent disturbances by wind, but mostly on a modest scale. Stands of trees on exposed sites may have suffered extensive wind damage at fairly frequent intervals.

Indians

Aboriginal populations were widespread throughout the Americas and must be considered as a potential agent affecting the composition of the lake sediments. Tsukada and Deevey (1967) found both pollen and charcoal fragments, which they related to pre-European swidden agriculture, in the cores of four Central American lakes.

As noted earlier, Indian populations were small in the Mirror Lake area (see *Chapter III.B.3*). Indians in the more mountainous regions kept to the river valleys where they spent the autumn, winter and spring months. They planted crops in early spring before migrating to the coast for the summer, at the end of which time they returned to carry out their harvest (Likens 1972c). There is no record of a village in the Woodstock region, but an Indian trail is known to have passed through, most probably along the Pemigewasset River (Proctor 1930) and Indian mounds of prehistoric origin are reported to have been found in both Woodstock and West Thornton (Kilbourne 1916). As discussed above, early explorers found small stands of white pine, which may have been related to Indian garden sites, in river valleys in the vicinity of Mirror Lake, but there is no evidence of widespread clearing or burning. Thus, while Indians were probably present in the Mirror Lake area from time to time, it is unlikely that they had a substantial effect on either the Hubbard Brook Valley or Mirror Lake.

Disturbance Intervals

Unlike the 100-yr period for natural fire rotations estimated by Heinselman (1973) for the

Boundary Water Canoe Area of northern Minnesota, the natural fire or cyclonic wind rotation for much of the White Mountain region seems to have been centuries or even a millennium. This conclusion is based on several lines of evidence. Indian populations, which were a significant factor in presettlement fires elsewhere, were sparse and had little incentive to burn forest areas deliberately; the lower elevations of the White Mountains are among the lowest wind-risk areas in New England (Smith 1946), and the northern hardwood forest is relatively resistant to fire. Although it is difficult to derive a meaningful index of fire susceptibility, extensive records of fires on national forest land from 1945 to 1976 (the period for which we have the most reliable documentation) suggest that forests in the White and Green Mountains are among the least burnable in the northern hardwood region (*Figure III.B.2–1*).

Many factors, including the recurrence of severe drought years, frequency of lightning, vigor of the vegetation, propensity to accumulate burnable matter and human activities, determine the susceptibility of a forested ecosystem to intense fires. Fires occur at the rate of 1.1 per 100,000 ha per year in the White Mountain National Forest, with about one-third of these ignited by lightning. This rate is the lowest of any of the eleven national forests located in the northern hardwood region. Of even more interest, an average of only 0.7 ha per 100,000 ha of the White Mountain National Forest actually burn each year—the smallest average area burned per year among the eleven forests. With the exception of the Green Mountain National Forest and Chequamegon National Forest in Wisconsin, the average area burned annually in the White Mountain National Forest is 15 to 40 times lower than that of other forests, including

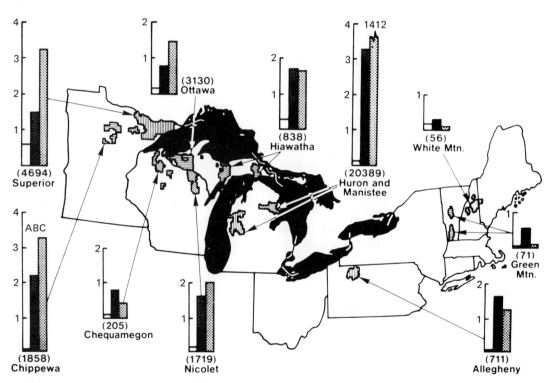

Figure III.B.2–1. Fire statistics for national forests in the northern hardwood region: Columns A and B (see Chippewa for key), average annual number of fires in units of 10, caused by lightning (A) and people (B); Column C, average acreage burned annually in units of 100 acres (1 acre = 0.405 ha). The number in parentheses is the largest acreage burned in a single year. (All data are per million acres of national forest land.) Data calculated from U.S. Forest Service Annual Fire Reports from the national forests over a 32-yr period, from 1945 to 1976. (Primary data supplied by Junius Baker, Jr., U.S. Forest Service.) (From Bormann and Likens 1979. Used with the permission of the publisher and the authors.)

B. Historical Considerations

the national forest in Minnesota that contains the Boundary Water Canoe Area. The highest rate of burning in the White Mountain National Forest in any year is 5.7 ha per 100,000 ha, which is from one to two orders of magnitude smaller than areas burned in the other national forests, again with the exception of the Green Mountain and Chequamegon National Forests (Bormann and Likens 1979).

The principal cause of fire in the northern hardwood region is human beings and their activities (Haines et al. 1975). The White Mountain National Forest receives more than twice as much recreational use per unit area as forest surrounding the Great Lakes (Bormann and Likens 1979). Despite this heavy use by people, the White Mountains still have low annual ignition and burn rates and the smallest single area burned in any one year, suggesting that a combination of meteorologic, biologic and topographic factors make the White Mountain forests less susceptible to fire.

Some of these factors can be tentatively identified. Vulnerability to fire or wind or both is determined by a number of physiographic and biologic characteristics, causing some areas to be more resistant than others to catastrophic damage. Given the well-watered, rugged landscape of the White Mountains, for instance, it seems reasonable that many areas would be protected from the easy sweep of fire or from intensive cyclonic winds.

The composition and developmental patterns of a forest-ecosystem also contribute to the vulnerability of the forest to fire and wind. Some species are more flammable than others, and the accumulation and spatial distribution of organic fuel is an important factor. Likewise, susceptibility to wind damage may be related to the structure of the stand, the size of individual trees and the degree of deciduousness. Although precise information on the susceptibility of the various developmental stages of northern hardwood forests to fire is not available, many ecologists and foresters are of the opinion that most northern hardwood forests are relatively resistant to fire (Chittendon 1905; Bromley 1935; Hough 1936; Egler 1940; Stearns 1949; Winer 1955; Bormann and Likens 1979). On the other hand, it would seem that as northern hardwood stands become composed of larger and older trees, their vulnerability to wind would increase. Toppling of large trees in an area could be coupled with an increase in the susceptibility of the area to fire.

The degree to which ecosystems are vulnerable to fire is also related to the rate at which biologic decomposition reduces the fuel load of dead wood or organic matter above the mineral soil (Wright and Heinselman 1973). In some ecosystems, litter production is so much greater than decomposition that fuel accumulation apparently reaches enormous proportions (Bloomberg 1950). In humid temperate and tropical systems, which are characterized by intense fungal, bacterial and detritivore activity, decomposition can reduce susceptibility to fire by a rapid turnover of litter with the result that the amount of dead wood and forest floor material present at most times tends to be minimal. Our studies indicate that the northern hardwood forests in the White Mountains follow this pattern (Bormann and Likens 1979).

European Settlement

The settlement of Europeans on and around the Mirror Lake watershed brought major changes in the relationship of the watershed to the dynamics of the lake. The activities of the settlers and their descendants, covered more fully in *Chapter III.B.3*, involved the clear cutting of most of the watershed, conversion of a substantial portion to pasture and arable land, a marked increase in forest fires in the vicinity of Mirror Lake, although we have no evidence of extensive burning on the watershed itself (see *Chapter III.B.1*), abandonment of agricultural land with replacement by natural succession and forest plantations, the establishment of permanent and seasonal domiciles within the watershed, diversion of Hubbard Brook through the lake, and recently the reduction of the area of the watershed following the construction of a major interstate highway. These dramatic land-use changes on the Mirror Lake watershed must have affected nutrient inputs into the lake and the quantity and contents of lake sediments.

Summary

Assuming only minor climatic fluctuations, the available evidence suggests to us that in the millennium preceding European settlement, catastrophic disturbance of the Mirror Lake watershed by fire or wind was not a major factor affecting the lake's limnology, at least on any regularly recurring basis (e.g. every century or two) such as has been postulated for lakes in the midwestern extension of the northern hardwood forest. This lack of periodic, catastrophic disturbance in the watershed, in turn, suggests that *regular* pulses of sediment, nutrient or charcoal are not to be expected in the sediment core of Mirror Lake. Indeed, from an analysis of charcoal in the Mirror Lake core (*Chapter III.B.1*), it was concluded that in the last millennium fire was not a dominant factor in the terrestrial ecosystems surrounding Mirror Lake. One is tempted to think that much of the period immediately preceding European settlement was characterized by hydrologic, nutrient and organic and inorganic sediment input not greatly different from that measured today in relatively undisturbed watersheds at higher elevations in the Hubbard Brook Valley. This picture would seem to have changed dramatically in the European period (see *Chapter VII.A.2*). Conversion of substantial portions of the watershed to agriculture, coupled with clear cutting of remaining forest resulted in changes in the sediment and nutrient input at least during those years when these activities were at their peak. Other human activities, not specifically related to the whole watershed, serve to further complicate the interpretation of the most recent sediments. These involve diversion of part of the flow of Hubbard Brook into Mirror Lake during the period from about 1850 to 1920, input of sewage, use of the lake as a regulated source of water power and recreation, and decreased watershed area resulting from highway construction.

3. Mirror Lake: Cultural History[1]

GENE E. LIKENS

On a crisp autumn day it is difficult to imagine that at one time the serene and colorful landscape around Mirror Lake was dotted with homes, a few farms and appreciable industry. Yet human use of this body of water and its watershed has changed considerably during the past, and is changing today. Even the name has changed: from Hobart's Pond to McLellan's Pond to Joberts Pond to Hubbard(s) Pond to Tannery Pond to Mirror Lake.

Currently West Thornton, at the junction of U.S. Route 3 and Mirror Lake Road, is a quiet village that attracts little more than glance as you pass through. However, in the nineteenth century West Thornton and Mirror Lake were important centers of human activity in northern New Hampshire. There were dance halls, a large "sugar bush," a soda bottling "factory," "stations" for the underground railroad, sawmills and a tannery in the vicinity of Mirror Lake. Several settlers' farms were located in the Mirror Lake watersheds. In fact, even nearby Hubbard Brook was diverted into the lake to provide a more constant supply of water for the activities associated with the lake. The lake provided a ready source of pure water as well as a pleasant setting for swimming, fishing and just plain contemplation as it does today.

What was the cultural history of this lake and its watershed? How have changes in the use of the watershed been reflected in changes in the lake? These are questions of much interest and ecologic importance, and the answers (*Chapter VIII*) provide valuable insights into environmental problems caused by increased use of aquatic resources by an expanding human population.

Indian Era

The earliest known Indian habitation in the northeastern woodlands dates from at least 10,000 to 8,000 B.C. and has been termed the Paleo-Indian period. For these people, hunting provided the primary source of food and the hunters ranged over large areas, but frequently returned to favorite camp sites. These Indians did a moderate amount of woodworking and were able to build fires (Griffin 1964). From

---

[1] This section is modified from an article published in *Appalachia* 39(2):23–41 (1972).

B. Historical Considerations

10,000 to 3,000 yr B.P., the Archaic period, and from 3,000 to 1,300 yr B.P., the early and middle Woodland periods, land was used primarily for hunting. Apparently the Indians fished, learned to fell and use trees, and had a seasonal cycle of activity that involved spending some of the year along the Atlantic coast. There was little use of the New England landscape during the Archaic period until approximately 5,000 yr B.P., apparently because the climate was cold and generally unfavorable. The period from 1,500 to 1,000 yr B.P. was characterized by semi-permanent villages, a large increase in population size, and a shift from food gathering to food production and storage. This latter activity apparently was related to the development of maize cultivation, starting in the middle Woodland period. Some evidence suggests that the ancient New England tribes were displaced by invading tribes from the south, just prior to the colonization by European settlers.

Indians of the Woodstock Area

The Indians in New England belonged to the Algonquin Family, and there were at least 12 tribes in New Hampshire at the time the Pilgrims landed at Plymouth Rock. Although there are no exact records, Indians of the Woodstock area probably belonged to the Pemigewasset Tribe and were members of the Pawtucket Confederation or Nation. Presumably an Indian with the name Pemigewasset (crooked mountain place of many pines) was selected as sagamore (chief) several centuries ago and thereafter the tribe bore his name (Speare 1964). The Pemigewassets were located near the present town of Plymouth (about 24 km south of Mirror Lake), but traversed much of the Pemigewasset River Valley into which both Mirror Lake and Hubbard Brook drain. The Pennacook Tribe, which ranged from Portsmouth, New Hampshire to Lowell, Massachusetts, led the Pawtucket Confederation until the Confederation became scattered after 1691.

All of the tribes in New England inhabited recognized tribal lands, but within each area they moved their dwelling place several times each year in accordance with season, demands for food or personal reasons, such as a death in the family (Vaughan 1965). They may have had several "permanent homes"—one for summer, one for winter and one for autumn hunting. Roger Williams wrote that the Indians migrated so often from one valley to another, it was difficult to remember which tribe was which (Child 1886). The Indians in Grafton County and the Woodstock area had practiced maize culture since 1,500 to 1,300 yr B.P., but this did not prevent frequent migration. In fact, the Indians probably made regular visits to plant small fields scattered at considerable distances throughout the river valleys of the area (Child 1886).

The population density of Indians in New Hampshire was never very great (*Figure III.B.3–1*) and since there are no records of large-scale forest clearing or burning activities it is safe to assume that the land-use practices of the Indians in northern New Hampshire, including the Hubbard Brook Valley, had minimal environmental impact (also see *Chapter III.B.2*).

Colonization and Land Use by White Men and Women

Towns were founded in New Hampshire by 1730, but it was not until 1763, at the end of the French and Indian Wars, that it was considered safe to settle north of Concord (W. R. Brown 1958). Both of the present-day towns of Woodstock and Thornton were granted in 1763. Woodstock, originally called Peeling, was granted to Eli Demerrit by Governor Wentworth. The Governor simply took a slice from the surrounding towns and made a new town, thus giving the town its first name. However, Peeling was not successfully settled and the area was regranted as Fairfield in 1771. Settlement followed in 1773 and the town name was changed to Woodstock in 1840. Nearby Thornton was settled in 1770. The first activities for these settlers must have been to clear some land and build homes to provide for their families. Obviously there are many advantages to doing this in river valleys and Mirror Lake provided a pleasant setting. However, the remaining Indians also were utilizing the river valleys and many early settlers felt it was safer to settle at higher elevations. Goldthwait (1927) believed that, "They (settlers) usually preferred the hillsides to the flat or terraced valleys, because the hills were better drained and because their soil,

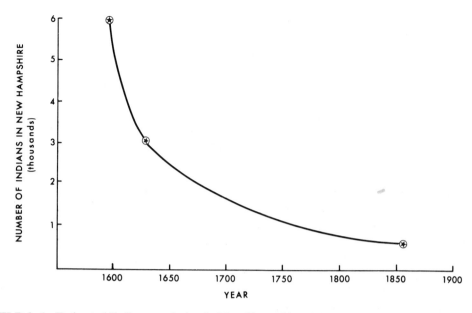

Figure III.B.3–1. Estimated Indian population in New Hampshire, 1600 to 1856. These estimates are based upon information from various sources, including Bouton (1856), Browne (1906), Child (1886), Driver and Massey (1957), Fogg (1874), Holmes (1915), Kilbourne (1916), Mooney (1928), Palfrey (1859) and Vaughan (1965), but reliable information is essentially lacking. Probably the most complete enumeration is given by Mooney (1928) who stated that the "original" (1600) population in New England was about 25,000. Palfrey (1859) and others believed that the number of permanent residents in northern New Hampshire was very low. The sharp decrease in Indian population by 1700 reflects the effects of an unknown fever or plagues that swept the southern New England coast in 1616–17, a smallpox epidemic in 1632–33 and numerous battles with the colonists and with other tribes.

though more stony, appeared less liable to become exhausted with use than the lighter loam of the valleys."

At first, Thornton grew much more rapidly in population than Woodstock and it was not until 1900 that the population of Woodstock surpassed that of Thornton (*Figure III.B.3–2*). These early settlers pastured sheep and cattle and raised crops on the cleared land. A few small farms were located in the northern part of the Mirror Lake watershed (*Figure III.B.3–3*) and in the vicinity of the present U.S. Forest Service Station within the Hubbard Brook Experimental Forest. In 1850 the principal crops in the Mirror Lake watershed were Indian corn, potatoes and oats. There was a Potato Hill district in Woodstock and eventually mills were built for manufacturing potato starch. As more settlers came into the area, less desirable land was cleared in an attempt to support families. In describing the settlement activities in central Grafton County, Goldthwait (1927) states,

"Abundant evidence remains of the vigor and hopefulness of these pioneers. With their ox teams they cleared incredible areas of hill country. Trees were chopped down or burned, stumps were uprooted, fieldstone was dragged off to make walls along every road and around every pasture; and crops sprang up on every sunny hillside." In the Hubbard Brook Valley, however, the main source of employment and income soon was derived from the forest itself in the form of timber products and by-products such as tannins from the bark of hemlock trees.

There are no accurate data on the amount of land that was cleared in New Hampshire, let alone the Mirror Lake watershed, and estimates vary. Old accounts in journals of the 1850s report that it was possible to stand in Lyme (central Grafton County), New Hampshire and look in any direction without seeing a tree. This condition is difficult to imagine since today 84% of the state is covered by forests. W. R. Brown (1958) estimates that by 1840 one-half of the

B. Historical Considerations

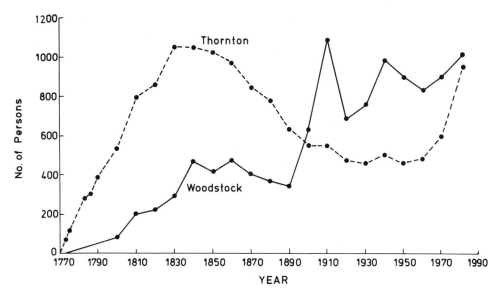

Figure III.B.3–2. Population in the towns of Woodstock and Thornton, 1770 to 1980.

state of New Hampshire was in cultivated land, but by 1860 this began to decline. But by 1840 the population was decreasing throughout the state, particularly in upland areas ". . . for these hill farms were not all productive, and they became less so as successive crops were harvested and fields were not sufficiently fertilized" (Goldthwait 1927).

Sawmills

Apparently Mirror Lake was previously called McLellan's Pond, probably because Aaron McLellan's Up and Down Sawmill was located on the outlet stream of the lake. McLellan built his sawmill on the lake around 1823, and about 1850 built a small cribwork dam filled with rock

Figure III.B.3–3. Mirror Lake, looking south toward swimming beach. Note the stone fence in the foreground, which indicates that clearing on the watershed was done at least in part for agricultural purposes (photograph published in *The Granite Monthly* in 1897).

and faced with wooden plank at the outlet to increase the water supply and facilitate its control. The dam raised the water level in the lake by approximately 1 m. The remains of this dam can be seen at the constriction some 70 m north of the present outlet of the lake. Also about 1850, a ditch was constructed so that a wooden baffle built in nearby Hubbard Brook would divert water from the Brook into Mirror Lake. The lake may have been used as a holding pond for logs, but its primary use was for waterpower to run the mill. McLellan's Mill burned, probably between 1856 and 1860.

Wilson Hill built another sawmill below the outlet of Mirror Lake in 1865. This mill was purchased by J. M. and W. N. Moulton in 1870 (*Figure III.B.3*–4). By 1886 it sawed one million board feet (1 board foot = 2,359 cm^3) and turned out 250,000 shingles each year (Child 1886).

About 1900 the lake level was raised an additional 0.5 m or so when John Emmons built another stone dam at the outlet. This dam was much larger and some 70 m downstream (south) of the original dam built by McLellan. Emmons' dam remains today as a portion of the reinforced structure built in 1964 by the New Hampshire State Water Resources Board.

In 1909 W. D. Veazey and Simeon C. Frye, partners in W. D. Veazey and Company of Laconia, New Hampshire built a large sawmill near the site of the Moulton Mill, a short distance downstream from the outlet of Mirror Lake (*Figure III.B.3*–5). It was moved from the Valley shortly after 1920.

Some white pine were logged on the lower portions of the Hubbard Brook Valley around 1879 (see Federer 1969) and hemlocks were used extensively as a source of tannins (see below), but prior to 1895, most of the forest in the Valley was nearly virgin, old-age northern hardwoods, with conifers particularly on north-facing slopes and at higher elevations. There has been some controversy concerning the predominance of conifers, particularly red spruce, in these early forests, but it is believed that the

Figure III.B.3–4. Moulton Sawmill at the outlet of Mirror Lake (photograph taken about 1890?).

B. Historical Considerations

Figure III.B.3–5. W. D. Veazey and Company Sawmill below outlet of Mirror Lake (photograph taken about 1910–16).

present-day composition of the forest approximates that which preceded it (see Bormann et al. 1970 and *Chapter III.B.1*).

Extensive cutting of the forest began around 1900, and by 1920 the whole valley was cut over (Bormann et al. 1970; *Figure III.B.3–6A* and *6B*). Salvage logging was done after the 1938 hurricane and experimental clear cutting of small areas was done in the HBEF during the 1960s and 1970s (see Likens and Bormann 1974a; Bormann and Likens 1979). Otherwise, the forest in the Valley has been regrowing since about 1920. A large area of the watershed along the northern shore of Mirror Lake had been cleared prior to 1900 (*Figure III.B.3–3*), presumably for agriculture. Much of this area was replanted in 1934 with white pine (*Figure III.B.3–7A* and *7B*) and is now heavily forested.

Tanneries

Richard Danforth came to Woodstock from Bristol, New Hampshire in 1852 and with Warren White built the first tannery building near the outlet of Mirror Lake (Child 1886). A brother, Almon H. Danforth, helped while the tannery was being built. A waterwheel installed in McLellan's dam provided power. The enterprise prospered and Charlton reported in 1856 that the tannery employed 20 persons. This first tannery burned in 1856 and Almon Danforth sold his interest in the business to Joseph Campbell. The second tannery was built on the same site and operated until 1888 when it burned. The tannery business continued to be prosperous and the buildings were large and roomy. The main building was 36 m long and 12 m wide, and joined to it were two wings— one 12 × 6 m, and the other 10 × 6 m. Bark peeled from hemlock trees in the area provided the tannins. The hemlocks were felled in the spring, and the bark was stripped and hauled to the tannery. Hides were brought to the tannery by horse- and oxen-drawn carts until near the end, when leather was shipped in on the new railroad (*Figure III.B.3–8*). Fogg (1874) reports that receipts from the sale of leather products from this tannery amounted to about $75,000

Figure III.B.3–6. Logging in the Hubbard Brook Valley in the winter about 1910. **A.** Looking northeast, from the Valley above the Mirror Lake watershed; a load of logs passing by several logging camps on the way to the mill. **B.** The Veazey Mill in the Valley, south of the Mirror Lake Road. The mostly deforested hill in the background is a part of the Mirror Lake watershed.

B. Historical Considerations

Figure III.B.3–7. A. Mirror Lake beach looking north (photograph taken prior to 1934). **B.** Mirror Lake, from town beach looking north, 4 June 1971. Note the increase in forest vegetation on the north shore since 1934.

Figure III.B.3–8. Junction of U.S. Route 3 and Mirror Lake (Tannery) Road. Mirror Lake Road extends into background, left and U.S. Route 3 curves to the right. Note railroad crossing signs and blacksmith shop in background, right on U.S. Route 3 (photograph taken about 1916–23).

per year. In 1886 Child indicated that the tannery had 12 employees and finished 12,000 sides of leather a year. The tannery so dominated the scene that Mirror Lake became known as Tannery Pond during this period (Hurd 1892). There were other ties between the logging and tanning industries. All of the lumber for the first and second tanneries was sawed at the McLellan Mill.

Bottling Works and Dance Halls

About 1892 a bottling works and dance pavilion was constructed by Joseph and Jesse Kendall along the north shoreline of Mirror Lake. Because there was no easy access to the north side of the lake, patrons would row across the lake in boats obtained in the vicinity of the present town beach on the south shore. The bottling works was moved about two years later to the southwest shoreline of the lake (present site of the Mirror Lake Hamlet) to provide better access for stagecoaches. Lemon, vanillin, root beer, birch beer, ginger beer and sarsaparilla were bottled and sold throughout New England. Part of the water for these soft drinks came from a well that is on a hill north of the lake, and was mixed with water from a bubbling spring on the side of the hill, which is now part of the HBEF. A mixture of these two waters produced the desired carbonation. The flavoring extracts were made at nearby Pleasant View Farm during the winter months. When extensive logging started about 1908 on nearby land, the bottling works and dance pavilion were moved to a new site east of U.S. Route 3 and out of the Mirror Lake watershed.

Camps

A variety of camps have been located along the shoreline of Mirror Lake since 1800. The most notable one was a camp for boys (Camp Musquaw, or black bear) founded by the Rev. J. Lincoln Thomas and Mr. Harold W. Lawrence in 1924. Land was cleared on the southern shore approximately 75 m west of the present public beach and tents were erected. As many as 50 boys attended camp each year at this location. This camp was moved in 1942 to the western shoreline of Mirror Lake (the present location of the Thomas cottage) and then closed in 1949. Currently, a rather large family camp (Camp Osceola) is located along the eastern shore.

Around 1894 the lake was renamed Mirror

B. Historical Considerations

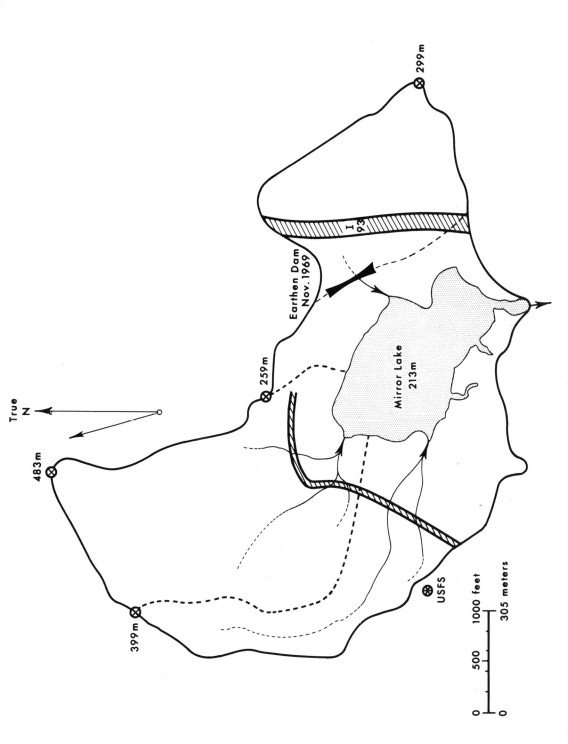

Figure III.B.3–9. Outline of Mirror Lake watershed, showing tributaries, subwatershed boundaries, access road, Interstate 93. The structure on the northeast tributary is an earthen dam built prior to the construction of I-93.

Lake by J. M. Kendall. Fogg (1874) describes the inhabitants of Woodstock as being employed in farming during the summer and lumbering in the winter. Thus after Campbell's tannery burned in 1888 and the heavy logging was completed in the Hubbard Brook Valley by 1920, the population around the lake declined and the land gradually became reforested.

Present Land Use

Although the number of buildings within the Mirror Lake watershed has increased during the last 120 years (Walling 1860; Hurd 1892) the year-round use has declined. There are five permanent residences in the Mirror Lake watershed and two small tourist camps with cottages and tents. There are three other small summer cottages and two mobile homes (three of these were established during 1970). One of the permanent residences has a tiny farm. The lake currently is used primarily for recreation—swimming, boating and fishing. Some water is used for drinking and there are no major inputs of sewage except for drainage from septic tanks and dry wells.

In 1969 the construction of Interstate 93 changed the watershed of Mirror Lake considerably (*Chapter III.A*). At one point on the eastern shoreline, the right-of-way of the road comes within approximately 110 m of the shoreline. And during 1970 an access road was added from the Mirror Lake Road to the northern side of the lake (*Figure III.B.3*–9). An earthen dam was built on the northeastern tributary to Mirror Lake prior to construction of I-93 to prevent the erosion and transport of silt and organic matter into the lake during road building. But the dam cuts off 68% of this subwatershed, or about 21% of the total lake watershed, thus decreasing the amount of water draining into the lake. Also, a drainage system was built to carry the road salt from I-93 away from the drainage into Mirror Lake, but salt (NaCl) content of the northeastern tributary is currently more than ten times higher than it was in 1969 (see *Chapter IV.D*).

Summary

Mirror Lake has reflected its surroundings in various ways since the formation of the lake some 14,000 yr ago. The diverse cultural activities associated with the lake and its watershed undoubtedly have had various effects on the ecology of the lake throughout this history.

Indians inhabited the area until the mid 1700s, but population density was low and the Indians were probably somewhat nomadic. Also, there are no records of large-scale forest clearing by fire or for agriculture, so human effects in the watershed were probably negligible until some 220 yr ago when white men and women settled in the area.

In the nineteenth century, Mirror Lake was an important center of human activity. A relatively large sawmill, a tannery, a dance hall, a soda bottling works, and camps all were located at one time or another within the Mirror Lake watershed. A dam was built on the outlet, and even nearby Hubbard Brook was diverted through the lake for water power.

Several small farms were located in the watershed during the nineteenth and early twentieth centuries. Some additional forest clearing was done prior to 1900, particularly the cutting of hemlock for bark used in the tanning operation, and heavy logging occurred during the period 1900–20.

After this intense logging activity of the early 1900s, the human population declined in the watershed and the land became reforested.

Today only a few houses and small seasonal camps remain, and the principal use of the area is for recreation. However, in 1969–71 an interstate highway (I-93) and access road were constructed through the watershed. The current population of the town of Woodstock (1,008) is close to its all-time high, and Thornton has nearly rebounded to its peak population size in 1830.

General Chronology for the Hubbard Brook Valley

Indian Era

12,000–10,000 yr B.P.	Paleo-Indian Period
10,000–3,000 yr B.P.	Archaic Period
3,000–2,500 yr B.P.	Early Woodland Period
2,500–1,300 yr B.P.	Late Woodland Period
1,300–1,000 yr B.P.	Pawtucket Confederation Scattered

White Men and Women Era

1763	Peeling (= Fairfield, = Woodstock) granted to Eli Demeritt by Governor Wentworth
1763	Thornton granted
1770	Thornton settled
1773	Fairfield (Woodstock) settled
early 1800 to 1920	Hubbard Brook diverted through Mirror Lake
1805	Hobart's Pond (= Mirror Lake)
1816	"Year Without a Summer"
1823	McLellan Pond (= Mirror Lake)
1823 to 1858?	McLellan's Up and Down Sawmill
1830	Joberts Pond (= Mirror Lake)
1840	Town name of Fairfield changed to Woodstock
1850	Upper stone dam constructed at outlet of lake; canal constructed to divert water from Hubbard Brook to lake
1850s	Present building of Osceola Camp built by J. Campbell
1852 to 1856	Danforth and White Tannery (burned 1856)
1852	Fifteen sawmills in Woodstock
1856 to 1888	Campbell Tannery (burned 1888)
1860 to 1910	Hubbard(s) Pond (= Mirror Lake)
1865 to 1908?	Hill (= Moulton) Sawmill
1883 to 1884	Railroad completed to North Woodstock
1892	Tannery Pond (= Mirror Lake)
1892 to 1908?	Mirror Lake Bottling Works and Dance Pavilion
1894	Mirror Lake named by J. M. Kendall
1895 to 1907?	J. Emmons built stone dam at present site on the outlet (lower dam)
1900 to 1920	Heavy logging in Hubbard Brook Valley
1909 to 1918	W. D. Veazey and Company Sawmill
1912 to 1921	Last school house in Hubbard Brook Valley (home of the late Henrietta K. Towers)
1920	Hubbard Brook watershed sold to White Mountain National Forest
1924 to 1942	Boys' camp (Camp Musquaw) along southern shoreline
1927 to 1937	Hubbard Brook State Game Refuge
1933	H. Emmons repaired and strengthened lower dam at outlet to Mirror Lake
1934	White pines planted in field on north side of Mirror Lake
1934	Construction of Mirror Lake Dance Hall (girls' camp)
1936	"Year of the Big Flood"
1938	"Year of the Hurricane"
1942 to 1949	Camp Musquaw moved to western shoreline
1955	Establishment of Hubbard Brook Experimental Forest, U.S.D.A. Forest Service
1963	Initiation of the Hubbard Brook Ecosystem Study
1964	Lower dam at Mirror Lake outlet repaired and strengthened by the New Hampshire Water Resources Board
1965	Town developed and opened beach on south side of Mirror Lake
1969 to 1971	Construction of Interstate 93 through watershed and access road around west side of Mirror Lake
1971	Town ordinance forbidding use of motor boats on Mirror Lake
1971	Mirror Lake Dance Hall destroyed by fire—arson

Chapter IV

Mirror Lake—Physical and Chemical Characteristics

A. Importance of Perspective in Limnology

W. John O'Brien

To the child enraptured by the stars on a clear night or delighted with the diminutive world within an ant hill, questions of scale come naturally. How big am I? How far away is the moon? How long until my next birthday? Physicists, too, contemplate the perspective of scale as they struggle to understand the forces controlling the expansion of the universe or compression of the atomic nucleus. Philosophers and writers such as Swift and Carroll, as well as Dr. Seuss (1958)* in more modern times, have dealt with the importance of scale and perspective. However, limnologists, with a few exceptions (Allen 1977; Harris 1980), have apparently been unconcerned about the vast differences in scale with which they deal.

Limnologists make measurements in an attempt to understand the workings of the world's largest lakes and also of such components of lakes as fish, phytoplankton, and phosphate molecules. Yet a phosphate molecule is 10^{25} times lighter than a 1-kg fish, which is 10^{16} times lighter than the mass of the world's lake of greatest volume, Lake Baikal. Thus the limnological scale ranges from a lake whose water weighs 23,000,000 megatons to a phosphate molecule weighing 1.6×10^{-16} μg; more than 41 orders of magnitude separate Lake Baikal from a phosphate molecule. We think of the sun as huge compared with the Earth, but the mass of our sun is only 3.32×10^5 times greater than that of the Earth. To have a mass 1×10^{41} times greater than the Earth, the sun would have to be considerably greater in diameter than our entire solar system. Put another way, our sun is only 1×10^{27} times the mass of an average person, while Mirror Lake is 6×10^{33} times the size of a phosphate molecule (*Table IV.A–1*).

Despite the magnitude of these scale differences, limnologists often use the very same instruments and methods to make inferences about how lake-ecosystems function and how phytoplankton cells behave. It is not uncommon for one limnologist interested in the workings of a lake and another interested in the environmental physiology of phytoplankton within the lake to sample the lake water or collect the phytoplankton with the same type of apparatus. It is important to remember that the size and scale of sampling devices have been most influenced in design by the average size of the limnologists who were to heft them.

The Scale of Measurement

Dramatic changes in scale within lakes are partly obscured by the common convention of shifting the units we use to express quantities and mass at each trophic level. Concentrations of plant nutrients are expressed in micrograms per liter, phytoplankton as number per milliliter, zooplankton as numbers per liter, and fish densities in kilograms per hectare (*Table IV.A–2*). Most of the numbers, then, that are used day by day are in units, tens, or hundreds. This convention makes it easier to remember and compute, but obscures vast changes in size and mass from one trophic level to another. A fish may weigh 1 kg but have a density of only 1

---

* This chapter is a continuation of a theme first presented in a paper entitled "Hearing Whoville: The importance of perspective in limnology," delivered at the 1980 meeting of the American Society of Limnology and Oceanography.

A. Importance of Perspective in Limnology

Table IV.A-1. Limnological Scale

Limnological Unit	Weight Expressed in Scientific Notation (g)	Weight Expressed in Fixed Notation (g)
Phosphate	1.6×10^{-22}	0.00000000000000000000016
Chlorella	5×10^{-13}	0.0000000000005
Daphnia	5×10^{-4}	0.0005
Bluegill	5×10^{1}	50.0
Largemouth bass	2.5×10^{3}	2,500.0
Mirror Lake	1×10^{12}	1,000,000,000,000.0
Lake Baikal	2.3×10^{19}	23,000,000,000,000,000,000.0

per 100 m³. By contrast, there may be 10^{16} phosphate molecules in 0.001 m³ of water from Mirror Lake; in mass, though, all the phosphate molecules in even 100 m³ will weigh but 0.16 g. Intriguingly, a 1-kg fish could well have 1 or 2 g of phosphate within its body.

Discontinuities in size and scale are difficult to comprehend. For example, a liter of water is as vast a space to a tiny algal cell as Yellowstone National Park is to a person. Phytoplankton that settle 0.2 to 0.4 m per day, a mere sweep of the human arm, are undergoing a considerable journey, comparable in human terms to more than 10 km. Likewise, vertically migrating zooplankton make similarly long journeys, equivalent to a commuter traveling 10 km in the morning and making another 10-km trip in the evening.

When zooplankton move quite rapidly, as some commonly do to evade their predators, they do so with unappreciated speed. A diaptomid copepod can travel at a speed of 145 cm/s (Swift and Fedorenko 1975). Of course, their jumps are not sustained over a full second, but a jump of even 0.1 s represents 14.5 cm or 145 body lengths for a 1-mm animal. A human running 145 body lengths in 0.1 s could win the 100-m dash in 0.034 s, which is much faster than the speed of sound (345 m/s) and would thus produce a sonic boom. Even the more modest measurements of copepods traveling 3 cm in less than 0.1 s would still be moving at the relative speed of a drag race car. One can appreciate how effective such an anti-predator jump might be by considering that a 3-cm jump of a copepod pursued by a fish is equivalent to a 2.5-m leap of a mouse away from an attacking wolf. As impressive as this is, such a 3-cm jump by a copepod still leaves it within the attack distance of most planktivorous fish.

We often recognize discrepancies in scale in one area or in the behavior of one animal group but fail to realize how common such differences are. A particularly good example is the popular amazement surrounding the eating habits of baleen whales, which strain small animals from sea water. Although these whales are several million times heavier than the prey upon which they feed, this situation is not particularly unusual in aquatic systems. A 2-mm long *Daphnia* for example, is ten million times heavier than the algal cells it eats. And consider the task of the *Daphnia*: it may have to handle 25 ml of water each day to extract its phytoplankton food, an amount equivalent to a person's raking leaves from four football fields every day.

Water in Perspective

It is not really possible to make all situations comparable in scale to humans. The tiny world of algae and zooplankton is fundamentally different from the one in which we exist. A profound difference lies in that nature of water, which causes it to be experienced differently by different-sized organisms. To humans, water is free flowing and easily manipulated, but to

Table IV.A-2. Common Densities or Weights at each Trophic Level

Phosphorus	10 μg/liter
Phytoplankton	100,000/ml
Zooplankton	50/liter
Fish	250/ha

small creatures it is viscous and enveloping. The effect of viscosity on objects moving through the water may be expressed by the Reynolds number (Re): $Re = ls/v$, where l is the length of the animal or plant, s is the average speed of the organism, and v is the kinematic viscosity, which in water is equal to 1×10^{-2} cm^2/s. If objects are large, move at high absolute speeds, or both, the Reynolds number is high, and water is experienced as we know it. A person swimming at 50 cm/s has a Reynolds number around 1×10^6. Dolphins, larger and faster than humans, have a still higher Re, while small goldfish and guppies have a Reynolds number of around 100. It is not until an organism's Reynolds number approaches 1 that water is experienced differently. A 1-mm *Daphnia* swimming 5 mm/s in Mirror Lake has a Reynolds number of 5, and a phytoplankton cell 10 μm in diameter, settling in the epilimnion of the lake at 0.2 m per day, has a Reynolds number of 2×10^{-4}.

At these low Reynolds numbers, water is sticky. As a human analogy, imagine swimming in a vat of molasses. The stickiness of water at low Reynolds numbers comes not from any fundamental change in the nature of water, but from the relationship between an organism's surface area and its volume. The surface area:mass ratio of phytoplankton and zooplankton is much greater than that of a fish or a person. Water, being a partially polar molecule, sticks to the surface of most objects placed in it, and thus forms a tiny microscopic film over all surfaces. This film of water is a trivial addition to the weight of a person in water, but represents a considerable shell when it surrounds a small phytoplankton cell. This water, which is literally stuck on the surfaces of small creatures such as phytoplankton and zooplankton, is shed only slowly and causes a substantial amount of drag on these organisms. This drag is so powerful that, in the world of low Reynolds numbers, momentum and inertia are totally irrelevant. When a zooplankton stops actively swimming it ceases moving almost instantly, with no residual glide.

Furthermore, while limnologists commonly refer to herbivorous zooplankton as being "filter feeders," zooplankton cannot filter algae from water the way we strain spaghetti from the cooking water with a colander. At low Reynolds numbers water is far too viscous to pass easily through the fine setae on zooplankton filtering appendages (Porter et al. 1982). If the water were, in fact, strained through these fine setae, it would have to be pumped through. Anyone who has tried to use a fork to strain a large crumb from a jar of honey will have an intuitive idea of the problem a hungry zooplankter faces in trying to obtain food. The process is considerably more strenuous than the word "filtering" suggests.

How Fast Is Slow?

There is another fundamental difference between the world of tiny creatures and that of humans: the speed of diffusion. At whole-body human scale, diffusion is slow. If our bodies used diffusion as the mechanism for obtaining oxygen, it would take days for new air to travel into our mouths and down to our lungs. Instead, we actively pump air in and out. By contrast, an oxygen molecule, and in fact most moderate-sized molecules, can diffuse across the diameter of an algal cell within seconds. It is not, of course, the speed of diffusion that changes, but the distances that are so vastly different. An algal cell may be but 10 μm long, whereas the human respiratory system is 500,000 μm long, a difference of 5×10^4. Thus, what is a thoroughly impractical method of obtaining oxygen or nutrients at the human scale is quite rapid and efficient on the plankton scale.

Worlds Within Worlds?

Another intriguing aspect of limnological scale is the vastness of even a small volume of water when compared with tiny phytoplankton cells. To a small algal cell, a liter of water is as large as the city of Los Angeles is to us. Consider the heterogeneity that exists in Los Angeles, and then imagine the micro-scale heterogeneity that might conceivably exist in a liter of Mirror Lake water.

Goldman et al. (1979) and Lehman and Scavia (1982) recently suggested that zooplankton excretion of plant nutrients such as nitrogen and phosphorus might produce micro-patches of concentrated nutrients. Such micro-patches might explain how ocean and lake phytoplankton can obtain nutrients even when average concentrations, which are what we measure, are very low. However, on the molecular scale, diffusion is so rapid that any micro-patches created by zooplankton would be short lived (Jackson 1980). Diffusion, a process that works at less than a snail's pace in human scale, operates with amazing rapidity in the scale at which such nutrients would be released by zooplankton. A phosphate concentration of 50 μg delivered into a volume of 5 μliter would diffuse to background levels within 10 s. Thus, even if phytoplankton came across micro-patches of nutrients left behind by zooplankton, their rate of uptake seems to be far too slow for them to obtain significant amounts of nutrients before the patch would diffuse away. If the concentration within a patch is assumed to be larger because of less frequent release by the zooplankton, it would, of course, diffuse away more slowly, but the opportunities for algae to encounter a patch would also be reduced since micro-patches would be less common. If algal uptake were faster than currently thought, diffusion still would be too slow to deliver more nutrients to the cell surface. Thus, it seems that no matter what the assumptions, phytoplankton might at best obtain only a small percentage of their daily nutrient needs from patches of concentrated nutrients. It seems that a liter of water is rather homogeneous to an alga. Thus, a more reasonable analogy to a person might be the large surface area of an ocean or a lake, rather than Los Angeles.

The Rate of Time

Limnologists also tend to consider time in human perspective, rather than adapting it to the short life spans of many lake organisms. An example is the expression of zooplankton filtering rates in milliliters of water processed per animal per hour, a measurement that assumes feeding is roughly constant over an hour. However, one hour in the life of a zooplankter might be 1% of its total life span, equivalent to 8.5 months in the life of a person who lives to be 70 yr old. Much can happen in an hour to a zooplankton, and one way in which zooplankton may selectively graze is by feeding more rapidly when they are in regions of preferred food and more slowly when surrounded by less preferred food (Richman et al. 1980).

Consideration of temporal scale is also important in phytoplankton experiments. It is common for phycologists to measure phytoplankton uptake of various nutrients (carbon, nitrogen and phosphorus) over a period of up to several hours. In fact, a fairly standard procedure for estimating inorganic carbon uptake is to incubate lake phytoplankton with a ^{14}C tracer for as much as 4 h. Such a procedure may be useful in describing the response of an entire phytoplankton community, but 4 h may represent half the life span of an individual algal cell. It is entirely possible that phytoplankton behave differently in the morning, when first exposed to light, compared with the afternoon, after almost half a lifetime's exposure to light. It does not seem strange to us that birds and mammals behave differently in the spring than they do in autumn, yet we generally disregard the possibility that phytoplankton may behave differently at different times of day.

Our perception of time may likewise influence our appreciation of the rate of nutrient uptake by phytoplankton. It is common to express nutrient uptake as the number of moles of nutrient assimilated per million algal cells per hour. As mentioned previously, an hour is a long time for an algal cell, particularly considering how rapidly the individual phytoplankton cell can obtain nutrients. Commonly reported uptake rates range from 1×10^{-7} to 1×10^{-9} micromoles per cell per hour, or between 1×10^8 and 1×10^{10} nutrient molecules per cell per hour. Even the lower rate represents approximately 30,000 nutrient molecules per cell *per second*. That is fast eating on any scale. However, the actual rate of eating cannot be calculated from such information as we now have, because we do not know how many *mouths* (uptake sites) are doing the eating. Are there 30,000 uptake sites, each taking up one nutrient molecule in each second? Or only 30 very rapid uptake sites each taking up one thousand molecules?

How Much Is Enough?

When we fail to consider discrepancies in scale, even our ideas of plenty and scarcity may be inappropriate. Consider the nutrient trophic status of lakes. A major theme in modern limnology has been to demonstrate that in many lakes overall productivity during the warm portions of the year may be regulated by the concentration of phosphorus in the lake water. Lakes such as Mirror Lake, with less than 5 μg PO_4-P per liter, are considered unproductive or oligotrophic, whereas lakes such as Lake Erie, with phosphorus levels ranging from 30 to 50 μg PO_4-P per liter, are eutrophic, often supporting massive blooms of phytoplankton (Vollenweider 1968; Likens 1975b). It is certainly true that the addition of phosphorus in large quantities can cause severe problems in water quality. However, even in an oligotrophic lake, 3.17×10^{16} molecules of phosphate exist in every liter of water, whereas a eutrophic lake has 3.17×10^{17} molecules per liter. The difference is one of perspective. Consider the task of the algal cell in finding phosphorus. Even in a eutrophic lake, each molecule of phosphate is embedded in a matrix of water, and there are 3.34×10^{25} molecules of water per liter. In other words, for every phosphate molecule there are one hundred million water molecules. Thus, even in a eutrophic lake a phytoplankton cell must endure a vast desert of non-nutritive water between each phosphate molecule that reaches it.

Then what makes a nutrient concentration of 3.17×10^{16} scarce and of 3.17×10^{17} abundant? The answer must lie with diffusion. At 5 μg PO_4-P, diffusion delivers phosphorus to the near-cell environment much more slowly than the rate of uptake; cell growth is therefore dictated by how rapidly diffusion moves nutrients close to the algal cell where they can be assimilated. When more phosphorus is present, diffusion delivers much more phosphorus to the near-cell environment, allowing phosphorus uptake to occur more rapidly than new cells can be built to use it.

Phytoplankton can do little to fight the tyranny of diffusion. At low Reynolds numbers they are surrounded by a sticky shell of *old* water that can be shed only slowly. Even if they settle or are flagellated and move actively, their nutrient uptake is still dominated by what diffusion can deliver (Purcell 1977).

Through an Elephant's Eyes?

Dr. Seuss saw Horton the elephant as the one to appreciate worlds too small to be easily seen or appreciated by others. If we are to study effectively and understand the workings and lives of creatures much smaller than ourselves, we too will need to develop Horton's empathy. We must constantly remind ourselves that to the tiny, water is sticky and diffusion speedy. We must remember lifetimes for phytoplankton and zooplankton come and go between weekends. Most importantly, we should strive to express our understanding of things as realistically as we can.

Summary

1. Limnologists work to understand systems as large as Lake Baikal and as small as nutrient uptake by a single algal cell, often using the same sampling apparatus, which has been designed primarily for human convenience.
2. The units of measurement used by limnologists are shifted with the scale of the organism or process being measured. This convention allows us to make computations in whole numbers but may obscure some important relationships.
3. Small aquatic organisms such as phytoplankton and zooplankton experience water and diffusion very differently than do human-sized creatures. To the very small, water is viscous and diffusion is very rapid. Our understanding of the life processes of such animals and plants will be enhanced by appreciating the way they experience the aquatic environment.
4. Most small aquatic organisms have such brief life spans that experiments extending for days, and in some cases just hours, may encompass most of the creature's life. Averaging data collected over extended periods of time may therefore be misleading when extrapolated to natural situations.

B. Physical and Chemical Environment

Gene E. Likens, John S. Eaton and Noye M. Johnson

Mirror Lake is typical of many of the lakes in northern New England. It is a small, clear, nutrient-poor lake surrounded by a watershed forested primarily by northern hardwood vegetation. There is some hemlock and white pine, mostly along the shoreline, a small white pine plantation along the northern shore and some spruce and fir, especially on ridgetops and near rock outcrops. The steep terrain of the current 85-ha watershed is drained by three primary subwatersheds identified as Northeast (NE; 10.4 ha), Northwest (NW; 42.6 ha) and West (W; 32.9 ha) (see *Figure III.B.3–9*). A small stream enters the lake from each of these subwatersheds. However, drainage water from some areas within these watersheds does not contribute to streamflow, but enters the lake directly as subsurface flow. Hydrologically these latter areas have been considered as separate from the subbasins drained by streams (see *Figure III.A–2*).

The sediments near shore are characterized by sand and gravel along the east, south and west, by boulders, cobbles and gravel along the north and by a bedrock outcrop on the northeast. The shallow pond between the present outlet dam and one constructed by early settlers has a mixture of sand and organic debris on the bottom. A facies map of the bottom sediments for the lake is given in *Figure III.A–6*. Rooted aquatic vegetation is very sparse throughout most of the lake (see *Chapter V.A.4*).

Currently, human use of the lake is limited to swimming (there is one small public beach), fishing and some boating (no gasoline-powered boats are allowed on the lake). Several small summer cottages, a camping area, a few year-round residences and small garden/pasture areas are present in the watershed near the lake (see *Chapter III.B.3*) for more information on the cultural history of the lake).

General Morphometry

Mirror Lake has an area of 15 ha, a maximum depth of 11.0 m and a mean depth of 5.75 m (*Figure IV.B–1*). The lake area to watershed area ratio is 1:5.7. Additional morphometric data are given in *Table III.A–1*. There is one surface outlet for the lake, which is currently dammed to about 1.5 m above the original outlet level (see *Chapter III.B.3* for the history of the dam). Area-depth and volume-depth curves for Mirror Lake are given in *Figures IV.B–2 and 3*. The original basin was thought to approximate a conic frustum (Lehman 1975), but a hyperboloid model may be more appropriate (see *Chapter VII.A* for sedimentation models). Mean depth divided by maximum depth of the present-day basin gives a value of 0.52, higher than for many kettle lakes (Wetzel 1975), and characteristic of ellipsoid basins (Carpenter 1983). Although useful in the abstract, these models do not account for irregularities, such as the conspicuous shelf along the southwest shore (*Figure IV.B–1*).

The Mirror Lake basin is presently filled with sediment to more than one-half of its original depth. Its current volume is 862×10^3 m^3 with 58.4% of the total volume contained in the top 4 m. In contrast, the bottom 4 m contains only 9.8% of the total volume. The volume below a depth of 10 m is 2,000 m^3; and because this volume receives organic matter settling from the much larger volumes of water above, it frequently becomes anaerobic. Thus, anaerobic bottom waters, usually typical of eutrophic lakes, are common below 10 m in oligotrophic Mirror Lake (see below) because of this volume effect (sediment focusing).

Thermal Characteristics

Mirror Lake shows many of the thermal characteristics of small lakes in the North Temperate Zone. It is thermally stratified with depth during the summer, inversely stratified during winter when covered by ice and snow, and usually has an isothermal period in spring and autumn when the entire lake mixes convectively (*Figure IV.B–4; Chapter IV.C*). Lakes with two complete mixing periods per year are referred to as dimictic (see Walker and Likens 1975, for discussion of mixing patterns and terminology in

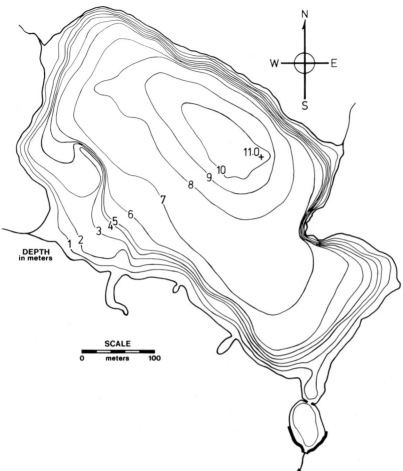

Figure IV.B–1. A depth-contour map of Mirror Lake. Contour intervals are in meters.

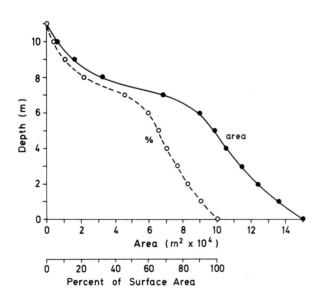

Figure IV.B–2. Hypsographic (depth-area) curves for Mirror Lake. Note different scales on the abscissa.

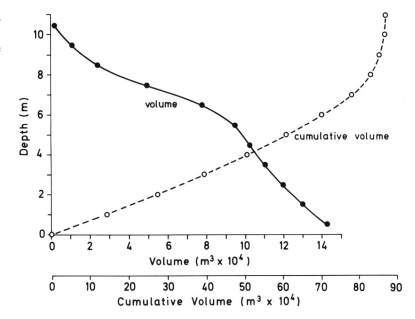

Figure IV.B–3. Depth-volume curves for Mirror Lake. Note different scales on the abscissa.

lakes). In summer the upper, relatively warm, isothermal wind-mixed layer is called the epilimnion. Below this layer the temperature decreases rapidly and the plane of maximum temperature change is called the thermocline (Hutchinson 1957). The bottom, relatively cold layer that is not mixed by the wind is called the hypolimnion. Complete mixing or *overturn* does not occur consistently during the spring in Mirror Lake. The lack of such mixing in the spring is referred to as spring meromixis (Åberg and Rodhe 1942).

During 1968 to 1970 we installed continuously recording thermographs in the deep water of the lake to examine the detailed thermal record of the hypolimnion. The record during each of the two years when the bottom water temperature was measured continuously was remarkably similar (*Figure IV.B–5*). Convective mixing reached the depth of the thermographs (~8.5 m) at approximately the same time in late October during both 1968 and 1969. At first the water warmed at this depth for several days and then cooled sharply at an average rate of 0.35°C/day from 9 to 21 November 1968 and 0.32°C/day from 13 November to 1 December 1969. The lake cooled, with continued mixing and loss of heat, until an ice-cover formed. Thereafter the bottom water warmed slowly throughout the winter. Note that the overall lake cooled well below 4°C before it froze over, immediately began to warm when the ice-cover formed, and that the initial rate (first 5 to 20 days) of heating in the bottom water (presumably by conduction of heat from the bottom sediments) was much more rapid than throughout the remainder of the ice-covered period. This variation in sediment heat flux throughout the ice-covered period had been postulated but not verified earlier (Likens and Johnson 1969). In late March and early April of 1969, some cooling in the bottom waters was observed, which was presumably related to a rather catastrophic event at the surface of the lake. On 11 March two water-level boards, and then on 25 March 1969 seven water-level boards, were removed from the dam at the outlet by personnel of the New Hampshire Water Resources Board. These boards were removed to lower the lake level as a flood-control measure. As a result, the surface water level in the lake on 27 March was suddenly lowered about 60 cm below what it had been on 11 March 1969. The reverberations caused by this massive outflow of water at the surface apparently were felt in the deepest regions of the lake, probably through baroclinic waves. However, mixing probably was slight and cooling was only on the order of 0.5°C at 7 m. Warming of the bottom waters during 1969–70 was much more regular and continuous until ice breakup. Spring overturn, or full convective mixing during both 1969 and 1970 was extremely weak and

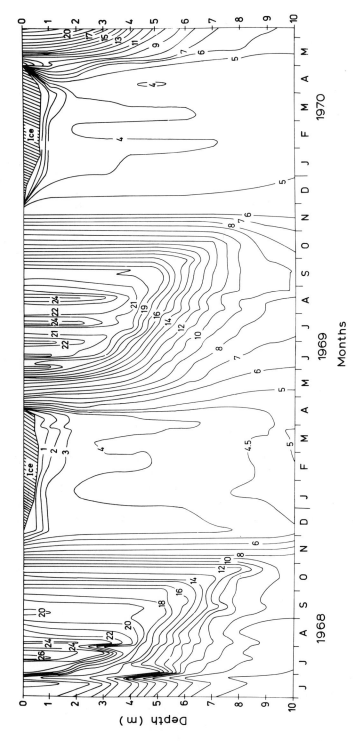

Figure IV.B-4. Depth-time temperature (°C) isopleths for Mirror Lake. (From Makarewicz 1974.)

B. Physical and Chemical Environment

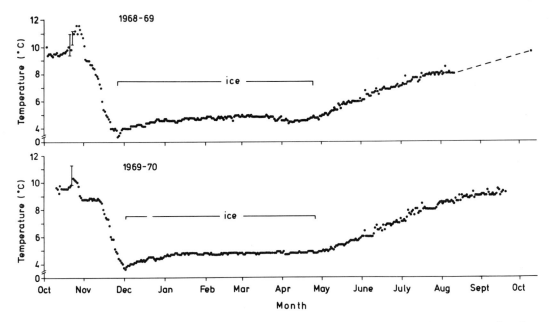

Figure IV.B-5. Bottom water temperatures in Mirror Lake measured with a continuous recording thermograph during 1968–69 and 1969–70. The bottom of the thermograph was located at a depth of 8.5 m; the total water depth at this location was 10.5 m. The vertical bars indicate the range of temperatures observed on specific days. Otherwise, each dot represents the mean daily temperature, where fluctuations during the day were negligible or small (<0.5°C). The period 31 October–12 November 1969 was characterized by cloudy, relatively calm weather.

of short duration, if it occurred at all (see *Figure IV.B–4*).

After the ice-cover disappeared, the bottom water warmed from solar radiation at a constant and similar rate during both 1969 and 1970. From 1 May 1969 to 15 July 1969 the temperature increased at an average rate of 0.037°C/day and from 1 May 1970 to 15 July 1970 the rate was 0.039°C/day. The overall rate from 1 May 1969 to 10 August 1969 was 0.030°C/day and from 1 May 1970 to 21 September 1970 it was 0.030°C/day. These bottom waters reached their maximum temperatures of approximately 9 to 10°C just prior to autumn overturn (*Figures IV.B–4 and 5*).

On a much shorter time scale, temperature variations occur within a given day as well as from day to day. For example, from 0900 h to 1900 h on 15 August 1974, temperatures in the upper 4 m increased by about 0.3 to 0.4°C above ambient (~23.8°C). The change at 5 m depth was even greater, from 21.5 to 22.2°C, but at 6 m the increase was only ~0.1°C. Overnight, from 1900 h on the 15th to 0900 h on the 16th, the surface 4-m layer cooled by 0.6 to 1.0°C; at 5 m the temperature increased by about 0.2°C and at 6 m fluctuations were largely damped. From 0900 h to 1330 h on 16 August, which was calm, the temperature increased by 1.2°C at 0 m, by 0.7°C at 1 m, by 0.6°C at 2 m, by 0.2°C at 3 m and there was no change at 4 m.

Small spatial variations in temperature occur horizontally as well as vertically in Mirror Lake. Synoptic surveys at 10 locations revealed that concurrent temperature measurements may vary by 1°C or more over the surface of the lake and by 0.5°C or so at depths of 4 to 6 m in midsummer.

On average, Mirror Lake is covered by ice and snow some 138 days or 38% of the year (*Table IV.C–3*). The ice may reach thicknesses of 60 to 75 cm during particularly cold or snow-free winters (see *Figure IV.B–4*). During periods of ice-cover, bottom-water temperatures are several degrees warmer than those under the ice and the lake is *inversely stratified* at least with regard to temperature (*Figure IV.B–4*). Theoretically, the ice and snow layer at the surface of the lake should dampen, but not eliminate environmental factors such as solar radiation and wind, which affect water temperatures within the lake. This effect is seen in *Figure IV.B–4*, where the 4° isotherm undergoes rather large oscillations in depth during this period. These oscillations probably result from the interplay between heating from the bottom sediments and intrusions of melt water into the lake (e.g. January thaw), although we find that runoff from the watersheds during spring snowmelt normally mixes to only shallow depths in the lake (see *Chapter IV.C*). Hydrodynamically, the lake is very unstable during this period.

The incoming streams drain forested watersheds. They are in deep shade during the summer and flow under a thick snowpack during winter. Thus, as with streams draining the Experimental Watersheds (see Likens et al. 1970), diurnal temperatures are relatively constant in these streams and the annual range of average monthly temperatures is only about 14°C (*Figure IV.B–6*). Small streams can warm or cool quickly (e.g. Burton and Likens 1973) and the average water temperature usually approximates the average air temperature (e.g. Macan 1958). The fact that the NW tributary is slightly warmer than the other two tributaries during spring and summer probably reflects its somewhat more open canopy, particularly near the mouth. The average temperature at the outlet represents the average surface water temperature of the lake and generally is warmer than the inlet streams throughout the year, particularly during May through September (*Figure IV.B–6*). The average monthly outlet temperature ranges from approximately 1.2°C in February to 24.6°C in July.

The average temperature of the lake, weighted for the entire volume of the lake, varies throughout the year in direct response to amount of solar radiation, but is displaced in phase (*Figure V.B.5–3*). The lag in time between the peak or minimum solar radiation and peak or minimum average water temperature is approximately two months.

Light

Solar radiation is the major source of energy for physical and metabolic activity in Mirror Lake. Solar radiation also is important for the formation of convective currents and in the heat budget (*Chapter IV.C*), it drives primary productivity of algae and macrophytes (*Chapters*

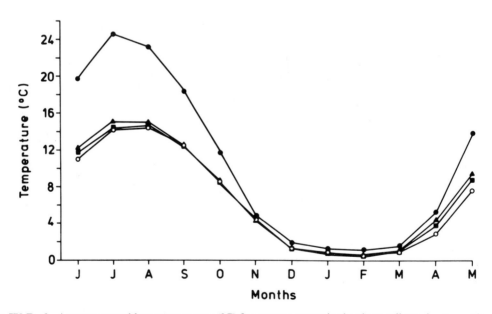

Figure IV.B–6. Average monthly temperatures (°C) for stream water in the three tributaries (near the mouth) and outlet discharge from Mirror Lake. Averages are based on at least seven years of data during the period 1968 to 1981. Monthly coefficient of variation values ($s_{\bar{x}}/\bar{x} \times 100$) varied from 1–4% during the summer months to 10–20% during the winter and spring months for stream water, and from 2–14% for the outlet. ○——○ is NE tributary; ▲——▲ is NW tributary; ■——■ is W tributary; and ●——● is outlet.

B. Physical and Chemical Environment

V.B.2, 3 and *4* and *V.C*) and affects the behavior of organisms (e.g. *Chapters V.A.5, 6* and *9*) in the lake. Because Mirror Lake contains very clear water, light can penetrate to the deepest regions of the lake, at least during certain seasons. We have measured the penetration of light in Mirror Lake in various ways.

A Secchi disc is an exceedingly simple, yet informative, limnological instrument for measuring relative transparency of water. Secchi disc values give a rough measure of the light penetration and/or clarity of water, and by implication, an index to the turbidity caused by suspended matter, both living and dead. Limnologists have proposed that the Secchi disc provides a quick and easy way to obtain synoptic information about the distribution or trend of eutrophication of lakes (e.g. Edmondson 1980).

The Secchi depth varies appreciably throughout the year in Mirror Lake (*Figure IV.B–7*). In summer, the Secchi depth may range from 5 to 7 m and in winter from 2.5 to 3.5 m (as viewed through a hole in the ice). The maximum value ever recorded at Mirror Lake was 8.5 m on 20 June 1979 and the minimum was 2.4 m on 22 December 1969. Obviously the presence, condition and depth of snow on a lake will greatly affect the penetration of light into the water. Values recorded only a few days apart in late June and early July of 1979 varied from 6.9 to 5.3 m. Secchi disc values range from a few centimeters in very turbid waters to over 40 m in a very few, extremely clear-water lakes (Wetzel 1975). Based on Secchi disc depths, it is reasonable to characterize Mirror Lake as a clear-water lake.

A Whitney Underwater Daylight Meter with or without various absorption filters has also been used to measure the transmission of solar radiation in Mirror Lake (see Kilham and Likens 1968). The Whitney Meter employs a Weston #1 photocell, which is most sensitive in the yellow region of the spectrum (~560 ηm), moderately sensitive in the green (~510 ηm)

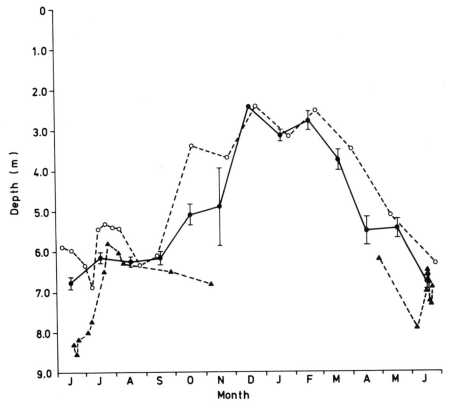

Figure IV.B–7. Secchi disc values in Mirror Lake. ○---○ are measurements in 1969–70; ▲---▲ are measurements in 1979–80; ●——● are monthly averages of all data from 1967 to 1981. Error bars indicate one standard deviation of the mean, $s_{\bar{x}}$.

and blue (~430 ηm) and very insensitive in the red region (~720 ηm). Relatively narrow-band absorption filters with transmission peaks at 435 ηm, 525 ηm and 610 ηm were used to measure the penetration of solar energy in three regions (blue, green and red) of the visible spectrum. The calibration and use of these filters is described in Kilham and Likens (1968).

Light intensity and quality can vary enormously with depth during the course of a single day or throughout the year. The amount of light at depth decreases sharply before sunset and then at all depths at sunset (*Figure IV.B*–8). Light returns to all depths within 2 to 3 h after sunrise. The 1% transmission level fluctuates from a depth of approximately 3.5 to 4.5 m under the ice and in early spring, to more than 9 m in June (*Figure IV.B*–9).

It is common for green wavelengths to penetrate most deeply in clear-water lakes (Wetzel 1975), and this also is the case in Mirror Lake (*Figure IV.B*–10). Transmission of green wavelengths fluctuates from 5 to 6% of surface values at 10 m in May and November to only 0.1% at 6 m in December. Similarly, red wavelengths vary from about 1 to 2% transmission of surface values at 10 m in November and June to only 0.1% at about 7 m in December; transmission of blue wavelengths varies from about 0.2% at 10 m in June through November to only 0.1% of surface values at 3 m in December under ice and snow. Mirror Lake is most transparent to total solar radiation in June (*Figure IV.B*–9).

The vertical extinction coefficient, k, is another index to the transparency (Clarke 1939),

$$k = \frac{\ln I_o - \ln I_z}{z} \quad (1)$$

where I_o is the intensity of light at the surface, I_z is the intensity at depth z. Traditionally in limnology, the thickness of the layer, z, is measured in meters. The extinction coefficient for the interval from 1- to 4-m depth in Mirror Lake during the summer is about 0.34. As such, Mirror Lake is the eighth clearest lake in Grafton County, New Hampshire (Kilham and Likens 1968).

Chemical Characteristics

Dissolved Substances

The chemistry of Mirror Lake has been determined by periodic sampling at 1- or 2-m intervals in the deepest part of the lake. The total amount of dissolved substances or particulate matter in each layer is calculated by multiplying the concentration times the volume of that particular layer of water. The total standing stock of an element in the lake is the sum of the layers, and the volume-weighted concentration for

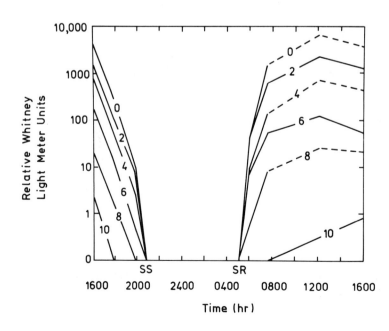

Figure IV.B–8. Light intensity in Whitney units at various meter depths (indicated on line) during 19–20 August 1969. Dotted line indicates values based on data from the previous day; SS = sunset; SR = sunrise. (Modified from Makarewicz 1974.)

B. Physical and Chemical Environment

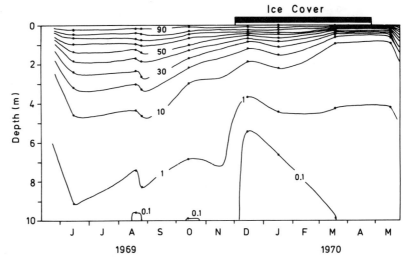

Figure IV.B–9. Depth-time isopleths of percent transmission of surface light in Mirror Lake during 1969–70.

the lake is this total divided by the total volume of the lake.

During periods of vertical stratification within the lake there is frequently an increase or decrease in the concentration of various substances in the deeper waters of the lake below the thermocline. This pattern is illustrated in August, and to a lesser degree, in January (*Figure IV.B*–11) and is most evident for dissolved inorganic carbon (DIC), dissolved O_2 and K^+. Such increases result in part from sedimentation of particulate matter from the epilimnion and subsequent mineralization in the hypolimnion. Some of the increases may be due to exchange with the sediments and its interstitial waters. The concentration of dissolved oxygen in the hypolimnion decreases near the bottom of the lake during the thermally-stratified season (*Figures IV.B*–11A and 17), often approaching or reaching zero in late summer or early autumn. This vertical pattern is evidence for aerobic decomposition in the bottom waters of the lake.

Also during summer stratification, there is often evidence of increased or decreased concentration of various dissolved substances near the thermocline. This pattern is most evident for dissolved oxygen (*Figure IV.B*–11A), which results from active photosynthesis by phytoplankton populations at this level (see *Chapter V.A.2*).

The fact that the concentration of some constituents may increase or decrease dramatically in the stable bottom portion of the lake is pertinent from the standpoint of vertical distribution

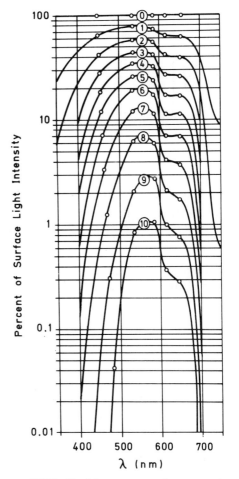

Figure IV.B–10. Mean spectral attenuation of downwelling light in 1975 (eight dates). Values are plotted at the optical centers of colored absorption filters and refer to the percentage of subsurface light penetrating to the depths indicated in meters. (From Moeller 1978b.)

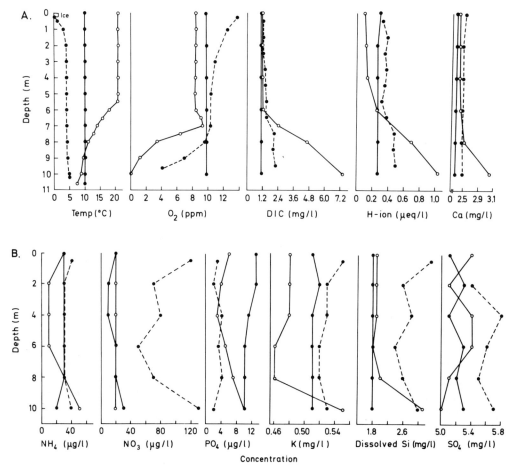

Figure IV.B–11. A and **B.** Vertical profiles of temperature and dissolved substances in Mirror Lake during 1975: ●---● is 8–10 January; ○——○ is 25–26 August; ●——● is 14–15 November 1975.

and migration of organisms, chemical reactions, etc. However, the contribution of these constituents to the total for the lake is minor. Not only are these bottom waters sealed off from the rest of the lake by the vertical density gradient (see *Chapter IV.C*), but because of the morphometry of the lake this bottom portion represents a small percentage of the total lake volume (or total standing stock of some chemical) (see *Figure IV.B*–3 and *Table III.A*–1). For example, in August of 1975 when the thermocline was at 5.75-m depth, the volume of water below the thermocline represented less than 20% of the total volume of the lake, and although the dissolved silica concentration at 10 m was 1.6 times greater than the concentration in the epilimnion, the standing stock of dissolved silica below the thermocline was only 21% of the total in the lake.

As shown in *Figure IV.B*–11, there may be some variation in the concentration of various ions during an annual cycle below the thermocline, primarily owing to the alternate periods of stratification and mixing of the lake. However, for the lake as a whole, the average, volume-weighted concentration of most elements (Ca^{2+}, Mg^{2+}, K^+, Na^+, NH_4^+, H^+, SO_4^{2-}, Cl^-, Al^{n+}) does not vary appreciably during the year. However, the concentration of nitrate and to a lesser degree dissolved silica does change significantly during the year (*Figure IV.B*–12). The control of the concentration of these two elements appears to lie primarily with the biologic community within the lake. Nitrate concentrations remain near the detection limit (0.01 mg/liter) from June through November, then rise dramatically to a maximum of 0.28 mg/liter in March or April. This pattern for nitrate in the

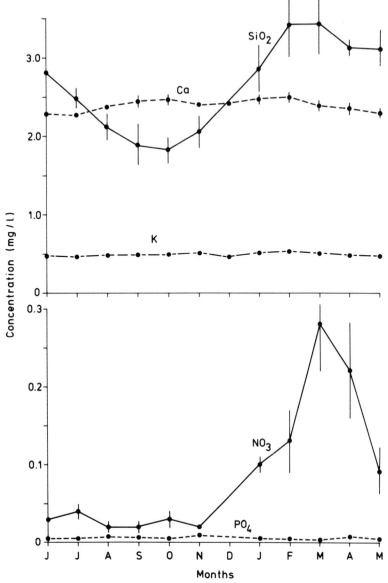

Figure IV.B-12. Average concentrations of various chemicals in Mirror Lake from 1967 to 1981. Error bars indicate one standard deviation of the mean, $s_{\bar{x}}$. Averages for each sampling date were volume weighted for the lake.

lake parallels that for nitrate in the forest-stream ecosystem (see *Figure II*–15). The action of the biologic community (diatoms in particular) on the dissolved silica concentration is less rapid and dramatic than on nitrate in the lake. The concentration of dissolved silica decreases steadily throughout the summer months reaching a minimum (approximately 1.8 mg/liter) in September or October. The concentration then increases, reaching its maximum value of about 3.4 mg/liter in February and March.

The only apparent long-term changes in the chemistry of Mirror Lake have been observed for Na^+ and Cl^-, and these changes occurred primarily since 1975. Volume-weighted sodium concentrations in Mirror Lake increased by 35% in four years from 1975–76 to 1979–80 (1.22 mg/liter to 1.65 mg/liter). During the same period, chloride concentrations increased 117% from 0.94 mg/liter to 2.04 mg/liter. These increases appear to be directly related to increased runoff of road salt as discussed in *Chapter IX* and/or to leaching of domestic sewage from septic tanks.

On an average annual basis, the ionic chemistry of Mirror Lake is clearly dominated by calcium and sulfate (*Table IV.B*–1). Bicarbonate and sodium are the next most important ions on both an equivalent and weight basis. On average, there is approximately 1,760 kg of dissolved substances in Mirror Lake, of which 27% is sulfate, 19% is bicarbonate, 13% is dissolved silica, 13% is dissolved organic carbon, 11% is calcium and 6% is sodium (*Table IV.B*–1).

The average chemistry of all of the tributaries and the outlet is also dominated by calcium and sulfate (*Table IV.B*–2). Sodium and chloride are much more abundant in the NE tributary than in the other tributaries, presumably because of the leaching of road salt from I-93 within this subwatershed (see *Chapter IX*). Also the NE tributary is more acid than the other two tributaries, the lake outlet or the lake itself. The water in the NE tributary is brown-colored and has much higher concentrations of dissolved organic carbon (DOC) and total dissolved solids. Presumably the higher acidity in this tributary is due to humic acids leached from decomposing organic debris in the watershed. The concentration of dissolved silica in the outlet is two to three times lower than that for the tributaries, presumably because diatoms in the lake utilize this element in their frustules. The siliceous frustules ultimately fall to the sediments. Diatoms are rare in the streams flowing into the lake, as well as in other shaded, headwater streams within the Hubbard Brook Valley.

The depth-time distributions of DOC and DIC are given in *Figures IV.B*–13 and 14. Note that the DOC concentrations are about an order of magnitude greater than the particulate organic carbon (POC) concentrations (*Figure IV.B*–15). The DIC consists of dissolved CO_2, H_2CO_3 and HCO_3^-, all of which are in relatively low concentrations in Mirror Lake.

Table IV.B–1. Average Volume-weighted Annual Concentrations and Standing Stock of Various Dissolved Substances in Mirror Lake during 1967 through 1981

Substance	Volume-Weighted Concentration		Standing Stock	
	mg/liter[a]	µEq/liter	kg/lake	Eq × 10³/lake
Ca^{2+}	2.39	119	1,986	99.1
Mg^{2+}	0.51	41.1	423.8	34.9
K^+	0.48	12.3	400.3	10.2
Na^+	1.32	57.4	1,099	47.8
H^+ [b]	0.00044	0.44	0.37	0.37
NH_4^+	0.035	1.8	27.8	1.54
NO_3^-	0.09	1.5	72.8	1.17
SO_4^{2-}	5.7	119	4,737	98.6
Cl^-	1.09	30.7	911	25.7
PO_4^{3-}	0.0056	0.18	4.66	0.147
HCO_3^- [c]	4.02	65.9	3,400	55.7
Dissolved Si	2.65	—	2,209	—
DOC[d]	2.75		2,313	
pH		6.36		
Total Dissolved Solids	21.0		17,577	
Σ+		232		194
Σ−		217		181

[a] Standard deviations of the means, $s_{\bar{x}}$, for concentrations are: Ca^{2+} ±0.022, Mg^{2+} ±0.004, K^+ ±0.066, Na^+ ±0.014, H^+ ±0.00009, NH_4^+ ±0.004, NO_3^- ±0.027, SO_4^{2-} ±0.08, Cl^- ±0.025, PO_4^{3-} ±0.001, HCO_3^- ±0.14, SiO_2 ±0.18.
[b] 1970 to 1981.
[c] 1970 to 1972.
[d] 1974, 1979–80.

Table IV.B-2. Average Volume-weighted Annual Concentrations of Dissolved Substances for the Inlet Tributaries and Outlet of Mirror Lake during 1970–71 through 1975–76

	Tributaries								Outlet	
	NE		NW			W				
Substance	mg/liter	μEq/liter	mg/liter	μEq/liter	mg/liter	μEq/liter	mg/liter	μEq/liter	mg/liter	μEq/liter
Ca^{2+}	3.77	188	2.64	132	2.60	130	2.28	114	114
Mg^{2+}	0.78	64	0.55	45	0.55	45	0.47	39	39
K^+	0.69	18	0.38	10	0.55	14	0.45	12
Na^+	2.46	107	1.05	46	1.38	60	1.17	51
H^+	0.0042	4.2	0.0008	0.8	0.0010	1.0	0.0010	1.0
NH_4^+	0.08	4.4	0.02	1.1	0.03	1.7	0.03	1.7
NO_3^-	0.05	0.8	0.07	1.1	0.21	3.4	0.25	4.0
SO_4^{2-}	7.96	166	6.77	141	6.25	130	6.09	127
Cl^-	4.46	126	0.64	18	1.04	29	0.90	25
PO_4^{3-} [a]	0.013	0.4	0.003	0.1	0.005	0.2	0.005	0.2
HCO_3^- [b]	5.6	92.4	4.5	74.7	5.4	89.1	3.8	62.5
Dissolved Si	10.1	—	6.5	—	7.5	—	2.9	—
DOC	5.2		1.5		2.3		2.1	
pH	5.38		6.10		6.0		6.0	
Total Dissolved Solids	41.2		24.6		27.8		20.4	
Σ+		386		235		252		219
Σ–		386		235		252		219

[a] 1972 to 1976.
[b] Only sporadic data for bicarbonate in stream water are available. On dates where bicarbonate was measured, the average sum of cation equivalents was 8% higher than the average sum of anion equivalents. Thus, bicarbonate data used here were obtained as a difference between the sum of + equivalents and − equivalents.

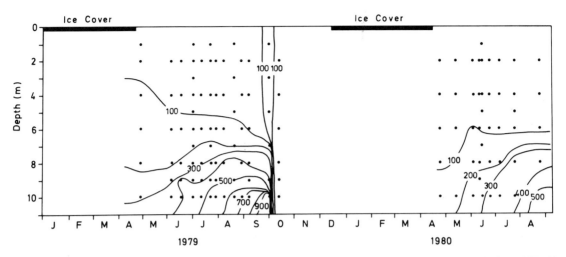

Figure IV.B–13. Depth-time isopleths for dissolved inorganic carbon (μM) in Mirror Lake during 1979–80. Dots indicate sampling locations.

There is a strong accumulation of DIC in the hypolimnion during the thermally-stratified summer period (June-October) (*Figures IV.B–11A and 13*). This pattern corresponds with the dissolved oxygen deficit that also forms during this period (*Figures IV.B–11A and 16*), both the result of increased decay of detrital organic carbon, which has settled from upper layers. Similarly, the increase in H^+ accompanying the increase in DIC (*Figure IV.B–11A*) would be anticipated as the reduced carbon (organic carbon) was transformed into CO_2 and carbonic acid. Because the lake is not convectively mixing at this time, the hypolimnion is isolated from the atmosphere and there is essentially no replenishment of O_2 or loss of CO_2 to the atmosphere. There is a weak vertical gradient of DIC until late summer or early autumn, when the concentration of DIC in the hypolimnion may exceed concentrations in the epilimnion by an order of magnitude. Concentrations of DIC are relatively well mixed to a depth of 5 to 6 m during June through October, but are stratified below this depth.

The pattern of DOC distribution throughout the year in Mirror Lake is weak and complicated (*Figure IV.B–14*). The concentrations in 1979 were generally higher than in 1980, but we

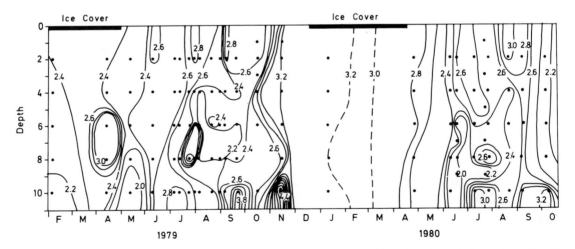

Figure IV.B–14. Depth-time isopleths for dissolved organic carbon (mg/liter) in Mirror Lake during 1979–80. Dots indicate sampling locations.

B. Physical and Chemical Environment

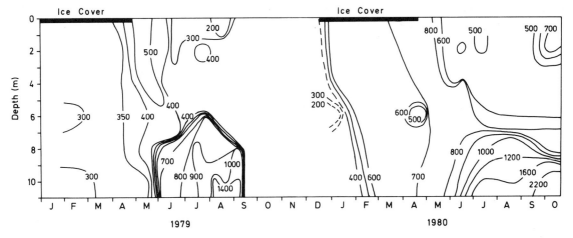

Figure IV.B–15. Depth-time isopleths for particulate organic carbon (μg/liter) in Mirror Lake during 1979–80.

have no data to assess long-term patterns. In general, the vertical stratification of DOC is weak and variable. Concentrations of DOC are relatively low in the hypolimnion during May and/or June, but small increases occur in the bottom waters during the remainder of the thermally-stratified period, particularly in early autumn. At a depth of 8 m, concentrations increase from June to a maximum in late July, followed by a decrease in concentration in August and then a buildup just before autumnal overturn (*Figure IV.B*–14). It is not clear why there should be more variability in DOC concentration at 8 m than at other depths throughout the year.

Particulate Matter (Seston)

There is a striking accumulation of POC in the hypolimnion of Mirror Lake during the thermally-stratified season (*Figure IV.B*–15), resulting in a strong vertical gradient in POC after

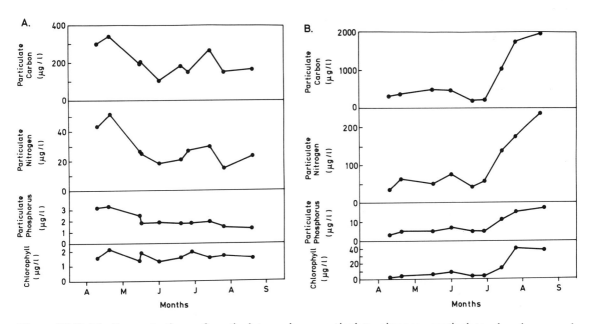

Figure IV.B–16. Concentrations of particulate carbon, particulate nitrogen, particulate phosphorus and Chlorophyll-*a* in **A**, surface waters (0.5 m) and **B**, bottom waters (9 m) in Mirror Lake during summer 1974.

June. The POC concentrations in the hypolimnion appear to be redistributed by some mechanism in September (*Figure IV.B*–15). However, this process does not correlate with the DOC profile (*Figure IV.B*–14) at this time of year (unless the 4.4 mg DOC/liter is an anomalous value) or with the DIC profile (*Figure IV.B*–13). Some accumulation of POC occurred just beneath the ice-cover, possibly as a result of increased phytoplankton concentrations, caused by more favorable light conditions at shallow depths under the ice.

The chemical composition of seston (particulate carbon, nitrogen, phosphorus and chlorophyll) was determined biweekly from late April to early September of 1974, at depths of 0.5, 3, 6 and 9 m. The results are summarized in *Table IV.B*–3, and trends during the summer for the 0.5- and 9-m depths are shown in *Figure IV.B*–16.

All of the parameters at the surface attained maximum values on 4 May, shortly after the ice melted and just as the lake was beginning to stratify thermally. A secondary peak of particulate carbon (PC) and particulate nitrogen (PN) concentrations was measured in late July. In contrast to the surface values, concentrations at 9 m increased dramatically during the thermally-stratified period (*Figure IV.B*–16).

The ratios (weight basis) between elements did not show any simple pattern with depth (*Table IV.B*–3) or month in the lake. Ratios of C:P tended to decline with depth, but this trend was not consistent on all dates. The ratio of C:Chl-*a* also declined with depth and was the most predictable pattern. However, values for epilimnetic samples were highly variable from week to week. In comparison with ratios for other aquatic ecosystems (*Table IV.B*–4), seston in Mirror Lake appears to be carbon rich or phosphorus poor. For example, the C:N, N:P and C:Chl-*a* ratios are 35, 60 and 45% higher than the respective ratios in Cayuga Lake, New York. The C:P ratio is more than twice the average ratio for Cayuga Lake or oceanic seston. This result was so striking that the method of oxidizing particulate matter for determination of phosphorus (Menzel and Corwin 1965) was compared with that of Stainton et al. (1974). On 17 June 1974 six replicate samples were taken from the 9-m depth in Mirror Lake. Three done by persulphate oxidation (Menzel and Corwin 1965) gave 7.69, 7.87 and 7.63 μg P/liter and three combusted in a Muffle furnace and then acidified (Stainton et al. 1974) gave 7.86, 7.10 and 7.20 μg P/liter. It appeared that P was not underestimated and the high C:P ratios stand unexplained. Much of this organic carbon must be in detritus or colloidal material that is depleted of phosphorus.

Based on these limited data, the volume-weighted concentration and standing stock for the whole lake were 0.27 mg/liter and 231 kg for PC, 0.038 mg/liter and 32.3 kg for PN, 0.0029 mg/liter and 2.48 kg for particulate phosphorus (PP) and 0.0027 mg/liter and 2.31 kg for Chl-*a*.

Table IV.B–3. Mean Values ± One Standard Deviation of the Mean of Particulate Carbon (PC), Particulate Nitrogen (PN), Particulate Phosphorus (PP) and Chlorophyll-*a* (Chl-*a*) in Seston Samples from Mirror Lake during 26 April 1974 through 2 September 1974[a]

	Depth (m)					
	0	1	3	5	6	9
PC	196 ± 24 (9)	300 (1)	206 ± 24 (9)	302 (1)	295 ± 49 (8)	750 ± 228 (9)
PN	26.4 ± 3.6 (9)	42.9 (1)	28.0 ± 3.2 (9)	38.0 (1)	40.5 ± 5.9 (9)	93.6 ± 21.5 (10)
PP	1.98 ± 0.18 (9)	3.16 (1)	2.01 ± 0.17 (9)	3.12 (1)	3.06 ± 0.31 (8)	8.23 ± 1.74 (9)
Chl-*a*	1.71 ± 0.09 (9)	1.52 (1)	1.87 ± 0.13 (9)	1.87 (1)	4.64 ± 0.84 (9)	12.86 ± 4.64 (10)
C:N	7.63 ± 0.54 (9)	6.99 (1)	7.57 ± 0.60 (9)	7.95 (1)	7.87 ± 0.66 (8)	7.17 ± 0.84 (9)
C:P	99 ± 8 (9)	95 (1)	103 ± 8 (9)	97 (1)	94 ± 9 (8)	80 ± 9 (9)
N:P	13.1 ± 0.9 (9)	13.6 (1)	14.1 ± 1.5 (9)	12.2 (1)	12.0 ± 0.9 (8)	11.6 ± 0.6 (9)
C:Chl	114 ± 11 (9)	197 (1)	111 ± 12 (9)	161 (1)	73 ± 6 (8)	72 ± 9 (9)

[a] Concentration values in μg/liter; ( ) is number of measurements.

B. Physical and Chemical Environment

Table IV.B–4. Weight Ratios of Mirror Lake Seston Compared with Values for Cayuga Lake, New York and Typical Oceanic Ratios

Sample	n	C:N	C:P	N:P	C:Chl-a	Reference
Mirror Lake (Mean of All Samples)	37	7.54	94.2	12.7	98	This work
Cayuga Lake (Epilimnetic Samples)	23	5.36	41.4	7.7	65	B. Peterson (1972, personal communication)
Ocean	?	5.64	40.6	7.2	variable depending on crop	Redfield et al. 1963

Average (volume-weighted) whole-lake ratios of 7.1:1 for C:N, 93:1 for C:P, 13:1 for N:P and 100:1 for C:Chl-a can be calculated for seston during April through September of 1974.

Dissolved Gases

Chemically and biologically, dissolved oxygen is one of the most important gases in a lake. Its presence or absence regulates a variety of oxidation-reduction reactions as well as the distribution and behavior of organisms. The average distribution of dissolved oxygen in Mirror Lake throughout the year is given in *Figure IV.B*–17. Complete vertical mixing during the autumn is obvious from these data. In contrast, replenishment of dissolved oxygen by convective mixing during the spring is complicated by the frequent occurrence of spring meromixis. Thus, these mean values show increased concentrations at 10 m in April, but not consistent overturn during this period. The highest concentrations occur just under the ice during December and February. Lowest surface values (<100% saturation) often occur during July through September when temperatures are maximal in the lake. A metalimnetic maximum is obvious at depths of 4 to 6 m during June, July and August, as the

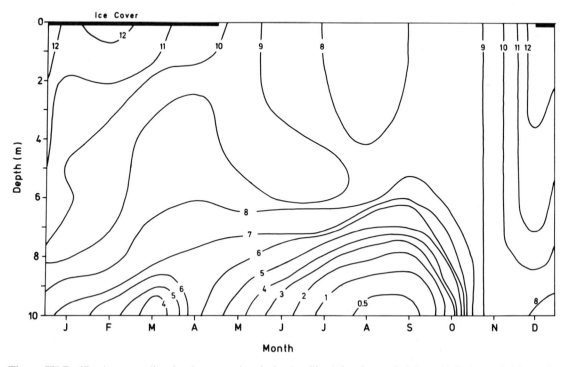

Figure IV.B–17. Average dissolved oxygen isopleths (mg/liter) for the period June 1967 through September 1978 in Mirror Lake.

water at these depths is supersaturated (>110% saturation at 4 to 5 m on *Figure IV.B*–18) with dissolved oxygen.

Anaerobic conditions frequently occur at or below a depth of 10 m in late summer or early autumn before overturn (e.g. *Figure IV.B*–11A). Zero dissolved oxygen has been observed at a depth of 9 m on 18 October 1969, 30 August 1971, 17 September 1974, 21 July 1977, and 0.1 mg/liter of dissolved oxygen was observed at a depth of 8 m on 5 September 1967. Commonly on these occasions H_2S is apparent in the samples. Low concentrations are also frequently observed in March near the end of the ice-covered period. The buildup of the dissolved oxygen deficit from the beginning of the ice-cover in December to complete circulation in October/November in the deepest portions of the lake is clear from the mean values (*Figure IV.B*–17). Unfortunately, we have little quantitative data on other gases in Mirror Lake.

Ice-Cover

Physical and chemical stratification within the ice-cover of a lake in winter gives some indication of how the ice was formed. In general, ice at the surface of the lake, accumulates as (1) clear *black ice* when water at the ice-water interface freezes, and (2) *white ice* when the deposition of snow on the surface of the ice is infiltrated with rain water or lake water, which comes to the surface through fissures in the ice-cover and then refreezes (see *Chapter IV.C*). This layering of black and white ice is usually quite apparent by late winter on Mirror Lake. In addition, lenses of interstitial melt water may form within the ice sheet as solar radiation is absorbed.

On 6 April 1972 we collected and chemically analyzed the various layers of the ice sheet at three locations on Mirror Lake. At the surface was a layer of "rotten," white ice averaging 5.2 ± 0.67 cm thick. Below this layer was a layer of grayish, white ice 28.0 ± 3.8 cm thick. There was internal melt water within this layer. At the bottom of the ice sheet was a layer of black ice 22.8 ± 3.0 cm thick. Thus, at this time the overall average thickness of the ice sheet was 56 cm.

The chemistry of the various layers was quite different (*Table IV.B*–5). The black ice contained only about one-third of the total chemical content, on a weight basis, as did the bottom layer of white ice, and less than 0.1 of the amount in the surface layer of white ice. There was less difference between the layers on an equivalence basis because of the smaller relative change in hydrogen-ion concentration between the three layers and because of the importance of this ion in the equivalence balance. The internal melt water contained much higher concentrations of dissolved salts than did the ice layer where it was found. This fractionation is the necessary consequence of the freeze-purification process, by which a growing crystal of

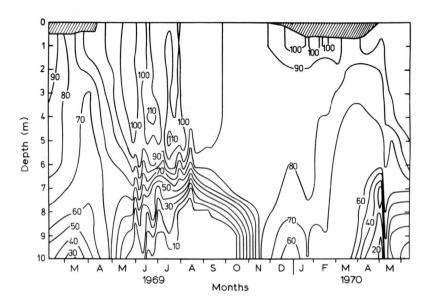

Figure IV.B–18. Isopleths of percent saturation for dissolved oxygen in Mirror Lake during 1969 and 1970. The shaded area indicates ice-cover. (From Makarewicz 1974.)

Table IV.B–5. Chemistry of Ice at the Surface of Mirror Lake on 6 April 1972[a]

Component	Layer A	Layer B	Layer C	Melt Water	Rain[b]	Lake Water[c]
Ca^{2+}	0.11 ± 0.012	0.01 ± 0.003	0.02 ± 0.003	0.04	0.04	2.37
Mg^{2+}	0.02 ± 0.003	0.002 ± 0.001	0.002 ± 0.001	0.012	0.02	0.50
Na^+	0.02 ± 0.007	0.01 ± 0.003	0.01 ± 0.003	0.06	0.06	1.19
K^+	0.03 ± 0.006	0.01 ± 0.006	0.03 ± 0.007	0.02	0.02	0.49
H^+ (μg/liter)	17.9 ± 3.78	6.0 ± 0.66	4.0 ± 1.5	62.0	100	0.3
NH_4^+	0.16 ± 0.026	0.06 ± 0.009	0.03 ± 0.003	0.18	0.27	0.05
NO_3^-	1.22 ± 0.17	0.37 ± 0.025	0.04 ± 0.010	1.78	2.5	0.25
SO_4^{2-}	0.8 ± 0.09	0.1 ± 0.10	<0.05	2.0	3.6	5.9
Cl^-	0.18 ± 0.017	0.18 ± 0.036	0.14 ± 0.060	0.24	0.23	0.60
PO_4^{3-} (μg/liter)	27 ± 7.3	21 ± 5.8	6 ± 3.2	31	5	15
Dissolved Si	0.32 ± 0.044	<0.05	0.09 ± 0.043	0.13	<0.05	2.3
Cond. at 25°C	11.9 ± 1.98	4.6 ± 0.45	2.9 ± 0.69	27.0	46.3	—
HCO_3^-	—	—	—	—	—	4[d]
Total mg/liter	2.91	0.77	0.28	4.43	6.84	11.37
$\Sigma+$ (μEq/liter)	35.4	10.6	8.0	77.6	101.4	226.5
$\Sigma-$ (μEq/liter)	42.3	13.8	5.8	78.1	122.0	210

[a] Values are in mg/liter ± one standard deviation of the mean except where noted. Layer A was the surface part of the white ice; B, the lower layer of white ice; C, the layer of black ice at the bottom of the ice sheet. The melt water was collected from Layer B.
[b] 28 March 1972.
[c] 9 May 1972.
[d] From *Table IV.B–1*.

ice excludes impurities from its lattice. The rejected impurities are left in concentrated form in the remaining fluid and are eventually incorporated into the last, most unstable crystals to form. Conversely, it is these impure, unstable crystals that are the first to melt when the temperature is raised. This process is illustrated by the fact that the melt water in layer B was ten times more acid than the ice of coexisting layer B. The chemistry of rain water, collected on 28 March (4.08 cm of precipitation), and of lake water on 9 May was appreciably different from that of the melt water within the white ice.

The chemical composition of the surface layer of white ice closely resembled the chemistry of average snow during the winter (see Table 13 in Likens et al. 1977; *Chapter II*). Sublimation and/or evaporation at the surface of the ice sheet could result in some increase in concentrations. In contrast to the surface layer, the chemical purity of the black ice argues that this ice is freeze-purified from lake water (see *Table IV.B–1*).

Summary

The physical and chemical attributes comprising Mirror Lake have been specified in considerable quantitative detail. These physical and chemical parameters are in turn determined by the interaction of the geologic, hydrologic, climatologic, vegetational, and more recently, the human setting of the ecosystem.

In size, shape and morphometry, Mirror Lake is typical of many lakes in New England; i.e. its basin was formed by glacial processes some 14,000 yr ago. Lake level is quite stable, being at sill depth for most of the year. Rocks, gravel and sand form the lake margin and shoaling waters. In contrast, the deeper basin is covered with gyttja. The relatively high organic matter of the sediments in the central portion of the basin exerts a crucial control over the chemical milieu of the hypolimnion during sustained periods of stratification.

Water and its dissolved substances are garnered, collected and fed into the lake by three streams as well as by direct flow of soil water from the forested watershed surrounding the lake. The tributaries are generally nutrient poor but vary in chemical composition; large increases in Na^+ and Cl^- concentrations have occurred in the NE tributary since 1969 because of contamination by road salt.

The chemistry of the lake water is dominated by Ca^{2+} and SO_4^{2-}. The average, volume-weighted concentration of total dissolved solids is approximately 21 mg/liter and the standing stock for the lake is about 17,600 kg. Of this, sulfate contributes 27%, bicarbonate 19%, dissolved silica 13%, DOC 13% and calcium 11%. DOC is ten times more abundant than POC. The amount and flux of such nutrients within the lake varies in both space and time. The largest variations, both seasonally and spatially, are shown by dissolved oxygen, NO_3^-, dissolved Si, DOC, H^+ and K^+. During periods of stratification, these chemicals manifest a strong gradient from hypolimnion to the surface mixing layer. Increased concentrations of DIC and H^+, and decreased concentrations of dissolved oxygen develop in the hypolimnion during summer stratification. The biologically-limiting nutrients are P and N and to a lesser extent Si, which, because of their low concentrations, impart the oligotrophic character to the lake. The bottom-most waters of Mirror Lake are subject to frequent episodes of anoxia after several months of stratification during the summer. Anaerobic bottom water, typical of eutrophic lakes, occurs because the volume of the hypolimnion is relatively quite small as compared with the epilimnion.

On average Mirror Lake contains approximately 230 kg of POC, 32 kg of PN, 2.5 kg of PP and 2.3 kg of Chl-*a* during the spring and summer period. These values give a ratio of C : N : P of 92 : 13 : 1, suggesting a strong depletion of phosphorus in the seston.

The chemistry of the surface white ice on the lake is similar to snow, whereas the chemistry of the deeper black ice results from purification (by freezing) of lake water.

The spatial and temporal distribution of heat within Mirror Lake is characteristic for a dimictic lake, i.e. a lake whose annual thermal state alternates between a warm and a cold stratification condition, separated by thermal convective periods. The cold stratification state (ice-covered period) is long (~138 days) and spring meromixis is common in Mirror Lake. At a given point within the lake, temperature changes on the order of 10^0 to 10^{1}°C are manifested during the course of the year. Over the time scale of hours to days, temperature fluctuations of 10^{-2} to 10^{-1}°C may be anticipated because of random advective phenomena.

The amount and the wavelength spectrum of light within Mirror Lake varies widely from season to season, hour to hour and depth to depth. The processes of reflection, especially during the ice-cover season, and refraction are sensitive functions of sun angle and, therefore, of season and hour of the day. Insolation, light pathlength and light wavelength are thus largely controlled by these factors. The quantity and quality of illumination in Mirror Lake is widely variable but quite predictable. The lake contains clear water (maximum Secchi disc value of 8.5 m) and green wavelengths penetrate the deepest.

In their composite effect, the heat, light and chemical milieu of Mirror Lake set and constrain the biologic system into its oligotrophic state. Although in detail every lake is unique, Mirror Lake exemplifies many of the smaller lakes of northern New England. It is clean, and functionally and aesthetically pleasing. Although Mirror Lake has demonstrated considerable stability in its behavior over the past decades, it is not immune to change. It would be most vulnerable to increased habitation and/or road-building in its watershed, which could induce radical shifts in its physical, chemical and biologic properties.

C. Stability, Circulation and Energy Flux in Mirror Lake

Noye M. Johnson, Gene E. Likens, and John S. Eaton

The physical status of the water in a lake basin is an important regulating factor for the kinds and rates of biogeochemical activity occurring within the lake. For example, in some lakes mechanical activity is vigorous while in others the waters are quiet; some are hot, others are cold;

C. Stability, Circulation and Energy Flux in Mirror Lake

some contain aged water, others are dominated by newly-added waters. These and related factors are determined largely by the energy exchanges, both thermal and mechanical, that take place between a lake and its ambient environment. These energy exchanges determine the timing, distribution and intensity of currents within the lake, both vertical (convective circulation) and horizontal (advective circulation).

The two major sources of energy driving such lake circulations are sun (insolation) and wind. Both must act initially through the upper surface of the lake, thus making the lake surface the maximum energy boundary for the system as a whole. Energy exchanges that occur at depth within the lake must first have affected the lake's surface. As a general rule, therefore, the flux of energy in a lake usually diminishes with depth, i.e. the deepest point in the lake is the least active, mechanically and thermally. The facts that (1) the sources of energy to a lake are few, and (2) that this energy must be applied through the lake surface, greatly simplify the physical understanding of lake circulation.

Under a given set of forces, just how the lake will actually respond in water motions is largely determined by the state of density stratification. That is, insolation or wind stress will induce different currents in different amounts as a function of the density stratification in the water-column. Density stratification readily lends itself to a quantitative assessment, which simplifies the understanding of the physical processes involved. In this chapter we analyze the components of density and the density distribution within Mirror Lake and, to the extent possible, explain specific circulation events on this basis.

Density Stratification

In conventional limnological terms, the circulation of Mirror Lake is described as dimictic; i.e. it convectively mixes or circulates twice a year, although occasionally spring meromixis (partial convective mixing) occurs (see Walker and Likens 1975). This concept of dimixis provides a qualitative abstraction that is useful for classifying lakes. There is no comparable concept in limnology to qualitatively assess the advective circulation of lakes.

Implicit in the concept of dimixis is the understanding that convective mixing in the whole volume of a lake can be induced by thermal energy exchanges alone. First, it is assumed that heating and/or cooling is accomplished *solely* at the lake surface. Second, for dimixis to occur, a lake theoretically must have water temperatures ranging from <4°C to >4°C at some time during its annual cycle. Thus, a dimictic lake will mix convectively in the spring when it is being heated from <4°C to 4°C, and again in the autumn when it is being cooled from >4°C to 4°C. This dimictic pattern and the implied thermal conditions do essentially occur in many lakes including the present Mirror Lake system. Theoretically, the dimictic pattern of convection will operate independently of any wind stress. Given enough time, conductive heat transfer between the lake surface and the overlying air will by itself bring about the predicted density inversions and the concomitant mixing. Under natural conditions, of course, wind stress will be present and will augment the convective mixing process.

It is significant that the value of 4°C is singled out as the most critical factor in the thermal convection process described above; pure water achieves its maximum density at 4°C. Thermal changes are generally the dominant factor controlling changes in density in lakes, including Mirror Lake. Thus, water temperature is conventionally used to gauge water density and the classification of lakes (Walker and Likens 1975). It must be emphasized, however, that other factors besides water temperature, such as salinity, suspended matter and pressure, also influence density, and on occasion may be dominant factors in density stratification, or the lack thereof (Chen and Millero 1977). Indeed, it is the overall change in density rather than the change in temperature that actually controls the mixing process in a dimictic lake. This point is graphically demonstrated in the case of Mirror Lake (*Figure IV.C–1*). The long-term average temperature distribution for Mirror Lake shows that the 4°C isotherm is rarely at the bottom of the lake, where it should be if it does in fact represent the heaviest water in the lake (*Figure IV.C–1*). The fact that Mirror Lake can be stratified with respect to density, yet simultaneously have 4°C water at mid-depths, affirms that other factors besides thermal state are acting to determine water density. We examine this problem in more detail later.

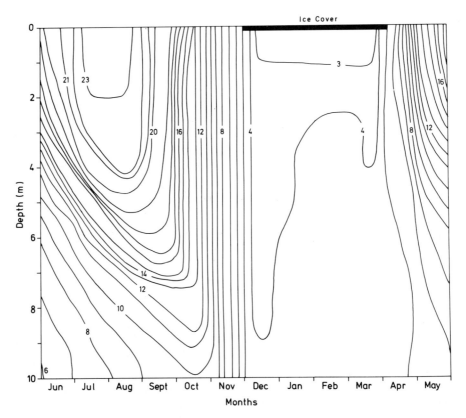

Figure IV.C–1. Average depth-time isotherms (°C) in Mirror Lake during 1967 to 1980. The dark area at the top represents the average ice-covered period for the lake.

Water temperatures, C, can be readily transformed into water density, ρ_T, by means of the empirical approximation of Tilton and Taylor (1937).

$$\rho_T = 1 - \left[\frac{(C - 3.9863)^2}{508929.2}\right]\left[\frac{C + 28.819414}{C + 68.12963}\right] \quad (1)$$

where ρ_T has units of g/cm^3 and C is degrees Celsius.

In sea water, where salinity differences between water masses commonly involve 10^3 ppm, temperature and salinity are of equal importance as density-determining factors. In fresh waters, however, salinity differences between water masses rarely exceed 10^2 ppm. In Mirror Lake the salinity distribution usually encompasses only a few parts per million of mass difference. The influence of salinity as a density-determining factor in fresh water, and particularly in Mirror Lake, is thus less important than in the sea. However, near 4°C in fresh water, changes of temperature cause diminishingly small changes in water density. For example, the change from 4°C to 3°C produces a difference in mass of only 8 mg/liter. It is near the 4°C point then in fresh water that a difference of a few mg/liter in salinity will be critical in density stratification. Just a few mg/liter of salinity distribution may be crucial under some conditions, say in ice-covered lakes for stabilizing a water-column (Hutchinson 1957).

The contribution of salinity to water density is additive and linear over relatively small ranges of salinity. That is, water at a constant temperature will become systematically heavier as salts are added to it.

The concentration of ionic salts in water can be estimated quickly from electrical conductivity measurements. For Mirror Lake the following relationship has been obtained:

$$TS(mg/liter) = 2.43 + 0.427846 \, K \quad (2)$$

where TS is total salinity concentration, K is electrical conductance in 10^{-6} cgs units. (The constants in the above equation are based on a

C. Stability, Circulation and Energy Flux in Mirror Lake

regression of 187 complete chemical analyses of Mirror Lake water from all depths and during all seasons from 1973 to 1976. The regression analyses on these data were significant at the 99.7% confidence level.) The intercept value (2.43) in equation (2) represents the non-ionized compounds in solution, essentially the mean dissolved silica content of Mirror Lake water (*Table IV.B-1*). When converting chemical mass units into density units we have assumed a molal contraction coefficient of 0.8 to account for the ligand formed between the water molecule and most ionic salts.

Liquid water also can be compressed by pressure with a corresponding increase in density. In calculating the increase of water density with depth, we used isothermal compression consants of 5.01×10^{-5} (unit volume/atm) at 0°C and 4.57×10^{-5} (unit volume/atm) at 25°C and interpolated accordingly.

We show in *Figure IV.C-2* the water density distribution in Mirror Lake over one annual cycle (1975–76). The density values portrayed are based on temperature, dissolved salt concentration and compressional effects. For any point in time (vertical axis, *Figure IV.C-2*), density stratification is indicated when one or more density contours are crossed. The intensity of the stratification is proportional to the number and frequency of density contours intersected by a vertical line. For example, the water-column was weakly stratified during the winter months, January through March, but strongly stratified in July–August, when a density contrast of some 3,000 ppm existed between surface and bottom waters (*Figure IV.C-2*).

As oriented in *Figure IV.C-2*, vertical contours indicate density homogeneity (instability) and implicitly a vertical mixing potential at the time. Using this criterion, it can be seen in *Figure IV.C-2* that during 1975–76 vertical mixing penetrated downward from August through October, culminating in complete mixing of the lake throughout November and December. In contrast, the spring mixing period was quite abbreviated, lasting only for a short time in April (*Figure IV.C-2*).

The distribution of mass as portrayed in *Figure IV.C-2* can be analyzed in quantitative terms. For each point in time (vertical axis, *Figure IV.C-2*) a center of mass could be calculated for the water-column. The relative position of the center of mass day by day would then specify in quantitative terms the relative potential energy of the water-column. This concept is essentially that contained in the Schmidt

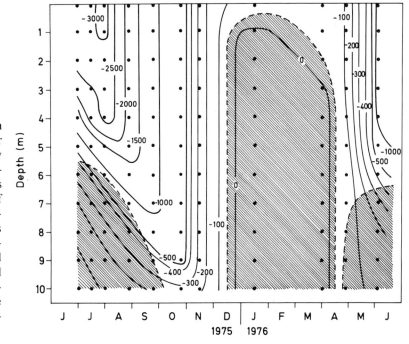

Figure IV.C-2. Distribution of water density in Mirror Lake during 1975–76. Density expressed as ppm mass difference from unity. Closed dots show time and location of water temperature and chemistry measurements. Densities were calculated from temperature and chemistry data and compensated for isothermal compression. Clear area represents mixing volume of the lake; patterned area is unmixed volume.

stability factor (Idso 1973):

$$S = A_o^{-1} \sum_{z_o}^{z_m} (z - z_{\bar{\rho}})(\rho_z - \bar{\rho})A_z\Delta z \quad (3)$$

where S is the stability in units of g cm/cm^2, A_o is the surface area of the lake, A_z is the area at any depth, z, $\bar{\rho}$ is the mean water density of the lake, ρ_z is the water density at depth z, $z_{\bar{\rho}}$ is the depth at which the mean density is found and z_m is the maximum depth of the lake. When S is multiplied by the acceleration of gravity, the units of S become ergs/cm^2, or work per unit area of lake surface. S then represents the amount of work that would be required to mix the lake into an isothermal, isohaline state. In another context, S describes how much the center of mass of the lake has been lowered by the stratification process, i.e. how much potential energy has been lost. We show in *Figure IV.C–3* the month-by-month status of Schmidt stability in Mirror Lake for 1969, reiterating in quantitative terms what we had previously inferred from *Figure IV.C–2*. That is, Mirror Lake is quite stable in summer, barely stable

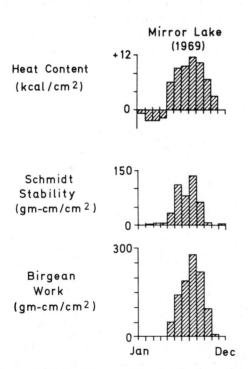

Figure IV.C–3. The monthly distribution of heat loading (see equation 5 in text), Schmidt stability (see equation 3 in text) and Birgean work (see equation 4 in text) for the 1969 annual cycle in Mirror Lake.

in winter, and convectively mixes between these times.

The data on distribution of mass shown in *Figure IV.C–2* can also be used to estimate the amount of mechanical work equivalent that has been expended through the surface of the lake to bring about the observed density distribution. This kind of energy expenditure is known as the Birgean work function (Idso 1973) and is frequently referred to as "wind work." The Birgean work function is given by

$$B = A_o^{-1} \sum_{z_o}^{z_m} z(\rho_i - \rho_z)A_z\Delta z \quad (4)$$

where B is the Birgean work function in units of g cm/cm^2 and ρ_i is the initial density at the depth concerned. Unlike the Schmidt stability function, the Birgean function requires a reference state, ρ_i, to be specified. In lakes of the Temperate Zone this reference state is commonly taken as $\rho_i = 1$, i.e. an isothermal lake at 4°C.

As defined in equation (2), positive work has been exerted when heavier water has been replaced by lighter water ($\rho_i > \rho_z$). (By way of analogy, this condition could be represented by the work done while pushing a cork underwater to a predetermined depth.) Negative work has been exerted when lighter water has been replaced by heavier water ($\rho_i < \rho_z$). (By similar analogy, the amount of work expended by a stone falling through the water-column to a predetermined depth.) The Birgean work function is a convenient measure of the mechanical work equivalent done on the lake through its surface. However, there is no reason why this work must come only from wind stress on the lake's surface. A solution to equation (2) requires that only the amount of density change be specified and is indifferent to the actual cause of this density change. In fact, the Birgean work function has been used in previous studies to assess the work done in deep water by density currents, currents that were not directly acting through the lake's surface (Johnson and Merritt 1979). In like fashion, if insolation penetrates the lake surface and heats water at depth, it will be manifested as Birgean work, even if no wind is acting.

We show in *Figure IV.C–3* the month-by-month values for Birgean work for Mirror Lake during 1969. Note that the magnitude of Birgean work (200 to 300 g cm/cm^2) is about double that

C. Stability, Circulation and Energy Flux in Mirror Lake

for Schmidt stability (150 to 200 g cm/cm^2) during the summer months. These energy differences suggest that more work has been expended on the lake to redistribute mass than would be used to break down the ambient stratification.

There is no necessary correlation between Schmidt stability and Birgean work; they involve completely separate concepts. Birgean work can be accomplished, in fact, in the absence of density stratification (Schmidt stability). An example of this would be the isothermal cooling of a lake during autumnal mixing. Under these conditions, the lake has no stability ($S = 0$), but negative Birgean work is being exerted. Mirror Lake commonly shows this situation during October–November. On the other hand, a lake may become stratified without having Birgean work accomplished. This situation is exemplified by the winter, ice-covered months in Mirror Lake.

It is sometimes instructive to portray the depth-distribution of Birgean work in a lake. Such profiles can be achieved by using the differential form of equation (4), which defines a *direct work* curve (e.g. *Figure IV.C*–4). We show in *Figure IV.C*–4 how Birgean work, both positive and negative, is accomplished in Mirror Lake during the autumnal cooling period for a typical year. Note that as the convective mixing process penetrates into the lake, the direct Birgean work curve is systematically displaced downward (left panel, *Figure IV.C*–4). The whole process may be viewed as a compound wave of Birgean work propagating downward with time and attenuating with depth (right panel, *Figure IV.C*–4). The leading edge of the mixing zone is preceded by a "front" of relatively warm water, which is driven downward in the form of positive Birgean work. Eventually much cooler, much denser water follows the front downward and is manifested as negative Birgean work. This phenomenon is not unique to Mirror Lake and has been noted previously in much larger lakes (Johnson and Merritt 1979) as well as other small lakes (Johnson et al. 1978). This sequence and disposition of Birgean work thus seems to characterize the convective mixing of many lakes during the breakdown of summer stratification.

Thermal Regimen

Up to this point the density stratification of Mirror Lake has been discussed mostly in terms of its mechanical energetics. The mechanical stability or instability of a lake is usually induced by or closely linked to the thermal energetics of the lake system.

Perhaps the most commonly used physical characterization of a lake is the temperature profile. The temperature profile describes the distribution of thermal energy within the water-column of the lake. As such, the temperature profile is quite analogous to the Birgean *direct*

Figure IV.C–4. Direct Birgean work curves (left) for Mirror Lake during the autumnal mixing period of 1969. The area under the direct work curve is the total Birgean work content of the water-column (see equation 4 in text). The monthly change in the distribution of Birgean work is shown in the right panel. Note how the mixing period is characterized by a sequence of negative and positive Birgean work propagating downward in the form of attenuating waves.

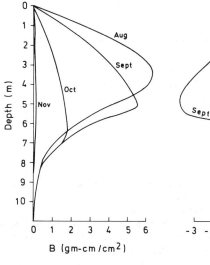
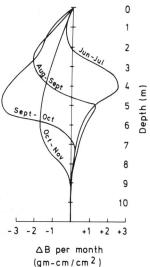

work curve (e.g. *Figure IV.C*–4), i.e. both show the distribution of certain forms of energy with depth. Like a Birgean *direct work* curve, a temperature curve can be transformed into an energy-loading term by integration. This integration is given by

$$\Theta_w = A_0^{-1} \sum_{z_o}^{z_m} c_z T_z A_z \Delta z \quad (5)$$

where Θ_w is heat content in units of calories/cm^2; T_z is the temperature at depth z and c is the specific heat of water in calories/cm^3-°C, usually taken as unity. If degrees Celsius are used, equation (5) yields results referenced to 0°C, the heat content at 0°C being zero by definition. If the lake is ice covered, the latent heat of fusion for water (-80 calories/g) must be substituted for the $c_z T_z$ product in equation (5) for each depth increment of ice. The ice layer represents negative calories with respect to the Celsius reference system. If the normalization factor used in equation (5) is lake volume (V), instead of lake surface area (A_0), then the energy loading will be expressed in terms of mean lake temperature, $\overline{T}_w$

$$\overline{T}_w = V^{-1} \sum_{z_o}^{z_m} T_z A_z \Delta z \quad (6)$$

In other words, $\overline{T}_w$ is the hypothetical temperature of the lake if it were rendered isothermal by complete mixing.

We show in *Figure IV.C*–3 the heat loading of Mirror Lake month by month for 1969. From *Figure IV.C*–3 it may be seen that some 14 kcal/cm^2 of heat energy are exchanged between Mirror Lake and its environment during the course of the year (*Table IV.C*–1). That is, the difference in heat content between the hottest and coldest conditions found in the lake during the year is 14 kcal/cm^2. This amount of heat exchange is defined as the annual Birgean heat budget (Wetzel and Likens 1979) and is a useful index for assessing a lake's potential for convective circulation.

It may be instructive here to compare the energy equivalent of the annual heat budget of Mirror Lake with its corresponding mechanical energy budget (*Figure IV.C*–3). The sum of Birgean work (equation 4) and Schmidt stability (equation 3) represents the minimal amount of mechanical work that has been exerted on or by the lake compared with its previous state (Idso 1973). In Mirror Lake the maximum sum for Birgean work and Schmidt stability occurs in summer and amounts to about 400 g cm/cm^2

Table IV.C–1. Mean Monthly Thermal and Stratification State of Mirror Lake (1967 to 1981)

	Q_w^a (kcal/cm^2)	S^b (g cm/cm^2)	B^c (g cm/cm^2)	z_{sm}^d (m)
June	8.97	91.5	133.8	4–5
July	10.81	116.2	226.6	4–5
August	11.63	99.1	292.9	4–5
September	10.48	42.6	253.4	5–6
October	7.76	6.5	128.6	7–8
November	4.37	0.4	20.4	9–11
December	0.40	2.0	0.3	<1
January	−0.86	2.0	0.2	<1
February	−1.61	2.3	0.3	<2
March	−2.36	2.7	0.4	<2
April	3.21	1.7	2.9	0–11
May	6.24	41.5	45.1	2–3

[a] Heat-loading (Q_w) includes heat of fusion of ice layer.
[b] S is stability.
[c] B is Birgean work function.
[d] Depth of the surface-mixing zone (z_{sm}) is essentially the mean depth to the top of the thermocline in the summer months and the depth of surface advection during the winter months.

C. Stability, Circulation and Energy Flux in Mirror Lake

(*Table IV.C–1*). This sum is equivalent to 4×10^5 ergs/cm²-yr of mechanical work. In contrast, the Birgean heat budget of Mirror Lake involves about 14 kcal/cm²-yr, which is equivalent to 5.9×10^{11} ergs/cm²-yr. It is evident from this comparison that the mechanical energy associated with density stratification in Mirror Lake is only one-millionth of that associated with the heating-cooling processes concurrently taking place. Thus, density stratification of a lake is an almost inconsequentital result of its thermal exchanges, at least with regard to the amount of energy involved. From the viewpoint of energy efficiency, the way in which density stratification is induced in lakes is extravagantly wasteful.

Because Mirror Lake is relatively shallow (11 m) and contains generally clear water (*Chapter IV.B*), sunlight commonly reaches the bottom of the lake. Insolation can then penetrate the whole body of Mirror Lake, heating it and causing density stratification. The wind-mixing zone in Mirror Lake in summer commonly extends to 5 m (*Table IV.C–1*). Thus, wind work *per se* is an important modifier of density distribution in Mirror Lake, but is most decidedly not the primary cause of the density contrasts. In terms of energy expenditure, insolation is the overwhelming force in creating density contrasts in the Mirror Lake system.

Bottom Heating

Another source of heat for the water in Mirror Lake is the heat contained in its bottom sediments. The bulk of this heat is solar in origin, but it also includes a small component of geothermal heat (Johnson and Likens 1967). The lake basin itself thus acts as a heat source (geothermal) and as a heat store (solar) for the waters of the lake. As a heat source, most lake bottoms will conduct <0.1 kcal/cm²-yr of geothermal heat from the Earth's interior (Johnson and Likens 1967). Nevertheless, this small amount of heat flux may still be crucial in certain isolated water bodies, such as meromictic lakes (Likens and Johnson 1969). On the other hand the amount of solar heat that passes in and out of lake bottom annually is substantial, frequently exceeding one kcal/cm²-yr for temperate lakes (Likens and Johnson 1969). For small lakes this amount of heat cannot be ignored, affecting as it does lake stratification and circulation. In small lakes, therefore, the bottom sediments become an integral part of the annual Birgean heat budget for the whole lake system.

The thermal variations in botton sediments are controlled by the heat conduction properties of the sediments and the pattern of temperature changes in the overlying water. The seasonal temperature variations in the sediments are approximated by a one-component Fourier wave (Likens and Johnson 1969). The temperature profile in the sediments at any depth and time, $T_{z,t}$ is thus given by

$$T_{z,t} = T_o + \beta z + Ae^{-\gamma z} \cos(\omega t - \gamma z) \quad (7)$$

where T_o is the mean annual temperature at the lake-sediment interface, β is the local geothermal gradient, z is depth below the lake-sediment interface, t is time, A is the amplitude of the annual temperature variation (half-range), ω is the angular velocity of the Fourier wave, $\gamma = (\omega/2\alpha)^{1/2}$ where α is the thermal diffusivity of the sediment. α is defined as $k/\rho c$ where k is the thermal conductivity of the sediment, ρ is density and c is heat capacity. In water-saturated sediments, the product ρc is commonly near unity, as is the case for water itself.

We show in *Figure IV.C–5* an example of how sediment temperatures in Mirror Lake lag behind seasonal temperature in the water. Note that 1 to 2 m below the sediment-water interface, temperatures are 180° out of phase with the overlying bottom water. That is, 1 to 2 m below the sediment-water interface the highest temperatures of the year are reached during the winter, and vice versa. Later we analyze the consequences of this ebb and flow of heat and temperature between sediments and water.

We show also in *Figures IV.C–5* and 6, that 4 m or more below the sediment-water interface no seasonal temperature changes take place. Instead, temperatures linearly increase with depth. This linear temperature gradient is in fact the local geothermal gradient, expressed as β in equation (7). As graphically shown in *Figures IV.C–5* and 6, it is the physical manifestation of the flow of heat from the interior of the Earth. The observed geothermal gradient (β) for the center of Mirror Lake is 0.1°C/m (*Figure IV.C–6*). The transient temperatures observed in the upper 4 m of sediment are those de-

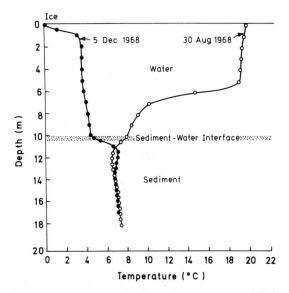

Figure IV.C–5. Distribution of temperature (°C) in Mirror Lake and underlying sediments during the summer and winter of 1968. Note how sediment temperatures tend to be out of phase with the seasonal temperatures of the overlying water.

scribed by the last term in equation (7) (*Figure IV.C–6*, right panel). Liddicoat (1970) has shown that equation (7) describes in adequate fashion the annual temperature fluctuation in a wide variety of New England and Adirondack mountain lakes.

Based on equation (7) and its premises, an expression for the annual heat budget of sediments, Θ_b, can be derived (Likens and Johnson 1969)

$$\Theta_b = 2Ak(\alpha\omega)^{-1/2} \qquad (8)$$

where Θ_b has units of calories/cm²-yr when ω is 2π/yr. Given equations (7) and (8), the heat flux, q_b, through the water-sediment interface also can be approximated (Likens and Johnson 1969)

$$q_b = \frac{\omega\Delta t}{2} \Theta_b \sin \omega t \qquad (9)$$

where q_b is the amount of heat conducted in the time interval Δt. Equation (9) shows that the heat flow in and out of the sediment varies si-

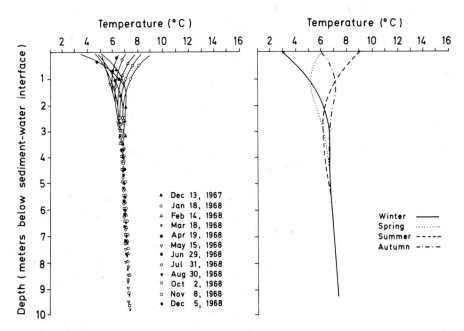

Figure IV.C–6. On the left is the observed distribution of temperature (°C) in the bottom sediments of Mirror Lake for one annual cycle (1968). Note the transient temperatures in the upper 4 m of sediment and the steady-state thermal condition below 4 m. On the right is the simulated distribution of temperatures calculated on the basis of heat conduction theory (see equation 7 in text) assuming a one-component annual wave, a mean annual sediment-water interface temperature of 6°C, an annual amplitude of 3°C, a geothermal gradient of 0.1°C/m and thermal conductivity of the sediments equal to thermal diffusivity (1.5 × 10⁻³ cgs). Note that the simulation results (right) closely duplicate the observed results (left).

nusoidally with time under the ideal condition presumed, and the magnitude of the heat flux is linearly related to Θ_b.

Likens and Johnson (1969) evaluated equation (9) for heat flux on a daily basis

$$q_b = 8.6 \times 10^{-4} \Theta_b \sin 2\pi \left(\frac{D}{365} - \frac{1}{8}\right) \quad (10)$$

where q_b has units of calories/cm²-day and D is the number of days elapsed since maximum water temperatures were reached at the lake bottom. Equation (10) explicitly shows a considerable phase difference (140 days) between the timing of the water temperature cycle and the timing of the subsequent heat flow into and out of the bottom. In other words the heating-cooling cycle of the water within the lake is almost, but not quite, 180 degrees out of phase with the heating-cooling cycle of the lake sediments.

It must be emphasized that the above equations are based on greatly simplified models, and the conclusions drawn from them are correspondingly idealized. Nevertheless, certain general patterns and useful insights may be gained by such analyses.

We show in *Table IV.C-2* the amount and distribution of stored heat in the bottom sediments of Mirror Lake. The magnitude of the sediment heat budget, Θ_b, is considerable; adding some 8% of the heat budget of the whole lake system. Note also the strong differential in bottom heating from the lake rim (2 kcal/cm²-yr) relative to the lake center (0.5 kcal/cm²-yr).

Ice-cover

The most conspicuous physical feature of Mirror Lake during the winter is its pervasive ice-cover. The onset of the ice-cover marks a profound change in the lake's albedo, its reaction to wind stress, its density stratification and advective currents. Similarly, the breakup of the ice-cover in spring brings about another drastic shift in its physical parameters.

Ice forms over Mirror Lake generally in late November–early December (*Table IV.C-3*). The initial ice sheet consists of clear, crystalline ice called "black ice" by Adams and Jones (1971) with no air inclusions or other optical defects. Once in place, such ice grows slowly on the bottom of the ice layer as the lake loses heat to the atmosphere. With the advent of the first major snowfall, however, a profound change occurs in the mode and rate of growth of the ice layer. The growth of black ice diminishes drastically and the site of major ice formation shifts to the top of the ice-cover. Ice forms in this situation by the successive accumulation and welding of snow to the top surface of the

Table IV.C–2. Average Annual Heat Budget of Bottom Sediments, Θ_b, in Mirror Lake

Depth Interval	Mean Annual Temp. Maximum	Mean Annual Temp. Minimum	Mean Annual Temp. Amplitude	$\Theta_b{}^a$ (cal/cm²-yr)	Area	$\Theta_b \times$ Area (cal/lake-yr)
0–1 m	23.5°C (Aug.)	1.0°C (Mar.)	11.3°C	1,950	1.4×10^8 cm²	2.73×10^{11}
1–2	23.3 (Aug.)	2.9 (Mar.)	10.2	1,770	1.2	2.12
2–3	23.1 (Aug.)	3.6 (Dec.)	9.8	1,700	0.9	1.53
3–4	22.7 (Aug.)	3.6 (Dec.)	9.6	1,660	1.0	1.66
4–5	21.0 (Aug.)	3.6 (Dec.)	8.7	1,510	0.6	0.91
5–6	18.6 (Aug.)	3.7 (Dec.)	7.5	1,290	0.9	1.16
6–7	16.2 (Sept.)	3.8 (Dec.)	6.2	1,076	2.2	2.37
7–8	13.7 (Sept.)	3.9 (Dec.)	4.9	850	3.6	3.06
8–9	11.8 (Oct.)	4.0 (Dec.)	3.9	677	1.6	1.08
9–10	10.3 (Oct.)	4.2 (Dec.)	3.1	530	1.0	0.52
10–11	9.6 (Oct.)	4.3 (Dec.)	2.7	460	0.6	0.28
					15.0×10^8 cm²	1.74×10^{12}
					mean $\Theta_b = 1.16$ kcal/cm²-yr	

[a] Θ_b calculated from equation (8), assuming $k = 1.5 \times 10^{-3}$ cgs, implicitly $\rho c = 1$.

Table IV.C–3. Mirror Lake Ice Data

Ice in	Ice Out
—	8 Apr. 1968
27 Nov. 1968[a]	25 Apr. 1969
2 Dec. 1969	27 Apr. 1970
7 Dec. 1970	26 Apr. 1971
29 Nov.–8 Dec. 1971	25 Apr.–1 May 1972
27 Nov.–5 Dec. 1972	16 Apr. 1973
10 Dec.–13 Dec. 1973	~24 Apr. 1974
1 Dec. 1974	28 Apr. 1975
9 Dec. 1975	18 Apr. 1976
22 Nov. 1976	14 Apr. 1977
8 Dec. 1977	28 Apr. 1978
26 Nov. 1978	23 Apr. 1979
14 Dec. 1979	13 Apr.–14 Apr. 1980
18 Nov.–2 Dec. 1980	2 Apr. 1981
Average: 3 December	Average: 20 April

Average Ice-Covered Period: 138 Days

[a] Estimated from thermal record (see *Chapter IV.B*).

ice-cover (Shaw 1965). Accumulated snow eventually depresses the ice layer enough so that lake water intrudes into and mixes throughout the snow. Lake water then spreads rapidly both vertically and horizontally through the dry snow by means of capillary flow, provided that a crack or fissure provides a supply of lake water. The resultant snow and lake-water slurry freezes and becomes attached to the top of the ice-cover. Winter rains and/or mixed rain-snow storms have the same ultimate effect, i.e. accumulation of ice on top of the ice sheet. Such "white ice" is distinctly stratified, cloudy with air bubbles and is finely crystalline (Adams and Jones 1971). It is invariably superimposed on a layer of clear black ice. The thickness of the annual ice layer on Mirror Lake is thus roughly proportional to the amount of snow that falls during the winter. Generally, the ice layer on Mirror Lake is 40 to 75 cm thick by late February.

The ice-cover is clearly an integral part of the lake system. It is held in place by buoyancy forces and contributes to the hydrostatic pressure at depth; it adds or subtracts water to the lake surface as freezing or thawing takes place. During the freezing process salts are locally concentrated in coexisting waters by exsolution (Ficken 1967); conversely, salts within the lake water are diluted by the thawing process. With respect to the lake's thermal regimen, the ice-cover is the dominant item in the winter heat budget. During the winter season, changes in heat content of the lake system are accomplished in the main by changing the thickness of the ice-cover. Heat added to the lake, say by insolation or a warm rain, will induce melting of ice. Conversely, heat losses from the lake will cause the ice layer to grow. Changes in the sensible temperature of the water-column during the winter are relatively small by comparison.

In our evaluation of Mirror Lake's annual heat budget, Θ_w, we have included the latent heat of fusion of the ice-cover as a necessary term. As seen in *Figure IV.C–3* and *Table IV.C–1*, the ice-cover accounts for several kcal/cm^2 of negative heat when referenced to water at 0°C. This amount of negative heat is significant when compared with the 10 to 12 kcal/cm^2 of positive heat manifested as summer heat income in the water-column (*Figure IV.C–3*). Some 17% of Mirror Lake's annual heat budget is thus attributed to the ice-cover alone. In some alpine lakes, which are ice covered for a longer time and with thicker ice, the annual heat budget is completely dominated by the ice freezing and thawing term (Johnson et al. 1978). In the extreme case where lakes have a perennial ice-cover, say those in polar regions, the importance of an ice-cover in the annual heat budget is maximized (Ragotzkie and Likens 1964). In such lakes the annual heat budget of the ice-cover is essentially one and the same as the annual heat budget for the lake itself.

The ice-cover on Mirror Lake usually breaks up during mid- to late April (*Table IV.C–3*). The breakup occurs as a sudden, discrete event, usually associated with a windy day. Residual ice may persist for a week or more as stranded blocks on the downwind shore. Prior to the actual breakup, however, the ice-cover has thinned in place and weakened. This thinning takes the form of melting from the top surface downward, so that the black ice layer is the last to decay. During this melting period, melt water puddles and flows over the ice surface until it drains into cracks and seams in the ice-cover. A moat of ice-free water develops along the lake shore throughout the melt period, usually March and April, as shore ice is preferentially melted for the most part by back radiation from the shoreline. Just before breakup, a raft of unanchored ice remains in the middle of the lake, separated from the shore by several meters of open water. The ice-cover is weakened by the differential melting of the contacts between the ice crystals making up the ice-cover. When the black ice layer loses its coherence by such differential melting, the whole ice-cover becomes essentially a structurally weak agglomerate of linear, upright ice crystals floating on the lake surface. This condition is referred to as *candle ice* and lasts for only a few days. Under such conditions, a mild wind can induce long-wavelength waves at the surface, with the candle ice behaving as a viscous, plastic sheet covering the lake. A stronger wind at this stage will effect an immediate breakup.

The existence of an ice-cover on a lake implies that the water-column on which the ice floats is theoretically 4°C or less. In this temperature range the effect of temperature change on the density of water is minimal. For example, in a lake filled with distilled water the maximum density contrast that could exist in the water below an ice-cover would be 132 ppm, i.e. the density difference between water at 0°C and 4°C. By comparison, the density contrast between top and bottom waters during summer thermal stratification can easily exceed 1,300 ppm (e.g. *Figure IV.C–2*) or ten times more than is possible during winter stratification. In an ice-covered lake, then, the impact of density-determining factors other than temperature, such as salt concentration and suspended matter concentration, is relatively enhanced. In Mirror Lake, and indeed in many ice-covered lakes (Hutchinson 1957), bottom water temperatures are commonly observed to increase throughout the period of ice-cover, presumably from the feedback of stored summer heat from the lake sediments. Temperatures well above 4°C are frequently observed at the bottom of Mirror Lake in late winter when it is ice covered. The density anomaly that results from this temperature anomaly is conventionally explained by a compensating diffusion of salts into the water-column from the bottom sediments (Hutchinson 1957).

A detailed chemical analysis of the salt in the Mirror Lake system reveals, however, that the inorganic compounds present are not sufficient by themselves to account for the observed density anomaly (*Figure IV.C–6*). We show in *Figure IV.C–6* that on this date (29 March 1976) even after the inorganic compounds are added to the water-column, a significant density inversion still exists below 4 m and especially at the bottom. It is evident, therefore, that inorganic salt by itself is not compensating for the entire thermal anomaly, so something else must be present. We speculate that dissolved organic matter and/or suspended materials in these bottom waters are abundant enough to stabilize the water-column. Our data suggest that only 4 to 8 mg/liter of dissolved organic matter are contained in these waters. However, just how these organic materials hydrate and affect water density is not known. Further work is necessary to resolve this problem.

It should also be noted that the apparent enhancement of density stratification caused by isothermal compression is, in fact, illusory (*Figure IV.C–7*), because no work is expended during an isothermal expansion or compression process accompanying a change in depth (pressure).

Advection and Advective Currents

Like all freshwater lakes, Mirror Lake is an open system with respect to its gains and losses of water. A consequence of these water transfers is that organized water movements, currents, are sometimes induced. These currents vary in size, velocity and trajectory from season to season, but certain predictable patterns seem to hold. In this section, we describe and explain seasonal current patterns in Mirror Lake.

As discussed in the preceding section, the period of ice-cover for Mirror Lake is associated with a constraining set of thermal and mechanical conditions. Density stratification is present, but mechanical stability is feeble (*Figure IV.C–3*; *Table IV.C–1*). The density stratification persists only because the ice-cover prevents wind stress from overpowering it. Also, in the Mirror Lake system the density of incoming stream water is very consistent during the winter; at this time, the temperature is constantly near 0°C. When such water intrudes into the lake it

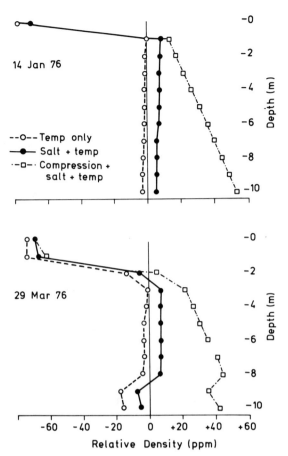

Figure IV.C–7. The density distribution within the water-column of Mirror Lake during the early (upper panel) and late (lower panel) ice-covered season. Density values are expressed in units of ppm departure from unit density. In early winter (upper panel) the density distribution based on temperature (see equation 1 in text) and observed salt content shows a slight inversion ($\simeq 2$ ppm). This instability would seem to disappear when isothermal compression of the water-column is accounted for. By late winter, however, the water-column shows an absolute instability of some 10 ppm even with the inclusion of isothermal compression in the analysis. The magnitude of this instability is greater than experimental error and argues that another density-determining factor (e.g. dissolved organic matter) must be present in Mirror Lake water besides temperature, dissolved inorganic salt and compression.

can only displace water of equal or lesser density than itself. Thus, the incoming stream water, which is at or near 0°C, can only penetrate the zone immediately below the ice-cover where temperatures are also near 0°C. Such water entering the lake will pass through the lake as an overflow density current (Wunderlich 1971), just beneath the ice-cover. The speed and the dimensions of this winter overflow current will be dictated by the mixing processes taking place at the boundaries of the current. Because the lake is density stratified, mixing will be much more prevalent in the horizontal direction than in the vertical direction. Dye tracer studies have shown that these currents may be only 1 to 2 cm thick but dispersed horizontally over tens of meters. The depth at which these overflow currents occur is variable from day to day, depending on the density of the incoming water. Throughout most of the winter, conditions are such that incoming water is confined to the first 50 cm or so beneath the ice-cover (*Figure IV.C*–2). In two separate years, stream water entering Mirror Lake in the spring was traced visually with dyes. On both occasions the incoming stream water was quickly organized into a thin ribbon at a depth of about 70 cm below the free-water surface, or some 30 cm beneath the ice-cover. The parcel of dyed water was observed to move as an integral unit across the lake, leaving a faint trail of dye behind in its wake.

As discussed previously, substantial amounts of summer heat are stored in bottom sediments (0.5 to 2 kcal/cm^2), and the feedback of this heat into the water at the lake bottom is delayed for a considerable time. From equation (10) and under idealized conditions, this feedback reaches its maximum about 140 days after the hottest water temperatures were reached in the lake. In Mirror Lake this maximum temperature usually occurs sometime in August or thereafter. For Mirror Lake, then, the time interval over which significant amounts of heat are being conducted from the lake bottom to overlying water extends from November through February; the maximum heat exchange occurs about the first of the year (*Table IV.C*–4). This time interval coincides well with the period of ice-cover for Mirror Lake.

On a daily basis the amount of heat actually exchanged is relatively minor, being about 17 calories/cm^2-day at its peak. However, during the course of the entire cooling period (November through March; *Table IV.C*–4) some 0.5 to 2 kcal/cm^2 will be added to the lake at its bottom interface. To put this into perspective, 1 kcal/cm^2 could theoretically raise the tempera-

Table IV.C–4. Comparison of Monthly Changes in Heat Storage in Mirror Lake with Atmospheric Energy Flux[a]

	Heat Storage Processes					Atmospheric Processes		
	$\Delta\Theta_b$	$\Delta\Theta_{water}$	$\Delta\Theta_{ice}$	Q_a^d	Σ	Q_s^b	Q_e^c	Σ
June	−0.26	+2.29	0	−0.33	+1.70	13.9	−4.6	9.3
July	−0.31	+1.33	0	−0.35	+0.67	14.8	−4.9	9.9
August	−0.27	−0.17	0	−0.39	−0.83	12.8	−5.3	7.5
September	−0.15	−1.94	0	−0.01	−2.10	9.7	−3.4	6.3
October	<0.01	−3.06	0	+0.38	−2.68	7.2	−2.1	5.1
November	+0.15	−3.19	−0.49	+0.19	−3.34	3.9	−0.8	3.1
December	+0.27	−1.17	−1.45	−0.04	−2.39	3.2	−0.5	2.7
January	+0.31	+0.50	−1.51	−0.02	−0.72	4.2	−0.4	3.8
February	+0.24	+0.12	−0.63	−0.02	−0.29	6.1	−0.4	5.7
March	+0.16	+0.42	+1.99	+0.01	+2.58	9.4	−0.4	9.0
April	<0.01	+2.22	+2.08	−0.009	+4.21	11.9	−0.7	11.2
May	−0.15	+2.88	0	−0.006	+2.67	13.9	−4.6	9.3

[a] Units: mean Δkcal/cm² for whole lake surface. Heat flux values derived from *Tables IV.C–1* and 2 and from unpublished U.S. Forest Service files.
[b] Q_s is the long-term average of solar radiation observed at the Hubbard Brook station.
[c] Q_e is observed and simulated evaporation.
[d] Q_a is advected heat including both ground-water and surface-water fluxes.

ture of the bottommost meter of water by 10°C. Significantly, this heat, although small in amount, is added to the bottom water of the lake, not the surface water. This condition is in marked contrast to the more powerful insolation and wind stress processes.

We show in *Table IV.C–2* that the bulk of the summer heat stored in the bottom sediments of Mirror Lake is concentrated in nearshore areas. At the deepest point in the lake, the bottom heat is only one-fourth of that in the nearshore areas (*Table IV.C–2*). The feedback of summer heat from sediments thus is differentially distributed over the lake bottom. That is, more bottom water heating takes place in the nearshore areas than in the deeper areas. The nearshore areas of Mirror Lake thus act like a *warm rim* relative to its center. This spatial difference in bottom heating can induce subtle currents in the otherwise quiet water of an ice-covered lake (Likens and Ragotzkie 1965; Likens and Johnson 1969).

These currents are thermally-induced density currents and are related to the unusual thermal expansion characteristics of water. When water <4°C is heated, it increases in density rather than decreasing as is the case with all other known liquids. For example, when water is heated from 0.0°C to 0.1°C its density increases by 6.7 mg/liter. When Mirror Lake is ice covered, heating of nearshore water, where ambient temperature must be <4°C, necessarily makes the affected water more dense. The tendency, then, is for this nearshore water, when heated, to slide downslope toward the deeper portions of the lake. If heavy enough, such density currents may assume the form of an underflow current that reaches the deepest parts of the lake (Hutchinson 1957), or it may mix away in part and become an interflow current, sometimes with a rotary trajectory (Likens and Hasler 1962; Likens and Ragotzkie 1965, 1966). Hutchinson (1957) invokes this type of density current to explain the transport of dissolved salt and heat into the deeper waters of ice-covered lakes. Hutchinson (1957) asserts that the diffusion of heat out of bottom sediments should be accompanied by the diffusion of salts out of the same sediments. We have in fact observed this joint accumulation of dissolved salt (*Figure IV.C–8*) and heat in the bottom water of Mirror Lake throughout the ice-cover period. The salts added to the water would increase its density

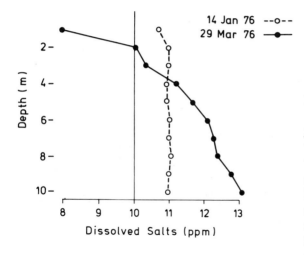

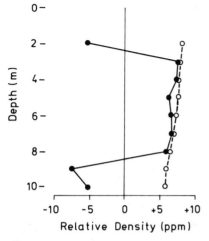

Figure IV.C-8. The dissolved salt and density distribution within the water-column of Mirror Lake during the early and late ice-covered season. We show in the upper panel how dissolved salt becomes less homogeneous in its distribution during the winter. Above a depth of 4 m, salinity is diluted by the advection of melt water from snow. Below 4 m, salinity is increased by diffusion of salt from the bottom muds and/or advection of salts from nearshore waters. In the lower panel we show how this differential salt distribution affects water density. (Density as portrayed is based only on temperature and salt distribution, ignoring isothermal compression.) Note that a significant density inversion persists throughout the winter but is especially conspicuous in the bottom waters. Evidently the increase in dissolved salt content of the bottom water is insufficient by itself to compensate for the decrease in density caused by warming of the bottom water. The density instability, some 10 ppm, may be the result of dissolved organic matter or suspended particles.

beyond any heat-induced density increases. The diffusion of salts out of bottom sediments will thus enhance any underflow density currents in ice-covered lakes.

In addition to the feedback of summer heat from bottom sediments, shallow waters and sediments in an ice-covered lake are also subject to heating by direct insolation. In mid-winter several factors mitigate against such insolation: (1) absorption by snow and ice, (2) the albedo of the ice-cover may be high at this time because of fresh snow, and (3) the sun angle is at its minimum. Nevertheless, on sunny days in mid-winter, insolation of the upper waters can be intense. This heating is especially pronounced where dark colored sediments lie immediately below the ice layer. By March and April the sun angle is higher and the ice-cover thinner and darker (lower albedo). Insolation of the upper layers of the lake is the principal heating process operating at this time, because the store of sediment-heat has been essentially exhausted by March and April.

Accompanying this insolation effect in early spring is the largest hydrologic event of the year for the Mirror Lake watershed—the spring flood. The snowpack generally melts during late March and early April and accounts for 30 to 40% of the annual runoff from the Mirror Lake watershed (see *Chapter IV.E*). Spring runoff is essentially water from melted snow, near 0°C and relatively salt free. As such, it passes through ice-covered Mirror Lake in the form of an overflow density current, actually an amplified version of the winter overflow current. The combination of this spring overflow current with the concurrent insolation of the upper water layers changes the thermal, chemical and mechanical states of the upper 2 m or so of the lake (*Figure IV.C-9*). It is significant, however, that these events near the surface do not affect the deeper parts of the lake (*Figure IV.C-9*); Hill 1967).

The breakup of the ice-cover signals a general change in the lake's physical state. As soon as the protective cover of ice is removed, the weak stability of the entire lake can be disrupted by wind and solar heating; i.e. convective overturn may proceed. The vigor, depth and duration of spring mixing is a sensitive function of the weather that immediately follows the ice breakup. A few days of mild temperatures and

Figure IV.C-9. The distribution of total dissolved salts with time in Mirror Lake during the ice-covered season. Closed dots represent water samples taken for complete chemical analyses. Isopleth units are mg/liter of inorganic salt. Note that during January-February the salinity of the water-column increases slightly in a homogeneous fashion. In the latter part of the ice-covered season, however, the homogeneous distribution of salt is disrupted (<4 m) by the spring flood advecting across the lake under the ice. Simultaneously, a slow diffusion or advection of salt into the bottom water takes place below the 4-m level. During this year (1976) the spring mixing season (late April-early May) did not render the lake isohaline (note the 11 mg/liter isopleth).

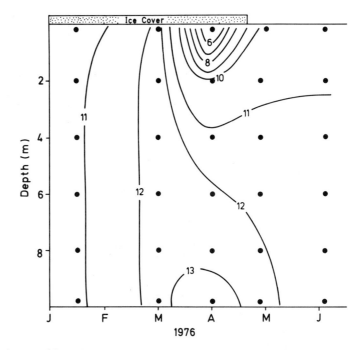

brisk wind will serve to mix the lake thoroughly to its bottom. As sometimes occurs, however, a few days of hot, quiet weather will attenuate or block convective mixing. We show in *Figure IV.C-10* such an occasion of *spring meromixis* (Åberg and Rodhe 1942). As seen in *Figure IV.C-10*, a weak mixing period during the spring of 1976 failed to completely oxygenate the bottom waters of Mirror Lake. In Mirror Lake spring meromixis is a common, but not invariable, occurrence.

Several sunny days after the breakup of the ice-cover are generally sufficient to form relatively stable density stratification in Mirror Lake, which shifts the circulation of the lake into yet a different mode. Density stratification inhibits downward mixing of incoming water and effectively seals off the deep waters of the lake for the duration of the spring and summer. This physical isolation of the deep waters is reflected in the systematic depletion with time of dissolved oxygen at the bottom of the lake (*Figure IV.C-10*).

In *Figure IV.C-2*, we have indicated the maximum depth (dashed line) to which ambient stream water entering the lake could potentially sink. Note that from May through July in 1975–76 this depth was approximately 6 m. It is the *maximum* depth of stream-water penetration because normal mixing at the surface would quickly attenuate any density contrast for the incoming stream water. Also during the summer months the volume of incoming water is typically small or occasionally nil. Surface outflow from Mirror Lake is exclusively from the mixing zone, the depth of which varies from season to season (*Table IV.C-1*).

The amount of streamflow into Mirror Lake during the summer rapidly declines with time, reaching its yearly low value by late summer or early autumn (*Chapter IV.E*). At the same time, however, evaporation from the lake surface is at its maximum (*Table IV.C-4*). The result is that by late summer generally there is no surface outflow at all from Mirror Lake, lake level being well below its outlet level (*Table IV.C-4*). Advection currents in Mirrow Lake during the summer stratification season are then nonexistent or are limited to mixing eddies at the mouths of the inlet streams. Below the mixing volume, vertical mixing in the lake is greatly restricted.

In *Figure IV.C-2*, we show how the mixing zone in Mirror Lake penetrates downward as the summer season progresses. The extent of this penetration can be quantitatively expressed as the loss in Schmidt stability with time (see *Figure IV.C-3; Table IV.C-1*). By October incoming stream water has a somewhat higher salt content (Likens et al. 1977) and has cooled to the point where its density is high enough to reach the lake bottom. By November the lake is

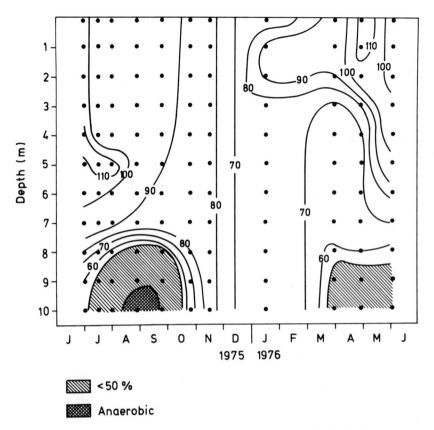

Figure IV.C–10. Distribution of dissolved oxygen in Mirror Lake during the 1975–76 annual cycle. Isopleths show percent saturation. Closed dots show time and location of oxygen measurements. Note the development of anaerobic bottom water during August-September, and the subsequent recharge of oxygen by convective mixing during November-December. The spring period of convective mixing (mid-April; see *Figure IV.C–1*) was insufficient to fully oxygenate the bottom water (see also *Figure IV.C–7*).

fully mixing, a condition that lasts until the lake freezes over. Simultaneously, increased autumn runoff (*Chapter IV.E*) brings the lake level up to its sill depth and outflow at the surface is resumed. Advection and convection during November are essentially one and the same, as the mixing volume of the lake is in fact the entire lake. Drogue studies at this time show a very complex system of water trajectories and water speeds (10^{-1} to 10^{-2} m/s). Countercurrents at depth compensate for the downwind drift of surface waters.

Water Residence Time

On an annual basis the mean, theoretical flushing time for the water in Mirror Lake is 1.02 years (or 12 months); i.e., a lake volume of 8.62×10^5 m^3 and a mean annual water flux of 8.45×10^5 m^3 (see *Chapter IV.E*). Implicit in this calculation is that incoming water is completely mixed throughout the lake basin. Because Mirror Lake is density stratified through most of the year, however, incoming waters do not mix through the whole lake volume at any given time (*Figure IV.C–2*; *Table IV.C–1*). The mean age of water within the mixing volume and the unmixed volume (bottom water) thus differ substantially. We show in *Figure IV.C–11* the month-by-month mean ages for the mixed layer and the unmixed layer as simulated for the conditions in Mirror Lake during 1975–76. Because of nonuniform mixing during most of the year, the simulated mean age of water in the lake (15 months) is 25% higher than that predicted by a theoretical flushing-time calculation (12 months). The principal factor here is the small size of the mixing volume in Mirror Lake during March and April when the maxi-

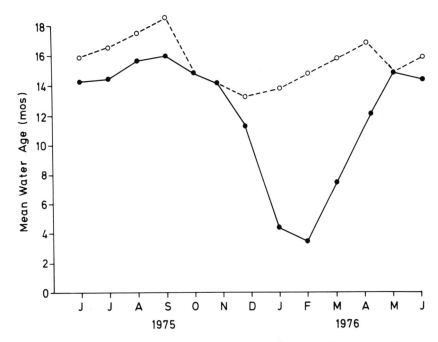

Figure IV.C–11. Simulated mean age of water in the mixing layer (closed circles) and bottom water (open circles) of Mirror Lake for 1975–76. The simulation was based on the amount and density of incoming water and the ambient density stratification of the lake. Note that the overflow density current, which occurs during the winter, ice-covered season, significantly freshens the waters of the mixing zone (closed circles), but does not affect the bottom waters (open circles) of the lake, which continue to "age". Primarily because of this under-ice mixing and advection mode, the mean age of water in the lake is substantially older (15 months) than what is conventionally calculated (12 months) by assuming complete mixing during the entire year. Also note the uneven water-age distribution with time and position; water age ranges from 4 to 18 months, depending on place within the lake and time of year.

mum input of water into the lake basin occurs. Even though large amounts of "new" water are entering the lake at this time, the bottom water in the lake is not affected; it continues to "age" unabated until the mixing process reaches it (May data, *Figure IV.C*–11).

An interesting question is posed by the location and extent of the currents, if any, which are associated with percolation water losses from the lake basin (*Chapter IV.E*). A large volume of water, approximately 3.6×10^5 m^3, leaves the lake (each year) by this route, mostly along the southern shore. We speculate that these water losses take place above the 4-m depth where in Mirror Lake the bottom is characterized by sand, gravel and boulders. The southern shore of the lake is in fact a gravel terrace deposit, which is water permeable (*Chapter III.A*), and probably acts as a general water sink. In contrast, below 4 m the lake bottom consists of gyttja, which is quite water impermeable. The transport mechanics and path of the water as it drains away from the lake basin by percolation are not really known. However, if the process is confined to the water above 4 m, it would be in the mixing zone during most of the year. Any currents, therefore, would meld with the normal turbulence of the mixing zone.

Seasonal Energy Flux

In this section, we analyze the amount of insolation flux reaching Mirror Lake month by month and assess how this thermal energy is dissipated by the lake system as a whole. In *Table IV.C*–4 we present the monthly heat fluxes associated with external heating (insolation) for the Mirror Lake system. Energy dissipation resulting from albedo, back-radiation and sensible heating of air are not considered in this analysis.

Seasonal patterns are clearly evident in the various components of internal heat storage

(*Table IV.C–4*). Of particular note is the dominance of the ice-cover term during the winter months both as a heat sink and heat source. It is also noteworthy that during the winter, the return flux of summer heat from bottom sediments (positive values of $\Delta\Theta_b$ is nearly sufficient by itself to account for the concurrent heating of the water-column (*Table IV.C–4*). During the summer months, advective heat transport assumes its major importance, being the dominant term (47%) in August. Over the year, advective heat loss is 21% greater than advective heat gain for the lake.

In the Mirror Lake system, the energy flux associated with internal heat storage clearly mimics the seasonal changes in insolation and evaporation (*Table IV.C–4*). The changes in heat storage are in fact directly proportional to insolation with a marked seasonal hysteresis (*Figure IV.C–12*). The seasonal changes in heating-cooling rate in Mirror Lake follow essentially a 4-component trajectory (*Figure IV.C–12*): (1) from July through November, when solar heating decreases with time, the lake loses heat proportionately; (2) from November through January, when solar heating is uniformly low ($\simeq 3$ kcal/cm^2), the lake continues to lose heat but at a diminishing rate; (3) from January through April, when solar heating increases with time, the lake gains heat proportionately; and (4) from April through July, when solar heating is uniformly high (~ 10 kcal/cm^2), the lake continues to gain heat but at a diminishing rate. Maximum heat exchanges in the Mirror Lake system occur during April and November and involve more than 3 kcal/cm^2-month. It is probably not just coincidence that April and November are also peak periods of convective mixing.

The correlation between insolation and heat exchange in the Mirror Lake system (r = 0.89, *Figure IV.C–12*) illustrates the obvious cause-and-effect relationship between them. Based on the slope of the relationship shown in *Figure IV.C–12*, it appears that a unit change in monthly insolation will induce a 0.78 change in heating-cooling of the lake.

On the average, complete convective mixing in Mirror Lake generally coincides with those intervals when heat exchange with the environment exceeds 3 kcal/cm^2-month, i.e. April and

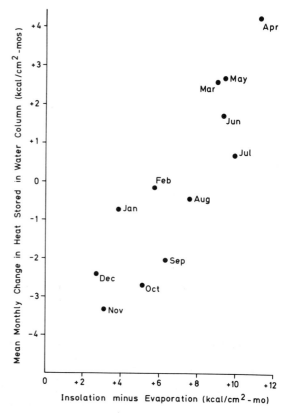

Figure IV.C–12. Changes in rate of heating and cooling in Mirror Lake as a function of insolation.

October-November (*Figure IV.C–12*; *Table IV.C–1*). For Mirror Lake it seems that 3 kcal/cm^2-month might reflect an energy threshold, which if exceeded predicates complete thermal convection. In other words, when meteorologic conditions are such that high energy exchange takes place, the lake tends to circulate. In this regard 1975–76 was a year for spring meromixis on Mirror Lake, i.e. spring convective overturn did not reach the bottom of the lake (*Figure IV.C–9*). It is significant that in April of that year (1976) only 2.51 kcal/cm^2 of heat exchange took place in the lake, well below the 4.30 kcal/cm^2 norm for April (*Table IV.C–4*; *Figure IV.C–12*).

In the above analysis, we have not considered the heat flux associated with direct precipitation on the lake. The temperature of rain in the summer months is not known, and the latent heat of snow in the winter months is incorporated in part into the ice-cover term. Assuming that the temperature of rain might approximate the wet-bulb air temperature, on the average the input of heat from rainfall during 15 April to 15 November might be ~0.8 kcal/cm² for Mirror Lake.

Summary

Mirror Lake has a rich variety of circulation pathways during the course of a year. The mechanisms for these circulation patterns can be explained generally from first principles of physics, and in specific instances by quantity and number. Seasonal changes coincide with repeatable circulation modes:

1. *Winter mode*—During the period of ice-cover on Mirror Lake a distinctive circulation system is maintained. Throughgoing water crosses the lake as a confined overflow density current just below the ice layer. Below this zone, the lake is density stratified, but weakly so (Schmidt stability <10 g cm/cm²). The return of summer heat from the bottom sediments (up to 17 calories/cm²-day) causes various thermal currents at depth. These underflow density currents are responsible for the advection of heat and salt into the deeper portions of the lake basin. In early spring especially, such density currents are augmented by insolation-heating at the sediment-water interface in nearshore waters. The latent heat of fusion of the ice-cover is the principal term in the winter heat budget, involving several thousand calories/cm².
2. *Spring-mixing mode*—Convective mixing in Mirror Lake takes place immediately after the breakup of the ice-cover. The period of mixing is short, spanning only a few days to a week. The intensity of the mixing is variable from year to year, but frequently fails to reach the deepest part of the lake (spring meromixis). Spring meromixis in Mirror Lake seems to occur in those years when April heat exchange is less than 3 kcal/cm². Advective currents are limited and diffused, being confined to the immediate mouths of inlet streams and the outlet.
3. *Summer stratification*—Mirror Lake attains about 100 g cm/cm² of density stability during its peak thermal stratification. The corresponding heat content of the lake is 10 to 12 kcal/cm². The stability of the bottom water is reflected by a systematic depletion of dissolved oxygen at depth during the summer. Throughgoing currents are generally absent as the lake surface is commonly below its outlet depth.
4. *Autumnal-mixing mode*—The mixing zone progressively penetrates downward in the lake from about August on. This mixing is caused by cooling of the lake surface and is shown as waves on the Birgean work curve. Both positive and negative Birgean work are exerted during the breakdown of summer stratification, so that total Birgean work is not an accurate account of what actually happens. By October, mixing has generally reached the bottom of Mirror Lake and complete mixing is sustained until the lake freezes over, usually in early December. Increased runoff in the autumn fills the lake to sill depth and throughgoing advection is resumed. Stream water entering the lake at this time is dispersed and no organized currents exist for long periods. The mixing volume of the lake is the entire lake.

The prime mover for all the physical activity described above for Mirror Lake is solar energy. The annual march of the sun dictates the *amount* of work that can be done season by season. On the other hand, the unusual density properties of water prescribe just what *kind* of work can actually take place.

The beautifully integrated ecosystem of a lake rests in large measure on this interaction between solar energy and water. Such factors as the supply and cycling of nutrients, the amount and distribution of dissolved oxygen, the timing and duration of metabolic environments, and the state of mechanical agitation in the water are all predicated by the thermal and mechanical energy exchanges taking place in the lake and its environment. The physical processes involved are simple and readily explained by first principles. A thorough understanding of these processes can provide insights into the much more complex processes associated with the biology of the ecosystem.

D. Approaches to the Study of Lake Hydrology

THOMAS C. WINTER

In many water-balance studies of lakes, one or more components of the hydrologic system are calculated as the residual. This approach is commonly used for economic reasons, but it also is considered acceptable by many workers because they are confident that the methods used to measure hydrologic components contain small errors, and therefore that the residual also contains small errors.

The question of accuracy in hydrologic-budget studies has recently been discussed by Winter (1981c). In that paper, errors associated with various methods of measuring and calculating all hydrologic components interacting with a lake were reviewed, and their effects on the accuracy of a residual term were evaluated. Winter also pointed out that many commonly used methods of estimating hydrologic components have appreciable error associated with the estimates, and that the residual, which contains all the accumulated errors, commonly has little meaning.

Hydrologic components that need to be considered in water-balance studies of lakes are shown schematically in *Figure IV.D–1*. The purpose of this chapter is to discuss comparative accuracy, advantages and disadvantages of selected methods of determining the quantity of water interacting with a lake from atmospheric, surface- and ground-water sources and sinks. Methods used to determine the water budget of Mirror Lake since 1968, as well as the more intensive studies that began during 1978, are discussed below. See *Chapter IV.E* for analysis of water budget during 1968 to 1980.

This section applies only to physical hydrology. Approaches to sampling and interpretation of chemical fluxes are not discussed here. Errors associated with chemical aspects of budget studies are added to those related to physical hydrologic aspects. A survey of approaches and associated error related to chemical budgets of lakes and reservoirs is presented by Winter (1981b).

Atmospheric Water

Precipitation

The choices available for measuring precipitation inputs to a lake include (in order of decreasing accuracy): (1) recording gauges at the lake, (2) nonrecording gauges read by observers living near the lake, and (3) gauges located a greater distance from the lake, such as at National Weather Service stations.

Instrument errors associated with gauging precipitation can be in the range of 1 to 5%. Errors are also related to the placement of gauges; for example, Neff (1977) found that a rain gauge whose orifice is 1 m above ground level catches 5 to 15% less water for long-term data, and as much as 75% less for individual storms than a gauge whose orifice is at ground level. Linsley et al. (1958) state that gauges equipped with windshields catch as much as 20% more water than those without windshields.

Errors associated with areal averaging of

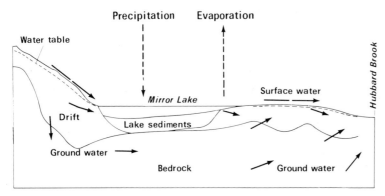

Figure IV.D–1. Schematic diagram of the hydrologic system interacting with Mirror Lake.

point-precipitation data can be greater than 60% for individual storms, depending on storm type, duration and gauge density. Concerning variations in rainfall estimates for different time periods at a given gauge density, daily mean rainfall can be in error as much as 60% at a gauge density approximately equivalent to the National Weather Service network of 648 km^2/gauge. As the time span for which values are averaged increases, however, the magnitude of error decreases. Errors in estimates of monthly precipitation are more in the range of 10 to 20%; errors in seasonal estimates are about 5% (Winter 1981c).

For comparison of rainfall estimates for various gauge densities for a given time, estimates of daily areal mean rainfall can be in error by about 60% with a gauge density of 725 km^2/gauge. The error decreases to about 10% with a gauge density of 21 km^2/gauge, and to 4% with a density of 2.6 km^2/gauge. Mathematical methods used to calculate quantities of precipitation over an area such as isohyetal (interpolation by use of contouring), average and weighted average, using identical precipitation amounts for each station, have differed by as much as 18%. These estimates of error are generally based on studies in relatively flat terrain.

Errors in estimates of precipitation related to density of gauges are considerably greater for mountainous settings than those given above (Molnau et al. 1980; Warnick 1951; Chang 1977). The study by Chang probably is the most pertinent to Mirror Lake because it was done in the Appalachian Mountains of West Virginia.

At Mirror Lake, precipitation data are obtained by a nonrecording gauge (no windshield) about 0.2 km east and a recording gauge (with windshield) about 0.4 km west of the lake (*Figure III.A–1*).

Evaporation

Evaporation commonly is not measured accurately in lake water-balance studies. In most studies, data from National Weather Service evaporation-pan stations are used, no matter how distant they are from the lake or how unrepresentative of the lake they might be. In fact, evaporation pans located immediately adjacent to a lake are only marginally acceptable as a measure of lake evaporation (Ficke et al. 1977).

Comparative studies that evaluated effects of differences in type, size and color of evaporation pans show variations in evaporation of as much as 30% between tanks and pans, 20% between different sizes (0.6-m vs. 3.6-m diameter) and 25% between different colors (Gangopadhyaya et al. 1966). Evaporation from a rinsed floating pan differed from a Class-A pan by 14 to 29% on a monthly basis, and 22% for a 6-month period (Neuwirth 1973).

To use any evaporation-pan data with confidence requires calibration of a pan-to-lake coefficient unique to the lake, and the only way to calibrate is to measure evaporation independently by some other method. Although the most commonly used pan-to-lake coefficient is 0.7, Kohler et al. (1955) point out that this value should be applied only to annual data, and only if effects of wind energy and heat transfer are considered. Pan-to-lake coefficients have been shown to range from about 0.4 to 2.0 for monthly data, and from 0.5 to 0.9 for annual data (Hounam 1973).

The two most common methods used to estimate evaporation with instrumentation at the lake are energy budget and mass transfer. Evaporation calculated by the energy-budget method is generally considered most accurate; with careful design of field installations and service of instruments, the error in annual estimates can be 10% or less, and in seasonal estimates can be about 13%. However, this method requires expensive instrumentation and large amounts of data must be processed.

The mass-transfer method is less accurate than the energy-budget method, but it requires fewer instruments and less computation. Errors in evaporation determined by the mass-transfer method (Harbeck et al. 1958) are related to estimates of the mass-transfer coefficient and measurement of wind speed and air and water temperature. Mean errors in estimating the mass-transfer coefficient are usually about 15%, when relating the mass-transfer product (wind speed times vapor-pressure gradient) to an accurate, independent measurement of evaporation, such as the energy budget. The error is slightly greater, perhaps about 20%, when relating the mass-transfer product to change in water level of the water body (Langbein et al. 1951), provided change in water level is measured accurately.

If this latter method were used, and if inflowing and outflowing streams were present, the change in lake level must be adjusted by distributing the volume of water entering and leaving by streamflow over the lake surface. The error introduced by making this adjustment can be large; for example, at Lake Michie, North Carolina, Turner (1966) showed that a 5% error in measured inflow of 0.76 m^3/s could cause an error of as much as 104% in change of lake level. In addition, errors in measurement of water temperature of only a few degrees can lead to errors as great as 40% in estimates of evaporation (Ficke 1972). Errors in estimating N, using the equation of Harbeck (1962), are expected to be greater than 16% one-third of the time and greater than 32% one-tenth of the time.

A number of empirical methods have been developed to estimate evaporation by using climatic data from National Weather Service stations, (e.g. Roberts and Stall 1967). Some of these equations require onsite sensors such as anemometers, thermometers and radiometers to measure various climatic parameters.

Whatever method is used to estimate evaporation, researchers should be aware of the possible effect of emergent aquatic plants on lake evaporation (Betson et al. 1978). Aquatic plants of all types, if they cover a significant portion of a lake, add complications to estimates of evaporation. Although the subject of transpiration is not discussed here because of the small number of emergent aquatic plants at Mirror Lake, a recent study by Shih (1980) is useful in pointing out the need to account for transpiration by aquatic plants in some studies.

For the hydrologic studies of Mirror Lake prior to 1978, evaporation losses from the lake were based on mean monthly evaporation-pan data during ice-free seasons from 1965 through 1970, and an estimate of snow sublimation during ice-covered periods. The pan was located at the U.S. Forest Service Headquarters, less than 0.5 km west of the lake. A pan-to-lake coefficient of 0.7 was used to relate pan data to the lake.

For recent intensive hydrologic studies of Mirror Lake, instrumentation for the mass-transfer method was installed during 1979 and for the energy-budget method during 1980 (*Figure IV.D–2A and B*). Because some of the instruments in place also are used for some of the more empirical methods (e.g. Penman 1948; Sverdrup 1937; Sutton 1949), it is planned that selected equations also will be evaluated.

Surface Water

Most surface water enters and leaves a lake by channelized streamflow. In some environments, however, overland runoff can be an important part of surface-water inflow.

Streamflow

For small streams, measurement of stream discharge can be done most accurately using devices through which flow is routed, such as weirs and flumes. Errors are generally about 5% if the devices are installed and used properly, and if recording instruments are used to monitor water level continuously.

For lakes that have large streams entering or leaving, it is usually necessary to establish a streamflow-gauging station. At such sites, stream stage is measured continuously, and many current-meter discharge measurements are made to establish a stage-to-discharge relationship.

The accuracy of individual current-meter discharge measurements is dependent on: (1) the velocity meter, (2) the number and distribution of velocity measurements, (3) the time of exposure of the meter, and (4) measurement of the cross-sectional area of the channel. Overall error in a current-meter discharge measurement can be <5%, if many velocity measurements are made per visit. If few velocity measurements are used and velocity-meter exposure times are short or flow rate is changing, the error in computed flow rates can be between 5 and 10%, or even greater.

Stage-to-discharge relationship curves can vary widely in quality; if a good relationship is developed, the error in estimating daily discharge is probably <10%. Volumes for longer periods should be even more reliable. If the stage-to-discharge relation is unstable, poorly defined, or both, daily discharge values may be in error by 30% or more.

In many lake studies, surface runoff is determined by establishing a correlation between occasional measurements of discharge at the site of interest to a continuous record of discharge from a nearby gauged site. Errors associated with this approach vary widely. It is also common to estimate streamflow by using a unit-runoff approach; that is, a value of runoff per unit area is determined from a gauged site, and the same value is then used to determine runoff from ungauged areas. This approach requires the assumption that the drainage basins are exactly alike.

In some studies, surface runoff is estimated without making any measurements of discharge in the basin of interest; again, the unit-runoff approach is used. The extrapolation of unit runoff from the gauged to the ungauged basin is made only on the basis of physical and landcover characteristics of the basins. Errors in the estimate of runoff for the ungauged basins can be as great as 70% using this approach (Scheider et al. 1978). Riggs (1973) found that use of multiple regression to estimate stream discharge based on basin characteristics resulted in errors of 10 to 53%, and errors greater than 100% were common in estimating low flows.

Nonchannelized Surface Runoff

All lakes have areas adjacent to them where surface runoff, if it occurs, moves directly to the lake. Areas of direct drainage are commonly overlooked or are considered insignificant in lake water-balance studies. An approach commonly used to determine discharge from such areas is to calculate a unit-area-runoff value from *similar* gauged drainage basins and apply the value to the ungauged areas. Because of differences in characteristics between gauged basins and ungauged basins, the unit-area-runoff approach is subject to appreciable error, as discussed in the previous paragraph. The error increases, of course, the more dissimilar the gauged and ungauged basins are.

The concept of overland flow is somewhat controversial; some hydrologists consider only water that flows as sheet wash over the land surface, while others include water that occurs in the soil zone and unsaturated zone. Because of the complexity of flow in the soil and unsaturated zones, and because it has been studied by scientists in a wide variety of disciplines, a number of concepts and terminologies have evolved.

In some environments, surface runoff from areas of direct drainage is treated as overland runoff, and appropriate equations have been developed to calculate discharge from these areas. However, it is believed by some hydrologists (Hewlett and Troendle 1975) that overland runoff is seldom, if ever, observed in certain environments, such as the forested areas in the eastern U.S. A number of concepts of flow to streams, and, by analogy, of direct flow to lakes, have been proposed for this type of environment. One of the concepts that was proposed during the 1960s and is widely accepted by forest hydrologists in the Appalachian Mountain region is that of a variable source area that contributes to subsurface stormflow. According to Hewlett and Troendle (1975), subsurface stormflow can be described as follows, ". . . infiltrated water in excess of retention storage is not confined to a vertical path to a water table, which in steep areas may not exist anyway. Instead water responds to changing hydraulic gradients and flows more or less parallel to the slope surface, depending on local moisture contents, soil conductivities, and the steepness of gradients." They state further that "A dynamic storage zone expands under the steady downslope movement of soil water. This pattern responds rapidly under intense rainfall, shrinks slowly afterward, and operates in such a way as to flush the lower slopes and shallow soiled areas more rapidly and more frequently than the upslope and deeper soiled areas." This concept has been questioned by some hydrologists (Dunne and Black 1970; Freeze 1972) and a lively debate has ensued.

At Mirror Lake, the unit-runoff method was used to estimate runoff to the lake. The unit-runoff value was determined from nearby, continuously gauged Watershed 3 (W3) of the Hubbard Brook Experimental Forest. Occasional measurements of discharge of the inlet streams to Mirror Lake have also been made as a check on this approach.

To measure inflow to Mirror Lake as accurately as possible, Parshall flumes (*Figure IV.D–2C*) were installed on streams W and NW in 1979, and another was installed on stream NE in 1981 (*Chapter III.A*). Information obtained from flume data, used in conjunction

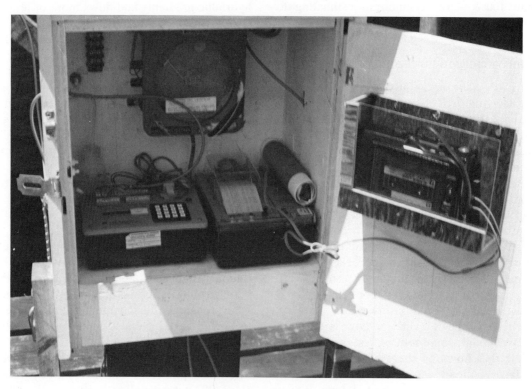

Figure IV.D-2. Instrumentation used in present hydrologic studies of Mirror Lake. **A.** Windspeed, wet- and dry-bulb air temperature, and water temperature sensors on raft near the middle of the lake. **B.** Data logger on raft.

D. Approaches to the Study of Lake Hydrology

C. Parshall flume on inlet stream. **D.** Group of piezometers near the shoreline.

with quantitative measurements of the drainage basins as discussed in *Chapter III.A*, permits comparisons of surface runoff between the Mirror Lake drainage subbasins, as well as with Hubbard Brook Experimental W3.

Ground Water

Interaction of lakes and ground water generally is not studied adequately in lake water-balance studies. Ground water is usually calculated as the residual, or it is sometimes considered unimportant relative to other hydrologic components. Yet, ground water and water flow in the unsaturated zone, as discussed in the previous section, even if small in quantity at some lakes, are commonly important with respect to water quality because they are carriers of constituents from nonpoint sources to a lake.

To understand ground-water flow, it is necessary to define the boundaries of the ground-water system interacting with the lake, for example, the thickness and extent of the geologic units comprising the system. It is also necessary to define certain internal characteristics of those geologic units, such as hydraulic conductivity, storage coefficient, porosity and secondary porosity, such as fracture patterns. In addition to those physical characteristics, it is necessary to define and monitor the distribution of hydraulic head within the ground-water system relative to lake level. To determine the distribution of head, the configuration of the water table (the upper surface of the saturated zone) in the drainage basin of the lake (Winter 1976, 1981a) needs to be defined using a network of water-table wells. Such wells are constructed so that they penetrate only the upper surface of the ground-water system. In addition, groups of piezometers are needed to define the vertical gradient of hydraulic head within the ground-water system at selected locations in the basin.

The geologic framework near a given lake is generally not known in sufficient detail for proper analysis of ground-water movement. Error analysis of geologic maps clearly shows the dependence of accurate contour maps on quality and quantity of control points. In areas of few geologic control points, errors in geologic contour maps can be large.

The dependence on quality and quantity of control points can also be observed in evaluating errors of hydraulic-gradient determinations. A study by Hanson (1972) shows that a water-table map drawn from few data points can be substantially in error, seriously affecting the calculation of hydraulic gradients.

Determination of values of aquifer coefficients, such as hydraulic conductivity, by laboratory analysis of samples may differ from values of *in situ* material by more than 100%. Error analysis of hydraulic-conductivity determinations by both sample analysis and aquifer tests is not commonly done. Aquifer-test data are generally considerably better than laboratory data, because the test is of *in situ* material and is representative of a much larger part of the aquifer. The onsite conditions for aquifer-test analysis, however, generally do not meet all the assumptions that must be made for proper use of analytical methods, and errors exceeding 100% in determination of aquifer coefficients are probably common. Provided hydrologic fluxes, such as recharge and evaporation, can be determined by independent methods, determination of hydraulic conductivity by statistical parameter-estimation techniques offers perhaps the best approach for aquifer units as a whole.

In the early studies of Mirror Lake, ground-water inflow and outflow were calculated from change in lake level during periods when the water surface was below the top of the spillways at the outlet and as the residual of the balance equation. Beginning in 1978, the work on ground-water flow near Mirror Lake has concentrated on defining the geometry of the ground-water system by test drilling and geophysical methods (see *Chapter III.A*), as well as monitoring hydraulic heads at selected sites.

Initially, it was believed that Mirror Lake was situated in a bedrock basin, and that ground water was moving primarily through fractures in crystalline bedrock. Following initial test drilling and surface-geophysical studies, it became apparent that the basin of the lake contained considerable thicknesses of glacial drift, and that, in many places, ground water was moving primarily through porous media. The general questions to be addressed then were: What is the relation of the lake to water movement in the drift, and in turn, what are the hydrodynamic relationships of water movement between bedrock and drift? Further, because the drift is relatively thick, even at higher altitudes in the drainage basin (*Figure III.A*–4), and wells have shown the existence of ground water in the drift at an altitude 61 m higher than the lake (*Figure III.A*–3), the same question of hydrodynamic interchange between the drift and bedrock needs to be addressed throughout most of the drainage basin of the lake.

Placement of wells and piezometers was designed to address the above questions. For example, theoretical studies (Winter 1976, 1978, 1981a) have indicated that the side of a lake toward a lower-altitude ground-water discharge area is the side most likely to have seepage losses to ground water. (See *Figure IV.D*–1 for a preliminary schematic interpretation of ground-water flow near Mirror Lake.) The studies also have shown that the hydraulic characteristics of the deposits and the shape of the water table on the down-gradient side of the lake are of greatest importance to understanding ground-water interchange with the lake. In the case of Mirror Lake, this would be the area between Mirror Lake and Hubbard Brook. Therefore, the initial ground-water work at Mirror Lake has concentrated on this side of the lake (see *Figure IV.D*–2D).

Because geophysical surveys on the north side of Mirror Lake have indicated that the ridge there consists entirely of drift, it is possible that Mirror Lake could have seepage to ground water on its north side (see *Figure III.A*–4). Although seepage from the lake on the north side seems unlikely, it will be necessary to define the configuration of the water table and distribution of hydraulic head within the ground-water system between Mirror Lake and the lower altitude area to the north of the ridge.

The ground-water study is designed to develop an understanding of basic hydrogeologic processes at Mirror Lake. A much simpler approach, such as using seepage meters (Lee 1977) could be used to determine ground-water inflow to and outflow from the lake. Although such direct-measurement devices are useful as reconnaissance tools to estimate quantities of flow, they do not permit study of the reasons *why* various patterns of inflow and outflow occur.

To fully understand subsurface-water movement, it is necessary to study flow in the unsaturated zone at selected sites in the drainage basin. It is not enough to define the shape and patterns of fluctuation of the water table; it is also important to understand how water moves through the unsaturated zone. Movement of water in this zone is dependent on hydrogeologic setting and climate.

Summary

Various methods are available for determining each component of the hydrologic system interacting with a lake. These range from the most accurate methods that require sensors and data loggers located at the lake to the least accurate methods that use data obtained an appreciable distance from the lake. If the most accurate methods are used, most hydrologic components, except ground water, can be determined with about 10% error. If the least accurate methods are used, errors can be greater than 100% for each component. Present (1981) studies of the water balance of Mirror Lake are designed to evaluate the cost, manpower and accuracy of several commonly-used approaches to determining each component of the hydrologic system.

E. Flux and Balance of Water and Chemicals

GENE E. LIKENS, JOHN S. EATON, NOYE M. JOHNSON and ROBERT S. PIERCE

Currently much research activity is being focused upon the biogeochemistry of natural ecosystems. Certainly a knowledge of the flux of water, nutrients and other chemicals, as well as internal cycling is vital to an understanding of the function (see *Chapter I*) and for management (see *Chapter IX*) of a lake-ecosystem. Much attention in the past has been given to instantaneous measurements of lake-water concentrations as a diagnostic procedure for determining or for predicting lake-ecosystem structure and function, but with relatively little success (cf. Schindler 1973); whereas the input and output fluxes for a lake-ecosystem are critical for evaluating the metabolic nature of the ecosystem (Vollenweider 1968; Likens 1975a; Likens and Bormann 1972, 1979) and for interpreting the history of sediments accumulated in the basin (*Chapter VII*). The runoff linkage between terrestrial watersheds and a lake-ecosystem is frequently the major control point for the management of the aquatic resource (see *Chapters VIII* and *IX*).

Large influxes of nutrients and other materials may completely overwhelm the normal functioning of a lake-ecosystem and accelerate eutrophication, or even produce polluted conditions (e.g. Hasler 1947; Likens 1972d, 1975a). To quantitatively evaluate human effects on aquatic ecosystems, detailed information on the input-output balance is required. Considering the magnitude and rate at which freshwater ecosystems are being subjected to accelerated cultural eutrophication (e.g. Anonymous 1969; Hasler and Ingersoll 1968), it is no longer appropriate to guess about these basic determinants of ecosystem function. Quantitative, long-term data are required.

Lake-Ecosystem Biogeochemical Model

We have employed a simple conceptual model for assessing the inputs and outputs of water and nutrients for an aquatic ecosystem (*Figure I–2*). Olsen and Chapman (1972) and many others have formulated much more complicated and detailed models. We have been able to use our model successfully in the application of real data for an understanding of overall input-output budgets related to ecosystem metabolism and biogeochemistry. This approach readily points up where the structure and function of the ecosystem must be studied in detail and, therefore, facilitates general ecosystem analysis. We believe that this approach has advantages over attempts to fit field data into a preconceived plan of how an ecosystem works. Use of the simplified basic model also provides a basis for experimentation at the ecosystem level, which can elucidate specific function and importance of the detailed structure. Our model is described in *Chapter I*.

Flux of Water

Precipitation

Estimates of direct precipitation inputs to Mirror Lake are based on standard precipitation gauge measurements made since 1970 at the U.S. Forest Service Headquarters, located less than 0.5 km west of Mirror Lake (*Figure II–1*). The area of the lake for this computation is assumed to be constant at 15 ha. In addition, a standard precipitation gauge, located about 0.5 km southeast of the lake at Pleasant View Farm, was established in 1974–75 and used for comparative measurements (see *Chapter II*).

Runoff

Estimates of the drainage to the lake from the three surrounding subwatersheds is calculated by using the gauged stream discharge (cm/month) from W3. Since precipitation is normally greater at W3 than at the U.S. Forest Service Headquarters (see *Chapter II*), these data are corrected proportionally for this difference in amount of precipitation. During the winter and early spring months (December through April) when most of the precipitation accumulates as a snowpack, this correction is based on a weighted mean value for the entire period rather than on a monthly basis as is done for the remainder of the year.

The precipitation corrected value for runoff from the Mirror Lake watershed, when calculated in this manner, makes no allowance for differences in terrestrial evapotranspiration (ET). In fact, the ET for the Mirror Lake watersheds as determined by difference is 43.5 cm, whereas the long-term ET value for the HBEF is 49.0 cm (range of 41 to 54 cm; see *Chapter II*).

At the present time we have no direct measurements of ET in the Mirror Lake watersheds. ET in these watersheds could be different than in the Experimental Watersheds. Factors contributing to a difference include: the presence of residential properties and roads in the Mirror Lake watershed, differences in vegetational types, differences in slope and aspect between the watersheds, and differences in the depth of soils, which affect both deep seepage and storage. It would have been possible to cal-

E. Flux and Balance of Water and Chemicals

culate runoff by difference (precipitation − ET) on an annual basis, assuming that the ET for the Experimental Watersheds was the same as for the lake watersheds. However, it is not possible to calculate runoff on a monthly basis by difference because of the accumulation of precipitation in a snowpack during the winter. We also might have used the BROOK model (Federer and Lash 1978) or some other empirical model to calculate ET (in fact, BROOK estimates ET to be 49.5 cm, or slightly higher than the long-term average for the Experimental Watersheds). Thus, we have simply corrected the runoff from W3 by the amount of precipitation at the U.S. Forest Service Headquarters. We realize that this approach may have introduced some error, since all components of the water budget are affected by this single percentage, whereas ET may not vary in this way.

Over a period varying from 16 to 27 yr, continuous measurements of precipitation and streamflow on eight watersheds have enabled the development of equations for predicting streamflow. Correlation coefficients as high as 0.98 for annual streamflow were found between the separate watersheds. These remarkable relationships support the contention that hydrologic characteristics such as precipitation, streamflow and evapotranspiration for nearby watersheds in the Hubbard Brook Valley (i.e. Mirror Lake) may be quite accurately predicted using data from the Experimental Watersheds.

The areas of the watersheds are based on planimeter measurements as follows: NE—32.6 ha; NW—42.6 ha; W—32.0 ha. An interstate highway (I-93) was constructed through subwatershed NE in 1969. Drainage associated with this highway was diverted away from the lake. Thus, the effective watershed area of NE after 15 November 1969 was only 10.4 ha. Measurements are reported for the period 1970–71 through 1979–80.

Outputs at the Dam

A concrete dam with two spillways is located on the outlet of the lake. Horizontal boards regulate the height of water in these spillways. In addition, we installed 120° V-notch weirs in the channel about 2 m below each spillway. Losses of water through the boards of the spillways have been estimated by periodic measurement of discharge through the calibrated 120° V-notch weirs. These losses were calculated to be 0.15 liter/s for the west weir and 0.037 liter/s for the east weir. Flow, Q, over the boards of the spillways was calculated using continuous measurements of lake level and the following formula for losses through a rectangular weir:

$$Q = CLH^{3/2} \qquad (1)$$

where

C = a constant, 3.0
L = width of the weir in cm
H = height of the water in cm

The relationship between the lake level and the top of the boards, H, was measured continuously with a water-level recorder and checked periodically (approximately every other day) by actual measurement; the relationship between the lake level and the flow rate through the weirs below the dam also was determined. Discharge measurements began in 1970, but during the period December 1970 through March 1971 there was a raft of frozen debris in front of the spillways. Estimates of flow during this period were determined from regular readings of the staff gauge in the weirs below the dam.

Deep Seepage

Estimates of water losses by deep seepage through the sediments of the lake were determined for 11 months during July 1970 through October 1972 by measuring the change in lake volume during months when the water level was continuously below the top of the outlet dam, corrected for fluvial inputs, evaporation losses and leakage losses through the dam.

Evaporation

Evaporation losses from Mirror Lake are based on mean monthly evaporation-pan data for the period 1965 through 1970. These data were collected at the U.S. Forest Service Headquarters. Data from the evaporation pan were multiplied by a coefficient of 0.7 to relate to lake surface area. See *Chapters II and IV.D* for additional information about measurements of evaporation in terrestrial and aquatic ecosystems of the Hubbard Brook Valley.

Precipitation Input

Overall, the amount of annual precipitation at the Pleasant View Farm site was not significantly ($p > 0.05$) different from that measured at the U.S. Forest Service site during 1974–75 to 1980–81. However, during two years—1974–75 and 1978–79—the amount was significantly ($p < 0.025$) greater at the U.S. Forest Service Headquarters site. Because of the length of record, measurements of amount of precipitation made at the U.S. Forest Service Headquarters site were used to estimate the input of water to the lake.

Annual precipitation averaged 121.5 cm during 1970–71 through 1980–81 (*Table IV.E–1*) at the U.S. Forest Service Headquarters site, or 93% of the average precipitation on the Experimental Watersheds (*Table II–2*). Average annual precipitation represented an input of 182,314 m^3 directly to Mirror Lake during this period (*Table IV.E–2*). As is the situation in the Experimental Watersheds, monthly precipitation is relatively constant throughout the year at Mirror Lake (*Table IV.E–1*). On the average, February was the driest month (6.6 cm) and December was the wettest (11.7 cm); the monthly average precipitation for the decade was 10.13 ± 0.43 cm.

Streamflow and Seepage Input

Annual streamflow and seepage averaged 663,063 m^3 (55,255 m^3/mo) or 78.0 cm/unit area of watershed for Mirror Lake during 1970–71 through 1979–80 (*Table IV.E–2*). The relative proportions of contribution from streamflow vs. seepage are not known quantitatively. However, an appreciable amount of drainage enters Mirror Lake directly from the watersheds along the shoreline without entering an inlet stream. Several methods have been used to estimate this seepage input.

During three months of the spring and early summer (April–June 1973) streamflow was measured in the inlet streams to Mirror Lake and compared with the computed total input of water from the watersheds (Jordan and Likens 1975). This short-term comparison would indicate that for the NE watershed only 14% of the drainage water entered the lake by way of the stream. The proportion entering as stream water for the NW watershed and W watershed was somewhat more, being 53% for each. For the lake as a whole it was estimated that 48% of the drainage water entered the lake by way of drainage streams.

Part of the watershed around the shore of Mirror Lake cannot be considered physiographically part of the drainage contributing to streamflow entering the lake (see *Chapter III.A*). Drainage from these areas which amount to approximately 22 ha or 26% of the total watershed area, obviously enters Mirror Lake directly. Seventy percent of the NE watershed has this type of area, which helps to explain the relatively small amount of water observed in that drainage stream.

Some 22% of the annual runoff occurred during April when the snowpack melted. Approximately one-half of the annual runoff to Mirror Lake occurred during the months of March, April and May. In contrast, only 14% of the total annual runoff occurred during the months of June through September (*Table IV.E–3*).

Lake Evapotranspiration Output

Based on the evaporation-pan data, annual evapotranspiration (transpiration was assumed to be negligible) from the lake surface was estimated at 50.7 cm/unit area, or a loss of about 76,000 m^3/yr. Our estimate of annual evaporation compares with a value of 63.5 cm determined for the region from an isopleth map of lake evaporation prepared by Kohler et al. (1959).

Monthly values of evaporation are given in *Table IV.E–3*. Some 63% of total annual evaporation occurs during the summer months of June through September. Evaporation was assumed to be 2.54 cm/mo during April and November when the lake was often partly covered with ice and snow, and 0.64 cm/mo from the snow and ice surface during the winter months of December through March (C. A. Federer, personal communication).

Fluvial Outputs

The fluvial outputs of water from the lake were of two types, deep seepage through the sediments of the lake and flow over the spillways of the dam at the outlet. Deep seepage was esti-

Table IV.E–1. Monthly Precipitation (cm) at the U.S. Forest Service Headquarters Site during 1970–71 through 1979–80

	1970–71	1971–72	1972–73	1973–74	1974–75	1975–76	1976–77	1977–78	1978–79	1979–80	$\bar{x}$	$s_{\bar{x}}$
June	7.42	5.84	10.16	19.51	9.86	14.48	7.52	12.07	17.04	4.80	10.87	1.54
July	6.30	12.52	13.67	18.52	8.92	14.78	10.87	5.79	4.34	6.88	10.26	1.45
August	13.08	9.75	12.83	10.57	12.88	8.81	13.51	9.73	7.80	11.10	11.01	0.63
September	13.00	8.00	3.58	14.15	16.81	14.78	11.71	16.92	2.69	9.35	11.10	1.61
October	8.38	10.01	9.70	8.97	4.11	13.77	19.15	19.66	8.05	13.56	11.54	1.57
November	7.85	6.93	13.08	11.68	9.88	12.60	3.96	6.76	6.22	10.19	8.92	0.96
December	11.94	9.30	14.27	26.37	8.38	9.58	9.22	12.85	8.46	6.58	11.70	1.79
January	6.60	7.85	13.97	10.52	10.16	13.31	5.74	18.42	20.05	4.29	11.09	1.68
February	10.06	7.16	6.32	7.90	7.47	11.02	5.69	2.67	4.75	2.87	6.59	0.87
March	7.67	14.07	7.77	9.86	11.23	9.63	12.89	5.41	11.61	13.16	10.33	0.88
April	3.12	7.37	7.75	11.48	5.87	7.32	9.32	7.70	11.99	11.96	8.39	0.90
May	9.47	6.38	14.91	14.96	2.06	16.66	3.76	9.47	14.83	5.13	9.76	1.68
Annual Total	104.89	105.18	128.01	164.49	107.63	146.74	113.34	127.45	117.83	99.87	121.54	6.54

Table IV.E-2. Water Budgets for Mirror Lake (m³)

	1970–71	1971–72	1972–73	1973–74	1974–75	1975–76	1976–77	1977–78	1978–79	1979–80	Mean ± $s_{\bar{x}}$ 1970–71 through 1979–80
Inputs											
Runoff[a]											
NE Watershed	67,433	63,797	85,196	124,129	64,828	109,588	71,764	83,738	77,590	63,983	81,205 ± 6,524
NW Watershed	275,885	261,002	348,551	507,831	265,226	448,889	293,601	342,585	317,434	261,766	332,277 ± 26,715
W Watershed	207,222	196,043	261,803	381,440	199,216	337,196	220,528	257,320	238,429	196,616	249,581 ± 20,067
Precipitation	157,335	157,770	192,015	246,735	161,445	220,110	170,010	191,175	176,745	149,805	182,314 ± 9,805
Total Input	707,875	678,612	887,565	1,260,135	690,715	1,115,783	755,903	874,818	810,198	672,170	845,377 ± 63,041
Losses											
Outlet	297,504	269,401	470,217	902,224	323,842	497,291	332,036	402,925	350,786	251,528	409,765 ± 60,383
Evaporation	76,000[b]	76,000[b]	76,000[b]	76,000[b]	76,000[b]	76,000[b]	76,000[b]	76,000[b]	76,000[b]	76,000[b]	76,000[b] —
Deep Seepage	310,000[c]	310,000[c]	310,000[c]	310,000[c]	310,000[c]	310,000[c]	310,000[c]	310,000[c]	310,000[c]	310,000[c]	310,000[c] —
Change in Volume	3,143	4,345	899	−1,948	−5,690	4,791	−7,247	10,395	899	−10,441	−85 ± 1,997
Total Losses	686,647	659,646	857,116	1,286,276	704,152	888,082	710,789	799,320	737,685	627,087	795,680 ± 60,624
Net Difference	+21,228	+18,966	+30,449	−26,141	−13,437	+227,701	+45,114	+75,498	+72,513	+45,083	49,697 ± 22,305
As % of Total Input	+3.0	+2.8	+3.4	−2.1	−1.9	+20.4	+6.0	+8.6	+8.9	+6.7	5.9

[a] Includes both streamflow and subsurface seepage.
[b] Based on monthly evaporation-pan data from 1965 through 1970.
[c] Estimated for an 11-month period during 1970 to 1972 when there was no outflow at the dam.

E. Flux and Balance of Water and Chemicals

Table IV.E-3. Monthly Fluvial Inputs and Outputs, and Evaporation for Mirror Lake during 1970-71 through 1979-80[a]

	Fluvial Input[b]			Fluvial Output[c]			Evaporation Output[d]	
	$\bar{x}$	$s_{\bar{x}}$	% of Total	$\bar{x}$	$s_{\bar{x}}$	% of Total	$\bar{x}$	% of Total
June	44,170	17,198	7	35,114	12,112	9	12,000	16
July	16,917	6,094	3	15,079	8,109	4	12,765	17
August	9,543	3,296	1	2,988	1,401	1	13,710	18
September	22,154	4,662	3	2,705	1,127	1	8,805	12
October	65,764	16,098	10	16,807	7,533	4	5,340	7
November	52,366	9,321	8	23,826	6,251	6	3,810	5
December	54,714	16,967	8	40,500	14,471	10	960	1
January	40,528	9,966	6	29,926	8,001	7	960	1
February	24,177	6,481	4	29,402	7,112	7	960	1
March	101,256	22,079	15	70,475	9,491	17	960	1
April	145,886	10,478	22	89,875	8,011	22	3,810	5
May	85,588	13,431	13	53,069	8,139	13	11,970	16
Annual Total	663,063	53,306	100	409,766	60,383	101	76,050	100
Annual $\bar{x}$	55,255	11,458		34,147	7,586		6,337	

[a] All inputs and outputs are in m^3.
[b] Inputs include both streamflow and subsurface seepage.
[c] Outputs include only surface losses ("water over the dam"). Estimate of average deep seepage from the lake is 30,000 m^3/mo.
[d] Estimated from evaporation-pan data—see text.

mated at 25,833 m^3/mo or 310,000 m^3/yr for the lake. This value, combined with the other hydrologic parameters of the annual hydrologic budget for the period 1970-71 to 1980-81, gave an annual net balance within ±5.9% (*Table IV.E-2*). In the final analysis, the variation in the water budget for the lake was assumed to occur within the deep seepage term, and using this assumption the mean annual value for deep seepage for the 10-yr period became approximately 360,000 m^3 or 30,000 m^3/mo. However, the majority of the difference between the two estimates of deep seepage occurred because of one year, 1975-76, when measured inputs exceeded outputs by about 228,000 m^3.

Deep seepage losses can be observed along the southeastern side of the lake. For example, in one location where the topography drops off sharply from the shoreline and below the Mirror Lake Road, the ground-water flow is obvious and channelized. It is assumed that the major portion of this seepage water originates as lake water, which seeps through sediments that are not covered by gyttja, i.e. between the surface of the lake and about 5-m depth (see *Figure III.A-6* and *Chapter VII.A*).

Fluvial losses of water over the dam at the outlet average about 409,800 m^3/yr or 34,150 m^3/mo (*Tables IV.E-2* and *3*). These losses have a distinct seasonal pattern, strongly reflecting the fluvial inputs (*Table IV.E-3*).

A budget of the annual fluxes of water is given in *Tables IV.E-2* and *4* for a 10-yr period, 1970-71 through 1979-80. Precipitation input in 1973-74, the wettest year, was about 1.6 times greater than in 1979-80, the driest year. The net difference values (*Table IV.E-2*) represent our error of closure on these budget estimates and we have algebraically added these values to deep seepage losses from the lake in the overall hydrologic balance (*Table IV.E-4*). The large net difference value in 1975-76 is due to the relatively small outlet loss in comparison to the inputs. We have verified our calculation and have no explanation for this discrepancy.

On average, the NW watershed contributed

Table IV.E–4. Average Annual Hydrologic Budget for Mirror Lake 1970–71 through 1979–80

	$m^3 \times 10^5$	% of Total
Inputs		
Runoff	6.631 ± 0.533	78
Precipitation	1.823 ± 0.098	22
Total	8.454	100
Outputs		
Outlet	4.098 ± 0.604	48
Deep Seepage	3.6[a]	43
Evapotranspiration	0.760	9
Total	8.458[b]	100

[a] Estimate of deep seepage plus average net difference value (see *Table IV.E–2*).
[b] Variable lake storage on an annual basis.

50% of the fluvial inputs to the lake; the Northeast inlet watershed contributed 12% and the West inlet watershed contributed 38% of the total fluvial inputs. Because of the way these inputs are calculated, this proportion directly reflects the relative areas of these three watersheds.

On average, drainage contributes 78% and precipitation 22% of the total input of water to Mirror Lake (*Table IV.E–4*). The losses are rather evenly distributed between outlet (48%) and deep seepage (43%) with evapotranspiration contributing only 9% of the total.

Flux of Nutrients

Methods

The chemistry of bulk precipitation was determined on weekly samples collected at the U.S. Forest Service Headquarters site, beginning in 1970. Samples of stream water from the watersheds surrounding the lake and from the outlet stream were collected and analyzed on a weekly basis. Samples for chemical analysis were collected at 1- or 2-m intervals within the lake. These lake samples were collected at variable intervals depending on the time of year, but were generally collected on a weekly or biweekly basis during the summer months and on a monthly basis during the remainder of the year. Seepage input was assumed to have the same dissolved chemical concentration as stream water in the corresponding watershed. Seepage output concentrations were assumed to be the same as those in the outlet stream or epilimnion of the lake (see *Chapter IV.B*). See *Chapters II* and *IV.B* and Likens et al. 1977 for additional information relative to sampling and procedures for determining concentration and input of dissolved substances.

Input of chemicals in particulate matter was estimated on the basis of watershed area for the lake and average annual gross output in kg/ha for W6 (Table 9 in Likens et al. 1977). Particulate outputs for the lake were determined on the basis of the chemistry and concentration of seston (B. Peterson, personal communication; *Chapter IV.B*) in the lake times the volume of discharge at the dam.

Input of nutrients in litter from the terrestrial watershed was calculated by multiplying the total litter input to the lake (see Jordan and Likens 1975; *Chapter V.C*) by the average element composition of litter (Table 5 in Gosz et al. 1976) at Hubbard Brook. Nutrient flux from insect emergence was calculated by multiplying the net insect emergence (*Chapter V.A.6*) times the average chemical concentration of adult chironomids in Norris Brook (R. Hall, personal communication; concentrations of nitrogen and sulfur were for immature insects in Norris Brook [Hall and Likens 1981]) within the Hubbard Brook Experimental Forest.

Flux to permanent sedimentation was calculated on the basis of deposition time (8 yr/cm) and chemical concentration of sediments at a depth of 25 cm from a core (Central) taken from the deepest portion of the lake (see *Chapters VII.A* and *VII.E*).

Inputs

The input of chemicals to the lake from direct precipitation closely resembles, in composition, seasonal pattern and amount, the input to the terrestrial, Experimental Watersheds (see e.g. Likens et al. 1977; *Table II–6* and *Table IV.E–5*). A study of paired samples of precipitation collected at the U.S. Forest Service Headquarters site (252 m MSL) and in the Experimental Watersheds (610 m MSL) in 1971–72 showed no significant difference in concentration of Ca^{2+}, Mg^{2+}, Na^+, K^+, NH_4^+ or Cl^- at

E. Flux and Balance of Water and Chemicals

Table IV.E-5. Average Monthly Input (kg) of Chemicals in Precipitation to Mirror Lake during 1970–71 through 1975–76[a]

Month	Ca^{2+}	Mg^{2+}	K^+	Na^+	H^+	NH_4^+	NO_3^-	SO_4^{2-}	Cl^-	SiO_2[b]	PO_4^{3-}[b]	DOC[c]
June	1.2	0.3	0.9	1.3	1.66	3.9	25.7	55.4	6.0	0.6	0.50	10.9
July	1.9	0.5	0.3	0.4	1.54	5.4	22.7	51.2	8.8	0.3	0.23	30.7
August	1.7	0.4	0.4	0.9	1.78	5.2	37.0	75.3	8.2	0.9	0.09	32.9
September	1.2	0.3	0.6	1.2	1.69	5.7	32.9	71.3	5.7	0.3	0.56	14.3
October	1.9	0.5	0.5	2.0	1.02	3.0	25.8	34.2	8.8	0.0	0.07	31.6
November	1.1	0.3	0.3	1.5	0.92	2.2	28.8	32.8	14.9	0.1	0.08	6.9
December	1.3	0.3	0.4	2.7	0.92	1.8	30.6	34.7	8.8	0.2	0.09	4.9
January	0.7	0.3	0.3	2.3	0.81	1.8	31.4	36.8	6.2	0.0	0.10	6.9
February	1.2	0.3	0.1	1.7	0.59	2.0	22.0	26.7	3.5	0.0	0.04	10.6
March	2.2	0.5	0.3	2.5	1.01	2.4	31.2	42.9	5.7	0.0	0.05	10.7
April	2.4	0.4	0.4	1.1	0.83	3.2	26.3	40.1	3.1	0.8	0.06	10.6
May	1.5	0.4	1.6	1.4	1.32	6.9	30.4	58.6	8.3	0.6	0.46	6.7
Annual Total	18.3	4.5	6.1	19.0	14.09	43.5	345	560	88.0	3.8	2.33	178
$\bar{x}$	1.52	0.37	0.51	1.58	1.17	3.62	28.7	46.5	7.3	0.3	0.19	14.8
$s_{\bar{x}}$	0.14	0.02	0.11	0.20	0.12	0.51	1.26	4.45	0.90	0.10	0.06	3.0

[a] Divide by 15 ha to calculate kg/ha of lake surface.
[b] 1972 to 1976.
[c] Dissolved organic carbon, 1976–77.

the two sites, but higher SO_4^{2-} (p < 1%) and NO_3^- (p < 5%) concentrations were recorded at the lower site (Likens et al. 1977). The difference in absolute input on a long-term basis was small, however. The average annual input to the lake during 1970 to 1976 was 23 kg NO_3^-/ha and 37 kg SO_4^{2-}/ha, whereas the average annual inputs to the Experimental Watersheds were 23 kg NO_3^-/ha and 38 kg SO_4^{2-}/ha for the same period. As was the case in the watersheds, appreciably more hydrogen ion, phosphate, ammonium, sulfate and dissolved organic carbon were added to the lake during the summer months than during winter (Table IV.E-5 and Figure IV.E-1). This finding is particularly interesting relative to lake metabolism, as these nutrient inputs may be important in supporting algal and bacterial growth in surface waters.

A disproportionally large amount of the total annual runoff input to Mirror Lake occurs during March, April and May (Table IV.E-6). This result occurs because of the large runoff of water during this period as the accumulated snowpack on the watershed melts. From 36% (PO_4^{3-}) to 64% (H^+) of the total runoff inputs occur during these three months. Input is greatest during April for all dissolved substances. In contrast, the smallest monthly input occurs during August, when a large proportion of the water from the watershed is lost by evapotranspiration.

Dissolved silica dominates the input of dissolved substances in runoff (Table IV.E-6). However, on an equivalent basis calcium (96.5×10^3 Eq/yr) and sulfate (98.1×10^3 Eq/yr) dominate, with sodium (41.1×10^3 Eq/yr), magnesium (33.4×10^3 Eq/yr) and chloride (24.9×10^3 Eq/yr) and presumably bicarbonate (57.6×10^3 Eq/yr; Table IV.E-7) added in much smaller amounts.

The average concentration of Ca^{2+}, Mg^{2+}, Na^+, H^+, NO_3^-, SO_4^{2-}, Cl^- and particularly of dissolved silica in runoff input is higher than the concentration in lake water (see e.g. Figure IV.E-1). In contrast, concentrations of K^+, NH_4^+ and PO_4^{3-} in runoff are approximately the same on an annual basis, as those in lake water (Figure IV.E-1).

The various subwatersheds are unique relative to their chemical inputs to Mirror Lake (Table IV.E-7 and Table IV.B-2). Runoff from the Northeast watershed is the most acid and contributes 40% of the total runoff input of hydrogen ion to the lake, even though it contributes only 12% of the water and represents only 12% of the total drainage area. The W subwa-

Table IV.E–6. Average Monthly Input (kg) of Various Dissolved Substances in Runoff to Mirror Lake during 1970–71 through 1975–76

Month	Ca²⁺	Mg²⁺	K⁺	Na⁺	H⁺	NH₄⁺	NO₃⁻	SO₄²⁻	Cl⁻	SiO₂	PO₄³⁻ ᵃ
June	150 ± 72	29.8 ± 14	22.4 ± 9.9	76.1 ± 36	0.062 ± 0.043	2.0 ± 0.8	3.7 ± 2.0	359 ± 180	55.6 ± 25	426 ± 188	0.45 ± 0.14
July	80.1 ± 25	16.1 ± 4.9	13.1 ± 3.6	39.0 ± 12	0.025 ± 0.016	1.1 ± 0.3	1.6 ± 0.4	159 ± 55	29.4 ± 10	228 ± 71	0.24 ± 0.06
August	37.9 ± 13	7.8 ± 2.7	7.9 ± 2.4	18.2 ± 6.1	0.007 ± 0.005	0.4 ± 0.1	1.6 ± 0.6	58.2 ± 22	10.6 ± 2.9	113 ± 42	0.11 ± 0.07
September	99.5 ± 25	20.4 ± 4.9	18.8 ± 3.9	47.3 ± 11	0.010 ± 0.003	0.7 ± 0.2	2.2 ± 0.9	166 ± 46	38.6 ± 14	285 ± 64	0.18 ± 0.06
October	175 ± 44	37.2 ± 8.9	37.6 ± 7.8	86.9 ± 22	0.031 ± 0.016	1.4 ± 0.2	3.6 ± 0.9	327 ± 100	78.4 ± 34	499 ± 110	0.27 ± 0.13
November	187 ± 31	40.3 ± 6.3	36.3 ± 5.4	94.4 ± 17	0.039 ± 0.011	1.8 ± 0.6	5.8 ± 1.0	416 ± 81	89.2 ± 24	532 ± 83	0.32 ± 0.10
December	180 ± 55	39.1 ± 12.1	29.2 ± 7.8	84.6 ± 25	0.077 ± 0.054	2.5 ± 1.8	10.8 ± 3.8	464 ± 177	74.8 ± 22	454 ± 113	0.32 ± 0.15
January	103 ± 24	22.0 ± 4.8	16.7 ± 3.8	50.9 ± 12	0.039 ± 0.019	0.8 ± 0.2	9.5 ± 2.0	237 ± 63	51.6 ± 18	258 ± 62	0.22 ± 0.09
February	97.0 ± 22	20.7 ± 4.5	15.5 ± 3.1	49.0 ± 13	0.032 ± 0.018	0.7 ± 0.2	9.3 ± 1.6	229 ± 62	54.3 ± 22	254 ± 55	0.12 ± 0.03
March	234 ± 44	49.1 ± 8.8	38.3 ± 7.3	115 ± 26	0.134 ± 0.075	2.4 ± 0.9	13.9 ± 3.1	589 ± 125	134 ± 47	593 ± 104	0.40 ± 0.09
April	335 ± 37	71.2 ± 8.4	58.4 ± 7.1	157 ± 14	0.285 ± 0.041	4.3 ± 1.2	15.6 ± 5.1	991 ± 106	154 ± 26	796 ± 71	0.51 ± 0.09
May	225 ± 38	52.1 ± 7.7	44.7 ± 6.8	127 ± 19	0.161 ± 0.062	3.0 ± 0.8	5.5 ± 1.9	716 ± 125	111 ± 27	698 ± 108	0.35 ± 0.10
Annual Total	1,934 ± 203	406 ± 41.3	339 ± 25.5	946.1 ± 123	0.903 ± 0.166	21.2 ± 6.0	83.1 ± 11	4,710 ± 571	881.6 ± 222	5,128 ± 524	3.51 ± 0.44
x̄	161.1	33.8	28.2	78.8	0.075	1.76	6.93	393	73.5	427	0.29
$s_{\bar{x}}$	24.4	5.18	4.35	11.7	0.023	0.33	1.39	76.9	12.3	59	0.036

ᵃ 1972 to 1976.

Table IV.E-7. Average Annual Runoff Inputs of Various Dissolved Substances to Mirror Lake during 1970–71 through 1975–76

Substance	NE Watershed kg	NE Watershed Eq × 10³	NE Watershed % of Total Runoff Inputs	NW Watershed kg	NW Watershed Eq × 10³	NW Watershed % of Total Runoff Inputs	W Watershed kg	W Watershed Eq × 10³	W Watershed % of Total Runoff Inputs
Ca^{2+}	323	16.1	17	926	46.2	48	685	34.2	35
Mg^{2+}	67.1	5.5	17	194	16.0	48	145	11.9	36
K^+	59.2	1.5	17	134	3.4	40	146	3.7	43
Na^+	211	9.2	22	370	16.1	39	365	15.9	39
H^+	0.36	0.36	40	0.29	0.29	32	0.25	0.25	28
NH_4^+	6.5	0.36	31	6.8	0.38	32	7.9	0.44	37
NO_3^-	4.6	0.074	6	23.3	0.38	28	55.2	0.89	66
SO_4^{2-}	683	14.2	15	2,377	49.5	50	1,649	34.3	35
Cl^-	383	10.8	43	225	6.3	26	274	7.7	31
PO_4^{3-}	1.14	0.036	32	1.05	0.033	30	1.32	0.042	38
HCO_3^- [a]	484	7.93	14	1,597	26.2	45	1,436	23.5	41
Dissolved Silica	870	—	17	2,291	—	45	1,967	—	38
DOC [b]	446	—	28	526	—	33	607	—	38
Water (m³ × 10³)	85.8		12	351		50	264		38
Total Dissolved Solids	3,539	—	18	8,671	—	44	7,339	—	38
$\Sigma+$		33.0			82.4			66.4	
$\Sigma-$		33.0			82.4			66.4	

[a] Calculated as difference between $\Sigma+$ minus $\Sigma-$.
[b] 1973–74.

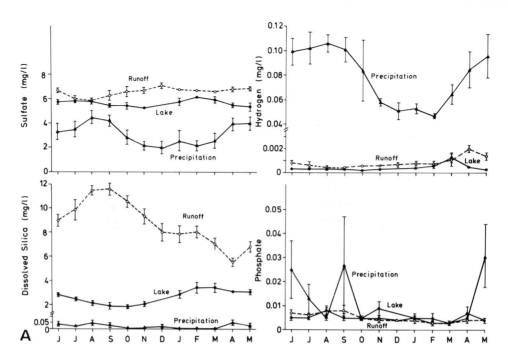
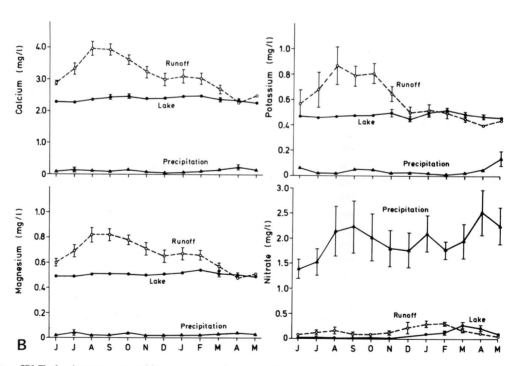

Figure IV.E-1. Average monthly concentrations (mg/liter) in direct precipitation (▲——▲), runoff (○- - -○) and lake water (●——●) for various dissolved substances in Mirror Lake. The averages for precipitation and runoff were for 1970 to 1976 (**A**), and for lake water was for 1967 to 1981 (**B**).

tershed contributes 66% of the total runoff input of NO_3^-, whereas the NE watershed contributes only 6%.

There are at least two anthropogenic sources of sodium and chloride in the drainage basin: (1) from road salt applied to I-93 during the winter (146 mt NaCl/km), and (2) from septic tank drainage. The increase in salt concentration in the NE tributary has been dramatic (*Figure IX-3* and *Table IV.E-7*) since the construction of I-93 (I-93 crosses the NE subwatershed). Several tourist cottages (Mirror Lake Hamlet) and three year-round residences are located close to the shoreline in the W subwatershed. These habitations are served by septic systems with drainage in porous soils. In comparison with the larger NW subwatershed, the outputs of Na^+ and Cl^- from the W subwatershed are elevated (*Table IV.E-7*). The same is true for PO_4^{3-}, NH_4^+ and K^+, which are usually increased in domestic sewage effluents (New Hampshire Water Supply and Pollution Control Commission 1975). Given the small size and runoff from the NE subwatershed, its relative contribution to the inputs (except for NO_3^-) for the lake is disproportionally large and unexpected (*Table IV.E-7*).

Most of the input of hydrogen ion and nitrate comes from precipitation falling directly on the surface of the lake (*Table IV.E-8*). Moreover, significant amounts of phosphorus, sulfur and chloride also are added to Mirror Lake each year in direct precipitation. In contrast, the vast majority of the input of dissolved Si, Ca^{2+}, Mg^{2+}, Na^+ and K^+ is contributed by runoff to the lake (*Table IV.E-8*). These relations demonstrate the influence of the terrestrial watershed in regulating fluvial outputs of nutrients. Nitrogen and phosphorus are readily assimilated and stored in the aggrading biomass of the terrestrial ecosystem and losses in drainage waters are small (Likens et al. 1977; Bormann and Likens 1979). Hydrogen ion in precipitation falling on the terrestrial ecosystem reacts with minerals, releasing Ca^{2+}, Mg^{2+}, etc. as a part of the weathering reaction. As a result, the acidity of the meteorologic input contributing to drainage water is reduced and the output of alkali and alkaline-earth metals is increased. Other processes within the terrestrial ecosystem, e.g. nitrification, may contribute to the acidity of drainage water (Likens et al. 1969). At Hubbard Brook we have found that external sources of hydrogen ion (meteorologic acid deposition) contribute 50 to 52% of the total protons used in chemical weathering reactions (Likens et al. 1977; Driscoll and Likens 1982).

Outputs

In general, the monthly pattern of fluvial outputs from Mirror Lake follows the pattern of fluvial inputs (*Tables IV.E-6* and *9*). Major fluvial losses occur in March, April and May during snowmelt, and minimal losses occur during August, September and October. Fluvial losses from the lake are low during late summer and early autumn (*Table IV.E-9*) when the lake level is lowest. These losses are not as small as might be expected, however, because seepage loss is 43% of the total water loss from the lake (*Table IV.E-10*) and seepage presumably occurs at a more or less constant rate throughout the year.

The chemistry of the fluvial outputs reflects closely the chemical composition of the surface waters in the lake. Calcium and sulfate dominate the concentration on an equivalent basis, followed by sodium, magnesium and bicarbonate (*Table IV.E-10*).

In general, approximately 57% of the chemi-

Table IV.E-8. Average Annual Inputs (kg) of Dissolved Substances and Water to Mirror Lake, 1970 to 1976

Element	Precipitation kg	%	Runoff kg	%
Hydrogen Ion	14.1	94	0.9	6
Nitrogen	119[b]	59	81[c]	41
Phosphorus[a]	0.76	40	1.14	60
Sulfur	187	11	1,572	89
Chloride	88.3	9	882	91
Potassium	6.2	2	339	98
Sodium	19.2	2	946	98
Calcium	18.3	1	1,934	99
Magnesium	4.6	1	406	99
Silica	1.9[b]	<1	2,397	>99
Water ($m^3 \times 10^5$)	1.892	21	7.009	79

[a] 1972 to 1976.
[b] 7 kg organic N.
[c] 46 kg organic N.

Table IV.E-9. Average Monthly Output (kg) of Various Dissolved Substances from Mirror Lake during 1970–71 through 1975–76[a]

Month	Ca^{2+}	Mg^{2+}	K^+	Na^+	H^+	NH_4^+	NO_3^-	SO_4^{2-}	Cl^-	SiO_2	PO_4^{3-} [b]
June	158 ± 42	32.6 ± 8.6	32.7 ± 8.4	83.6 ± 23.8	0.025 ± 0.010	2.2 ± 0.6	5.89 ± 3.61	413 ± 112	58.0 ± 17.1	174 ± 40	0.99 ± 0.56
July	120 ± 27	25.2 ± 5.3	24.9 ± 5.0	63.7 ± 14.7	0.018 ± 0.006	2.1 ± 0.5	2.62 ± 0.91	314 ± 71	45.8 ± 10.9	111 ± 22	0.35 ± 0.08
August	74 ± 8	15.4 ± 1.5	15.5 ± 1.3	38.8 ± 3.6	0.010 ± 0.001	1.1 ± 0.1	1.20 ± 0.53	192 ± 15	26.6 ± 2.7	67 ± 6	0.17 ± 0.05
September	78 ± 9	16.7 ± 1.7	16.2 ± 1.0	40.3 ± 3.5	0.013 ± 0.003	1.4 ± 0.05	0.81 ± 0.09	200 ± 15	27.8 ± 3.3	64 ± 3	0.19 ± 0.06
October	94 ± 23	20.0 ± 4.8	19.9 ± 3.6	48.2 ± 11.0	0.018 ± 0.008	1.5 ± 0.2	0.66 ± 0.19	218 ± 46	34.1 ± 9.4	77 ± 15	0.21 ± 0.07
November	132 ± 27	27.8 ± 5.6	28.4 ± 5.4	69.4 ± 14.4	0.022 ± 0.007	2.0 ± 0.5	1.97 ± 0.56	317 ± 60	51.1 ± 12.4	120 ± 22	0.22 ± 0.08
December	205 ± 54	43.3 ± 12.0	41.9 ± 11.3	105.7 ± 28.9	0.040 ± 0.018	3.5 ± 1.4	8.75 ± 3.48	507 ± 147	79.3 ± 24.1	201 ± 57	0.44 ± 0.12
January	168 ± 25	34.6 ± 4.9	33.7 ± 4.6	86.1 ± 13.5	0.039 ± 0.012	2.3 ± 0.5	13.83 ± 3.80	415 ± 73	63.4 ± 11.1	195 ± 37	0.33 ± 0.08
February	173 ± 23	36.4 ± 5.2	33.5 ± 3.8	87.6 ± 12.3	0.055 ± 0.018	2.5 ± 0.5	23.26 ± 5.23	452 ± 78	67.5 ± 11.6	253 ± 55	0.24 ± 0.04
March	230 ± 14	47.3 ± 2.8	42.7 ± 1.5	115.0 ± 8.2	0.162 ± 0.068	3.4 ± 0.4	43.07 ± 8.35	599 ± 61	92.6 ± 11.3	360 ± 43	0.29 ± 0.01
April	241 ± 19	50.6 ± 4.1	48.2 ± 3.7	124.3 ± 7.7	0.236 ± 0.063	3.9 ± 0.8	62.46 ± 16.53	721 ± 87	103.0 ± 9.4	367 ± 40	0.38 ± 0.02
May	207 ± 24	43.2 ± 5.1	42.1 ± 4.8	106.9 ± 13.4	0.037 ± 0.004	2.7 ± 0.7	5.35 ± 1.41	586 ± 87	85.3 ± 14.0	282 ± 35	0.29 ± 0.06
Annual Total	1,881 ± 221	393.2 ± 46.1	379.7 ± 39.6	969.7 ± 120.3	0.675 ± 0.153	28.5 ± 5.1	169.87 ± 21.42	4,934 ± 587	734.6 ± 107.6	2,270 ± 279	4.11 ± 0.80
$\bar{x}$	157	32.8	31.6	80.8	0.056	2.4	14.16	411	61.2	189	0.34
$s_{\bar{x}}$	17	3.4	3.2	8.4	0.020	0.3	5.67	49	7.3	31	0.06

[a] Losses include outlet plus seepage.
[b] 1972 to 1976.

E. Flux and Balance of Water and Chemicals

Table IV.E–10. Average Annual Fluvial Outputs of Various Dissolved Substances from Mirror Lake during 1970–71 through 1975–76

Substance	Outlet			Seepage		
	kg	Eq × 10^3	% of Total Fluvial Outputs	kg	Eq × 10^3	% of Total Fluvial Outputs
Ca^{2+}	1,048	52.3	56	834	41.6	44
Mg^{2+}	218	17.9	55	175	14.4	45
K^+	208	5.3	55	172	4.4	45
Na^+	539	23.4	56	431	18.7	44
H^+	0.456	0.45	68	0.219	0.22	32
NH_4^+	16.1	0.89	56	12.4	0.69	44
NO_3^-	115	1.9	68	55.2	0.89	32
SO_4^{2-}	2,803	58.4	57	2,131	44.4	43
Cl^-	416	11.7	57	318	9.0	43
PO_4^{3-}	2.30	0.073	56	1.81	0.057	44
HCO_3^{a}	1,748	28.7	57	1,341	22	43
Dissolved Silica	1,342	—	59	928	—	41
DOC^b	957	—	57	734	—	43
Water ($m^3 \times 10^3$)	460	—	57	353	—	43
Total Dissolved Solids	9,413	—	57	7,134	—	43
$\Sigma+$		100.2			80.0	
$\Sigma-$		100.8			76.3	

[a] 1970 to 1972.
[b] 1973–74.

cal outputs from the lake are carried by water flowing over the dam and 43% are lost in seepage through the sediments (*Table IV.E–10*). Outlet losses of hydrogen ion and nitrate are somewhat greater than this general relationship because concentrations of these two ions are highest during the spring melt runoff period when fluvial losses from the lake are proportionally greater.

Input-Output Balance

Annual inputs, outputs and accumulations or losses of dissolved substances for Mirror Lake are given in *Tables IV.E–11, 12 and 13*. Based on a net balance of inputs in precipitation and runoff, and outputs in outlet discharge and seepage losses, some elements (H^+, dissolved Si, NH_4^+, NO_3^- and PO_4^{3-}) show a statistically significant accumulation within the lake on an annual basis (*Table IV.E–14*). No element shows a statistically significant net loss from the lake (*Table IV.E–14*). Some 95% of the hydrogen-ion and 59% of the inorganic nitrogen inputs are retained, on average, in the lake each year. For some elements the balance from year to year is highly variable (*Table IV.E–13*). The unusually large fluvial losses at the outlet during 1973–74 (*Table IV.E–2*) were responsible for net losses of Ca^{2+}, Mg^{2+}, K^+, Na^+ and Cl^- during that year (*Table IV.E–13*).

The average monthly inputs of various dissolved substances, relative to the standing stock in the lake, are shown in *Figure IV.E–2*. Monthly inputs represent a small percentage of the total standing stock in the lake, except for hydrogen ion, and during spring runoff (i.e. calcium, sulfate and dissolved silica).

The theoretical turnover time (year) for dissolved substances can be calculated as the quotient derived from dividing the standing stock in the lake by the total input. On this basis the turnover times per year are 1.16 for K^+, 1.14 for Na^+, 1.03 for Mg^{2+}, 1.02 for Ca^{2+}, 0.93 for Cl^-, 0.90 for SO_4^{2-}, 0.80 for PO_4^{3-}, 0.43 for dissolved Si, 0.43 for NH_4^+, 0.17 for NO_3^- and

Table IV.E–11. Annual Inputs of Dissolved Substances (kg/lake-yr) in Direct Precipitation plus Runoff to Mirror Lake during 1970 to 1976

Dissolved Substance	1970–71	1971–72	1972–73	1973–74	1974–75	1975–76	$\bar{x} \pm s_{\bar{x}}$
Ca	1,560	1,531	1,960	2,544	1,540	2,581	1,953 ± 204
Mg	327	332	399	538	331	535	410 ± 41
K	341	288	360	421	260	401	345 ± 26
Na	715	690	967	1,243	765	1,412	965 ± 123
H	14.4	13.8	15.8	18.4	13.0	14.8	15.0 ± 0.78
NH_4-N	46	52	43	81	37	42	50 ± 6.4
NO_3-N	79	107	111	129	77	77	97 ± 9.1
SO_4-S	1,291	1,512	1,710	2,485	1,342	2,210	1,759 ± 199
Cl^-	586	571	852	990	844	1,976	970 ± 212
PO_4-P	—	—	2.23	2.32	1.05	2.03	1.91 ± 0.29
SiO_2-Si	1,768[a]	1,960[a]	2,604	3,097	1,896	3,052	2,397 ± 245

[a] Runoff only.

Table IV.E–12. Annual Outputs (outlet plus seepage) of Dissolved Substances (kg/lake-yr) from Mirror Lake during 1970 to 1976

Dissolved Substance	1970–71	1971–72	1972–73	1973–74	1974–75	1975–76	$\bar{x} \pm s_{\bar{x}}$
Ca	1,503	1,355	1,932	2,662	1,456	2,379	1,881 ± 221
Mg	314	285	391	560	310	499	393 ± 46
K	323	282	398	524	293	456	380 ± 40
Na	742	673	1,001	1,375	763	1,264	970 ± 120
H	0.297	0.428	0.949	1.176	0.315	0.887	0.675 ± 0.153
NH_4-N	16	14	21	41	17	23	22 ± 4.0
NO_3-N	33	47	40	56	24	29	38 ± 4.8
SO_4-S	1,430	1,192	1,641	2,377	1,194	2,044	1,646 ± 196
Cl	489	488	785	1,007	560	1,079	735 ± 108
PO_4-P	—	—	1.94	1.27	0.68	1.49	1.34 ± 0.26
SiO_2-Si	871	763	1,174	1,520	737	1,297	1,061 ± 130

Table IV.E–13. Annual Net Accumulation (+) or Loss (−) of Dissolved Substances (kg/lake-yr) for Mirror Lake during 1970 to 1976

Dissolved Substance	1970–71	1971–72	1972–73	1973–74	1974–75	1975–76
Ca	57	176	28	−118	84	202
Mg	13	47	8	−22	21	36
K	18	6	−38	−103	−33	−55
Na	−27	17	−34	−132	2	148
H	14.1	13.4	14.9	17.2	12.7	13.9
NH_4-N	30	38	22	40	20	19
NO_3-N	46	60	71	73	53	48
SO_4-S	−139	320	69	108	148	166
Cl	97	83	67	−17	284	897
PO_4-P	—	—	0.29	1.05	0.37	0.54
SiO_2-Si	897	1,197	1,430	1,577	1,159	1,755

E. Flux and Balance of Water and Chemicals

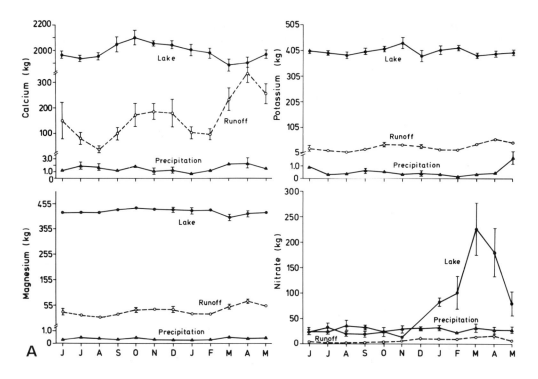

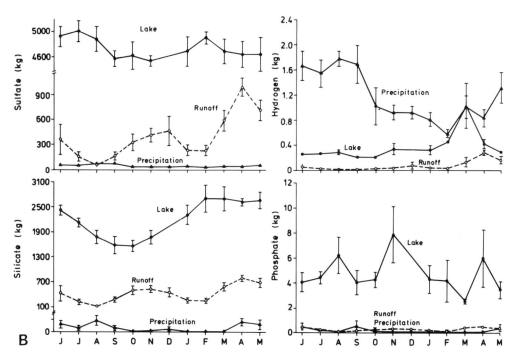

Figure IV.E–2. Average monthly inputs (kg/mo) in direct precipitation (▲——▲), runoff (○---○) and the standing stock (kg) in lake water (●——●) for various dissolved substances in Mirror Lake. The averages for precipitation and runoff were for 1970 to 1976 (**A**), and lake water was for 1967 to 1981 (**B**).

Table IV.E–14. Average Annual Input-output Balance for Dissolved Substances (kg ± One Standard Deviation of the Mean) for Mirror Lake during 1970–71 through 1975–76

Element	Precipitation + Runoff Input	Outlet + Seepage Loss	Net	Net Retention as % of Input
SiO_2-Si	2,399 ± 245	1,061 ± 130	+1,338 ± 127	56
Cl	970 ± 212	735 ± 108	+235 ± 138	ns
SO_4-S	1,759 ± 199	1,646 ± 196	+112 ± 61	ns
Ca	1,953 ± 204	1,881 ± 221	+72 ± 47	ns
NO_3-N	97 ± 9.1	38 ± 4.8	+58 ± 5	60
Na	965 ± 123	970 ± 120	−5 ± 37	ns
Mg	410 ± 41	393 ± 46	+17 ± 9.8	ns
NH_4-N	50 ± 6.4	22 ± 4.0	+28 ± 4	56
H	15.0 ± 0.78	0.68 ± 0.15	+14.2 ± 0.6	95
PO_4-P[a]	1.91 ± 0.29	1.34 ± 0.26	+0.55 ± 0.17	29
K	345 ± 26	380 ± 40	−35 ± 18	ns

[a] 1972–76.

0.025 for H^+. Clearly the theoretical residence time for hydrogen ion and to a lesser extent for nitrogen and dissolved silica in Mirror Lake is very short; i.e. the standing stock of hydrogen ion is replaced about 40 times per year, whereas K^+, Na^+, Mg^{2+} and Ca^{2+} require somewhat longer than a year to flush through the lake on the average. The average theoretical residence time for water in the lake is about 1.02 yr, although the actual residence time for epilimnetic and hypolimnetic water may be appreciably different than the theoretical value (see *Chapter IV.C*).

Dissolved nutrients are assimilated by organisms in the lake, and through production of feces or death of the organisms themselves, the nutrients eventually accumulate in the sediments (see *Chapter VII.E*). The large net retention of inorganic N, Si and P attests to the role these critical nutrients play in the metabolism of the lake, and to the fact that the sediments can be major sinks for limiting nutrients moving within the landscape (*Table IV.E*–15). The mechanisms whereby relatively large amounts of chloride are retained in the lake are unknown.

There are very few studies of nutrient balances for lake-ecosystems, and none that we know of where litter inputs, insect emergence, etc. are considered. A detailed analysis of this sort may be very informative, because for some elements these poorly known inputs or losses may be important to the overall balance. For

Table IV.E–15. Input-output Balance for Dissolved Substances in Various Lakes

Element	Input (mt/yr)	Output (mt/yr)	% of Input Retained
Mirror Lake, New Hampshire			
P	0.0019	0.0013	32
N	0.147	0.060	59
S	1.759	1.646	6
K	0.345	0.380	−10
Rawson Lake, Ontario[a]			
P	0.032	0.005	84
N	0.635	0.094	85
S	0.302	0.317	−5
K	0.110	0.124	−13
Cayuga Lake, New York[b]			
P	170	61	64
N	2,744	513	81
S	24,984	31,983	−28
K	3,499	3,969	−13
Clear Lake, Ontario[a]			
P	0.035	0.009	74
N	0.995	0.126	87
S	—	—	—
K	0.096	0.082	15

[a] Schindler et al. 1976.
[b] Likens 1972a.

E. Flux and Balance of Water and Chemicals

Table IV.E-16. Average Annual Nutrient Budgets for Mirror Lake 1970 to 1976[a]

	Chemical Substance				
	Ca	Mg	Na	K	H-ion
Inputs					
Precipitation (bulk)	18.3 ± 2.4	4.5 ± 0.5	19.0 ± 2.0	6.1 ± 0.4	14.1 ± 0.64
Litter	10	1	<1	4.6	—
Dry Deposition or Fixation	—	—	—	—	—
Fluvial					
Dissolved	1,934 ± 203	406 ± 41	946 ± 123	339 ± 25	0.90 ± 0.17
Particulate	18	16	21	44	—
Domestic Sewage/					
Road Salt Seepage	?	?	>4[b]	?	—
Total	1,980	427	990	394	15.0
Outputs					
Gaseous Flux	—	—	—	—	—
Fluvial					
Dissolved	1,881 ± 221	393 ± 46	970 ± 120	380 ± 39.6	0.675 ± 0.153
Particulate	0	0	0	—	—
Net Insect Emergence	0.35	0.33	0.35	2.0	—
Permanent Sedimentation	42[c]–117[d]	16[c]–68[d]	32[c]–221[d]	39[c]–205[d]	—
Total	1,923–1,998	409–461	1,002–1,191	421–587	0.67
Change in Lake Storage	0	0	+4	0	0

	Cl	N	S	P	Si
Inputs					
Precipitation (bulk)	88 ± 14	119[e]	187 ± 20.1	0.76 ± 0.15[f]	1.8 ± 0.2[f]
Litter	<1	13.6	1.5	1.0	72[g]
Dry Deposition or Fixation	—	~15	~59	—	—
Fluvial					
Dissolved	882 ± 222	81[h]	1,571 ± 190	1.14 ± 0.14[f]	2,397 ± 245
Particulate	<1	9[i]	2.6	10.6 ± 5.5[j]	526
Domestic Sewage/					
Road Salt Seepage	>6[k]	?	?	8.5[b]	?
Total	976	238	1,821	22	2,997
Outputs					
Gaseous Flux	—	?	?	—	—
Fluvial					
Dissolved	735 ± 108	115[l]	1,646 ± 196	1.34 ± 0.26	1,060 ± 130
Particulate	<1	11[i]	0	0.9	?
Net Insect Emergence	0.2	13	0.9	2.4	?
Permanent Sedimentation	207[b]	127[c]–139[d]	39	16[c]–18[d]	2,400[c]–4,000[d]
Total	942	266–278	1,686	21–23	3,460–5,060
Change in Lake Storage	+34	−6[m]	0	0	0

[a] Values in kg/yr ± $s_{\bar{x}}$.
[b] By difference.
[c] Spatially integrated sedimentation since about 1840 A.D., from *Table VII.A.2–1*.
[d] Extrapolation of precultural deposition rate (*Table VII.A.2–4*) and chemical concentration from a sediment depth of 25 cm (Central core) to 50% of the lake area.
[e] Organic 7; inorganic 112 ± 10.7
[f] 1972 to 1976.
[g] Estimated.
[h] Organic 46; inorganic 35 ± 7.2.
[i] All organic.
[j] Organic 3.0; inorganic 7.6.
[k] Based on assumption of NaCl, using Na value.
[l] Organic 54; inorganic 61 ± 8.8.
[m] Decrease in NO_3^- storage.

example, the annual input of phosphorus in litter to Mirror Lake is greater than the input in direct precipitation, and loss of phosphorus through net insect emergence is larger than fluvial outputs (*Table IV.E–16*).

Net loss of nutrients by means of emergence and dispersal of aquatic insects has not been thought to be a significant flux by some authors (e.g. Vallentyne 1952); however, some 11% of the annual input of phosphorus, 5% of nitrogen and 0.5% of potassium to Mirror Lake during 1970 to 1976 were lost as adult insects emerged from the lake and then died somewhere in the surrounding watershed (*Table IV.E–16*). Aquatic insects accumulate the entire chemical content of their protoplasm from the aquatic ecosystem and then can carry it away from the lake as flying adults. This flux is one of the few avenues whereby nutrients can be transported "uphill" (*sensu* Leopold 1941) from aquatic to terrestrial ecosystems by organisms (see *Chapter I;* Likens and Loucks 1978).

Conversely, terrestrial litter accounted for 5% of the total annual input of phosphorus, 6% of the nitrogen and 1% of the potassium to Mirror Lake during 1970 to 1976.

Overall, the evaluation of all known inputs and outputs to Mirror Lake showed a remarkable balance for calcium and magnesium, and reasonable balance for potassium, sodium, nitrogen and silica. Appreciably more sulfur and especially more hydrogen ion were added in inputs than lost in outputs, but gaseous losses of sulfur occur (*Table IV.E–16*). All of the other elements had relatively large values in the balance that were estimated by difference.

Because of a lack of data, we assumed that stream water and seepage runoff concentrations were the same. However, this assumption was probably not correct for phosphorus, nitrogen, sodium and chloride, which may seep directly into the lake from local septic tanks and from road salt, but may not contaminate stream water to the same extent. Likewise, we assumed that the concentrations of dissolved substances in the outlet water and in deep seepage were the same. This assumption is probably reasonable for Mirror Lake.

Estimated inputs in direct seepage of P, Na^+ and Cl^- are significantly large as determined by difference from the balance of other terms in the budget (*Table IV.E–16*). Unknown amounts of NH_4-N may seep into the lake from septic tank drainage, which would tend to balance the nitrogen budget.

Unfortunately, chloride concentrations have not been measured in the sediments of Mirror Lake, thus the sedimentation rate had to be estimated by difference (*Table IV.E–16*). This rate for chloride sedimentation seems too high relative to the rate for the other elements. Nevertheless, the measured input (direct precipitation and runoff) of chloride also exceeds the fluvial outputs from the lake during most years (*Table IV.E–13*), indicating an overall annual accumulation of 235 kg within the basin during the period 1970 to 1976 (*Table IV.E–14*). An increase of 235 kg Cl/yr in the water of the lake would raise the concentration by 0.27 mg/liter-yr, or significantly above the increase actually observed (0.04 mg/liter-yr during 1970 to 1976). Because most of the increase in lake-water concentration has occurred since 1975–76 (see *Chapter IV.B*), it is likely that a large part of the average sedimentation rate for chloride should be reapportioned into the change-in-storage term for the lake budget and is a function of the averaging process in constructing these budgets.

The actual amount of gaseous exchange of nitrogen and sulfur at the air-water interface of Mirror Lake is unknown. We have crudely estimated that benthic fixation of nitrogen may contribute about 15 kg N/yr to the lake (see *Chapter V.B.4*), although fixation could be up to twice this value. Flett et al. (1980) detected no nitrogen-fixing algae or fixation of nitrogen within the water-column of similar oligotrophic lakes within the ELA. Although we did not make measurements, we would not expect large amounts of nitrogen fixation within the water-column of Mirror Lake, even though algal N-fixation was commonly found in ELA lakes where the total N:P ratio (weight basis) for inputs was <~10. The input ratio of N:P in Mirror Lake is 6.6.

Deposition of gaseous and particulate sulfur directly on Mirror Lake has been estimated using deposition velocities of Sheih et al. (1979) over water and measured concentrations of SO_2 and particulate sulfur in 1973 to 1976 (Eaton et al. 1978). These estimates show a total input of

47 kg/yr and 12 kg/yr of sulfur to the lake for SO_2 and particulate S, respectively. We have no estimate of the possible loss of gaseous sulfur (e.g. H_2S) from the lake-ecosystem.

Obviously, more than just inputs in direct precipitation and runoff and losses in outflow and deep seepage must be known to characterize the nutrient budget for a lake-ecosystem. Moreover, long-term data are necessary to describe "average" conditions for natural ecosystems (Likens 1975a; Likens et al. 1977; Likens and Bormann 1979; Schindler et al. 1976; Likens 1983). Detailed and quantitative data on the hydrology of a lake and its watershed are crucial for determining the flux of nutrients in humid regions such as at Mirror Lake. Precipitation varied from 99.9 cm to 164.5 cm during 1970–71 through 1979–80 (*Table IV.E*–1), and total input of water (precipitation *plus* runoff) was 1.9 times greater in 1973–74 than in 1979–80 (*Table IV.E*–2). Such fluctuations similarly affect the net accumulation or loss of dissolved substances for the lake (*Table IV.E*–13).

Annual averages may obscure seasonal or other shorter-term relationships. For example, some 55% of the NO_3^-, 56% of the dissolved silica and 57% of the K^+ in the annual precipitation and runoff input to Mirror Lake occurred during November through April when the lake was metabolically less active (*Tables IV.E*–5 and 6). Thus temporal variations should be considered when evaluating or predicting the effect of nutrient flux on the metabolism or biogeochemistry of a lake. Fluctuations in climate, summer droughts, lack of snow during the winter, or floods may accentuate these effects. Unfortunately, limnologists tend to think primarily about short-term averages, or only about "the year the lake was studied," rather than about the long-term variations or about the effects of episodic extremes in hydrologic and nutrient flux. Quantitative, long-term data provide a reliable baseline against which perturbations caused by human activities can be evaluated (Likens 1983).

Summary

Detailed hydrologic data during 1970 to 1980 and nutrient flux measurements during 1970 to 1976 for precipitation, evaporation, runoff, outflow and deep seepage are reported for Mirror Lake. Annual precipitation averaged 121.5 cm/unit area or 182,314 m^3/yr. Runoff into the lake averaged 663,063 m^3/yr (78 cm/unit area of watershed), lake evaporation was 76,000 m^3/yr (51 cm/unit area), deep seepage losses from the lake were 360,000 m^3/yr (240 cm/unit area) and outlet losses averaged 409,800 m^3/yr (273 cm/unit area) during 1970 to 1980. Annual inputs varied by about twofold during the period of study.

A relatively large percentage of the runoff of water and dissolved substances occurred during March, April and May when the snowpack on the watershed melts. In contrast, small runoff occurred during late summer because of evaporative water loss. Dissolved silica dominates the runoff on a weight basis, but Ca^{2+} and SO_4^{2-} dominate on an equivalent basis. The runoff for each subwatershed is chemically unique. For example, runoff from the NE subwatershed is the most acid and the W subwatershed contributes most of the nitrate to the lake.

Direct precipitation contributes most of the H^+ and NO_3^- input to the lake and significant amounts of PO_4^{3-}, SO_4^{2-} and Cl^-. In contrast, most of the dissolved Si, Ca^{2+}, Mg^{2+}, Na^+ and K^+ comes into the lake in runoff. Concentrations of NO_3^- and SO_4^{2-} in precipitation were significantly higher at the elevation of Mirror Lake than at the Experimental Watersheds.

Hydrogen ion, N and dissolved Si have short residence times within the lake, whereas K^+, Na^+, Mg^{2+}, Ca^{2+} and water have theoretical residence times greater than one year. Seasonal or episodic inputs of water and nutrients may be more important to the metabolism or biogeochemistry of a lake than average annual fluxes.

Detailed average nutrient budgets, including litter input and net insect emergence are given for the major nutrients in Mirror Lake. Significant amounts of H^+, nitrogen, phosphorus and dissolved silica accumulate annually within the lake basin. Thus, lacustrine sediments can serve as major sinks for nutrients (e.g. N and P) moving in the landscape.

Because of temporal variability, quantitative, long-term data on hydrologic and nutrient flux are required to characterize the metabolic, biogeochemical and developmental state of an aquatic ecosystem.

CHAPTER V

Mirror Lake—Biologic Considerations

A. Species Composition, Distribution, Population, Biomass and Behavior

1. Bacteria

MARILYN J. JORDAN

Classification and Energetics

Bacteria are ubiquitous organisms, distinguishable from all other living organisms by their structural organization. They are difficult to study because of their small size (0.2 to 50 μm long). There are only a few observable features, such as cellular size and shape or presence or absence of spores, which are useful for identification. Therefore, bacteria are classified primarily on the basis of their metabolic capabilities, a time-consuming procedure requiring isolation and culture (not always possible) in numerous different media. Ecologically, the taxonomy of bacteria is less important than the trophic positions and metabolic activities of broad groups of bacteria in natural systems. From such perspective, bacteria may be grouped according to their sources of energy and of carbon (*Table V.A.1*–1). The information in *Table V.A.1*–1 is somewhat oversimplified because some bacteria can simultaneously utilize both inorganic and organic compounds as energy sources, and bacteria utilizing any given energy source (inorganic, organic or mixed) can utilize almost any carbon (CO_2, organic compounds or a mixture of both) source (Rittenberg 1969).

If the energy source and primary C source for bacterial production is organic matter, the production is considered secondary (heterotrophic) production. If the C source is CO_2 and the energy source is sunlight, the production is considered primary (photoautotrophic) production. Such terminology is difficult to apply to the chemoautotrophic bacteria because of the dichotomy in their energy and C sources. Although the chemoautotrophs obtain their C from CO_2, from a trophic viewpoint they may be considered secondary producers since their energy is derived from the reduced products of anaerobic decomposition of photosynthetically-produced organic matter.

The heterotrophic and chemoautotrophic bacteria, collectively known as the chemotrophs, obtain energy by mediating reduction-oxidation (redox) reactions. As organic matter is oxidized, the e^- removed are transferred to the most reducible e^- acceptor (oxidant) available. The degree of reducibility can be expressed by the term "relative electron activity," or ρ_ε, defined as $-\log \{e^-\}$. The ρ_ε is measured by the E_h, or redox potential, expressed in volts. Large positive values of E_h indicate low e^- activity, or oxidizing conditions. Small positive or large negative values of E_h indicate reducing conditions.

The most reducible e^- acceptor available in aerated waters is O_2. The process of oxidation of organic matter using O_2 as the final e^- acceptor is carried out by the aerobic heterotrophic bacteria. The end products are CO_2 and H_2O, plus release of all of the chemical bond energy originally stored in the organic molecules. Bac-

A. Species Composition, Distribution, Population, Biomass and Behavior

Table V.A.1-1. Types of Bacterial Energy Metabolism[a]

	Energy Source	Primary C Source	Electron Donor (ED) or Acceptor (EA)
Autotrophy			
Photoautotrophy	Light	CO_2	H_2, H_2S[b] (ED)
Chemoautotrophy[d]	Inorganic Compounds	CO_2	O_2 (EA)[c]
Heterotrophy			
(1) Aerobic	Complex Organic Compounds	90–100% of total C from organic compounds	O_2 (EA)
(2) Anaerobic	(a) Complex Organic Compounds (fermentation)	70–93% of total C from organic compounds	Organic Compounds (EA)
	(b) Simple Organic Compounds (CH_4, CH_3OH, etc.)	10–70% of total C from organic compounds and 30–90% from CO_2	Inorganic Compounds (EA)

[a] From Sorokin (1966), Doetsch and Cook (1973), Lechevalier and Pramer (1971).
[b] Photolithotrophy.
[c] Photoorganotrophy.
[d] Chemolithotrophy or chemosynthesis.

teria are not 100% efficient, but they do utilize a large proportion of the available energy. The aerobic heterotrophs also obtain most of the C needed for biosynthesis from organic molecules. As the O_2 is depleted, the E_h falls and different taxa of bacteria become involved in utilizing progressively less easily reduced e^- acceptors in the sequence shown in *Table V.A.1-2*. When the E_h becomes sufficiently negative, fermentation and reduction of SO_4^{2-} and CO_2 may occur nearly simultaneously. In fermentation reactions organic compounds are the final e^- acceptors. The by-products of fermentation include various simple organic acids, alcohols and acetone.

The reactions shown in *Table V.A.1-2* may occur sequentially in time in a closed system (e.g. sewage digester) or may occur vertically distributed in anaerobic waters and/or in sediments (Stumm and Morgan 1981). Correspondingly, some of the different types of bacteria carrying out the reactions listed in *Table V.A.1-2* may also be stratified in the sediments. Other bacteria are facultative aerobes; that is, they can produce the enzymes necessary for either aerobic or anaerobic respiration and can be found throughout the water and sediment.

The position of the various redox layers shifts upward or downward as the O_2 supply de-

Table V.A.1-2. The Sequence of the Principal Microbially-mediated Oxidation and Reduction Reactions Occurring in the Presence of Excess Organic Matter and Limited Oxygen[a]

1. Aerobic Respiration
   $(CH_2O) + O_2 \rightarrow CO_2 + H_2O$
2. Denitrification
   $5(CH_2O) + 4NO_3^- + 4H^+ \rightarrow$
   $\qquad 5CO_2 + 2N_2 + 7H_2O$
3. Nitrate Reduction
   $2(CH_2O) + NO_3^- + 2H^+ \rightarrow$
   $\qquad 2CO_2 + NH_4^+ + H_2O$
4. Fermentation
   $2(CH_2O) + H_2O \rightarrow HCOO^- + CH_3OH + H^+$
5. Sulfate Reduction
   $2(CH_2O) + SO_4^{2-} + H^+ \rightarrow$
   $\qquad 2CO_2 + HS^- + 2H_2O$
6. Methane Fermentation
   $4CH_3OH \rightarrow 3CH_4 + CO_2 + 2H_2O$
   $4H_2 + CO_2 \rightarrow CH_4 + 2H_2O$

[a] Modified from Stumm and Morgan 1981, *Aquatic Chemistry*. Copyright © 1970 by John Wiley & Sons, Inc. Used with permission.

creases or increases. Even the deepest water of Mirror Lake is well oxygenated throughout much of the year, with the boundary between the aerobic and anaerobic zoning occurring in the deepest waters or in the sediments.

Fermentation is much less efficient than aerobic respiration. For example, the fermentation of glucose to CO_2 and alcohol releases only about 8% of the energy available from its complete oxidation (Lechevalier and Pramer 1971). However, there are bacteria that are able to utilize the energy remaining in the simple organic molecules produced by fermentation. These bacteria obtain 30 to 90% of their biosynthetic C from CO_2 (Table V.A.1–1), even though their energy is obtained heterotrophically (Sorokin 1966).

Additional energy also remains in the various end products of anaerobic decomposition, such as H_2S, NO_2^-, NH_4^+, CH_4 and Fe^{2+}. These products are soluble and tend to diffuse upwards through the anaerobic zone. At the boundary between the anaerobic and aerobic zones, the chemoautotrophic bacteria utilize the energy in these compounds, using O_2 as the oxidant (Table V.A.1–1). The chemoautotrophic bacteria are considered autotrophs because the source of their biosynthetic C is CO_2 and the source of their energy is inorganic molecules. However, the chemoautotrophic bacteria should be thought of as *secondary producers* since they obtain their energy from the reduced products of anaerobiosis that originated from photosynthetically-produced organic matter (Sorokin 1966; Rich and Wetzel 1978). Thus, the entire decomposition process is complex but well integrated, involving many components that may be separated from each other in time and/or space.

Aerobic heterotrophs are the most important bacteria in the waters of Mirror Lake. Anaerobic bacteria occur in the sediments, in thick littoral leaf packs and possibly also in the deepest hypolimnetic water in late summer or in the late winter/early spring. In all of these cases where anaerobic bacteria are found, organic matter is abundant but the supply of oxygen is limited. Under the most highly anaerobic conditions methane may be formed. Low levels of methane have been detected in the hypolimnion of Mirror Lake, presumably released by methanogenic bacteria. Methanogenic bacteria also appear to be active in the thick leaf packs found in shallow littoral water, since the abundant gas bubbles released when the leaf packs are disturbed are rich in CO_2 and CH_4. Chemoautotrophic bacteria, which utilize as an energy source the reduced inorganic compounds produced by anaerobic respiration, would be found at the boundary between aerobic and anaerobic conditions near the sediment surface or possibly in the deepest part of the hypolimnion in the summer.

The remaining bacteria in Table V.A.1–1, which have not been discussed, are the photoautotrophs. In the photosynthetic process in eukaryotes and blue-green algae, CO_2 is reduced by electrons from H_2O, with the resultant release of O_2. Bacterial photosynthesis is an anaerobic process that uses compounds other than water as the reductant for CO_2. The green and purple sulfur bacteria, for example, oxidize H_2S and produce elemental sulfur. Photosynthetic bacteria occur abundantly only in situations where sufficient light reaches anaerobic waters. Photosynthetic bacteria are unlikely to be important in Mirror Lake since less than 1% of incident sunlight reaches the hypolimnion, which in any case is anoxic only during a few weeks or months of the year.

Enzymes and Structures

Aquatic bacteria live in an environment that generally has a very low concentration of dissolved nutrients and organic compounds. For example, the concentration of dissolved organic matter in Mirror Lake is usually less than 4 mg/liter, and the concentration of glucose in most aquatic systems is just a few micrograms/liter (Wetzel 1975). The bacteria are able to actively transport solutes across the cell membrane. Since many potential bacterial carbon sources are particulate or are molecules too large to be transported across the cell membrane, bacteria use enzymes to degrade these substrates into smaller molecules.

In the water of lakes, bacteria may be free-floating, attached to floating particles or attached to surfaces such as rocks, macrophytes and other organisms. Most of the bacteria suspended in the water-column of Mirror Lake are free-floating, and only a few percent are attached to detrital particles. Only within the last ten years has the mode of bacterial attachment to particles become known. Bacteria produce fibers of polysaccharides, or branching sugar molecules that extend from the bacterial surface. These tangled fibers form a feltlike *glyco-*

calyx that surrounds the bacterial cell, or group of cells, enabling the cell to adhere to surfaces (*Figure V.A.1*–1; Costerton et al. 1978). The glycocalyx may aid bacterial nutrition by binding certain molecules and also may keep bacterial digestive enzymes and digestive products close to the cell. The glycocalyx also may physically group bacteria into something like an organized community. For example, the bacteria that release H_2 from organic compounds may be bound closely with other bacteria that use H_2 to reduce CO_2 to methane (CH_4) (Costerton et al. 1978).

Numbers and Biomass

There are several direct and indirect methods of determining bacterial numbers and/or biomass. The same techniques may be used for water and sediment, except that sediment samples must first be diluted with bacteria-free water. An indirect method that has been used for a long time is the plate-count method. Another method, the most probable number (MPN) technique, requires the inoculation of a serial dilution of the sample into tubes of media. Both methods underestimate by as much as several orders of magnitude the total number of bacteria present in the sample, for only a few types of bacteria are able to grow in the media provided. Although these methods are inadequate for determining total numbers of bacteria, the use of appropriate media may enable the selective enumeration of particular physiological types of bacteria in rich laboratory media.

A good method for determining total numbers of bacteria is direct counting using a fluorescence microscope. Samples to be counted by the fluorescence method are first stained with a dye such as acridine orange, which binds with the DNA of all living and dead organisms in the sample. The sample is filtered through a black filter, preferably polycarbonate Nuclepore filters since they have a flat surface and uniform pore size (Hobbie et al. 1977). When viewed with the proper wavelength illumination, bacteria on the filter fluoresce yellow-green or red. With this method, the bacteria are easy to see and to distinguish from detritus because the bacteria appear as glowing points of light against the black background, and the detritus is barely visible. Most bacteria are too small (~ 0.6 μm) to exhibit any definite shape.

Nearly all lakes contain approximately one to ten million bacteria per milliliter of water. Total numbers of bacteria in the epilimnion of Mirror Lake range between one and four million per milliliter, but numbers in the hypolimnion may

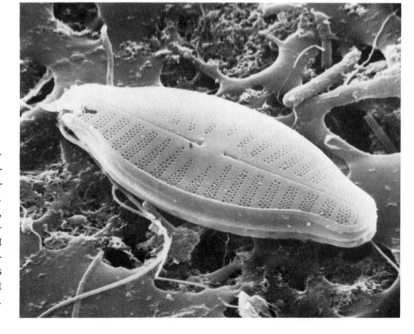

Figure V.A.1–1. Diatom frustule surrounded by filamentous bacteria and other bacterial organic matter on filter. Sample from Lake Tahoe, California/Nevada. *Pseudomonas* sp. in upper right about 1 μ long. Two hundred ml water from a depth of 20 m was filtered through the solvinert filter with pore size 1.2 mm. (Photo by Hans Paerl.)

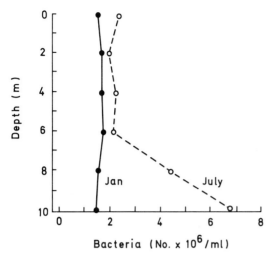

Figure V.A.1–2. Profiles of bacterial numbers in the water-column of Mirror Lake in January 1976 (●———●) and July 1976 (○- - -○).

be as high as eight million by the end of the summer stratification period (*Figure V.A.1–2*). These bacteria are the normal heterotrophic flora. The top 10 cm of gyttja sediments contain an average of 2 to 10×10^9 bacteria/cm^3 of sediment. In many lakes, numbers of bacteria are highest at the sediment surface and decrease with depth, but such stratified counts have not been done for Mirror Lake.

Bacterial biomass may be estimated from bacterial counts by measuring cell sizes, calculating the weighted-mean cell volume, and then calculating the wet weight assuming an average specific gravity of approximately 1.0. Dry weight is about 20% of wet weight, and carbon content is 45% to 50% of dry weight (Sorokin and Kadota 1972). Bacteria in the water and sediment of Mirror Lake range from about 0.3 to 1.4 μm in length, and the average cell contains about 9×10^{-9} μg of carbon. On an area basis, the mean annual standing biomass of sestonic bacteria in Mirror Lake is 80 mg C/m^2, and of benthic bacteria is 4,000 mg C/m^2. Sestonic bacterial C comprises approximately 4% of the total sestonic C; benthic bacterial C comprises less than 1% of the total benthic C (*Table V.C–1*). Carbon contained in the glycocalyx is not included in these estimates.

Summary

Bacteria are ubiquitous prokaryotic organisms of very small size (0.2 to 50 μm long). Since observable features of bacteria are few, bacterial classification is based primarily upon their metabolic capabilities. Autotrophic bacteria obtain carbon for protoplasmic synthesis from CO_2 and energy from light (photoautotrophy) or from the oxidation of reduced inorganic compounds (chemoautotrophy). Heterotrophic bacteria obtain carbon primarily from organic molecules (aerobic heterotrophy) or from both organic molecules and CO_2 (anaerobic heterotrophy), and obtain energy from the oxidation of organic molecules. The bacteria in the water of Mirror Lake are primarily aerobic heterotrophs, most of which are suspended as single cells, but some are attached to detrital particles. The vertical distribution of bacteria in the sediment shifts upward and downward seasonally, along with the redox potential, indirectly controlled by changes in the degree of oxygenation. Heterotrophic bacteria metabolizing aerobically are found at the oxygenated surface of the sediment, and heterotrophic bacteria metabolizing anaerobically are found deeper in the sediments. Anaerobic bacteria, including methanogens, also may be found in dense littoral leaf packs. Chemoautotrophic bacteria are found at the boundary between the aerobic and anaerobic zones, where they oxidize the reduced inorganic compounds that diffuse upward from the anaerobic zone below.

Total numbers of bacteria in water and sediment may be accurately and directly counted using fluorescence microscopy. Plate counts or most probable number techniques underestimate total numbers of bacteria, but the use of selective culture media can be valuable for studying specific physiological bacterial types. Mirror Lake is typical of most freshwater lakes in containing between 1 and 8×10^6 bacteria per milliliter of water (varying seasonally and with depth), and between 2 and 10×10^9 bacteria/cm^3 of sediment. Since the average cell contains about 9×10^{-9} μg of carbon, the mean annual standing crop of bacteria in Mirror Lake is 80 mg C/m^2 in the water (4% of total sestonic C) and 4,000 mg C/m^2 (to a depth of 10 cm) in the sediment (<1% of total sediment C).

2. Phytoplankton

F. deNoyelles and Gene E. Likens

The suspended algae in the pelagic region of a lake are referred to as phytoplankton. These algae are exceedingly diverse in form, size, pigments, taxonomy and ecologic niche (see Hutchinson 1967). Oligotrophic lakes in particular, may contain hundreds of species, but usually only a few predominate. Many of the common species are small (a few μm in diameter), fragile or poorly preserved and have largely been ignored in limnological studies. Mirror Lake contains an abundance of these species and special attention has been given to them in our studies.

The chemistry and biology of Mirror Lake are similar to many other small oligotrophic lakes in North America. Such data from the intensively studied lakes in the Experimental Lakes Area (ELA) of Ontario, Canada (e.g. Armstrong and Schindler 1971; Schindler and Holmgren 1971; Findlay and Kling 1975; Findlay 1978; Fee 1979; Prokopowich 1979), in particular, provide important comparisons and are used to elaborate on the findings from our studies of phytoplankton in Mirror Lake. Thus, in our description of the phytoplankton of Mirror Lake, basic characteristics are presented along with comparisons to other similar oligotrophic lakes, particularly those in the ELA. Where the ecology of the lake may be changing from what is considered to be natural, the abundance and composition of phytoplankton are used to detect such changes and perhaps explain their origin.

Species Composition and Distribution

The composition of phytoplankton species in Mirror Lake was determined for two periods during 1968 to 1975. From July 1968 through August 1971, a number of the more abundant species were identified and counted by F. de-Noyelles on 46 sampling dates (*Table V.A.2–1*). These samples included all seasons of the year (*Figure V.A.2–1*). Four depths (0, 3, 6, 9 m) were considered on each date. From a few of these samples another investigator (S. K. Holmgren, personal communication in 1968) prepared a more detailed species list. A species list was also compiled by P. Godfrey (personal communication) in 1975. Species recorded by all investigators are listed in *Table V.A.2–2*. In the 1975 study, counts were included from

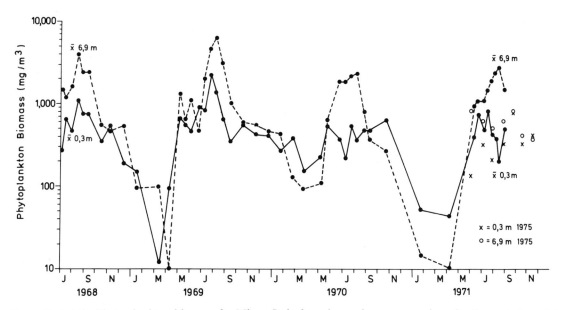

Figure V.A.2-1. Phytoplankton biomass for Mirror Lake based on volume conversion of cell counts from July 1968 to August 1971 and June to November 1975. The biomass of the epilimnetic and hypolimnetic phytoplankton are represented by the means of the 0- and 3-m values and the 6- and 9-m values, respectively.

Table V.A.2–1. Species of Phytoplankton Recorded in Mirror Lake from July 1968 through August 1971[a]

	0 m		3 m		6 m		9 m	
	S	W	S	W	S	W	S	W
Anabaena spiroides	8– 6	0– 0	3– 10	0– 0	1– 0	0– 0	3– 2	0– 0
Ankistrodesmus falcatus (S-0, 3)	12– 3	0– 0	9– 2	0– 0	4– 3	0– 0	1– 1	0– 0
Asterionella formosa (W)	2– 6	2– 12	1– 1	3– 31	10– 2	4– 15	16– 7	5– 7
Bitrichia chodatii (Y)	17– 3	2– 0	16– 2	4– 1	13– 1	3– 4	8– 1	2– 0
Botryococcus braunii	0– 0	0– 0	3– 10	0– 0	3– 10	0– 0	1– 10	0– 0
Ceratium hirundinella	2– 75	0– 0	8–112	0– 0	6– 100	0– 0	0– 0	0– 0
Chlamydomonas sp. (S-9)	4– 17	0– 0	0– 0	0– 0	4– 3	0– 0	11– 36	2– 4
Chromulina sp. (S-0, 3)	13– 30	2– 79	19– 22	1– 8	2– 6	0– 0	0– 0	0– 0
Chroococcus limneticus (S)	21–182	0– 0	19–403	1– 6	20– 253	0– 0	14– 107	0– 0
Chrysidiastrum catenatum	4– 49	0– 0	0– 0	0– 0	0– 0	0– 0	0– 0	0– 0
Chrysochromulina parva (Y-0, 3, 6)	27– 6	4– 2	28– 7	6– 9	18– 4	4– 10	6– 2	4– 6
Chrysococcus sp. (Y-0, 3, 6)	28– 17	9– 4	24– 7	6– 1	27– 9	9– 2	11– 6	4– 1
Chrysolykos planctonicus (Y)	8– 1	5– 2	5– 2	2– 3	7– 1	1– 0	1– 2	2– 0
Chrysosphaerella longispina (S-6, 9)	15– 18	0– 0	20– 35	0– 0	18– 744	0– 0	5– 232	0– 0
Coccomyxa sp. (Y-6, 9)	34– 9	3– 3	19– 9	1– 1	35– 30	5– 2	19– 24	6– 6
Cosmarium depressum	4– 3	0– 0	3– 55	1– 3	5– 3	0– 0	2– 3	0– 0
Crucigenia tetrapedia	0– 0	0– 0	1– 0	0– 0	5– 6	0– 0	2– 0	0– 0
Cryptomonas erosa (Y-9)	23– 34	7– 61	28– 87	8– 56	35– 72	8–135	35– 205	6–120
Cryptomonas marssonii (Y)	35– 43	8– 31	32– 54	6– 14	36– 33	6– 6	34– 56	5– 11
Cryptomonas platyuris (S-9)	1– 21	0– 0	0– 0	0– 0	0– 0	0– 0	22– 82	1– 21
Cryptomonas sp. (S-9)	4– 20	0– 0	1– 3	0– 0	3– 39	1– 3	30– 440	5– 39
Cyclotella comensis (Y)	5– 1	4– 0	8– 4	5– 3	7– 7	8– 2	7– 6	5– 1
Cyclotella comta (Y)	23– 70	2– 42	27– 74	5– 64	22– 77	5– 79	11– 39	5– 22
Dictyosphaerium pulchellum	0– 0	0– 0	0– 0	0– 0	0– 0	0– 0	1– 26	1– 0
Dinobryon bavaricum (S-3, 6)	4– 9	1– 1	7– 25	0– 0	11– 20	0– 0	8– 2	0– 0
Dinobryon borgei (S)	10– 1	2– 0	7– 1	1– 0	13– 2	0– 0	6– 5	0– 0
Dinobryon cylindricum (S-6)	1– 28	0– 0	1– 16	0– 0	10– 15	1– 1	6– 3	0– 0
Dinobryon divergens (S)	13– 43	2– 36	13– 54	2– 6	11– 42	3– 3	5– 37	2– 1
Dinobryon divergens v. angulatum (S)	19– 4	1– 0	14– 5	1– 0	15– 5	3– 0	8– 8	0– 0
Dinobryon divergens v. schauinslandii (S)	24– 5	3– 3	23– 4	1– 0	15– 8	3– 0	8– 8	0– 0
Dinobryon seucicum	0– 0	0– 0	1– 0	0– 0	0– 0	0– 0	1– 0	0– 0
Dinobryon tabellariae (S)	11– 3	1– 0	5– 0	1– 2	10– 3	0– 0	3– 2	1– 0
Elakatothrix gelatinosa (S)	16– 0	1– 0	5– 0	2– 0	19– 0	2– 1	9– 0	2– 0
Euglena sp.	0– 0	0– 0	0– 0	0– 0	0– 0	0– 0	5– 6	0– 0
Gloeocystis planctonica (S)	13– 4	0– 0	14– 4	0– 0	23– 5	0– 0	16– 3	1– 0
Gonium pectorale	0– 0	0– 0	1– 1	0– 0	0– 0	0– 0	0– 0	0– 0
Gonyostomum semen (S-6)	5– 23	1– 5	16– 82	0– 0	26– 227	0– 0	19– 88	0– 0
Gymnodinium mirabile (S-6, 9)	11–100	0– 0	21– 93	1– 50	27– 980	0– 0	21– 235	0– 0
Gymnodinium sp. (S-0, 3, 6)	15– 13	0– 0	14– 5	1– 3	11– 9	3– 3	2– 9	2– 9
Katablepharis ovalis (Y)	36– 6	8– 6	27– 5	7– 8	33– 6	9– 4	13– 4	9– 5
Mallomonas akrokomos v. parvula (Y)	9– 1	3– 5	10– 2	2– 0	28– 3	2– 6	12– 5	2– 0
Mallomonas caudata (S-6, 9)	5– 38	1– 16	4– 16	0– 0	11– 26	1– 16	13– 139	4– 43
Mallomonas globosa (S)	6– 4	1– 2	8– 7	2– 0	10– 7	3– 1	8– 4	0– 0
Mallomonas tonsurata (S)	25– 85	3– 13	33–101	2– 13	32– 141	1– 13	18– 232	2– 13
Melosira ambigua	4– 48	0– 0	2– 73	1– 20	4– 7	1– 20	5– 30	1– 41
Merismopedia tenuissima (S)	32– 3	0– 0	29– 4	1– 0	29– 5	2– 0	26– 1	0– 0
Monochrysis aphanaster (Y)	33– 43	9– 3	34– 10	3– 2	35– 16	4– 5	18– 8	3– 4
Ochromonas sp.	2– 8	0– 0	8– 2	1– 2	3– 3	0– 0	0– 0	0– 0
Oocystis sp. (Y)	35– 5	3– 1	30– 7	6– 1	33– 10	5– 2	25– 4	6– 2
Peridinium inconspicuum (S-0)	18– 46	0– 0	3– 21	0– 0	1– 6	0– 0	1– 49	0– 0
Peridinium polonicum	3– 91	0– 0	0– 0	0– 0	0– 0	0– 0	0– 0	0– 0
Peridinium willei (Y-6, 9)	1– 75	2–150	1– 75	1–300	14– 118	1– 75	6– 100	0– 0
Phacus longicaudata	0– 0	0– 0	0– 0	0– 0	0– 0	0– 0	2– 120	0– 0
Pseudokephyrion elegans (S)	11– 1	1– 0	5– 1	1– 0	4– 5	1– 2	4– 2	1– 0
Pseudokephyrion entzii (S)	21– 2	1– 0	8– 1	1– 0	9– 5	0– 0	4– 20	2– 0
Pseudokephyrion sp. (Y-0, 3, 6)	23– 1	3– 1	21– 1	1– 0	19– 1	2– 0	2– 0	2– 0
Quadrigula pfitzeri (S)	13– 1	1– 0	15– 2	1– 3	15– 6	0– 0	16– 3	0– 0
Rhabdomonas incurva	0– 0	0– 0	0– 0	0– 0	0– 0	0– 0	14– 6	0– 0
Rhizosolenia eriensis (S)	6– 25	0– 0	6– 24	0– 0	6– 17	1– 3	6– 17	1– 3
Rhodomonas minuta v. nannoplanctica (S)	10– 0	0– 0	16– 1	1– 0	26– 1	4– 0	17– 1	0– 0
Rhodomonas pusilla (Y)	34– 6	9– 34	36– 16	10– 25	34– 7	8– 21	25– 2	6– 24
Scenedesmus abundans	0– 0	0– 0	0– 0	0– 0	0– 0	0– 0	1– 1	0– 0
Scenedesmus bijuga	7– 4	0– 0	2– 1	0– 0	3– 1	0– 0	0– 0	0– 0
Staurastrum sp.	5– 3	0– 0	4– 3	0– 0	5– 3	0– 0	3– 3	1– 3
Staurodesmus sp. (S)	4– 5	1– 5	4– 5	1– 5	10– 9	1– 5	2– 5	0– 0
Synedra radians (Y)	3– 2	1– 2	2– 14	2– 4	2– 14	1– 2	7– 4	4– 2
Synedra sp. (Y-6, 9)	3– 4	0– 0	4– 7	0– 0	10– 4	4– 4	13– 4	3– 1
Synura adamsii	2– 2	0– 0	4– 2	0– 0	7– 13	0– 0	2– 2	0– 0
Synura uvella (S-9)	4– 52	0– 0	7– 18	1– 86	8– 35	1– 4	18– 445	0– 0
Tabellaria fenestrata (Y)	16– 16	2– 14	17– 43	5– 7	21– 23	4– 26	13– 6	7– 26
Tetraëdron minimum	1– 0	0– 0	0– 0	0– 0	0– 0	0– 0	0– 0	0– 0

A: Species Composition, Distribution, Population, Biomass and Behavior

Table V.A.2–1. Continued

	0 m		3 m		6 m		9 m	
	S	W	S	W	S	W	S	W
Tetraëdron minimum v. *tetralobulatum* (Y)	5– 2	4– 2	4– 7	5– 0	4– 4	4– 0	5– 0	4– 0
Trachelomonas spp. (S-9)	0– 0	0– 0	0– 0	0– 0	0– 0	1– 2	21– 15	1– 17
Uroglena volvox	0– 0	0– 0	0– 0	0– 0	1– 2	0– 0	0– 0	0– 0
Uroglenopsis americana	2– 7	0– 0	2– 24	0– 0	7– 87	1– 2	0– 0	0– 0
TOTAL PHYTOPLANKTON[b]	36–635	10–170	36–753	10–184	36–1,923	10–199	35–1,424	10–171

[a] Recorded for each species is its frequency of occurrence per number of sample dates available, followed by the mean biomass (live weight mg/m^3) for four depths (0, 3, 6, 9 m) and two periods (S, May–November; W, December–April). For example, *A. spiroides* was recorded on 8 of the 36 summer sample dates at 0 m with an average biomass of 6 mg/m^3 for those 8 dates. Total sampling dates and mean biomass are given for total phytoplankton. Summary distribution follows each species name (when recorded for more than 20 sampling dates or depths) with season of greatest abundance as S, W, or Y (year-round) followed by depths of greatest abundance if not uniform at all depths.

[b] Detection limit for organisms included in total phytoplankton biomass was 23 cells/ml, but for individual species considered here detection limit was lowered to one cell/ml.

Table V.A.2–2. Species of Phytoplankton Recorded in Mirror Lake from 1968 to 1975 by three Investigators Working Independently[a]

Cyanophyceae
  Anabaena circinalis Rabenhorst (3)
  Anabaena spiroides Klebahn (1)
  Anabaena sp. (3)
  Aphanizomenon flos-aquae (L.) Ralfs (3)
  Aphanocapsa delicatissima West & West (2, 3)
  Aphanotheca clathrata G. S. West (2)
  Chroococcus dispersus (Keissl.) Lemmermann
    v. *minor* G. M. Smith (3)
  Chroococcus limneticus Lemmermann (1, 2, 3)
  Chroococcus turgidus (Kütz.) Nägeli (2)
  Gloeocystis vesiculosa Nägeli (3)
  Gomphosphaeria lacustris Chodat (2, 3)
  Lyngbya nordgaardii Wille (3)
  Merismopedia elegans A. Braun (2)
  Merismopedia glauca (Ehr.) Nägeli (2, 3)
  Merismopedia tenuissima Lemmermann (1, 2, 3)
  Microcystis aeruginosa Kützing (2, 3)
  Oscillatoria agardhii Gomont (3)
  Oscillatoria agardhii v. *isothrix* Skuja (2)
  Oscillatoria limnetica Lemmermann (3)
  Oscillatoria limosa (Roth) C. A. Agardh (2)
  Oscillatoria rubescens De Candolle (3)
  Oscillatoria subbrevis Schmidle (3)
  Oscillatoria tenuis C. A. Agardh (3)
Chlorophyceae
  Ankistrodesmus braunii (Näg.) Brunnthaler (3)
  Ankistrodesmus falcatus (Corda) Ralfs (1, 2, 3)
  Ankistrodesmus falcatus v. *acicularis* (A.
    Braun) G. S. West (2)
  Ankistrodesmus setigerus (Schröd.) G. S. West (2)
  Ankistrodesmus spiralis (Turner) Lemmermann (2)
  Arthrodesmus incus (Bréb.) Hassall (2, 3)
  Arthrodesmus phimus Turner (3)
  Botryococcus braunii Kützing (1, 2, 3)

  Botryococcus protuberans West and West v.
    minor G. M. Smith (3)
  Carteria sp. (2)
  Chlamydomonas sp. (1)
  Chlamydomonas spp. (2)
  Chlamydomonas sp. (3)
  Chlorella vulgaris Beyerinck (2, 3)
  Closterium incurvum deBrébisson (3)
  Coccomyxa minor Skuja (2)
  Coccomyxa sp. (1)
  Coelastrum microporum Nägeli (3)
  Cosmarium depressum (Näg.) Lund (1, 3)
  Cosmarium depressum v. *achondrum* (Boldt)
    West and West (2)
  Cosmarium impressulum Elfving (3)
  Cosmarium nitidulum DeNotaris (3)
  Crucigenia tetrapedia (Kirch.) West & West (1, 2)
  Dictyosphaerium pulchellum Wood (1, 2, 3)
  Dispora crucigenioides Printz (3)
  Elakatothrix gelatinosa Wille (1, 2)
  Euastrum abruptum Nordstedt v. *lagoense*
    (Norst.) Krieger (3)
  Euastrum binale (Turp.) Ehrenberg (3)
  Eudorina elegans Ehrenberg (2)
  Gemellicystis neglecta Teiling and Skuja (2)
  Geminella minor (Näg.) Heering (2, 3)
  Gloeococcus schroeteri (Chod.) Lemmermann (2)
  Gloeocystis planktonica (West & West) Lem-
    mermann (1, 2, 3)
  Gonium pectorale Müller (1, 2)
  Gymnozyga moniliformis (Ehr.) v. *maxima*
    Irenee-Marie (2, 3)
  Kirchneriella contorta (Schmidle) Bohlin (3)
  Kirchneriella elongata G. M. Smith (3)
  Nephrocytium agardhianum Nägeli (3)

Table V.A.2–2. *Continued*

Nephrocytium limneticum (G. M. Smith) Skuja (2, 3)
Oocystis borgei Snow (2, 3)
Oocystis crassa Wittrock (3)
Oocystis lacustris Chodat (2)
Oocystis parva West & West (3)
Oocystis pusilla Hansgirg (3)
Oocystis submarina Lagerheim v. *variabilis* Skuja (2)
Oocystis sp. (1)
Pandorina morum (Müll.) Bory (2, 3)
Paulschulzia pseudovolvox (Schulz & Teiling) Skuja (2)
Pediastrum boryanum (Turp.) Meneghini (2)
Pediastrum tetras (Ehr.) Ralfs (2)
Pyramimonas tetrarhynchus Schmarda (2)
Quadrigula closterioides (Bohlin) Printz (2)
Quadrigula lacustris (Chod.) G. M. Smith (3)
Quadrigula pfitzeri (Schröd.) Printz (1, 2)
Scenedesmus abundans (Kirch.) Chodat (1)
Scenedesmus apiculatus (West & West) Chodat (2)
Scenedesmus bijuga (Turp.) Lagerheim (1, 3)
Scenedesmus opoliensis P. Richter (3)
Scenedesmus quadricauda (Turp.) deBrébisson (2, 3)
Selenastrum minutum (Näg.) Collins (3)
Sphaerozosma excavata Ralfs v. *subquadratum* West & West (3)
Spondylosium planum (Wolle) West & West (2, 3)
Staurastrum arachne Ralfs (2)
Staurastrum cuspidatum Bréb. v. *divergens* Nordstedt (3)
Staurastrum dejectum deBrébisson (2)
Staurastrum lacustre G. M. Smith (3)
Staurastrum limneticum Schmidle (2, 3)
Staurastrum lunatum Ralfs (3)
Staurastrum megecanthum Lundell (3)
Staurastrum muticum deBrébisson (3)
Staurastrum natator W. West (3)
Staurastrum paradoxum Meyen (3)
Staurastrum pentacerum (Wolle) G. M. Smith (3)
Staurastrum spiculiferum G. M. Smith (3)
Staurastrum sp. (1)
Staurodesmus sp. (1)
Stichococcus minor Nägeli (2)
Tetradesmus smithii Prescott (3)
Tetraëdron asymmretricum Prescott (3)
Tetraëdron minimum (A. Braun) Hansgirg (1, 2, 3)
Tetraëdron minimum v. *tetralobulatum* Reinsch (1)
Volvox aureus Ehrenberg (2)
Euglenophyceae
Euglena gracilis Klebs (3)
Euglena viridis Ehrenberg (2)
Euglena sp. (1)
Euglena sp. (3)
Lepocinclis sp. (2)
Menoidium costatum Korschik (2)
Phacus longicaudata (Ehr.) Dujardin (1, 2)
Rhabdomonas incurva Fresenius (1)
Trachelomonas spp. (1)
Trachelomonas sp. (2)
Chrysophyceae
Bitrichia chodatii (Reverdin) Chodat (1, 2, 3)
Chlorochromonas polymorpha Gavaudan (3)
Chromulina mikroplankton Pascher (3)
Chromulina minima Doflein (3)
Chromulina nebulosa Cienkowsky (3)
Chromulina ovalis Klebs (3)
Chromulina woroniniana Fisch
Chromulina sp. (1)
Chromulina spp. (2)
Chromulina sp. (3)
Chrysidiastrum catenatum Lauterborn (1, 2)
Chrysidalis peritaphrena Schiller (3)
Chrysochromulina parva Lackey (1, 2)
Chrysococcus rufescens Klebs (2, 3)
Chrysococcus sp. (1)
Chrysococcus spp. (2)
Chrysoikos skujai (Nauwerck) Willen (2)
Chrysolykos planctonicus Mack (1)
Chrysosphaerella longispina Lauterborn (1, 2)
Dinobryon acuminatum Ruttner (2, 3)
Dinobryon bavaricum Imhof (1, 2)
Dinobryon borgei Lemmermann (1)
Dinobryon crenulatum West & West (2)
Dinobryon cylindricum Imhof (1, 2)
Dinobryon divergens Imhof (1, 2, 3)
Dinobryon divergens v. *angulatum* (Sel.) Brunthaler (1)
Dinobryon divergens v. *schauinslandii* (Lemm.) Brunnthaler (1)
Dinobryon sociale Ehrenberg (2, 3)
Dinobryon sertularia Ehrenberg (2, 3)
Dinobryon suecicum Lemmermann (1)
Dinobryon tabellariae (Lemm.) Pascher (1, 3)
Kephyrion boreale Skuja (2)
Kephyrion littorale Lund (2)
Kephyrion ovum Pascher (3)
Mallomonas acaroides Perty (2, 3)
Mallomonas akrokomos Ruttner (2)
Mallomonas akrokomas v. *parvula* Conrad (1, 2)
Mallomonas caudata Conrad (1, 2, 3)
Mallomonas elongata Reverdin (2)
Mallomonas globosa Schiller (1, 2)
Mallomonas pseudocoronata Prescott (3)
Mallomonas reginae Teiling (2)
Mallomonas tonsurata Teiling (1, 2, 3)
Monochrysis aphanaster Skuja (1, 2)

Table V.A.2–2. *Continued*

Monochrysis spp. (2)
Monosiga brevicollis Ruin. (3)
Ochromonas sp. (1)
Ochromonas spp. (2)
Ochromonas spp. (3)
Pseudokephyrion elegans (Conr.) Conrad (1)
Pseudokephyrion entzii Conrad (1, 2)
Pseudokephyrion spirale Schmid (2)
Pseudokephyrion sp. (1)
Synura adamsii G. M. Smith (1, 3)
Synura uvella Ehrenberg (1, 2, 3)
Uroglena volvox Ehrenberg (1)
Uroglenopsis americana (Calkins) Lemmermann (1, 2, 3)

Diatomeae
Asterionella formosa Hassall (1, 2, 3)
Coscinodiscus sp. (3)
Cyclotella bodanica Eulenst. (3)
Cyclotella comensis Grunow (1, 2)
Cyclotella comta (Ehr.) Kützing (1, 2, 3)
Cyclotella glomerata Bachmann (3)
Cymbella sp. (3)
Cymbella angustata (W. Sm.) Cleve (3)
Cymbella lanceolata (Ag.) Ag. (3)
Cymbella tumidula Grun. ex A.S. (3)
Diatoma vulgare Bory (3)
Fragilaria capucina Desm. (2)
Fragilaria construens (Ehr.) Grunow (3)
Fragilaria crotonensis Kitton (2, 3)
Melosira ambigua (Grun.) Müller (1, 2, 3)
Melosira distans (Ehr.) Kützing v. alpigena Grunow (2)
Melosira granulata (Ehr.) Ralfs (3)
Melosira italica (Ehr.) Kützing v. subarctica Müller (2)
Melosira varians C. A. Agardh (3)
Neidium iridis (Ehr.) Cl. v. amphigomphus (Ehr.) A. Mayer (3)
Nitzschia acicularis W. Smith (3)
Rhizosolenia eriensis H. L. Smith (1, 3)
Stephanodiscus dubius (Fricke) Hust. (3)
Synedra delicatissima W. Sm. v. angustissima Grunow (2)
Synedra nana Meister (3)

Synedra radians Kützing (1, 2, 3)
Synedra rumpens Kützing (3)
Synedra sp. (1)
Tabellaria fenestrata (Lyngb.) Kützing (1, 2, 3)
Tabellaria flocculosa (Roth) Kützing (2)

Cryptophyceae
Chroomonas coerulea (Geitler) Skuja (2)
Cryptomonas compressa Pascher (3)
Cryptomonas erosa Ehrenberg (1, 2, 3)
Cryptomonas marssonii Skuja (1, 2)
Cryptomonas obovata Skuja (2)
Cryptomonas ovata Ehrenberg (2, 3)
Cryptomonas platyuris Skuja (1)
Cryptomonas rostratiformis Skuja (2, 3)
Cryptomonas rufescens Skuja (2)
Cryptomonas sp. (1)
Gonystomum semen (Ehr.) Diesing (1, 2)
Katablepharis ovalis Skuja (1, 2)
Rhodomonas lacustris Pascher and Ruttner (3)
Rhodomonas minuta Skuja (2, 3)
Rhodomonas minuta v. nannoplanctica Skuja (1)
Rhodomonas pusilla (Bachm.) Javorn. (1, 2, 3)
Rhodomonas tenuis Skuja (2, 3)
Sennia parvula Skuja (2)

Peridineae
Amphidinium sp. (2)
Ceratium hirundinella (Müll.) Schrank (1, 2, 3)
Cryptaulax vulgaris Skuja (2)
Glenodinium gymnodinium Penard (3)
Glenodinium palustre (Lemm.) Schiller (2, 3)
Glenodinium pulvisculus (Ehr.) Stein (3)
Gymnodinium fuscum (Ehr.) Stein (2)
Gymnodinium helveticum Penard (2)
Gymnodinium lacustre Schiller (2)
Gymnodinium mirabile Penard (1, 2)
Gymnodinium uberrimum (Allman) Kofoid and Swezy (2)
Gymnodinium sp. (1)
Peridinium cinctum Ehrenberg (2, 3)
Peridinium inconspicuum Lemmermann (1)
Peridinium palustre (Lindemann) Lefèvre (2)
Peridinium polonicum Wolle (1)
Peridinium willei Huitfeld-Kaas (1, 2)
Peridinium wisconsinense Eddy (2, 3)

[a] A given species occurring in the lake may, therefore, be listed here under more than one species name inflating the total number of entries (where obvious, these were eliminated by the authors). Each species listed is followed by a notation identifying the contributing investigator: F. deNoyelles (1); S. K. Holmgren (2); and P. Godfrey (3).

seven dates between 2 June and 11 November, with at least four depths (0, 3, 6, 9 m) considered on each date. All phytoplankton samples prepared for counting were preserved in acid Lugol's solution and settled in counting chambers. Identifications and counts were made with a Wild inverted microscope. Organisms were counted by F. deNoyelles and P. Godfrey at detection limits of 1 to 23 cells/ml.

The species of phytoplankton recorded in Mirror Lake at all depths are commonly included in the species lists from other oligotrophic lakes (Järnefelt 1956; Hutchinson 1967; Bozniak and Kennedy 1968; Johnson et al. 1970; Schindler and Nighswander 1970), including the Class-B lakes (see *Chapter V.B.2*) at the ELA (Kling and Holmgren 1972; Findlay and Kling 1975; Findlay 1978). Because the taxonomy of the phytoplankton is rarely, if ever, completely mastered, such lists will certainly differ in detail, in part because of preparation by different investigators. It can be valuable also to compare more general characteristics of the algal communities, such as relative abundance of major taxonomic groups. It appears typical for these small (<20 ha) clear oligotrophic lakes of moderate depth (>6 m) and summer thermal stratification that their algal communities are dominated by Chrysophyceae, Diatomeae, Peridineae and Cryptophyceae with

Table V.A.2–3. Distribution with Depth of Some Common Species during July and August from 1968 through 1971[a]

Species	0 m		3 m		6 m		9 m	
Asterionella formosa	0	(0)	0	(0)	11	(1)	22	(2)
Chromulina sp.[b]	16	(18)	19	(17)	7	(10)	2	(1)
Chroococcus limneticus	371	(10)	918	(8)	457	(10)	271	(5)
Chrysochromulina parva[b]	7.7	(13)	5.1	(15)	3.5	(9)	0	(0)
Chrysococcus sp.[b]	3.5	(19)	7.4	(17)	5.2	(19)	1.3	(6)
Chrysosphaerella longispina[b]	19	(4)	88	(5)	1,013	(11)	572	(2)
Cryptomonas erosa[b]	88	(2)	110	(8)	83	(7)	251	(16)
Cryptomonas marssonii[b]	61	(13)	55	(16)	58	(12)	69	(14)
Cryptomonas platyuris[b]	0	(0)	0	(0)	0	(0)	213	(5)
Cryptomonas sp.[b]	0	(0)	0	(0)	0	(0)	524	(19)
Cyclotella comta	137	(6)	144	(8)	139	(7)	74	(2)
Dinobryon spp.[b]	14	(14)	28	(12)	29	(14)	4	(8)
Euglenophyta[b]	0	(0)	0	(0)	0	(0)	63	(11)
Gloeocystis planktonica	7.0	(1)	15	(2)	8.6	(9)	8.0	(4)
Gonyostomum semen[b]	50	(1)	83	(6)	202	(16)	119	(11)
Gymnodinium mirabile[b]	60	(5)	65	(10)	1,366	(18)	306	(14)
Katablepharis ovalis[b]	7.1	(17)	6.8	(10)	7.4	(12)	3.3	(4)
Mallomonas caudata[b]	0	(0)	0	(0)	124	(1)	330	(4)
Mallomonas tonsurata[b]	175	(5)	272	(5)	262	(13)	194	(5)
Merismopedia tenuissima	3.8	(18)	5.1	(17)	7.2	(18)	1.6	(13)
Monochrysis aphanaster[b]	13	(14)	11	(15)	10	(14)	4	(5)
Oocystis sp.	5	(11)	18	(8)	12	(12)	5	(7)
Peridinium willei[b]	0	(0)	0	(0)	128	(10)	113	(4)
Pseudokephryon spp.[b]	3.1	(10)	1.8	(8)	2.0	(9)	0.4	(1)
Rhodomonas pusilla[b]	5.7	(12)	10.6	(16)	8.2	(14)	1.5	(4)
Synura uvella[b]	0	(0)	38	(3)	101	(2)	1,088	(7)
Tabellaria fenestrata	74	(2)	41	(3)	41	(3)	24	(2)
Total Phytoplankton[c]	506	(19)	861	(19)	2,884	(19)	1,892	(19)

[a] Summary of 19 sampling dates with $\bar{x}$ abundance in mg/m^3 determined for each species only when it was present. The number of dates when each species was present is recorded in parentheses.
[b] Flagellate.
[c] Includes all species recorded on the 19 dates.

only minor appearances by the Chlorophyceae and Cyanophyceae. This pattern is evident for Mirror Lake based on the distribution of biomass among the more common species from 1968 to 1971 (*Tables V.A.2–1* and 3). A listing of all species recorded in Mirror Lake arranged according to their taxonomic groupings is given in *Table V.A.2–2*. This same pattern was evident in 1975, although not presented here. From the other oligotrophic lakes cited, the same groups dominated the summer biomass. The dominance of these groups also persists during the winter months in Mirror Lake (*Table V.A.2–1*) and in the other lakes.

The seasonal distribution of species can be considered from the study during 1968 to 1971, which extended through all seasons. In *Table V.A.2–1*, seventy-five of the more common species are listed according to season and depth, with frequency of occurrence and mean biomass at times of occurrence. Of the 53 species recorded in 20 or more samples (out of a possible 183), 52 species were most abundant in terms of biomass and frequency of occurrence during the May through November ice-free season though 20 of these were commonly present year round. Twenty-nine of these 52 species were recorded in less than seven of the 40 samples taken from under the ice.

The species diversity, considering all species recorded on a given date and combining all depths, also varied with season. For the 36 dates sampled from four periods during May to November, the total number of species ranged from 33 to 52, averaging 43. For 10 dates from three periods during December to April, the range was 12 to 36, with an average of 24. This latter season of ice-cover is also a time of lower phytoplankton biomass (*Figure V.A.2–1*), which may be due in part to the greatly reduced illumination under the ice (*Chapter IV.B*).

Species diversity in terms of mean number of species present for individual dates and depths is summarized in *Table V.A.2–4* for the July 1968 through August 1971 period. This diversity was highest at 6 m for all periods, i.e. the entire interval, along with ice-free (May–November), summer stratification (July–August) and mixing (October, November, May) periods. Diversity was lowest at 9 m during all of these periods. Also included in *Table V.A.2–4* is the mean number of species unique to a particular depth on a given date. Mean numbers of unique species were highest at 9 m for each of the seasonal periods. Based on this pattern of diversity it can be hypothesized that the 6-m depth is within a zone of transition, an ecotone of sorts, between the two layers of the water-column of greatest physical and chemical difference. This transition zone is most obvious during stratification when 6 m is within the metalimnion (*Figures V.A.2–2* and *V.B.2–1* and 2). Ecotones share some conditions of both bordering zones, are generally less extreme than the extremes on ei-

Table V.A.2–4. Mean Number of Species Recorded in Mirror Lake from July 1968 through August 1971[a]

Period	0 m	3 m	6 m	9 m	0–3 m[b]	6–9 m[b]
July 1968–August 1971	21.1	21.6	23.2	18.4		
(46)	3.4	3.1	4.3	4.9	8.8	11.4
December–April	11.6	12.3	13.2	11.8		
(10)	2.6	2.8	2.8	2.8	6.6	7.0
May–November	23.8	24.1	26.0	20.3		
(36)	3.7	3.2	4.7	5.5	9.4	12.8
July–August	24.3	24.1	26.5	20.2		
(19)	4.4	3.0	5.4	6.3	10.4	14.0
October, November, May	23.8	24.9	25.1	20.8		
(9)	2.7	3.3	3.4	4.1	7.1	9.3

[a] Presented for four annual periods and four depths. The mean number of unique species (for a given date recorded at only one depth is also presented). The number of dates (samples) used for each mean is recorded in ( ).
[b] Unique to the stratum recorded at one or both depths.

ther side and as a result support a greater variety of species. Certainly settling or upward migration from below of some species into this transition zone at 6 m adds to the diversity, but is not solely responsible for it, because at 9 m, where settling also should have strong influence, diversity is lowest. The largest number of unique species are recorded at 9 m, perhaps because fewer species can tolerate the lower dissolved oxygen and light conditions throughout the year and low temperatures during the summer.

The relative contribution that each species makes to the structure of the phytoplankton community can be considered further from *Tables V.A.2–1 and 3*. For a given species, frequency of occurrence can be compared with the highest possible frequency, and mean biomass to the total for that depth. For example, *Chrysosphaerella longispina,* although present less often than *Rhodomonas pusilla* at 6 m during the ice-free (*Table V.A.2–1*) and summer stratification (*Table V.A.2–3*) seasons, when present comprised a considerably larger portion of the total biomass. Like *R. pusilla,* many other species occurred frequently but each represented only a small fraction of the total biomass (e.g. *Gloeocystis planktonica, Katablepharis ovalis, Merismopedia tenuissima, Pseudokephyrion* spp. and *Chrysochromulina parva*). It also is evident that during the summer in the epilimnion (0, 3 m) the community biomass is more evenly distributed among many species, while in the metalimnion and hypolimnion (6, 9 m) fewer species comprise a larger proportion of the total biomass. The total biomass in these deeper waters is often much greater than that of the epilimnion which, as is discussed later, is typical for lakes of this type.

Vertical Distribution

There also are similarities in the vertical distribution of the phytoplankton species between Mirror Lake and Class-B oligotrophic lakes in the ELA. One shared feature is the greater biomass of phytoplankton below the epilimnion. At the ELA, bands of high biomass occurred in deep water, generally where photosynthetically available light was 1 to 3% of surface irradiance (Fee 1976). These bands of high biomass, generally 0.5 to 2 m thick (Fee et al. 1977), varied in

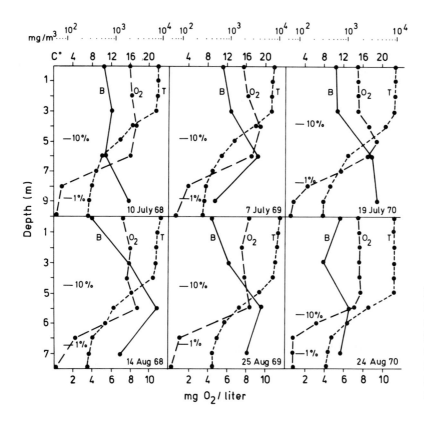

Figure V.A.2–2. Depth profiles in Mirror Lake during the summer for biomass (mg/m^3 from volume-converted cell counts), dissolved oxygen (mg/liter), temperature (°C) and irradiance (1 and 10% depths).

depth in different lakes or in the same lake over time, partly owing to differences in light penetration. Most often the bands occurred in the metalimnion and upper hypolimnion at depths of 4 to 8 m. In Mirror Lake, 10 and 1% of surface irradiance generally occurred at 4 to 5 m and 8 to 9 m, respectively (*Figure V.A.2–2* and *Chapter IV.B*), with the metalimnion beginning at 4 to 6 m. Samples from 6 and 9 m in Mirror Lake contained the greatest biomass during the thermal stratification period, which usually extended from June into September (*Table V.A.2–3* and *Figures V.A.2–1* and 2 for 1968 through 1971). A similar distribution occurred in 1972, 1974 and 1979, based on chlorophyll profiles (*Figures V.A.2–3* and *V.B.2–2*) and in 1975 on phytoplankton counts (*Figure V.A.2–1*). The presence of biomass peaks below the epilimnion has been recorded in other oligotrophic lakes (Juday 1934; Gessner 1948; Findenegg 1966; Baker and Brook 1971; Kiefer et al. 1972; Tilzer 1973; Kerekes 1974), although their presence has not been sought in most studies or has been overlooked because of the sampling method and the narrowness of the bands.

Another similarity in the vertical distribution of phytoplankton between Mirror Lake and many ELA lakes is the composition of the species that dominate the biomass peaks below the epilimnion. Fee (1976, 1978b) described some of the dominant species for the ELA lakes as Chrysophycean flagellates of the genera *Dinobryon*, *Synura*, *Uroglena* and *Chrysosphaerella*. In one of these ELA lakes, deNoyelles et al. (1980) found that species from three of these genera were dominant. In Mirror Lake, *Chrysosphaerella longispina* and *Synura uvella* are

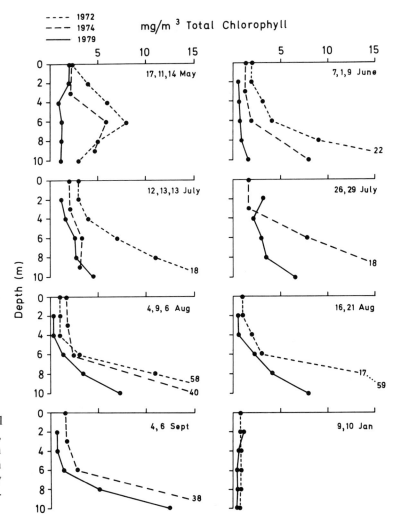

Figure V.A.2–3. Chlorophyll depth profiles from 1972 (----), 1974 (– – –) and 1979 (———) in Mirror Lake. Concentrations in mg Chl-*a*/m^3 were determined by extraction followed by fluorescence readings.

among the dominant species (*Tables V.A.2*–1 and 3) and are the same species reported by deNoyelles et al. (1980) as dominant in the ELA lake. Other common species in Mirror Lake (*Tables V.A.2*–1 and 3) include members from the Peridineae (*Gymnodinium mirabile, Peridinium willei*), the Cryptophyceae (*Gonystomum semen, Cryptomonas erosa, C. platyuris*) and the Chrysophyceae (*Mallomonas caudata*). Species of these genera are also common in the meta- and hypolimnetic waters of the ELA lakes (Fee et al. 1977; Findlay 1978; deNoyelles et al. 1980) and other oligotrophic lakes (Findenegg 1966; Tilzer 1973; Johnson et al. 1970).

The species dominating the biomass peaks below the epilimnion share certain characteristics in common in both Mirror Lake and the ELA Class-B lakes. Beyond their taxonomic affinities discussed above, they are mostly flagellates (Fee 1976; Fee et al. 1977; *Table V.A.2*–3). Also, while maintaining large numbers below the epilimnion, they are often rare or absent within the epilimnion (Fee 1976; Fee et al. 1977; *Tables V.A.2*–1 and 3). This pattern suggests that the peaks can arise from growth rather than from mere settling of organisms. Fee (1978b) has described several types of peaks in the ELA lakes, including ones arising from settling organisms and from photosynthetic bacteria. Peaks composed of the latter obligate anaerobes are restricted to anoxic bottom waters, generally a meter or so below phytoplankton peaks, and with illumination as low as 0.1% of the surface value. Some phytoplankton peaks arise from growth *in situ,* but are not dominated by the species noted earlier. Instead, such peaks contain filamentous Cyanophyceae (Lake 230, Fee et al. 1977) as has been noted elsewhere (Klemer 1976; Konopka 1981). The factors responsible for the development of these less common species have not been determined. Phytoplankton peaks produced in place or from settling from above can occur separately in the same lake or can be mixed together as a single peak (Fee 1978b). In this case, the species composition of the peak compared with that of the epilimnion can be used as an indicator of the origin of a peak.

The distribution of *Chrysosphaerella longispina* in Mirror Lake (*Figure V.A.2*–4) is suggestive of a peak produced in place. Periods when *Chrysosphaerella* did increase in the epilimnion generally coincided with maximum convective mixing (*Chapter IV.C*) and thus entrainment of the biomass produced below the epilimnion. A similar type of distribution can be seen for several other species including those of *Cryptomonas* and *Euglena* and *Gonystomum semen, Gymnodinium mirabile, Mallomonas caudata, Peridinium willei* and *Synura uvella*. Some of these species were never recorded at 0 and 3 m and, as a group, comprise most of the phytoplankton biomass at the 6- and 9-m depths (*Tables V.A.2*–1 and 3). Fee (1976) speculated that because of their motility these species can maintain their position in a zone of low turbulence, perhaps along a gradient of light. All of the species listed above are motile flagellates. Although the growth of these species may be slow in these low light and low temperature zones, the large biomass can, perhaps, develop in part because of the large cell and colony sizes of many of these species, which minimize respiratory and grazing losses, respectively. All of the species listed above have cells larger than 1,000 μm^3 in volume or are colonial. A further consideration of behavioral similarities among these species is presented later.

The characteristics discussed above for the phytoplankton in Mirror Lake and many ELA lakes seem to represent the usual conditions for these organisms in such oligotrophic lakes. Any departure from these conditions may indicate changes in the natural environment, perhaps related to unnatural disturbances. The phytoplankton community in Mirror Lake during 1975 did seem to show some changes though not to such a degree that the community was no longer characteristic of an oligotrophic lake. As stated above, phytoplankton biomass during summer stratification in 1975 was still greater below the epilimnion (*Figure V.A.2*–1; *Table V.A.2*–5). However, there were some changes in the distribution of particular species.

Before noting these changes, a word of caution is necessary concerning their interpretation. First, only three July and August dates are available from 1975, compared to 19 dates from 1968 to 1971. More importantly, different investigators counting the 1968 to 1971 samples vs. the 1975 samples may account for some of the discrepancies in the identification of these organisms. For example, *Chrysosphaerella long-*

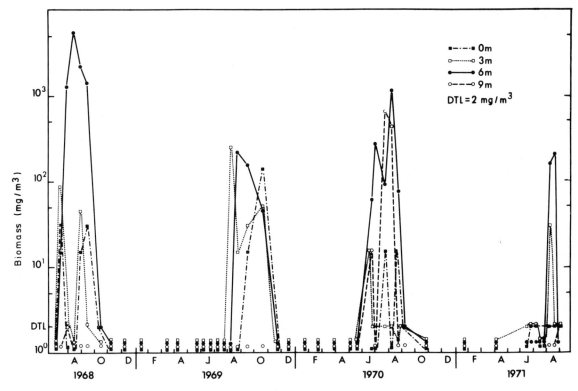

Figure V.A.2–4. Distribution of *Chrysosphaerella longispina* in Mirror Lake from July 1968 through August 1971 at four depths in biomass as mg/m^3 (from volume conversion of cell counts).

ispina and *Synura uvella* may have been overlooked in the 1975 samples because these colonial Chrysophyceae separate into single cells soon after preservation with acid Lugol's solution. When these samples were examined again several years later by the investigator who counted the 1968 to 1971 samples, both species were recorded. Accurate counts could not be made at this time because of some deterioration of the cells with age. It did appear, however, that both species were lower in abundance than in 1968 to 1971. In recent years through the summer of 1980, *Chrysosphaerella* has frequently been observed below the epilimnion, but abundances have not been determined (J. Cole, personal communication). The particular species or group of species selected for comparison in *Table V.A.2–5* were chosen to minimize problems of identification.

There is some evidence of changes in phytoplankton species distribution in 1975, especially in the abundance of certain species in the deepwater assemblages. Besides the possible declines of *Chrysosphaerella* and *Synura*, others also declined, including *Gonystomum, Gymnodinium* and *Mallomonas caudata* (*Table V.A.2–5*). Recent reexamination of these samples confirmed these declines. Other species increased in abundance, including those of *Gloeocystis, Cyclotella* and *Dinobryon* (*Table V.A.2–5*). It appears that species common to the epilimnion were contributing more to the biomass below the epilimnion than in previous years. This change may be an indication that settling organisms were replacing those produced in place. The distributions of the other species in *Table V.A.2–5* generally remained the same. This possible evidence for changing environmental conditions in Mirror Lake is discussed below when factors regulating the structure and behavior of the phytoplankton communities are considered.

Phytoplankton Biomass

The biomass of the phytoplankton community in Mirror Lake at various times throughout the year was approximated by converting the cell

Table V.A.2-5. Comparison of the July–August Distributions of Some Common Species from 1968 to 1971 and 1975[a]

	0 m		3 m		6 m		9 m	
Species	1968 to 1971	1975	1968 to 1971	1975	1968 to 1971	1975	1968 to 1971	1975
Asterionella formosa	0	2	0	3	11	7	22	2
Chrysosphaerella longispina[b]	19	—	88	—	1,013	—	572	—
Cyclotella spp.	137	37	144	49	122	125	75	41
Cryptomonas spp.	75	42	97	20	92	33	842	319
Dinobryon spp.	14	4	28	3	29	16	4	65
Gloeocystis planktonica	7	6	15	41	9	90	8	204
Gonyostomum semen	50	0	83	0	202	0	119	0
Gymnodinium spp.	50	76	65	0	1,366	0	306	0
Mallomonas caudata	0	0	0	0	124	0	330	0
Mallomonas spp.	148	34	195	3	232	50	165	56
Merismopedia spp.	4	39	5	29	7	19	2	10
Oocystis spp.	5	13	18	4	12	16	5	11
Peridinium spp.	0	0	0	0	128	56	113	280
Rhodomonas pusilla	6	2	11	3	8	4	2	3
Synura uvella[b]	0	—	38	—	101	—	1,088	—
Tabellaria fenestrata	74	6	41	26	41	11	24	28
Total Phytoplankton[c]	506	351	861	230	2,884	422	1,892	817

[a] Summary of 19 sampling dates for 1968 to 1971 and three for 1975 with x̄ abundance in mg/m^3 determined for each species only when it was present.
[b] Not counted in 1975, see text for explanation.
[c] Includes all species recorded on the 19 dates.

counts to volume, using individual size correction factors for each species. Assuming the weight of an algal cell to be equal to the weight of an equal volume of water, live weight in mg/m^3 was determined. These values can be converted to cell carbon by multiplying by 0.12 (Strickland 1960). The biomass estimates for 1968 to 1971 (*Table V.A.2-6*) are presented for four periods of the year: summer stratification (July, August) autumn circulation (October, November), winter ice-cover (January, February, March) and spring circulation (May). Other months are omitted as transitional periods. The highest biomass at each depth was recorded during summer stratification, with the typical peaks below the epilimnion. Biomass during autumn circulation became more evenly distributed and later during ice-cover declined to the lowest values for the year. During spring circulation, biomass increased at all depths to levels and uniform distributions similar to those during autumn circulation. The biomass levels in 1975 were lower, perhaps in part because of the real or artificial absence of the species noted above (*Figure V.A.2-1; Table V.A.2-5*). These ranges in biomass (live weight) with depth and season in Mirror Lake are similar to those in the Class-B oligotrophic lakes at the ELA (Schindler and Holmgren 1971; Kling and Holmgren 1972; Findlay and Kling 1975; Findlay 1978) and elsewhere (Kalff and Knoechel 1978).

The biomass of phytoplankton in Mirror Lake has also been estimated by measuring chlorophyll in the water-column by pigment extraction, followed by fluorescence measurements. Chlorophyll profiles are shown for various times of the year during 1972, 1974 and 1979 in *Figure V.A.2-3*. These chlorophyll concentrations with depth and season are similar to

Table V.A.2–6. Distribution of Phytoplankton Biomass with Depth and Season from 1968 to 1971[a]

Depth	Spring Circulation May	Summer Stratification Jul.–Aug.	Autumn Circulation Oct.–Nov.	Winter Ice-cover Jan.–Mar.
0 m	418	506	399	167
3 m	577	861	607	174
6 m	821	2,884	546	162
9 m	542	1,892	412	124

[a] Biomass (mg/m^3) calculated from cell counts converted to volume.

those in the oligotrophic lakes in the ELA. There, epilimnion values range from <1 to 5 mg/m^3, and values in deeper waters may be many times higher (Armstrong and Schindler 1971; Fee 1976, 1979; Fee et al. 1977). Similar epilimnetic values are reported for oligotrophic lakes elsewhere (Sakamoto 1966; Macan 1970; Kerekes 1974; Carlson 1977; Oglesby and Schaffner 1978; Smith 1979).

The carbon content of phytoplankton can be estimated by multiplying live weight by 0.12 or Chlorophyll-a by 45 (Strickland 1960). B. Peterson (personal communication) in a series of enrichment studies with Mirror Lake phytoplankton to obtain carbon-to-chlorophyll ratios verified the accuracy of these conversions for Mirror Lake.

Structure and Behavior

In discussing the structure together with the behavior of the phytoplankton community of Mirror Lake, we continue our consideration of the vertical distribution of the phytoplankton. In the intensively studied lakes in the ELA, which are chemically and physically similar to Mirror Lake, the phytoplankton below the epilimnion sometimes differ from the phytoplankton within the epilimnion in composition and relative abundance of the species (Schindler and Holmgren 1971; Fee 1976, 1978b; Fee et al. 1977; Findlay 1978). The species dominating the deeper assemblage, particularly those absent from the upper waters, are adapted to (perhaps even dependent upon) the different quality and lower quantity of light, lower temperature or higher nutrient conditions normally found there (Fee 1976; deNoyelles et al. 1980; Knoechel and deNoyelles 1980). Such physical and chemical conditions occur in the hypolimnetic waters of Mirror Lake as well (*Chapter IV.B*). The phytoplankton of these deeper waters include photosynthesizing organisms, not just moribund cells settling out of the epilimnion or heterotrophs capable of survival in the absence of light (Schindler and Holmgren 1971; deNoyelles et al. 1980; Fee 1980; Knoechel and deNoyelles 1980). Therefore, the small amount of light, generally less than 10% of the surface illumination (Fee 1976; Shearer and DeClercq 1977; Fee et al. 1977) that reaches these organisms is necessary for their growth. To remain as a characteristic component of these lakes, the organisms must also be capable of surviving some aphotic periods such as during prolonged cloud cover. Such viability in the absence of light has been reported (Tilzer et al. 1977) and may be dependent upon slow respiration, facilitated by low temperatures, and upon the use of carbohydrate reserves rather than heterotrophy (Rodhe et al. 1966; Morgan and Kalff 1975).

The upward migration of these algae in response to the loss of light would not greatly enhance their survival since within a meter or so they would leave the hypolimnion or lower metalimnion and be increasingly mixed throughout the epilimnion. Above the lower metalimnion, these algae would be exposed to environmental conditions that would not likely support their survival, as indicated by their general absence from these waters during the summer.

It appears that these deep assemblages may lead a fragile existence, particularly with respect to the low light conditions and their dependence on light for growth. Any loss of this major component of the phytoplankton community in terms of biomass and production may alter conditions elsewhere in the lake. However, little is known of this since we have not traditionally looked for either responses of these organisms to environmental conditions or ways their responses may influence other conditions in the lake. As a possible example of the latter, *Chapter V.D* discusses how algal biomass produced in the hypolimnion, compared with the epilimnion, may undergo less mineralization before settling to the sediment surface. Upon reaching the sediment surface, less of what is mineralized will be released to the water-column. Thus, if phytoplankton production were to partially shift to the epilimnetic waters, more nutrients would be cycled to these waters, perhaps further intensifying the eutrophication of these upper waters.

As mentioned elsewhere (*Chapters III.B.2 and 3, VIII and IX*), Mirror Lake may be increasingly affected by human activities in the watershed, including land clearing, home and road-building and sewage disposal. Changes in the structure and behavior of the phytoplankton, particularly below the epilimnion, may provide an early indicator of ecologic changes following such disturbances. Considering that the light penetrating below the epilimnion of an oligotrophic lake is already close to minimal for the requirements of the phytoplankton (Fee 1976; Fee et al. 1977; deNoyelles et al. 1980; Knoechel and deNoyelles 1980), any further light reduction will likely evoke changes in the phytoplankton community. Such reductions in amount of light would occur in connection with declining clarity of the epilimnetic waters. This decline might result from an increase in the input of suspended matter from increased erosion. Nutrient inputs may create the same condition by increasing algal biomass in the epilimnion, thus increasing the suspended matter. Lakes at the ELA that were artificially enriched with phosphorus in the epilimnion showed reduced phytoplankton production below the epilimnion, generally to less than 10% of the annual total, compared with untreated lakes where rates remained higher (Fee 1980). In these eutrophic lakes, the euphotic zone also became shallower (Fee 1980). Annual production in Mirror Lake has not been determined for the epilimnion vs. hypolimnion.

Stimulation of phytoplankton growth in the hypolimnion can also occur as a consequence of eutrophication. Such stimulation could occur from deep injection of nutrient sources such as from sewage outflow into the hypolimnion. This response was demonstrated in an ELA lake in which nutrients were added to the hypolimnion (Schindler and Fee 1974; Fee 1976). With the nutrients remaining more within the hypolimnion during the summer, phytoplankton growth in the epilimnion was little affected and the transparency of the epilimnion remained unchanged. However, production increased below the epilimnion and remained a significant portion of the total (Fee 1976; Schindler et al. 1980). The overall disturbance of the lake as a consequence of eutrophication was, however, much less when compared with that in other lakes receiving similar quantities of nutrients, but where they were added to the epilimnion (Schindler et al. 1980).

Actual eutrophication trends have not yet been verified by both chemical and biologic evidence in Mirror Lake, but there are indications of changes in algal community structure as discussed earlier. Some species once dominant in the late 1960s and early 1970s had become less abundant by 1975. Other species like the Chlorophycean *Gloeocystis planktonica* noticeably increased by 1975 (*Table V.A.2–5*). There has also been an increase in a filamentous Chlorophyceae, *Spirogyra* sp., occupying the inlets and surface of the littoral sediments. This increase may also indicate the beginning of eutrophication, but at the same time could be an early sign of the effects of another form of pollution, acid deposition. In the past few years, increasing concern has been focused on the effects of acid deposition on lakes, especially in this area of North America (Schofield 1976; *Chapter II*). An experimentally acidified oligotrophic lake in the ELA developed an extensive growth of a filamentous Chlorophycean, *Mougeotia* sp., over the littoral sediments (Schindler 1980b). *Spirogyra* sp. has also been abundant on occasion in acidified experimental enclosures in the same lake (Müller 1980). Species composition of the phytoplankton in this lake also shifted from the Chrysophyceae to the Chlorophyceae. The clarity of the epilimnion increased not from any reduction of phytoplankton biomass, since this did not occur, but perhaps because of an alteration of the color of the dissolved organic substances (Schindler 1980b). The increased light available to the sediments may be one factor enhancing the growth of *Mougeotia*. *Spirogyra* and *Mougeotia* both share close taxonomic affinity (Chlorophycean family Zygnemataceae). The ecologic implications suggested by the increase of *Spirogyra* in Mirror Lake remain to be clarified.

Summary

Past characteristics of the structure of the phytoplankton community in Mirror Lake have been established. The species present, their depth and seasonal distributions, and those dominating the phytoplankton by their biomass or frequency of appearance, are similar to conditions in other oligotrophic lakes of similar size and chemical characteristics. The Chrysophyceae, Bacillariophyceae, Diatomeae, Peridineae and Cryptophyceae dominate the phytoplankton in such lakes but could be replaced by Chlorophyceae or Cyanophyceae from the effects of eutrophication or acid deposition. The hypolimnetic peaks of algal biomass in Mirror Lake, as in many other oligotrophic lakes, are now recognized as common during summer stratification. They often develop from species growing in place rather than from settling. The peaks are composed of species that are rare or absent elsewhere in the water-column. This pattern too can change with different types of human disturbances as altered physical and chemical conditions favor the growth of new species within the surface or deep-water phytoplankton or over the littoral sediments. Changes in the algae of Mirror Lake may already be providing the first indicators of increasing human influence on the ecology of the lake.

3. Periphyton

GENE E. LIKENS, BRUCE J. PETERSON and JONATHAN J. COLE

The algae that grow on submerged benthic surfaces are referred to collectively as periphyton. Periphyton is typically found in the shallow regions of lakes, attached to surfaces such as rocks, mud or macrophytes.

The periphyton community is difficult to study quantitatively and has often been overlooked altogether. Nevertheless, attached algae in lakes may be both floristically diverse and abundant (Wetzel 1975). Probably more than 90% of all algal species grow in the attached habitat (Round 1964b) and periphyton frequently has high, sustained rates of productivity (see *Chapter V.B.4*). Quantitative measurements of abundance and biomass of periphyton are difficult for several reasons. First, it is tedious and technically difficult to separate the organisms from sediment particles. Second, the innumerable microhabitats that are available in the littoral zone often contribute to an extremely heterogeneous spatial distribution. Third, the littoral zone is characterized by a highly variable environment (e.g. with respect to light, temperature, wave action, erosion and deposition, settling zone for allochthonous materials, etc.) contributing to appreciable temporal and spatial heterogeneity.

Wetzel (1975) has summarized the elaborate and often confusing terminology associated with periphyton; the following terms are used here:

Epipelic algae—algae growing on fine, organic sediments;
Epiphytic algae—algae growing on the surfaces of macrophytes;
Epilithic algae—algae growing on rock surfaces;
Epipsammic algae—algae growing on or in sand.

Knowledge of periphyton and its role in the Mirror Lake ecosystem is quite limited. There have been no synoptic or quantitatively detailed studies of this biotic component in the lake. However, based on numerous observations it is apparent that benthic algae are present everywhere in Mirror Lake (*Table V.A.3*–1). Periphyton is common on the sediment surface,

Table V.A.3-1. Some Genera of Epipelic Algae found in Mirror Lake during 1979 to 1981[a]

Chlorophyceae
  Closterium[b]
  Colacium (epizoic on cyclopoids)
  Cosmarium
  Closterium
  Desmidium[b]
  Euastrum
  Gonium
  Micrasterias[b]
  Pediastrum
  Phacus
  Pleurotaenium[b]
  Spinocosmarium
  Spirogyra
  Staurastrum[b]
  Staurodesmus
Chrysophaceae
  Many species, including pennate, centric and filamentous species[b]
Bacteriophyceae
  Calothrix
  Merismopedia
  Nostoc[b]
  Oscillatoria[b]

[a] D. Strayer, personal communication.
[b] Abundant.

on rocks and on macrophytes in the littoral zone of Mirror Lake and is generally dominated by diatoms and blue-green algae. Scattered patches or mats of *Anabaena* sp. occur locally in wind-protected areas during the autumn where the water is 1 to 3 m deep, and more widespread mats of *Oscillatoria* sp., together with other genera of the Oscillatoriales, develop at depths greater than 5.5 m during the autumn or winter. These mats disintegrate during the spring (Moeller and Roskoski 1978). Desmids (e.g. *Desmidium* sp. and *Closterium* sp.) and diatoms also occur abundantly on the epipelic sediments between 6 to 8 m deep and commonly on deeper sediments of the profundal region. In addition, colonies of *Nostoc* sp. are common on cobbles and other inorganic surfaces, while colonies of *Gleotrichia* sp. are commonly attached to surfaces of the macrophytes, *Isoetes* and *Lobelia,* in Mirror Lake.

A noticeable increase in the abundance of epipelic *Spirogyra* sp. has developed in the inlet areas and on littoral sediments since 1969, and most dramatically during the past few years in Mirror Lake. Such biotic changes may be part of a general eutrophication trend in the lake (see *Chapters V.A.2* and *IX*).

It is relevant here to summarize Round's (1975a,b,c; 1960, 1961a,b) detailed studies of epipelic algae in the English Lake District. Round found that the main growth period of epipelic diatoms and blue-green algae was in the spring, essentially following the increasing values of light and temperature. In highly productive lakes benthic algae may be shaded by dense growths of phytoplankton or macrophytes as the season progresses. However, internal shading of light should not be a factor in the littoral zone of Mirror Lake because the water is clear and macrophyte development is low (*Chapter V.A.4*). Round further found that epipelic algae were absent below a depth of 8 m. Stockner and Armstrong (1971) observed that epilithic algae were negligible in abundance below a depth of 10 m in lakes of the Experimental Lakes Area (ELA), Ontario. We have not observed any cutoff depth for periphyton growth on the sediments in Mirror Lake. When the lake is clear, periphyton can grow and are found in the deepest parts of the lake. Mats of *Oscillatoria* have been observed as deep as 10.5 m in Mirror Lake, especially in late summer (D. Strayer, personal communication).

Oscillatoria appears to be a characteristic deep-water form in Mirror Lake. Mats 3 to 5 mm thick in places, are abundant on the deepest sediments of the lake from July through autumn overturn (corresponding with periods of hypolimnetic anoxia). The occurrence of these mats may affect sediment-water exchanges by taking up substances released from the sediments and by changing the mechanical properties of the sediment surface.

Diatoms usually dominated the epipelic communities in the English Lake District. Large diatom biomass corresponded to high organic matter content of the sediments, whereas moderate organic matter content favored development of blue-green algae. In Mirror Lake, cores of recent sediments from the center of the lake (*Chapter VII.C*; Sherman 1976) show that an overwhelming percentage of the diatom frustules accumulated were of planktonic origin, but most of the species were benthic forms. Many of these benthic diatoms may have been transported from the littoral region to the central portion of the lake by water currents, however.

Epiphytic populations obviously depend on the presence of macrophytes and on the seasonal cycle of surface area that these plants provide. Epiphytic algae occur in Mirror Lake but the macrophyte population is sparse (see *Chapter V.A.4*).

Some data on the biomass (as estimated from Chlorophyll-*a*) and the area colonized by epilithic algae in Mirror Lake are available (*Table V.B.4*–1). The mean of eight determinations of total chlorophyll on the rock surfaces in the cobble portion of the littoral zone during July and August was 40 mg Chl-a/m^2. This value is very similar to that obtained for epilithic algae in Rawson Lake (#239) in the ELA (Stockner and Armstrong 1971). A conversion of Chlorophyll-*a* to algal carbon was made assuming a carbon-to-chlorophyll weight ratio of 50:1 (Strickland 1960). On this basis then, we estimate a mean biomass of 2,000 mg C/m^2 in the rock littoral zone. Since this zone occupies only 19% of the total benthic area, the summer epilithic algal biomass is approximately 380 mg C/m^2 on a whole lake basis. Thus, the standing stock of epilithic algae is approximately equivalent to the standing stock of carbon in phytoplankton (see *Chapter V.C*). The biomass of epipelic and epiphytic algae is unknown.

Stockner and Armstrong (1971) found that epilithic algae in Rawson Lake contained about 2.6% N and 0.06% P of dry organic matter. Assuming that Mirror Lake is similar to Rawson Lake, then the epilithic algae in Mirror Lake would have standing stocks of about 20 mg N and 0.5 mg P/m^2.

It has been suggested that the occurrence and metabolic importance of attached, benthic blue-green algae in clear-water, oligotrophic lakes, such as Mirror Lake, may be greater than in eutrophic lakes (Round 1957c; Moore 1974; Kann 1940). However, much more detailed study is required to establish this relation, particularly in Mirror Lake.

Summary

Diatoms, desmids and blue-green algae are common on submerged surfaces in Mirror Lake. Epipelic blue-green algae frequently form widespread mats in wind-protected and profundal areas of the lake. The biomass of epilithic algae may be 380 mg C/m^2 (entire lake surface) in Mirror Lake.

4. Macrophytes

ROBERT E. MOELLER

Everywhere around Mirror Lake the shore is coarsely sandy or boulder strewn. Accumulations of terrestrial leaves locally form a soft, peaty debris among boulders, but silty sediments are nearly absent, occurring only in the mouths of the two western inlets. Small patches of floating-leaved waterlilies and *Sparganium* (bur reed) and the flowering stalks of submersed plants such as *Eriocaulon* (pipewort) and *Lobelia* (water lobelia) provide a summertime hint of the community of large benthic plants growing in the sandy lake bottom. This vegetation is mainly submergent, firmly rooted in the substrate, short in stature, and generally inconspicuous. Longer strands of *Utricularia* (bladderwort) rise toward the surface, but deeper meadows of *Potamogeton* (pondweed) and *Nitella* (stonewort) are invisible to observers above the surface of the lake.

The "aquatic macrophytes," as limnologists call them, are important to many aquatic ecosystems as primary producers. Of course this function is shared with the planktonic and the periphytonic algae in Mirror Lake. The large aquatic plants were surveyed with the aim of establishing their relative importance in comparison with the phytoplanktonic algae. In this ecologic treatment, all large plants growing in the lake are included among the "hydrophytes," as botanists traditionally refer to them, disregarding the preference of some specialists to restrict the term to plants that can reproduce sexually while submersed (e.g. Den Hartog and Segal 1964). In fact some of the common species in Mirror Lake do not flower and set fruit unless emersed, a rare development caused by a greater than usual drop of water level during some summers. Among the questions to be considered in this chapter are the taxonomic composition of the hydrophyte community, the absolute amounts of each species in the lake and the vegetational zonation according to water depth.

The transition from the successional forest covering most of the drainage basin to the lake itself is very abrupt. During particularly dry summers the lake surface falls 15 to 25 cm below the high-water level established by the dam at the outlet (*Chapter IV.E*). Shoreline boulders record the high-water level as a conspicuous algal band and the terrestrial soil as a narrow vertical band of *Sphagnum* with other mosses. The summer decline of water level exposes some of the submersed hydrophytes, which are mostly adapted to such seasonal emersion.

A closer examination of the terrestrial vegetation within 1 m of the lake margin has shown an ecotonal mixture of terrestrial and aquatic communities. Plants from five ecologic groups occur together near the shoreline:

1. The forest trees and shrubs. These root near the shore and frequently overhang the margin of the littoral zone.
2. The herbs of the forest floor. These occur in association with a forest canopy, often rooting within a few centimeters of the high-water level.
3. Shrubs and herbs that are specialized for wet, even water-logged soils. These are frequent at the lake margin, but rarely comprise more than an irregular fringe around the lake. Included are shrubs like speckled alder (*Alnus rugosa*), high-bush blueberry (*Vaccinium corymbosum*), and witch-hazel (*Hamamelis virginiana*). Other shrubs such as leather-leaf (*Chamaedaphne calyculata*) and sweet Gale (*Myrica gale*) are well known from acid bogs as well as from the shores of softwater lakes. St. John's wort (*Hypericum ellipticum*) is a conspicuous herb around much of the lake's margin, and the tiny insectivorous sundew (*Drosera rotundifolia*) is common in the *Sphagnum* band at the high-water mark.
4. Weeds of old fields and roadsides. These take advantage of light and, where human impact is greatest, the disturbed soils of the lake shore, but they are not common. Representatives are goldenrod (*Solidago graminifolia*), aster (*Aster novi-belgii*), meadowsweet (*Spirea latifolia*) and bracken (*Pteridium aquilinum*).
5. Terrestrial forms of several of the submersed angiosperms, notably *Gratiola aurea*, *Juncus pelocarpus* and *Eleocharis acicularis*. These are uncommon, but may be demographically significant as a sexually reproducing component of the predominantly submersed, and asexual, populations.

Conspicuously missing are any stands of emergent aquatic plants (such as large sedges, rushes, cattails, etc.). The stony, generally steep lake bottom probably inhibits their colonization. The absence of emergent vegetation, together with the only sparing representation of floating-leaved species, produces the sharp physiognomic discontinuity at the lake margin.

The 23 species of hydrophytes in Mirror Lake illustrate the varied evolutionary history of the freshwater flora (Sculthorpe 1967; Hutchinson 1975). The 14 angiosperms may all be distant-to-recent descendants of earlier terrestrial plants, and their specialization to the aquatic habitat varies greatly. Some species of *Potamogeton* flower underwater in such a way that sexual recombination among flowers and individuals is possible, but other plants flowering submersed (e.g. *Elatine minima* and the deeper individuals of *Lobelia dortmanna*) have closed flowers that would seem to preclude that possibility. Most of the angiosperms bear their flowers above the water's surface, like the familiar waterlilies. During most summers *Utricularia purpurea* produces no flowers, because few if any shoots reach the surface. On the other hand the two species of *Isoetes*, primitive cryptogamic vascular plants related to ferns, produce spores copiously underwater. The local flora includes three bryophytes (two mosses and one liverwort), three charophytes (large, highly specialized green algae) of the genus *Nitella*, which also produce many spores underwater, and one red alga of the genus *Batrachospermum* (which is actually blue-green in color).

The species occurring in Mirror Lake represent only a small fraction of the submersed and floating-leaved species of eastern North America. Many others are not adapted to the low concentrations of dissolved nutrients that prevail in the water and sandy sediments of Mirror Lake. Vegetational surveys of the lakes of Wisconsin (Fassett 1930; Swindale and Curtis 1957) and Minnesota (Moyle 1945) have established a floristic gradient that reflects the gradient of water chemistry extending from acidic lakes of low alkalinity to strongly alkaline or even saline lakes. Few hydrophytes tolerate the full range of conditions. Many of the softwater species from Wisconsin and Minnesota occur in Mirror Lake or two nearby softwater New Hampshire lakes (Table V.A.4–1), whereas the hardwater species are generally absent. This pattern is evident even within the single genus *Potamogeton* (Hellquist 1980), so that an experienced limnologist could tell from the list of species that these three New Hampshire lakes were softwater, oligotrophic (or perhaps dystrophic) lakes.

The same floristic pattern has been repeatedly and extensively described from northern Europe (Iversen 1929; Samuelsson 1934; Lohammer 1938; Maristo 1941; Olsen 1950; Spence 1964; Seddon 1972), where the climate, surficial geology, glacial history, and plant floristics create a close parallel to the conditions of northeastern North America. Only a few of the American species are known from Europe as well, but the presence of one of these, *Lobelia dortmanna*, highlights the broad analogy between the New Hampshire lakes and the clear "*Lobelia*-lakes" that figured prominently in Naumann's conception of aquatic "oligotrophy" (Blomgren and Naumann 1925; Thunmark 1931; Naumann 1932).

The Rosette Growth-Form in Softwater Lakes

Botanists long ago noticed that many of the common submersed plants of softwater lakes produced a rosette of short, stiff leaves (2 to 15 cm long) from a basal, usually perennial stem that was firmly rooted in the substrate. The rosette growth-form represents an evolutionary convergence in the morphology of plants that differ greatly in their taxonomic affinities, including dicotyledonous angiosperms (e.g. *Lobelia dortmanna* of the family Campanulaceae, *Subularia aquatica* of the Crucifereae), monocots (e.g. *Eleocharis acicularis* of the Cyperaceae, *Sagittaria graminea* of the Alismataceae) and vascular cryptogams of the Isoëtaceae (e.g. *Isoetes tuckermani*). To these true rosette species (or "isoetids" in the strictest sense) must be added a few other small, stiff-stemmed species that occur with them in shallow water, often forming a multispecific matlike vegetation. Among these latter plants are *Gratiola aurea* and *Elatine minima* in Mirror Lake, but the most interesting is an easily overlooked submersed form of *Utricularia cornuta*. In nearby Lower Baker Pond this tiny plant sends up short, virtually leafless photosynthetic stems from a delicately rooted rhizome that creeps through the sandy substrate. Only the upper centimeter of the 2- to 3-cm tall stems protrudes above the sand. Like its more familiar nonrooting relatives (e.g. *U. purpurea* in Mirror Lake), *U. cornuta* is carnivorous: tiny bladders attached to the lower part of the stems trap invertebrates moving among the sand grains.

One interesting adaptation of most rosette species is their toleration of seasonal emersion. Sometimes they can be found growing at wet sites (bog mats, outlet streams, etc.) where submersion occurs only infrequently. Associated with this "amphibious" character is a narrow depth range. Some species of *Isoetes* do grow at moderate depths—like the *I. tuckermani* at 5 m in Mirror Lake and *I. macrospora* as deep as 9 m in Lake George, New York (Sheldon and Boylen 1977); but most of the angiosperm isoetids are found shallower than 3 m, even in very clear water. As the water level falls during dry summers, the shallowest individuals in populations of *Gratiola aurea*, *Juncus pelocarpus*, *Eleocharis acicularis*, *Sagittaria graminea* and *Lobelia dortmanna* gradually emerge from Mirror Lake. Many plants do better than merely tolerate this emersion; they flourish, increasing rapidly in stature and biomass, and flowering abundantly (see also Potzger and Van Engel 1942). So perhaps it is really the toleration of submersion that is an ancillary adaptation of these amphibious plants. Perhaps their scarcity on the lake shore is due to an inability to resist prolonged desiccation, freezing during New Hampshire's severe winters, or competition from other terrestrial species. Submersed, their growth may be slow and stunted (as it clearly is in *Gratiola aurea*, at least), but the populations persist and proliferate vegetatively across the infertile sandy bottom to which few plants are adapted at all.

The submersed vegetation of hardwater lakes

Table V.A.4–1. Large Aquatic Plants of Three Softwater Lakes in Grafton County, New Hampshire[a]

Species	Mirror Lake	Russell Pond	Lower Baker Pond[b]
1. Submersed vascular plants of the mat community (*isoetid* growth-form):			
a. Rosette plants:			
Eleocharis acicularis L. (R. & S.)	+++	+	++
Eriocaulon septangulare With.	+++	+++	+++
Isoetes macrospora Dur.		+++	
Isoetes muricata Dur.[c]	++	+	
Isoetes tuckermani A. Br.	+++	++	+++
Juncus pelocarpus Mey.	+++	+++	+++
Lobelia dortmanna L.	+++	++	++
Sagittaria graminea Michx.	+++		++
Subularia aquatica L.		++	++
b. Plants with short, stiff stems:			
Elatine minima (Nutt.) Fish & Mey.	++		++
Gratiola aurea Muhl.	+++	+	
Myriophyllum tenellum Bigel.		+++	+
Utricularia cornuta Michx.		+	++
2. Submersed plants with long, flexuous stems or leaves (*elodeid* growth-form):			
a. Vascular plants:			
Eleocharis robbinsii Oakes			+
Najas flexilis (Willd.) Rostk. & Schmidt			++
Potamogeton berchtoldii Fieber[d]	+++		++
Potamogeton sp.		+	
Potamogeton spirillus Tuckerm.	+		++
Utricularia inflata Walt. var. *minor* Chapm.		++	
Utricularia minor L.		++	+
Utricularia purpurea Walt.	+++		+
b. Bryophytes:			
Chiloscyphus fragilis (Roth) Schiffn.	++	?	?
Drepanocladus fluitans (Hedw.) Warnst.	++	+++	
Fontinalis novae-angliae Sull.	++	++	++
c. Charophytes and Rhodophytes:			
Batrachospermum sp.	++		
Chara braunii Gm.			+
Nitella flexilis (L.) Ag.	+++	+++	
Nitella furcata (Roxb. ex Bruz.) Ag.			+++
Nitella gracilis (Sm.) Ag.	+		
Nitella tenuissima (Desv.) Kütz.	+++		+
3. Floating-leaved vascular plants (*nymphaeid* growth-form):			
Nuphar microphyllum (Pers.) Fern.			+
Nuphar variegatum Engelm.	++	+	+
Nymphaea odorata Ait.	++		+
Potamogeton epihydrus Raf. var. *ramosus*	++	+	++
Potamogeton natans L.			+
Sparganium angustifolium Michx.	++	+	
Sparganium fluctuans (Morong) Robins.			+

[a] Relative abundances are indicated subjectively for all lakes, according to the following scale: + = infrequent, nowhere abundant; ++ = frequent, or locally abundant; +++ = frequent and locally abundant. Nomenclature of vascular plants follows Fernald (1950), charophytes follow Wood (1967) and mosses follow Grout (1931).
[b] The flora listed from Lower Baker Pond does not include emergent and floating-leaved species in a marsh near the outlet.
[c] Synonymous with *I. braunii* Dur. and *I. echinospora* Dur. var. *braunii* (Dur.) Engelm.
[d] Synonymous with *P. pusillus* L. var. *tenuissimus* Mert. & Koch.

is dominated by flexuous-stemmed species (the "elodeids") like *Myriophyllum spicatum* and *Elodea canadensis* (which have weakly rooting stems of indeterminate growth, whose buds are borne above the substrate) or *Vallisneria americana* (which has long, flexuous leaves borne from a subterannean rhizome). Waters of intermediate hardness often contain a mixture of isoetid and elodeid growth-forms (Wilson 1935, 1939). Emergent plants and the floating-leaved plants (or "nymphaeids") are present in many soft- as well as hardwater lakes, especially at sites that are protected from the wind where fine sediment can accumulate. The waterlilies (e.g. *Nuphar variegatum* and *Nymphaea odorata*) seem to be relatively indifferent to water chemistry, although the other Mirror Lake nymphaeids (*Sparganium angustifolium* and *Potamogeton epihydrus*) are known principally from softwater habitats.

The predominance of the isoetid growth-form in the submersed vegetation of softwater lakes and the elodeid form in hardwater lakes is a striking ecologic difference. Notwithstanding the increasingly detailed subdivision of the major hydrophyte growth-forms into finer and finer categories (Hutchinson 1975, 118), there is little real understanding of the adaptive mechanisms behind these morphological differences. The observed distributions of plants among waters of differing alkalinity, conductivity, pH, or whatever cannot reveal what particular environmental factors actually influence the distribution of a particular species. Many chemical and biologic factors are at least broadly correlated with such composite chemical properties of the littoral sediments. Growth studies and tracer experiments with several rooted hydrophytes have demonstrated that the substrate is an important source of such plant nutrients as N and P (Bristow and Whitcombe 1971; Nichols and Keeney 1976; Carignan and Kalff 1979), especially when these elements are strongly depleted in the water surrounding the shoots, as is commonly the case because of uptake of nutrients by phytoplankton.

The below-ground portion of isoetids, which is mostly roots, comprises about 20 to 60% of the plant weight (*Table V.A.4*–2). The importance of roots is further accentuated by their surface area, which probably greatly exceeds that of the leaves in most species. Although many isoetids lack finely divided roots and some appear to lack fine root-hairs, they have been reported to harbor the mycorrhizal fungi that are responsible for much nutrient absorption among terrestrial plants (Søndergaard and Laegaard 1977). The well-developed root systems of isoetids may be a key adaptation to life in nutrient-poor waters, where they are often found rooted in sandy, infertile substrates. The roots absorb not only nutrients like N, P and K but also much of the CO_2 needed for photosynthesis (Wium-Andersen 1971; Sand-Jensen and Søndergaard 1978).

Seddon (1972) has speculated that the characteristic occurrence of isoetids in softwater lakes is a consequence of their competitive exclusion from hardwater sites. By this argument, only the isoetids—and a few specialized elodeids—absorb nutrients efficiently enough at low concentrations to persist, submersed, in infertile softwater lakes. Notable among the softwater elodeids are the individuals of the carnivorous species of *Utricularia*, which apparently cope with the poor availability of dissolved nutrients by utilizing nutrients concentrated in their prey. In alkaline waters of potentially higher fertility, elodeids may outcompete the diminutive isoetids, excluding them spatially from the littoral zone. A few isoetids do occur in some hardwater lakes (Spence 1964; Lang 1967; Seddon 1972), but are restricted to shallow, waveswept shoals and strands to which their robust, strongly rooted form and their tolerance of emersion are suited. *Ceratophyllum demersum,* an abundant elodeid in many hardwater lakes, requires relatively high concentrations of inorganic carbon for optimal photosynthesis. These concentrations exist in hardwater lakes, since *C. demersum* utilizes bicarbonate as well as dissolved free CO_2, but soft waters are suboptimal for this species (Steemann Nielsen 1944). Phosphorus, N, and possibly other nutrients are likely to play similar roles, perhaps simultaneously with CO_2, in keeping most elodeids from becoming established in softwater lakes.

This theory of competitive exclusion by faster growing hydrophytes cannot explain the impoverishment or absence of the isoetid flora at some sites where the macrovegetation of elodeids seems to be too sparse to constitute an effective competition. An example of this situation in New Hampshire is Partridge Lake,

Table V.A.4–2. Biomass Characteristics of Large Aquatic Plants in Mirror Lake[a]

Species	Biomass		
	In Lake (kg)	Below Ground (%)	Maximal (g dry wt/m^2)
1. Submersed vascular plants of the mat community (*isoetids*):			
a. Rosette plants			
Eleocharis acicularis	20	30	30
Eriocaulon septangulare	43	50	100
Isoetes muricata	1.9	50	—[b]
Isoetes tuckermani	27	35	20
Juncus pelocarpus	30	25	?[c]
Lobelia dortmanna	55	55	80
Sagittaria graminea	40	35	50
b. Plants with short, stiff stems:			
Elatine minima	3.2	35	—
Gratiola aurea	68	35	100
2. Submersed plants with long, flexuous stems or leaves (*elodeids*):			
a. Vascular plants:			
Potamogeton berchtoldii	2.9	10	60
Potamogeton spirillus	<0.5	15	—
Utricularia purpurea	49	0	20
b. Charophytes and Rhodophytes:			
Batrachospermum sp.	<0.5	0	—
Nitella flexilis	140	15	60
Nitella gracilis	<0.5	?	—
Nitella tenuissima	2.1	10	?
c. Bryophytes:			
Chiloscyphus fragilis	<0.5	0	—
Drepanocladus fluitans	2.3	0	?
Fontinalis novae-angliae	8.6	0	—
3. Floating-leaved vascular plants (*nymphaeids*):			
Nuphar variegatum	17	70	200
Nymphaea odorata	16	25	300
Potamogeton epihydrus	1.1	30	30
Sparganium angustifolium	0.8	15	20

[a] The total biomass in the lake is given as kg dry wt, the belowground biomass as percentage of the total. The belowground percentage is based on a few collections from sites in Mirror Lake where the species is relatively abundant; variations with depth, substrate, etc., have not been studied. The maximal biomass for each species refers to small quadrats of 0.05 or 0.5 m^2 where that species is strongly dominant.

[b] Dash (—) means that the species always occurs in Mirror Lake as scattered individuals or clumps of ≤0.05 m^2.

[c] A question mark (?) means that the data were not obtained.

where *Eriocaulon septangulare* occurs as scattered clonal patches in very shallow water (mostly less than 0.5 m). Aside from some *Isoetes muricata* reported by Collins and Likens (1969), the isoetids are otherwise absent, yet most of the littoral sediment lies unvegetated.

Most of the species characteristic of softwater lakes—whether isoetids, vascular elodeids, or mosses—cannot use the bicarbonate dissolved in lake water for photosynthesis (Gessner 1959; Raven 1970; Wium-Andersen 1971; Moeller 1978a). These plants may suffer carbon starvation in quiet waters if the combined productivity of phytoplankton and benthic plants is high enough to cause a diurnal rise in pH and a concomitant reduction in the concentration of free CO_2. This condition arises in productive hardwater lakes, and eutrophied softwater lakes (Schindler 1971). The availability of atmospheric CO_2 in wave-swept or emersed habitats within alkaline lakes might favor the local occurrence of isoetids. Since the concentration of free CO_2 in a lake's epilimnion is set by atmospheric and benthic exchange, and is little different in softwater as opposed to unproductive hardwater lakes, the unavailability of free CO_2 is probably not a general explanation for the rarity of isoetids in hardwater lakes as a whole.

Hydrophyte Remains in the Lake Sediments

Mirror Lake is geologically young, much younger than the species of hydrophytes that now colonize the lake bottom. Such plants could not have survived the latest glaciation in northern New England. They must have migrated into the region sometime during the late-glacial or postglacial period. Preserved in the lake mud is a fragmentary record of the hydrophyte community since late-glacial times. Few aquatic species leave much record among the microscopic pollen preserved in the sediment, except for the frequent microspores of *Isoetes*. The aquatic angiosperms flower rather sparingly, and the few pollen grains they release become virtually lost in the much greater density of terrestrial grains. A better record of the macrovegetation comes from the sporadic occurrence of angiosperm seeds, and the abundance of *Isoetes* megaspores and large oospores of charophytes in sediments near the margin of the deep-water zone of steady gyttja accumulation. A sediment core that was taken by Margaret B. Davis and coworkers at the northwestern end of Mirror Lake, in 6.2 m of water (core #783), has revealed a continuous history of the hydrophyte flora (see *Figure VII.F–4*).

The oldest lacustrine sediments containing remains of freshwater organisms are gray, sandy silts deposited more than 12,000 yr ago (*Chapter VII.A*), before boreal forests had colonized the deglaciated landscape. Hydrophytes are represented by many oospores of two charophytes. One of these, *Nitella flexilis,* has persisted to the present. In fact the core was taken beneath a sparse stand of *N. flexilis*. The other, a species of *Chara*, occurred continuously until 4,500 yr ago, then disappeared completely. The earliest vascular plants appeared about 11,000 to 12,000 yr ago, as the boreal forest became established and the sediments changed to an organic gyttja. *Isoetes tuckermani* and *Eleocharis acicularis* grew within the lake at that time. Both have persisted to the present. An unidentified *Potamogeton* was also in the lake. A peak in the abundance of *Isoetes* megaspores (about $2/cm^3$) 10,000 yr ago is contemporaneous with a peak of the microspores in the deep-water core (M. B. Davis, unpublished data). Megaspores deposited before 10,000 yr ago all lack the outer siliceous cover. Possibly these dissolved sometime after burial.

Several members of the modern vegetation first left fossil proof of their presence in sediments 7,500 to 8,000 yr old: *Isoetes muricata, Lobelia* cf. *dortmanna, Elatine minima, Potamogeton spirillus* and *Nuphar* cf. *variegatum*. The spores or seeds of these shallow-water species are uncommon and occur sporadically. We cannot be sure they were not present somewhat earlier, and that their presence has been continuous. *Najas flexilis* was present about 7,500 to 8,000 yr ago, too, but it has since disappeared from the lake. Changes in abundance of the various aquatic macrofossils within the core may reflect changes in plant densities near the site—perhaps in response to changing water depth or light attenuation—or changes in depositional patterns. The gyttja in the core from 7,500 to 8,000 yr ago is sandy and rich in fine wood fragments. In addition to the earliest records of several shallow-water species, it contains seeds from shoreline plants, like *Hypericum*. Perhaps shallow-water debris was being redeposited at the coring site.

The continuous presence of *Isoetes tuckermani* since almost 12,000 yr ago, and the absence of a distinctive hardwater flora in the early post-glacial period, suggest that Mirror Lake has had relatively soft water throughout most, if not all, of its history. To be sure, most *Chara* are hardwater species, and *Najas flexilis* is mostly found in lakes of moderate hardness (Swindale and Curtis 1957). Nonetheless, both *Najas flexilis* and *Chara braunii* (whose oospores closely resemble those from Mirror Lake) occur today with numerous rosette species in Lower Baker Pond (*Table V.A.4–1*). That lake has a slightly higher conductivity than Mirror Lake, but would certainly be judged oligotrophic on the basis of its hydrophytes. The long persistence of *Nitella flexilis* and *Isoetes tuckermani* argues for a long history as a relatively softwater lake, but the fossil diatoms provide stronger evidence for a more alkaline phase early in the lake's history (*Chapter VII.B and F*).

There are few published records of softwater species in lake sediments dating from the glacial periods. A pertinent study by Watts (1970) proves that several of these species occurred, together, hundreds or even a thousand kilometers south of their present ranges. About 20,000 yr ago two small ponds in northern Georgia contained such species as *Eriocaulon septangulare, Lobelia dortmanna, Elatine minima, Isoetes* cf. *macrospora, Juncus* cf. *pelocarpus* and *Potamogeton spirillus*—a striking floristic resemblance to the softwater lakes of New England and the western Great Lakes region today. Migrating waterfowl probably carried seeds and spores north into the newly deglaciated landscape. The reconstitution of the isoetid-dominated softwater vegetation might have had to wait, at some sites, for glacial tills to be leached of the carbonates and other compounds that cause the alkalinity of fresh waters (e.g. Vasari and Vasari 1968). It remains to be established whether the delay in appearance of some species in Mirror Lake was due to inhospitable chemical conditions within the lake, or to a delay imposed by the migration route of these plants from their glacial refuges.

Within the last 150 years the overall increase, and probable fluctuation, in lake level resulting from human activities (*Chapter III.B.3*; Likens 1972c) has presumably caused changes in the spatial distribution of the hydrophytes. Closer study of recent sediments might reveal these changes. The longer-term perspective provided by the analyzed core indicates that no major changes have occurred to the flora or to the type of vegetation.

The future of the macrovegetation in Mirror Lake is uncertain because of human activities at the lake shore, within the terrestrial drainage basin, and hundreds of kilometers away. The dumping of sand to create beaches, combined with the trampling from human feet, has reduced or eliminated the mat vegetation at several sites around the lake periphery. A further threat exists in the form of eutrophication, as more cottages and vacation homes are built within the drainage basin. Increased supplies of nutrients and total dissolved solids could change the ecologic composition of the lake enough so that species now absent could become established, and even be a nuisance. Rosette species apparently have disappeared from some European lakes subjected to cultural eutrophication. The very chemical conditions that favor the softwater species, the poverty of surrounding soils in leachable cations and carbonate, expose this vegetation to the additional threat of acidification (Likens and Bormann 1974b). *Lobelia dortmanna* and *Isoetes lacustris* apparently have declined in abundance in some Swedish lakes because of the acidification of rain and snow (Grahn et al. 1974). Because most softwater species cannot tolerate ever more pervasive human disturbance—that is, they are not "weeds" preadapted to thrive under human disturbance—they are likely to be eliminated from many parts of their present-day range before lake infilling or reglaciation accomplish the same end.

Quantitative Survey of Hydrophytes

Freshwater habitats may support a floristically rich vegetation, but most of these species are microscopic algae. Compared with the phytoplankton or the surrounding terrestrial vegetation, hydrophyte communities are strikingly impoverished (Curtis 1959). The submersed community in Mirror Lake (19 species) is relatively rich, as submersed freshwater communities go. The vegetational cover over the lake bottom is, however, very sparse.

During a systematic survey of the lake bottom in July and August, 1973–74, a diver noted the density of each species, according to a subjective eleven-point cover scale, in almost one thousand quadrats located throughout the littoral zone. The biomass of each species was estimated from an intercalibration between measured dry weight and units of the cover scale, which was based on special plant collections (Moeller 1978b, 1975). This indirect method

A: Species Composition, Distribution, Population, Biomass and Behavior   185

could be wildly inaccurate for a particular species in a particular quadrat. A test of the method in 39 quadrats where the mat vegetation was surveyed visually, then collected and weighed, indicated that the method was accurate overall, despite its imprecision. The problem confronting the survey was that of extreme heterogeneity, or small-scale patchiness, of the macrovegetation. Most of the quadrats (77%) contained plants, but the biomass ranged from 0.01 to 100 g dry wt/m^2, with a variance-to-mean ratio of 24:1. The indirect determination of biomass allowed a much more extensive collection of data than would have been possible if all plants had been dug up, washed, sorted to species, and weighed.

The mean biomass of hydrophytes over the entire littoral zone (0 to 7.2 m) was 6.1 g dry wt/m^2. This value is approximately equivalent to 5.4 g organic (ash-free) matter or 2.5 g C/m^2. Floating-leaved species make up only 7% of the biomass; the remainder is submersed. The deepest species, *Nitella flexilis,* is also the most abundant (26% of the total biomass). The elodeid growth-form in its broadest sense, including charophytes and mosses, contributes 39% of the total biomass, and is dominant below a depth of 2 m. Above this depth the isoetids form an irregular, only locally well-developed mat vegetation (54% of the total biomass).

The estimated 530 kg of dry plant mass in Mirror Lake (*Table V.A.4–2*) is not uniformly distributed throughout the littoral zone (*Figure V.A.4–1*). Biomass is greater in the region between 0- to 2-m depth, where most of the isoetids occur, than at greater depths (*Figure*

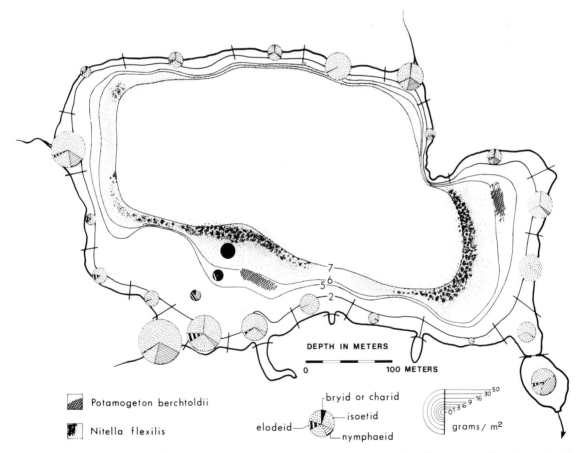

Figure V.A.4–1. Spatial pattern of hydrophyte biomass in Mirror Lake in mid-summer. The size of each circle indicates the mean biomass in each of 21 arbitrarily delimited subsections of the 0- to 2-m zone, and in the 2- to 5-m, 5-to 6-m, and 6- to 7.2-m zones for the lake as a whole. The relative contributions of different growth-forms are represented diagrammatically by shading patterns within each circle. Note that mosses ("bryid" growth-form) and charophytes ("charid" growth-form) are separated from the vascular elodeids.

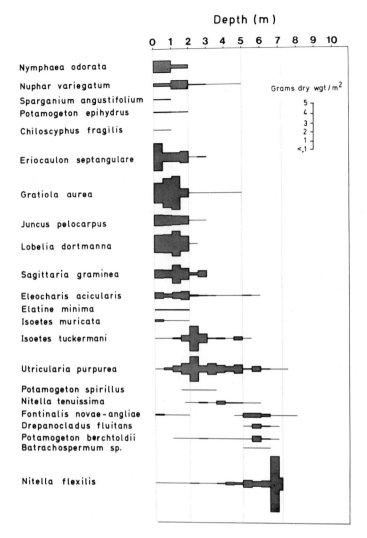

Figure V.A.4–2. Depth-distributions of 22 hydrophytes in Mirror Lake. The vegetational continuum has been divided somewhat arbitrarily into three zones: a 0- to 2-m zone dominated by isoetids, a 2- to 5-m zone characterized by *Utricularia purpurea* and *Isoetes tuckermani,* and a 5- to 7.2-m zone, lying in and below the mid-summer thermocline, dominated by *Nitella flexilis.* The 23rd species, *Nitella gracilis,* is infrequent at depths of about 2 to 5 m, but was not consistently distinguished from other species of *Nitella* during the survey.

V.A.4–2) and is greatest near the two western inlets (about 50 and 30 g dry wt/m² in the West and Northwest inlets, respectively). The favorability of inlets for the development of aquatic vegetation has been repeatedly observed and discussed (Pearsall 1920, 1921; Wilson 1935, 1939; Spence 1964, 1967). An obvious explanation involves greater access to inflowing nutrients. The western inlets to Mirror Lake are sedimentary environments, although most of the fine material settling there is subject to eventual resuspension and transport to adjacent areas of the littoral zone, or to the deeper sediments. Newly sedimented material is likely to be richer in adsorbed nutrients than the coarse, well-leached substrates that prevail elsewhere in the littoral zone. Neither the chemistry nor the fertility *per se* of littoral substrates has been investigated in Mirror Lake, but the greater vegetational development near the inlets must be a clue to the factors that dictate the sparse vegetation elsewhere. This clue points to nutrient availability. It is possible that factors such as the physical nature of the substrate or protection of these sites from wave activity also may be involved. The mouths of the western inlets are somewhat unstable habitats. A major flood during late June 1973 deposited enough sand and particulate organic debris near the western inlets to smother some of the rosette vegetation.

Juday (1942) called attention to the greater vegetational biomass in the productive, hardwater lakes of southern Wisconsin than in the northern softwater lakes. Compared with Green Lake and Lake Mendota, Wisconsin, with biomass exceeding 150 g organic matter/m², the isoetids of Weber Lake, Wisconsin, formed a vegetation of only 8.8 g/m² (*Table V.A.4–3*). Mirror Lake, with 5.4 g organic matter/m² fits this pattern, but elsewhere much higher densities occur in some softwater lakes. The dense vegetation of isoetids in Lake Kalgaard, Denmark is 20 times the density in Mirror Lake. Thus isoetids are not necessarily to be associated with a sparse vegetation. Submersed aquatic macrovegetation reaches its greatest abundance in some eutrophic, hardwater lakes (200 to 300 g organic matter/m²), but it is interesting to note that intermediate levels of about 100 g/m² can be found in oligotrophic softwater lakes, eutrophic hardwater lakes, and relatively unproductive marl lakes ("alkalitrophic" lakes).

The variability of vegetational abundance in softwater lakes (*Table V.A.4–3*) is a limnologically important factor that determines the relative importance of hydrophytes in the metabolism of these aquatic ecosystems. In clear, softwater lakes the littoral zone tends to comprise a large proportion of the lake area—at least in many small lakes—owing to the extensive range of depths at which the bottom receives enough light for plant growth. Where conditions are especially favorable (for reasons that have not yet been rigorously explained), hydrophytes may contribute the major portion of the lake's primary production (Søndergaard

Table V.A.4–3. Biomass of Littoral Vegetation in Lakes of Different Trophic Status[a]

Lake	Littoral Depth (m)	Biomass and Dominant Growth-Form (g organic matter/m²)		Reference
Oligotrophic Lakes				
Trout	6.5	0.06	elodeids	Wilson (1941)
Mirror	7.2	5.4	isoetids & elodeids	Moeller (1978b)
Weber[b]	>4.5	8.8	isoetids	Potzger and Van Engel (1942)
Port-Bielh[c]	17	22	elodeid (*Nitella*)	Capblancq (1973)
Latnjajaure[c]	35	40	elodeid (moss)	Bodin and Nauwerck (1968)
Kalgaard	4.5	107	isoetids	Sand-Jensen and Søndergaard (1979)
Øvre Heimdalsvatn[c]	7	115	isoetid & moss	Brettum (1971)
Dystrophic Lakes				
Suomunjärvi	2	30	emergent	Toivonen and Lappalainen (1980)
Vitalampa	3	63	nymphaeid & isoetids	Eriksson (1974)
Punnus-yarvi	2	150	emergent	Andronikova et al. (1972)
Pääjärvi	2	275	emergent & isoetids	Kansanen et al. (1974)
Eutrophic Lakes				
Batorin[d]	3.5	46	emergent & submersed	Winberg et al. (1972)
Myastro[d]	6.0	128	emergent & submersed	Winberg et al. (1972)
Green	8	151	elodeids	Rickett (1924)
Mendota	5	172	elodeids	Rickett (1922)
Wingra	2.7	220	elodeids	Adams and McCracken (1974)
Dgał Maly[e]	2.5	266	emergent & elodeids	Gerlaczyńska (1973)
Naroch[d]	9.5	314	elodeid (*Nitella*)	Winberg et al. (1972)
Alkalitrophic Lakes				
Lawrence	7	140	elodeids	Rich et al. (1971)

[a] Values refer to the mean biomass compiled over the entire littoral zone, to the indicated depth. These are lakes of northern Europe and northeastern North America.
[b] Does not include the deeper moss zone.
[c] Arctic or alpine lakes of granitic terrane.
[d] Assuming 5 kcal = 1 g organic matter.
[e] Aboveground biomass only.

and Sand-Jensen 1978). The dramatic differences in vegetational abundance among softwater lakes might be related to chemical properties of the substrate, perhaps even to subsurface supplies of CO_2 or other plant nutrients reaching hydrophyte roots by means of water seeping into the lake.

Depending on the morphology of a lake basin, emergent or floating-leaved species sometimes constitute a major part of the hydrophyte vegetation, regardless of the transparency, fertility, or hardness of the lake water. The humic-stained softwater lakes cited in *Table V.A.4–3* (Naumann's dystrophic lakes) have a rather sparse submersed vegetation of isoetids, whose role in lake metabolism is reduced further by the narrow range of depths that receive enough light to support them. However, the hydrophyte production in many dystrophic lakes is bolstered by moderate biomass of emergent plants (Pääjärvi and Suomunjärvi, Finland) or floating-leaved plants (Vitalampa, Sweden), so that the hydrophytes as a whole play a more important role than they do in Mirror Lake.

Marl lakes like Lawrence Lake, Michigan (Naumann's alkalitrophic lakes), represent an extreme case among hardwater lakes. In these lakes, the production and biomass of phytoplankton are suppressed owing to the precipitation of limiting nutrients, notably P, with calcium carbonate (Otsuki and Wetzel 1972; Wetzel 1975, 645). Owing to suppression of the phytoplankton, alkalitrophic lakes are relatively clear, and the littoral zone is consequently extensive. Within the littoral zone of Lawrence Lake, hydrophytes are abundant. Their relatively high biomass may reflect nutrients stored in the sediments. Nutrients lost to the phytoplankton may be available, at least in part, to hydrophytes rooted in littoral marl.

To place Mirror Lake in broader limnological perspective, it is important to keep in mind the striking differences of hydrophyte colonization among lakes with similarly unproductive or "oligotrophic" phytoplankton, as exemplified by lakes Mirror, Kalgaard, and Lawrence. Mirror Lake lies near the poorly vegetated end of the spectrum, although some small lakes of the Canadian Shield appear to be even more sparsely vegetated (Schindler et al. 1973).

Vegetational Zonation According to Depth

It is not unusual for the submersed macrovegetation of small soft- and hardwater lakes to be dominated by only one to two really abundant species (e.g. Rich et al. 1971; Adams and McCracken 1974; Sand-Jensen and Søndergaard 1979), but such is not the case in Mirror Lake (*Table V.A.4–1*). The shallow-water isoetid community is a mosaic of nine species, seven of which are of roughly equal importance (20 to 68 kg). As depth increases within the littoral zone of many lakes, the number of hydrophytes present also decreases until finally only one to two species grow as monospecific stands (Spence 1964, 1967). This pattern is well illustrated in Mirror Lake (*Figure V.A.4–2*). The relatively diverse isoetid community of the upper littoral zone (0 to 2 m) gives way to an intermediate zone dominated by *Isoetes tuckermani* and *Utricularia purpurea* (2 to 5 m), which in turn gives way to a deep-water zone (5 to 7.2 m) dominated by *Nitella flexilis*. The deepest species occur within or just below the mid-summer thermocline (*Chapter IV.B*). *Nitella flexilis* and *Potamogeton berchtoldii* form mutually exclusive stands on the gyttja that are strikingly parallel to the depth contours (*Figure V.A.4–1*). Other cryptogams occur in these cool, dimly illuminated waters. *Fontinalis novae-angliae* and *Batrachospermum* are restricted to cobbles and boulders, whereas *Drepanocladus fluitans* grows on the gyttja surface.

The vegetational zonation that emerges from the species distributions of *Figure V.A.4–2* represents the response of individual species to a microclimatic gradient of temperature and light (*Figure V.A.4–3*). The upper littoral is a relatively distinct zone, with its multiplicity of rosette plants and a scattering of floating-leaved species. Most of the rosette plants reach a depth-limit somewhere between 2 and 3 m. The deeper zones are merely arbitrary groupings of species. Plants in the deepest zone are rarely exposed to temperatures exceeding 20°C. In 1975 they received no more than 5 to 12% of the surface illuminance during the summer. Unlike some elodeids of hardwater lakes whose rapidly growing stems carry the young shoots several meters upward into the water-column during the summer, and thus toward more favorable

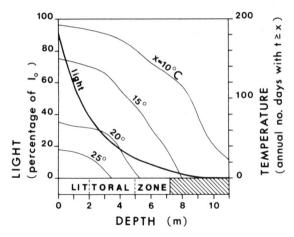

Figure V.A.4–3. The depth-gradient as a microclimatic gradient of light and temperature. The light curve represents the attenuation of downwelling radiation, computed by assuming an equal energy intensity at all wavelengths (400 to 700 nm) above the lake surface, a 10% reflection of all wavelengths at the lake surface, and depth-attenuation as measured with a photometer equipped with colored absorption filters (May-September data, 1975). The temperature gradient is depicted as a family of curves. Each curve presents, as a function of depth, the annual number of days (in 1975) on which temperature exceeded the indicated temperature.

illuminance (e.g. *Myriophyllum spicatum:* Adams et al. 1974), the tallest species in Mirror Lake, *Potamogeton berchtoldii*, extends barely 60 cm above the sediment at its maximal development. Water temperature in the upper littoral zone exceeds 20°C for at least 60 days of the year, and light is 40 to 90% of surface intensity—quite a different physical environment from that prevailing at a depth of 6 to 7 m.

The abundant light in the upper littoral zone does not mean that plant distributions there—most notably their deep limits—are uninfluenced by light. At least one of the shallow-water isoetids, *Lobelia dortmanna*, responds to experimental shading with a reduction in growth. Reducing the illuminance by 48% during the 1976 growing season caused a 30% decline in leaf production by plants at a depth of 1.6 m (Moeller 1978c). It is not surprising that this species occurs no deeper than 3 m, even in very clear lakes. More surprising is the bimodal distribution of *Fontinalis novae-angliae*. It grows as clumps adhering to rocks both in the thermocline and at the lake's edge, especially where shaded by overhanging trees. Somewhat perplexing are the distributions of *Nitella flexilis* and *Potamogeton berchtoldii*. In addition to their sharply delimited stands in deeper water, scattered individuals of both species can be found growing well in as little as 1 m of water (*Figure V.A.4–2*).

Potamogeton berchtoldii is well known to tolerate the low illuminance and cool waters in the lower epilimnion of temperate lakes (Pearsall 1920; Wilson 1935, 1941; Hutchinson 1975). Its depth limit is probably set by light requirements, or a combination of light and temperature requirements. The depth limit of *Nitella flexilis* almost certainly depends on light, since this species can grow deep in the hypolimnion of clear temperate lakes (as in Russell Pond). This species reaches its limit at about 6 to 8% of surface light intensity (green-yellow region of the spectrum) in northern New Hampshire, corresponding to depths of 5.6, 7.2 and 16 m in Partridge Lake, Mirror Lake and Russell Pond, respectively (Moeller 1978b). But what could account for the equally sharp, shallow limit of the *N. flexilis* zone and of the two stands of *P. berchtoldii*? There are no other hydrophytes present to invoke as competitors, and no obvious change in the sediment.

Vascular plants in general do not grow below the thermocline, even in exceptionally clear lakes (Hutchinson 1975). Observations of the seasonal growth patterns in *Utricularia purpurea* at depths of 2, 4 and 6 m in Mirror Lake have demonstrated how light attenuation and thermal stratification cause a reduction of growth in the lower epilimnion (Moeller 1980). Below a depth of 3 m, the growing season begins only as the epilimnion accretes downward (*Chapter IV.B*). Thus the onset of the growing season is retarded past the time of maximal solar irradiance in late spring and early summer. Light is not limiting at 2 or 4 m during the growing season, but may be limiting during the short growing season at 6 m. There is little experimental or, for that matter, observational data to resolve the relative importance of light, temperature and pressure in truncating the depth-distribution of vascular plants (cf. Hutchinson 1975). Their importance probably varies from one lake to another, depending on the species present, the clarity of the water and the depth of the thermocline.

Patchiness of the Mat Vegetation

The submersed vegetation of the upper and intermediate littoral zone (0–5 m) might be considered a "pioneering" plant community, because it leaves much of the largely inorganic, asedimentary and perhaps somewhat unstable substrates uncolonized. Single individuals of each species, and small clumps of individuals, are widely dispersed across the upper littoral zone, but dense vegetation on a scale of 0.05 m² or greater is very infrequent. The densest stands, or low mats in the case of most rosette species, reach barely 100 g dry wt/m² (*Figure V.A.4-4*). These local patches occur throughout the littoral zone, although there is usually no obvious reason why similar substrate within a spatial extent of at most a few meters should support such heterogeneous vegetation. The densest patches contribute heavily toward the mean biomass of the littoral zone. An accurate assessment of their frequency requires an intensive sampling, which has been attained in Mirror Lake at the expense of the precision of plant weights in individual quadrats.

One biologic factor that encourages the formation of local patches is vegetative reproduction. Most species can proliferate as asexual clones and form dense patches as new plants develop from lateral offsets of established stems. Some species rely almost exclusively on vegetative reproduction (*Eleocharis acicularis, Gratiola aurea, Juncus pelocarpus, Utricularia purpurea*), but rare sexual reproduction, for instance by emersed plants, may provide embryos capable of being dispersed both within Mirror Lake and across the landscape. Other species reproduce sexually on a routine basis, although vegetative proliferation is mainly responsible for producing such locally dense patches (*Eriocaulon septangulare, Lobelia dortmanna, Nitella flexilis, Nuphar variegatum, Nymphaea odorata, Potamogeton berchtoldii, P. epihydrus, Sagittaria graminea* and *Sparganium angustifolium*). A few species are, however, exclusively sexual. These include the annual plants (*Elatine minima, Nitella tenuissima* and *Potamogeton spirillus*) and some perennials (*Isoetes muricata* and *I. tuckermani*). *Isoetes tuckermani* and *Nitella tenuissima* form locally dense clumps or meadows. These species grow below the zone of strongest wave action. Thus their spores tend to sink near the parent plants, where they germinate later.

Along with vegetative reproduction comes a strong degree of dominance within individual quadrat-sized plots (0.05 to 0.5 m²). Excepting stands of *Nitella flexilis* below 6.5 m, at depths where other macrophytes do not venture, plots usually are not monospecific. In fact, species richness increases with increasing plot biomass up to the maximal biomass attained in the lake (*Figure V.A.4-4*). One explanation for this pattern is that the establishment of one species may improve the chance of another species invading and surviving. Severe competition must ultimately regulate the overall density within developing patches and determine which species—if not the first to become established—predominates. The unknown factors that retard the coalescence of patches into a uniformly dense vegetation also may prevent

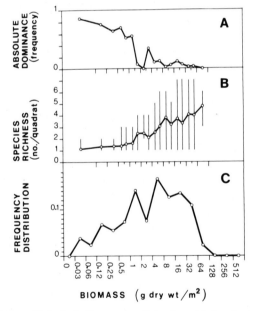

Figure V.A.4-4. Synopsis of the isoetid-dominated vegetation in the 0- to 2-m zone. **A.** The proportion of vegetated quadrats having only one species; **B.** The mean number of species per quadrat (with range); and **C.** The frequency distribution of biomass among vegetated quadrats, all plotted against a logarithmic scale of increasing biomass. Note that 87% of the 694 quadrats, each 0.05 m² in area, were vegetated.

the competitive elimination of some species from the lake, or at least prevent a lakewide dominance of only two to three species. In contrast to Mirror Lake, such dominance and lower species richness is evident in two small Danish lakes where isoetid species colonize the sandy littoral zone much more thickly (Nygaard 1958; Sand-Jensen and Søndergaard 1979).

Why are the hydrophytes so sparse in Mirror Lake? Attempts to answer that question through further research should take into account two complementary aspects of the problem: (1) the relatively low biomass attained even in the densest patches of most species, and (2) the rarity of dense patches.

In dense patches, intraspecific competition likely regulates the total biomass at levels sustained by major plant growth factors. Rosette plants in other lakes may reach a biomass of 250 g dry wt/m^2 (Brettum 1971; Sand-Jensen and Søndergaard 1979), but in Mirror Lake only two isoetid species reach 100 g/m^2 (Table V.A.4–2). Primary nutrient elements such as N, P, and inorganic CO_2 must be suspected as limiting elements. Light is abundant in the upper part of the littoral zone, though self-shading may play a role in limiting the biomass of the deeper elodeid stands (*Potamogeton berchtoldii* and *Nitella flexilis*) to their low maximum of 60 g dry wt/m^2.

The patchiness of the vegetation, on the other hand, results from a balance among such demographic factors as reproduction, dispersal, growth rates and mortality. The "maxima" attained by some species in Mirror Lake are so low that one wonders if densities limited by nutrient resources have even been reached. Many factors in addition to primary plant nutrients could be involved. For example, in the *Lobelia dortmanna* stand on the south shore of Mirror Lake, tiny seedlings are abundant in shallower water, though their sparsity nearer the depth-limit of the population suggests that decreased reproduction may contribute to the dramatic thinning of the stand at about 1.8 m (Moeller 1978c). There was no obvious change in the density of *L. dortmanna* from 1973 to 1980 at that site, so a substantial mortality must have balanced the evident recruitment of young plants into the population. The only mortality factor that has been identified specifically is an aquatic caterpillar that feeds on leaves of this plant, killing some of them (Fiance and Moeller 1977). Such herbivory may not even be the major cause of mortality within the population, but is no doubt significant. A fertilization experiment in 1978 and 1979 revealed that increased supplies of N, P and K to the roots of adult plants significantly increased the rate at which new leaves appeared. But would continued fertilization over a period of several years bring about a denser vegetation? Only long-term experiments can provide the same type of data that can be obtained with algal enrichments in a week or two.

This survey represents a description of a possibly dynamic vegetation at only one point in time. The present patchiness of the macrovegetation encourages its depiction as a pioneering community. We do not know, however, if there are changes taking place that would be detectable on a scale of decades or longer. There is probably no true vegetational succession taking place. The lake's shores are in (or very near) an asedimentary equilibrium; infilling at the lake margins is negligible. The present-day vegetation has undergone major spatial shifts during the last 150 years owing to the probably irregular increase in lake level. The vegetation may not be in equilibrium with its habitat, yet it is even less certain that there is any inherent trend of development toward the nearly complete, relatively dense colonization found in some softwater oligotrophic lakes. In Lake Fiolen, Sweden, Thunmark (1931) noted sporadic holes in the nearly uniform benthic vegetation, but in Mirror Lake the scattered patches of well-developed vegetation are the curiosities.

Summary

Mirror Lake's hydrophytes establish the lake as a North American analog of the clear *Lobelia*-lakes in northern Europe. The total biomass of large aquatic plants in midsummer has been estimated as 530 kg dry wt (470 kg organic matter, or 215 kg C). The vegetation includes:

1. Four floating-leaved species, which make up 7% of the biomass;
2. Ten submersed species of "elodeid" growth-form, equal to 39% of the biomass;
3. Nine submersed species of "isoetid" growth-form, which contribute 54% of the biomass.

The vegetation extends, rather sparsely, across 58% of the lake area, from depths of 0 to 7.2 m. Isoetids predominate in the 0- to 2-m zone, where the vegetation is most dense (up to 100 g/m^2 for isoetids, up to 300 g/m^2 for floating-leaved species, with an overall mean of 11 g dry wt/m^2). Biomass declines in the 2- to 5-m zone (1 to 3 g/m^2), but rises somewhat in the 5- to 7.2-m zone (3.5 g/m^2) because of stands of *Nitella flexilis* between 6.5 and 7.2 m that locally reach 60 g/m^2. The extreme patchiness of the vegetation suggests that demographic factors must be investigated for clues to the low vegetational density in this softwater oligotrophic lake.

Seeds and spores preserved in sediment demonstrate that the most abundant plant today, *Nitella flexilis* (26% of the total biomass), has existed in the lake since late-glacial times. *Isoetes tuckermani* arrived early in the postglacial, and several of the rosette species, including *Lobelia dortmanna*, arrived by 7,500 yr ago.

5. Zooplankton

Joseph C. Makarewicz

As the primary consumers of algae in a lake-ecosystem, zooplankton play an integral role in transferring energy to other consumers (such as fish and salamanders), in affecting phytoplankton diversity and succession, and in nutrient cycling; in short, in the metabolism and biogeochemistry of a lake. An intensive study was undertaken from October 1968 to September 1972 to investigate these aspects of the limnetic zooplankton community of Mirror Lake. The zooplankton of Mirror Lake is characterized by a diverse community of Rotatoria, Cladocera and Copepoda (*Table V.A.5–1*) similar to the zooplankton communities of many nutrient-poor lakes of the North Temperate Zone, such as those of the Experimental Lakes Area (ELA) (Schindler 1972; Schindler and Noven 1971; Patalas 1971). On the basis of productivity, the cladocerans are the dominant zooplankton group, while on a biomass basis the copepods are the dominant group within the limnetic community of Mirror Lake.

Rotatoria

Classified as a Class of the Ashchelminthes or as a separate phylum, the Rotatoria (*Figure V.A.5–1*) are pseudocoelomate animals generally possessing an elongated body shape 100 to 500 μm in length with a head, trunk and foot region. A thin and flexible cuticle covers the body surface of rotifers (e.g. *Polyarthra*), but in some species (e.g. *Keratella*) the body surface is thick and rigid and is termed a *lorica*. The ciliated corona at the anterior end serves for locomotion and for gathering food into the mouth. Below the mouth lies the *mastax,* a structure unique to rotifers. The mastax consists of muscles and a set of jaws or trophi that function in seizing, tearing, grinding or macerating food. Most rotifers are herbivores, but some, such as *Asplanchna* in Mirror Lake, are predators of other animals such as rotifers and protozoa.

Rotifer reproduction cycles between 1 to 2 sexual (mictic) generations and 20 to 40 asexual (amictic) generations per year (*Figure V.A.5–2*). In the amictic generations, reproduction is parthenogenetic with the amictic females and eggs being diploid in chromosome number. Crowding of amictic females, pheromones produced by females, changes in water temperature (Wetzel 1975) and ingestion of algae containing vitamin E (Gilbert 1968; Gilbert and Thompson 1968) will induce sexuality in rotifers with the production of mictic females. The mictic female is indistinguishable morphologically from the amictic female. However, the mictic egg produced by the mictic female is haploid in chromosome number. If a mictic female is not fertilized, the diploid fertilized or resting egg will become thick walled, an adaptation to adverse environmental conditions. Copulation, in most rotifers, is by hypodermic impregnation. Sperm are deposited directly into the pseudocoel by the male penis, which can penetrate any part of the female body wall (Barnes 1963). Hatching of the resting egg is triggered by

Table V.A.5–1. Species of Zooplankton in Mirror Lake[a]

Limnetic	
Cladocera[b]	
Herbivores	
Holopedium gibberum	
Daphnia catawba	
Daphnia ambigua[*c]	
*Scapholeberis kingi**	
Bosmina longirostris	
Copepods	
Herbivores	Omnivores/Predators
Diaptomus minutus	*Cyclops scutifer*
*Epischura lacustris**	*Mesocyclops edax*
	*Macrocyclops albidas**
	Tropocyclops prasinus
	*Cyclops vernalis**
Rotatoria	
Herbivores	Omnivores/Predators
Epilimnion	*Asplanchna priodonta*
Keratella taurocephala	
Keratella crassa	
Metalimnion	
Conochilus unicornis	
Keratella cochlearis	
Polyarthra vulgaris	
Kellicottia longispina	
Hypolimnion	Miscellaneous
Kellicottia bostoniensis	*Keratella quadrata**
Conochiloides dossuarius	*Ascomorpha* sp.*
Littoral	
Cladocera	
Herbivores	Predators
Sida crystallina	*Polyphemus pediculus*
Latona setifera	
Alonella excisa	
Ophryoxus gracilis	
Eurycercus lamellatus?	

[a] Historically, both *Bosmina longirostris* and *Eubosmina tubicen* have been observed in Mirror Lake sediments (*Chapter VII.D*). The current Bosminidae assemblage discussed in *Chapter V.A.5* was classified according to Edmondson's (1959) system, which did not distinguish between *B. longirostris* and *E. tubicen* (Deevey and Deevey 1971).
[b] Nomenclature follows Brooks (1957), Ahlstrom (1943) and Edmondson (1959).
[c] Organisms marked with an asterisk (*) are extremely rare in the limnetic zone.

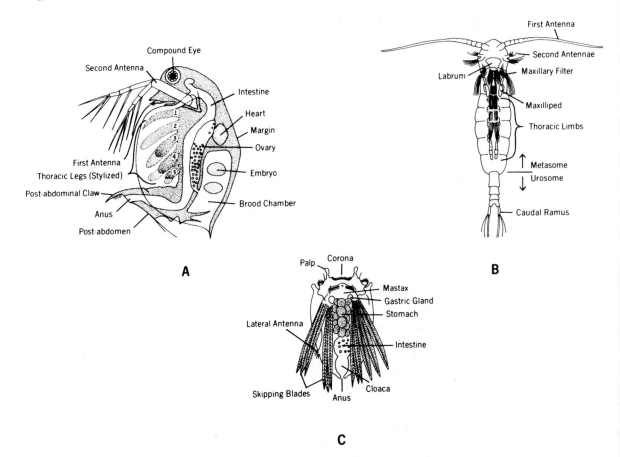

Figure V.A.5–1. Anatomy of a cladoceran (**A**), a copepod (**B**), and a rotifer (**C**).

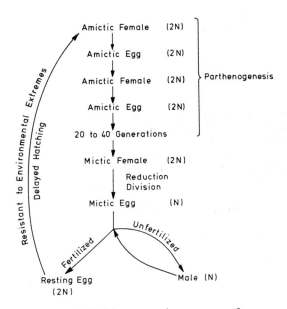

Figure V.A.5–2. Diagrammatic sequence of generations in a typical ploimate rotifer.

changes in temperature, osmotic pressure, water chemistry and oxygen content (Pennak 1978). The resting egg will develop into an amictic female and the cycle repeats itself.

Based on studies in the laboratory, the life span of limnetic rotifers in Mirror Lake varies from 7.5 to 14.1 days at a temperature of 16.8°C. However, food quality and quantity and temperature are known to affect the generation time of rotifers. Increasing food levels will decrease the generation time (*Table V.A.5–2*) and may increase the brood size. Also, changes in temperature generally change rate processes in poikilotherms such as the Rotatoria. Thus the annual fluctuation in lake temperature (0–28°C, see *Chapter IV.B*) will affect the physiological life span, generation time, rate of molting and rate of brood production.

During the 3-yr zooplankton study, rotifer biomass of Mirror Lake was 0.32 kg C/ha-yr or 16.5% of the total zooplankton biomass. Four

Table V.A.5–2. Generation Time (in days) for Rotifers at Varying Food Levels, Culture Temperature 16.8°C[a]

		Development Time (days)		
Species	Food Level	Egg	Post-Embryonic	Mean Generation Time ± S.E.
Polyarthra vulgaris	Low	3.5	8.5	12.0 ± 0.71
	High	2.7	6.0	8.7 ± 0.88
Keratella cochlearis	Low	2.1	8.3	10.4 ± 0.20
	High	2.2	7.0	9.2 ± 0.17
Keratella taurocephala	Low	2.4	10.1	12.5 ± 0.41
	High	2.0	8.0	10.0 ± 0.29
Kellicottia longispina	Low	2.8	12.3	15.1
	High	2.5	8.3	10.8 ± 0.57

[a] Modified from "Structure and function of the zooplankton community of Mirror Lake, New Hampshire," by J. C. Makarewicz and G. E. Likens, *Ecological Monographs*, 1979, 49(1):109–127. Copyright © 1979 by The Ecological Society of America. Used with permission. See Makarewicz and Likens (1979) for more information on food levels and culturing techniques.

rotifers—*Polyarthra vulgaris, Keratella cochlearis, Keratella crassa* and *Keratella taurocephala*—accounted for 70% of the rotifer biomass. *P. vulgaris*, a monacmic (i.e. a species exhibiting one maximum per year) species, reached a maximum density in the metalimnion or hypolimnion during the spring and comprised 22.1% of the rotifer biomass. In some years, a population pulse of this rotifer was evident in the autumn.

K. taurocephala characteristically inhabits somewhat acid waters such as Mirror Lake (pH 5.5 to 6.8). Although this species accounts for 11.7% of the rotifer biomass annually, its seasonal distribution is quite variable in Mirror Lake. Generally, there is a mid-summer population peak that may extend into the autumn. In some years, a population maximum occurred in the late spring.

K. crassa, an epilimnetic species with a population maximum in June–July and then again in the autumn, is an example of a diacmic species.

K. cochlearis comprised 26.0% of the rotifer biomass and in some years was the dominant rotifer. A major population pulse occurs in the 3- to 6-m region during the summer and throughout the water-column during most winters.

A number of other rotifers, *Kellicottia bostoniensis, K. longispina, Asplanchna priodonta, Conochilus unicornis* and *Conochiloides dossuarius* are found in Mirror Lake. Together, comprising only 26.0% of the rotifer biomass, they are apparently of relatively little metabolic importance in the lake. However, these organisms are a reservoir of species that may become more prevalent in the lake with changing chemical and biologic conditions. For example, *K. bostoniensis* appears to be increasing in abundance within the Mirror Lake ecosystem (*Figure V.A.5–3*). This increase may be related to "improvements" in the Mirror Lake watershed, such as the construction of a major highway, of an access road partially around the lake, and of an increase in use of summer homes and subsequent nutrient additions from septic field drainage.

Seasonally, *C. unicornis* and *A. priodonta* generally have a spring peak and *K. bostoniensis, K. longispina* and *C. dossuarius* a summer peak with an accompanying autumn or winter pulse. The vertical distribution of the two cold stenotherms, *K. bostoniensis* and *C. dossuarius*, is characterized by their restriction to the hypolimnion (5.5 to 8.5°C) during the summer. However, during the autumnal turnover when the lake temperature is between 4 and

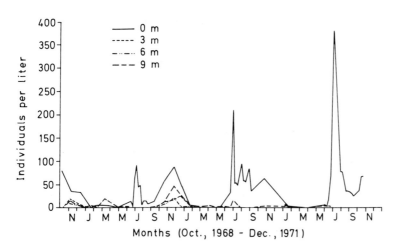

Figure V.A.5–3. Seasonal distribution of *Kellicottia bostoniensis*.

10.5°C, these two species are distributed throughout the water-column.

Cladocera

The Cladocera are an Order of the phylum Arthropoda, commonly known as water fleas (*Figure V.A.5–1*), and range in size from 0.2 to 4.0 mm (Dumont et al. 1975). Most Cladocera are herbivorous, feeding on algae, bacteria and organic detritus, but some, such as *Polyphemus* in Mirror Lake, are predators feeding on protozoa, rotifers and crustaceans. Limnetic species are usually light-colored or translucent (Pennak 1978).

A single compound eye and often a small eyespot, or *ocellus*, are located on the head (*Figure V.A.5–1*). Locomotion is provided by strokes of the second antennae, a large biramous structure. Movement in Cladocera, such as *Daphnia*, is characterized by a hopping motion. However, *Holopedium gibberum*, a species found in Mirror Lake, typically swims upside down. A posterior extension of the abdomen termed the post-abdomen or *abreptor* is important in taxonomic identifications of this group. Eggs develop in a brood chamber along the dorsal part of the body.

For a greater part of the year, development of eggs into a new individual occurs without fertilization, that is by parthenogenesis as in the rotifers. After a variable number of parthenogenetic generations, some of the eggs develop into males. The production of males is induced by unfavorable physical and biotic conditions, as in the rotifers. If unfavorable conditions continue, sexual eggs appear. Upon fertilization, the eggs move into the brood chamber, called an *ephippium*, whose walls become thickened and darkened to withstand severe environmental conditions.

Development of the individual is affected by quantity and quality of food and by temperature. For example, the effect of increasing food supply on *B. longirostris* and *H. gibberum* in Mirror Lake increased the brood size and the length of the organism (*Figure V.A.5–4*), while generation time decreased, and the life span appeared not to change significantly. As in the Rotatoria and Copepoda, increasing temperature will decrease metabolic processes such as generation time.

Annual biomass of Cladocera in Mirror Lake was 0.53 kg C/ha or 26.5% of the total zooplankton biomass. Two species, *Daphnia catawba* and *H. gibberum*, are the dominant Cladocera in the lake. *D. catawba*, comprising 73.5% of the Cladocera biomass, reaches a population peak during the late spring and summer. In July the population declines until autumn when a second peak occurs (*Figure V.A.5–5*). An overwintering population is not evident in Mirror Lake. *H. gibberum*, having the second largest Cladocera biomass, generally has population peaks during the spring in the upper 6 m of the lake, with an occasional second pulse in the late summer.

The studies of Mirror Lake zooplankton have concentrated on the limnetic region with little attention given to the littoral or inshore zones. However, we do know that a different assemblage of zooplankton occurs in the littoral re-

A: Species Composition, Distribution, Population, Biomass and Behavior

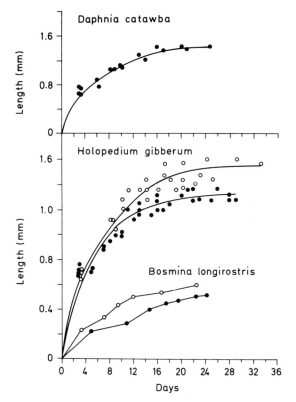

Figure V.A.5–4. Growth curve of Cladocera at higher (open circles) and lower (closed circles) food concentrations.

gion of the lake. For example, *Polyphemus pediculus* is rarely found in the limnoplankton but occurs in increasing numbers toward the shore (*Figure V.A.5–6*). Also, the presence of *Sida crystallina* and *Latona setifera* in the littoral zone of Mirror Lake is indicated by their high density in the stomachs of *Nothopthalmus viridescens,* an aquatic newt living in the littoral region of Mirror Lake (see *Chapter V.A.8*).

Copepoda

An elongated cylindrical body, 0.3 to 3.2 mm in length, is common to the Calanoida and Cyclopoida, suborders of the Class Crustacea (*Figure V.A.5–1*). The Calanoida feed by propelling water past the body by moving four pairs of feeding appendages (Koehl and Strickler 1981). The second maxillae captures parcels of water containing food particles (phytoplankton and organic debris), which are pushed into the mouth by the first maxillae. The major genera of the Cyclopoida are mainly carnivores with mouth parts modified for seizing and biting. The five pairs of swimming legs and the first antennae function in locomotion. Reproduction is sexual except for one European species (*Attheyella*)

Holopedium gibberum
  High:
  Chlorophyll-*a* = 4.8 parts per billion (ppb)
  Mean life span (MLS) = 21.8 days
  Mean brood size (MBS) = 3.93
  n = 5
  Temperature (T, °C) = 18.0 ± 0.7

  Low:
  Chlorophyll-*a* = 1.7 ppb
  MLS = 24.8 days
  MBS = 1.80
  n = 5
  T = 18.8 ± 1.2

Bosmina longirostris
  High:
  Chlorophyll-*a* = 4.3 ppb
  MLS = 23.2 days
  MBS = 2.0
  n = 2

  Low:
  Chlorophyll-*a* = 1.53 ppb
  MLS = 24.2 days
  MBS = 1.25
  T = 18.0 ± 0.7

Daphnia catawba

  Low:
  Chlorophyll-*a* = 1.7 ppb
  MLS = 17.0 days
  MBS = 2.83
  T = 18.3 ± 1.2
  n = 4

The line fitted to *H. gibberum* and *D. catawba* is a significant fit ($p < 0.05$). (Modified from "Structure and function of the Zooplankton community of Mirror Lake, New Hampshire," by J. C. Makarewicz and G. E. Likens, *Ecological Monographs*, 1979, 49(1):109–127. Copyright © 1979 by the Ecological Society of America. Used with permission.)

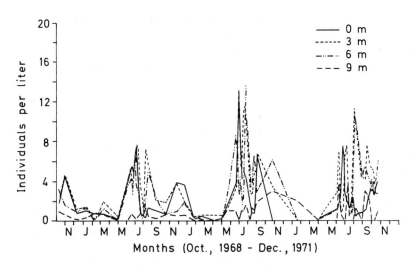

Figure V.A.5–5. Seasonal distribution of *Daphnia catawba*.

that reproduces by parthenogenesis (Pennak 1978). The structure of the first antennae, urosome and fifth leg is used to distinguish these two copepod groups.

The free-swimming larva or nauplius developed from a copepod egg has three small appendages. Before the adult stage is reached, a series of molts takes place. The nauplius goes through six molts, getting larger and adding appendages with each molt. The next molt results in the first of the five copepodite stages. The final molt results in an adult animal capable of

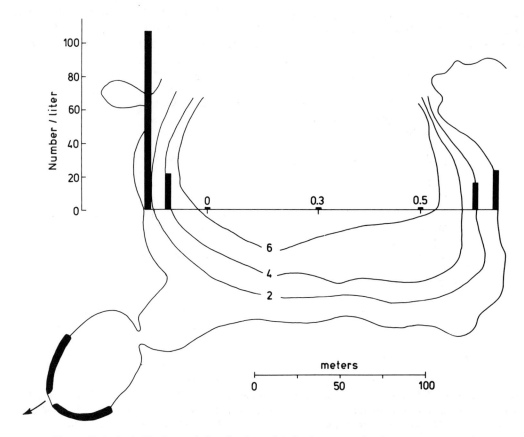

Figure V.A.5–6. Horizontal distribution of *Polyphemus pediculus* in Mirror Lake.

reproduction. Life span is variable in the group in Mirror Lake, ranging from three months in *Tropocyclops* to one year in *Diaptomus* and *Mesocyclops*.

Accounting for 57.0% of the total zooplankton biomass in Mirror Lake, Copepoda biomass (1.14 kg C/ha-yr) is greater than the combined biomass of the Rotatoria and Cladocera. Three species, *Diaptomus minutus*, *Cyclops scutifer* and *Mesocyclops edax*, are the important limnetic copepods. *D. minutus*, a small calanoid copepod, is the dominant copepod comprising 48.8% of the copepod biomass. Possessing a single generation per year with egg production in the winter, this species is found in the water-column throughout the year.

M. edax, a cyclopoid copepod with one generation per year, undergoes diapause (i.e. a period of suspended development or growth) in the sediments during winter. Adults of *C. scutifer* occur during the spring while adults of *M. edax* occur in the summer. Similarly, when the nauplii of *C. scutifer* are present in the water-column, the nauplii of *M. edax* are in diapause in the sediments (*Figure V.A.5–7*). Since they have similar feeding requirements, this alternation of life history reduces competition between these two ecologically similar species.

Sampling Methods

Quantitative rather than qualitative data are required to evaluate the role of organisms in a lake-ecosystem. Zooplankton are commonly sampled with devices such as the Clarke-Bumpus and Isaac-Kidd samplers, the ½-m vertical tow net and the Likens-Gilbert filter. With the Likens-Gilbert filtering system (Likens and Gilbert 1970), a point sample is taken with a water bottle and filtered through a filter in the boat. Closing the water bottle immediately is imperative if a quantitative sample of the larger, more motile zooplankton is desired (Smyly 1968).

The towed metered devices (i.e. Clarke-Bumpus, Isaac-Kidd and ½-m vertical tow; see Wetzel and Likens 1979 for pictures) provide large-volume samples over a relatively large horizontal or vertical vector. However, as the net is towed through the water, the meshes of the net clog with animals and debris often mak-

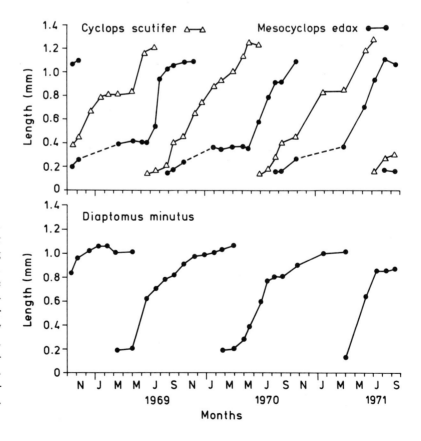

Figure V.A.5–7. Growth curves of *Mesocyclops edax* and *Cyclops scutifer*, showing an alternation of life history. (Modified from "Structure and function of the zooplankton community of Mirror Lake, New Hampshire," by J. C. Makarewicz and G. E. Likens, *Ecological Monographs*, 1979, 49(1):109–127. Copyright © 1979 by the Ecological Society of America. Used with permission.)

ing the volume measurements unreliable. McNaught (1971) recommends not using a net finer than 363 μm (#2 net) in productive waters because clogging is more pronounced with finer meshes, which may cause inaccurate measurement of the volume filtered (Tranter and Smith 1968). Although the towed metered devices provide a more representative sample of the lake because of the large volume of water sampled, they may not provide a quantitative sample, particularly for small organisms such as rotifers. This problem was demonstrated in the following study in Mirror Lake.

Within a period of 90 minutes, three tows or hauls were made with a Clarke-Bumpus and Isaac-Kidd sampler, a ½-m vertical tow net and a water bottle. All species of zooplankton collected in the towed devices were found in the water-bottle sample (*Table V.A.5–3*). In all cases except one, the water bottle caught more of the larger, more motile Cladocera and Copepoda than the other devices. More *D. catawba* were caught by the Clarke-Bumpus apparatus than the water bottle; however, the means were not significantly different (p = 0.05). The number of rotifers and nauplii missed by the towed devices equipped with a #10 mesh net is particularly striking. In our study of the zooplankton of Mirror Lake, we, therefore, elected to use the Likens-Gilbert technique, for a quantitative sample at a given point in the lake, even though the sample may not have been representative of a larger area.

Sampling and Counting Errors

The total error involved with taking a sample, counting it and expanding it to the remainder of the lake can be divided into three components: counting error, site sampling error and lake sampling error. By counting one sample by both the Sedgewick-Rafter and settling technique (see Wetzel and Likens 1979 for a description of these techniques), an estimate of the counting error is obtained. With Mirror Lake samples (*Table V.A.5–4*), counting errors for both the inverted microscope method and the Sedgewick-Rafter approach are generally low, with the settling technique providing a more precise count. At least three samples taken at the same station will provide an estimate of the site sampling error (*Table V.A.5–5a*). Because we were consistently able to catch all the rotifers and probably most of the crustaceans and because

Table V.A.5–3. Comparison of the Catching Efficiencies of a Clarke-Bumpus Sampler (#10 net, 158 μ), Isaac-Kidd Sampler (#10 net), ½-m Vertical Tow (#10 net) and a Likens-Gilbert Sampler (35-μ net)[a]

	Mean ± S.E./m³				
	Likens-Gilbert	Isaac-Kidd	Clarke-Bumpus	½-m Vertical Tow	Likens-Gilbert
				(for the entire water-column)	
Bosmina longirostris	505 ± 84	81 ± 34	231 ± 34	378 ± 27	682 ± 84
Holopedium gibberum	1,052 ± 126	72 ± 72	872 ± 92	599 ± 95	1,043 ± 115
Daphnia catawba	5,598 ± 295	1,910 ± 230	8,243 ± 1,008	2,465 ± 628	5,416 ± 390
Cyclops scutifer	2,315 ± 547	562 ± 463	1,750 ± 390	775 ± 144	2,858 ± 339
Diaptomus minutus	6,397 ± 673	1,828 ± 93	3,416 ± 1,377	4,588 ± 181	6,906 ± 421
Copepod nauplii	20,059 ± 1,667	87 ± 87	1,015 ± 34	4,598 ± 1,180	44,792 ± 483
Polyarthra vulgaris	62,062 ± 4,024	0.0	0.0	116 ± 21	76,850 ± 2,811
Keratella cochlearis	9,705 ± 572	0.0	88 ± 88	26 ± 19	35,608 ± 807
Kellicottia longispina	11,599 ± 1,035	1,336 ± 873	8,029 ± 509	2,676 ± 457	12,183 ± 620
Asplanchna priodonta	1,052 ± 126	0.0	0.0	249 ± 12	1,139 ± 137
Kellicottia bostoniensis	0.0	0.0	0.0	2,615 ± 425	11,106 ± 295

[a] From "Structure and function of the zooplankton community of Mirror Lake, New Hampshire," by J. C. Makarewicz and G. E. Likens, *Ecological Monographs*, 1979, 49(1):109–127. Copyright © 1979 by The Ecological Society of America. Used with permission.

Table V.A.5–4. Counting Error Associated with the Settling Technique and Sedgewick-Rafter Cell for Zooplankton in Mirror Lake[a]

	Mean	S.E. of the Mean
Settling Technique (n = 3)		
Keratella taurocephala	17.0	0.00
Keratella cochlearis	304.7	1.33
Keratella crassa	95.0	1.53
Polyarthra vulgaris	259.0	6.66
Kellicottia longispina	148.0	1.00
Copepod nauplii	130.7	0.67
Daphnia catawba	24.7	0.33
Diaptomus minutus	66.7	0.67
Cyclops scutifer	12.0	0.00
Sedgewick-Rafter Cell (n = 5)		
Keratella cochlearis	55.2	4.0
	15.2	0.9
	15.6	3.3
Kellicottia longispina	2.4	0.7
Polyarthra vulgaris	90.4	6.0
	18.6	2.8
Conochiloides dossuarius	19.6	5.7
	44.2	4.6
Keratella taurocephala	5.6	0.9
	12.6	1.4
Kellicottia bostoniensis	12.6	1.8

[a] From "Structure and function of the zooplankton community of Mirror Lake, New Hampshire," by J. C. Makarewicz and G. E. Likens, *Ecological Monographs*, 1979, 49(1):109–127. Copyright © 1979 by The Ecological Society of America. Used with permission.

the site sampling error was generally <10%, we believe that our samples from Mirror Lake are quantitative at one station (Central) in the lake. However, because of the contagious dispersion of many organisms (*Table V.A.5–6*) and the small number (one) of stations sampled in the seasonal studies, the error statement for the total number of organisms in the lake is large (*Table V.A.5–5b*).

Vertical Migration

Daily cycles of plankton activity are prominent in most lakes. Fluctuation in ambient light is often or always one of the controlling factors for zooplankton migrations toward the surface during the night and downward in the water-column during the day (i.e. nocturnal migration). Although light is the stimulus for the migration, the adaptive significance of this behavior is not fully understood. Migration into a zone of greater food availability, the metabolic and demographic advantages of residing in cooler water, and/or lower predation may provide a selective advantage by enhancing the ability of the population to survive.

Vertical migrations as large as 60 m have been observed for the cladoceran *Daphnia longispina* and 46 m for the copepod *Cyclops strenuus* with descent and ascent rate as fast as 3.1 min/m and 4.3 min/m (Allee et al. 1949). With their smaller body size, the amplitude of migration in rotifers is smaller but can be surprisingly large. For example, George and Fernando (1970) observed a daily amplitude in vertical migration of 5.6 m in *Polyarthra vulgaris* in Sunfish Lake, Ontario.

K. longispina provides an example of a nocturnal migration pattern in Mirror Lake (*Figure V.A.5–8*). At 1600 h the depth of the average individual (Worthington 1931) had moved up in the water-column from 1.9 to 3.9 m. With sunrise, the population began to move downward in the lake.

When maximum numbers of plankton occur at dawn and dusk in surface layers of the lake, it is known as a twilight migration. In Mirror Lake *K. crassa* displays this type of migration pattern (*Figure V.A.5–8*). As the sun sets, there is an upward movement of the population through the metalimnion. During the night, the population drops back into deeper water. With increasing light intensity, *K. crassa* once again moves upward in the water-column and maintains this position until noon. By 1600 h the population is moving downward.

A migratory movement downward during the night and upward to the surface during the morning by an organism is termed a *reverse migration* (Hutchinson 1967). Reverse migrations are thought to be rare (Cunningham 1972; Hutchinson 1967), yet in Mirror Lake four species of rotifers—*K. cochlearis, C. unicornis, P. vulgaris* (*Figure V.A.5–7*) and *C. dossuarius*—displayed this type of behavior.

The phenomenon of reverse migration in rotifers might be due to convection currents. For example, some convection currents do occur when the surface waters are cooled during the night, become more dense and sink, possibly

Table V.A.5-5. Site Sampling (a) and Lake Sampling (b) Error for Selected Zooplankton in Mirror Lake[a,b]

(a) Site Sampling Error (n = 5)		
Species	Mean	S.E. of the Mean
Keratella taurocephala	91.8	10.9
Keratella cochlearis	85.2	8.3
Polyarthra vulgaris	447.2	16.3
Copepod nauplii	88.4	4.2
Daphnia catawba	31.0	4.4
Mesocyclops edax	14.8	1.1
Diaptomus minutus	6.8	0.4

(b) Lake Sampling Error—Mean for eight selected stations scattered throughout the lake to one station (Central) in the lake

Species	Strata	Number of Stations (n)	Mean ± 95% Confidence Level	Mean for Central ± 95% Confidence Level (n = 3)
Polyarthra vulgaris	0	8	49.9 ± 54.3	29.0 ± 28.4
Polyarthra vulgaris	3	7	314.8 ± 317.8	213.3 ± 46.1
Polyarthra vulgaris	6	3	78.7 ± 697.1	88.7 ± 12.4
Kellicottia longispina	0	8	1.0 ± 1.0	1.3 ± 1.4
Kellicottia taurocephala	0	8	34.2 ± 11.4	43.5 ± 16.3
Cyclopoidae	0	8	5.5 ± 2.9	3.5 ± 2.2
Diaptomus minutus	0	8	1.9 ± 1.7	2.7 ± 6.6
Daphnia catawba	3	7	3.0 ± 2.0	7.0 ± 7.4

[a] From "Structure and function of the zooplankton community of Mirror Lake, New Hampshire," by J. C. Makarewicz and G. E. Likens, *Ecological Monographs*, 1979, 49(1):109–127. Copyright © 1979 by the Ecological Society of America. Used with permission.
[b] All samples were taken with a Likens-Gilbert filter.

carrying organisms downward. Yet in Mirror Lake reverse migration is apparent in organisms found in the hypolimnion, an area normally isolated from strong currents. In the metalimnion, the population of *P. vulgaris* displayed a reverse migration while in the epilimnion a nocturnal migration was evident (Makarewicz 1974). If a downward-moving convectional current were causing the reverse migration, the top quarter of the population should have been carried down. Also, reverse migrations are found in strong-swimming organisms, such as copepods and cladocerans (Cunningham 1972; Thienemann 1919; Bayly 1963), which should be able to counteract the force of a weak convectional current by swimming.

The gradients of temperature and oxygen present during the summer in Mirror Lake may have affected the amplitude of migration (Russell 1927), but it is difficult to conceive how they would trigger a positive phototaxis in some rotifers and not in others. Small changes in pH and cation concentrations with depth exist in Mirror Lake, but do not vary greatly over a 24-h cycle. Predator avoidance might be a factor affecting vertical migration (Zaret and Suffern 1976). Not much information is available on when predatory zooplankton are actively feeding. Many freshwater fish actively feed at dusk and dawn. Do predatory zooplankton feed at the same time, causing the twilight migrations of rotifers? Bayly's (1963) pH hypothesis is not applicable,

Table V.A.5-6. Horizontal Distribution of Zooplankton[a,b]

Species	Depth	x̄	s^2	χ^2	Dispersion[c]
Polyarthra vulgaris	0	59.4	1,014	119.6	C
	3	362.0	32,168	533.0	C
	6	93.0	3,845	84.0	C
Kellicottia longispina	0	1.52	1.59	2.09	R
	3	7.45	14.77	6.00	R
	6	10.3	11.5	4.47	R
Conochilus unicornis	0	49.6	67.9	95.7	C
	3	25.6	186.6	51.02	C
	6	2.37	0.72	1.53	R
Keratella taurocephala	0	36.0	144.8	28.2	C
	3	24.3	266.0	65.7	C
	6	2.31	0.49	0.85	R
Keratella cochlearis	0	2.00	2.7	9.5	R
	3	8.51	62.0	43.7	C
	6	125.3	369.1	11.8	C
Copepod nauplii	0	26.2	210.3	562	C
	3	58.3	144.2	14.8	C
	6	42.4	1,162.0	27.0	C
Cyclopoidae	0	5.52	11.8	14.9	R
	3	23.1	27.6	7.2	R
	6	16.9	41.0	9.7	R
Diaptomus minutus	0	2.9	14.5	34.0	C
	3	4.7	4.8	6.1	R
	6	26.1	735.0	112.5	C
Daphnia catawba	0	1.2	2.8	19.0	C
	3	3.0	4.8	9.6	R
	6	3.5	1.8	2.0	R

[a] From "Structure and function of the zooplankton community of Mirror Lake, New Hampshire," by J. C. Makarewicz and G. E. Likens, *Ecological Monographs*, 1979, 49(1):109–127. Copyright © 1979 by The Ecological Society of America. Used with permission.
[b] See Elliot (1971) for methods.
[c] C = contagion; R = random.

as the pH of Mirror Lake is generally between 5.5 and 6.8, much lower than the 8.2 believed by Bayly to cause reverse migrations. Rotifers in Mirror Lake, however, might be sensitive to small changes at different parts of the pH scale. When the daily light cycle involves variation over low light intensities, such as in the winter under the ice, Cunningham's (1972) hypothesis that reverse migrations in some species may be determined by a positive phototaxis may be true. However, the hypothesis is not universal and, in essence, is refuted by the reverse migrations of rotifers in Mirror Lake under a daily light cycle with large variations in light intensities. At present, the causes of a reverse migration are not understood from field studies. Experimental manipulations of these organisms in controlled environments should be done.

Summary

At least 29 species of zooplankton occur in Mirror Lake. Twenty-six of these species were found in the limnetic zone of the lake.

The average annual zooplankton biomass during 1968–1971 was about 2 kg C/ha. Rotifers represented about 16%, cladocerans about 26% and copepods about 57% of the total zooplankton biomass. *P. vulgaris, K. cochlearis, K. crassa* and *K. taurocephala* comprised 70% of the rotifer biomass. *D. catawba* and *H. gib-*

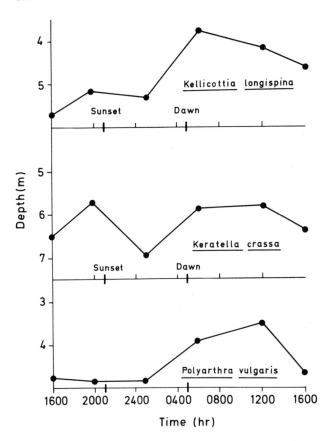

Figure V.A.5–8. Vertical migration of *Polyarthra vulgaris, Keratella crassa* and *Kellicottia longispina*. Values are reported as the depth of the average individual.

berum are the dominant cladocerans, with *D. catawba* accounting for about 73% of the cladoceran biomass. *D. minutus, C. scutifer* and *M. edax* are dominant copepods, with *D. minutus* comprising about 49% of the total copepod biomass in Mirror Lake.

Vertical migration, including reverse migration, of zooplankton was observed in Mirror Lake, although the magnitude of movement was smaller than observed in many other lakes.

Many of the zooplankton had a contagious horizontal distribution in the limnetic zone, which contributes significantly to lake sampling error. A Likens-Gilbert sampler was found to have an equal or better catching efficiency than other devices tested.

6. Benthic Macroinvertebrates

RHODA A. WALTER

The benthic environment of a lake can be divided into three regions: the upper littoral, the lower littoral (often called the sublittoral) and the profundal. The upper littoral region roughly subtends the epilimnion. Temperatures, light and dissolved oxygen concentrations are relatively high; the sediments are heterogeneous and aquatic macrophytes are often abundant. The lower littoral region encompasses the metalimnion and is characterized by steep gradients in light, temperature, dissolved oxygen and sediment type. The profundal region corresponds roughly to the hypolimnion and is devoid of vascular macrophytes; it is characterized by lower temperatures, less light and moderate to low levels of dissolved oxygen. The substrate is often a soft, semi-gelatinous detrital mud, called *gyttja*. Many invertebrate species are restricted to one of these three regions of a lake, where environmental conditions are most suitable for their existence. Because of the more homogeneous habitat and harsher environmental conditions existing in the profundal region of most lakes, the species diversity is often less than in the littoral regions.

In Mirror Lake, three main taxa of macroinvertebrates—chaoborids, chironomids and oligochaetes—dominate the profundal region (7.2 to 11 m), and few other groups are even represented. The littoral regions—upper littoral (0 to 3 m) and lower littoral (3 to 7.2 m)—are dominated by chironomids, amphipods, oligochaetes and sphaeriids; but at least a dozen other groups are represented. The benthic invertebrates of Mirror Lake include representatives from most of the common invertebrate phyla (*Table V.A.6–1*). For many groups, especially the Oligochaeta and Chironomidae, the lists of species are far from complete.

Biology of the Benthic Invertebrates in Mirror Lake

Chaoboridae

The larva of the phantom midge, *Chaoborus* (*Figure V.A.6–1A*), is a common inhabitant of the profundal region of lakes. *Chaoborus* undergoes complete metamorphosis; the larval and pupal stages are aquatic, and the short-lived adult is terrestrial. In Mirror Lake, the emergence period for *C. punctipennis* from the profundal region is long, lasting from mid-June through mid-August. In other lakes, emergence may take place as a swarm over short periods.

Chaoborus spends most of its life in four larval instars. The first two instars are positively phototactic and remain in the surface water (Berg 1937) or in the metalimnion or hypolimnion above the sediment (Juday 1921; LaRow 1969). Most third and fourth instars are bottom inhabitants during the day (LaRow 1969) except in fishless lakes (von Ende 1979). They burrow into the sediment, and their abundance is positively correlated with the softness of the sediment (Parma 1971). Since softness of the sediment tends to increase with water depth, *Chaoborus* are most common in profundal regions.

Chaoborus larvae, especially *C. punctipennis* (Hilsenhoff and Narf 1968), are often exposed to anaerobic conditions since they commonly inhabit lakes whose hypolimnia are anoxic for long periods during the summer. They can survive long periods under anaerobic conditions but may often escape continuous anoxia by nocturnal migrations to the oxygen-rich epilimnion.

Chaoborus larvae feed, as well as replenish oxygen supplies, during vertical migration. The nocturnal nature of these migrations may be a mechanism for predator avoidance (Zaret and Suffern 1976; von Ende 1979).

Only a portion of the *Chaoborus* population migrates on any one night (Berg 1937; Parma 1971), and the older larvae and pupae make up a large proportion of all migrations (Parma 1971). The factors controlling occurrence and timing of migration are not well understood. Field observations and laboratory experiments suggest that light intensity controls the timing of migration (Teraguchi and Northcote 1966; Chaston 1969), possibly by determining whether or not larvae will emerge from the sediment (LaRow 1969). Conditions in the water-column, such as oxygen concentrations and availability of food, may determine whether or not the larvae will actually migrate after they emerge from the sediment (LaRow 1970). LaRow (1970) hypothesized that oxygen concentration is the major regulatory factor controlling migration since migration occurs mainly under low-oxygen conditions. However, Fedorenko and Swift (1972) have reported diurnal migrations of *Chaoborus* larvae in a lake with consistently high oxygen concentrations. Although extensive diurnal migration of *Chaoborus* does not usually occur during winter under the ice (Wood 1956; Malueg 1966; LaRow 1968; Parma 1971), it has been observed in some lakes (Eggleton 1931a; Northcote 1964).

All *Chaoborus* species are nocturnal carnivores (Kajak and Ranke-Rybicka 1970; Parma 1971). They prefer Copepoda over Cladocera and Cladocera over Ostracoda (Parma 1971). In theory, predation by *Chaoborus* on benthic macrofauna is negligible, since they are usually in the water-column at night; however, *Chaoborus* larvae have been known to feed on benthic chironomid larvae (Tubb and Dorris 1965) and tubificid oligochaetes (Swuste et al. 1973).

In summer, *Chaoborus* larvae are usually distributed in moderate numbers throughout the

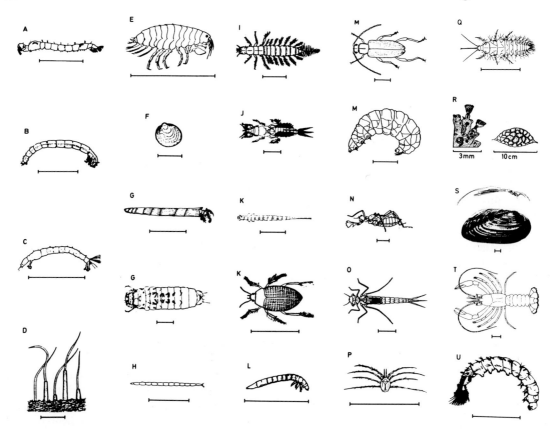

Figure V.A.6–1. The benthic micro- and macroinvertebrates of Mirror Lake. The scale bar represents 5mm, unless otherwise specified.

A. *Chaoborus* larva, Diptera (American Public Health Association 1971, 815)
B. *Chironomus* larva, Diptera (American Public Health Association 1971, 815)
C. *Ablabesmyia,* tanypod larva, Diptera (American Public Health Association 1971, 815)
D. *Tubifex* worms, Oligochaeta (Pennak 1978, 2nd ed., 278)
E. *Hyalella,* Amphipoda (Morgan 1930)
F. *Pisidium,* sphaeriid clam, Mollusca (American Public Health Association 1971, 820)
G. Caddisfly larva (*Limnephilus*) and case (*Triaenodes*), Trichoptera (case = American Public Health Association 1971, 814; larva = Pennak 1978, 605)
H. *Palpomyia,* heleid larva, Diptera (Pennak 1978, 698)
I. *Sialis* larva, Megaloptera (American Public Health Association 1971, 813)
J. *Ephemera,* mayfly nymph, Ephemeroptera (Pennak 1978, 537)
K. *Haliplus,* haliplid larva and adult, Coleoptera (Pennak 1978, 626)
L. *Narpus,* elmid larva, Coleoptera (American Public Health Association 1971, 817)
M. *Donacia,* Chrysomelid larva and adult, Coleoptera (Pennak 1978, 659)
N. *Macromia,* dragonfly nymph, Odonata (American Public Health Association 1971, 812)
O. *Enallagma,* damselfly nymph, Odonata (Hutchinson 1967, 80)
P. *Unionicola,* mite, Acarina (Hutchinson 1967, 88)
Q. *Sisyra* larva, Neuroptera (Pennak 1978, 591)
R. *Pectinatella,* individuals and colony, Ectoprocta (American Public Health Association 1971, 804)
S. *Elliptio,* Mollusca (Burch 1975)
T. *Orconectes,* crayfish, Decapoda (Pennak 1978, 476)
U. *Eoparargyractis plevie* larva, Lepidoptera (Fiance and Moeller 1977)

Table V.A.6–1. The Benthic Micro- and Macroinvertebrates of Mirror Lake[a]

P[b]—Protozoa
  Amoeba sp.
  Colacium sp.
  Coleps sp.
  Difflugia sp.
  Dileptus sp.
  Metopus sp.
  Ophrydium sp.
  Paramecium sp.
  Stentor spp.
  Vorticella sp.
P—Porifera
  Anheteromeyenia argyrosperma (Potts)
  Ephydatia mulleri (Lieberkuhn)
  Eunapius fragilis (Leidy)
  Spongilla lacustris Linnaeus
P—Coelenterata
  Chlorohydra sp.
  Hydra americana Hyman
P—Platyhelminthes
  Dugesia sp.
  Gyratrix hermaphroditus Ehrenberg
  Macrostomum sp.
  Mesostoma spp.
  Microdalyellia sp.
  Microstomum lineare (O. F. Müller)
  Myostenostomum sp.
  Opisthocystis goetti (Bresslau)
  Otomesostoma auditivum (Du Plessis)
  Phagocata sp. (?)
  Prorhynchella minuta Ruebush
  Prorhynchus stagnalis Schultze
  Rhynchomesostoma sp.
  Rhynchoscolex simplex Leidy
  Stenostomum leucops (Duges)
  Stenostomum unicolor O. Schmidt
  Stenostomum spp.
  4 undetermined species
P—Gastrotricha
  Aspidiophorus sp.
  Chaetonotus spp.
  Ichthydium spp.
  Lepidodermella spp.
  Polymerurus spp.
P—Rotatoria
  C[c]—Digononta
    Dissotrocha aculeata Ehrenberg
    Dissotrocha macrostyla (Ehrenberg)
    Habrotrocha spp.
    Macrotrachela sp.
    Rotaria cf. *macrura* Ehrenberg
    Rotaria rotatoria (Pallas)
    Rotaria tridens Montet

  C—Monogononta
    Ascomorpha sp.
    Cephalodella forficula (Ehrenberg)
    Cephalodella spp.
    Collotheca sp.
    Colurella spp.
    Dicranophorus spp.
    Erignatha clastopsis (Gosse)
    Gastropus minor (Rousselet)
    Lecane spp.
    Lepadella sp.
    Macrochaetus spp.
    Microcodon clavus Ehrenberg
    Monommata astia Myers
    Monommata spp.
    Notommata spp.
    Polyarthra spp.
    Proales sp.
    Proalinopsis spp.
    Scaridium longicaudatum (O. F. Müller)
    Stephanoceras fimbriatus (Goldfuss)
    Synchaeta sp.
    Taphrocampa selenura (Gosse)
    Taphrocampa sp.
    Testudinella sp.
    Tetrasiphon hydracora Ehrenberg
    Trichocerca spp.
    Trichotria sp.
    Wierzejskiella sp.
    5 undetermined genera
P—Nematoda
  Achromadora sp.
  Alaimus sp.
  Anonchus sp.
  Aphanolaimus sp.
  Criconemoides sp.
  Chromadorita sp.
  Dorylaimoidea
  Ethmolaimus sp.
  Ironus sp.
  Mermithidae
  Monhystera spp.
  Monhystrella sp.
  Mononchus sp.
  Prismatolaimus sp.
  Prodesmodora sp.
  Rhabdolaimus sp.
  Tobrilus sp.
  3 undetermined genera
P—Tardigrada
  Dactylobiotus grandipes Schuster et al.
  Macrobiotus sp.
  Pseudobiotus sp.

Table V.A.6–1. *Continued*

P—Ectoprocta
  Fredericella sultana (Blumenbach)
  Pectinatella magnifica Leidy
  Plumatella fruticosa Allman
P—Annelida
  Aeolosoma spp.
  Amphichaeta americana Chen
  Aulodrilus pigueti Kowalewski
  Chaetogaster diastrophus (Gruithuisen)
  Dero digitata (O. F. Müller)
  Dero obtusa d'Udekem
  Enchytraeidae
  Ilyodrilus templetoni (Southern)
  Limnodrilus hoffmeisteri Claparède
  Lumbriculidae
  Nais communis Piguet
  Nais simplex Piguet
  Piguetiella blanci (Piguet)
  Pristina aequiseta Bourne
  Pristina leidyi Smith
  Quistodrilus multisetosus (Smith)
  Rhyacodrilus montana (Brinkhurst)
  Slavina appendiculata (d'Udekem)
  Specaria josinae (Vejdovsky)
  Spirosperma nikolskyi (Lastockin and Sokolskaya)
  Stylaria lacustris (Linnaeus)
  Tubifex tubifex (O. F. Müller)
  Uncinais uncinata (Ørsted)
  Vejdovskyella comata (Vejdovsky)
P—Mollusca
  C—Gastropoda
    Campeloma sp.
    Helisoma sp.
    Lymnaea sp.
  C—Pelecypoda
    Elliptio complanata (Lightfoot)
    Pisidium sp.
    Sphaerium sp.
P—Arthropoda
  C—Arachnida
    Parasitengona
      Arrenurus sp.
      Forelia sp.
      Hygrobates sp.
      Lebertia sp.
      Limnesia sp.
      Piona sp.
      Unionicola sp.
    Halacaridae
      Lobohalacarus sp.
      Porohalacarus sp.
      Porolohmanella sp.
      Soldanellonyx sp.
    Oribatida
      Hydrozetes lacustris (Michael)
      Trimalaconothrus novus (Sellnick)

  C—Crustacea
    Amphipoda
      Hyalella azteca Saussure
    Cladocera
      Alonella excisa (Fischer)
      Alonella nana (Baird)
      Alona affinis (Leydig)
      Alona cf. *barbulata* Megard
      Alona quadrangularis (O. F. Müller)
      Alona rustica Scott
      Chydorus bicornutus Doolittle
      Chydorus nr. *brevilabris* Frey
      Chydorus piger Sars
      Chydorus sp.
      Disparalona acutirostris (Koch)
      Eurycercus longirostris Hann
      Ilyocryptus sordidus (Liéven)
      Ilyocryptus spinifer Herrick
      Latona parviremis Birge
      Latona setifera (O. F. Müller)
      Macrothrix laticornis (Jurine)
      Monospilus dispar Sars
      Ophryoxus gracilis Sars
      Polyphemus pediculus Linnaeus
      Rhynchotalona falcata (Sars)
      Scapholeberis kingi Sars
      Sida crystallina (O. F. Müller)
    Copepoda
      Acanthocyclops vernalis (Fischer)
      Attheyella obatogamensis Coker
      Bryocamptus minutus group
      Bryocamptus zschokkei (Schmeil)
      Canthocamptus assimilis
      Diacyclops nanus (Sars)
      Eucyclops agilis (Koch)
      Macrocyclops albidus (Jurine)
      Mesocyclops edax Forbes
      Microcyclops rubellus (Lilljeborg)
      Microcyclops varicans (Sars)
      Paracyclops affinis (Sars)
      Paracyclops sp.
      Parastenocaris brevipes Kessler
    Decapoda
      Orconectes sp.
    Ostracoda
      Candona sp.
      Cypria turneri Hoff
      Darwinula stevensoni (Brady and Robertson)
  C—Insecta
    O[d]—Collembola
      Isotomurus palustris Müller
    O—Ephemeroptera
      Caenis sp.
      Ephemera sp.
      Ephemerella sp.

Table V.A.6–1. *Continued*

Heptagenia sp.
Hexagenia sp.
O—Odonata
  Argia sp.
  Basiaeschna sp.
  Cordulia sp.
  Dromogomphus sp.
  Enallagma sp.
  Epiaeschna sp.
  Ischnura sp.
  Macromia sp.
  Neurocordulia sp.
O—Megaloptera
  Chauliodes sp.
  Sialis sp.
O—Neuroptera
  Sisyra sp.
O—Coleoptera
  Bidessus sp.
  Dineutus sp.
  Donacia sp.
  Elmidae
  Galerucella sp.
  Haliplus sp.
  Helodidae
O—Trichoptera
  Agrypnia sp.
  Banksiola sp.
  Limnephilus sp.
  Mystacides sp.
  Oecetis sp.
  Oxyethira sp.
  Phylocentropus sp.
  Polycentropus sp.
  Triaenodes sp.
O—Lepidoptera
  Eoparargyractis plevie
O—Hemiptera
  Cymatia sp.
  Mesovelia sp.
  Notonecta sp.
  Rheumatobates sp.
O—Diptera
  F[e]—Tabanidae
    Chrysops sp.
  F—Heleidae
  F—Chaoboridae
    Chaoborus flavicans Meigen
    Chaoborus punctipennis Say

sub-F[f]—Chironominae
Tribe—Chironomini
  Chironomus anthracinus Zetterstedt
  Chironomus sp.
  Cladopelma sp.
  Cryptochironomus sp.
  Cryptotendipes sp.
  Dicrotendipes spp.
  Glyptotendipes lobiferus Say
  Lauterborniella sp.
  Microtendipes sp.
  Pagastiella sp.
  Parachironomus sp.
  Paracladopelma sp.
  Paralauterborniella sp.
  Phaenopsectra sp.
  Polypedilum haltere Coq.
  Polypedilum tritum Walker
  Stichtochironomus sp.
Tribe—Pseudochironomini
  Pseudochironomus sp.
Tribe—Tanytarsini
  Cladotanytarsus sp.
  Micropsectra sp.
  Paratanytarsus sp.
  Stempellina sp.
  Tanytarsus spp.
sub-F—Orthocladiinae
  Brillia sp.
  Heterotanytarsus sp.
  Heterotrissocladius sp.
  Parakiefferiella sp. (?)
  Psectrocladius spp.
  Rheocricotopus sp.
  Zalutschia zalutschicola
  Zalutschia spp.
  2 undetermined species
sub-F—Diamesinae
  Protanypus sp.
sub-F—Prodiamesinae
  Monodiamesa sp.
sub-F—Tanypodinae
  Ablabesmyia mallochi Walley
  Clinotanypus sp.
  Conchapelopia sp.
  Labrundinia sp.
  Larsia sp.
  Procladius denticulatus Sublette
  Procladius sp.

---

[a] Contributions by D. Strayer, R. Moeller and D. Mazsa.
[b] P = phylum.
[c] C = class.
[d] O = order.
[e] F = family.
[f] sub-F = subfamily.

profundal regions in Mirror Lake, up to 5,000 larvae per m². However, in winter, densities of *Chaoborus* often increase dramatically in the deepest areas of lakes (*Table V.A.6–2*). For example, at 10 m in Mirror Lake, *Chaoborus* densities increased from 1,600 larvae per m² in August 1973 to 38,000 larvae per m² by January 1974. Reasons for these high densities are discussed in "Depth-Distribution," below.

Chironomidae

Chironomids (*Figure V.A.6–*1B), or nonbiting midges, undergo complete metamorphosis. They spend most of their lives as benthic larvae, which exhibit four discrete instars.

Many larvae are free-living, while others inhabit tubes constructed from reworked sediments. In Mirror Lake, chironomid tubes are particularly abundant in the 6- to 8-m depth interval where the tubes are long (2 to 8 cm), thin and often aggregated to form tufts approximately 10 cm in diameter on the sediment surface. Some chironomid species reach their highest abundance in regions of dense macrophyte growth, while other species are restricted to deeper regions where competition and predation may be lower. Although chironomid larvae are usually benthic, the first instar chironomids can be planktonic (Hilsenhoff 1966; Oliver 1971; Potter and Learner 1974) and older instars and pupae may undergo nocturnal migrations (Hamilton 1965; Davies 1974).

The length of the life cycle is highly variable; for example, smaller species and those exposed to favorable growth conditions, such as warm water or high food densities, have shorter life cycles (Merritt and Cummins 1978). Emergence to adults may take place as a swarm (over several days) or over a longer period of time (several weeks or months), depending on the species and lake. Chironomids emerge throughout the summer from all depths in Mirror Lake, with greatest numbers emerging from the upper littoral region. Adults are terrestrial and have a short life span (several days).

The chironomid fauna are a very important component of most lakes in terms of both biomass and numbers of species and individuals. A natural temperate lake usually has at least 50 species, which often account for more than half of the macroinvertebrate species (Merritt and Cummins 1978). The abundance and biomass of chironomids varies widely from lake to lake (*Tables V.A.6–3a* and *3b*), but they usually comprise >50% of the total numbers and biomass of benthic macrofauna.

In the littoral regions of Mirror Lake, chironomids are relatively small in size (typically <4 mm) and the number of abundant species is high (>10); whereas in the profundal region, chironomids are large (typically 5 to 13 mm) and the number of abundant species is low (<5). The species list for Mirror Lake is too incomplete to determine the percent contributed by chironomid species. However, chironomids as a group dominate the macrobenthic fauna of Mirror Lake in terms of numbers (49%) and biomass (65%) of organisms throughout the year. Mean annual abundance appears to be high when compared with other oligotrophic lakes, while mean annual biomass appears to be moderate (*Table V.A.6–3b*). Peak abundance, which can reach 23,000 chironomids per m², occurs in the littoral regions in July (*Figure V.A.6–*2A) and in the profundal region in January (*Figure V.A.6–*3A). Peak biomass (up to 5.3 g dry wt/m²) occurs in the profundal region at all times of the year (*Figures V.A.6–*2B and 3B).

Table V.A.6–2. Densities of *Chaoborus* in Various Lakes during the Winter

Lake	No./m²	Source
Lake Vechten, Netherlands	5,000–14,000	Parma 1971
Cedar Bog Lake, Minnesota	20,000	Lindeman 1942
Little McCauley Lake, Ontario	27,339	Wood 1956
Mirror Lake, New Hampshire	38,000	Walter 1976
Myers Lake, Indiana	43,456	Stahl 1966
Clear Lake, Kansas	65,200	Lindequist and Deonier 1943
Third Sister Lake, Michigan	71,000	Eggleton 1931a
Linsley Pond, Connecticut	97,000	Deevey 1941

Table V.A.6–3a. Mean Annual Abundance of Chironomids in Various Lakes

Lake	Sieve Size (μm)	No./m^2	% of Total Benthic Macrofauna	Source
Entire Lake				
3 Rotorua Lakes, New Zealand[a,b]	—	103	21	Forsyth 1978
2 Rotorua Lakes, New Zealand[c]	—	568	40	Forsyth 1978
Simcoe, Ontario[b,c]	—	590[e]	62	Rawson 1930
Char, N.W.T., Canada[a]	67	7,800	—	Welch 1976
Mirror, New Hampshire[a]	250	9,800	49	revised from Walter 1976
Loch Leven, Scotland[c]	175; 500	14,000	—	Charles et al. 1974, 1976; Maitland and Hudspith 1974
Eglwys Nunydd, South Wales[c]	150	28,800	83	Potter and Learner 1974
Werowrap, Australia[c,d]	300	40,000–140,000	99	Paterson and Walker 1974
Profundal Region Only				
Cayuga, New York[a,b]	750	284	5	Henson 1954
Loch Lomond, Scotland[b]	200	470	63	Slack 1965
Lizard, Iowa[c]	425	1,210[f]	79	Tebo 1955
Mirror, New Hampshire[a]	250	6,000	38	revised from Walter 1976
Esrom, Denmark[c]	200–510	9,250	45	Jonasson 1972
Loch Leven, Scotland[c]	175	14,300	—	Charles et al. 1974, 1976
Myvatn, Iceland[c]	60; 112	58,800	91	Lindegaard and Jonasson 1979

[a] Oligotrophic.
[b] Mesotrophic.
[c] Eutrophic.
[d] Saline.
[e] Averaged over the ice-free season only.
[f] Averaged over the summer only.

Chironomini

Species within the genus *Chironomus* are difficult to identify. Chromosome analysis (J. Sublette and J. Martin, personal communication) shows that the largest chironomid and the dominant species in the profundal region of Mirror Lake is *C. anthracinus*. It represents a large species complex in North America, and Mirror Lake specimens resemble those from northern Europe (J. Sublette, personal communication). *C. anthracinus* feeds on organic detritus (Izvekova 1971) and algae, especially planktonic diatoms that have settled to the bottom (Jonasson 1972). In Lake Esrom, Denmark, growth of *C. anthracinus* is limited to two short periods in the spring and autumn, which correlate with peaks in primary production as well as presence of oxygen in the hypolimnion (Jonasson 1972).

Chironomus anthracinus dominates the biomass of Mirror Lake at all times of the year, primarily because of its large size. Abundance is relatively low in summer (up to 2,300/m^2) compared with other chironomids (*Figure V.A.6-2C*); and peak abundance (9,600/m^2) is reached in the profundal region in January (*Figure V.A.6-3A*). Larvae often overwinter as third and fourth instars and probably emerge as a swarm in late May and early June in Mirror Lake. A second generation may emerge in late summer. *C. anthracinus* larvae occasionally undergo nocturnal vertical migration in Mirror Lake. The most active migrators are probably the fourth instar larvae, especially those ready to pupate (Dugdale 1956; Davies 1974).

Tanytarsini

Larvae of the tribe Tanytarsini are often bright red but are relatively intolerant of poor oxygen conditions compared with the Chironomini (Bryce and Hobart 1972). The Tanytarsini are

Table V.A.6–3b. Mean Annual Biomass of Chironomids in Various Lakes

Lake	g dry wt/m^2	% of Total Benthic Macrofauna	Source
Entire Lake			
Char, N.W.T., Canada[a]	0.43	—	Welch 1976
Simcoe, Ontario[b,c]	0.81[e]	65	Rawson 1930
Mirror, New Hampshire[a]	1.1	65	revised from Walter 1976
Eglwys Nunydd, South Wales[c]	3.44	79	Potter and Learner 1974
Tundra Pond J, Alaska[a]	5.42	>90	Butler et al. 1980
Parvin, Colorado[c]	6.08	61	Buscemi 1961
Werowrap, Australia[c,d]	8.07	99	Paterson and Walker 1974
Loch Leven, Scotland[c]	8–9	—	Charles et al. 1974, 1976; Maitland and Hudspith 1974
Profundal Region Only			
Cayuga, New York[a,b]	0.07	9[f]	Henson 1954
Loch Lomond, Scotland[b]	0.30	—	Slack 1965
Lizard, Iowa[c]	1.3[g]	93	Tebo 1955
Mirror, New Hampshire[a]	1.9	68	revised from Walter 1976
Loch Leven, Scotland[c]	8.34	—	Charles et al. 1974, 1976
Esrom, Denmark[c]	10.43	63[f]	Jonasson 1972

[a] Oligotrophic.
[b] Mesotrophic.
[c] Eutrophic.
[d] Saline.
[e] Averaged over the ice-free season only.
[f] Includes shell weight for *Pisidium* sp.
[g] Averaged over the summer only.

common in the oxygen-rich littoral regions of oligotrophic and eutrophic lakes, and the genus *Tanytarsus* is typical of the oxygen-rich profundal regions of oligotrophic lakes. They usually feed on detritus, plant particles and benthic algae (Mundie 1957; Hamilton 1965; Armitage 1968); but some species are filter feeders (Bryce and Hobart 1972).

In Mirror Lake, Tanytarsini reach their maximum abundance (17,000/m^2 in July) in the lower littoral region (*Figure V.A.6*–2C), where they seem to be associated with a secondary peak in macrophyte biomass (predominantly *Nitella flexilis*) (*Figure V.A.6*–4C). They probably build the tubes that are so common in this region; perhaps Tanytarsini are so abundant here because this region is the shallowest (hence, most oxygen-rich) area of the lake where suitable material for tube-building (gyttja) occurs. Despite their abundance, Tanytarsini make a minor contribution to biomass compared with *Chironomus anthracinus* (*Figure V.A.6*–2D). In Mirror Lake, Tanytarsini emerge throughout the summer.

Orthocladiinae

The subfamily Orthocladiinae is large, but poorly known. Since the larvae lack hemoglobin, orthoclads are the chironomids most sensitive to low-oxygen conditions (Mundie 1957) and are restricted to oxygen-rich water (Bryce and Hobart 1972). They are often the dominant chironomid larvae in cold northern lakes (Bryce and Hobart 1972). Orthoclad larvae feed on detritus and algae in the sediment and on algae scraped from stones and from stems and leaves of aquatic macrophytes (Hamilton 1965; Bryce and Hobart 1972). The larvae are found principally in areas of vegetation (Morgan and Waddell 1961; Sandberg 1969; Bryce and Hobart 1972), where their green color provides camouflage (Bryce and Hobart 1972). In Mirror Lake, orthoclads are most common in the littoral regions, reaching peak abundance (6,800/m^2 in

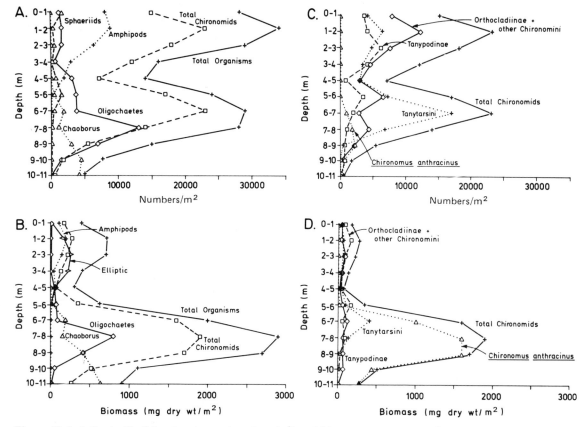

Figure V.A.6–2. A–D. July abundance (numbers/m²) and biomass (mg dry wt/m²) of the dominant benthic macroinvertebrates (**A** and **B**) and the dominant taxa of chironomids (**C** and **D**) at each water-depth interval in Mirror Lake.

July) in the 1- to 2-m depth interval (*Figure V.A.6*–2C), possibly associated with the peak in macrophyte biomass (*Figure V.A.6*–4C).

Diamesinae

The Diamesinae closely resemble the Orthocladiinae in morphology and are most abundant in oligotrophic lakes, especially those situated in cold climates (Bryce and Hobart 1972). The larvae are detritivores or herbivores and live on the surface sediments or burrow into them (Merritt and Cummins 1978). The Diamesinae in Mirror Lake have been lumped with the Orthocladiinae because it is difficult to separate the two groups taxonomically.

Tanypodinae

The distinctive head capsule and stiltlike prolegs of the tanypod larvae are adaptations for the free-living, carnivorous existence that makes this subfamily unique among the chironomids (*Figure V.A.6*–1C). They are extremely motile and seem to be widely distributed in most lakes (Miller 1941), both above and below the thermocline. They lack hemoglobin and tend to be less resistant to anoxic conditions than *Chironomus* or tubificids (Eggleton 1931b). Many species are absent from anoxic waters (Berg and Petersen 1956), although the common species *Procladius choreus* is characteristic of the profundal region of eutrophic lakes (Mundie 1957).

Although Tanypodinae are known to feed on tubificids (Brinkhurst 1964; Saether 1970), crustaceans (Kajak and Dusoge 1970) and other chironomids (Hamilton 1965) by sucking out the gut contents of the prey, very few are thought to be obligate carnivores (MacDonald 1956; Oliver 1971). Their guts are often filled with algae, especially large-celled taxa such as *Closterium*, *Micrasterias* and *Pinnularia* (MacDonald 1956;

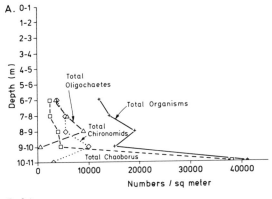

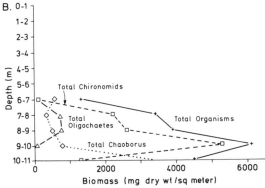

Figure V.A.6–3. A–B. January abundance (numbers/m^2) and biomass (mg dry wt/m^2) of the dominant benthic macroinvertebrates at each water-depth interval in Mirror Lake.

Davies and McCauley 1970; Izvekova 1971), which may be selected directly from the mud rather than by carnivory (Armitage 1968). Such alternate modes of feeding may be used more in winter and by early instars of some species (Merritt and Cummins 1978).

In Mirror Lake, tanypod larvae reach their maximum abundance (6,100/m^2 in July) in the upper littoral region, but are found throughout the lake (*Figure V.A.6*–2C). Adults emerge from the profundal region in early summer and from shallower depths in mid- and late summer.

Oligochaeta

Aquatic oligochaetes of the family Tubificidae (*Figure V.A.6*–1D) are often found in anaerobic environments such as the profundal regions of eutrophic lakes, because they are very resistant to lack of oxygen (Juday 1908; Eggleton 1931b); they also may be common in well-oxygenated oligotrophic lakes. The common species *Tubifex tubifex* is ecologically tolerant and has been found at all depths and in all substrate types occurring in lakes (Brinkhurst 1964).

Tubificids burrow into the sediment head first, enabling their tails, which contain the respiratory appendages, to undulate in the water above the sediment-water interface. Tubificids are deposit-feeders, their main food being bacteria (Wavre and Brinkhurst 1971). Since their preferred food is often bacteria growing on the feces of other species of tubificids, many individuals of two or more species (especially *Tubifex tubifex* and *Limnoldrilus hoffmeisteri,* the two dominant oligochaetes in Mirror Lake) are often clumped together, resulting in lower respiration rates since less time must be spent searching for food (Chua and Brinkhurst 1973).

Concentrations in excess of 200,000 oligochaetes per m^2 may be found in some lakes (e.g. Toronto Harbour, Lake Ontario; see Brinkhurst 1970). In Mirror Lake, they reach maximum numbers (8,900/m^2 in January) in the profundal region (*Figures V.A.6*–2A and 3A). Although oligochaetes are common throughout most of Mirror Lake, they are absent from depths greater than 10 m at all times of the year, possibily owing to competition from *Chaoborus* (see *Chapter VI.B*).

Amphipoda

Amphipods or scuds (*Figure V.A.6*–1E) are laterally compressed crustaceans. Since amphipods require abundant dissolved oxygen (Pennak 1978) they are usually restricted to littoral regions; however, several species are found in the oxygen-rich profundal regions of deep oligotrophic lakes. *Hyalella azteca,* the species found in Mirror Lake, is a common littoral species associated with vegetation (Wohlschlag 1950). In Mirror Lake, *Hyalella azteca* reaches its maximum abundance (8,800/m^2 in July) in the same depth interval (1 to 2 m) where macrophyte biomass reaches its peak (*Figure V.A.6*–2A).

Sphaeriidae

Sphaeriids or fingernail clams (*Figure V.A.6–1F*) are common benthic inhabitants; compared with most clams, they are tiny—smaller than a fingernail. Sphaeriids are adapted to a wide range of conditions, and occur on all types of substrate except clay and rock (Pennak 1978).

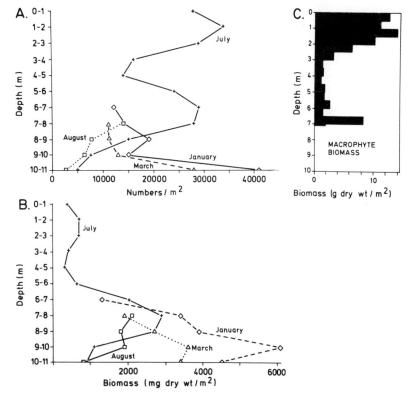

Figure V.A.6–4. A–C. (**A** and **B**) Abundance and biomass of total benthic macroinvertebrates at each water-depth interval in Mirror Lake in various months; (**C**) distribution of macrophyte biomass with water depth in Mirror Lake (Moeller 1978b).

They burrow into the sediment, feeding on bacteria (Jonasson 1972), detritus and benthic algae. In Mirror Lake, peaks in abundance (1,600/m^2 in July) are reached in the upper littoral (0 to 1 m) and lower littoral (5 to 6 m) regions (*Figure V.A.6–2A*).

Unionidae

The freshwater mussel *Elliptio complanata* (*Figure V.A.6–1S*) is one of the largest and most conspicuous members of the benthic macrofauna of Mirror Lake. This filter-feeding mollusc reaches a length of 80 mm in Mirror Lake and may live for 15 to 20 years (Strayer et al. 1981). The life cycle of unionid mussels such as *E. complanata* is an unusual one: the tiny (0.3 mm) larvae, called *glochidia*, are not free-swimming, as are most marine bivalve larvae, but are parasitic on fish for a short time before beginning their benthic existence (Pennak 1978). Because of this, the distribution and abundance of a unionid mussel may be affected by the distribution and abundance of its host fish, which in the case of *E. complanata* is the yellow perch *Perca flavescens* (Matteson 1948).

Elliptio complanata is most abundant on gentle sandy slopes between 1 and 4 m depth in Mirror Lake, where it forms a large portion (38%) of total macrobenthic biomass (*Figure V.A.6–2B*). However, lakewide densities are low (0.032 adults/m^2 totaling 52 mg/m^2 of dry organic matter), and the turnover time of the mussel population is so long (about 9 yr) that its contribution to secondary production in the lake is almost negligible—6.4 mg/m^2-yr as dry organic matter (Strayer et al. 1981). In lakes having much denser mussel populations than Mirror Lake, mussel biomass may be on the order of 10 g/m^2 of dry organic matter (Okland 1964; Magnin and Stanczykowska 1971), and unionids may assume important roles in benthic secondary production, calcium cycling (Green 1980) and the mixing of sediments (McCall et al. 1979).

Other Organisms

Other organisms that are abundant in the littoral regions of Mirror Lake, but are seldom present in the oxygen-poor profundal regions, include members of the Trichoptera, Heleidae, Sialidae, Ephemeroptera, Coleoptera, Odonata and Parasitengona (*Figure V.A.6–1G–P*).

Trichoptera or caddisflies (*Figure V.A.6–1G*) undergo complete metamorphosis and all stages except the adult are aquatic. Most lake species spin nets or build stationary or portable cases of stones, sand grains or parts of leaves or twigs, which they inhabit. Most species are omnivorous. In Mirror Lake, they are most abundant in the upper littoral region with maximum numbers in July reaching $680/m^2$.

The Heleidae or biting midges (*Figure V.A.6–1H*) are a family of dipterans; the adults (called no-see-ums) are terrestrial and the larvae are aquatic. The larvae live in diverse habitats, but may be particularly abundant in floating masses of algae (Pennak 1978). The larvae are predators or scavengers. In Mirror Lake in July, they are present in low numbers (up to $400/m^2$) in the littoral regions.

The larvae of Sialidae or alderflies (*Figure V.A.6–1I*) are predatory and live in diverse, shallow-water habitats. Pupation takes place on shore. In Mirror Lake, sialid larvae occur in low numbers (up to $400/m^2$ in July) in the littoral regions.

The Ephemeroptera, or mayflies, undergo gradual metamorphosis; the adults are short-lived terrestrial inhabitants, while the nymphs (*Figure V.A.6–1J*) inhabit well-oxygenated regions of lakes. Two large burrowers possessing tusklike projections from their mandibles are present in Mirror Lake—*Ephemera* and *Hexagenia*. They dig U-shaped tubes and obtain oxygen by setting up a current of water through the tube by body undulations. They usually feed on organic detritus. In Mirror Lake, their distribution is apparently so patchy that they have never been found in benthic samples, although they have been observed emerging. *Ephemera* emerge in mid-June, *Hexagenia* in mid-July.

Two families of Coleoptera are common benthic inhabitants in Mirror Lake—Haliplidae and Elmidae. All life stages of these holometabolous insects are aquatic. The Haliplidae or crawling water beetles (*Figure V.A.6–1K*) inhabit the shallow upper littoral regions, crawling and climbing among filamentous algae and submersed vegetation—their chief sources of food. They are poor swimmers, although the adults must swim to the surface at intervals to obtain air. Elmid beetles (*Figure V.A.6–1L*) crawl among vegetation and feed on algae and detritus. They are unable to swim, and adults do not need to surface for air as do the haliplids. Coleoptera inhabit the upper littoral region of Mirror Lake; their maximum abundance ($250/m^2$ in July) seems to be correlated with maximum biomass of macrophytes.

Two genera of chrysomelid beetles (*Figure V.A.6–1M*)—*Donacia* and *Galerucella*—infest the yellow waterlily, *Nuphar variegata*, in Mirror Lake. *Donacia* larvae feed on the submersed bases of the leaves and on the rhizome, breathing through tubes inserted into the airspaces of the plant. *Galerucella* larvae feed and pupate on the surfaces of the floating leaves. Adult chrysomelids feed on these floating leaves also.

The Odonata, or dragonflies (*Figure V.A.6–1N*) and damselflies (*Figure V.A.6–1O*), undergo gradual metamorphosis; adults are terrestrial and the nymphs are aquatic. The nymphs are large compared with most other benthic invertebrates and are carnivorous, feeding on a variety of other benthic organisms. Of the species present in Mirror Lake, some climb among vegetation, others sprawl on the bottom, and others burrow into sand or silt. In Mirror Lake, they inhabit the upper littoral region in low numbers (up to $100/m^2$) and are associated with macrophytes.

The Parasitengona or water mites (*Figure V.A.6–1P*) are completely aquatic, spiderlike organisms with globular, brightly colored (often red) bodies. They swim or crawl and are most abundant among aquatic macrophytes. They are carnivorous or parasitic, ingesting fluid from their victims. In Mirror Lake, their distribution is patchy in the littoral and upper profundal regions. Peak numbers (up to $580/m^2$) may be correlated with macrophyte abundance.

Many other benthic invertebrate groups represented in Mirror Lake were neglected in routine studies because of their patchy distributions. These groups are interesting, though, and add diversity to the benthic environment of lakes. They are briefly described below.

Spongilla lacustris is a freshwater sponge common throughout the upper littoral, and occasionally the lower littoral, regions of Mirror Lake. Its fingerlike shoots are green owing to the presence of symbiotic green algae, or zoochlorellae. Sponges are filter-feeders. They form gemmules—specialized reproductive structures that serve as highly resistant resting stages. The green larvae of the Sisyridae or spongilla-flies (*Figure V.A.6*–1Q) parasitize sponges, sucking sponge cells for nourishment.

Ectoprocta (a type of bryozoan; *Figure V.A.6*–1R) are sessile colonies of zooids that form gelatinous masses or tubular branches on cobbles, boulders or submerged debris. Ectoprocts are filter-feeders and require clear water. In Mirror Lake, they occur from the upper littoral to the upper profundal regions, with maximum abundance in the upper littoral region. *Pectinatella magnifica* forms carpets encrusting boulders or sticks in Mirror Lake and reaches its maximal development in late summer or early autumn. It then forms specialized sesso- and statoblasts, which are released from the colony and overwinter in the water-column or on the sediments.

The crayfish *Orconectes* (*Figure V.A.6*–1T) is also a large (up to 13 cm) invertebrate that inhabits the upper littoral region of Mirror Lake. It is active mainly at night, remaining hidden under stones, debris, or in the substrate during the day. It is omnivorous, feeding mainly on algae and larger aquatic plants.

The aquatic caterpillar (Lepidoptera), *Eoparargyractis plevie* (*Figure V.A.6*–1U), is common in the upper littoral region of Mirror Lake, where it feeds on the leaves of large aquatic plants—particularly the rosette plants *Lobelia, Isoetes, Gratiola* and *Eriocaulon* (Fiance and Moeller 1977). After extensive feeding in August and September, these moth larvae retire to silken hammocks constructed at the base of the plants. They resume feeding late in the following spring before pupating within silken cocoons, also constructed on the food plants. Adults emerge in June (R. Moeller and S. Fiance, unpublished observations).

Other benthic forms include protozoa, tardigrades, rotifers, cladocerans, copepods and nematodes. Most of these organisms are too small to be considered macrofauna, but are discussed in the section on benthic microinvertebrates (*Chapter V.A.7*).

Benthic Macroinvertebrate Sampling Methods

Since quantitative information on abundance, biomass and species composition of benthic invertebrates and on emergence of adult insects depends so much on the adequacy of sampling and sorting methods, a brief description of those used in Mirror Lake follows. Details concerning these procedures have been given by Walter (1976).

Ideally, the number of samples taken in each depth interval should be proportional to the area of that interval and should be determined by the size of the sampler (Ranta and Sarvala 1978; Downing 1979). In Mirror Lake, the number of random samples taken at each depth at various seasons (*Table V.A.6*–4) was limited by practical considerations. Samples from the gyttja were taken with a standard Ekman dredge (0.023 m^2 area) and an FRB (Fisheries Research Board of Canada, see Hamilton et al. 1970) multiple corer (0.0016 m^2 area for each of four cores). Samples from substrates other than gyttja (i.e. sand, gravel, detritus and beds of macrophytes) were taken by SCUBA divers using corers made from polyvinylchloride tubing (0.0025 m^2 area).

All samples except those taken in winter were sieved with a 250-μm sieve. Although this sieve size is small, it still represents a compromise between efficiency and accuracy since early instar chironomids still may be lost (Jonasson 1955, 1958; Hamilton 1965; Hilsenhoff 1966). Jonasson (1958) determined that for Lake Esrom, a larger sieve size was adequate in winter when most of the larvae were in third and fourth instars. Therefore, a 500-μm sieve was used for Mirror Lake samples in winter.

All samples were preserved in 10% formalin and stained with rose bengal (0.1 g/liter of 10% formalin). To sort the organisms, samples were floated in a sugar solution (1.125 sp. gr.) using the procedure of Anderson (1959), examined for organisms that did not float and then floated again. Organisms were preserved in 70% alcohol. Since length is easier to measure than the dry weight of each organism, a length-dry weight relationship was determined for each major taxon. These curves, shown on the same graph for comparison purposes (*Figure V.A.6*–5), were used to estimate dry weight for each organism. For chironomids and *Chaoborus*—the two taxa that dominated the biomass in Mirror Lake—the length-dry weight curves were corrected for the loss in both length and weight that occurs during preservation in formalin and alcohol (Walter 1976).

The supplementary census of the freshwater

Table V.A.6–4. Summary of the Number of Benthic Samples Analyzed at each Depth and Season in Mirror Lake

Depth Interval	% of Total Lake Area	16–27 Aug. 1973 (Ekman)	5–7 Jan. 1974 (Ekman)	20–22 Mar. 1974 (Ekman)	July–Aug. 1974 (corers)
Gyttja					
>10 m	4.1	5	1	2	3
9–10 m	6.7	9	1	2	3
8–9 m	10.7	13	1	1	5
7–8 m	23.9	12	3	2	13
6–7 m	14.5	—	1	—	7
5–6 m	6.0	—	—	—	4
Nongyttja					
4–5 m	4.3	—	—	—	2
3–4 m	6.7	—	—	—	7
2–3 m	6.0	—	—	—	13
1–2 m	8.0	—	—	—	8
0–1 m	9.3	—	—	—	11
Total	100.2	39	7	7	76

mussel population was carried out by SCUBA divers, who collected by hand all adult mussels along 23 1-m wide transects extending from shore to the edge of the profundal region (see Strayer et al. 1981).

A cursory study of insect emergence from the lake was made by monitoring one funnel emergence trap (submerged 1 m below the lake surface) at each of three locations (water depth 2.4, 7.6 and 9.1 m) for one-third of the total emergence period for the ice-free season. Traps were emptied every 1 to 3 days and organisms preserved in 70% alcohol.

Horizontal and Vertical Distribution

Even within a small area of sediment, there is variation in the abundance of organisms. By taking many benthic samples at one station (i.e. over a small area) it is possible to compare the within-station variation with the variation between many stations within the same depth interval. In Mirror Lake, within-station variation is usually as great as, or greater than, between-station variation (*Table V.A.6–5*). Within any station, this variation may be due to two major factors—the spatial and vertical distribution of organisms in the sediment.

In most aquatic systems the benthic organisms tend to be aggregated or clumped (Elliott 1971). This clumping may be due to a number of factors; the most important for benthic forms probably are (1) the environmental conditions (e.g. temperature and oxygen stratification) and the sediment-type distribution, (2) oviposition patterns and methods of reproduction (oligochaetes, for example, may reproduce by budding), and (3) the benefits associated with living in a group, either monospecific (Konstantinov 1971) or mixed (Chua and Brinkhurst 1973). The

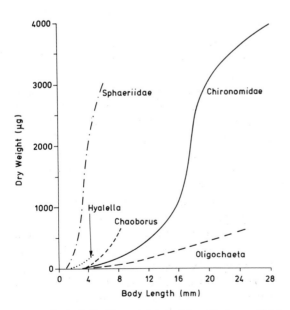

Figure V.A.6–5. Comparison of length-dry weight curves for different taxa in Mirror Lake.

Table V.A.6–5. Comparison of 95% Confidence Intervals for Geometric Means (medians) of Within-station and Between-station Samples from Mirror Lake

	Numbers/m^2							
	Total Organisms		Chaoborus		Chironomids		Oligochaetes	
	Median	Maximum Minimum	Median	Maximum Minimum	Median	Maximum Minimum	Median	Maximum Minimum
5 Ekman samples from 1 station at 8.5 m	7,223	9,879 5,281	723	1,136 460	3,356	5,153 2,186	3,003	4,211 2,141
Mean of 13 stations in the 8–9-m depth interval	7,200	9,100 5,800	2,300	3,400 1,600	2,700	3,800 1,900	1,600	2,500 1,000
9 samples (Ekman dredge and corers) from 1 station at 9.1 m	5,364	6,337 4,208	184	894 38	2,328	3,568 1,519	1,864	3,027 1,147
Mean of 9 stations in the 9–10-m depth interval	5,800	7,700 4,400	2,200	3,600 1,400	2,100	3,200 1,400	950	1,700 530

spatial patterns may change seasonally or with age of the organisms and may depend on whether an entire group or one species is examined (Paterson and Fernando 1971).

Small-, medium- and large-scale dispersion of benthic invertebrates was examined in Mirror Lake by calculating the index of dispersion for each taxon from samples taken over small (<0.1 m^2), medium (<100 m^2) and large (>6,000 m^2) areas of substrate. Taxa agreeing with the Poisson series were assumed to be randomly distributed, while taxa showing a significant departure from the Poisson series were assumed to be clumped (Table V.A.6–6).

Results indicated that over a small (<0.1 m^2) area, organisms may be randomly distributed. However, since the degrees of freedom and the number of organisms per sample were very small, the numbers must have a large range to show a significant departure from randomness. Over medium and large areas, where sample sizes were larger and results more meaningful, most taxa showed clumped distributions. In the deeper profundal region, where substrate and environment are most homogeneous, organisms were least clumped. All organisms in the shallower water were clumped. This clumping probably contributed the largest amount of variation to samples from a given station or depth-interval in Mirror Lake. Parametric statistical analyses cannot be performed on data obtained from such benthic samples unless an appropriate transformation such as log (x + 1) or a fourth-root transformation can be found that will normalize the data (Elliott 1971; Downing 1979).

Benthic organisms are also vertically distributed in lake sediment and may in some cases burrow deeper than the sampler can penetrate. Usually, the number of organisms below a depth of 15 cm in the sediment is negligible (e.g. Berg 1938; Cole 1953; Mundie 1957); however, nonfeeding *Chironomus plumosus* and *Spaniotoma* (*Orthocladius*) *akanusi* can burrow to 51 and 70 cm, respectively (Hilsenhoff 1966; Yamagishi and Fukuhara 1972). Since large larvae tend to burrow deeper than small larvae (Berg 1938; Birkett 1958), deep burrowing may be more common in the winter when the larvae are generally in the third and fourth instar and are not feeding (Hilsenhoff 1966). This factor may represent an important source of error for biomass and secondary production estimates because these estimates are influenced most by the oldest—hence, largest—larvae.

In Mirror Lake, the samplers usually penetrated to 12 cm, and often to 14 cm. Studies of

Table V.A.6-6. The Spatial Dispersion[a] of each Taxon for Different Scales of Dispersion and Different Depth Intervals in Mirror Lake

	Small-Scale Dispersion (<0.1 m^2)						Medium-Scale Dispersion (<100 m^2)	
	(July 1974, multiple corer)						(Ekman)	(Ekman and corer)
	Depth Interval (m)							
	>10	9–10	8–9	7–8	6–7	5–6	8.5 m	9.1 m
Total Organisms	R[b]	R-C[c]	R	C[d]	C	C	C	C
Total *Chaoborus*	R	R-C	—	—	—	—	R	C
Chaoborus punctipennis	R	R-C	R	R-C	R	R	—	—
Chaoborus flavicans	R	R	—	—	—	—	—	—
Total Chironomidae	R	R	—	—	—	—	C	C
Chironomus anthracinus	R	R	R-C	R	R	—	—	—
Other Chironomini	—	—	—	—	—	—	—	—
Orthocladiinae	—	R	R	R	C	C	—	—
Tanytarsini	—	R	R-C	R-C	R	R	—	—
Tanypodinae	R	R	R	R	R	R	—	—
Oligochaeta	—	R	R-C	R	R	R	C	C
Hyalella azteca	—	—	—	—	—	—	—	—
Sphaeriidae	—	—	—	—	—	—	—	—

	Large-Scale Dispersion (>6,000 m^2)									
	(August 1973 Ekman)						(August 1974 corer)			
	Depth Interval (m)						Depth Interval (m)			
	>10	9–10	8–9	7–8	6–7	5–6	4–5	3–4	2–3	1–2	0–1
Total Organisms	C	C	C	C	C	C	C	C	C	C	C
Total *Chaoborus*	R	C	C	C	C	C	C	C	C	C	C
Chaoborus punctipennis	R	C	C	C	C	C	C	C	C	C	C
Chaoborus flavicans	R-C	C	R	C	C	C	C	C	C	C	C
Total Chironomidae	C	C	C	C	C	C	C	C	C	C	C
Chironomus anthracinus	C	C	C	C	C	C	C	C	C	C	C
Other Chironomini	—	—	—	—	—	—	—	—	—	—	—
Orthocladiinae	R	R	C	C	C	C	C	C	C	C	C
Tanytarsini	—	R	C	C	C	C	C	C	C	C	C
Tanypodinae	R	R	C	C	C	C	C	C	C	C	C
Oligochaeta	—	C	C	C	C	C	C	C	C	C	C
Hyalella azteca	—	C	C	C	C	C	C	C	C	C	C
Sphaeriidae	—	C	C	C	C	C	C	C	C	C	C

[a] Determined by testing the index of dispersion $\frac{s^2}{\bar{x}}$ [n − 1] for agreement with the Poisson distribution.

[b] R = random.

[c] R-C = borderline.

[d] C = clumped.

vertical distribution of chironomids, oligochaetes and *Chaoborus* in Mirror Lake sediments (Strayer 1984) suggest that most (95 to 99%) chironomids and oligochaetes live above 4 cm in the sediment and a few penetrate to 12 cm. However, *Chaoborus* burrows into the sediment to at least 12 cm in the deepest regions of the lake where the gyttja is the least consolidated.

Abundance and Biomass

Mean numbers and biomass for the dominant taxa in Mirror Lake are given in *Table V.A.6–7* for the profundal region and for the entire lake at different seasons and on an annual basis. The annual mean biomass of 1.8 g dry weight per m^2 is moderate when compared with other oligotrophic and eutrophic lakes (*Table V.A.6–8*). Conversely, the annual mean abundance of 20,000 organisms per m^2 for the entire lake seems to be high when compared with other oligotrophic lakes, and even ranks high among the eutrophic lakes (*Table V.A.6–9*). Abundance in the profundal region alone is also similar to that for eutrophic lakes (*Table V.A.6–9*). Such high abundance seems unusual for an oligotrophic lake such as Mirror Lake. However, the high abundance in the profundal region of Mirror Lake results from the fact that the hypolimnion is eutrophic since the deeper part becomes anaerobic in late summer or autumn because of the comparatively large amount of organic matter that settles into the small volume of hypolimnetic water—approximately 9% of the total lake volume (see *Chapters V.B.6* and *VI.A* for more discussion). Perhaps the high abundance in the upper littoral region is not as unusual as it appears either. In Mirror Lake, benthic samples in the littoral region were taken with deep-penetrating (15 cm) diver-operated corers, and the samples were processed with a comparatively small sieve size (250 μm), resulting in a good estimate of actual numbers of or-

Table V.A.6–7. Mean Numbers and Biomass for the Dominant Taxa in Mirror Lake

	Total Organisms	Total *Chaoborus* Larvae	Total Chironomid Larvae	Total Oligochaetes	Total Sphaeriidae	Total Amphipoda
Numbers (No./m^2)						
Entire Lake						
Annual $\bar{x}^a$	20,000	2,700	9,800	3,200	600	2,100
July	24,000	1,300	14,000	5,200	420	2,100
Profundal Region (>7 m)						
Annual $\bar{x}^a$	16,000	4,400	6,000	—[b]	—	—
January	18,000	6,300	5,900	5,200	140	0
March	13,000	5,800	4,100	3,000	58	0
July	20,000	2,300	8,900	8,700	90	0
August	10,000	1,800	5,900	2,200	200	0
Biomass (mg dry wt/m^2)						
Entire Lake						
Annual $\bar{x}^a$	1,800	280	1,100	180	150	40
July	1,600	170	980	260	54	40
Profundal Region (>7 m)						
Annual $\bar{x}^a$	2,800	520	1,900	—	—	—
January	4,000	710	2,700	550	80	0
March	2,500	850	1,300	200	43	0
July	2,400	310	1,500	520	21	0
August	1,900	170	1,400	150	81	0

[a] Extrapolated values based on means for January, March, July and August and on the July depth-distribution for the littoral region.
[b] — Not determined.

Table V.A.6–8. Mean Annual Biomass of Benthic Macroinvertebrates in Various Lakes

Lake	g dry wt/m^2	Source
Keyhole, N.W.T., Canada[a]	0.1[i]	Hunter 1970
4 Large Canadian Lakes[a]	0.1–0.4[i]	Rawson 1955
5 Alpine Lakes, Canada[a]	0.2–0.4[i]	Rawson 1941
Wingra, Wisconsin[c]	0.4[g]	Wissmar and Wetzel 1978
Findley, Washington[a]	0.7[g]	Wissmar and Wetzel 1978
4 Lakes, N.W.T., Canada[a]	0.4–2.2[h]	Healey 1977
Washington, Washington[c]	0.8	Thut 1969
Cayuga, New York[a,b]	0.9	Henson 1954
Char, N.W.T., Canada[a]	1.2[g]	Rigler 1978
Simcoe, Ontario[b,c]	1.2[h]	Rawson 1930
Memphremagog, North Basin, Quebec/Vermont[b]	1.2	Dermott et al. 1977
Mirror, New Hampshire[a]	1.8[f]	Walter 1976
Waskesiu, Saskatchewan, Canada[c]	2.5[i]	Rawson 1941
Plöner See, Germany[c]	3.3	Lundbeck 1926 as cited in Rawson 1930
Memphremagog, South Basin, Quebec/Vermont[c]	3.4	Dermott et al. 1977
Paul, B.C., Canada[e,a]	3.6[i]	Rawson 1941
Eglwys Nunydd, South Wales[c]	4.4	Potter and Learner 1974
Mendota, Wisconsin—0 to 7 m[c]	5.4[i]	Juday 1942
57 North German lakes	5.6[i]	Lundbeck 1926 as cited in Rawson 1930
Marion, B.C., Canada[a]	5.9[i]	Hargrave 1969
Mendota, Wisconsin—>20 m[c]	7.7	Juday 1942
Werowrap, Australia[c,d]	8.2	Paterson and Walker 1974
Parvin, Colorado[c]	10.0	Buscemi 1961
Esrom, Denmark[c]	13.8[f]	Jonasson 1972

[a] Oligotrophic.
[b] Mesotrophic.
[c] Eutrophic.
[d] Saline.
[e] Alpine.
[f] Not including mollusc shells.
[g] Assuming dry weight = 45% carbon.
[h] Averaged over ice-free season only.
[i] Averaged over the summer only.

ganisms present. Perhaps if deep-penetrating samplers and, particularly, smaller sieve sizes were used in more benthic studies of the littoral regions of other lakes, the quantity of smaller littoral organisms would not be underestimated, and abundances of benthic macrofauna more similar to those in Mirror Lake would be found.

The littoral region of Mirror Lake is inhabited by a large number of organisms of low individual weight, while the profundal region is inhabited by a moderate number of organisms of high individual weight. In the littoral region, organisms are either very large (*Elliptio*) or small as a result of the particular sediment (sand and gravel), oxygen, temperature and food conditions present in the littoral region. As the sediment becomes finer toward the profundal region, the individuals become larger. In the profundal region, where oxygen is often low or absent, only large organisms can create a strong enough respiratory current to bring in oxygen-enriched water (Brundin 1951).

Table V.A.6–9. Mean Annual Abundance of Total Benthic Macroinvertebrates in Various Lakes

Lake	Sieve Size (μm)	No./m^2	Source
Entire Lake			
3 Rotorua Lakes, New Zealand[a,b]	—	510	Forsyth 1978
Simcoe, Ontario, Canada[b,c]	—	820[e]	Rawson 1930
4 Lakes, N.W.T., Canada[a]	600	800–4,200[e]	Healey 1977
Great Slave, N.W.T., Canada[a]	—	1,600[f]	Rawson 1953
2 Rotorua Lakes, New Zealand[c]	—	1,700	Forsyth 1978
Borrevann, Norway[c]	650	1,900	Okland 1964
Mirror, New Hampshire[a]	250	20,000	Walter 1976
Eglwys Nunydd, South Wales[c]	150	36,000	Potter and Learner 1974
Werowrap, Australia[c,d]	300	40,000–140,000	Paterson and Walker 1974
Profundal Region Only			
Cayuga, New York[a,b]	750	5,800	Henson 1954
Mirror, New Hampshire[a]	250	16,000	Walter 1976
Linsley Pond, Connecticut[b,c]	670	19,000	Deevey 1941
Esrom, Denmark[c]	200–510	20,440	Jonasson 1972
Myvatn, Iceland[c]	60–112	64,500[e]	Lindegaard and Jonasson 1979

[a] Oligotrophic.
[b] Mesotrophic.
[c] Eutrophic.
[d] Saline.
[e] Averaged over ice-free season only.
[f] Averaged over the summer only.

Seasonal Distribution

Abundance and biomass of benthic invertebrates in lakes exhibit marked seasonal variations. Numbers and biomass often reach maximum values in early winter. This pattern is true for Mirror Lake, where high abundance and maximum biomass occur in the profundal region in January (*Table V.A.6–7, Figures V.A.6–3 and 4*). At this time of year, emergence has ceased and, with renewal of oxygen throughout the benthic regions at autumnal overturn, large numbers of larvae have hatched from eggs laid earlier in the summer and autumn (Eggleton 1931b). In addition, *Chaoborus* larvae, whose young instars are planktonic, often reach third instar after autumnal overturn and become benthic. Fourth instar larvae, which will be ready to emerge in early spring, also add significantly to the biomass of overwintering larvae.

In most lakes, abundance tends to decrease throughout the winter owing to mortality from starvation, predation or anaerobiosis. *Chaoborus* mortality over this time period may be as high as 55 to 70% (Borutzky 1939; Parma 1971). Winter mortality for the dominant profundal organisms in Mirror Lake was estimated from changes in abundance between January and March within the profundal region. Mortality of *Chaoborus* was only 3.4%, while that for *Chironomus anthracinus* was 46.5%, and for Oligochaeta, 42.5%. This mortality was probably not due to anaerobiosis because the oxygen concentrations were high—at least 40% saturation at 10 m in an 11-m water-column. The high mortality of chironomids and oligochaetes could be due to (1) predation by *Chaoborus*, which show a growth of 33 μg dry wt per organism between January and March and have been known to feed on benthic macrofauna; (2) predation by yellow perch, since they are the most abundant fish in the lake (*Chapter V.A.9*) and are known to feed in the profundal region in winter (see Pearse and Achtenberg 1918, for Lakes Mendota and Wingra in Wisconsin); or

(3) natural mortality resulting from food reserves that are insufficient for organisms to survive the winter (possibly resulting from low phytoplankton production the previous autumn).

During the winter, decreases in biomass from the causes listed above may be offset somewhat by increases in the weight of individual organisms. However, in Mirror Lake, biomass did show a marked decrease between January and March (*Figure V.A.6–4B* and *Table V.A.6–7*) owing to high mortality of *C. anthracinus* larvae, which are the major contributors to macrofaunal biomass in the lake.

In early summer, abundance decreases as emergence begins. Biomass lost as a result of emergence is somewhat offset by growth of fourth instar larvae, which have not yet emerged.

Abundance and biomass tend to decrease throughout the summer, often reaching minima in late summer before autumnal overturn. Such minima occur in many lakes (Deevey 1941; Craven 1969; Eggleton 1931b); and the August data from Mirror Lake appear to show such minima (*Table V.A.6–7*). By this time, emergence has depleted numbers and especially biomass, and growth in the profundal region may be slow because of low oxygen levels. In addition, newly laid eggs may not have hatched, and many new larvae are still in their planktonic stages.

Depth-Distribution

Depth-distributions for abundance and biomass of benthic invertebrates in Mirror Lake in July, August, January and March are shown in *Figures V.A.6–4A and 4B*. Comparisons can be made with the depth-distributions for macrophyte biomass (*Figure V.A.6–4C*) and for percent saturation of oxygen and rate of change of temperature for July (*Figure V.A.6–6A*) and January (*Figure V.A.6–6B*).

July

In July, a maximum in abundance (34,000 organisms/m²) occurs in the upper littoral region at water depths of 1 to 2 m. Chironomids dominate and amphipods are also abundant (*Figure V.A.6–2A*). This peak in abundance is undoubtedly due to the habitat and food provided by the

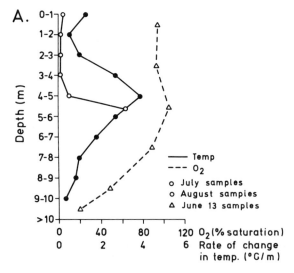

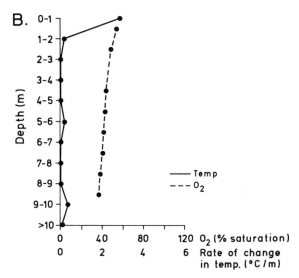

Figure V.A.6–6. A–B. Distribution of percent saturation of oxygen, temperature and rate of change of temperature with water depth in Mirror Lake for July (**A**) and January (**B**).

peak in macrophyte biomass (and its associated epiphytic algae), which also occurs at 1 to 2 m. Biomass shows only a small peak in the upper littoral region (1- to 3-m water depth), since the dominant organisms (*Figure V.A.6–2B*) are small chironomids.

A marked minimum in abundance and biomass occurs in the lower littoral region of Mirror Lake in July (*Figures V.A.6–4A and 4B*). Many other lakes show a minimum in abundance and biomass of benthic organisms in this region (Miller 1941; Berg 1938; Rawson 1930;

Okland 1964). Although this region is well oxygenated, it is occupied by the thermocline and is characterized by rapidly decreasing temperatures with increasing depth (*Figure V.A.6–6A*). In addition to this, internal seiches may expose sediments throughout the region to rapid changes in water temperature and oxygen concentration. This minimum corresponds to a minimum in macrophyte biomass in Mirror Lake (*Figure V.A.6–4C*), and it is often just beyond the lower limit of the rooted vegetation in other lakes (Okland 1964). In some lakes, predation by fish is heavy in this region because of the cooler, yet still oxygenated waters (Deevey 1941); but in Mirror Lake, fish predation in this region is probably insignificant (G. Helfman, personal communication). Rawson (1930) and Miller (1941) attributed this minimum to the fact that conditions in this transitional zone are unfavorable for truly littoral or truly profundal species.

In Mirror Lake in July, a second peak in abundance (29,000 organisms/m^2) occurs at water depths of 6 to 8 m in the lower littoral and upper profundal regions, associated with high abundance of Tanytarsini. At 7 to 9 m in the upper profundal region, the major peak in biomass occurs, resulting primarily from large numbers of *Chironomus anthracinus*, which is by far the largest larva in the lake. Such a zone of high concentration of numbers and biomass often occurs just below the thermocline and has been observed in many other lakes (Curry 1952; Eggleton 1934, 1936; Cole and Underhill 1965; Timms 1974; Berg 1938).

Apparently in the summer, the upper profundal region provides the most favorable habitat for strictly profundal species; and organisms present in the lower profundal region early in the summer migrate toward the upper profundal region as oxygen concentrations decrease (Eggleton 1931b), resulting in a zone of high density in the upper profundal region.

Numbers and biomass decrease with increasing depth in the profundal region of Mirror Lake in July, mainly owing to low temperatures and low oxygen concentrations; and minima in numbers and biomass occur in the deepest depth interval (>10 m). Numbers are lower than in the thermocline region; but biomass is higher because of the much larger size of the organisms inhabiting the profundal region.

January

By January, the depth-distributions of abundance and biomass of benthic macrofauna in Mirror Lake have changed markedly in the profundal region (*Figures V.A.6–3A* and *3B*).[1] The zone of highest concentration in the profundal region has shifted from the upper to the lower profundal region. Peak numbers of 41,000 organisms per m^2 occur in the >10-m zone and result mainly from extremely high concentrations of *Chaoborus* larvae. Peak biomass occurs at the 9- to 10-m zone, mainly owing to moderate numbers of *Chaoborus* and *Chironomus anthracinus* larvae in third and fourth instars.

A similar descent of the concentration zone after autumnal turnover has been documented in two Michigan lakes (Eggleton 1934). High concentrations of organisms in the deepest regions in winter are common in many other lakes and, as in Mirror Lake, are due to high numbers of *Chaoborus* sometimes accompanied by high biomass of a *Chironomus* species.

Several phenomena probably contribute to this tremendous increase in abundance and biomass in winter (often soon after autumnal overturn) in the deepest zones of a lake, with the most important factor depending on the particular lake. These factors include the following:

1. After autumnal overturn, oxygen conditions in the lower profundal region become more favorable for the existence of benthic invertebrates (Deevey 1941; Eggleton 1931b, 1934).
2. After ice formation on the lake, temperatures are highest in the deepest regions of the lake (*Chapter IV.B*).
3. Abundance of organisms throughout the lake is high at this time of year because no insect emergence is occurring.
4. As organisms migrate to the lower profundal region, their concentrations increase owing to the smaller amount of area available for them to inhabit.

Emergence of Adult Insects

Information on emergence of adult insects whose larvae inhabit the benthic regions in Mirror Lake was obtained by monitoring 32%

---
[1] No data available for the littoral region in January.

(from 6 June to 14 August) of the total emergence period. Daily rates of emergence of all insects over the time period monitored (*Figure V.A.6–7*) showed that emergence tended to be highest in the upper littoral region. This pattern was expected since the density of benthic larvae was higher in the upper littoral region than in the lower littoral or profundal regions. Also the number of life cycles per summer would be higher in the upper littoral region because of the more rapid growth rates caused by the higher water temperatures and the small size of the organisms (Crisp 1971). These factors must outweigh losses attributable to fish predation, which should be maximal in the upper littoral region. Maximum emergence rates often occur in the upper littoral regions of other lakes also (e.g. Scott and Opdyke 1941; Sandberg 1969).

Chaoborus dominated total emergence at the two profundal stations in Mirror Lake, while the Tanypodinae and Tanytarsini dominated at the upper littoral station (*Table V.A.6–10*). Of the Chironomidae, the Tanypodinae were by far the dominant emergers (43 to 80% of the total chironomids emerging), both in the upper littoral and in the profundal regions. Since the tanypods were seldom dominant in the benthic fauna at any depth, their importance may have been overemphasized by the particular time period when emergence was monitored. In addition, the two dominant chironomids in the lake—*C. anthracinus* and Tanytarsini—appear to have emerged before or after the time period when emergence was monitored. To obtain truly quantitative data, emergence should be monitored throughout the entire emergence season (see Rosenberg et al. 1980 for optimum sampling scheme).

The total annual emergence from Mirror Lake was predicted to be approximately 4,500 adults/m^2-yr, assuming that (1) the rate of emergence was constant over the entire emergence period; (2) the 9.1-m trap was representative of the depth zone greater than 8 m; (3) the 7.6-m trap was representative of the 5- to 8-m zone; and (4) the 2.4-m trap was representative of the

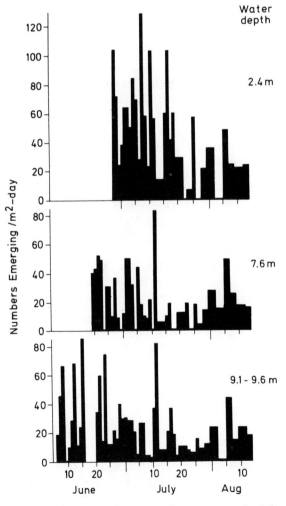

Figure V.A.6–7. Daily rates of emergence of adult insects from three locations in Mirror Lake.

Table V.A.6–10. Percent Contribution of Various Taxa to Total Adults Emerging per Square Meter between Late June and mid-August[a]

	% of Total Adults		
	2.4 m	7.6 m	9.1 m
Chaoboridae	12	51	70
Chironomidae	77	47	29
Tanypodinae	[33]	[38]	[18]
Orthocladiinae	[5]	[0]	[1]
Tanytarsini	[21]	[0]	[0]
Chironomini	[7]	[6]	[5]
Other	[11]	[3]	[5]
Trichoptera	7	1	<1
Unidentified	3	1	<1
Total Adults	100	100	100

[a] 23% of the entire emergence season.

0- to 5-m zone. Annual emergence is not as high as would be expected from the high density of benthic insect larvae. Annual emergence from Mirror Lake is only moderate compared with other lakes (*Table V.A.6–11*); however, few estimates are available for other oligotrophic lakes.

Losses of energy, carbon and nutrients as a result of insect emergence from Mirror Lake and other lakes are summarized in *Table V.B.6–3*. Losses of energy and carbon from Mirror Lake were estimated using values from the literature (see Walter 1976 for details). Losses of nutrients were determined using the average percent composition in adult and immature insects from Norris Brook of the Hubbard Brook Experimental Forest (R. J. Hall, personal communication; Hall and Likens 1981). Since adult females return to the lake to oviposit, it was assumed, based on the observations of several investigators (Tubb and Dorris 1965; Hilsenhoff 1966; Parma 1971) that the females die over the water and thus do not represent a net output of energy or nutrients from the lake. Since female adults are substantially heavier than males, the total emergence value was divided by three to give the predicted annual amount of each nutrient removed from the lake-ecosystem. The total amount of energy lost from Mirror Lake by means of insect emergence—5.8 Kcal/m^2-yr—represents <1% of net primary production (Makarewicz 1974) and is probably insignificant compared with other modes of energy transfer in Mirror Lake. The annual amount of organic carbon lost from the lake by way of emergence is only one-sixth that lost by respiration of the macrofauna. However, the loss of N and P as a result of emergence is substantial, especially when compared with annual inputs to Mirror Lake (*Table IV.E-13* and *16*). The implications of this loss are discussed in *Chapter VI.B*.

Summary

The dominant benthic macroinvertebrates in the littoral regions of Mirror Lake (0 to 7.2 m) are chironomid midge larvae, amphipods, oligochaetes and sphaeriid clams. The profundal region (7.2 to 11 m) is dominated by organisms adapted to living in anaerobic environments—the phantom midge larva *Chaoborus,* the chironomid midge larva *Chironomus anthracinus* and oligochaetes. The Tanytarsini, a tribe of chironomid midge that is sensitive to low-oxygen concentrations, lives only in the uppermost reaches of the profundal region where it is associated with the macrophyte *Nitella*. Chironomids

Table V.A.6–11. Annual Emergence of Adult Insects in Various Lakes

Lake	Annual Emergence (No./m^2-yr)	Reference
Char, N.W.T., Canada[a]	690[e]	Welch 1973
Erken, Sweden[c]	420–2,200[e]	Sandberg 1969
Tundra Ponds, Alaska[a]	500–5,000	Butler et al. 1980
Mirror, New Hampshire[a]	4,500	Walter 1976
Eglwys Nunydd, South Wales[c]	5,000	Potter and Learner 1974
4 Ponds, Southwest Ontario	2,200–7,200	Judd 1953, 1960, 1961, 1964
Kempton Park E. Res., England[c]	2,800–7,300[e]	Mundie 1957
Clear, California	>6,500[f]	Lindequist and Deonier 1942
2 ponds, Hertfordshire, England[b,c]	2,600–16,700	Learner and Potter 1974
Loch Dunmore, Scotland[d]	6,000–21,000	Morgan and Waddell 1961
2 Sewage Lagoons, Oregon[c]	8,400–35,000[e]	Kimerle and Anderson 1971

[a] Oligotrophic.
[b] Mesotrophic.
[c] Eutrophic.
[d] Small trout loch.
[e] Chironomids only.
[f] *Chaoborus* only.

are the dominant group of benthic macroinvertebrates in Mirror Lake in terms of numbers (up to 23,000/m^2) and biomass (up to 5.3 g dry wt/m^2). Common insect inhabitants of the littoral regions of Mirror Lake include caddisflies, biting midges (no-see-ums), alderflies, mayflies, beetles, dragon- and damselflies and moths. Other common inhabitants of the littoral regions include the mussel *Elliptio,* water mites, sponges, bryozoans and crayfish.

Benthic samples were taken in the summer (July and August) and winter (January and March). An Ekman dredge and multiple corer were used to sample the gyttja. Substrates other than gyttja were sampled with diver-operated hand corers. Mussels were sampled along transects by SCUBA divers. Three traps were used to monitor adult insect emergence over one-third of the total emergence period.

Most benthic invertebrate taxa in Mirror Lake show a clumped distribution. Within a given depth interval, within-station variance is usually as great as, or greater than, between-station variance.

Mean annual biomass was 2.8 g dry wt/m^2 for the profundal region and 1.8 g dry wt/m^2 for the entire lake. Mean annual abundance of total benthic macroinvertebrates for the profundal region was 16,000 organisms/m^2 and for the entire lake 20,000 organisms/m^2. Such high abundance for an oligotrophic lake may be explained by the eutrophic nature of the profundal environment and its macrofauna, and by the small sieve size used to process the benthic samples (which is important in the littoral region where many of the organisms are small).

Abundance and biomass reach a peak in winter, decline over the summer because of emergence and predation, and reach a low prior to autumnal overturn. In July, the depth-distribution of benthic macrofauna closely follows the distribution of macrophyte biomass in the littoral regions, with peaks in the upper littoral region at 1 to 2 m and just below the thermocline at 6 to 8 m. A minimum in abundance and biomass occurs in the region of the thermocline. Abundance of benthic macrofauna declines with depth in the profundal region as a result of low oxygen concentrations. In January, the depth-distribution in the profundal region is reversed, and peak abundance (up to 41,000 organisms/m^2) occurs in the deepest part of the lake, mainly because of high concentrations of *Chaoborus*.

Total annual emergence was estimated to be 4,500 adult insects/m^2, with highest emergence rates occurring in the littoral regions. Annual losses of energy and nutrients from Mirror Lake to its watershed as a result of adult insect emergence were estimated to be 5.8 kcal/m^2, 506 mg organic C/m^2, 84 mg total N/m^2 and 16 mg total P/m^2. Losses of energy and organic carbon appear to be negligible; however, losses of N and P are substantial when compared with annual inputs to Mirror Lake.

7. Benthic Microinvertebrates

David L. Strayer

As was pointed out in *Chapter V.A.6,* many benthic animals are too small to be retained on the sieves used in routine macrobenthic surveys. As a result, these animals have been largely ignored by freshwater biologists, in spite of the fact that they vastly outnumber macrobenthic animals in both density and species diversity. They have received considerable attention, though, from marine ecologists (see reviews by McIntyre 1969 and Fenchel 1978), who have divided this fauna into the *meiofauna:* animals passing the coarse sieves (250- to 1,000-μm mesh) used in macrobenthic work, but retained on fine sieves (40- to 100-μm mesh); and the *microfauna:* animals, especially protozoans, that pass even these fine sieves. This classification is one of convenience; size fractions do not generally correspond to any natural ecologic or taxonomic groupings. Meio- and microbenthic animals serve as competitors, predators and prey of macrobenthic animals, and may perform similar functions in lake-ecosystems.

A detailed survey of the benthic microinvertebrates of Mirror Lake is presently in progress, but sufficient preliminary data are available to give a general overview of the fauna. Because protozoans were excluded from this study (see below), the term *benthic micrometazoans* is used to refer to the fauna considered in this chapter.

Methods

Sampling methods devised for the collection of *macro*benthic animals are generally acceptable for the study of smaller animals, except that size-selective procedures such as sieving must not be used in sorting animals from sediment. A variety of methods have been used to separate meio- and microbenthic animals from sediments (see the review of Hulings and Gray 1971); in this study, sediment samples were carefully examined under a dissecting microscope and animals were removed using a pipette. Sorting was done within a few days of sample collection, because the animals are much easier to find when they are alive and moving. It took 25 to 40 h to sort a sample in this way. Because of time limitations, therefore, it was not possible to consider the benthic Protozoa, which were sometimes very abundant in the Mirror Lake sediments.

Sediment samples were taken by using the multiple corer described in *Chapter V.A.6*, but only one of the four cores in a set was analyzed. During 1980–81, 20 cores were collected from each of four depths on the gyttja (6, 7.5, 9 and 10.5 m). Twelve additional samples were taken by diver from shallower water in the summer and autumn of 1981.

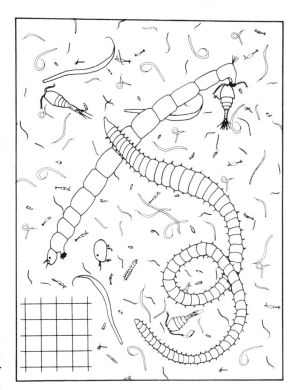

Figure V.A.7–2. Size structure of the Mirror Lake zoobenthic community. This sketch represents 1 cm² of lake bottom with its associated metazoan community. The animals, the dimensions of the figure and the 0.5-mm mesh sieve are all drawn to the same scale. Population densities are based on the means for the area of gyttja bottom in Mirror Lake.

Abundance and Biomass

Total density of benthic micrometazoans (*Figure V.A.7–1*) ranges from 500,000 animals/m² to 3,500,000 animals/m² and does not show a strong dependence on depth or season. There is, however, a weak minimum at 7.5 m, and densities are somewhat lower during periods of stratification than during turnover. The mean micrometazoan density for all samples (1,200,000/m²) is about 60 times higher than the macrofaunal densities reported by Walter in *Chapter V.A.6*, even though she used a relatively fine-meshed sieve (250 μm) in her study. The great majority of benthic animals in Mirror Lake are small (*Figure V.A.7–2*).

It is not possible to make a direct comparison between Mirror Lake and the few other lakes in which meio- and microbenthic animals have been studied because of substantial method-

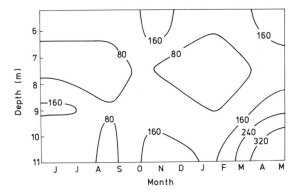

Figure V.A.7–1. Abundance (number/cm²) of benthic micrometazoa on the gyttja in Mirror Lake, 1980–81.

ological differences among the studies (in, for example, sorting procedures and extent of sampling). Nonetheless, the densities found in Mirror Lake are consistent with those reported for other lakes (Stanczykowska and Przytocka-Jusiak 1968; Holopainen and Paasivirta 1977; Oden 1979; Anderson and de Henau 1980; Nalepa and Quigley 1983), as well as those given for marine faunas (McIntyre 1969; Fenchel 1978).

The lakewide average biomass of benthic micrometazoans in Mirror Lake is 0.6 g dry weight/m^2, or about 30% of macrobenthic biomass. In other lakes for which similar estimates have been made, meiobenthic biomass was 10 to 300% of macrobenthic biomass (Stanczykowska and Przytocka-Jusiak 1968; Holopainen and Paasivirta 1977; Paasivirta and Sarkka 1978; Oden 1979; Anderson and de Henau 1980; Nalepa and Quigley 1983).

Diversity

Members of about 60 families of micrometazoans have been found in samples from the gyttja (Table V.A.6–1; Figure V.A.7–3), representing 10 of the 14 phyla known from fresh water. Taxonomic diversity (Figure V.A.7–4) declines sharply with increasing depth, but never falls below 7 families per sample, even during periods of anoxia. Seasonal variation is less obvious, but there is a small increase in diversity during turnover. This increase is caused by an influx of littoral species that are carried into deeper water by water movements. These species (e.g., *Hydra,* planarians and

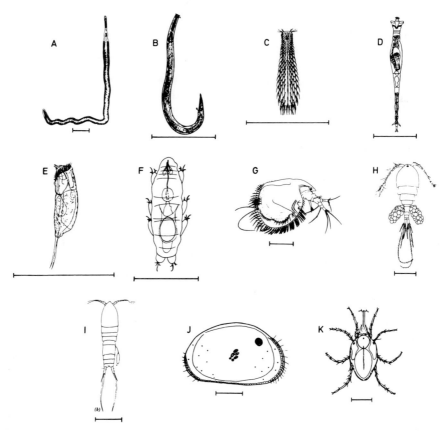

Figure V.A.7–3. Some benthic micrometazoans: **A.** The microturbellarian *Rhynchoscolex simplex;* **B.** A benthic nematode; **C.** *Chaetonotus,* a gastrotrich; **D.** the bdelloid rotifer *Rotaria tridens*; **E.** the ploimate rotifer *Cephalodella*; **F.** A tardigrade; *G. Ilyocryptus sordidus,* a benthic cladoceran; **H.** A cyclopoid copepod; **I.** A harpacticoid copepod; **J.** An ostracod; **K.** a halacarid mite. The scale bar represents 200 μm. (From Luther 1960; Hyman 1951; Brunson 1959; Bartos 1951; Pennak 1978; Hutchinson 1967; Wilson and Yeatman 1959.)

A: Species Composition, Distribution, Population, Biomass and Behavior

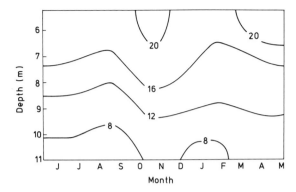

Figure V.A.7–4. Diversity of benthic micrometazoa on the gyttja in Mirror Lake, 1980–81. The numbers plotted are mean number of taxonomic families per sample (each sample contains about 200 individuals). Because they have not yet been identified to family, it was assumed that all nematodes belonged to one family, so this figure underestimates true familial diversity.

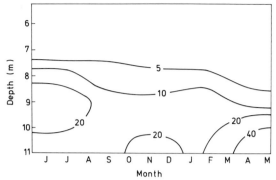

Figure V.A.7–5. Abundance (number/cm^2) of the bdelloid rotifer *Rotaria* cf. *tridens* on the gyttja in Mirror Lake, 1980–81.

mayflies) clearly do not belong in the profundal zone and disappear from the fauna with the re-establishment of thermal stratification.

Total species diversity is not yet known, but there are probably about 150 species of micrometazoans living on the gyttja in Mirror Lake. In other lakes where investigators used appropriate techniques, similarly high diversities were found (Monard 1920; Moore 1939; Cole 1955). Lacustrine benthic invertebrate communities are more diverse than is generally realized (see the incomplete *Table V.A.6–1*).

Effects of Depth and Season on Abundance

The distributions of the two most abundant species of bdelloid rotifers are shown in *Figures V.A.7–5* and *6*. These two species have markedly different distributions, but in both cases, population density is clearly related to water depth. Seasonal variation is much less pronounced. Available data suggest that, in general, population densities of benthic micrometazoans are much more a function of depth than of season. In this respect, the benthic micrometazoans may resemble the benthic macroinvertebrates more closely than the planktonic micrometazoans, which often show great seasonal variation in population density (cf. *Chapter V.A.5*).

It is not surprising that the distributions of various benthic micrometazoan species are correlated with water depth; a number of important environmental factors vary with water depth, especially during stratification (see *Chapters VI.A* and *B*). It is not clear, however, which of these factors are responsible for restricting the distributions of the micrometazoans.

Biology of the Benthic Micrometazoa

Biologic features of some of the benthic microinvertebrates are presented in the following section, along with brief remarks on their distributions in Mirror Lake. Unless noted otherwise, the general biologic information is drawn from Hyman (1949, 1951) and Pennak (1978).

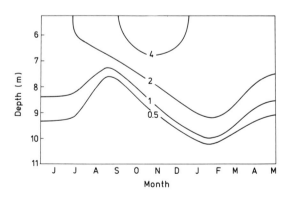

Figure V.A.7–6. Abundance (number/cm^2) of the bdelloid rotifer *Dissotrocha macrostyla* on the gyttja in Mirror Lake, 1980–81.

Microturbellarians

These small (0.3 to 6 mm long), free-living flatworms (known to many American biologists as rhabdocoels) are common inhabitants of freshwater benthic environments. Little is known of their distribution and importance in lakes. Most microturbellarians are voracious predators of protozoans, rotifers and microarthropods (e.g. Case and Washino 1979; Maly et al. 1980), although some species feed on algae. Most species are associated with vegetation or oxygenated sediments, but some worms of the genus *Mesostoma* are regularly found in the pelagic zone, both in Mirror Lake and elsewhere (Hutchinson 1967). Because microturbellarians contract into a nondescript ball of jelly on routine preservation, they are undoubtedly overlooked in many studies of the zooplankton and macrobenthic invertebrates.

Microturbellarians have been found in most samples taken from oxygenated sediments in Mirror Lake, from 1-m to 9-m water depth. Densities are typically 5,000 to 50,000/m^2. The most common and widespread species in Mirror Lake is *Rhynchoscolex simplex* (*Figure V.A.7-3A*), which is believed to feed on dipteran larvae and oligochaetes (Luther 1960). About 20 species of microturbellarians have been identified from Mirror Lake, but more are certainly present. The few extensive studies of lacustrine microturbellarian communities found 15 to 50 species (e.g. Chodorowski 1959; Rixen 1968; Kolasa 1977, 1979).

Nematodes

Free-living nematodes are probably the most abundant benthic metazoans in lakes and may be expected from almost every benthic habitat. Freshwater nematodes are usually 0.5 to 5 mm long and like their relatives, the parasitic roundworms, are usually slender and colorless (*Figure V.A.7-3B*). Various species feed on detritus, bacteria, plants or small animals.

Nematodes have been found in every sediment sample taken from Mirror Lake, from the shallow littoral to the anoxic profundal. They are very abundant (200,000 to 2,500,000/m^2); about 70% of the metazoans living on the gyttja in Mirror Lake are nematodes. The relatively low densities of nematodes reported in other lakes are probably underestimates, as most other workers used sieves in their work (e.g. Prejs 1970; Bretschko 1973). Very little work has been done on the roles of nematodes in lake-ecosystems, but it is likely that they are similar to those of the more familiar soil nematodes in terrestrial ecosystems (Yeates 1979).

Gastrotrichs

Gastrotrichs are tiny, ciliated metazoans, common in many benthic environments. The Mirror Lake species are 100 to 500 μm long, and many are strikingly ornamented with spines or scales (*Figure V.A.7-3C*). They apparently feed primarily on bacteria, but may eat algae and detritus (Bennett 1979).

Gastrotrichs are found throughout Mirror Lake, but are most abundant on the gyttja, where they may be found even under anaerobic conditions. Densities on the gyttja are usually 10,000 to 600,000/m^2. *Chaetonotus* and *Lepidodermella* are the most common of the five genera found.

Rotifers

Although rotifers are well known in zooplankton, most rotifer species are benthic. The benthic rotifers of a lake may be almost unbelievably diverse, especially if the water is nutrient-poor. For example, Oden (1979) found 372 species of benthic rotifers living in the shallow littoral zone of Par Pond, South Carolina, and similarly large faunas have been reported from nutrient-poor lakes in other parts of North America (e.g. Harring and Myers 1928). This taxonomic diversity corresponds to an equally great ecologic diversity in the benthic rotifers, and various species feed on detritus, bacteria, algae and small animals; there are even parasitic species.

Members of two very different orders of benthic rotifers live in Mirror Lake.

Bdelloid rotifers (*Figure V.A.7-3D*) creep about among the sediment particles, where they feed on fine particulate matter. Eight species of bdelloids have been found in Mirror Lake, where they occur at densities of 40,000 to 500,000/m^2. Most species are restricted to oxygenated sediments, but one species, *Rotaria* cf. *tridens,* is commonly found in great numbers (up to 250,000/m^2) in anoxic profundal sediments (*Figure V.A.7-5*).

The order Ploima contains a wide variety of free-swimming benthic rotifers (e.g. *Figure V.A.7–3E*) as well as most of the familiar planktonic rotifers. Ploimates are abundant in Mirror Lake (usually 20,000 to 100,000/m^2), but are scarce in the seasonally anoxic profundal sediments. Twenty-five genera of ploimates, some of them containing several species, have been identified from Mirror Lake. More genera are certainly present.

Tardigrades

These animals are known as *water-bears* because of their bearlike shape and ponderous locomotion. They are usually herbivorous and are found in a variety of benthic habitats in lakes. In Mirror Lake, tardigrades (*Figure V.A.7–3F*) are found on oxygenated sediments from the shoreline to about 9 m, but are nowhere abundant. Densities are usually less than 10,000/m^2.

Microarthropods

Tiny arthropods are abundant in benthic environments as well as in the pelagic zone. The benthic microarthropod community of Mirror Lake includes cladocerans, cyclopoid and harpacticoid copepods and halacarid mites.

Benthic cladocerans are known to limnologists mainly because the remains of one family, the Chydoridae, are common in lake sediments and are important in paleolimnological investigations (*Chapter VII.C*). While living, these animals creep or swim about in vegetation or along the substratum, where they feed primarily on algae and detritus. In addition to the chydorids, lesser numbers of macrothricid and sidid cladocerans have been found in the sediments of Mirror Lake. Most of the cladoceran species in Mirror Lake have population maxima at water depths of less than 6 m, but one species, the macrothricid *Ilyocryptus sordidus* (*Figure V.A.7–3G*), is common on oxygenated parts of the gyttja. Total cladoceran density rarely exceeds 20,000/m^2.

Cyclopoid copepods (*Figure V.A.7–3H*) are common in benthic environments as well as in the pelagic zone (the planktonic species are discussed in *Chapter V.A.5*). In fact, many planktonic species spend a part of their life on the sediments (Hutchinson 1967, 639–642; Sarvala 1979), so almost all cyclopoids are more or less benthic. Adult cyclopoids are 0.3 to 3 mm long and are good swimmers. Cyclopoids seize and chew their food, which may be either animal or plant. These animals are abundant in most benthic samples from Mirror Lake and are frequently found even under anoxic conditions. Densities are usually 5,000 to 75,000/m^2.

Harpacticoid copepods are characteristic inhabitants of freshwater and marine benthic environments, but have received very little attention from North American limnologists. The adults (*Figure V.A.7–3I*) are 0.3 to 1.4 mm long and crawl or dart about on the bottom. Harpacticoids are abundant on sandy sediments in Mirror Lake (up to 100,000/m^2) and are found in lesser numbers on the oxygenated gyttja.

Ostracods (*Figure V.A.7–3J*) are small bivalved crustaceans that burrow into the sediments or swim and crawl about at the sediment surface. Little is known of the distribution and importance of benthic ostracods in lakes, although they are abundant in many lakes and their valves are preserved in lake sediments. Ostracods feed on detritus, bacteria and algae. They are abundant on oxygenated sediments in Mirror Lake and reach their highest densities (more than 25,000/m^2) at water depths of 7.5 m on the gyttja.

Mites of the family Halacaridae are found in the Mirror Lake sediments, along with the familiar water mites of the suborder Parasitengona (which are discussed in *Chapter V.A.6*). Halacarids (*Figure V.A.7–3K*) are tiny (less than 1 mm long), flattened mites, sometimes of striking morphology, that creep among the sediment particles. They are a characteristic part of the marine meiofauna (e.g. Newell 1947), but little is known of their distribution and abundance in fresh water (but see Paasivirta 1975a). They are probably predators. Halacarids are found from the shoreline to about 6 m depth in Mirror Lake but are nowhere abundant. Typical densities are 1,000 to 5,000/m^2.

Chironomids and Oligochaetes as Meiobenthos

It is well known that some chironomids and oligochaetes pass the sieves used in studies of the benthic macrofauna. The number of such meiobenthic chironomids and oligochaetes may be very large (e.g. Jonasson 1958), especially if small juveniles are present and if a relatively coarse sieve is being used. It is not possible to calculate directly how many of the chironomids and oligochaetes in Mirror Lake passed the 250-μm sieve used for the work of *Chapter V.A.6*, because seven years passed between that study and the work reported in this chapter. Nonetheless, a rough estimate can be made by comparing the densities of chironomids and oligochaetes reported in *Chapter V.A.6* with those found in the present study. I found three times as many chironomids and nine times as many oligochaetes as did Walter, suggesting that many of these animals pass even a 250-μm mesh sieve. Thus, even these "macrobenthic" animals are not retained quantitatively on most sieves. This result serves as another reminder that any classification of benthic animals by size is to some extent arbitrary.

Roles of the Benthic Micrometazoa in the Lake-Ecosystem

The paucity of information about the benthic micrometazoans and their distribution in lakes makes it difficult to estimate their importance in the ecosystem. These animals are small, so presumably they have shorter turnover times than macrobenthic animals (although this remains to be demonstrated). If this were so, then production and respiration of meio- and microbenthic communities would equal or exceed that of macrobenthic communities, since meiofaunal biomass has been found to approximately equal macrofaunal biomass in lakes (Holopainen and Paasivirta 1977; Anderson and de Henau 1980). In fact, Holopainen and Paasivirta (1977) estimated that meiobenthic secondary production was about 100 to 400% of macrobenthic secondary production in Lake Paajarvi, Finland. Furthermore, it is likely that other functions attributed to macrobenthic animals (e.g. mixing sediments, grazing and controlling bacterial populations—see *Chapter V.B*) are performed by animals classified as meio- or microbenthic as well. Much more research is needed before we will be able to identify and quantify the roles of these potentially important animals.

Summary

Most benthic animals are too small to be retained on the sieves (250- to 1,000-μm mesh) used in routine studies of macrobenthic animals. These animals, constituting the meio- and microfauna, are diverse, abundant and widespread in lakes, although they have been studied only infrequently and their existence is scarcely known to many limnologists. In Mirror Lake, 98 to 99% of the benthic animals (excluding the Protozoa) pass a 250-μm screen. The species diversity of the Mirror Lake fauna has not yet been determined, but members of about 60 taxonomic families have been found. There is considerable variation in community composition with water depth, but seasonal variation appears to be much less pronounced. It is likely that meio- and microbenthic animals play important roles in lake-ecosystems.

8. Salamanders

THOMAS M. BURTON

Mirror Lake contains a small population of the red-spotted newt, *Notophthalmus v. viridescens*. Commonly, this salamander has a 2- to 3-month aquatic larval stage, a 2- to 4-yr terrestrial (eft) stage and an aquatic, adult (newt) stage. However, this highly adaptable species may have a completely aquatic life cycle, as has been reported for several neotenic populations in Massachusetts and New York (Healy 1970). In Mirror Lake, courtship and mating occur from late May to late June. Based on dissection of gravid females, the female deposits from 55 to 200 (mean of 143 ± 56) eggs. These eggs are "glued" in pockets in leaves of aquatic plants according to observations elsewhere (Bishop 1941). Hatching in Mirror Lake starts about the end of the first week of July. Larvae are approximately 7.5 to 8.5 mm long at hatching and have undeveloped legs and well-developed gills and balancers. No estimate of the larval population size is available from Mirror Lake. Adult newts are cannibalistic on their own larvae (*Figure*

Figure V.A.8–1. Food habits of the newt, *Notophthalmus viridescens*, in Mirror Lake.

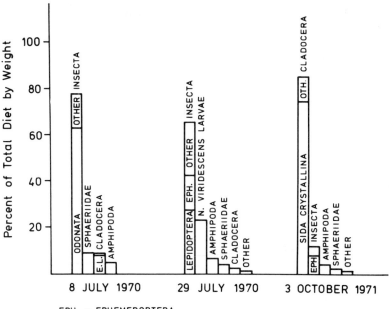

V.A.8–1) with these larvae making up about 20% of the diet of the adults in July and August. The larvae hide in the vegetation in the very shallow areas (water less than 0.5 m deep) at the margin of the lake. They grow rapidly and transform into the terrestrial eft stage at a total length of about 36 mm and a weight of 0.5 to 0.8 g wet weight in mid-September. Most efts remain on the land for three years and return to Mirror Lake during both a spring and autumn migration period. Some efts remain on the land for as long as five years. The efts transform into the aquatic or newt stage after entering the water and remain completely aquatic for the rest of their lives.

The adult male newt can be aged by counting testes lobes. About 3% of the summer population of adults are three years old, about 25% are four years old, about 57% are five years old, and the remainder are six years old or older. The 3-yr-olds are probably the newly arrived spring migrants, while the 4-yr-olds include a combination of spring and autumn migrants from the year before. There is additional recruitment in the 4- and 5-yr-old groups from older migrants from the land.

The adult newts are found in very limited areas of the littoral zone of Mirror Lake. Their distribution is highly correlated with areas of vegetation in the shallow littoral areas of the lake near the inlets and outlet. Indeed, about 90% of the population is found in the area of the West inlet and in the outlet pond. These areas of Mirror Lake have more dense populations of waterlilies and other aquatic plants that can serve both as a refuge from predators and as areas of dense prey populations for the newt. Some adults will remain in or near floating masses of *Utricularia purpurea* and can be found with such masses along any part of the shoreline, especially along the southern shore between the West inlet and the outlet. Occasionally, a few individuals will stray into the deeper vegetated zones, but the population is essentially found in water less than 2 m deep.

Predation is normally not a major problem for newts because of the residual effect of toxic skin secretions (tetradotoxin) produced by the eft stage prior to returning to the water (Brodie 1968); but smallmouth bass in Mirror Lake do prey on newts, at least to a limited extent according to stomach analyses (Mazsa 1973). Bullfrogs, turtles, older raccoons and some snakes can eat the efts and adults with few ill effects (Hurlbert 1970). Small populations of *N. viridescens* adults (50 to 150) occurred in the Northwest and Northeast inlet areas in 1970 and 1971 but not in 1972. Thus, small subpopulations do occur at different areas in different years, but the West inlet and outlet pond areas are the main centers of the population.

The population size for the West and Northwest inlet areas in 1970 was estimated using mark-recapture techniques. Adults were captured while they were in courtship aggregations, were marked by toe clipping and released. The population for these two inlets was estimated to be 2,600 ± 1,400 (estimate ± 95% error limit) in 1970. In 1971 and 1972, the population for the whole lake was estimated using a SCUBA censused quadrat technique (Burton 1977). The population in 1971 was estimated to be 2,600, while the population in 1972 was estimated as 1,950. Thus, the population of adult *N. viridescens* in Mirror Lake appears to be between 2,000 and 3,000. Since the mean weight of the adult newt population is 0.62 g dry weight per salamander, the biomass in Mirror Lake is 1,550 g dry weight.

Adult *N. viridescens* prey on a wide variety of invertebrates (*Figure V.A.8-1*). The most common prey items vary throughout the summer. While the data are not conclusive, it appears that newts take whatever prey is available in the proper size class with seasonal differences reflecting the availability and catchability of the prey (*Figure V.A.8-1*). For example, large mayfly larvae were the most important prey item in early July 1970; terrestrial Lepidoptera larvae falling into the lake from overhanging trees owing to an outbreak of defoliating *Heterocampa* caterpillars were most important in late July 1970; while the cladoceran, *Sida crystallina*, was most important in the October 1971 sample (*Figure V.A.8-1*). It is possible that newts form *search images* for particular prey. The stomachs of a few specimens collected in early July 1970, were engorged with the cladoceran, *Eurycercus lamellatus*, while others contained only insects. Chironomidae (midge) larvae, Sphaeriidae (fingernail clams) and Amphipoda (*Hyalella azteca*) were always small but significant components of the diet.

The larvae of *N. viridescens* prey primarily on the amphipod, *Hyalella azteca* (55% of total weight of prey); Chironomidae larvae (16% of prey); Cladocera, especially the small littoral species *Ophryoxus gracilis* and *Alonella excisa* (13% of prey); Sphaeriidae (9% of prey); and Ostracods (5% of prey).

Summary

There are about 2,500 adult newts (*N. viridescens*) found in a very restricted area of the littoral zone of Mirror Lake. About 90% of the population is found in the West inlet and outlet pond areas of the lake in or near masses of vegetation such as waterlilies, floating masses of *Utricularia purpurea*, etc. In these limited areas, this salamander is one of the top predators and preys on a wide variety of invertebrates. The total biomass is about 1,550 g dry weight. Migrations out of and into the lake by the eft stage probably represent a small net flux of energy and nutrients for the lake. Data on this question are lacking, but some discussion is included in *Chapter V.B.7*.

Other amphibians using the shallow areas of the lake for breeding and larval development include the northern frog, *Rana p. pipiens,* and the American toad, *Bufo americanus*. Numbers of these organisms can be substantial in restricted, very shallow areas of the lake, but are unlikely to be an important component of animal biomass for the entire lake.

9. Fishes

GENE S. HELFMAN

Mirror Lake contains five species of fish that are common throughout much of temperate North America: smallmouth bass (*Figure V.A.9-1*), yellow perch, chain pickerel, brown bullhead and white sucker. A community of such low diversity immediately raises a number of ecologic questions. Why are there so few species? What characteristics of the environment contribute to or cause such low diversity? Why do we find these particular species? What characteristics of the species are responsible for their successful colonization of the lake? In this chapter I review aspects of the life history of the fishes of Mirror Lake and relate this infor-

Figure V.A.9–1. A smallmouth bass in Mirror Lake. (Photograph by R. E. Moeller. Used with permission.)

mation to the question of limited diversity in the fish fauna and also to the role of fishes in the Mirror Lake ecosystem.

Although Mirror Lake fishes *per se* have received comparatively little investigation (see Mazsa 1973), the species that occur there have been well studied elsewhere and this information can be applied with reasonable confidence to events and relationships in Mirror Lake.

Species

The smallmouth bass, *Micropterus dolomieui*, is a predator that ranks "second to none among all the game fishes of the fresh waters of North America" (Hubbs and Bailey 1938; see also Robbins and MacCrimmon 1974 for general review). It occurs naturally or has been introduced into lakes and streams throughout most of eastern North America. Its natural range overlaps with that of the four native species in Mirror Lake and, although it is introduced in Mirror Lake, its behavior and relationships in the lake are probably representative of natural situations. The smallmouth bass prefers cooler, clearer water than its close relative, the largemouth bass. The most common habitat is large, clear lakes and clear streams with moderate current. In both situations it is most often found in association with rocky shoals or fallen logs where vegetation is relatively scarce and where water varies from 2 to 6 m depth. Adult fish are roving predators, frequently patrolling a small area of shoreline in a lake or moving between a few *home pools* in a stream (Munther 1970). They are inquisitive fish and it is not uncommon for a large individual to swim up to the faceplate of a SCUBA diver and peer inside.

As a fish grows, its food preferences shift from plankton to immature aquatic insects to crayfish and fishes. As much as 90% of the diet of adult fish may be crayfish, followed by fishes and aquatic insects (Tester 1932). Food is taken from the bottom, midwater and the surface. I have seen an individual hunting for food in Mirror Lake move from surface to bottom and back to the surface repeatedly in water as deep as 5 m. Among the fishes eaten, yellow perch appear to be a preferred species (Tester 1932), although the diet includes most other associated fishes, including smallmouth bass. These predators are most active and successful during crepuscular (dawn and dusk twilight) periods, when "they may move so far inshore that their backs are exposed, and splash noisily as they lunge about after fishy morsels" (Hubbs and Bailey 1938). They also feed during the day and are relatively inactive at night, with younger fish ceasing activity earlier in the evening and resuming activity later in the morning than larger individuals (Helfman 1981). As water temperatures decrease in the autumn, the fish move to deeper water, where they may aggregate and become torpid. This winter resting site is generally in regions of low light penetration and little current, such as crevices between rocks, holes, submarine caves, hollow logs (Coble 1975) or even sediment traps in Mirror Lake that are being used in paleoecologic studies (*Chapter VII.A*).

Smallmouth bass were first introduced into Mirror Lake in 1941–42, and then again in 1951 and 1953 (Newell in Mazsa 1973). Mazsa studied aspects of the biology of smallmouth in Mirror Lake and his findings are in general agreement with information from other areas where the fish is a native or an introduced species. He found that large fish ate less zooplankton and small insect larvae (e.g. chironomids, *Chaoborus*, Ephemeroptera nymphs) than smaller fish did. Small and medium smallmouth bass ate larval fish, while large individuals consumed primarily fish and crayfish. Allochthonous food sources were important and at times constituted as much as half of the volume of food consumed by some fish. Allochthonous foods included caterpillars of the genus *Heterocampa*, which were eaten by larger bass, and formicid flying ants, which were consumed by all size classes.

Smallmouth bass occurred around the entire periphery of the lake, concentrating where solid structures such as large rocks and logs were present, but they were also found in pelagic regions in midlake. Growth (measured as change in total length) occurred at a similar, although slightly reduced, rate compared with other northern lakes, particularly after the first two years. However, bass survivorship in Mirror Lake was estimated at between 60 and 70% annually for age IV fish and older, which is at the high end of the range given in Carlander (1977) for more heavily fished populations at various locales. The slope of the log length (cm) to log weight (g) regression calculated by Mazsa (1973) is 2.995, which appears to be slightly less than the mean slope of 3.282 calculated from 23 other North American populations (Carlander 1977). This comparison would suggest that Mirror Lake fish are gaining less weight with increasing size than the average smallmouth bass.

Mazsa (1973) estimated that there were approximately 600 to 700 smallmouth bass in Mirror Lake, based on a tag-recapture study that was admittedly selective because it was based largely on fish caught while angling. This number is probably an underestimation of actual population size because only two-year-old and older fish were used in the calculations. One-year-old fish constituted 43% of the individuals Mazsa captured, which is similar to the percentage of one-year-olds found in other smallmouth bass populations (e.g. 51% in Fajen 1975). If one-year-olds are included in the calculations, the estimated population size is nearly doubled, which could be a significant change in terms of potential ecologic relationships. With one-year-olds included, the standing stock of smallmouth bass in Mirror Lake would be approximately 80 to 90 fish/ha, or at the upper end of the density range (0.25 to 138 fish/ha) given by Carlander (1977) for large lakes.

The yellow perch, *Perca flavescens*, is probably one of the best known of freshwater fishes (see Colby 1977; Thorpe 1977a,b; Ney 1978). It is a highly variable fish, plastic in many attributes, including habitat preferences, feeding patterns, growth, schooling, coloration, developmental trends and diel and seasonal activity cycles (e.g. Helfman 1979). The yellow perch, combined with the possibly conspecific Eurasian perch, *Perca fluviatilis* (see Thorpe 1977a), has a circumpolar distribution that includes most of northern North America and Europe. Although it may occur in any number of lake, river, and stream habitats, the preferred combination of habitat characteristics is found in relatively clear, mesotrophic lakes with muddy, sandy or gravel bottoms, with abundant open water, modest amounts of littoral vegetation and substantial populations of forage fishes (Thorpe 1977a,b). Larval stages aggregate in open water and then move into littoral areas as individuals grow, although schools of both juveniles and adults may be found in shoreline as well as a variety of deeper littoral and sublittoral regions. Most feeding, both on pelagic and on benthic organisms, is done by fish in schools. In many lakes, schools of perch make regular late afternoon and early morning migrations between deep water and shallow areas, although other schools in shallow water may not exhibit such migrations (Scott 1955; Helfman 1979). Feeding is restricted to daylight hours. A general progression in diet occurs as fish grow, beginning with rotifers and planktonic crustaceans, then including benthic invertebrates and finally developing into piscivory

(Scott and Crossman 1973). This shift in food preferences may or may not occur in different water bodies and at different fish sizes. Consequently, all sizes may feed on plankton, or different individuals from a single spawning of one adult pair may specialize on plankton, benthic invertebrates, or fish fry (Il'ina 1973). "Thus the differences are primarily a response to availability, perch making use of any prey of a size appropriate to its gape" (Thorpe 1977b). Cannibalism is common. Feeding occurs throughout the year and in some lakes fish under ice may feed solely on benthic invertebrates (Hartman 1974 in Thorpe 1977b), or primarily on Cladocera (personal observation, Cazenovia Lake, New York).

In Mirror Lake, most physical and biotic factors of depth, temperature, bottom type, light penetration and forage types fall within the preferred range for yellow perch, particularly for the younger stages. Growth rates of perch in Mirror Lake (*Figure V.A.9–2*) exceed or follow closely the "standard" or moderate growth rate proposed by Tesch (1955 as cited in Thorpe 1977a) for *Perca fluviatilis* through the first six years. Growth rates then fall behind the standard rate in years VII and VIII. Perch in Mirror Lake also fare well in comparisons with various North American populations of yellow perch, attaining lengths at all ages that are in the middle of the ranges reported by Carlander (1950 in Thorpe 1977a). Mazsa (1973) found that perch in Mirror Lake were growing as fast as or faster than perch in nearby populations in Massachusetts and Pennsylvania.

Food habits of yellow perch in Mirror Lake were also typical for North American populations. Young of the year fed largely on *Bosmina* and *Daphnia*, plus other small zooplankters and some *Hyalella*. Cladocerans remained important in the diets of all size classes, although they were utilized most heavily by smaller fish. Aquatic invertebrates (Anisoptera, chironomids, *Hexagenia*, Trichoptera, *Chaoborus* and *Hyalella*) became increasingly important for larger fish. Unlike populations in other lakes, piscivory was limited to the small and medium size classes and consisted of feeding on fish fry. Very little use was made of allochthonous terrestrial foods such as ants or caterpillars.

Perch are the most abundant fish in Mirror Lake, with a population of about 3,000 individ-

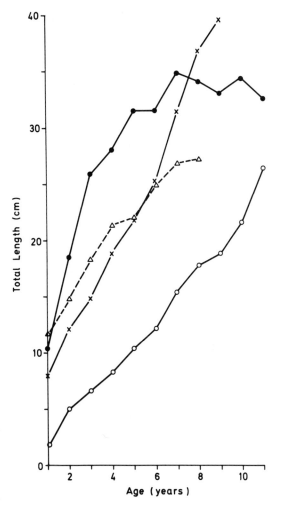

Figure V.A.9–2. Growth rates of yellow perch in Mirror Lake as compared with growth in other regions. △ = Mirror Lake (data from Mazsa 1973); × = "standard" or moderate growth rate proposed by Tesch (1955 in Thorpe 1977a) for the congeneric European perch, *Perca fluviatilis;* ○ and ● = minimum and maximum growth rates for North American populations of yellow perch (Carlander 1950 in Thorpe 1977a).

uals one year old and older, although "large confidence intervals make the estimate . . . an approximation to order of magnitude" (Mazsa 1973). Mazsa's estimates did not include fish smaller than 9 cm (less than one year old). It is, however, impractical to attempt to calculate the increase in the population estimate that would occur if these fish had been counted because year class strength typically varies greatly within yellow perch populations (e.g. LeCren 1955). My own qualitative diving observations confirm that perch are much more abundant than other species in the lake.

The chain pickerel, *Esox niger,* is a relative of the northern pike and muskellunge (see Wich and Mullan 1958; Carlander 1969; Crossman and Lewis 1973). This predator is native to sluggish streams and clear, heavily vegetated lakes of eastern and southcentral North America from Nova Scotia to central Florida. It is a solitary, diurnally foraging predator that hovers motionless in the water-column amid vegetation for prolonged periods. Its irregular coloration blends well with the dappling light that filters through the branches and leaves of macrophytes. Adult fish can occasionally be observed moving slowly through vegetation as they take up a lurking position near a group of actively foraging prey such as yellow perch. Although large pickerel are primarily piscivorous, food habits change several times during the growth of an individual. Recently hatched young feed initially on zooplankton, then the diet diversifies to include insect larvae and some small fish. At about 100 to 150 mm total length, fish become the principal food, with some larger invertebrates such as crayfish also being eaten. As with many piscine predators, little selectivity is shown in food choice (e.g. Larkin 1979); the chain pickerel's diet includes minnows, sunfishes, catfishes, chain pickerel and "almost anything alive and small enough to be engulfed" (Scott and Crossman 1973). Larger chain pickerel take larger food items, being able to swallow a forage fish whose depth is equal to the pickerel's depth (Carlander 1969). At night, pickerel rest motionless but easily disturbed on the bottom, generally in shallower water than they occupy by day (Helfman 1981). They are active and take food during the winter under the ice, feeding on essentially the same food as during summer (Scott and Crossman 1973).

Chain pickerel in Mirror Lake appear to be growing at a considerably slower rate and attaining smaller maximum sizes than pickerel in other regions (Mazsa 1973). Food habits of pickerel in Mirror Lake, however, conform to those in other regions; they shift from zooplankton to small invertebrates and fish (including white suckers), then to crayfish, and finally to larger fish, mostly yellow perch (Mazsa 1973). All food taken by pickerel in Mirror Lake is autochthonous in origin.

Ideal pickerel habitat in Mirror Lake is not abundant, being limited to areas of less than 3-m depth where vegetation with vertical stems, such as lily pads or fallen logs with attached branches, is dense (*Chapter V.A.4*). Pickerel hunt by lurking unnoticed amid such structure and surprising passing prey. A shortage of preferred habitat would presumably result in lowered prey-capture success and, therefore, support a relatively small pickerel population. Using angling data (Mazsa 1973) estimated that there were less than 200 adult fish in Mirror Lake. Standing crop calculations for chain pickerel in Massachusetts ponds range from 1.2 to 11.5 kg/ha (Carlander 1969). By taking the mean Massachusetts value of 5.6 kg/ha for Mirror Lake (15 ha) and a median pickerel size of 25 cm total length (equals 75 g total weight) (Mazsa 1973), a population of approximately 1,000 fish could be expected. With limitations imposed by a lack of both preferred habitat and prey fishes such as minnows, pickerel density in Mirror Lake is probably at the lower end of the range found in Massachusetts. Hence Mazsa's estimate of 200 fish is probably justified.

The brown bullhead, *Ictalurus nebulosus* is a moderate-sized catfish and is known as the hornpout in New Hampshire (see Trautman 1957; Emig 1966). It has a natural range that includes nearly half of the U.S., occurring from the Maritime Provinces of Canada to Florida, west to central Alabama, and then northwest across the midwest to Saskatchewan; it has been introduced into Idaho and California, as well as Germany, England and several other countries. The preferred habitat of the brown bullhead is relatively shallow ponds, small lakes and shallow bays of larger lakes and slow-moving streams. The fish is found most often in association with abundant aquatic vegetation and sand to mud bottoms. Brown bullhead are extremely tolerant of adverse temperature, oxygen and pollution conditions. At winter temperatures, they can survive at oxygen levels as low as 0.2 ppm (Scott and Crossman 1973).

This fish is a nocturnally foraging, benthic omnivore, feeding on or near the bottom and eating "offal, waste, molluscs, immature insects, terrestrial insects, leeches, crustaceans (crayfish and plankton), worms, algae, plant material, fishes and fish eggs" (Scott and Crossman 1973). Fry and fingerlings up to 75 mm long eat zooplankton and chironomids, while adults shift to larger foods (Carlander 1969). Algae may constitute up to 60% of the diet. Recent studies have demonstrated that brown bullhead can extract significant amounts of carbon from algae, including blue-greens (Gunn et al. 1977). Most food is apparently found by the senses of taste and smell, the chin barbels and entire body being densely covered with taste buds (Atema 1971). By day, these fish often rest on the bottom among dense weeds, sometimes with only their tails visible (Helfman 1981). They bury themselves in the bottom muds for extended periods (Carlander 1969) and are apparently inactive during the winter.

There is little specific information available on the brown bullhead population of Mirror Lake. Mazsa (1973) analyzed the stomach contents of five adults. He found a variety of food items, with benthic invertebrates and detrital plant material predominating. The two most abundant invertebrates in the stomachs were Anisoptera and *Hexagenia* nymphs. My own visual observations of brown bullhead in Mirror Lake suggest that they are fairly common in the shallow (1 to 3 m) regions during the night, feeding along the bottom and particularly concentrating their activities at the bases of large boulders. In the morning, they move to slightly deeper water and come to rest in *Utricularia* mats a few minutes before or after sunrise. Their abundance appears to be similar to that of chain pickerel.

"The white sucker [*Catostomus commersoni*] seems to be a rather plastic animal and characters vary from area to area" (Scott and Crossman 1973; see Geen et al. 1966). Also known as the common sucker, its distribution includes most of the northern ⅔ of North America. Fish are usually found in schools in warm, shallow bays and lakes in the upper 10 m of water. They are moderately active during the daytime, but active feeding is usually restricted to near sunrise and sunset when they move into shallower water (Scott and Crossman 1973). As with the other species in Mirror Lake, white suckers go through a series of dietary shifts with increasing age. These shifts also correspond to anatomical changes. Fry smaller than 20 mm have terminal mouths and feed in the water-column on small zooplankters. When the fish reaches about 16 to 18 mm long, the mouth begins to move from a terminal to a ventral position and the young fish assume the benthic foraging habits of older individuals, feeding on chironomids, trichopterans, molluscs and entomostracans in the bottom ooze. "Adults have rather generalized food habits" (Pflieger 1975). There is some argument as to whether they feed on fish eggs (Carlander 1969; Scott and Crossman 1973).

White suckers are the least abundant fish in Mirror Lake. Mazsa (1973) captured only four individuals during 2.5 months of sampling, and I saw only one individual during 10 dives. The suckers that do occur in the lake are apparently quite large for their ages (Mazsa 1973), although age determinations based on the examination of scales are suspect (Beamish and Harvey 1969). If the age determinations are correct, the results agree well with studies that have shown that individual growth is inversely related to population density (Carlander 1969). The rarity of white suckers in Mirror Lake is somewhat puzzling, given that they are the most abundant fish in the Merrimack drainage system (Bailey 1938). Low numbers may relate to the absence of tributary streams flowing into Mirror Lake. Such streams are the usual spawning place of white suckers. It is unlikely that white suckers play a significant role in the trophic relationships of Mirror Lake, beyond providing a short-lived springtime input of demersal eggs to be eaten by benthic feeding brown bullheads, yellow perch and smallmouth bass.

Discussion

Determinants of Diversity

Mirror Lake was formed after the last glacial retreat, approximately 14,000 yr ago (*Chapter VII.A*). The fish community was, therefore, constituted since that time, probably from colonizers moving out of the Pemigewasset River, up Hubbard Brook, and into the lake. There are 24 fish species in the Pemigewasset River and in other water bodies of Grafton County; 14 of these species are commonly found in lentic situations and are native to the area (Bailey 1938). Mirror Lake has only four of these fishes (excluding the introduced smallmouth bass). The question, again, is why so few and why the particular species that occur in the lake.

It should be noted that it is not surprising that only five fish species occur in a lake only 15 ha in area. Other small lakes in the vicinity of Mirror Lake have similarly low numbers; larger lakes in Grafton County have more fish species (Bailey 1938). An areal value of 15 ha is actually misleading because the littoral zone above 3-m depth includes only 3.5 ha (see *Table III.A.1*). This narrow region contains the habitat types most used by the five species of fishes. The observation of five species in Mirror Lake is surprisingly close to 7.4, the number that would be predicted from the species/area regression developed for North American lakes by Barbour and Brown (1974). However, size in itself is a correlate of low species diversity, not necessarily a cause.

Low productivity of plankton may explain, in part, why there are so few fish species. Mirror Lake is an oligotrophic lake of comparatively low annual production (see *Chapters V.B.2* and *V.C*); it has apparently been so since its formation (Likens and Davis 1975). This low production is very marked in the segments of the ecosystem that directly affect fish success. Zooplankton production, when compared with phytoplankton or net primary production, falls at the lower end of the range of values for several lakes compared by Makarewicz and Likens (1979; *Chapters V.B.5*). The zooplankton production-to-biomass (P:B) ratio, a measure of the turnover and, hence, replacement rate of the resource, is also comparatively low in Mirror Lake (Makarewicz and Likens 1979). Secondary production of benthic invertebrates is also relatively low in Mirror Lake (Walter 1976; *Chapter V.B.6*). Certainly a limited food resource plays some role in maintaining the low diversity of the community.

Can anything be concluded from the preceding species accounts concerning the determinants of the actual species composition in the Mirror Lake fish community? One characteristic emerges from most of the accounts—an opportunistic feeding pattern. This characteristic is emphasized through a comparison of the various taxa consumed by the different fishes (*Figure V.A.9–3*). Each species of fish utilizes most major food types to some degree at some time in its life, with the possible exception of plant material. This observation suggests that the fishes are trophic generalists. How might a generalist strategy relate to the question of low diversity? What are the advantages of being a trophic generalist in an environment such as Mirror Lake?

The question of generalist feeders among fishes has recently been treated by Keast (1979). A dietary generalist, according to Keast, is one with a diet "composed of many prey types in moderate amounts" (p. 243), whereas a specialist has a diet made up of one or a few prey types. Keast investigated the diet diversity of 17 species of fish in Lake Opinicon, Ontario, which is at the same latitude as Mirror Lake. Five of the fishes were generalists, six specialists, and the remaining species fell into an intermediate category. Two species in Mirror Lake were considered in Keast's study, brown bullhead and yellow perch. Both had very high diet diversity indices and were classified as generalists.

It is not possible to calculate diet diversity indices for the other Mirror Lake species and compare them directly with Keast's findings because the available data on feeding in Mirror Lake (Mazsa 1973) were based on a "percent points" rather than the "percent volume" method used by Keast. However, using Keast's more general criterion of a generalist as a fish with eight or more taxa constituting more than 3% of the mean diet, year I and II smallmouth bass might also be categorized as generalist (10 taxa), while larger bass are intermediate feeders (seven taxa for Age III and IV, six taxa for Age V+, data from Mazsa 1973). Smallmouth in Mirror Lake are, in fact, more generalized feeders than the generalist yellow perch in Mirror Lake, which has a diet diversity by the same criterion of only five taxa (Mazsa 1973). Chain pickerel, as befits a piscivore with a body shape specialized for lurking predation, has a mean diet diversity of five in young fish and three in adults; it would probably fall into the relative specialist category of Keast (1979).

By Keast's arguments, generalist feeders should be more common in lakes where re-

A: Species Composition, Distribution, Population, Biomass and Behavior

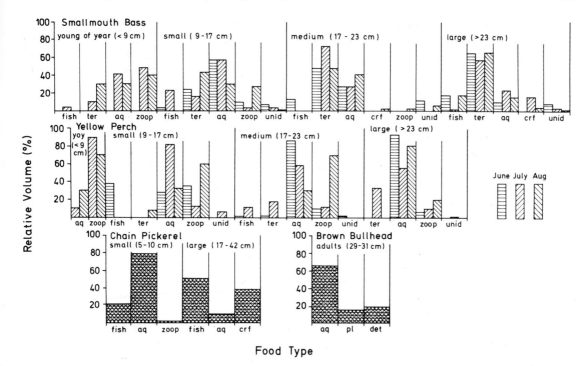

Figure V.A.9–3. Foods of Mirror Lake fishes, summer 1970. Food habits are summarized by resource type, fish size and month. Data for chain pickerel and brown bullhead are summer means and should be considered preliminary. Values for yellow perch young of the year (yoy) are approximate. Data from Mazsa (1973). Fish sizes are cm total length. Food types: fish = fish, ter = terrestrial invertebrates, aq = aquatic invertebrates, zoop = zooplankton, unid = unidentified, crf = crayfish, pl = plant material and det = detritus.

sources fluctuate widely. "The generalist feeding strategy is advantageous in highly seasonal cold temperate lakes where different prey types peak in numbers at different times" (Keast 1979). Food supplies in such lakes fluctuate seasonally and the different prey organisms peak in abundance at different times. There would, therefore, be an advantage to shifting rapidly from one food type to another. Fluctuations in numbers of Zygoptera nymphs were tenfold in Lake Opinicon, fivefold in Anisoptera and Ephemeroptera nymphs and threefold in amphipods and isopods. The biomass of chironomid larvae in September was one-fifth of what it was earlier in the season. By relying on aquatic invertebrates which "in the course of producing several generations during a summer undergo large fluctuations in population size, lake fishes are forced into being able to switch opportunistically, rather than specialize" (Keast 1979).

The availability of food resources in Mirror Lake follows a similar pattern. Walter (1976,

Chapter V.A.6) collected data on the abundance of benthic invertebrates in water deeper than 7 m on four occasions during one year. Benthic invertebrates were the food resource that all five fish species in the lake used most commonly (Mazsa 1973; see also *Figure V.A.9–3*). The abundance of eight different invertebrate taxa (measured as numbers of individuals) varied between 2- and 24-fold, and biomass varied from 2- to 38-fold during the year. Assuming that similar trends exist in shallower water, fishes in Mirror Lake are also experiencing notable fluctuations in resource availability and could consequently be expected to adopt the generalist strategy predicted by Keast.

The advantage of generalizing is shown dramatically when total production of resource type, rather than of individual taxa, is considered. The maximum variation in total number of benthic invertebrates between any two of Walter's sampling dates was 2.1. This variation indicates that a fish capable of capturing any of the eight taxa would experience at most an approximately twofold difference in food abundance at different times of the year. In contrast, the average maximum difference between sampling periods for the eight individual food taxa was 6.1. Hence a specialist would experience on the order of three times more variability in its food supply than would a generalist. Similarly, the greatest variation in biomass between two dates for all invertebrates was 2.2, whereas the average value of the maximum differences between sampling dates in the eight individual taxa was 8.0. As a consequence, a specialist feeding on a limited number of taxa would face much more variability than a generalist that "ignores" taxonomic considerations and feeds from the spectrum of organisms making up the food resource. The generalist species in Mirror Lake are apparently capable of exploiting the entire resource spectrum, including both benthic invertebrates and zooplankton. This nonspecificity in feeding may be an important attribute that has allowed these fishes to exist in Mirror Lake whereas other species may have failed.

Despite opportunistic feeding by the various species, important differences in feeding preferences still exist and may play a role in determining the species composition of the community. Perch show a preference for zooplankton, smallmouth for crayfish and terrestrially-derived foods, pickerel for small fish, crayfish and odonate nymphs, bullheads for plant material, detritus and benthic prey captured at night, white suckers for benthic prey captured by day. By exploiting the same resources when abundant, but also by specializing on fairly exclusive, complementary food types at other times, coexistence of the species in the community may be enhanced (see also Zaret and Rand 1971).

The Role of Fishes in the Mirror Lake Ecosystem

Chapter V.B.7 uses the available data on population sizes and feeding to estimate biomass, secondary production, energy flow, and nutrient cycling in the fish community in Mirror Lake. Two other, more qualitative characteristics of the fish community also merit further discussion. These characteristics are (1) the general dependence of all species on autochthonous food sources, which results in unusual feeding habits; and (2) the relative *tightness* of the internal cycling of nutrients and energy within the fish community itself.

The food eaten by all Mirror Lake fishes except smallmouth bass is produced in the lake, either as aquatic invertebrates, insect larvae, plants or fishes. Consequently, the fishes are involved in comparatively little exchange of nutrients and energy between terrestrial and aquatic systems. Given this dependence on autochthonous materials and the relatively low productivity of the lake, it is not surprising that some of the food habits of the different fishes are somewhat unusual. The largest factor appears to be the absence of abundant prey fishes for piscivorous smallmouth bass, yellow perch and pickerel. In many temperate freshwater systems, schooling minnows (Cyprinidae) provide the major food type for such predators. Minnows are absent from Mirror Lake, and thus the only available prey items for the predators are the young of the species present. From Mazsa's (1973) data, it appears that the yellow perch is the only species that might occur regularly in substantial numbers, and this species is notorious for undergoing large fluctuations in year class strength and abundance. As a result, the predators rely on alternative sources. "Bass as large as 40 cm total length feed quite heavily on insects . . . [although] insects are considered to be food of only small and medium-sized bass" (Mazsa 1973). Similarly, while the diet of large yellow perch elsewhere may include up to 75% small fishes (Scott and Crossman 1973), fish were absent from the stomachs of larger yellow perch in Mirror Lake. Half of the diet of large chain pickerel, a lurking predator, consisted of aquatic invertebrates (see *Figure V.A.9–3*).

Dependence on autochthonously-produced materials results in relatively tight internal cycling within the fish community. A food web diagram (*Figure V.A.9–4*) of the fish community has several bidirectional arrows; each species is connected to every other fish species by a two-headed arrow, plus a loop to account for cannibalism. It appears that every species of fish in the lake eats and falls prey to every other

A: Species Composition, Distribution, Population, Biomass and Behavior

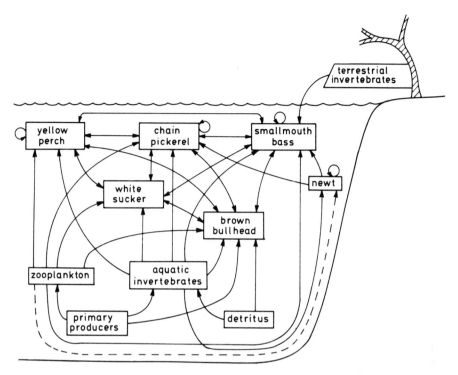

Figure V.A.9–4. Relationships of fishes in the food web of Mirror Lake. Arrows indicate the direction in which energy is transferred. Interactions of all life history stages are depicted, including consumption of eggs and fry. Circular arrows indicate cannibalism, which has not been verified for the white sucker and brown bullhead. No quantitative relationships are implied by different compartment sizes. Additional details on the nonfish components of this food web are given in *Chapter VI.D.1*. (Based on data in Mazsa 1973, Scott and Crossman 1973 and Burton 1977 and *Chapter V.B.7*.)

species and itself at some time during its life. Bullhead and perch (and perhaps sucker) are more likely to take eggs and fry; juvenile bass and pickerel and adult perch eat juvenile fishes when available, whereas adult bass and pickerel feed on larger fishes. The opportunistic switching between prey types, the overlap in foods eaten, the generalist patterns shown by most species, and the mutual piscivory that takes place, all highlight the complexity of this apparently simple community of five species.

Summary

Mirror Lake contains only five species of fish: smallmouth bass, yellow perch, chain pickerel, brown bullhead and white sucker. Only smallmouth bass and yellow perch are abundant. Several factors may have contributed to this low diversity and abundance, including the small size of Mirror Lake, a lack of appropriate habitats, and the low productivity of the plankton and aquatic invertebrate resources of the lake. Low productivity may be intimately linked with both low diversity and the species composition of the fish community. Each fish species eats progressively larger food as it grows, without necessarily dropping smaller items from its diet. Hence the fishes are opportunistic feeders, using a wide variety of autochthonous food resources, including the eggs and young of their own and other fish species. Because they feed on a diversity of food types, the fishes are less affected by the seasonal fluctuations in numbers and biomass that characterize the life histories of their prey. This *generalist strategy* may explain why these particular species have successfully colonized Mirror Lake.

B. Production and Limiting Factors

1. Bacteria

MARILYN J. JORDAN

Bacterial productivity is a measure of the rate of formation of bacterial protoplasm, expressed either in terms of energy or carbon. The following discussion deals with the heterotrophic bacteria, which appear to be the most important (and are certainly the most thoroughly studied) group of bacteria in Mirror Lake.

The heterotrophic bacteria obtain both energy and carbon from the decomposition of organic matter. Decomposition includes the physical disintegration and dissolution of dead organic structures, as well as microbial breakdown of complex organic molecules ("waste" materials) into smaller organic molecules, chemical nutrient ions, water and CO_2. The major biologically-mediated pathways of decomposition in aquatic systems are by microbial (bacterial and fungal) breakdown of organic matter, and degradation by means of the intermediary metabolism of the animal community (*Chapter V.B.7*; Saunders 1976). It is not clear which, if either, of these pathways predominates in planktonic systems. Although zooplankton may be the primary, direct mediators of nutrient regeneration in aquatic systems (Johannes 1968), only bacteria can assimilate the varied mix of dissolved organic compounds that occur at very low concentrations in natural waters (Hobbie and Wright 1966). By converting these dissolved compounds into a particulate form that can be consumed by grazers, heterotrophic bacteria return energy that might otherwise be unavailable to the systems (Paerl 1978). In water bodies receiving large amounts of allochthonous organic matter, bacterial biosynthesis could be of comparable or greater magnitude than plant photosynthesis (e.g. Kuznetsov 1968; Overbeck 1979). Fungi apparently are not important decomposers of small detrital particles suspended in fresh waters, but may be important in the benthic decomposition of plant litter (Willoughby 1974). Part of the C assimilated by heterotrophic bacteria is respired (R) yielding reducing power (NADH), heat and CO_2, and the rest of the C is incorporated into bacterial protoplasm (P = production). The *assimilation efficiency,* or growth yield, is defined as the ratio of the C incorporated into protoplasm to the total C assimilated, or [P/(P + R)]. A growth yield may vary widely depending on the substrates being metabolized and the growth rate of the bacteria (Forrest 1969; Postgate 1973; Williams 1973). An average growth yield of 50% for aquatic bacteria appears to be reasonable based on presently available information (Hobbie and Crawford 1969; Payne 1970; Williams 1970; Crawford et al. 1974; Cole 1982a,b).

Methods

The productivity of algae and photoautotrophic bacteria, and at least part of the productivity of chemoautotrophic bacteria, can be estimated (at least in theory) by measuring the rate of uptake of ^{14}C-labeled CO_2. An analagous method is not possible for heterotrophic bacteria since it probably is not feasible to identify and measure the uptake rates of all the diverse substrates assimilated by heterotrophic bacteria in the natural environment. Among the direct methods that have been suggested for measuring heterotrophic bacterial production are the following:

1. Decreases in O_2 in incubated water samples prefiltered to remove phyto- and zooplankton (Vollenweider 1969). This method is subject to filtration effects as well as to *bottle effects* resulting from the long incubation times (days) required to obtain observable O_2 changes.
2. Increases in bacterial biomass (measured by direct counting) in incubated water samples prefiltered to remove zooplankton (Vollenweider 1969). This method suffers from the same limitations as the preceding method.
3. Uptake of $(^{14}C)CO_2$ in water samples incubated in the dark. Sorokin (1961) and Romanenko (1964) estimated that CO_2-C uptake comprised about 6% of the total C

uptake by heterotrophic bacteria. If only bacteria take up CO_2 in the dark, then total heterotrophic bacterial C assimilation could be estimated by multiplying the rate of dark CO_2-C uptake by 100/6. Algal uptake of CO_2 in the dark and variability in bacterial CO_2 assimilation seriously limit the usefulness of this method, as discussed below.

4. Uptake of $(^{35}S)SO_4^{2-}$. Sulfur is an essential element, apparently assimilated by bacteria (and phytoplankton) in direct proportion to C (Roberts et al. 1955). By multiplying the rate of uptake of S (as SO_4^{2-}) by the cellular C:S ratio, the rate of uptake of C by the plankton can be estimated (Monheimer 1972, 1974; Jordan and Peterson 1978; Jassby 1975; Campbell and Baker 1978). The fraction of the total planktonic SO_4^{2-} uptake that was due to bacteria must then be determined (Jordan and Likens 1980). Sulfate is the primary source of SO_4^{2-} in most aerobic waters. The SO_4^{2-} method is limited to aerobic environments since under anaerobic conditions SO_4^{2-} may be reduced to H_2S. Also, the total SO_4^{2-} concentration must be low enough (several mg/liter or less) so that a measurable fraction of the added $(^{35}S)SO_4^{2-}$ can be taken up in a short-term incubation. Thus, the straightforward filtration procedure used to measure SO_4^{2-} uptake in fresh waters would not be sensitive enough for use in saline lakes or sea water.

5. Carbon budget. It is also possible to estimate bacterial production in systems for which all the other major organic C fluxes are known. For example, if all the significant inputs of organic C are known, and all the outputs of organic C except for those losses resulting from bacterial respiration are also known, the rate of bacterial respiration can be estimated by difference (Jordan and Likens 1980). If the growth yield = 50%, then production = respiration.

Production in Mirror Lake

Mirror Lake is one of the few lakes for which a complete annual organic C budget has been prepared (see *Chapter V.C*). Production of both benthic and pelagic bacteria in Mirror Lake has been estimated based on the organic C budget and has also been estimated directly for pelagic bacteria by measuring uptake of $(^{14}C)CO_2$ in the dark, uptake of $(^{35}S)SO_4^{2-}$, and *in situ* biomass changes based on direct counts by fluorescence microscopy. Details of the methods are given in Jordan and Likens (1975, 1980).

Based on the carbon budget, respiration of bacteria was estimated as 3.3 g C/m^2-yr in the epilithibenthos, 12.7 g C/m^2-yr in the gyttja and 6.5 g C/m^2-yr in the water for a total of 22.5 g C/m^2-yr (*Chapter V.C*). Assuming a growth yield of 50%, production would equal respiration. The upper limit for total bacterial production was set by the total annual organic C inputs to the water and sediments of the lake. The net input of C to the lake was 60 g C/m^2-yr, calculated by subtracting plant respiration from total gross inputs (*Tables V.C–2* and *V.C–3*). If every atom of organic C entering the lake were ultimately taken up by bacteria just once, and with an average bacterial growth yield of 50%, maximum total bacterial production would have been 30 g C/m^2-yr, 38% higher than the actual estimate of 21.7 g C/m^2-yr.

Pelagic bacterial production also can be estimated from the DOC component of the C budget. Inputs of DOC (g C/m^2-yr) were:

Algal excretion	8.7
Drainage water	8.4
Precipitation	2.9
Total	20.0

Outputs of DOC (g C/m^2-yr) were:

Outflow	6.6
Ground-water losses	2.1
Total	8.7

Thus the net input of DOC was 11.3 g C/m^2-yr, which if assimilated by pelagic bacteria with a 50% growth yield would result in bacterial production of 5.6 g C/m^2-yr. However, algal excretion as measured by $(^{14}C)DOC$ release is affected by assimilation of labeled DOC by bacteria during incubation with $(^{14}C)HCO_3$ (Derenbach and Williams 1974) and by cell breakage during filtration. The total DOC excretion by algae in Mirror Lake could be as high as 15 g C/m^2-yr, since bacteria apparently take up nearly half of the $(^{14}C)DOC$ released by algae during photosynthesis incubations in Mirror Lake and since decomposing algal cells leach additional DOC (J. Cole, unpublished data). If so, pelagic bacterial production could be as high as 9 g C/m²-yr.

Another way to estimate pelagic bacterial production based on the C budget is to add up the various rates at which bacteria are lost from the water of the lake. In a steady-state system where there is no net change in biomass (ΔB) from year to year, net annual bacterial production (P) should equal the sum of the losses:

$$P = \Delta B + L - I + G + S + D \quad (1)$$

where L = losses of bacteria by way of the lake outlet, I = inputs of bacteria in precipitation and streamflow, G = grazing by zooplankton and protozoa (*Chapter VI.C*), S = sinking and D = DOC excretion and lysis. These various losses have been estimated (revised from Jordan and Likens 1980), yielding a production rate of approximately 0.58 g C/m²-yr:

$$P = 0 + 0.05 - 0.016 + 0.55 + 0 + 0$$
$$= 0.58 \text{ g C/m}^2\text{-yr} \quad (2)$$

Although bacterial biomass remains approximately constant from year to year, small short-term fluctuations may be measured (*Chapter V.A.1*). The sum of these increases in biomass was 0.2 g C/m²-yr. This value is a lower limit, because most of the bacterial biomass produced was removed by the loss factors discussed above and did not result in increased biomass.

These estimates of production based on the C budget are subject to error because the various components of the budget were investigated in different years (1969 to 1976). Also, there is variability in the measurements themselves and in bacterial growth yields. Even more important, estimating a small number (bacterial production) by the difference between two large numbers (inputs and outputs) is subject to large variance. Accurate, direct measurements would be preferable. Measurements of (^{35}S)SO$_4^{2-}$ uptake and dark (^{14}C)CO$_2$ uptake show promise, but these methods suffer from limitations discussed below.

The total uptake of S by the plankton of Mirror Lake, measured using (^{35}S)SO$_4^{2-}$, was about 600 mg S/m² in 1975 and 1,000 mg S/m² in 1976. This uptake was presumed to be due to bacteria plus phytoplankton, because the S needs of the zooplankton should be met in particulate food consumed. To separate bacterial uptake of SO$_4^{2-}$ from algal uptake, samples were filtered through 1- and 3-μm pore size Nucleopore filters following incubation with (^{35}S)SO$_4^{2-}$. The filters retained most of the algae and passed most of the bacteria in the filtrate (Jordan and Likens 1980). Thus the ^{35}S incorporated by algae (84% of the total) could be distinguished from ^{35}S incorporated by bacteria (16% of the total). These estimates were corrected for any retention of bacteria on, and passage of phytoplankton through, the filters based on the retention and passage of chlorophyll and incorporated (^{14}C)-glucose in separately-filtered samples (Derenbach and Williams 1974; Jordan and Likens 1980).

To estimate planktonic bacterial production, the mean annual bacterial S uptake of 817 mg S/m² must be multiplied by the ratio of C to S in bacterial protoplasm. The C:S ratio of bacteria in laboratory cultures varies from about 50:1 to 200:1 (Buchanan and Fulner 1928; Roberts et al. 1955) or possibly as high as 500:1 (ZoBell 1963; Monheimer 1972, 1974). The mean C:S ratio of five bacterial isolates from Mirror Lake grown in glucose medium was 103 (Jordan and Peterson 1978), but was as low as 70 when the medium was supplemented with yeast extract.

The C:S ratio of bacteria in the natural environment is not known, but most likely ranges between 50:1 and 100:1 for Mirror Lake (Jordan and Likens 1980). Bacterial production estimated using a C:S ratio of 50 (6.5 g C/m²-yr) agrees fairly closely with estimates based on the C budget (*Table V.B.1–1*), but more data are needed before a generally applicable C:S ratio for natural plankton can be assigned. Uptake of (^{35}S)SO$_4^{2-}$ appears to be a potentially accurate method for measuring planktonic bacterial production in aerobic fresh waters, although methodological problems do remain.

The other direct method used was based on CO$_2$ uptake. Multiplying total CO$_2$ uptake in the dark by 100/6 yields an estimate of 34 g C/m²-yr for heterotrophic bacterial production in the water of Mirror Lake. This value appears to be a serious overestimate when compared with the results of all other methods (*Table V.B.1–1*), especially since net C inputs to the water-column were only 52. The production value was calculated by assuming that only bacteria take up CO$_2$ in the dark and then by assuming this C uptake was 6% of total heterotrophic bacterial C assimilation (Sorokin 1961; Romanenko

Table V.B.1-1. Planktonic Bacterial Production in Mirror Lake ($\bar{x} \pm$ S.E.)[a]

Method	g C/m²-yr
1. Carbon fluxes, 1970 to 1978	
DOC budget	5.6 ± 1.9
19% of phytoplankton POC production[b]	5.4 ± 1.7
Sum of bacterial losses[c]	0.6 ± 0.3
2. Biomass increases, 1974 to 1976	>0.2
3. SO_4^{2-} uptake,[d] 1975–76 (C:S = 50)	6.5 ± 1.8
4. Dark CO_2 uptake,[e] 1975–76	
If 56% bacterial	20
If 100% bacterial	34

[a] Revised from Jordan and Likens 1980.
[b] Assumes a C:S ratio of 50 for bacteria (see text).
[c] Drainage, zooplankton grazing, sinking.
[d] Bacterial SO_4^{2-} uptake estimated as 16% of total SO_4 uptake.
[e] Bacterial dark CO_2-C uptake taken as 6% of total bacterial C assimilation.

1964). Both assumptions are questionable. First, phytoplankton probably do take up significant quantities of CO_2 in the dark (Morris et al. 1971). Algal uptake of CO_2 in the dark was measured for Mirror Lake water using the same differential filtration technique described above for SO_4^{2-}, and was found to account for about 42% of the total dark CO_2 uptake (Jordan and Likens 1980). Second, the heterotrophic bacterial assimilation ratio for dark CO_2-C may range from 0.4 to 30% (Overbeck and Daley 1973; Overbeck 1979). The ratio for natural populations could be much higher owing to the inclusion of CO_2 uptake by chemosynthetic bacteria. Even after subtraction of algal dark CO_2 uptake, bacterial production in Mirror Lake was estimated as 20 g C/m²-yr using an assimilation ratio of 6%, a value significantly higher than all other estimates (*Table V.B.1*–1). Dark CO_2 uptake may have accounted for more than 6% of total bacterial C assimilation, or the fraction of dark CO_2-C uptake attributable to bacteria might have been overestimated because it was based on only six summer measurements.

The extent to which the CO_2 taken up by heterotrophic bacteria is actually used for biosynthesis is an unknown and often overlooked issue. Overbeck and Daley (1973) postulated that CO_2 fixation in most heterotrophs is not synthetic, that is, "CO_2 fixation serves only to regenerate intermediates which are 'lost' from the tricarboxylic acid (TCA) cycle . . . ." Overbeck (1979) found that cultured planktonic bacteria took up $^{14}CO_2$ in the dark only when supplied with a utilizable substrate, and that the incorporated $^{14}CO_2$ appeared in compounds of the TCA cycle.

Bacterial Growth Rates

Based on the results of all methods except dark CO_2 uptake (*Table V.B.1*–1), production (P) of the planktonic heterotrophic bacteria in Mirror Lake most likely ranged between 3 and 8 g C/m²-yr. This growth rate can be expressed as a doubling time (t_d), the average time that would be required for bacterial biomass to double if not harvested:

$$t_d = \ln 2 \div (P/B) \quad (3)$$

Since the average annual standing biomass (B) of planktonic bacteria in Mirror Lake is 80 mg C/m², t_d would be between 2.5 and 7 days with a seasonal range from 1.2 in September to >100 in February. Many published values for bacterial doubling times in lakes and reservoirs range from just 1.7 h to 2.1 days (Abdirov et al. 1968; JIBP-PF Lake Biwa 1975; Sorokin and Donato 1975), or occasionally as high as 9 days (Drabkova 1965; Vinberg 1969; Niewolak 1974), and may be too low because of overestimates of production (P) based on dark CO_2 uptake, underestimates of biomass (B), or both. Until recently, bacterial biomass was estimated from direct counts by using Millipore-type filters, which give counts severalfold lower than Nucleopore filters (Daley and Hobbie 1975).

Similar calculations for sediment bacteria are more complicated. Production resulting from bacteria in the gyttja in Mirror Lake has been estimated as 12.6 g C/m²-yr (or 21 g/m² of gyttja, which covers 60% of the lake bottom; see *Chapter VII*), but the depth of sediment in which significant bacterial growth occurs is unknown. Using the biomass of bacteria in the top 10 cm of gyttja (4.0 g C/m²), t_d would be 80 days. If only the top 1 cm of sediment contained active bacteria, t_d would be 8 days. Total benthic (epilithibenthic + gyttja) bacterial production was 17.3 g C/m²-yr, or 2 to 6 times greater than planktonic bacterial production. Planktonic bacterial production could be a greater proportion of total decomposition in deeper lakes, where detritus would have a longer residence time in the water-column.

Microbiologists work with laboratory cultures of bacteria, which have doubling times of minutes or hours, under optimum conditions. The doubling times of days to weeks for bacteria in lakes seems very long in comparison. Aquatic bacteria are thought to be C limited rather than N or P limited in most lakes since labile, easily metabolized compounds like glucose are present in very low concentrations (a few µg/liter) and most of the DOC pool is believed to be composed of large refractory molecules that are difficult to metabolize (*Chapter V.D*). The aquatic bacterial population may consist of cells that are all growing very slowly. Alternatively, most cells may not be growing at all, while a small fraction of the population could be growing rapidly because of a locally rich input of substrate such as a dead plankter, fecal pellet, etc. (Stevenson 1978). A significant fraction of the bacterial flora of lakes is soil bacteria washed into the lakes by runoff (Monheimer 1974). These soil bacteria are probably adapted to the higher nutrient concentrations found in the soil solution and may grow slowly, if at all, in low-nutrient lake water.

Summary

The heterotrophic bacteria are the most important and most intensively studied group of bacteria in Mirror Lake. Only bacteria can assimilate the varied mix of dissolved organic compounds that occur at very low concentrations in natural waters. Part of the C assimilated by heterotrophic bacteria is respired (R) and the rest is incorporated into bacterial protoplasm (P = production). An average growth yield [P/(P + R)] of 50% is assumed for aquatic bacteria.

Heterotrophic bacterial production can be estimated by a variety of methods, none of which is generally applicable in all water bodies: decreases in O_2 or increases in bacterial biomass in prefiltered, incubated water samples, biomass changes *in situ*, dark uptake of $(^{14}C)CO_2$, uptake of $(^{35}S)SO_4^{2-}$, and indirect estimates based upon a carbon budget. Planktonic heterotrophic bacterial production in Mirror Lake has been estimated using all but the first two of the above methods, and values most likely range between 3 and 8 g C/m^2-yr. Since the average standing biomass (B) of planktonic bacteria in Mirror Lake is 80 mg C/m^2, the average time required for planktonic bacterial biomass to double $[t_d = \ln 2 \div (P/B)]$ would be between 2.5 and 7 days. Actual rates could vary seasonally from as little as 1.2 days in September to >100 days in February.

Production by benthic bacteria was estimated as 3.3 g C/m^2-yr in the epilithibenthos and 12.7 in gyttja sediments. Thus, total benthic bacterial production was two to five times greater than planktonic bacterial production.

2. Phytoplankton

F. deNoyelles and Gene E. Likens

The microscopic suspended algae, phytoplankton, make up a major portion of the autotrophic component of the aquatic food web in lakes. In this respect they can both support and influence the metabolism of other levels of the food web. In Mirror Lake the phytoplankton account for 88% of the primary production, with production by benthic algae and macrophytes being of minor importance (*Chapter V.C*; Jordan and Likens 1975). The phytoplankton provide an energy link between the physical environment (solar energy) and the animal and microbial consumers. They are also a nutrient link between the chemical environment and other levels of the food web (*Chapter I*). Being dependent upon absorption of dissolved nutrients for growth, the phytoplankton build their biomass directly from the chemical environment. The consumer levels of the food web receive most nutrients from the ingestion of particulate food. The phytoplankton also strongly influence the structure of the chemical environment by the absorption of nutrients and by the release of various inorganic and organic substances (*Chapter IV.B*).

It is evident that much can be learned about

the ecology of a lake through studies of the structure and behavior of the phytoplankton community. Such studies can also reveal changes in the ecology of the lake that may be a departure from prior, noncultural relationships. We know from many studies that lakes under the early influences of pollution display marked changes in both the structure and behavior of their phytoplankton communities. Such changes have been demonstrated for oligotrophic waters when their naturally low concentrations of plant nutrients begin to increase at an unnatural rate (Lund 1969; Schindler et al. 1971, 1974a; Schindler and Fee 1974; deNoyelles and O'Brien 1978; Kalff and Knoechel 1978). This response has been termed "cultural eutrophication" to distinguish it from natural eutrophication, which involves slower increases in nutrients over many years (Hasler 1947). Cultural eutrophication may involve a variety of human activities, culminating in nutrients entering the lake from water or airborne wastes, from fertilizers applied to adjacent lands, or from soils washed in by erosion. These nutrient increases may bring about changes in a phytoplankton community, followed by changes throughout the ecosystem over just a few months or years, which many generations of ecologists would not observe in the much slower natural eutrophication of the lake.

Most studies of phytoplankton in oligotrophic lakes have focused on the epilimnion during the summer, where it was assumed that the most influential activity of the phytoplankton was occurring. It has only been based on findings of a few recent studies detailing the vertical distribution of phytoplankton in oligotrophic lakes that this assumption could be challenged (Schindler and Holmgren 1971; Kiefer et al. 1972; Tilzer 1973; Fee 1976, 1980; Fee et al. 1978; and *Chapter V.A.2*). From lakes at the Experimental Lakes Area (ELA), it has been demonstrated that considerable biomass and productivity of phytoplankton is often found below the epilimnion (Schindler and Holmgren 1971; Fee 1976, 1980). Such conditions are known to develop from algae settling through the epilimnion and accumulating in the more dense, deeper waters, or from algae, which are rarely found in the upper waters, growing in place (Fee 1978b). The mere stratification of a lake can lead to deep biomass accumulations, but for active growth to occur in such zones they must receive sufficient light. Thus the presence of both deep biomass and production requires specific lake conditions beyond just stratification.

Concurrent biomass and production peaks below the epilimnion seem to be typical for small (<20 ha), protected lakes of moderate depth (>6 m max), where summer stratification occurs and the clarity of the water permits sufficient light for algal photosynthesis to occur below the epilimnion (Fee 1976, 1978b, 1980). The hypolimnion is often shallow and relatively nutrient rich. Lakes of this type at the ELA are termed Class-B lakes (Schindler and Holmgren 1971; Kling and Holmgren 1972). Mirror Lake has these physical and chemical characteristics as described earlier (Chaper IV.B) and is similar in the structure of the phytoplankton community (*Chapter V.A.2*); additional characteristics of the phytoplankton in terms of production and behavior now remain to be described.

Phytoplankton production in Mirror Lake was studied by Gerhart (1973), Jordan and Likens (1975) and J. Cole (personal communication). During 1973–74, phytoplankton production in Mirror Lake accounted for about 90% of total plant production on an annual basis. During this time, phytoplankton production for the entire water-column, as estimated from ^{14}C uptake measurements at 1-m intervals two to six times a month, ranged from a low of 1.4 mg C/m^2-day in February to a high of 595 mg C/m^2-day in July. The annual net phytoplankton production was 29 g C/m^2-yr for the ice-free season (Jordan and Likens 1980). Gerhart (1973) in 1972 recorded a high of 470 mg C/m^2-day and a total for the ice-free season of 37 g C/m^2.

In the Class-B oligotrophic lakes at the ELA, highs of 400 to 700 mg C/m^2-day are generally recorded (Schindler and Holmgren 1971), with totals for the ice-free season of 20 to 40 g C/m^2 (Schindler 1972; Fee 1979, 1980). Similar values have been reported for other oligotrophic lakes in North America and elsewhere (Rodhe 1958; Findenegg 1964; Sakamoto 1966; Vollenweider 1968; Allen 1972; Kalff and Knoechel 1978; Smith 1979).

The distribution of phytoplankton production with depth in Mirror Lake has been determined at various times from 1971 to 1979. During periods of summer stratification in 1971–72, maximum production occurred below the epilimnion between 6 and 8 m on sunny days but not on cloudy days (*Figure V.B.2–1*). Production profiles from 1979 (*Figure V.B.2–2*) also show some occurrence of peaks below the epilimnion, however, during August such peaks are less apparent. With the sparse production data available before 1979, this alone cannot be used as an indicator of changing lake conditions. Production peaks coincided with the higher chlorophyll levels below the epilimnion but not with the highest levels, which were always recorded at 10 m (*Figures V.A.2–3* and *V.B.2–2*). Considering the effect of cloudy days on the deep phytoplankton production and the rapidly declining production below 8 m where irradiance generally fell below 1%, it is evident that light penetration strongly regulates production in the deeper waters. As discussed earlier, changing conditions in the epilimnion affecting light penetration can thus drastically affect these deep phytoplankton assemblages. Production peaks below the epilimnion have been recorded in Class-B lakes at the ELA (Schindler and Holmgren 1971; Fee 1976, 1978b, 1980) and in other oligotrophic lakes (Findenegg 1964, 1966; Larson 1972; Tilzer 1973). A similar influence of irradiance on the production of these deep assemblages has been described by Fee (1976, 1978a, 1980), deNoyelles et al. (1980) and Knoechel and deNoyelles (1980).

It should be noted that although the highest chlorophyll values in Mirror Lake always occur at 10 m during summer stratification, production was always near zero there (*Figure V.B.2–2*). It would seem that the peaks of chlorophyll concentration at the greatest depths are not produced there but may accumulate by settling from above or are made up of a different contingent of species less dependent upon light for growth, perhaps capable of heterotrophy as reported elsewhere (Maeda and Schimura 1973). Though chlorophyll levels are highest at 10 m, the biomass of intact organisms (from 9-m samples counted by microscope) often is not (*Tables V.A.2–2, 3, 5 and 6*), suggesting that settling dead material may add to these chlorophyll peaks. On the other hand, from *Tables V.A.2–3,*

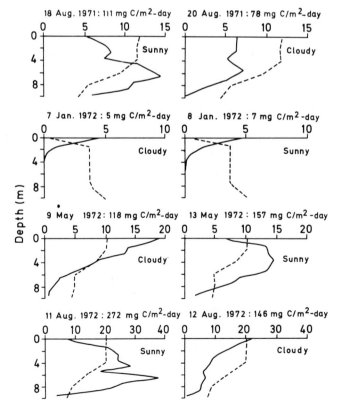

Figure V.B.2–1. Phytoplankton production (——) and temperature (----) profiles in Mirror Lake. Values plotted are for 4-h exposure periods in May and August and 5-h exposure periods in January.

B: Production and Limiting Factors

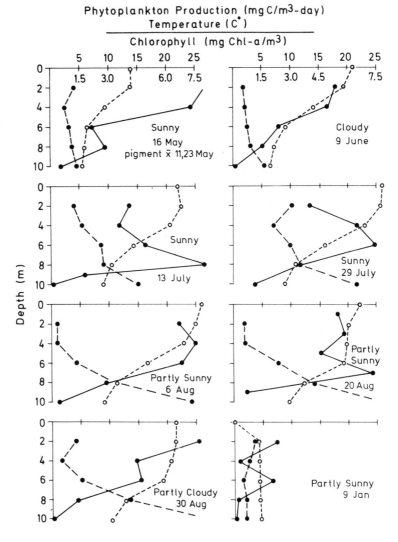

Figure V.B.2–2. Phytoplankton production (●——●), chlorophyll (●--●), and temperature (○--○) profiles for 1979 in Mirror Lake. Production values plotted are mg C/m³-day in net particulate production and the chlorophyll values are mg Chl-a/m³ from extraction followed by fluorescence readings.

4, and 5 it appears that species exist at 9 m that are not found at shallower depths in the lake. Therefore, both conditions may contribute to this situation, showing that below the epilimnion, characteristics of the peaks may differ with depth. Fee (1978b) has described this condition for the subepilimnion assemblages in the ELA lakes as discussed in *Chapter V.A.2*.

Regulatory Factors

Environmental factors regulating the structure and behavior of freshwater phytoplankton communities from oligotrophic waters have been reviewed by many authors (e.g. Lund 1965; Hutchinson 1967; Lehman et al. 1975; deNoyelles and O'Brien 1978; Kalff and Knoechel 1978; Schindler 1978). Factors of particular importance include the availability of nutrients (particularly phosphorus, nitrogen and carbon) and light (quality and quantity) for regulating growth, along with settling and grazing for regulating loss.

In this chapter we stress the importance of studying the vertical distribution of phytoplankton in oligotrophic lakes. Certainly factors influential in the epilimnion may differ in type or degree from those influential below the epilimnion. For example, nutrient availability may be particularly important in regulating growth of phytoplankton in the epilimnion where light is in excess, but with increasing depth, light becomes more important and nutrients may be in excess. Evidence for this shift in regulating factors can be found in studies done on Mirror Lake.

Growth stimulation was recorded in experiments where phytoplankton from the epilimnion of Mirror Lake were exposed to increased levels of nitrogen and phosphorus (*Figure V.B.2–3*). Three different bioassay methods were used to evaluate nutrient limitation of phytoplankton in Mirror Lake (Gerhart 1973). These were (1) *in situ* incubations in 130-ml bottles with ^{14}C, (2) nutrient enrichments of natural phytoplankton communities in 8-liter continuous cultures in the laboratory, and (3) enrichment of large tubes (0.8 m diameter by 4.3 m long; 1,700 liters) suspended in the lake. The results of the continuous culture, tube and long-term (several days) bottle (^{14}C) bioassays showed that growth of phytoplankton was limited by nitrogen and phosphorus (Gerhart and Likens 1975). Additions of either nitrogen or phosphorus alone, however, failed to stimulate phytoplankton growth. A study during 1974 by B. Peterson (personal communication) also demonstrated response to combined N and P addition (single additions were not used) as increases in particulate C and Chlorophyll-*a* over three- to four-days with *in situ* bottle incubations. No stimulation of growth has been observed in any short-term (4 to 30 h) ^{14}C bioassays done in Mirror Lake, which suggests that the incubation times were too short for the positive effects of nutrient enrichment to overcome the negative effects of enclosure in 130-ml bottles (Gerhart and Likens 1975) or to overcome any inherent lag in the increase in ^{14}C uptake following uptake of added nutrients (O'Brien and deNoyelles 1976). Others also have suggested that short-term ^{14}C bioassays may be misleading relative to actual growth responses over intervals of several days (e.g. Barlow et al. 1973; Knoechel and deNoyelles 1980). The natural concentrations of nitrogen and phosphorus in Mirror Lake water were consistently low in the epilimnion and higher in deeper waters (*Chapter IV.B*).

Biomass and successional responses were monitored in the tubes and in the continuous cultures. The species responding to enrichment with nitrogen and phosphorus differed considerably between the continuous culture and tube experiments (Gerhart 1973). *Synedra, Rhizoso-*

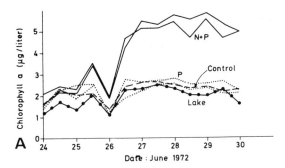

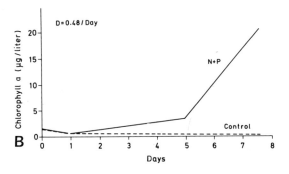

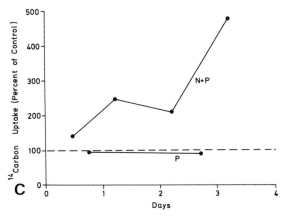

Figure V.B.2–3. A. Chlorophyll-*a* changes in enriched and control polyethylene tubes and in open lake water. Enrichment began on 24 June and was done daily at 70 µg N/liter-day and 35 µg P/liter-day. **B.** Chlorophyll-*a* changes in enriched and control continuous cultures operated at a constant dilution rate (D). Enrichment beginning at time 0 continued at rates of 106 µg N/liter-day and 43 µg P/liter-day. **C.** ^{14}C uptake for 4-h incubations of subsamples drawn from enriched continuous cultures during their first three days of operation. P was added alone to subsamples at 3 µg P/liter-day and N and P combined at 5 µg N/liter-day and 1 µg P/liter-day. From Gerhart and Likens 1975, *Limnology and Oceanography* 20(4):649–653. Copyright © 1975 by The American Society of Limnology and Oceanography. Used with permission.

B: Production and Limiting Factors

lenia, Dinobryon, Hyalobryon and *Anabaena* increased consistently in continuous cultures, whereas the cryptomonads and *Chrysochromulina parva* increased in tubes. Such species responses and associated grazing responses by zooplankton complicate the interpretation of nutrient bioassay experiments relative to the management of cultural eutrophication of natural lakes.

These responses to nutrient enrichment differ somewhat from those recorded in ELA lakes using similar methods. Phosphorus alone caused a stimulation of growth in the ELA lakes (Schindler 1971). This difference may be explained by differences in nitrogen availability for phytoplankton in Mirror Lake compared with the ELA lakes. From studies done on nutrient budgets in these lakes (Schindler 1976; Likens and Loucks 1978; *Chapter IV.E*) there is an indication from the lower nitrate concentrations and lower N/P in Mirror Lake water that for P to stimulate algal growth, N must also increase as it would quickly be exhausted with an increase in P. With short-term exposure (hours to days) to P in the experiments described above, the resident community would likely not show an increase in growth. This response does not mean that longer-term P exposure could not stimulate algal growth in the lake. Under such conditions and with time, nitrogen-fixing algae could become more abundant in the community in response to greater P availability and using a natural source of N not available to other algae. The newly structured phytoplankton community would then also increase in production as the nitrogen fixers became more abundant. Such successional changes have occurred with long-term P additions to ELA lakes. Even though N may not be initially limiting there, sources other than N_2 are soon exhausted with P increase (Schindler 1976, 1977).

The effects of exposing the phytoplankton from below the epilimnion to nutrient increases have not been determined in Mirror Lake. However, from the stimulation of growth observed with increases in irradiance, there is evidence for nutrients being available in excess of demands. J. Cole (personal communication), on five occasions during the summer of 1979, collected phytoplankton at 8 m and incubated them for 24 h at that depth and at 2 m. On the average there was a 16-fold increase in net primary production at 2 m and, therefore, enough nutrients available (from cell storage, absorption, or active decomposition) to accommodate this increase. Although such an experiment does not separate the effects of light and temperature, a comparable experiment for an ELA lake did and showed similar results. There phytoplankton collected from a biomass peak in the upper hypolimnion were transported to laboratory continuous cultures. Temperature was maintained for that depth (6 m, 10°C) and the irradiance was doubled over the natural levels (15 μEi/m^2-s) and those of control cultures. Although this irradiance change was considerably less ($\sim$1/20) than that created in the Mirror Lake experiment, biomass still increased over the controls by 21% after 11 days (deNoyelles et al. 1980; Knoechel and deNoyelles 1980). Additional evidence for the controlling influence of light on these deep assemblages in Mirror Lake can be seen in the comparison of deep production on successive sunny and cloudy days (*Figure V.B.2–1*).

Some studies of nutrient limitation for deep-dwelling phytoplankton in ELA lakes have been conducted. In the experiment by deNoyelles et al. (1980) described above, one culture received P addition causing biomass and production increases for some species (Knoechel and deNoyelles 1980). Healey and Hendzel (1980), using physiological indicators such as chemical composition ratios and metabolic parameters, have found evidence for N and P deficiency in such species in several lakes. However, N and P deficiency was less than that determined for epilimnetic populations in the same lakes. One lake at the ELA has received direct nutrient additions to the hypolimnion (Schindler and Fee 1974). In this lake, two basins were separated by a partition and one basin received the nutrient additions, leaving the other basin as a control. The total phytoplankton growth in this fertilized basin was considerably less than that of other lakes receiving similar additions of nutrients to the epilimnion (Fee 1976; Schindler et al. 1980). The peaks of biomass and production persisted below the epilimnion in this basin since the minimal growth in the epilimnion only slightly reduced light penetration. However, the hypolimnion peaks were greater than in other lakes in both biomass and production, indicating some stimulation by nutrients.

It appears from the above discussion that phytoplankton growth in Mirror Lake is regulated to a large degree by N, P and light availability but the nature of the regulation differs with depth. Knowing this, we can conclude this chapter with a consideration of how the composition of the species and their spatial distribution as described earlier (*Chapter V.A.2*) may be maintained for this type of oligotrophic lake. It has been suggested by some limnologists that the species dominating an oligotrophic lake do so not because they thrive there, but because they do less poorly than other species (Järnefelt 1952; Kalff and Knoechel 1978). With the potential for N and P limitation, these species likely have certain adaptations to facilitate nutrient uptake at low concentrations as suggested by Lehman et al. (1975) and deNoyelles and O'Brien (1978). Since light may become limiting in deeper waters, the species residing there may possess certain characteristics of terrestrial shade plants such as larger amounts of the light-harvesting pigments (Boardman 1977). Also, as discussed earlier (*Chapter V.A.2*), with perhaps only relatively slow growth possible under these low light conditions, it also would be advantageous for them to reduce losses to respiration and grazing. The species comprising the deep peaks that originate from growth seem to reduce these losses as a function of the cold temperatures, their large cell size, and their colonial structure. These species may be adapted for deep growth in part to take advantage of the somewhat higher N and P concentrations there. They further demonstrate their specialization for existing below the epilimnion by being motile and presumably phototactic, so as not to settle out of the restricted euphotic zone. Other tactic responses also may be involved, because although light-limited, these species do not ascend into shallower waters when that action would take them into the epilimnion. In contrast, species common to the epilimnion, most of which do not possess these adaptations, derive more advantage by remaining in the upper waters and thus may possess certain structural features that increase their buoyancy. With inevitable settling, benefit still can be derived because passage through the water may increase nutrient uptake (Titman and Kilham 1976) in this zone where nutrients are most limiting. This limitation, of course, has to be counterbalanced by rapid growth, enough to offset these losses.

It would be possible to describe other features of the oligotrophic lake that regulate the phytoplankton communities, but this is better left for reviews such as those by Lund (1965) and Kalff and Knoechel (1978). Rather, we conclude by emphasizing that the species of phytoplankton in an oligotrophic lake do seem to be adapted for doing less poorly there, rather than thriving, because the prevailing nutrient and light limitations would preclude the latter. Their existence, though supported by the oligotrophic condition, is fragile, as evidenced by their rapid displacement when this condition is even slightly altered. With their loss, the lake loses not only whatever influence their unique behavior provides but also a vertical distribution of biomass and production whose influence on other components of the ecosystem has yet to be assessed. The early signs of such changes may have been detected in Mirror Lake with changes in the structure and production of the deep component of the phytoplankton community (see also *Chapter V.A.2*). It remains for other signs to either support or refute this before our perception of the lake's future will enable or convince us to better control those destructive influences that humans impose.

Summary

It is again evident, considering production and limiting factors, as it was in *Chapter V.A.2* considering species composition and biomass, that Mirror Lake shares many characteristics in common with other oligotrophic lakes. Daily highs of net primary production during the summer of less than 1,000 mg C/m^2 and annual totals for the ice-free season of 20 to 40 g C/m^2 are consistent with those of many oligotrophic lakes of similar size and chemical characteristics. The presence of considerable net primary production below the epilimnion in Mirror Lake is compatible with the measurements in ELA lakes. Nutrient and light limitation has been observed for the phytoplankton of Mirror Lake and, as in ELA lakes, depth in the water-column is a determining factor as to which predominates. With nutrient limitation in Mirror Lake both P and N can be in short supply, while in other lakes it may be one or the other, or both. In ELA lakes, P is most often limiting (Schindler 1976, 1977; Schindler et al. 1978).

In most oligotrophic lakes, including Mirror Lake, P is most often the most threatening nutrient for eutrophication since a phytoplankton community can compensate for minimal nitrogen supplies by succession to an assemblage dominated by species capable of fixing atmospheric nitrogen (N_2). From both chapters dealing with the phytoplankton of Mirror Lake we have considerable evidence that communities here are structured and function like phytoplankton communities in other oligotrophic lakes and also share the same vulnerabilities to human disturbances.

3. Macrophytes

ROBERT E. MOELLER

The underwater survey of the littoral zone indicated that the hydrophytes could not be a very important component in the lake's metabolism. This conclusion was based on an understanding of the annual pattern of growth in hydrophytes as a whole, and in three species from Mirror Lake in particular. To be sure, the biomass of large aquatic plants in Mirror Lake in midsummer (530 kg dry wt) greatly exceeds that of the planktonic algae, but the hydrophytes play a much less dynamic role in lake metabolism than these algae. Hydrophyte biomass turns over slowly during the year, older tissue decaying as new tissue is formed. Generally only a small amount of the yearly production of new hydrophyte tissue (annual net production) is consumed directly by herbivores (Straskraba 1968). Tissue becomes senescent, then dies on the plant, leading to a large release of soluble organic and inorganic compounds (Godshalk and Wetzel 1978; Carpenter 1980). Most of the dissolved and particulate organic matter formed by hydrophytes is ultimately respired by bacteria and fungi, or by animals that feed, in turn, on these microbial decomposers. A few resistant particles finally end up in the sediments.

Animals do feed directly on some of the plants. Muskrats occasionally visit Mirror Lake in the autumn or spring and have been seen digging up and eating shallow-water plants like *Lobelia dortmanna* and *Eriocaulon septangulare*. An aquatic caterpillar feeds on shoots of several shallow-water species, including *L. dortmanna*, *E. septangulare*, *Gratiola aurea*, *Isoetes muricata* and *I. tuckermani* (Fiance and Moeller 1977). Some plants are killed by the caterpillars, apparently because they lack adequate reserves to recover from defoliation.

Measuring Hydrophyte Production

The determination of hydrophyte biomass in the lake was a necessary step toward the calculation of annual net production, but there are at least three different levels of sophistication at which the calculation can be refined beyond that step. The simplest method is to measure the increase in biomass during the growing season. The resulting value is an underestimate of net production, because some tissue decays even while new tissue is accumulating. This approach usually involves repetitive sampling of an entire stand of vegetation (Westlake 1965), but one variation is to sample the mean weight of individual plants during the growing season and to correct for changes in plant density, or assume density constant (Kansanen and Niemi 1974). This approach would have been difficult to apply to Mirror Lake owing to the great spatial variability of biomass per unit area (*Chapter V.A.4*) and, in plants like *Lobelia dortmanna*, the variability of plant weights (Moeller 1978a).

A more elaborate approach is periodically to mark individual plant parts and compute annual net production as the sum of new tissues produced during the year (e.g. Sand-Jensen 1975). To be most accurate this approach must be combined with a demographic analysis, so that biomass accruing as new plants and as increase in size of plants is included. This approach is especially suited to the mostly perennial, even evergreen nature of the Mirror Lake vegetation, in which the carry-over of tissue from one year into the next obscures the relationship between mid-summer biomass and annual net production.

A rather different approach is to measure the photosynthetic activity of plant tissue as the rate of inorganic carbon uptake (Wetzel 1964, 1969) or oxygen release (e.g. Ikusima 1966). Because photosynthesis is expressed per unit of plant tissue, biomass per unit area of lake bottom must also be measured periodically, and photosynthetic rates measured on different parts of the shoot must be integrated into a composite value applicable to the entire plant (e.g. Adams et al. 1974; Adams and McCracken 1974). Annual net production is then calculated as an integration over the growing season of the product: $Ps_n \cdot B$, where Ps_n is net photosynthesis (corrected for nighttime respiration) and B is biomass. Concurrent decay is equivalent to the amount of newly synthesized matter that fails to appear in subsequent biomass determinations. Over an annual cycle, decay equals the net production for stable populations.

Two difficulties hindered application of the photosynthetic approach to the Mirror Lake plants. First, the isoetid species like *Lobelia dortmanna* apparently absorb most of their photosynthetic carbon by way of the roots (Wium-Andersen 1971; Sand-Jensen and Søndergaard 1978). Consequently, measurements of gas exchange must be made from the rhizosphere, where CO_2 levels are higher than in the lake water but where turbulence is reduced (Sand-Jensen and Søndergaard 1978). Second, short-term photosynthesis by submersed aquatic plants often is not carbon saturated; that is, any increase or decrease in the carbon supply to the plant surface will cause the photosynthetic rate of enclosed plants to deviate from that of naturally growing plants. This problem is especially severe in poorly buffered soft waters like Mirror Lake, because the progressive uptake of CO_2 or HCO_3^- results in increasing pH and precipitously decreasing concentrations of free CO_2—the form of inorganic carbon used preferentially by submersed plants, and exclusively by some of them (Raven 1970). For example, *Utricularia purpurea* in Mirror Lake takes up only free CO_2. When this plant was illuminated in sealed chambers, net uptake of inorganic carbon ceased when approximately 20% of the total inorganic carbon had been absorbed. At that point the free CO_2 concentration had dropped to 0.1 to 0.2 mg CO_2/liter, the compensation point at which uptake was reduced to a level supported by respired CO_2 (Moeller 1978c). Accurate values of net production depend on overcoming these difficulties. The biomass-marking method was instead selected for studies in Mirror Lake, as it provided concurrent estimates of organic matter synthesis and decay, and hence of the biomass turnover in the perennial plants, yet avoided the problems associated with carbon-limited photosynthesis and gas exchange within the rhizosphere.

Seasonal Growth and Net Production of Three Species in Mirror Lake

Three species were selected among the hydrophytes in Mirror Lake to illustrate the seasonal growth dynamics of contrasting growth-forms. Insights gained from these species into the relationship between annual net production (P) and the maximal biomass (B) have allowed an estimation of the annual net production in the lake as a whole. The P/B quotients presented here represent correction factors by which biomass, as determined in the mid-summer survey of the littoral zone, is converted to the dynamic unit of production. In practice, a random (*Nuphar variegatum*) or arbitrary (*Lobelia dortmanna*, *Utricularia purpurea*) selection of individual stems was monitored for tissue production and decay over a one-year period (Moeller 1978a, 1980). The study plants were part of localized populations where each species was relatively abundant and strongly dominant.

The final data as summarized here are based only on plants that survived the entire period of observation without experimental accident or natural mortality (factors that are difficult to distinguish). The implicit assumption that natural mortality is zero results in an underestimate of the population-wide P/B quotient. Natural mortality appears to be very low in *N. variegatum* and among the larger stems of *U. purpurea*. Major losses of leaf tissue and some mortality of rosettes occur in *L. dortmanna* owing to the depredations of an aquatic caterpillar (Fiance and Moeller 1977). The gradual replacement of large adult plants through the net growth of smaller plants must increase the P/B values reported here by perhaps 10% (*N. variegatum* and *U. purpurea*) to 20% (*L. dortmanna*). Comprehensive demographic studies of survivorship, reproduction and size-specific P/B quotients are required to refine these values.

Lobelia dortmanna (water lobelia) and other isoetids are both perennial and evergreen (or "wintergreen"). They all retain a large proportion of their midsummer biomass through the following winter—60% in the case of *L. dortmanna*. There is a good deal of variation within the growth-form, from plants like *L. dortmanna* in which each rosette may live several to many years, to stoloniferous species like *Eleocharis acicularis* and *Juncus pelocarpus,* in which the individual plants persist indefinitely as a series of shorter-lived rosettes.

The biomass-marking study of *L. dortmanna* revealed the seasonal pulse of leaf production, followed by a pulse of leaf senescence, that are obscured by the rather small (40%) fluctuation in biomass during the year (*Figure V.B.3*–1). The seasonal increase in biomass underestimates annual net production by 42%, owing to the partial overlapping of leaf production and decay. Although *L. dortmanna* produces no new leaves during the late autumn and winter, there is conceivably enough photosynthesis taking place to balance winter respiratory demands. At the study site, a diffuse stand of 10 g dry wt/m^2, annual net production is approximately 7 g (thus P/B = 0.7).

Utricularia purpurea (purple-flowered bladderwort) is a perennial stem of indeterminate growth. It is rootless, yet stays more or less anchored to the bottom by the weight of detritus encrusting the older sections of stem. The seasonal variation of biomass is only 48% of the mid-summer maximum, and underestimates the annual net production by 52%. *Utricularia purpurea* has a broad depth-range in Mirror Lake (0 to 6.5 m), and its growing season is determined by the seasonal dynamics of thermal stratification (Moeller 1980). The deeper plants only begin to grow as warm epilimnial water is mixed progressively deeper during the summer (*Figure V.B.3*–2). Plants growing at a depth of 6 m have a shortened growing season whose onset is delayed past the time of maximal illuminance. Plants of equivalent size produce only 25 to 40% as much biomass at 6 m as at 2 or 4 m, and their growth seems to be limited by both

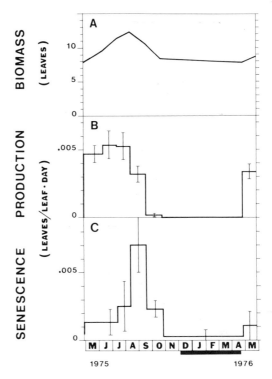

Figure V.B.3–1. Seasonal growth pattern of the perennial hydrophyte, *Lobelia dortmanna*. **A.** The mean number of leaves per plant; **B.** the rate of leaf production; and **C.** the rate of leaf senescence. These parameters are good estimates of the biomass dynamics of whole plants, since leaves are relatively uniform in size, roots are produced only at the base of new leaves, and belowground biomass fluctuates in seasonal synchrony with leaf biomass. (Redrawn from Moeller 1978a, *Canadian Journal of Botany* 56(12):1425–1433. Copyright © 1978 by the National Research Council of Canada. Used with permission.)

temperature and light during the short growing season. At all depths, growth slows in September, as thickened winter buds form at the apex of all stems. Photosynthesis declines to less than 0.7% of the mid-summer amount by the time the water temperature reaches 5°C. Overwintering stems are effectively dormant (Moeller 1978c). At the study site, a very diffuse stand of 5 g dry wt/m^2 at a depth of 2 m, the annual net production is 5 g/m^2 (thus P/B = 1.0).

Nuphar variegatum (yellow waterlily or spadderdock) is a perennial floating-leaved plant that overwinters as a thick subterranean rhizome tipped with dormant buds. Delicate,

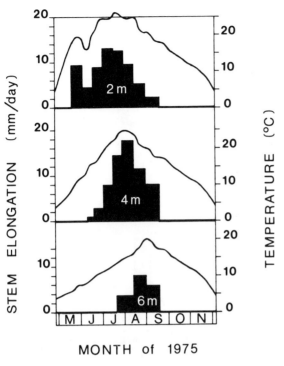

Figure V.B.3–2. Seasonal growth pattern of the rootless hydrophyte, *Utricularia purpurea*, at three depths (histograms), compared with trends of water temperature (smooth curves). Measurements from plants of different stem diameter were standardized to rates expected of plants 0.9 mm in diameter. (Redrawn from Moeller 1980, *Freshwater Biology* 10:391–400. Copyright © 1980 by Blackwell Scientific Publications Ltd. Used with permission.)

short-stemmed submersed leaves are produced in late spring and again in late summer. The familiar long-stemmed floating leaves and yellow flowers develop during June–September in Mirror Lake. Even during mid-summer, photosynthetic tissues, including flowers and fruits, make up only 32% of the total biomass, but these tissues turn over relatively rapidly during the growing season. The annual net production of green tissues is 2.6 times their maximum biomass, compared with turnover rates of about 0.33 per yr for roots and 0.14 for rhizomes. It turns out that the annual net production of the plants as a whole almost equals the mid-summer biomass (P/B = 0.9 in the studied stand of 200 g dry wt/m^2), but this equivalence obscures the very different rates of turnover in above- and belowground tissue. New biomass produced by the shoots is allocated to floating leaves (58%), flowers and fruits (22%), submersed leaves (7.6%), roots (6.1%) and rhizomes (6.1%).

In all three species a large proportion of the maximum biomass is retained through the winter (*Figure V.B.3–3*). Such plants of "obscured annual regrowth" (Westlake 1965) store structural tissue, carbohydrates and nutrient elements that may protect the populations from environmental hazards to which tiny plants, growing from seeds, spores or abscised buds, would be vulnerable. The storage function is performed by belowground organs in *N. variegatum*, whose floating leaves are expendable structures persisting barely one month. In *L. dortmanna* both leaves and belowground tissues share the storage function. Leaves persist about 10 to 11 months on the average. *Utricularia purpurea* is all stem. The winter buds do not abscise as in *U. vulgaris* (Maier 1973), so the old stem acts as the storage organ. Extensive dieback does occur in *Potamogeton berchtoldii* and the three annual species in Mirror Lake, but the evergreen characteristic of most species creates a distinctive vegetational physiognomy in Mirror Lake and other softwater lakes of Europe and northeastern North America.

Judging from the observations on *L. dortmanna* and *U. purpurea*, the maintenance of green shoot tissues during the winter does not necessarily mean that the growing season extends appreciably into the late autumn or winter. On the other hand, detectable increases in biomass (e.g. Borutskii 1950) and measured photosynthetic rates (Sheldon and Boylen 1976) suggest that some other vascular plants of temperate lakes couple the evergreen habit with an extended growing season.

Biomass Turnover Related to Nutrient Cycling

Perennial plants that retain much of their tissues past the main growing season have an opportunity to absorb and store nutrients for the subsequent formation of new tissue. Boyd (1970, 1971) noted a disproportionate uptake of critical elements such as N and P early in the growing season by some emergent aquatic plants, and suggested that such accumulation of excess nutrients could be an important adaptation for growth in oligotrophic lakes. Thus, there are potentially three components to the

B: Production and Limiting Factors

Figure V.B.3–3. Seasonal changes in biomass and in the relationship of biomass to annual net production for three species of contrasting growth-form. The biomass present on any date (subdivided into tissues, as indicated) is represented by the thickness of the solid part of each figure, scaled in units of annual net production (left) and in units of local stand biomass (right). Shading indicates the cumulative senescence of biomass (subdivided into tissues, as indicated) during the 1975 growing season. Width of field in photographs: 6 cm (*Lobelia* and *Utricularia*) or 50 cm (*Nuphar*).

conservatism of perennial plants within nutrient cycles:

1. The conservation of a large fraction of the mid-summer biomass along with the nutrient elements it contains (*biomass holdover*). This conservation means that plants need not grow each year from tiny embryos or buds that are presumably more severely subject to environmental hazards, such as herbivory, than adult plants.
2. The ability to store preferentially such mobile elements as N, P and K that can be resorbed from senescing tissues (*nutrient resorption*). This component of conservation would be associated with a detectable increase of elemental concentrations in overwintering tissue.
3. The suggested ability to absorb and store excess elements at times when the plants are not actively synthesizing new biomass (*nutrient hoarding*). This component of conservation also implies an increase of elemental concentrations in overwintering tissue. In the absence of actual measurements of nutrient uptake, only an absolute increase in the quantity of an element present in a population would allow this net uptake to be recognized above and beyond increases in concentration attributable to resorption.

Periodical chemical analyses of *L. dortmanna* and *U. purpurea* in conjunction with the growth studies allow an evaluation of these components of nutrient conservation. In *Figure V.B.3–4* are plotted the seasonal trends of elemental masses of N, P and K compared with the trends of biomass in the two populations. All curves are normalized to that amount of element or biomass present in August, the time of maximal biomass. Throughout most of the year, concentrations of N, P and K in *L. dortmanna* exceed those in midsummer. Resorption from senescent tissue probably contributes part of the increase, but the absolute increases in the amount of each element in the population during mid-autumn 1975 demonstrate that a net uptake also occurred. Autumn 1975 was warmer than average, and autumn 1976 colder than average; so the absence of a similar absolute increase in 1976 suggests that net uptake occurs under relatively warm conditions immediately following the cessation of leaf production and the pulse of leaf senescence. The excess N stored in overwintering *L. dortmanna* (mainly in the leaves) may have contributed roughly 25% of the following year's growth requirement (Moeller 1978a). But the excess P stored in late autumn was gradually leached away before the next growing season began. In *U. purpurea* no net autumnal uptake could be recognized, nor was there even an increase in concentration of N. Slight increases in P and K concentration could represent resorption alone. As in *L. dortmanna,* P apparently was lost during the winter.

Thus in *L. dortmanna* components (2), nutrient resorption and (3), nutrient hoarding, both occur to some extent, but in *U. purpurea* even resorption appears minimal. Obviously component (1), biomass holdover, is predominant and impressive in its magnitude in both species. In both species the seasonal uptake of N, P and K is closely regulated by the seasonal pattern of biomass synthesis. When the plants are not growing actively, absorption of nutrients does little more than balance losses to secretion or leaching.

The mid-summer biomass serves as a rough approximation of the annual net production of the large aquatic plants in Mirror Lake. This implicit P/B quotient of 1.0 is consistent with values determined for the three species that were actually studied. Although P/B quotients exceed 2.5 in some submersed temperate species from freshwater and marine sites (e.g. Adams and McCracken 1974, Sand-Jensen 1975), values of 1 to 1.5 are probably more common. The low value found for *Lobelia dortmanna* (P/B = 0.7) agrees with a similar study of this species in Lake Kalgaard, Denmark (P/B = 0.8; Sand-Jensen and Søndergaard 1978), but not all rosette plants turn over their tissues this slowly. Net production of *Littorella uniflora* in Lake

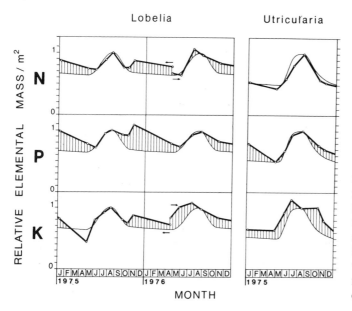

Figure V.B.3–4. Seasonal changes in biomass and biomass-bound pools of N, P and K in two perennial hydrophytes. Amounts of biomass (thin line) and nutrients (thick line showing data points) are relativized to the amounts of each present in mid-August. Shading between the lines indicates elemental enrichment above the mid-August concentrations. In *Lobelia dortmanna* the concentration of N was higher in August 1976 than in August 1975, whereas K was lower in 1976 than in 1975, necessitating the rescaling of N and K curves in April 1976.

B: Production and Limiting Factors

Kalgaard is 1.5 times the maximum biomass (Sand-Jensen and Søndergaard 1978), equivalent to that of several plants of elodeid growth-form from more fertile sites (Ikusima 1966; Raspopov 1972; Dawson 1976). If we accept a onefold turnover of mid-summer biomass annually, then the annual net production of hydrophytes in Mirror Lake amounts to 2.5 g C/m² of littoral zone, or 1.4 g/m² of lake surface. This is only 4% of phytoplankton production (*Chapter V.C*).

The minor contribution of hydrophytes to the carbon budget of Mirror Lake is not a surprise to anyone who has noted the sparsity of plants on the sandy or rocky bottom. It is misleading to think that the littoral zone itself is small in proportion to the lake area. Hydrophytes thinly colonize a broad zone from 0 to 7.2 m in depth, equal to 58% of the lake area. In some other lakes with similar morphometry and water clarity, and similarly "oligotrophic" phytoplankton densities, hydrophyte production exceeds that of the phytoplankton. In these cases the littoral zone is more uniformly, and densely, colonized. Examples are a marl lake in Michigan (Lawrence Lake: Rich et al. 1971) and a softwater lake in Denmark (Lake Kalgaard: Søndergaard and Sand-Jensen 1978; Sand-Jensen and Søndergaard 1979).

It should be a major aim of comparative limnology to establish what factors are responsible for the great differences in abundance of macrovegetation among small lakes. Mirror Lake's phytoplankton appears to be limited by inadequate supplies of both N and P during the summer (*Chapter V.B.2*; Gerhart 1975), but it is not certain that these elements also control the density of hydrophyte populations. The emplacement of commercial plant food pellets, providing N, P and K, within the rhizosphere of *Lobelia dortmanna* did cause the rate of leaf production, and hence of net production, to increase by 50 to 80% over periods of 7 to 12 weeks (R. Moeller, unpublished data). Long-term studies are required to test whether nutrient enrichment can lead to increases in the vegetational density as well. Any factor that affects the reproduction or survival of these mostly perennial plants acts to limit their contribution to lake metabolism.

The measured hydrophyte biomass in mid-summer, 530 kg dry matter (*Chapter V.A.4*), contains approximately 215 kg organic C (46% of the organic matter) as well as various amounts of other elements (*Table V.B.3–1*). These estimates come from two series of chemical analyses: first, the August chemical analyses of tissues collected in conjunction with biomass turnover studies of *Lobelia dortmanna, Utricularia purpurea* and *Nuphar variegatum*; second, the analyses of triplicate collections of seven other species collected from one to three sites on one to three dates during July or August of several years. The mean values reported in *Table V.B.3–1* are subject to a large variance, especially for the seven species collected from a variety of sites. Replicate values often differed by amounts as large as the mean itself; consequently, the data do not provide precise means for the individual species. Spatial heterogeneity is the main source of variation, judging from the smaller variance associated with triplicate collections of plants at the sites where studies of biomass turnover were carried out (e.g. *Lobelia dortmanna*: Moeller 1978a).

Although the interspecific differences evident in *Table V.B.3–1* are, in many cases, no larger than intraspecific variability, some conspicuous contrasts deserve mention. The carnivorous vascular plant, *Utricularia purpurea,* is distinguished from all other Mirror Lake species analyzed by high concentrations of Ca and Zn. The macroalga, *Nitella flexilis,* is indistinguishable from other submersed plants except for somewhat lower levels of Na. All species are rich in Na—usually 1 to 2% of organic weight. As discussed by Hutchinson (1975), such high levels of Na are typical of submersed freshwater plants in marked contrast to most terrestrial species, but the physiological basis for this difference has not been resolved.

Because the large aquatic plants are relatively sparse in Mirror Lake, their role in nutrient cycles also must be very small. Elemental pools bound as hydrophyte biomass (e.g. the 14 kg of N and the 0.8 kg of P) exceed pools tied up in the living phytoplankton, but the rapid turnover of plankton populations reduces the biogeochemical significance of hydrophyte growth and decay. The hydrophytes, except for the rootless *Utricularia purpurea*, probably derive most of their nutrients from the littoral sediments. Their growth cycles thus would result in a very modest transfer of nutrients across the

Table V.B.3–1. Tissue Chemistry of Large Aquatic Plants in Mirror Lake[a]

Species		Ash (% dry wt)	N	P	Ca	Mg	Na	K	Zn (µg/g-OM)
					(% Organic Matter)				
A. Elemental concentrations in selected species:									
Nuphar variegatum		7.9	1.4	0.12	0.63	0.06	0.40	2.7	30
	(a)[b]	7.5	2.3	0.24	1.05	0.11	0.35	2.4	33
	(b)	8.1	0.9	0.06	0.40	0.04	0.43	2.8	29
Eriocaulon septangulare		11.0	2.6	0.08	0.07	0.06	1.2	1.7	180
	(a)	9.0	3.3	0.12	0.08	0.06	1.9	2.6	80
	(b)	13.0	2.0	0.05	0.07	0.05	0.6	0.8	280
Gratiola aurea		11.0	2.6	0.20	0.67	0.22	1.9	2.1	410
	(a)	15.0	2.8	0.23	0.82	0.30	2.2	2.6	290
	(b)	4.0	2.3	0.16	0.41	0.09	1.2	1.4	610
Isoetes tuckermani		15.0	2.9	0.20	0.47	0.15	1.4	3.0	260
	(a)	16.0	3.1	0.21	0.63	0.19	1.7	4.0	250
	(b)	13.0	2.5	0.18	0.19	0.09	1.0	1.2	290
Juncus pelocarpus		6.5	2.2	0.13	0.31	0.18	1.2	1.8	140
	(a)	7.1	2.4	0.14	0.32	0.20	1.4	2.0	130
	(b)	4.8	1.8	0.12	0.27	0.11	0.7	1.2	170
Lobelia dortmanna		9.3	1.8	0.11	0.55	0.21	1.3	2.7	160
	(a)	7.8	1.7	0.10	0.56	0.13	1.0	3.1	140
	(b)	11.0	1.9	0.13	0.55	0.28	1.4	2.2	170
Sagittaria graminea		19.0	3.4	0.20	0.50	0.16	2.1	4.3	150
	(a)	19.0	3.5	0.20	0.51	0.13	2.1	4.3	140
	(b)	19.0	3.2	0.21	0.48	0.15	2.1	4.3	170
Potamogeton berchtoldii		13.0	3.8	0.45	0.73	0.13	1.9	2.7	210
Nitella flexilis		11.0	2.6	0.15	0.84	0.35	0.7	2.4	190
Utricularia purpurea		16.0	3.5	0.11	3.2	0.80	1.3	1.7	1,500
B. Mean concentrations for the species reported above (units as above):		12.0	2.7	0.18	0.80	0.23	1.3	2.5	320
C. Estimated elemental mass bound in hydrophyte tissue during mid-summer (kg/lake)[c]		63.0	14.0	0.8	4.6	1.5	6.5	13.0	0.2

[a] Values represent means of three plant collections from mid-summer (mid-July to early September).
[b] Aboveground (a) and belowground (b) concentrations are presented separately for plants with appreciable belowground tissue, along with composite values for the whole plant.
[c] Calculated by applying species chemistry to biomass values reported in *Chapter V.A.4*, and assuming the mean concentrations for all species that were not analyzed (amounting to 57 of the 530 kg dry wt in the lake).

lake bottom, probably on the order of a onefold turnover of the elemental pools bound in the mid-summer biomass.

Summary

By marking individual plant parts—notably leaves—and following the formation and senescence of tissue over an annual period, the relationship between annual net production (P) and mid-summer biomass (B) was determined for three perennial plants in Mirror Lake. The P/B quotients ranged from 0.7 (the isoetid, *Lobelia dortmanna*) and 0.9 (the waterlily, *Nuphar variegatum*) and 1.0 (the rootless elodeid, *Utricularia purpurea*). Overwintering tissues amounted to 52% of the mid-summer biomass in *U. purpurea* (green stems overwinter), 60% in *L. dortmanna* (both leaves and belowground tissues overwinter), and 75% in *N. variegatum* (belowground rhizomes and roots overwinter). Overwintering tissues conserve nutrients as well as carbohydrates in the structural tissue, but serve to accumulate little (*L. dortmanna*) or no (*U. purpurea*) excess nutrients during this period when competition from short-lived algae and bacteria might be reduced.

Annual net production of the hydrophyte community in Mirror Lake can be estimated as a onefold turnover of the mid-summer biomass, or 2.4 g C/m^2 of littoral zone. This value is very low in comparison with some other small lakes and also when compared with phytoplankton production, and results more from the very sparse colonization of the littoral zone than from low rates of biomass turnover. The observed rates of biomass turnover are not atypical of temperate hydrophytes.

4. Periphyton Production

BRUCE J. PETERSON and GENE E. LIKENS

Photosynthesis by periphyton ranges from a major or even a dominant source of autochthonous organic carbon for many lake ecosystems (e.g. Wetzel 1964; Hargrave 1969; Rich et al. 1971) to a very minor source in others (Schindler et al. 1973). Apparently, growth conditions can be optimal in the littoral habitat where there are surfaces for attachment and proximity to high light intensities and rich nutrient sources. The overall metabolic significance of the periphyton community to the lake ecosystem depends in large part on the morphometry (area of littoral zone) and substrate characteristics of the lake.

The measurements of production, respiration and biomass of periphyton in Mirror Lake are limited in scope. They provide preliminary estimates only for the cobbled benthic areas of the lake. Some additional conservative calculations of production are made for the epipelic and epiphytic communities based on rates of nitrogen fixation (Moeller and Roskoski 1978).

Methods

Measurements of oxygen and CO$_2$ exchanges by the epilithic algae on rocks and cobbles were obtained by transferring rocks from the lake bottom into transparent plastic chambers placed at the 1-m depth in the rocky littoral zone. Dark chambers were covered with foil and black plastic. Care was taken not to disturb the flocculent layer of debris that covered some of the rocks. In experiments 2 and 3 (*Table V.B.4*–1) the chamber covers were attached 24 hours after the rocks were placed on the chamber bottoms to ensure minimum disturbance of the epilithic community. Changes in oxygen and total dissolved inorganic carbon (DIC) were measured at intervals of from 4 to 8 h in both light and dark chambers during the day and during the following night. Chambers were fitted with manual stirrers, which allowed complete mixing of water within the chambers before each set of samples was taken.

Samples for oxygen and DIC were withdrawn from the chambers through serum-stopper ports with 50-ml plastic disposable syringes. Oxygen samples were fixed immediately by injecting 0.5 ml each of MnSO$_4$ and alkaline iodide into the syringes. On return to the laboratory the samples were acidified with sulfuric acid and titrated with N/20 sodium thiosulfate (American Public Health Association 1965). DIC samples were acidified at the laboratory and equilibrated with helium in the sampling syringe. The head-

Table V.B.4–1. Epilithiphyton Production, Respiration and Chlorophyll-a at the 1-m Depth in Mirror Lake

Date	Site	Chamber Nos. Light (L) Dark (D)	Net Daylight Community Production (mg C/m²)	Net 24-h Community Production in Light Chamber	Total 24-h Dark Respiration	Gross 24-h Primary Production	RQ $\Delta CO_2/\Delta O_2$ (mole/mole)	Chl-a (mg/m²)
				(mg C/m²-day)				
10 Jul. 74	1[a]	L1	−35[b]	—	—	—	—	—
		L2	−55	−21	82[c]	+103[d]	—	—
		D1	—	—	+ 97	—	1.11	—
6 Aug. 74	1	L1	−31	+ 3	+ 82	+ 79	0.63	54
		L2	−12	+17	+ 71	+ 54	0.55	64
		D1	—	—	+128	—	0.68	63
	2[e]	L3	− 9	+41	+122	+ 81	1.43	15
		L4	−13	+37	+123	+ 86	0.62	23
		D2	—	—	+106	—	0.84	27
17 Aug. 74	1	L1	−14	+12	+ 63	+ 51	0.68	—
		L2	−11	+18	+ 71	+ 54	1.04	55
		D1	—	—	+113	—	0.66	—
	3[e]	L3	+37	+89	+128	+ 39	1.25	—
		L4	+36	+90	+132	+ 42	1.76	—
		D2	—	—	+166	—	1.22	20
		x(n)	−10.7 (10)	+31.8 (9)	+106 (14)	+65.4 (9)	0.96 (13)	40 (8)
		s_x	9.1	12.5	7.9	7.4	0.10	7.3

[a] Site 1 was a relatively fertile site on the east shore near the outlet.
[b] Negative value means DIC concentration decreased in the light chamber during the daylight period (photosynthetic CO_2 uptake exceeded respiratory CO_2 release). Positive values indicate DIC increase.
[c] 24-h dark respiration in the light chambers was computed by extrapolating the hourly respiration rate measured either by darkening the chamber or at night.
[d] Calculated for light chambers only using the assumption that the hourly respiration rate in the light period (14 h) equaled the respiration rate in the dark period (10 h).
[e] Sites 2 and 3 were less productive and were located on the south shore.

space gas was analyzed for CO_2 on a gas chromatograph following the procedure of Stainton (1973).

Estimates of community metabolism were calculated as follows. Net daily (light period only) community production was calculated from the absolute (positive or negative) change in CO_2 in the light chamber extrapolated to the whole-light period based on the percent of the daily radiation actually occurring during the hours of incubation. Net 24-h community production was calculated from the absolute change in DIC in a single chamber over one entire day/night cycle. This calculation could only be done for the light chambers where gas exchange for both light and dark conditions was measured (N = 9). Gross 24-h production was calculated by a procedure analogous to the light-dark, O_2-bottle method. Net 24-h change in CO_2 in the light chamber plus total change in CO_2 in the dark chamber equals gross 24-h epilithic algal production. This calculation was made for each light chamber individually by assuming that the measured rate of dark respiration could be extrapolated to 24 hours. Net 24-h primary production was assumed to be equal to 60% of gross production. Respiratory quotients (RQ) on a molar basis were computed for each chamber as $\Delta CO_2/\Delta O_2$ in the dark period only.

After each experiment, the rocks in each chamber were scrubbed with a stiff bristle brush to remove attached algae. A subsample of the material removed by scrubbing was filtered onto a glass fiber filter. The filter was homoge-

B: Production and Limiting Factors

nized in 90% acetone with a teflon-glass pestle and mortar. The chlorophyll was extracted in the refrigerator for 24 hours and the concentration of chlorophyll was determined with a Turner fluorometer. The fluorometer was standardized against duplicate samples done on a Beckman DU spectrophotometer by the method of Lorenzen (1967). The colonized surface area of the rocks in each experiment was determined by fitting foil to the contour of each rock and cutting to the outline of the algal colonization, which was quite evident after scrubbing. The foil was dried and weighted to obtain colonized area. The surface area of rocks in the chambers averaged 1.4 times the area of the chambers. We assumed the lake bottom had the same area of colonized surface per square meter as did the chambers. The displacement volume of the rocks in each experiment was determined to calculate the residual water volume in the chambers by difference, so calculations of metabolic rate could be made.

Results

The rates of production and respiration measured in each chamber are summarized in Table V.B.4–1. Production and standing stocks of Chlorophyll-a were higher at the east-shore site but respiration per square meter was not significantly different at the three locations. Community respiration (24 h) exceeded daylight net production and usually exceeded 24-h gross production as well. Net carbon uptake and Chlorophyll-a (uncorrected for phaeopigments, which made up as much as 50% of the total pigment) occurred in the same range as those reported by Schindler et al. (1973) for similar lakes (#239, Rawson and #240) in the Experimental Lakes Area of Ontario. Schindler et al. reported net carbon uptake of 0 to 100 mg/m^2-day and Chlorophyll-a from 10 to 60 mg/m^2, while we found values from less than 0 to 55 mg/m^2-day and 15 to 64 mg/m^2.

The third experiment included a very dark afternoon, and there was no detectable production of dissolved oxygen in any of the light chambers. The two light chambers on the more productive east shore showed a slight uptake of CO_2 while CO_2 increased in both light chambers at site 3.

Rates of respiration were surprisingly constant from day to day and chamber to chamber (Table V.B.4–1). The narrow range from 62 to 166 mg C/m^2-day is surprisingly small in view of the obviously heterogeneous nature of the substrate and biomass. These values represent the most reliable measurements of the study. The mean respiration quotient, RQ, of 0.96 is not significantly different from 1.0.

Annual Values

Some cautious extrapolations of these results are of interest for comparison with other carbon fluxes in Mirror Lake (see Chapter V.C). The cobbled substrate represents 19%, sand and gravel 13% and gyttja 68% of the bottom area in Mirror Lake. If we assume that the average rate of gross algal production on cobbles and rocks (Table V.B.4–1) continues for 200 days, gross annual production is 13.1 g C/m^2-yr. Net epiphytic production is assumed to be 60% of gross production or 7.8 g C/m^2-yr. Respiration by heterotrophs averaged 71 mg C/m^2-day in these experiments. If we assume this rate continues for 200 days and that the rate is one-fourth as great (assuming a respiratory Q_{10} of 2.0) during the winter months, annual respiration is 17.1 g C/m^2-yr in the rocky littoral zone. On a whole-lake basis, net epilithic algal production is approximately 1.5 g C/m^2-yr and heterotrophic respiration is 3.25 g C/m^2-yr. While the data base for these calculations is limited, it is clear that epilithic algal carbon fixation is a small part of the total organic carbon input to the Mirror Lake ecosystem. Based on these experiments, the rocky littoral community appears to consume more organic carbon than it produces and is probably supported by organic carbon inputs from allochthonous and planktonic sources (see Chapter V.C).

Some idea of the magnitude of epipelic and epiphytic production can be obtained from the estimates of nitrogen fixation by the littoral benthic algae (Moeller and Roskoski 1978). On a whole-lake basis, annual N-fixation by algae in these habitats was estimated to be about 0.1 g N/m^2. Assuming a C:N weight ratio of 6:1, a minimum estimate of net epipelic and epiphytic algal production would be 0.6 g C/m^2-yr. Thus, the lakewide total for epilithic, epipelic and epiphytic net primary production would be 2.1 g C/m^2-yr. Schindler et al. (1973) estimated an annual net productivity for periphyton of 810 and 900 mg C/m², based on the entire lake surface in Rawson Lake and Lake #240 of the Experimental Lakes Area.

Nutrient Flux and Cycling

Not only are periphyton involved in the flux and cycling of carbon in Mirror Lake, but they also incorporate and cycle all of the other nutrients that are physiologically important to algae. In addition, because heterocyst-bearing blue-green algae are common to abundant on plant and inorganic surfaces in the littoral zone (see *Chapter V.A.3*), periphyton also may play a significant role in nitrogen flux and cycling in Mirror Lake. Moeller and Roskoski (1978) found that epiphytic blue-green algae fixed inorganic nitrogen at rates ranging from 0.1 to 1.2 µg N/g macrophyte dry wt/h, and epipelic algae fixed from 0.5 to 400 µg N/m² and epilithic-epipsammic algae fixed from 40 to 207 µg N/m² of littoral lake bottom per hour during the late summer and autumn of 1975 in Mirror Lake. Extrapolating the highest values to an arbitrary 1500-h growing season, annual N-fixation would be about 1.8 mg N/g dry weight of macrophytes/yr for epiphytic algae, 600 mg N/m²-yr for epipelic *Anabaena* mats and 300 mg N/m²-yr for *Nostoc*-colonized cobbles. We estimate a total fixation of about 100 mg N/m²-yr on an entire lake basis. Such extrapolation must be treated with much caution, however, because spatially the pattern of fixation activity was highly variable and nothing is known about seasonal variations. Nevertheless, it appears that the periphyton may be a significant source of inorganic nitrogen for the lake.

Summary

Biomass (Chlorophyll-*a*), production and respiration were measured for epilithic algae on the cobbled substrate of Mirror Lake. Gross 24-h primary production ranged from 39 to 103 mg C/m²-day and Chlorophyll-*a* from 15 to 64 mg/m². Community respiration ranged from 63 to 166 mg C/m²-day. Production was greater at the east-shore site with high chlorophyll, but respiration was not significantly different at the various sites.

Annual epilithic net production was estimated at about 7.8 g C/m² for the cobbled substrate area or 1.5 g C/m² on a whole-lake basis. Annual epilithic heterotrophic respiration was estimated at 3.25 g C/m² for the entire lake. Epipelic plus epiphytic net production was estimated at about 0.6 g C/m²-yr for the entire lake.

Nitrogen fixation by epiphytic, epipelic and epilithic algae might be as high as 1.8 mg N/g dry wt-yr, 600 mg N/m²-yr and 300 N/m²-yr, respectively, for macrophyte, soft and cobbled sediment substrates in the lake, or totaling about 0.1 g N/m²-yr on an entire lake basis.

5. Zooplankton

JOSEPH C. MAKAREWICZ

Methodology

Production in animals is defined as that part of the assimilated food that is retained and incorporated into the biomass of the organism. For a specific population of zooplankton over a given period of time, production is the sum of the growth increments of all the individuals in the population. The increase in numbers and size of somatic cells, gametes, exuviae and other separable parts of the body should be considered as part of growth (Winberg 1971a).

Zooplankton production may be regarded simply as growth and differs conceptually from primary production. That is, gross primary production is the rate at which radiant energy is stored by photosynthetic and chemosynthetic organisms as organic substances. For primary producers like algae, the rate of storage of organic matter in excess of that amount used in respiration is termed net primary production (see Odum 1959). In zooplankton, the assimilated energy utilized in respiration is energy derived from organic matter, which was produced by primary producers, not synthesized by the consumer. Because consumers only utilize food, which has already been produced, secondary production should not be divided into "gross" or "net" secondary production. By analogy, gross primary production can be compared with zooplankton assimilation, and net primary production can be compared with zooplankton production without the term "net" (Edmondson 1974).

The measurement of zooplankton production is technically very different from the measurement of primary production. Primary production of the phytoplankton community can be measured directly, as with the ^{14}C procedure, or indirectly with the light-dark bottle oxygen method. In theory at least, a few measurements conveniently integrate the production of all algal species in the ecosystem. No similar procedure exists for integrating the production of all species of zooplankton.

Before an attempt can be made to estimate the production of any zooplankton species, it is necessary to have some knowledge of the timing and nature of reproduction. In zooplankton populations, two modes of reproduction are evident: continuous and noncontinuous. In species with noncontinuous reproduction (e.g. most copepods), the organism usually has a long life cycle and a short period of reproduction. The group of newly born individuals can be considered as a cohort (i.e. individuals of the same age).

Because it is often possible to identify cohorts by differences in body size, appendages and other morphological features, changes in cohort abundance with time and average growth of the individuals in the population can be measured directly from the field data, greatly simplifying the production measurement. A procedure that can be used to measure production in populations with noncontinuous reproduction is discussed later in this section.

With continuous reproduction, eggs may be produced by the population on any day of the year. Because new recruits (i.e. juveniles) are potentially entering the population every day, it is difficult to recognize and separate cohorts. This situation applies to most zooplankton populations, thus, it is necessary to use techniques that do not require the recognition and following of a cohort.

Most methods of estimating production of a population with continuous reproduction are based on the finite rate of growth of individuals (Winberg 1971a). The doubling-time method (Galkovskaya 1965), which is based upon the inverse relationship of the daily individual growth increment to generation time, is suitable for rotifiers because only a small amount of growth occurs after hatching of the egg (*Figure V.B.5–1*):

$$P = \frac{\overline{NW}}{GT}, \qquad (1)$$

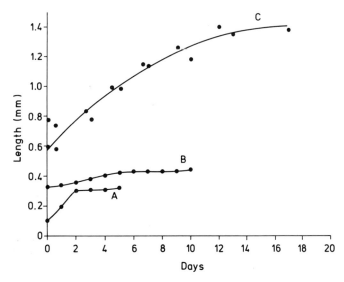

Figure V.B.5–1. Growth of the rotifers **(A)** *Rotaria rotatoria* and **(B)** *Adineta vaga* (Modified from Edmondson 1946) and **(C)** the cladoceran *Daphnia catawba* from Mirror Lake.

where

- P = production in weight (unit area or volume of lake)/unit time;
- $\overline{N}$ = mean density of a species per unit area or volume of lake;
- GT = generation time, the development time in days from hatching of the parent rotifer to the hatching of its first offspring;
- $\overline{W}$ = average weight.

For example, if we consider a rotifer population at time zero consisting of two individuals/liter with a generation time of half a day, the number of individuals produced in one day will be four. If the mean weight of the individuals in the population is 2 μg, then production will be 8 μg/liter-day.

Unlike the situation in rotifers, the addition of new tissue through the growth of an individual is often significant in cladocerans and copepods (*Figure V.B.5–1*). By dividing each species population into size classes or recognizable developmental stages (e.g. egg, nauplii, copepodid, adult), we see that production of copepods and cladocerans is the sum of the weight increments of each stage and may be calculated in the following manner:

$$P = \sum_{i=1}^{n} \frac{N_n \Delta W_n}{DT_n}, \quad (2)$$

where

- P = production in weight (unit area or unit volume)/unit time;
- N_n = the number in a size class or developmental stage;
- DT_n = development time, duration of time in a size class or developmental stage;
- W_n = the change in weight during time DT, where
- n = the number of size classes or developmental stages.

Information on the growth of individuals in the population or changes in weight between developmental stages is required for this procedure. This method can be used for populations with continuous or noncontinuous reproduction in which individuals possess a significant growth in size.

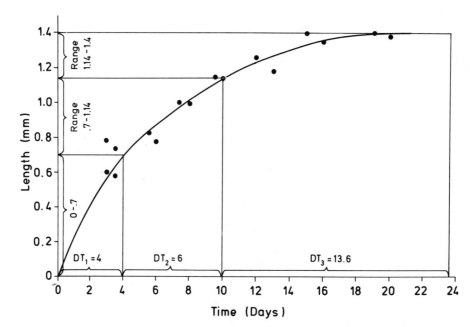

Figure V.B.5–2. Growth curve of *Daphnia catawba* in Mirror Lake (DT = development time of a size class). Modified from "Structure and function of the zooplankton community of Mirror Lake, New Hampshire," by J. C. Makarewicz and G. E. Likens, *Ecological Monographs*, 1979, 49(1):109–127. Copyright © 1979 by The Ecological Society of America. Used with permission.

B: Production and Limiting Factors

Example

The growth curve for *Daphnia catawba* was obtained by culturing the organism in water from Mirror Lake (see Makarewicz and Likens 1979 for details) and was arbitrarily divided into size classes (*Figure V.B.5–2*). The length-weight relationship of *Daphnia catawba* was previously established by weighing organisms of different length on a sensitive balance. The number (no./m^3) of *D. catawba* in each size class was estimated from samples taken with a Likens-Gilbert filter (Likens and Gilbert 1970) and counted by the settling procedure of Utermöhl (1958). The data in *Figure V.B.5–2* and *Table V.B.5–1* give the necessary information and provide an example for calculating production of *D. catawba* over a 24-h period.

The production for a monthly period (P_m) with four sampling dates will be equal to the average daily production times the number of days in the time period:

$$P_m = \frac{P_1 + P_2 + P_3 + P_4}{4} \times (28, 30 \text{ or } 31 \text{ days}). \quad (3)$$

The time interval between P_n and P_{n+1} (i.e. sampling dates) should be no longer than the shortest development time. This time interval between sampling dates can be flexible, changing with season as the water temperature changes. With increasing temperatures in the summer, a shorter development time will occur and thus a shorter time span is necessary as compared with the winter.

Important aspects of the production equation (i.e. growth rate, development and generation times) are affected by temperature and nutrition. In general, rate processes in poikilotherms, such as zooplankton, are increased with increasing temperature.

Nutrition also affects some of the parameters of the production equation. With higher food levels, a decrease in generation time for rotifers occurs. Cladocerans become larger faster, affecting development time, and have a greater maximum length (see also *Chapter V.A.5*). Because food levels change with season (*Figure V.B.5–3*) and depth (*Figure V.B.5–4*), nutritional effects on production parameters need to be considered. The best procedure is to derive the growth curve, from which the development time can be obtained, from direct observations of lengths from the natural lake population. This method is often possible with crustacean populations with noncontinuous reproduction. The growth curves of *Mesocyclops edax* and *Cyclops scutifer* in Mirror Lake over a 3-yr period were obtained in this manner (see *Figure V.A.5–7*). The effects of food and temperature are considered simultaneously.

When the growth curve cannot be obtained by direct observation of a lake population because of overlapping cohorts (i.e. in a population with continuous reproduction), organisms have to be cultured at different food levels (e.g. *Figure V.B.5–5*). Although this can be done, a problem exists relative to applying the various growth curves, obtained by culturing at different food levels, to the natural population.

Table V.B.5–1. Data Required for Estimating Production of *Daphnia catawba*[a]

Size Class	N No./m^3	L_1 mm	L_2 mm	W_1 mg	W_2 mg	ΔW mg	DT days	P mg/m^3-day
1	1,679	0.00	0.70	0.0	4.1	4.1	4.0	1,721
2	1,095	0.70	1.14	4.1	7.3	3.2	6.0	584
3	638	1.14	1.40	7.3	8.2	0.9	13.6	42
							=	2,347

or

$$P = \frac{1,679 \cdot 4.1}{4.0} + \frac{1,095 \cdot 3.2}{6.3} + \frac{638 \cdot 0.9}{13.6} = 2,347 \text{ mg/m}^3\text{-day}$$

[a] W_1 and W_2 are determined from length-weight relationships.

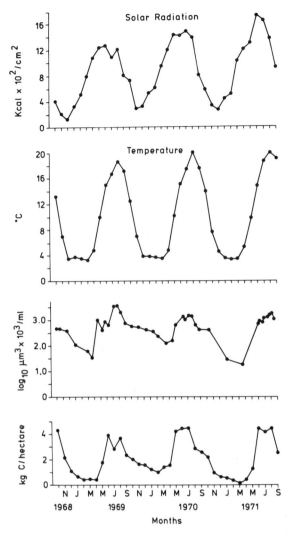

Figure V.B.5–3. Seasonal variation in solar radiation, temperature, phytoplankton biomass and zooplankton production in Mirror Lake. Phytoplankton biomass values are daily means weighted for depth at station Central. Temperature values are monthly means weighted by volume of each 1-m stratum of the entire lake. Solar radiation values represent the total amount of solar radiation impinging on the lake per month. Zooplankton production values are monthly means weighted for depth. Modified from "Structure and function of the zooplankton community of Mirror Lake, New Hampshire," by J. C. Makarewicz and G. E. Likens, *Ecological Monographs*, 1979, 49(1):109–127. Copyright © 1979 by The Ecological Society of America. Used with permission.

One approach used by Makarewicz and Likens (1979) in solving this problem is based on the effect of an increased food supply on a cladoceran to increase the size of the organism and to increase the number of eggs in the brood pouch (*Figure V.B.5–5*). Thus, the maximum length and the brood size of individuals in a natural population are indirect measures of food supply. The decision on which growth curve should be used for the production estimate can be achieved by comparing brood size and maximum length of the organism in the field with laboratory-generated growth curves and brood sizes at different food levels. *Table V.B.5–2* provides an example of this procedure.

To measure zooplankton production, the following data are required: (1) the density of each species or each size class of a species with depth and season; (2) the effect of food and temperature on growth rates of species and on generation and development times; and (3) the weight of each species or size class of each species. Furthermore, since temperature and food levels vary with season and depth, all of these factors need to be integrated over a year. As compared with primary production, the measurement of zooplankton production is a monumental task that has been attempted in only a handful of the lakes of the world. Further information on measuring zooplankton productivity, including other methods, may be obtained from Makarewicz and Likens (1979), Edmondson (1974), Edmondson and Winberg (1971) and Winberg (1971b).

Zooplankton Production and Nutrient Cycling in Mirror Lake

In spite of a voluminous literature on zooplankton, the role of zooplankton, particularly rotifers, is inadequately understood in aquatic ecosystems. For example, do rotifers, with their short generation times as compared with crustaceans, contribute a significant proportion to the total community zooplankton production and, thus, play an important role in energy transfer in an oligotrophic ecosystem such as Mirror Lake? Does the rotifer contribution to community productivity change in lakes of different trophic status? Rotifers possess fragile, easily-decomposed bodies. Could they be important in

B: Production and Limiting Factors

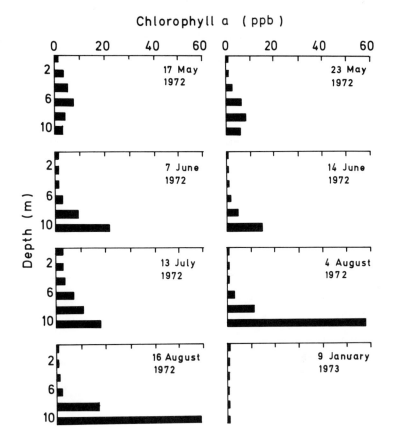

Figure V.B.5–4. Chlorophyll-depth profiles in Mirror Lake. (From Gerhart 1973.)

Table V.B.5–2. Mean Maximum Length and Brood Size of Selected Monthly Field Samples and Laboratory Specimens of *Holopedium gibberum*[a,b]

	Maximum Length (mm)	Brood Size	Growth Curve Employed
Laboratory Culture (enriched)	1.55	3.93 ± 0.26	—
Laboratory Culture (unenriched)	1.34	1.80 ± 0.37	—
Oct. 1968	1.37 ± 0.02	2.20 ± 0.20	unenriched
Oct. 1969	1.52 ± 0.05	4.00 ± 0.07	enriched
Nov. 1968	1.55 ± 0.03	3.25 ± 0.95	enriched
Mar. 1969	1.30[c]	2.00[c]	unenriched
May 1972	1.30[c]	2.00[c]	unenriched
June 1971	1.33 ± 0.01	1.00 ± 0.00	unenriched
Aug. 1971	1.37 ± 0.03	1.75 ± 0.25	unenriched

[a] From Makarewicz and Likens (1979)
[b] Values given are the mean ±S.E.
[c] One individual.

Holopedium gibberum—Enriched	Unenriched
Chl-a = 4.8 ppb (parts per billion)	1.7 ppb
MBS (mean brood size) = 3.93	1.80
T (temperature) = 18.0 ± 0.7°C	18.2 ± 1.2°C

Bosmina longirostris—Enriched	Unenriched
Chl-a = 4.3 ppb	1.53 ppb
MBS = 2.0	1.25
T = 18.0 ± 0.7°C	18.0 ± 0.7°C

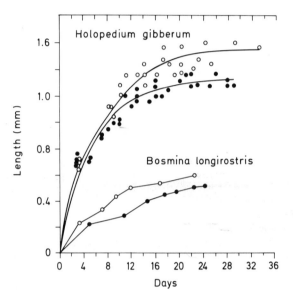

Figure V.B.5-5. Growth curves of *Holopedium gibberum* and *Bosmina longirostris* in enriched (open circles) and unenriched (closed circles) cultures. Modified from "Structure and function of the zooplankton community of Mirror Lake, New Hampshire," by J. C. Makarewicz and G. E. Likens, *Ecological Monographs*, 1979, 49(1):109–127. Copyright © 1979 by The Ecological Society of America. Used with permission.

cycling nutrients within the ecosystem? Our objective in the study of the zooplankton community of Mirror Lake was to integrate structure and function at the ecosystem level in answering these questions.

In Mirror Lake the 0- to 4.5-m strata (~epilimnion) contributed 68.5% of the annual zooplankton production. The 4.5- to 7.5-m strata, an area that approximates the metalimnion during summer stratification, accounted for 26.3% of the annual zooplankton production, while the 7.5- to 10.9-m region (~hypolimnion) contributed only 5.1% of the zooplankton production between October 1968 and September 1971. If production is compared on a concentration basis (mg C/m^3) and not weighted for the volume of each strata, 16.3% of the zooplankton production occurred in the hypolimnion.

Vertically, zooplankton production generally reaches a maximum at a 3-m depth through the year. The exception is the 9-m strata during the month of July when a second peak in production is evident (*Figure V.B.5-6*). This latter hypolimnetic production peak is related to the accumulation of algae and organic detrital particles in the hypolimnion of Mirror Lake (*Figure V.B.5-4*).

In August and September, the 9-m peak is not evident even though the high food levels still exist. However, unlike July, the hypolimnetic water in August and September often has less than 3 ppm of dissolved oxygen, suggesting that zooplankton productivity is inhibited by low oxygen concentration.

An ordination of species rank against percent of total zooplankton production describes a dominance-diversity curve with four distinct groups (*Figure V.B.5-7*). The first group, consisting only of the dominant species *Daphnia catawba*, represents 29% of the total zooplankton production. Those organisms that have consistently high densities each year and are gener-

B: Production and Limiting Factors

Figure V.B.5–6. Vertical distribution of zooplankton community production; 3-yr mean.

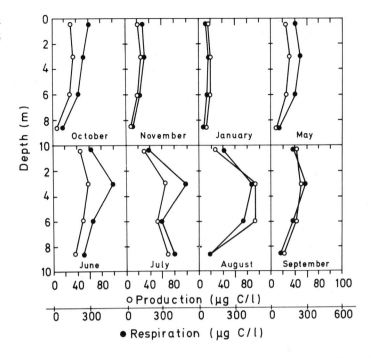

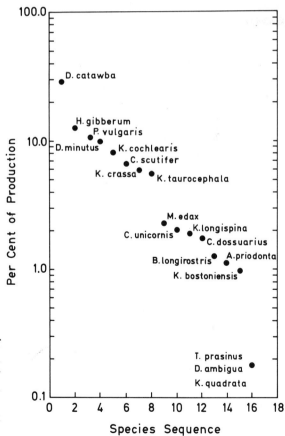

Figure V.B.5–7. Rank-species abundance curve of the zooplankton community. From "Structure and function of the zooplankton community of Mirror Lake, New Hampshire," by J. C. Makarewicz and G. E. Likens, *Ecological Monographs,* 1979, 49(1):109–127. Copyright © 1979 by The Ecological Society of America. Used with permission.

ally found in each season of the year comprise the second group. This group of seven species accounts for 59.5% of the zooplankton production. Thus, eight zooplankton species (Groups 1 and 2) represent 88.5% of the annual zooplankton production. Four of the species are rotifers.

Unlike the second group, the third group (seven species) consists of organisms present in only one year in relatively high densities or relegated to areas of the lake where temperature and oxygen were low. For example, *Kellicottia longispina* and *Bosmina longirostris* were prevalent only in the second year of the study, while *Kellicottia bostoniensis* and *Conochiloides dossuarius* are found only in the cold hypolimnion during summer stratification.

The last group contains rare organisms. These organisms, together with those in Group 3, represent a reservoir of species that could become prevalent in the lake with eutrophication or with changes in predation pressure.

Zooplankton community production ranged from 22.3 kg C/ha-yr to 29.3 kg C/ha-yr with a 3-yr mean of 25.2 kg C/ha-yr. Seasonally, a cyclic periodicity of zooplankton community production is evident and is correlated with the seasonal pattern of phytoplankton biomass, solar radiation and the volume-weighted, mean lake temperature (heat content) (*Figure V.B.5–3*). Because of the high specific heat of water, the heat content of the lake lags behind the total amount of solar energy impinging on the lake's surface by at least 1 to 2 months. The effect of solar energy on physical and biologic processes, such as summer stratification of the lake and phytoplankton succession from a cold-water to a warm-water assemblage, is well known (Hutchinson 1967; Wetzel 1975).

A question of concern to ecologists has been the effect of temperature vs. food supply on zooplankton productivity. Both temperature and food supply will affect rate processes, such as development and generation times (see *Chapter V.A.5*), and thus productivity. Food supply (quality and quantity) will differ with season in a lake and between lakes. The ratio of production to biomass is an estimate of the turnover rate of the community and is a function of temperature, and quality and quantity of food. For a given unit of biomass, the P/B ratio provides an indication of how much new organic matter was synthesized. In lakes of differing trophic status, there is a tendency for the P/B ratio to increase with the productivity of the lake (*Table V.B.5–3*; Patalas 1970). With eutrophication one would expect an increase in the availability of food (quantity) to zooplankton. An increase in food should increase brood size and growth rate and decrease generation time. These factors, accompanied with the warmer temperatures generally expected in a eutrophic lake relative to an oligotrophic lake,

Table V.B.5–3. Percentage of Production of Cladocera, Copepoda and Rotatoria and Production/Biomass (P/B) Ratios in Various Lakes during the Vegetative Period (Ice-free Period)[a,b]

	Increasing Primary Production					
	Krugloe[c]	Krivoe[c]	Mirror	Naroch[d]	Myastro[d]	Batorin[d]
Primary Production	3.2	12.0	37.0	43.0	140.0	156.0
Rotatoria	19.2	15.2	39.8	43.5	23.9	29.3
Cladocera	71.8	36.1	40.9	31.2	47.6	45.5
Copepoda	8.9	48.7	19.3	25.2	28.6	25.1
Zooplankton Production / Zooplankton Biomass	12.7	10.1	10.1	18.4	16.2	19.1

[a] Modified from "Structure and function of the zooplankton community of Mirror Lake, New Hampshire," by J. C. Makarewicz and G. E. Likens, *Ecological Monographs,* 1979, 49(1):109–127. Copyright © 1979 by The Ecological Society of America. Used with permission.
[b] Primary production values are in g C/m^2-vegetative period.
[c] Alimov et al. (1972).
[d] Winberg et al. (1972).

should increase the turnover rate of the biomass, i.e. increase the P/B ratio.

Although the P/B ratio increases with increasing primary productivity (food supply), the relationship presented is not a strong one. At least two factors may be the cause of it: quality of food and total food available. With eutrophication there is a tendency for the phytoplankton community to shift to large "net" phytoplankton dominated by colonial and filamentous blue-green and green algae that are often too large or unmanageable in size or toxic for zooplankton to assimilate. Even though a greater amount of food is present in a eutrophic lake, not all of it is available to herbivorous zooplankton because many species of blue-greens and green represent dead ends in the planktonic food chain (Porter 1977). This condition is reflected in increased generation times, which tend to lower P/B ratios in zooplankton cultured in lake water containing blue-green algae (Makarewicz 1976).

Lake Naroch, U.S.S.R., has a relatively high P/B ratio considering its low phytoplankton production. However, when macrophyte and periphyton production are included, the total primary production of Lake Naroch is 3 times the value given (Winberg 1972). Considering the total reduced carbon produced and potentially available to zooplankton through detritivore food chains, the P/B ratio for this lake becomes more reasonable.

Seasonally, in Mirror Lake the cyclic pattern of high zooplankton productivity (*Figure V.B.5–3*) in the spring and summer and the gradual decline in the winter follows the seasonal cycle of phytoplankton biomass (see also *Chapter V.B.2*). As expected, the initial increase in zooplankton productivity is closely tied to the onset of the spring phytoplankton bloom and moderating lake temperatures.

The role of food availability in affecting zooplankton production is further demonstrated by data from the winter of 1969–70. Phytoplankton biomass was higher during this winter than during the other winters when the lake was studied, and this higher biomass was reflected in a higher zooplankton biomass and production. Several species of phytoplankton in a size range small enough to be eaten by zooplankton were present in high densities (*Figure V.B.5–8*), unlike during other winters, and these species ap-

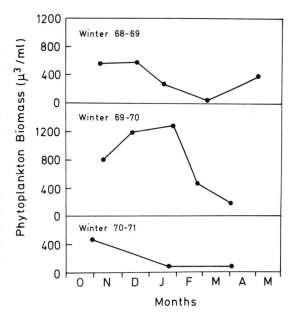

Figure V.B.5–8. Biomass of selected algal species during the winters of 1968 to 1971. Biomass includes the following species: *Cryptomonas erosa, Cryptomonas marssonii, Rhodomonas pulsilla, Katablepharis ovalis, Chrysococcus* spp., *Erkenia subaequiciliata* and *Monochrysis aphanagter*. From "Structure and function of the zooplankton community of Mirror Lake, New Hampshire," by J. C. Makarewicz and G. E. Likens, *Ecological Monographs*, 1979, 49(1):109–127. Copyright © 1979 by The Ecological Society of America. Used with permission.

parently accounted for the increase in zooplankton production. Indirect evidence (i.e. the occurrence of clear, hard ice; the lack of a snow cover on the lake; and the occurrences of oxygen supersaturation just below the ice, presumably resulting from photosynthesis) suggests that greater amounts of solar radiation were available to the phytoplankton community during the winter of 1969–70.

Temperature could also limit zooplankton production by controlling critical metabolic processes. Would two lakes of similar primary productivity, one with consistent year-round higher temperatures (i.e. a tropical lake) and the other with warm summer and cold winter water (i.e. a temperate lake), have a similar secondary productivity? One might expect that the effect of a higher year-round lake temperature on zooplankton would be an increase in turnover rates and, thus, a higher productivity in the tropical than in the temperate lake. The question can be partially answered if restated as: Would ecologic efficiency (zooplankton production/net primary production) increase from temperate to tropical lakes? The answer is apparently no, as ecologic efficiency in tropical Lake Lanao is 4.4% (Lewis 1979), while in temperate Mirror Lake it is 5.6%. Unfortunately, the comparison of the two lakes is not a valid one because it is not possible to isolate the effect of quality of food from temperature on zooplankton production in each lake. However, within a lake this question may be asked: How much of the seasonal variation that occurs in zooplankton productivity is due to both temperature and food supply? Multiple regression analysis of the Mirror Lake data indicates that 70% of the variation in zooplankton productivity is accounted for by quantity of food and by temperature.

Production-Respiration Ratios

Community zooplankton respiration varied from a high of 122 kg C/ha-yr to a low of 96 kg C/ha-yr with respiration being 79.7% of assimilation. There is a seasonal periodicity of zooplankton community respiration described annually by a unimodal curve (*Figure V.B.5–9*). Respiration increases during the spring and reaches a peak in the summer. As zooplankton biomass and temperature decrease, respiration falls to a minimum in winter. With spring the cycle begins again.

Information on the functioning of a zooplankton community may be obtained by observing seasonal production-respiration ratios. In the spring when the lake is cool, the P/R ratio increases, indicating that more energy from primary producers and bacteria is diverted into the synthesis of new organic matter. In the summer the ratio decreases, with a large fraction of the energy used in respiration in the warmer water. With cooler water temperature in the autumn, there is an increase in the ratio with more energy again being utilized in the production of new organic matter. As the winter progresses, the decrease in the ratio reflects the larger decrease in production relative to respiration. Because of the minimal thermal and food conditions imposed on the community, the synthesis of new matter almost ceases during the winter with the available energy being utilized to maintain the extant organisms.

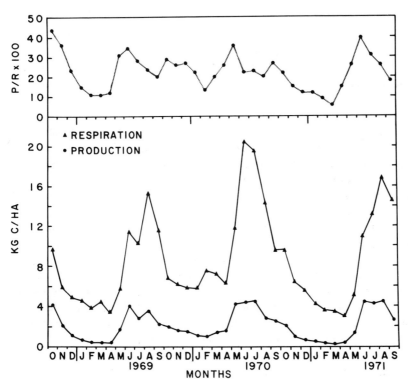

Figure V.B.5–9. Mean monthly community zooplankton production and respiration. From "Structure and function of the zooplankton community of Mirror Lake, New Hampshire," by J. C. Makarewicz and G. E. Likens, *Ecological Monographs,* 1979, 49(1)109–127. Copyright © 1979 by The Ecological Society of America. Used with permission.

Importance of Rotatoria

Rotifers, because of their small size, alleged low abundance and thus their low biomass, were thought to play a minor role in the transfer of energy in lake-ecosystems (Brooks 1969; Porter 1977). In 1971, however, research on the ELA lakes in Canada indicated that rotifers comprised 12% of the zooplankton biomass (Schindler 1972), a value considerably higher than the 1 to 4% generally attributed to this group. Schindler (1972) believed that the high biomass of rotifers he observed was due to the fine mesh net (28 μm) used on his sampling device. In fact, a #20 (75-μm mesh) net commonly used in limnological studies may give significantly low numbers of rotifers for natural populations (Likens and Gilbert 1970; *Chapter V.A.5*). Considering that the biomass of rotifers was higher than previously thought and combining this with their higher turnover rates, as compared with the cladocerans and copepods, an important role in energy transfer seemed likely for rotifers.

It is commonly thought that the copepods are important in oligotrophic lakes such as Mirror Lake. In fact, copepods do comprise ~57% of the total zooplankton biomass in Mirror Lake. However, the copepods, with their slow growth over an entire year, represent only ~20% of the zooplankton production while rotifers account for ~40% of the annual zooplankton production in Mirror Lake (*Table V.B.5–4*).

The high production rates of rotifers in Mirror Lake are due to high turnover rates, as compared with the crustaceans, and to the relatively high biomass of rotifers. The high biomass value in Mirror Lake is due to the use of sampling equipment with fine mesh nets (48 μm or less). Any sampling scheme not employing nets

Table V.B.5–4. Annual Biomass and Production of Rotifera, Cladocera and Copepoda for a 3-yr Period in Mirror Lake, New Hampshire[a]

	Rotifera	Cladocera	Copepoda	Total
Biomass				
Oct. 1968–Sept. 1969				
kg C/ha	0.26	0.51	1.26	2.03
%	12.8	25.1	62.1	
Oct. 1969–Sept. 1970				
kg C/ha	0.48	0.56	1.54	2.58
%	18.6	21.7	59.7	
Oct. 1970–Sept. 1971				
kg C/ha	0.24	0.52	0.61	1.37
%	17.5	38.0	44.5	
Mean %	16.3	28.3	55.4	
Production				
Oct. 1968–Sept. 1969				
kg C/ha-yr	8.26	10.09	5.69	24.04
%	34.4	42.0	23.7	
Oct. 1969–Sept. 1970				
kg C/ha-yr	13.59	9.08	6.60	29.27
%	46.4	31.0	22.5	
Oct. 1970–Sept. 1971				
kg C/ha-yr	8.60	11.08	2.59	22.27
%	38.6	49.8	11.6	
Mean %	39.8	40.9	19.3	

[a] From "Structure and function of the zooplankton community of Mirror Lake, New Hampshire," by J. C. Makarewicz and G. E. Likens, *Ecological Monographs*, 1979, 49(1):109–127. Copyright © 1979 by The Ecological Society of America. Used with permission.

as fine as 48-μm mesh is of doubtful value as a quantitative measure of rotifer density and thus of zooplankton biomass and production in aquatic ecosystems.

It is now known that rotifers play a major role in energy transfer in many lakes of the world (*Table V.B.5–3*). In oligotrophic Lakes Krivoe and Krugloe (U.S.S.R.), rotifers account for 15.2 and 19.2% of the zooplankton community production, while in eutrophic Lakes Naroch, Myastro and Batorin (U.S.S.R.), the rotifer contribution to community production ranges from 23.9 to 43.5%. Because these lakes range in trophic status from extreme oligotrophy (Krugloe) to eutrophy (Batorin), the important role of rotifers in lake metabolism may be a much more common phenomenon than previously thought.

Nutrient Cycling

Zooplankton are not only important in the transfer of energy within the lake ecosystem but also in nutrient cycling. Excretion of P by zooplankton and remineralization of P by decomposition of zooplankton tissue undoubtedly play a major role in maintaining the metabolism of lake-ecosystems by making P available to primary producers whose productivity could not be maintained by allochthonous inputs alone (see *Chapter VI.C* for detailed discussion).

Summary

A relatively simple procedure, such as the ^{14}C procedure for primary production, integrating the production of all zooplankton species in a lake does not exist. The measurement of community zooplankton production requires data on the density of each species or size class of a species with depth and season, on the effect of food and temperature on growth of species and on generation and development times, and on the weight of each species or size class of each species.

Copepods comprise 55.4% of the total zooplankton biomass. However, the copepods, with their slow growth over an entire year, represent only 19.3% of the zooplankton production, while rotifers account for 39.8% of the zooplankton production annually in Mirror Lake. Although the cladocerans are the dominant group in Mirror Lake, accounting for 40.9% of the annual zooplankton production, it is evident that rotifers play a much more important role in energy transfer than previously thought in temperate lakes of varying trophic status.

A seasonal cycle in excretion of phosphorus from the Mirror Lake zooplankton community is evident with a peak in June and a minimum in late winter. Zooplankton excretion of phosphorus ranged from 102 to 477 mg P/m^2-month in March and June, respectively. Excretion of phosphorus by zooplankton is an order of magnitude higher than allochthonous inputs for each season of the year. Functionally within Mirror Lake, copepods account for only 17% of the zooplankton production but do contribute 47.7% of the P excreted by the zooplankton community. Cladocerans, on the other hand, are important in the lake-ecosystem considering their high productivity (41% of the total zooplankton community production), but are not major contributors to phosphorus cycling within the lake. Rotifers account for over 40% of the phosphorus excreted by zooplankton in Mirror Lake. Excretion of phosphorus by zooplankton and remineralization of phosphorus by decomposition of zooplankton tissue undoubtedly play a major role in maintaining the metabolism of lake-ecosystems by making phosphorus available to primary producers whose productivity could not be maintained by allochthonous inputs alone.

6. Benthic Macroinvertebrates

Rhoda A. Walter

The rates of production and respiration of the benthic macrofauna in lakes are dependent on many interrelated factors, including (but not limited to) the trophy, depth, morphometry, temperature and oxygen regimes of each lake. Rates are highly variable, even among lakes of similar trophy; but, in general, production and respiration increase with increasing trophy (Tables V.B.6–1 and 2). Loss of energy and nutrients because of emergence of insects also shows this trend (Table V.B.6–3). The actual importance of the benthic macrofauna in a given lake depends on the relation between these rates of production, respiration and loss resulting from emergence and the fluxes in energy and nutrients in other components of the lake-ecosystem. The Mirror Lake studies provide an opportunity to examine this relationship because fluxes through all major components of the ecosystem have been measured or estimated.

Production

A direct calculation of benthic secondary production by the Hynes-Coleman method (Hynes and Coleman 1968) was not possible for Mirror Lake because the time intervals between benthic samples were too long for estimating the number of generations per year. Therefore, secondary production of the benthic macrofauna in Mirror Lake was estimated by four independent methods that were designed to establish its upper and lower limits (Table V.B.6–4). Benthic production in Mirror Lake is 2 to 10 g dry wt/m^2-yr (1 to 5 g C/m^2-yr) and is similar to zooplankton production and net macrophyte pro-

Table V.B.6–1. Comparison of Benthic Macrofaunal Production in Various Lakes

Lake	Total Benthic Macrofaunal Production		Source
	g dry wt/m^2-yr	kcal/m^2-yr	
Lake Flosek, Poland[d]		0.5[g]	Kajak et al. 1972
Char Lake, N. Canada[a]	1.6[f]		Rigler 1978
Findley Lake, Washington[a]	1.6[f]		Wissmar and Wetzel 1978
3 tundra ponds, Barrow, Alaska[a]	1.6–7.0		Butler et al. 1980
Lake Memphremagog, Quebec/Vermont			Dermott et al. 1977
North Basin[b]	>2.8		
South Basin[c]	>7.4		
Lake Wingra, Wisconsin[c]	4.9[f]		Wissmar and Wetzel 1978
Marion Lake, British Columbia[a]	7.6[f]		Wissmar and Wetzel 1978
Mirror Lake, New Hampshire[a]	2–10	18–34	revised from Walter 1976
Lake Mendota, Wisconsin[c]		26	Juday 1940
Lake Sniardwy, Poland[c]		50[g]	Kajak et al. 1972
Lake Ontario[a]		59	Johnson and Brinkhurst 1971
Lake Esrom, Denmark[c]		100	Jonasson 1972
Lake Mikolajskie, Poland[c]		130	Kajak et al. 1972
Lake Warniak, Poland		140[g]	Kajak et al. 1972
Eglwys Nunydd Reservoir, South Wales[c]	37	150	Potter and Learner 1974
Lake Myvatn, Iceland[c]		160	Jonasson 1979
Loch Leven, Scotland[c]	38	187	Maitland and Hudspith 1974; Charles et al. 1974, 1976
Lake Taltowisko, Poland[b]		200[g]	Kajak et al. 1972
Bay of Quinte, Lake Ontario[c]		63–254	Johnson and Brinkhurst 1971
Lake Werowrap, Australia[c,e]	68	323	Paterson and Walker 1974
Sewage lagoons, Oregon[c]		18–459	Kimerle and Anderson 1971

[a] Oligotrophic.
[b] Mesotrophic.
[c] Eutrophic.
[d] Dystrophic.
[e] Saline.
[f] Assuming dry weight = 45% carbon.
[g] Nonpredatory benthic organisms only.

Table V.B.6–2. Annual Flux of Energy and Carbon by Means of Respiration of Benthic Macrofauna in Mirror Lake and Other Lakes

Lake	Benthic Macrofaunal Respiration		Source
	Energy (kcal/m²-yr)	Organic C (mg/m²-yr)	
Char Lake, N. Canada[a]		1,300	Rigler 1978
Mirror Lake, New Hampshire[a]	25[c]	2,800	revised from Walter 1976
Marion Lake, British Columbia[a]		18,000	Hargrave 1969
Lake Ontario[a]	102		Johnson and Brinkhurst 1971
Bay of Quinte, Lake Ontario[b]	39–135		Johnson and Brinkhurst 1971
Loch Leven, Scotland[b]	189–225		McLusky and McFarlane 1974
Lake Esrom, Denmark[b]	432		Jonasson 1972
Sewage lagoons, Oregon[b]	44–520		Kimerle and Anderson 1971

[a] Oligotrophic.
[b] Eutrophic.
[c] Using an oxy-caloric coefficient of 3.38 kcal/g O_2 (Crisp 1971).

duction in the lake (*Chapters V.B.3* and *V.B.5*). It is within the range of benthic production values from other oligotrophic lakes, and low compared with eutrophic lakes (*Table V.B.6–1*).

The primary factor controlling this productivity, as well as abundance and biomass (*Chapter V.A.6*), of the bottom fauna in Mirror Lake and other lakes is probably the amount of food available to the benthic environment (Rowe 1971; Kajak et al. 1972; Jonasson 1972). This availability depends on the amount of total ecosystem source carbon (Likens 1972b and *Chap-*

Table V.B.6–3. Annual Flux of Energy, Carbon and Nutrients by Means of Emergence of Adult Insects from Mirror Lake and Other Lakes

Lake	Adult Emergence				Source
	(kcal/m²-yr)	Organic C	Total N	Total P	
		(mg/m²-yr)			
Char Lake, N. Canada[a]	0.8		13.1	0.95	Welch 1973
Lake Pääjärvi, Finland[a]	1.1		13.7	0.9	Paasivirta 1975b
Findley Lake, Washington[a]			80	11	Likens and Loucks 1978
Mirror Lake, New Hampshire[a,c]	5.8	506	84	16	revised from Walter 1976
Lake Wingra, Wisconsin[b]			140	10	Likens and Loucks 1978
Marion Lake, British Columbia[a]		max 1,300			Hargrave 1969
Lake Esrom, Denmark[b]	53				Jonasson 1972
Sewage lagoons, Oregon[b]	9–131				Kimerle and Anderson 1971

[a] Oligotrophic.
[b] Eutrophic.
[c] Includes only the portion of emergence that is estimated to leave the lake (one-third of total emergence).

Table V.B.6–4. Comparison of Secondary Production Estimates for the Benthic Macrofauna of Mirror Lake

Method	g dry wt/m²-yr	g C/m²-yr[e]	kcal/m²-yr
P:E ratio[a]	10.2	5.1	20
C available to benthic organisms[b]	6.0–9.6	3.0–4.8	21–34
Waters' average turnover rate[c]	6.0	3.0	34
Maximum biomass[d]	2.5	1.2	18

[a] Based on (1) the mean P:E ratio for the benthic species in Lake Esrom (Jonasson 1972) that are dominant in Mirror Lake and (2) an estimated emergence of 1,654 mg C/m²-yr for Mirror Lake.
[b] Based on 10% of the total amount of C available to the benthic environment—30 to 48 g C/m²-yr (*Chapters V.C* and *VII.A*).
[c] Based on Waters' (1969) average turnover rate of 3.5, assuming only one generation per year.
[d] Based on maximum benthic macrofaunal biomass of 2.5 g dry wt/m² for the entire lake in January.
[e] C is assumed to be 50% of dry weight.

ter *VI.D*), the secondary production occurring in the water-column itself, the length of the column of water (which relates to the bacterial mineralization occurring in the water-column [*Chapter V.D*]) and the amount of food leaving by way of the lake's outlet(s). The quality of the food—i.e. particle size, nutritional value, etc.—must also be important in determining productivity. Sediment type, macrophyte distribution, the seasonal oxygen and temperature regime, the seasonal distribution of the food available to the bottom (determined by the seasonal pattern of phytoplankton production) and the rate of decomposition in the sediment all determine whether the potential macrobenthic production will result from a few organisms with a high turnover rate or a large number of organisms with a lower turnover rate.

Approximately 30% of net primary production reaches the benthic environment in Mirror Lake (*Chapter V.D*). A larger percent of net primary production and total ecosystem source carbon may be available to the benthic environment in Mirror Lake than in many other lakes because zooplankton production (2.5 g C/m²-yr) is small compared with other lakes even as a percent of total net primary production (*Chapter V.B.5*). But when absolute amounts of organic carbon (or energy) reaching the benthic environment in the form of tripton (all of the nonliving suspended material) were examined, it was found that the amount of energy available to the benthic environment in oligotrophic Mirror Lake is 45 to 72% of that available in a mesotrophic lake, and only 12 to 34% of that available in two eutrophic lakes (*Table V.B.6–5*). This relation is what would be expected for a typical oligotrophic lake because net primary production would be low. Perhaps, though, utilization of tripton is more efficient in oligotrophic lakes where the tripton is composed mainly of flagellates and diatoms, than in eutrophic lakes where blue-green algae predominate.

In most oligotrophic lakes, the benthic community and the tripton reaching the hypolimnion and the profundal sediments exert a low oxygen demand; the profundal region remains oxygenated throughout the year, and is inhabited by typical oligotrophic organisms requiring high oxygen levels.

Because of the basin configuration in Mirror Lake, the volume of water in the hypolimnion is only about 9% of the total lake volume, and the volume below 9 m is only 1.2% of the total lake volume. The focusing effect of sediment (*Chapter VII.A*) is very pronounced because the lake's tripton and resuspended sediment from the littoral regions are funneled into this small volume. This concentration of dead, sinking material creates an oxygen demand in the hypolimnion which, when combined with the respiratory demands of the profundal benthic environment, is substantial enough to create anaerobic conditions below 9 m in the water-column by the end of the summer (and some-

Table V.B.6–5. Quantity of Tripton Reaching the Sediment-water Interface (and thus Available to the Benthic Organisms) on an Annual Basis for an Oligotrophic and a Eutrophic Lake

Lake	Trophy	Tripton Reaching the Sediment-Water Interface	
		(g C/m^2-yr)	(kcal/m^2-yr)
Mirror Lake, New Hampshire	oligotrophic	31^a–46^b	215^c–335^c
Lake Mikolajskie, Poland^d	eutrophic		1,100^e–1,800^f
Horw Bay, Lake of Lucerne, Switzerland	mesotrophic	67^g	
Rotsee, Switzerland	eutrophic	142^g	

^{a} Chapter V.D.
^{b} Chapter VII.A.
^{c} Using 6.96 kcal/g C based on mean caloric content of aquatic primary producers from Cummins and Wuycheck (1971).
^{d} 460-ha area; 11-m mean depth.
^{e} Kajak et al. 1972.
^{f} Lawacz 1969.
^{g} Bloesch et al. 1977.

times the winter) stratification period. Anoxic conditions probably occur at much shallower depths at the sediment-water interface, because the oxygen gradient is very steep close to the sediment (see *Figure VI.B*–1). Thus, the physical and chemical environment in the profundal region of Mirror Lake is eutrophic in nature, and the benthic invertebrates inhabiting it are adapted to this eutrophic environment and to the relatively high concentrations of tripton and redeposited sediment reaching the sediment surface. Mirror Lake, then, is an oligotrophic lake with a eutrophic profundal region—the profundal region having a eutrophic benthic fauna with higher numbers, biomass and production than in most oligotrophic lakes.

In the littoral regions in Mirror Lake, almost all of the tripton reaching the sediment surface is either utilized quickly by the benthic community or is resuspended during autumnal or spring overturn or during storms and redeposited in the profundal region. The net result is little or no sediment accumulation in most of the littoral area over the past 14,000 yr (*Figure VII.A.1*–1), so food derived from sedimentation of tripton must be scarce in these regions. In view of this scarcity, the high numbers of organisms in this area (*Chapter V.A.6*) seem anomalous, because even though individual size is small, turnover of biomass—hence, production—must be high because the warm water in summer contributes to high rates of matura-

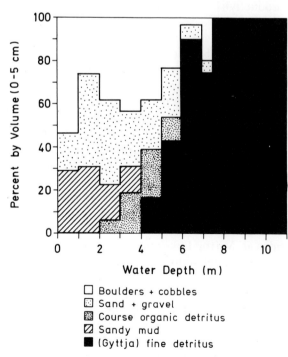

Figure V.B.6–1. Percent by volume of five substrate types in the 0- to 5-cm layer assessed visually during quadrat surveys (Moeller 1978b).

tion, enabling more than one generation per year to be produced. Food must be available from some other source in addition to tripton. The peak abundance in the 0- to 3-m region appears correlated with occurrence of peak macrophyte biomass and particulate organic debris (occurring either as thick deposits or as thin layers on top of gravel or sand; *Figure V.B.6–1*). The peak abundance in the lower littoral and upper profundal regions (*Chapter V.A.6*) appears to be associated with occurrence of a second peak in macrophyte biomass (*Figure V.A.6–4C*) and with the shallowest occurrence of gyttja, which is rich in nutrients (*Chapter VII.A.2*). In the 3- to 5-m zone between these two areas, the macrophyte biomass is low and the substrate is predominantly sand, boulders, cobbles and sandy mud. So it appears that the areas where high abundance of macrofauna occurs also have moderate to high levels of food available in the form of detritus, probably from aquatic macrophytes and leaves shed from deciduous trees around the lake's perimeter. In addition, periphyton (*Chapter V.B.4*) must provide food in these areas. Thus, in the littoral regions, the benthic macrofauna are able to utilize food sources—macrophyte and leaf detritus and algae—that are not available to the profundal macrofauna, which must depend on tripton alone (except where blue-green algal mats are present).

Respiration

Respiration of the benthic macrofauna in Mirror Lake ranged from 0.5 mg C/m^2-h in late summer to 0.2 mg C/m^2-h in winter, based on weighted mean respiration for each depth interval, calculated from published respiration rates for each taxon, and assuming a respiratory quotient of one (see Walter 1976 for details). In July, respiration was lowest in the upper littoral region (0.3 mg C/m^2-h) and highest in the lower littoral and profundal regions (0.5 mg C/m^2-h) where biomass was highest.

Mean annual respiration of the benthic macrofauna for the entire lake was estimated to be 2.8 g C/m^2-yr (*Table V.B.6–2*), with chironomids being the dominant contributors (*Table V.B.6–6*). Organic C lost by way of respiration of benthic macrofauna is small compared with the total output of organic C from the lake

Table V.B.6–6. Mean Annual Respiration of the Benthic Macrofauna in Mirror Lake, Partitioned Among the Dominant Taxa

	Mean Annual Respiration (Entire Lake)
	(g C/m^2-yr)
Total Organisms	2.8
	% of Total Organism Respiration
Total *Chaoborus*	21
Total Chironomids	65
Total Oligochaetes	4
Other	10

(*Chapter V.C*). The amount of energy lost by the benthic macrofauna through respiration (25 kcal/m^2-yr) represents only 4.1% of net primary production (*Chapter V.C*). On an annual basis, the percent of total benthic respiration attributed to the benthic macrofauna in Mirror Lake is 12 to 25% depending on which value for annual respiration of the total benthic community is used (see *Chapters V.C* and *V.D*).

Hourly respiration of the total benthic community, including the benthic bacteria, algae and other microorganisms as well as the macrofauna, was estimated by measuring changes in dissolved inorganic carbon or dissolved oxygen inside darkened Plexiglas chambers containing undisturbed cobbles, sand or gyttja and incubated *in situ* at various water depths and seasons in Mirror Lake (*Table V.B.6–7*). The macrofauna accounted for an estimated 2 to 12% of this respiration (*Table V.B.6–7*), which is lower than the 12 to 25% estimated on an annual basis. This discrepancy has not been resolved. When this percent is compared with the benthic macrofaunal biomass in Mirror Lake and lakes of different trophic status, no trend is apparent (*Table V.B.6–8*).

Benthic bacteria, protozoa and algae must play a major role in respiration in Mirror Lake and other lakes where macrofauna respiration is relatively small. In Mirror Lake, the ratio of benthic macrofaunal biomass to benthic microbial biomass is only 28%. This ratio must have to be large in order for respiration of the benthic

Table V.B.6–7. Total Benthic Community Respiration and Benthic Macrofauna Respiration at Various Times of the Year in Mirror Lake[a]

Water Depth (m)	Substrate	Month	Respiration		Macrofauna as % of Total Community	mg O$_2$/liter[c]	Temp (°C)[c]
			Total Benthic Community	Macrofauna[b]			
			(mg C/m²-hr)	(mg C/m²-hr)	%		
2	cobbles	Jul./Aug.	4.4[d]	—	—	high	23
3	sand	Jul.	7.9[e]	0.2	2	high	22
7	gyttja	Aug.	9.5[e]	0.8	8	medium	13
9	gyttja	Oct.	3.3[f]	0.4	12	high (8.6)	12
10	gyttja	Nov.	16.1[f]	0.4	2	high (saturated)	7
10	gyttja	Jan.	3.7[f]	0.4	11	<4.2	5
10	gyttja	Mar.	4.0[f]	0.3	8	<1.7	4
10	gyttja	May	2.8[f]	0.3	11	high (8.7)	5
			(mg C/m²-yr)	(g C/m²-yr)	%		
Entire Lake		Annual	21[g]	2.8[h]	13		

[a] Revised from Walter 1976.
[b] Respiration estimated using mean biomass estimates from Mirror Lake and respiration rates from the literature.
[c] Taken at sediment-water interface where possible; otherwise, taken from the appropriate depth at Central (11.0-m water-column).
[d] M. Jordan and B. Peterson (unpublished data).
[e] Derived from O$_2$ uptake (including chemical uptake by the sediment), assuming RQ = 1 (R. Walter, unpublished data).
[f] M. Jordan (unpublished data).
[g] Chapter V.C.
[h] Based on mean annual biomass for Mirror Lake macrofauna and respiration values from the literature.

TABLE V.B.6–8. Comparison for Different Lakes of Percent of Benthic Oxygen Uptake Attributable to Macrofauna

Lake	Benthic O$_2$ Uptake due to Benthic Macrofauna	Biomass of Benthic Macrofauna	Source
	% of total	(g dry wt/m²)	
Castle Lake, California[b]	<5	0.9	Neame 1975
Lake Mendota, Wisconsin[c]	6.5	4.8–7.7	Juday 1940
Mirror Lake, New Hampshire[a]	2–12	1.7	Walter 1976
Marion Lake, British Columbia[a]	33	5.9	Hargrave 1969
Lake Esrom, Denmark[c]	67	13.8[d]	Jonasson 1972

[a] Oligotrophic.
[b] Mesotrophic.
[c] Eutrophic.
[d] Does not include *Pisidium casertanum*.

Table V.B.6–9. Estimated Annual Mean Standing Stocks for Calories, Organic C, N and P in the Benthic Macrofauna of Mirror Lake, Partitioned Among the Dominant Taxa[a]

	Annual Mean (Entire Lake)			
	kcal/m^2	mg Organic C/m^2	mg N/m^2	mg P/m^2
Total Organisms	9.6	850	170	20
	% of Total Organisms			
	kcal	Organic C	N	P
Total *Chaoborus*	17	16	14	23
Total Chironomids	69	65	70	54
Total Oligochaetes	9	11	10	7
Other	5	8	6	16

[a] Does not include *Elliptio*.

macrofauna to exceed that of the benthic microbes.

Standing Stocks

Annual mean standing stocks for calories, organic carbon, nitrogen and phosphorus in Mirror Lake (*Table V.B.6–9*) were estimated using values from the literature (see Walter 1976 for details). In relation to other living components of the Mirror Lake ecosystem, the benthic macrofauna represent major storage reservoirs for energy, organic carbon and nutrients (*Table VI.D–4*). The storage of nutrients in the benthic macrofauna may be an important mechanism for recycling N and P to the lake by preventing these nutrients from being permanently sedimented (*Chapter VI.B*).

The depth-distributions for standing stocks of calories, organic C, N and P in the benthic macrofauna of Mirror Lake (*Table V.B.6–10*) basically correspond with the depth-distribution for macrofauna biomass (*Figure V.A.6–4*). Total standing stocks on an areal basis are highest in the profundal region and lowest in the upper littoral region (*Table V.B.6–10*) because of the dominance of chironomids (mainly *Chironomus anthracinus*) and *Chaoborus* in the profundal region. These two taxa account for 77 to 86% of the standing crop of calories, organic C, N and P in the benthic macrofauna (*Table V.B.6–9*).

Summary

Production of benthic macrofauna in Mirror Lake was estimated to be 2 to 10 g dry wt/m^2-yr (1 to 5 g C/m^2-yr), which appears to be moderate for an oligotrophic lake. The hypolimnion of Mirror Lake is eutrophic in nature, because of its relatively small volume and the funneling of tripton and resuspended sediment from the littoral regions; and the benthic invertebrates of the profundal region have higher numbers, biomass and production than in most oligotrophic lakes because of this readily available food source. The oligotrophic littoral regions also have higher numbers of benthic macroinvertebrates than would be expected, although their biomass is low. Most of these littoral organisms are associated with the macrophytes and are

Table V.B.6–10. Depth Distribution of Standing Stocks of Calories, Organic C, N and P in the Benthic Macrofauna of Mirror Lake in July[a]

Water Depth (m)	Calories (kcal/m^2)	Organic C (mg/m^2)	N (mg/m^2)	P (mg/m^2)
0–3	2.1	230	37	5.1
3–7	6.0	550	92	12.0
>7	14.0	1,200	200	24.0

[a] Does not include *Elliptio*.

probably obtaining food from macrophyte and terrestrial leaf detritus and periphyton, in addition to whatever tripton is available before it is resuspended and funneled into the profundal region.

Mean annual respiration of the benthic macrofauna for the entire lake was estimated to be 2.8 g C/m²-yr. This amounts to 12 to 25% of total benthic community respiration, but is an insignificant component of the organic carbon budget for the Mirror Lake ecosystem.

The benthic macrofauna represent major storage reservoirs for energy, organic carbon and nutrients; chironomids (mainly *Chironomus anthracinus*) and *Chaoborus* account for 77 to 86% of this storage.

7. Vertebrates

T. M. BURTON

Information on vertebrate production in Mirror Lake is limited. No actual measurement of production has been conducted. Thus, the approach taken in this section is to derive estimates of production based on measured population estimates and biomass for Mirror Lake and literature values for utilization of energy by vertebrates. Vertebrate production in the lake is dominated by fish in the pelagic and most of the littoral zone and by salamanders (the newt, *Notophthalmus viridescens*) in some restricted areas of the littoral zone near the inlets and outlet.

To estimate or measure population energy flow, the parameters in equation (1) must be known where:

$$A = I - E, \qquad (1)$$

A = assimilated energy
I = ingested energy
E = egested energy.

In other words, the feeding rate and energy content of the food ingested and the efficiency of uptake of this ingested energy (i.e. food energy ingested minus energy not assimilated and lost in such egestion processes as defecation, urination and gaseous losses) must be known. Conversely, assimilation (A) may be expressed as energy expended in metabolism and production as in equation (2) where:

$$A = R + P, \qquad (2)$$

R = respiration
P = production of new biomass through growth and reproduction.

A complete energy budget for a population is represented by equation (3).

$$I - E = R + P \qquad (3)$$

In terms of trophic dynamics within a lake, the important processes to know are ingestion, because this measures the impact on the previous trophic level, and production, because this measures the amount of energy that can be transferred to the next trophic level.

Salamanders

The biomass of newts in Mirror Lake is about 1,550 g (*Chapter V.A.8*; Burton 1977). Biomass for the red-backed salamander (*Plethodon cinereus*) in the nearby northern hardwood forest was 20% of annual ingestion (Burton and Likens 1975). Based on this relationship, ingestion by the newt population is 7,750 g dry wt/yr. Of this ingested energy, 19% (based on data for *P. cinereus*) or 1,472 g dry wt/yr is lost in egestion, resulting in assimilation of 6,278 g dry wt/yr. Approximately 39% or 2,448 g dry wt/yr of assimilated energy is lost by respiration, while 61% or 3,830 g dry wt/yr goes into production. Thus, the estimated energy budget for salamanders in Mirror Lake is summarized as:

(I) 7,750 g dry wt/yr
− (E) 1,472 g dry wt/yr
= (R) 2,448 g dry wt/yr
+ (P) 3,830 g dry wt/yr.   (4)

Production is estimated as 3,830 g dry wt/yr. Assuming that one-half of dry matter is organic carbon, then production is 1,915 g C/yr for the entire lake or 12.8 mg C/m²-yr. As such, the salamander production is equivalent to 0.03% of net primary production (*Table V.B.7*–1). Actually, nearly all the salamanders are found in areas near the inlets and outlet (approximately

B: Production and Limiting Factors

Table V.B.7–1. Biomass and Nutrient Standing Crops in the *N. viridescens* in Mirror Lake, New Hampshire[a,b]

Element	% Dry Weight ($\bar{x} \pm 1$ S.E.)	Grams in Lake	g/ha
Newt biomass (dry wt)		1,550	103.3
Protein (N × 6.25)	52.64	815.7	54.4
Calcium	6.19 ± 0.22	96.0	6.4
Magnesium	0.16 ± 0.004	2.5	0.2
Potassium	0.91 ± 0.019	14.1	0.9
Sodium	0.45 ± 0.017	7.0	0.5
Phosphorus	3.54 ± 0.12	54.9	3.7
Nitrogen	8.42 ± 0.25	130.5	8.7
Sulfur	0.55 ± 0.14	8.5	0.6
Zinc	0.016 ± 0.001	0.24	0.02

[a] Modified from Burton 1977, *Copeia* 1977(1):139–143.
[b] Based on a population of 2,500 adults.

1 ha). Thus, production averages about 190 mg C/m²-yr in these areas.

The nutrient content of adult newts was analyzed and standing crops of each nutrient were calculated for the Mirror Lake population (*Table V.B.7*–1). Standing crop of phosphorus in salamander biomass was 2.9% of annual input of dissolved P to Mirror Lake, while standing crop of nitrogen was only 0.09% of annual dissolved input. Thus, it appears that salamanders do not represent a major sink for these nutrients in Mirror Lake. The calcium content of both the terrestrial efts and the adult newts is rather high (*Table V.B.7*–1), compared with populations of terrestrial and streamside salamanders in nearby forests (Burton and Likens 1975). This apparent high calcium requirement seems to be met by a tendency for the terrestrial eft stage to selectively eat snails. Data on this selectivity are not conclusive, but certainly the terrestrial eft stage does eat a much higher percentage of snails than any other salamander in the area (Burton 1976). The high nutrient content of efts returning to Mirror Lake could represent a significant input of nutrients into this system. If one-half of the adult population reenters the lake from the land each year (85% of the adult population is in the 3- to 5-yr-old range so this estimate appears to be reasonable—see *Chapter V.A.7*), they would represent an input of about 42 g nitrogen, 31 g calcium, and 18 g phosphorus (1,250 efts × 0.4 g dry wt × % dry wt of nutrient) with all other constituents representing lesser inputs. These amounts of nutrients represent insignificant inputs when compared with total inputs from other sources (*Chapter IV.E*). These efts would also represent an energy input of 2,300 kcal [4,598 cals/g dry wt (see Burton and Likens 1975) × 500 g of efts migrating into a lake].

Fish

Data on fish production are essentially nonexistent for Mirror Lake. The five species present include the smallmouth bass, *Micropterus dolomieui;* the yellow perch, *Perca flavescens;* the chain pickerel, *Esox niger;* the white sucker, *Catostomus commersoni;* and the brown bullhead, *Ictalurus nebulosus* (*Chapter V.A.9*). Of these five species, population estimates are only available for the smallmouth bass and the yellow perch (Mazsa 1973). These population estimates are based on limited capture-recapture data from the summer of 1970 and are subject to large potential error because there were relatively few recaptures. The large error is especially applicable for the yellow perch estimate. Mazsa (1973) suggested on the basis of his observations that the population of chain pickerel was less than 200 individuals. Populations of the other two species are assumed to be very low (<200) and are ignored in the following calculations. Using these crude population estimates, proportional catch data, and conversion to weight based on length-weight curves for Mirror Lake, the biomass estimates in *Table V.B.7*–2 have been derived. These data are only order-of-magnitude estimates.

On a whole-lake basis, biomass is 480 mg dry wt/m²-yr or 240 mg C m²-yr. Waters (1977) compiled data on annual production/biomass (P/B) ratios for numerous species of fish and concluded that the P/B ratio was between 0.5 and 1.0 for most species. Using these P/B ratios for fish in Mirror Lake would suggest that annual production is between 240 and 480 mg dry wt/m²-yr (120 to 240 mg C/m²-yr). Trout consume 6.4 to 7.8 times their own weight in food in a year (Dadikyan 1955 cited in Winberg 1971b). Using the assumption that fish in Mirror Lake consume 7.0 times their own biomass would

Table V.B.7–2. Population and Biomass Estimates for Fish in Mirror Lake[a]

Species	Population Size	Biomass g dry wt[b]	Biomass g carbon[c]	% of Total Biomass
Micropterus dolomieui	660	18,960	9,480	26.4
Perca flavescens	2,800	50,540	25,270	70.4
Esox niger	200	2,300	1,150	3.2
Total	3,660	71,800	35,900	100.0

[a] Derived from limited capture/recapture data from Mazsa (1973) and are order of magnitude estimates only.
[b] 20% of wet weight.
[c] 50% of dry weight.

suggest that ingestion is about 3.4 g dry wt/m²-yr (1.7 g C/m²-yr). If assimilation is 80% of consumption as it was for salamanders, then 2.7 g dry wt/m²-yr (1.3 g C/m²-yr) is assimilated by fish populations. Respiration can be calculated by difference as 2.2 g dry wt/m²-yr. Thus, a complete energy budget for the fish populations would be:

$$\begin{aligned}\text{(I) } & 3.4 \text{ g dry wt/m}^2\text{-yr} \\ - \text{ (E) } & 0.7 \text{ g dry wt/m}^2\text{-yr} \\ = \text{ (R) } & 2.2 \text{ g dry wt/m}^2\text{-yr} \\ + \text{ (P) } & 0.5 \text{ g dry wt/m}^2\text{-yr.} \end{aligned} \quad (5)$$

An estimate of respiration also can be derived by using the "basic equation" of Winberg (1960) relating metabolism to weight. This equation (6) relates mean metabolic rate (Q):

$$Q = 0.3 \, W^{0.8} \quad (6)$$

in ml O_2/h to the weight (W) of the fish in grams. Using this equation and the mean weight of each species and converting to g dry wt (ml O_2 = 4.86 calories; 5,000 calories = 1 g of fish tissue) yields an estimate of respiration of 2.4 g dry wt/m²-yr (1.2 g C/m²-yr). This estimate is remarkably close to the respiration value derived by difference in equation (5) and suggests that the population energy budget is at least the right order of magnitude. Respiration calculated by this technique also suggests the relative role of each species in the energy budget with 72% of respiration from yellow perch, 24.5% from smallmouth bass and 3.5% from chain pickerel.

Fish and salamanders are the top predators in Mirror Lake. Together they consume about 3.4 g dry wt/m²-yr (1.7 g C/m²-yr) and produce about 0.5 g dry wt/m²-yr (0.25 g C/m²-yr). This consumption is high compared with net zooplankton and benthic production (5 and 2 to 10 g dry wt/m²-yr, respectively; *Chapters V.B.5* and *V.B.6*). If we use maximal values of benthic production (10 g dry wt/m²-yr), 23% of benthic and zooplankton production is being consumed by vertebrates in Mirror Lake. The smallmouth bass do take some of their diet (25 to 50% for older individuals) from terrestrial insects near shore. Corrections for this input would suggest that fish and salamanders are consuming about 20% of net zooplankton and benthic production. If we use minimum benthic production values (2 g dry wt/m²-yr), vertebrates are consuming about 40% of net zooplankton and benthic production. Thus, vertebrates, which can be thought of as secondary/tertiary consumers, appear to be taking from 20 to 40% of the energy produced by the primary to secondary invertebrate consumers in Mirror Lake. These values seem reasonable from a theoretical standpoint. Trophic level efficiency [the amount of energy produced (respiration plus yield) by one level that is assimilated by the next higher trophic level] tends to increase at the higher trophic levels with herbivores assimilating 10% or less of primary production, primary consumers assimilating 15% or less of herbivore production, etc. (cf. Whittaker 1975). Trophic level efficiency is between 12 and 17% for the vertebrate populations. Thus, the estimates of vertebrate production seem reasonable even though they are based on very poor population data for fish in Mirror Lake.

The chemical composition of fishes in Mirror Lake has not been analyzed. The population could serve as a potentially large sink for nutrients such as N, P and Ca because of the relatively large amounts of N tied up in proteins and the relatively large amounts of Ca and P tied up in bone and scales. Using a P concentration of 4.75% dry wt (Kitchell et al. 1975), the fish population would contain about 3,400 g P (71,800 g fish biomass × 4.75%). If a P concentration of 3.81% dry wt were used (Goodyear and Boyd 1972), the fish population would contain 2,736 g of P. Thus, the P immobilized in fish biomass represents from 7.6 to 9.4% of total annual input (36 kg; Table IV.E–13) with about half of this P (1,370 to 1,700 g) in the refractory bones and scales that are not likely to be remineralized (Kitchell et al. 1975). If a production to biomass (P/B) ratio of 1.0 is used (Kitchell et al. 1975), the fish population processes 12 to 15% (2,740 to 3,400 g) of the annual input of P each year and converts 6 to 7.5% (1,320 to 1,700 g) of this input into refractory P in bones and scales that is lost in permanent sedimentation. Thus, the fish population alone may account for as much as 4 to 5% of the total P lost to permanent sedimentation each year in Mirror Lake (31 kg; Table IV.E–13). This calculation illustrates how important a relatively small fish population can be to nutrient cycling within a lake. Using the order of magnitude estimate of fish biomass (Table V.B.7–2) and the elemental composition of largemouth bass (Goodyear and Boyd 1972), order of magnitude estimates of the standing stocks of major nutrients in the Mirror Lake fish population can be calculated (Table V.B.7–3). These estimates and a P/B ratio of 1.0 can be compared with nutrient budget data for Mirror Lake (Chapter IV.E) to assess the importance of fish populations in nutrient cycling in Mirror Lake.

Even the limited data available on vertebrate populations illustrates the importance of these animals in both energy flow and nutrient cycling dynamics within Mirror Lake. When this role is coupled with their possible effect on zooplankton and phytoplankton populations (Chapters V.B.5 and V.B.2), it is apparent that vertebrates (primarily fish) do play a major role in the lake much beyond their limited numbers and biomass.

Table V.B.7–3. Standing Crop of Major Nutrients in Fish Biomass in Mirror Lake, New Hampshire Based on a Fish Biomass of 71.8 kg[a]

Element	Average Composition[b] (% dry wt)	Standing Crop (g in lake)	Standing Crop (g/m^2)
Nitrogen	11.46 ± 0.14	8,200	0.055
Phosphorus	3.81 ± 0.10	2,700	0.018
Calcium	6.51 ± 0.18	4,700	0.031
Magnesium	0.17 ± 0.01	120	0.0008
Potassium	1.37 ± 0.02	980	0.007
Sodium	0.44 ± 0.01	320	0.002
Sulfur	0.95 ± 0.02	680	0.005

[a] See Table V.B.7–2.
[b] Average composition of *Micropterus salmoides* from Goodyear and Boyd (1972).

Summary

Vertebrates in Mirror Lake ingest 3.5 g dry wt/m^2-yr and assimilate 80% of this energy. Of the assimilated energy, 81% is lost as respiration while 19% goes into new production (0.5 g dry wt/m^2-yr). Fish populations account for 98.5% of ingested energy and 95.1% of production while the salamander population accounts for the remainder. Salamanders are found primarily in shallow inlet and outlet areas and are more important in those areas. Frogs and toads represent an additional, but unknown, source of vertebrate production in Mirror Lake. Because these forms are restricted to shallow areas of the lake, their role in production is likely to be small.

Vertebrates consume from 20 to 40% of zoobenthic plus zooplanktonic production annually. They also represent a sink for 8 to 9% of total annual phosphorus input. Thus, vertebrates are an integral part of energy flow and nutrient cycling in Mirror Lake.

INTRODUCTION TO CHAPTERS V.C AND V.D

The next two sections of Chapter V concern the production and destruction of organic carbon in Mirror Lake. Initially, we took a broad approach in an attempt to quantify the inputs to and outputs from the lake and, subsequently, developed a budget of organic carbon for Mirror Lake (Chapter V.C). Further studies were focused on the metabolism of organic carbon within the lake itself (*Chapter V.D*). In fact, these studies are strongly interrelated, but because of the different approaches involved, some differences in interpretation were inevitable. The source and metabolism of organic carbon in lake-ecosystems have been studied for nearly a century, but we are still far from a complete quantitative understanding. The following two chapters represent our current understanding of the flux and cycling of organic carbon for Mirror Lake.

C. Organic Carbon Budget

MARILYN J. JORDAN, GENE E. LIKENS and BRUCE J. PETERSON

Energy flow and nutrient cycling are closely coupled (see *Chapter I*), so that studying the cycling of organic carbon provides a sound conceptual framework for understanding the production and control processes in ecosystems. A holistic, ecosystem approach is particularly valuable because systems often have unexpected and important properties distinct from their components. Understanding the relationship between nutrient fluxes across system boundaries and nutrient cycles within the system enables integration of an aquatic system with the surrounding atmosphere and terrestrial landscape. For example, understanding the relative importance of allochthonous and autochthonous organic carbon inputs to the productivity of a lake could provide a basis for predicting effects of disturbance on the lake's trophic status.

The complexity of most ecosystems makes comprehensive, holistic studies very difficult and expensive in time and money. We have been fortunate in the Mirror Lake study, since the lake's proximity to the HBEF made available a large amount of applicable ecologic information regarding hydrology, nutrient fluxes, phytosociology and geology (see *Chapter II*). These and basic limnological studies provided the relevant physical, chemical and biologic data required to formulate an organic carbon budget for the lake.

To do a complete organic carbon budget for a lake, all inputs of organic carbon must be estimated, including primary production, material carried in the inlet stream(s), direct litterfall from surrounding terrestrial vegetation, and carbon in precipitation. Similarly, losses of organic carbon must be measured, including respiration (CO_2), material flushed out the outlet stream, ground-water losses, permanent sedimentation and insect emergence. We do not deal in this section with internal carbon transfers such as zooplankton grazing and respiration, carbon fluxes from algae to bacteria, etc. Such within-system fluxes are discussed in *Chapter V.D* on decomposition, based on data gathered after the study of the overall lake C budget had been completed, and in *Chapter VI*. Many of the budget values presented here have been revised from those published in 1975 by Jordan and Likens. The new values reflect more extensive data.

Biomass

Although biomass measurements are not a necessary part of an organic carbon budget, when combined with separate productivity measurements biomass data may be used to estimate growth rates or turnover times for populations. This information is useful in understanding the trophic relationships of aquatic ecosystems.

Only 7% of the organic carbon in the water of Mirror Lake, and 0.6% in the top 10 cm of the sediments, occurs in living organisms. The rest of the organic carbon is in detritus, including both particulate and dissolved carbon (*Table V.C–1*). Dissolved organic carbon (DOC) is about 11 times greater than particulate organic carbon (POC), a distribution typical for fresh waters (Wetzel and Rich 1973). Concentrations of DOC ranged from 1 to 4 mg C/liter, and POC from 0.2 to 1.5 mg C/liter. Concentrations tended to be highest in the deepest water, but there were no consistent patterns of seasonal change (see *Chapter IV.B*).

In most fresh waters the concentration of dis-

C. Organic Carbon Budget

Table V.C–1. Mean Annual Standing Stocks in Mirror Lake in mg C/m² [a]

Seston	
Phytoplankton	375[b]
Zooplankton	200[b]
Bacteria	80[d]
Detritus (particulate)	1,300[c]
Bacteria (benthic)	4,000[d]
Benthic invertebrates	700[c]
Macrophytes	1,350[c]
Fish	250[d]
Epilithic algae	380[d]
Epipelic algae	?
Epiphytic algae	?
Salamanders	5[c]
DIC	8,600[b]
DOC	13,800[b]
Benthic detritus (10 cm)	1,000,000[b]

[a] Revised from Jordan and Likens 1975.
[b] ±20%
[c] ±50%
[d] Right order of magnitude.

solved inorganic carbon (DIC = CO_2 + HCO_3^- + CO_3^{2-}) is greater than the concentration of DOC, the opposite of the case for the soft water of Mirror Lake (*Table V.C*–1). Direct atmospheric diffusion is the main source of DIC because the acidic inlet streams contribute little HCO_3^-, and the even more acidic rain and snow contribute essentially none (see *Chapter IV.B*). The concentration of DIC in the epilimnion ranged from about 0.6 to 2.8 mg C/liter. In the hypolimnion concentrations of 3 to 8 mg C/liter were common from June through autumn overturn in October (see *Chapter IV.B*).

Methods used in measuring the biomass of living organisms are described in *Chapter V.A.1–9*. Dissolved organic carbon and POC were determined using an infrared gas analyzer (Menzel and Vaccaro 1964). Dissolved inorganic carbon was determined by gas chromatography (Stainton 1973). The lake was sampled at least once a month; weekly samples were taken from the inlets and outlet, from August 1973 to August 1974. These sampling intervals were adequate since concentrations of DIC, DOC and POC in Mirror Lake do not change rapidly. When possible, the more rapid changes in DOC and POC occurring in the inlet streams during storm events were more closely monitored.

Inputs

There are two categories of organic carbon input for any aquatic ecosystem: autochthonous and allochthonous inputs. Autochthonous inputs are due almost entirely to the photosynthesis of phytoplankton, periphyton, macrophytes and photoautotrophic bacteria. Allochthonous inputs are transported into the system by meteorologic and geologic vectors: precipitation, blowing of shoreline litter (see page 297) and streamflow. Biologic vectors may be important in some lakes.

Autochthonous Inputs

The methods used to estimate production by the macrophytes, epilithic, epipelic and epiphytic algae and bacteria are described in *Chapter V.B*. The remaining autochthonous carbon source, production by phytoplankton, is the largest single input of organic carbon in Mirror Lake (*Table V.C*–2). Currently, the best method for measuring phytoplankton production in oligotrophic water is based on fixation of (^{14}C)CO_2. This method is deceptively easy to use but is accompanied by many problems in both procedures and interpretation (Peterson 1980). We discuss some of the problems below, because the reliability of the ^{14}C method affects the accuracy of the Mirror Lake organic carbon budget.

Samples are incubated *in situ* with (^{14}C)HCO_3^- for several hours, processed, then the amount of ^{14}C fixed (as POC, or POC + DOC) is usually counted by liquid scintillation. The fraction of the added ^{14}C that was fixed is multiplied by the total DIC concentration in the sample, correcting for isotopic discrimination, to estimate the total carbon fixed. The importance of contaminant-free isotope and the use of proper sample blanks cannot be overemphasized. Care should be taken to reduce losses of (^{14}C)CO_2 from the (buffered) stock of (^{14}C)HCO_3^- during preparation of samples for incubation. We found that this loss was minimized by preparing stock solutions in 5-ml or greater portions, rather than 1-ml portions, and

Table V.C–2. Inputs of Organic Carbon for Mirror Lake in mg C/m²-yr[a]

Autochthonous (gross)		
Phytoplankton (POC and DOC)		56,500[f]
Epilithic algae		2,500[d]
Epipelic and epiphytic algae		>1,000[d]
Macrophytes		2,500[d]
Dark CO_2 fixation		2,100[d]
	Sum	64,600
Allochthonous		
Precipitation		1,400[b]
Shoreline litter		4,300[b]
Fluvial:[e] DOC		10,500[c]
FPOC (0.45 μm–1 mm)		300[c]
FPOC (>1 mm)		50[c]
CPOC		800[c]
	Sum	17,350
Total Inputs		81,950

[a] Revised from Jordan and Likens 1975.
[b] ±20%.
[c] ±50%.
[d] Right order of magnitude.
[e] DOC = dissolved organic carbon, FPOC = fine particulate organic carbon, CPOC = coarse particulate organic carbon.
[f] Daytime ^{14}C fixation = 47,000 (POC = 38,300; DOC = 8,700); gross POC fixation = 47,800; net POC fixation = 28,700 (60% × 47,800); day and night respiration = 19,100 (40% × 47,800). Net POC = 0.75 × [^{14}C]POC. The 0.75 is a correction for nighttime respiration based on net = 60% gross + R = 10% P_{max} (Steemann Nielsen 1958).

the results agreed well, although some ^{14}C losses occurred in the Schindler method if the insides of the purging tubes were not rinsed down prior to withdrawal of aliquots for counting.

We followed the common procedure of incubating samples during 1000 to 1400 hours EST. Incubation length has an effect on measurement of ^{14}C fixation. The longer the incubation, the lower the measured hourly carbon fixation rate, perhaps owing to respiration of an increasingly large fraction of the total fixed ^{14}C (Rodhe 1958; Vollenweider and Nauwerck 1961). A 4-h incubation probably is a good compromise for measuring net fixation, because overly long incubations may cause *bottle effects*. That is, containment in a bottle may alter the physiology of the phytoplankton through disruption of fluxes of DIC and nutrients, altered rates of zooplankton predation, etc.

Full-day photosynthesis was calculated by the routinely used procedure of multiplying 4-h production values by the ratio of total daily surface solar radiation to the amount of solar radiation during the 4-h incubation period. Solar radiation was measured with a recording pyrheliometer at the U.S. Forest Service Station 0.5 km from Mirror Lake. Samples were incubated at 1-m depths; total integral production per m² was calculated using a computer program that accounted for the effect of lake level on the hypsographic curve.

The daily ^{14}C measurements are represented in *Figure V.C*–1. Estimates of average daily production for January through icemelt in April are the means of the actual measurements. Estimates of mean daily production for each remaining month of the year were based on regressions of measured productivity vs. solar radiation, done separately for spring, summer and autumn. In addition, average production was estimated graphically and the results differed by only 8%.

The data plotted in *Figure V.C*–1 represent net *daylight* fixation of carbon (POC + DOC). Total annual carbon fixation during daylight measured by (^{14}C)CO_2 uptake was 47 g/m², of which 38.3 g was POC and 8.7 g (18.5%) DOC. To estimate the actual net carbon fixation, which theoretically could be measured as harvestable particulate carbon after 24 h, nighttime respiration must be subtracted from daylight

by keeping the stock solution on ice during use. Following incubation, the samples may be processed by two basically different procedures. The more common procedure is to filter subsamples through membrane filters, which are then subjected to acid fumes to drive off any (^{14}C)CO_3^{2-}. The filters must be placed in the liquid scintillation fluid before they dry, or some fixed ^{14}C may be lost by volatilization. The filtrate may be acidified and bubbled to purge the solution of (^{14}C)DIC; any ^{14}C remaining in solution is presumed to be DOC lost from the phytoplankton during incubation. The second procedure developed by Schindler et al. (1972) is simpler, faster and potentially less subject to error. Replicate subsamples are acidified and bubbled to purge all (^{14}C)DIC, then aliquots are mixed with liquid scintillation fluid and counted for total fixed ^{14}C (particulate plus dissolved). We used both procedures and found

C. Organic Carbon Budget

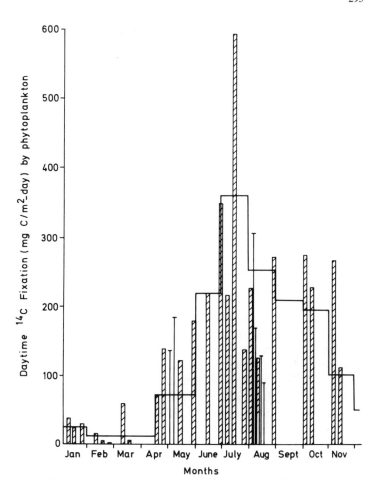

Figure V.C–1. Daytime ^{14}C fixation (mg C/m²-day) by phytoplankton. Hatched bars = 1973–74 (M. Jordan, personal communication); vertical lines = 1971–72 (D. Gerhart 1973); horizontal lines represent monthly mean values.

fixation. This step is often overlooked when ^{14}C is used to measure phytoplankton production. We assume that the rate of nighttime respiration = daytime respiration = χ, so net 24-h C production + 2χ = gross C production. We assume daytime ^{14}C (POC) fixation is *net* daytime production (i.e. does not include daytime respiration); therefore, ^{14}C (POC) fixation – χ = net 24-h C production. We further assume that net integral (POC) production is 60% of gross integral production based on the work of Steemann Nielsen (1958) (R = 10% P_{max}). No better assumption is as yet available. Based on these assumptions the following relationships exist:

net 24-h C production
$\quad\quad\quad\quad\quad$ = (0.6) (gross production)
$\quad$ total respiration = 2χ
$\quad\quad\quad\quad\quad$ = (0.4) (gross production)
$\quad\quad\quad\quad\chi$ = (0.2) (gross production).

It therefore follows (for POC only) that:

^{14}C fixation – χ = net production
$\quad\quad\quad\quad\quad$ = (0.6) (gross production)
^{14}C fixation = net production + χ
$\quad\quad\quad\quad\quad$ = (0.8) (gross production)
(1.25) (^{14}C fixation) = gross production
(0.75) (^{14}C fixation) = net production.

Applying the above relationships to the Mirror Lake daylight ^{14}C (POC) fixation of 38.3 g C/m²-yr yields the following estimates:

	POC
Net (POC) production	28.7
Respiration	19.1
Gross (POC) production	47.8

Algal excretion of DOC measured using ^{14}C uptake (8.7 g C/m²-yr) is not directly affected by algal respiration. Thus, the total gross production including POC and DOC is 47.8 + 8.7 = 56.5 g C/m²-yr. The amount of algal DOC excretion may have been underestimated by approximately 2.7 g C because of rapid assimilation and respiration of DOC by bacteria during the 4-h incubations (*Chapter V.D*).

The above reasoning is based on the assumption that ^{14}C fixation is *net* daytime C production, when in actuality ^{14}C fixation is probably intermediate between net and gross production. The degree to which ^{14}C fixation deviates from net production depends on the growth rate of the phytoplankton assemblage (Peterson 1978).

Production by phytoplankton, epilithic algae and macrophytes is given as gross, rather than net, production in *Table V.C-2*. However, the concept of gross photosynthetic production is only of theoretical value in an organic carbon budget because net carbon production is the actual input to the system. Net production for the epilithic algae and macrophytes was estimated first, then gross production was calculated as 1.67 + net production (net = 60% of gross). Total net photosynthetic production by phytoplankton (POC + DOC), epilithic algae, epipelic algae, epiphytic algae and macrophytes was $37.4 + 1.32 + 0.36 + 0.04 + 1.50 = 40.6$ g C/m^2-yr. Phytoplankton contributed 92% of the net photosynthetic carbon inputs to Mirror Lake. The relative importance of production by phytoplankton vs. attached algae vs. macrophytes varies among different lakes. In general, macrophytes tend to be more important in shallow, fertile lakes and phytoplankton more important in low-nutrient lakes, with attached algae and submersed macrophytes abundant only in moderately fertile lakes of sufficient clarity (Wetzel and Rich 1973). Photosynthetic bacterial production may be significant in lakes with anaerobic water within the euphotic zone, as may occur in meromictic lakes and in reservoirs receiving large amounts of allochthonous carbon (e.g. Brunskill and Ludlam 1969).

It is difficult to interpret the significance of CO_2 fixation in the dark. This fixation has generally been measured using opaque bottles incubated with (^{14}C)CO_2 during the daytime, and may not occur at the same rate as in a nighttime incubation. "Dark CO_2 uptake" values are often subtracted from (^{14}C)CO_2 uptake in transparent bottles before calculation of phytoplankton production. This procedure was followed for Mirror Lake. Dark uptake of CO_2 usually has been attributed solely to bacteria by many microbial ecologists (Romanenko 1964; Sorokin 1961) and solely to the algae by many algal physiologists (Morris et al. 1971). Dark CO_2 uptake was examined by differential filtration for Mirror Lake; 63% of uptake in the epilimnion and 15% of uptake in the hypolimnion was due to bacteria (Jordan and Likens 1980). About half of the total dark CO_2 uptake of 2.1 g C/m^2-yr (*Table V.C-2*) would have been due to phytoplankton and presumably was a net input of carbon. Of the remaining dark CO_2 uptake an unknown, but probably small, fraction could have been due to chemosynthetic bacteria. The rest would have been due to the heterotrophic bacteria and may or may not have been a net input, as was discussed in *Chapter V.B.1*. We are considering all the dark CO_2 uptake to be a net input for lack of a conclusive interpretation.

Allochthonous Inputs

Drainage into Mirror Lake was estimated from a precipitation-corrected, areal relationship based on continuous hydrologic measurements in the HBEF. An areal basis provides a better estimate of total runoff into the lake because appreciable drainage may enter directly at the shoreline (see *Chapters IV.D* and *IV.E*).

The most important allochthonous inputs were fluvial, particularly fluvial DOC (*Table V.C-2*). The concentration of DOC was assumed to be identical in both streamflow and direct runoff, and not strongly related to discharge (Fisher and Likens 1973; McDowell 1982). Thus, DOC inputs were estimated by multiplying the mean DOC concentration for each inlet by the annual total drainage for that watershed. Stream DOC concentrations normally range from 0.1 to 7.0 mg C/liter, but may increase 4 times during unusually heavy rains; runoff during the year varies by four orders of magnitude. Thus accurate hydrologic measurements are far more critical than DOC concentrations for calculating total fluvial DOC inputs.

Streamflow is the major source of fine particulate organic carbon (FPOC), and the concentration of FPOC (<1 mm to 0.45 μm) is dependent on discharge (Bormann et al. 1969). Concentrations predicted by regressions of Bormann et al. (1974) for the HBEF agreed well with measured values and were used to calculate total input.

C. Organic Carbon Budget

An average concentration of 0.048 mg/liter for fine organic matter (>1 mm) was used since there was no correlation between concentration and flow rate in the HBEF streams (Bormann et al. 1974). Coarse particulate organic matter inputs were estimated as 7 kg/ha (Bormann et al. 1974). Fluvial organic matter (dry weight) was assumed to be 50% carbon.

Litterfall is also an important allochthonous input. The ratio between forest litterfall and litter that would fall or blow into the lake per meter of shoreline was estimated from measurements on transects at nearby forest edges. Annual litterfall in the woodland around Mirror Lake was assumed to be the same as in the HBEF (5,702 kg/ha; Gosz et al. 1972), and was approximately 45% carbon (dry weight). Litterfall contributed annually about 354 g C/m of Mirror Lake shoreline. Dry leaves were 62% of the total litterfall (219 g C/m) and the remainder was twigs, branches, flowers, seeds, pollen, etc., which agrees closely with an estimate for Polish lakes of 500 g/m organic matter in dry leaves (about 225 g C/m) (Szczepanski 1965) and also for Lake Wingra (320 g C/m shoreline; Gasith and Hasler 1976).

Amounts of rain and snow were continuously measured at the U.S. Forest Service Station 0.5 km from Mirror Lake (*Chapters II* and *IV.E*). Mean hydrologic values for 1974–75 were used (precipitation 1.6×10^5 m^3 and runoff 5.3×10^5 m^3).

Precipitation provides 1,400 mg C/m^2-yr, or 8% of total allochthonous C inputs to Mirror Lake. A volume-weighted average total organic C (TOC) concentration of 2.38 mg C/liter in precipitation (rain and snow) had been measured in a limited number of samples collected in 1973–74 (Jordan and Likens 1975) but extensive samples taken in 1976–77 using collectors modified to avoid contamination with organic matter contained just 1.28 mg C/liter (*Chapter II*; Likens et al. 1983). This lower value is assumed to be more accurate and is used here. Similar concentrations of organic carbon in rain samples have been found elsewhere: Ithaca, New York = 1.4 to 9.0; Davis, California = 1.1; Lake Tahoe, California = 14.5; Christchurch, New Zealand = 0.9; Canberra, Australia = 8.1 mg C/liter (G. Likens, personal communication). Mean values of 3.98 mg C/liter in rain and 2.31 mg C/liter in snow were reported for precipitation in Wisconsin (Gasith 1975). A mean concentration of about 3.7 mg dissolved organic C/liter has been reported for precipitation in northwestern Ontario (Schindler et al. 1976). Dissolved organic carbon contributed by precipitation (84% of TOC, or 1,200 mg C/m^2-yr) is less than net annual DOC release by phytoplankton (8,700 mg C/m^2-yr) or mean standing stock of DOC in Mirror Lake (13,600 mg C/m^2).

Gross autochthonous production provides 79% of the carbon inputs to Mirror Lake and allochthonous sources provide the remainder. If respiration by plants is discounted, net autochthonous sources (43.2 g/m^2-yr) provide 71% of the actual fixed carbon inputs of 60.5 g/m^2-yr. The relative importance of autochthonous vs. allochthonous inputs may depend in part on the flushing rate. Annual hydrologic inputs to Mirror Lake are almost equal to the volume of the lake, so the theoretical flushing time is about one year. Lakes with a short flushing time tend to have a lower biomass and productivity attributable to phytoplankton, and autochthonous inputs of carbon would be relatively less important. The effective flushing time for Mirror Lake is actually greater than one year, since the spring runoff tends to flow over the surface (~1-m depth) of the lake to the outlet (see *Chapter IV.C*).

Outputs

Respiration

Respiration converts organic carbon to CO_2, which becomes part of the DIC pool, and ultimately may be lost to the atmosphere. Respiration of a whole ecosystem is exceedingly difficult to measure, because of environmental heterogeneity and temporal variation. In Mirror Lake there are at least three major sediment types, including gyttja (68% of the total area), cobbles and boulders (19%), and sand and/or gravel (13%) and these provide innumerable microhabitats. Similarly, plankton biomass and presumably respiration vary with depth in the water-column and with season of the year. To actually measure total lake respiration, component by component, during a single year would be a major undertaking. Many of the components of respiration in Mirror Lake have been estimated in various years from 1969 to 1976. We have ignored year-to-year variability and included all these estimates in our budget for a *typical year*.

Respiration of phytoplankton and macrophytes was estimated as 40% of gross POC production (see above). Methods of measuring the respiration of zooplankton, benthic invertebrates and salamanders are described elsewhere (*Chapters V.B.5–7;* Makarewicz 1974; Walter 1976; Burton 1973). Fish respiration was estimated using the data for O_2 consumption by perch (a major fish species in Mirror Lake; *Chapter V.A.9*) given in Figure 22 of Fry (1957, page 47). We calculated the value for α in the equation (ml O_2/h) = $\alpha(W)^{0.8}$ where W is the weight in grams of the fish. The total O_2 consumption in each size class of each species of fish for each of the four major annual temperature intervals was calculated. Total annual O_2 consumption was converted to carbon, assuming a respiratory quotient of 1.0. We measured the respiration of epilithic organisms using benthic chambers incubated *in situ* as described in *Chapter V.B.4*. Of the total epilithibenthic respiration, 3.2 g C/m^2-yr was due to bacteria (and protozoa) and 1.0 g C/m^2-yr was due to epilithic algae (see *Chapter V.B.4*).

The only remaining potentially significant respiratory losses of organic carbon would be those due to bacteria in the gyttja sediments and lake water. Respiration of the gyttja bacteria was estimated using two different approaches. First, total benthic respiration was estimated as 24.6 g C/m^2-yr based on rates of accumulation of CO_2 in cores of gyttja sediment incubated *in situ* and on rates of accumulation of CO_2 in the hypolimnion and under ice-cover. Part of this benthic respiration was due to the gyttja bacteria, as calculated below (in g C/m^2-yr):

24.6	Total benthic respiration
−2.8	Respiration of benthic invertebrates (Walter 1976; *Chapter V.B.6*)
−3.2	Respiration of bacteria on rocks and sand
−1.4	Respiration of epilithic and epipelic algae
17.2	Respiration of bacteria in the gyttja

The second approach was based on sedimentation rates. The mean (three years) sedimentation rate as measured with sedimentation traps was 35.3 g C/m^2-yr over the gyttja sediments, which cover 68% of the lake area (*Chapter VII.A.2*), or 24.0 g C/m^2 for the total lake area. Of this amount, only 12.8 g C/m^2-yr (lake area) accumulates in permanent gyttja sediments (*Chapter VII.A.2*). The disappearance of the remaining 11.2 g C/m^2 from the gyttja is thought to be primarily due to respiration by bacteria and benthic invertebrates and to insect emergence. Subtracting the 2.8 g C/m^2-yr respired by gyttja invertebrates and the 0.2 g C/m^2-yr lost from the gyttja through insect emergence leaves 8.2 g C/m^2-yr attributable to respiration of gyttja bacteria. Because we have no way of determining which of the above two approaches is more accurate we will use the mean value of 12.7 g C/m^2-yr. A maximum estimate of 16 g C/m^2-yr could be derived from Cole's estimates of sedimentation (*Chapter V.D*).

Respiration of bacteria in the lake water was the only significant loss of organic carbon that could not be measured either directly or indirectly. However, this respiration may be estimated by difference. Since total organic carbon inputs exceeded measured outputs by 6.5 g C/m^2-yr, and since inputs and outputs must balance, the difference may be attributed to the planktonic bacteria. Total bacterial respiration (epilithibenthic + gyttja + water) thus, was 3.3 + 12.7 + 6.5 = 22.5 g C/m^2-yr. Some unknown but probably small fraction of this total was due to respiration of protozoa, which could not be readily separated from bacterial respiration in the field measurements.

Exports

The rate of permanent sedimentation of organic carbon in accumulated gyttja was estimated at 1,920 kg C/yr for the period 1840 to 1978 in Mirror Lake (*Chapter VII.A.2; see Table VII.A–1*). This recent, cultural horizon of the sediments is marked by a major rise in abundance of herbaceous pollen and varies in thickness from about 12 to 25 cm. The 10-cm depth in the sediments, chosen as a functional boundary for nutrient flux in the lake-ecosystem, corresponds then to approximately 1900 A.D. This rate of accumulation represents 12.8 g C/m^2-yr over the entire lake bottom, although in reality the sediment is focused into the deeper 68% of the basin (*Chapter VII.A*), where the annual rate since 1840 has averaged 18.8 g C/m^2.

C. Organic Carbon Budget

The mean annual water loss from the lake outlet in 1970 to 1980 was 4.1×10^5 m^3, based on continuous measurements. An additional 3.6×10^5 m^3 was lost annually through deep seepage (see *Table IV.E–4*). The mean concentration of DOC was assumed to be the same in the seepage water as that measured in the outlet stream, 2.08 mg C/liter. Thus, DOC losses through the outlet and in deep seepage were 6 and 5 g C/m^2-yr, respectively. The mean concentration of FPOC measured in lake outlet water was 0.272 mg C/liter, for a total FPOC loss of approximately 0.8 g C/m^2-yr (*Table V.C–3*).

A small amount of organic carbon, 0.5 g C/m^2-yr, was lost through insect emergence (*Chapter V.B.6*).

Comparison of Outputs

Respiration accounts for 69% of the carbon outputs from Mirror Lake. Of the remainder, 16% is lost through permanent sedimentation, 14% in fluvial and ground-water exports and less than 1% in insect emergence. These values include plant respiration. If we exclude estimated plant respiration from the budget and deal only with net carbon outputs, respiration would account for 59% of carbon outputs, sedimentation 21%, fluvial and ground-water export 19% and insect emergence 1%.

Carbon Cycling

In addition to Mirror Lake, essentially complete annual carbon budgets have been prepared for several other lakes in temperate and arctic climates (*Table V.C–4*). Since primary production was reported as net (or net daylight ^{14}C fixation) in these other studies, we have removed plant respiration from the Mirror Lake budget to permit comparison. Net carbon fluxes may be more appropriate for use in interbudgetary comparisons because they are based on "real" (harvestable) carbon inputs and outputs, which can be utilized by the lake biota.

It is difficult to draw ecologic generalizations from the data in *Table V.C–4*, because the few lakes that are included differ in many critical factors such as latitude, geologic substrate, terrestrial vegetation and land use, nutrient loading, anthropogenic carbon inputs, etc. For example, eutrophic Lake Wingra receives high inputs of N and P in urban runoff (Gasith 1975; Likens and Loucks 1978), alkaline Lawrence Lake is partially bordered by extensive areas of marsh (Wetzel et al. 1972), high-altitude, oligotrophic Findley Lake has a very short growing season (Wissmar et al. 1977) and arctic Char Lake has an extensive carpet of benthic moss (Kalff and Welch 1974). Nevertheless, it is apparent that flushing time is a major factor influencing organic carbon cycling in these lakes. In the more rapidly flushed lakes the ratio of allochthonous to autochthonous carbon inputs (Findley 1.5, Marion 7.9) is greater than in the other more slowly flushed lakes (0.1 to 0.9), the percent of total organic carbon inputs lost through respiration (Findley 38, Marion 10) is lower than in the more slowly flushed lakes (62 to 90) and the internal recycling of organic carbon is approximately one-sixth (Findley 0.108, Marion 0.031) of the values observed in the more slowly flushed Lakes Wingra (0.572) and Mirror (0.661) (Richey et al. 1978). Both flushing rate and P loading control phytoplankton

Table V.C–3. Outputs of Organic Carbon for Mirror Lake in mg C/m^2-yr[a]

Respiration		
Phytoplankton		19,100[c]
Zooplankton		10,000[b]
Epilithic algae		>1,000[d]
Epipelic and epiphytic algae		>400[d]
Macrophytes		1,000[c]
Benthic invertebrates		2,800[c]
Fish		200[d]
Salamanders		0
Bacteria on rocks and sand		3,250[d]
Bacteria in gyttja		12,700[c]
Bacteria in plankton		6,550[c]
	Sum	57,000
Exports		
Permanent sedimentation		12,800[c]
Ground water (DOC)		4,800[c]
Fluvial: (DOC)		5,990[b]
(FPOC)		780[b]
Insect emergence		500
	Sum	24,950
Total Outputs		81,950

[a] Revised from Jordan and Likens 1975.
[b] ±20%.
[c] ±50%.
[d] Right order of magnitude.
[e] Estimated by difference.

Table V.C-4. Carbon Budgets of Temperate and Arctic Lakes[a]

	Char	Mirror	Lawrence	Wingra	Findley	Marion
Location	N.W.T., Canada	New Hampshire	Michigan	Wisconsin	Washington	B.C., Canada
Land use and vegetation	tundra	forest/rec.	ag/marsh	urb/for	forest	forest
Flushing time (years)	9–14	1.0	0.75	0.48	0.14	0.015
Depth (m) $\bar{x}$/max	10.2/27	5.8/11	6.0/13	2.5/6	7.8/27	2.4/6
Area (ha)	52.6	15.0	5.0	137	11.4	13.3
Autochthonous (Net) — Inputs						
Phytoplankton	4.1	37.4	58.1	390	5.1	8
Macrophytes	—	1.5	91.4	117	—	18
Algae—epiphytic	—	—	39.8	—	—	—
—epipelic + epilithic	16.9	2.1	2.1	—	—	42
Bacterial dark CO_2 uptake	—	2.1	17.3	—	—	—
Sum =	21.0	43.2	208.7	507	5.1	68
Allochthonous						
Precipitation	0.4	1.4	<0.2	2.5	—	—
Litterfall	—	4.3	<0.1	1.2	6.6	17
Fluvial—DOC	2.1	10.5	21.0	17.1	0.9	467
—POC	—	1.1	4.1	25.4	—	54
Sum =	2.5	17.3	25.4	46.2	7.5	538
Total Inputs =	23.6	60.5	234.1	553	12.6	606
Outputs						
Respiration[b]—planktonic	~0	16.8	29.2	300	2.5	—
—benthic	14.6	18.8	117.5	148	1.6	57
Permanent sedimentation	1.0	12.8	16.8	16	4.6	8
Outflow—fluvial	2.1	6.7	38.6	34	1.5	474
—ground water	—	4.9	0.5	—	—	—
Insect emergence	—	0.5	—	—	0.65	1.3
Total Outputs =	17.7	60.5	202.6	498	10.8	540

[a] References: Char—deMarch 1975, 1978; Kalff and Welch 1974; Welch and Kalff 1974. Lawrence—Wetzel and Rich 1973; Wetzel et al. 1972. Wingra—Gasith 1975; Gasith and Hasler 1976. Findley—Wissmar, et al. 1977. Marion—Hall and Hyatt 1974; Hargrave 1969.
[b] *Not* including plant respiration.

C. Organic Carbon Budget

production (e.g. Vollenweider 1968, 1969; Devol and Wissmar 1978). Zooplankton production is in turn directly proportional to phytoplankton production, being lowest in Findley, Char and Marion Lakes (0.5, 0.13, 1.0 g C/m^2-yr), intermediate in Mirror Lake (2.1 g C/m^2-yr) and highest in eutrophic Lake Wingra (23.0 g C/m^2-yr), whereas insect production tends to be more closely related to allochthonous carbon inputs (Richey et al. 1978). Makarewicz and Likens (1979; also see *Chapter VI.D*) reported a direct relationship between zooplankton production and net phytoplankton production in eleven lakes, including Mirror Lake. The relationship was strengthened when macrophyte and periphyton production were added to net phytoplankton production.

Wissmar and Wetzel (1978) suggested that the efficiency with which energy is transferred from primary production (phytoplankton, macrophytes and algae) to secondary consumers (zooplankton, fish and insects) tends to be *higher* in less productive lakes, as revealed by ratios of secondary production to total primary carbon inputs in lakes Findley, Mirror, Marion and Wingra. They assumed that less efficient energy transfer resulted in greater inputs of energy to benthic communities. In contrast, based on data from eight lakes Hargrave (1973) suggested that "primary production in oligotrophic waters is *less* efficiently utilized by pelagic communities," because benthic respiration (oxygen consumption) expressed as a percentage of primary production is greater in less productive lakes. Resolution of this disagreement is hindered by variability in other influential factors such as theoretical vs. actual flushing time, allochthonous organic carbon inputs, mixing depth, ratio of planktonic to benthic production and the relative importance of aerobic to anaerobic respiration, plus small sample size. For example, although for Lakes Findley, Mirror, Marion and Wingra the ratio of secondary to primary production (0.23, 0.13, 0.08 and 0.05, respectively) is indeed inversely proportional to total primary production (5, 43, 68, 507 g C/m^2-yr) as noted by Wissmar and Wetzel, this apparent relationship is destroyed by the addition of just one data point: Char Lake, with a total annual autochthonous production of 21 g C/m^2 (only five of which are due to phytoplankton) has a ratio of secondary to primary production of just 0.006. Before this and other aspects of whole-lake organic carbon cycling can be clarified, data from a greater number of lakes will be needed, preferably including lakes similar in such important variables as flushing time, nutrient inputs, depth, etc. Moreover, from an ecosystem point of view the organic carbon that provides energy for consumers may originate from allochthonous as well as autochthonous sources (the Ecosystem Source Carbon of Likens 1972b) and efficiencies of utilization are obscured by complexities and reuse of organic carbon in detritivore-consumer food webs (see *Chapter VI.D* for further discussion).

Summary

Biomass

Only 7% of planktonic and 0.6% of benthic organic carbon occurs in living organisms; the bulk of organic carbon is in particulate and dissolved detritus.

Inputs

A net total of 60.5 g C/m^2-yr enter Mirror Lake annually, of which autochthonous sources provide 71%. Phytoplankton are the major producers (37.4 g C/m^2-yr, net), accounting for 88% of autochthonous inputs. Fluvial inputs are the most important allochthonous carbon source (19% of total net inputs), followed by litterfall (7%) and precipitation (2%).

Outputs

The major loss of organic carbon is through respiration: 57.0 g C/m^2-yr including plant respiration (69% of gross inputs) or 35.6 g C/m^2-yr excluding plant respiration (59% of net inputs). Planktonic and benthic respiration are of roughly equal magnitude, with most of the planktonic respiration attributable to phytoplankton and zooplankton, and most of the benthic respiration attributable to bacteria. Twenty-one percent (12.8 g C/m^2-yr) of net carbon inputs are permanently buried in the sediments and 19% of net inputs (11.6 g C/m^2-yr) are lost through outflowing waters. Insect emergence accounts for 1% of carbon losses.

D. Decomposition

JONATHAN J. COLE

Approximately 80% of the annual income of organic carbon to Mirror Lake comes from the primary production of phytoplankton; biologically mediated oxidation of this organic carbon to CO_2 is a major loss of organic carbon from the lake (see *Chapter V.C*). This oxidation is largely the result of microbial metabolism. An evaluation of the organic carbon budget suggested that the transfer of C from algae to bacteria is one of the larger fluxes of organic C within the Mirror Lake ecosystem and attempts were made to directly measure several aspects of this transfer. New insights about algal decomposition and other internal fluxes of organic C in the lake have refined our interpretation of the organic C budget presented in *Chapter V.C*. Specifically, the release of soluble organic carbon to the lake from algal growth and decomposition is quite large, nearly twice as large as the allochthonous input of DOC. Microbial metabolism in the lake consumes this algal DOC but probably does not depend upon the DOC supplied from fluvial inputs.

The direct measurement of the decomposition of planktonic algae in oligotrophic water is a difficult problem. The concentrations of particulate and dissolved materials are low and the rates of reaction slow. Traditional methods (e.g. oxygen uptake, CO_2 production, etc.) proved ineffective. Because an extremely sensitive and realistic approach was needed, we chose to use radiotracers. Thus, the natural phytoplankton community was grown in bottles suspended *in situ* in the presence of ^{14}C-bicarbonate. Growth was continued for several algal generations to allow for a reasonably uniform distribution of the radioactive label. We then measured, under a variety of field and experimental conditions, the rate at which that material was converted to $(^{14}C)CO_2$ and $(^{14}C)DOC$. Methods were developed to (1) perform *in situ* radiorespirometry under both aerobic and anaerobic conditions; (2) obtain data on the decomposition process over the course of several days without using long-term incubations in bottles; (3) measure heterotrophic metabolism of the compounds actually released during algal growth and death; and (4) perform long-term measurements (months) of algal decomposition on the sediment surface in permeable incubation vessels.

Algal organic matter has multiple fates in a lake (*Figure V.D-1*). Some soluble compounds are released during algal growth and death; other compounds dissolve from algal detritus after cell death. Some cells are consumed by zooplankton; others sink to the bottom either as free cells or in aggregates such as zooplankton feces. Dissolution and decomposition continue on the sediments and some of the material is consumed by benthic invertebrates. Finally the material that resists decomposition may be exported from the lake or interred into the permanent sediments.

Photosynthetically-Produced DOC

In traditional measurements of primary production in which $(^{14}C)HCO_3$ is added to lake water, some $(^{14}C)DOC$ is always found in the water after the incubation. Researchers in various environments have found between 5 and 70% (Hellebust 1974) of net primary production in the form of this photosynthetically-produced dissolved organic carbon (PDOC; *sensu* Wiebe and Smith 1977). In Mirror Lake we find that about 20% of measured net primary production is in the form of PDOC. During cell growth and death a number of processes including autolysis (Golterman 1964), lysis by parasites (Daft and Stewart 1973; Burnham et al. 1976), mechanical damage by grazers (Pourriot 1963) or active excretion of metabolites all may cause the release of PDOC. Clearly the PDOC that is measured is the net production of PDOC because microbes metabolize some fraction of the gross PDOC during the incubation (Nalewajko and Lean 1972; Larsson and Hagstrom 1979). To measure both the gross release and microbial metabolism of PDOC we used a serial filtration technique in which the ^{14}C that had been incorporated into particulate matter was fractioned into nine or ten size classes (Cole et al. 1982). We calculate that as much as 50% of the gross production of PDOC is metabolized (incorporated plus re-

D. Decomposition

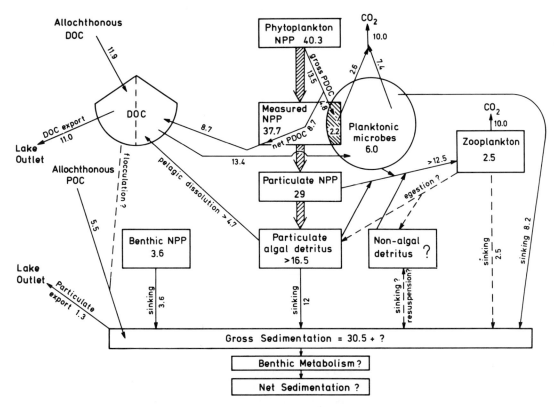

Figure V.D–1. Diagram of some aspects of the carbon cycle in Mirror Lake to show the role of decomposition of algal organic matter. All units are g C/m²-yr. The numbers within the boxes represent production (primary or secondary); the arrows represent a transfer between components. The estimates of these fluxes are explained in the text. (NPP—net primary production; PDOC—photosynthetically-produced dissolved organic C; DOC—dissolved organic C; and POC—particulate organic C.) Measured NPP is smaller than actual NPP by the amount of PDOC respired by microbes during ^{14}C primary production measurements. A portion of measured primary production consists of bacteria that have assimilated PDOC (cross-hatched portion). Sedimentation is equal to the sum of the estimated inputs to the sediments plus an unestimated input from resuspended particles.

spired) by microorganisms during a 24-h incubation (*Figure V.D–2*).

Thus, in the epilimnion in summer the gross production of PDOC reaches 40% of net primary production. The ^{14}C in microbial biomass would be tallied as "algal production" in traditional measurements (*Figure V.D–2*); the ^{14}C respired by the microbes, of course, would be overlooked altogether. Over the course of the year we estimate that the gross production of PDOC was about 13.5 g C/m² and microbes metabolize about 4.8 g C/m²-yr, of which about 2.7 g C/m² is respired (*Figure V.D–1*). Thus, total net primary production would have been 2.7 g C/m² larger than measured primary production (*Figure V.D–1*) and the gross release of PDOC would have been 180% of measured (or net) PDOC. The net PDOC is probably utilized by microorganisms but on a slower time scale; we are not certain of the fate of the net PDOC.

Zooplankton Grazing

The production of zooplankton (rotifers and crustaceans) was studied during three annual cycles in Mirror Lake (Makarewicz and Likens 1979; *Chapter V.B.5*). Zooplankton assimilation averaged 12.5 g C/m²-yr, of which 10 g C/m²-yr was respired and 2.5 g C/m²-yr became new biomass (production). We do not have quantitative estimates on ingestion, egestion or excretion, nor do we know precisely what comprises the ingested material, but will assume here that all the egested material is in particu-

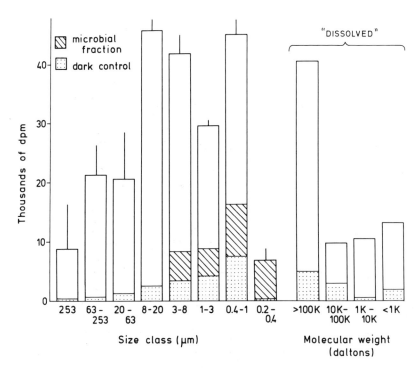

Figure V.D–2. The fate of photoassimilated C in the epilimnion of Mirror Lake. Lake water from 3 m plus ^{14}C-bicarbonate was incubated in clear (open bars) and opaque (stippled bars) BOD bottles, which were suspended *in situ*. The ^{14}C that had been assimilated during 24 h was fractioned into a series of particulate- and dissolved-size classes by serial differential filtration (Cole et al. 1982) and ultrafiltration (Cole et al. 1984). For the particulate-size classes, mean and standard deviation are shown for triplicate incubations; for the dissolved-size classes means of pooled samples are shown.

The cross-hatched portion represents bacterial net incorporation of PDOC according to Cole et al. (1982); bacteria respire about 1.2 times the amount of the PDOC, that they incorporate in 24 h (not shown).

late form and that the amount excreted is included in our measurements of PDOC. Zooplankton consume, then, at least 12.5 g C/m²-yr of particulate organic C. If the efficiency of assimilation by zooplankton were about 30% (cf. Wetzel 1975), it is possible that all the algal production in Mirror Lake passes through the guts of these animals. *Figure V.D–1* is constructed as if zooplankton consume only algal C, but it should be kept in mind that zooplankton, in fact, consume some mixture of phytoplankton, bacterioplankton and detritus.

Mineralization of Algal Organic Matter by Bacteria

When dead algal cells are added to lake water at low concentrations (<50 µg C/liter) they are mineralized to CO_2 almost entirely by bacteria. If, for example, a lake-water sample is autoclaved or poisoned with formalin prior to the addition of the ^{14}C-labeled algal substrate, (^{14}C)CO_2 is not produced (Cole and Likens 1979). Mineralization is also inhibited by antibiotics (Cole and Likens 1979). If lake water is filtered at various pore sizes prior to the addition of the algal substrate, it is possible to get an idea of the size-class of particle causing the observed mineralization (*Figure V.D–3*). Prefiltration at 0.1 or 0.2 µm removes more than 90% of the *mineralization activity* in an unfiltered sample (*Figure V.D–3*). Mineralization is not measurably reduced by prefiltration at 5 µm and is only reduced by half by prefiltration at 0.4 µm (*Figure V.D–3*). Clearly, the mineralization of algal detritus (under these experimental conditions) is associated with very small particles (0.2 to 1 µm).

In summary, the agent of mineralization is a

D. Decomposition

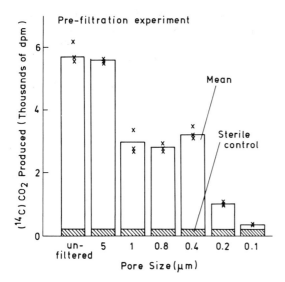

Figure V.D–3. The size-classes of particles that are active in the mineralization of ^{14}C-labeled algal cells. Lake water from 3-m depth was filtered through sterile Nuclepore filters at the indicated pore size prior to the addition of heat-killed, ^{14}C-labeled algal cells (final concentration of added material was 50 μg C/liter). The lake-water samples were then incubated *in situ* for 24 h and the amount of (^{14}C)CO$_2$ produced from the oxidation of the ^{14}C-labeled algal cells was measured.

Samples of lake water were autoclaved and then filtered (at 5 μm) prior to the addition of the ^{14}C-labeled algal cells (cross-hatched area = sterile controls).

small particle that is destroyed by heat or strong toxins and is inhibited by antibiotics. Such an agent is probably bacterial.

Decomposition of Particulate Algal C

After accounting for the above losses to PDOC and grazing, at least 16.5 g C/m^2-yr of algal origin remain in particulate form (*Figure V.D–1*). The decomposition of recently killed phytoplankton may be divided into two distinct processes, dissolution and mineralization (to CO$_2$) of the dissolved carbon. Dissolution, at least over the course of several days, is largely a chemical process, which is not necessarily mediated by microorganisms (Cole 1982b). Dissolution will proceed at equal rates in sterilized and nonsterile lake water. The rate of dissolution is relatively insensitive to changes in temperature below about 12°C; above this temperature dissolution is markedly temperature-dependent increasing by about 0.01/day (expressed as a first-order reaction) for each degree Celsius. During the first day following cell death there is a large initial release (about 25%) of the soluble components of the cell. On subsequent days the rate of dissolution is slower. This pattern of algal dissolution after cell death has been observed by other researchers in both marine and fresh waters (Waksman et al. 1937; Grill and Richard 1961; Golterman 1964; Fallon and Brock 1980; and many others).

The mineralization of the soluble C that leaches from algal cells depends on several factors including temperature, the concentration of bacteria as well as the age of the algal detritus (i.e. how long the algal cell has been decomposing). The soluble compounds that leach from dead algal cells are mineralized relatively quickly in both epilimnetic and hypolimnetic water (*Figure V.D–4*). For example, the leachate from algal detritus that was four days old (i.e. the algae were alive four days earlier and had been decomposing for three days) was converted to CO$_2$ at about 30%/day at a depth of 3 m in July, and at about 12%/day at a depth of 10 m in July or at either depth in the winter (*Figure V.D–4*). The complete conversion of four-day-old particulate detritus to CO$_2$ (that is, the product of the leaching rate and the mineralization rate) is extremely slow in comparison (*Figure V.D–4*): 6.5%/day at 3 m in July; 0.5%/day at 10 m in July; and even more slowly in winter. Clearly the limiting step in the complete conversion of algal detritus to CO$_2$ is the slow rate of dissolution, especially in cold water (*Figure V.D–4*).

Long-Term Decomposition

By incubating labeled algal detritus in permeable vessels (pore size 0.8 μm) on the sediments of the lake we found that at 4- and 8-m depth, respectively, 16 and 30% of the initial particulate material remained undecomposed after a one-year exposure. The undecomposed fraction was not affected by the inclusion or exclusion of meiobenthic animals (Cole 1982b). The long-term average rate of loss from particulate algal detritus is thus about 0.2% per day. A time se-

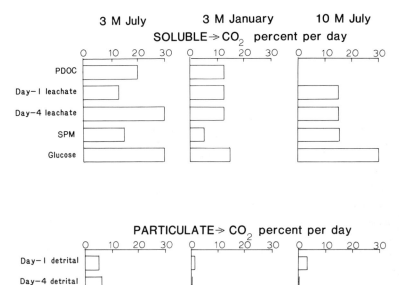

Figure V.D–4. Mineralization rates for soluble and particulate C of algal origin in Mirror Lake. *Upper panel:* The percent of soluble C converted to CO_2 in 24 h at 3-m depth in July (left), 3 m in January (center) and 10 m in July (right). Day 1 and Day 4 leachate refer to the soluble compounds that leach from particulate algal detritus one and four days after cell death. SPM (soluble plankton material) is a warm-water extract (50°C) from recently killed phytoplankton. *Lower panel:* The percent of particulate C converted to CO_2 during 24 h in Mirror Lake (as above). Day 1, 4 and 30 detrital refers to the C that remained in particulate form 1, 4 and 30 days after cell death.

ries of similar incubations revealed that most of the loss occurred within the first 10 to 20 days; after about 150 days little further loss occurred. The rate of decomposition did not fit a first-order kinetic rate law. Conservatively, then, at least 70% of the algal detritus will decompose within one year.

Benthic and Pelagic Decomposition

Although we have detailed information on the decomposition process and have a good idea of the total amount of decomposition, it is difficult to estimate how much decomposition is benthic and how much occurs in the water-column. It is especially difficult to estimate benthic decomposition because we do not have direct measurements of microbial metabolism on the surface of the sediment.

The release of PDOC is restricted to the photic zone in Mirror Lake. While PDOC is clearly important for the organic nutrition of bacterioplankton, it accounts for only about one-half to one-fourth of the organic carbon used by the bacterioplankton (Cole et al. 1982). The decomposition of particulate algal detritus in the water-column supports the remainder of bacterial production (Cole 1982b).

The extent to which particulate algal detritus decomposes in the water-column or on the sediments depends on a number of factors. Most critical are the location of cell death (i.e. in the water or on the bottom), the residence time of the dead cell in the water-column and the water temperature. Because we do not know the exact location of cell death or, for that matter the natural causes of algal mortality, some assumptions will be made. We assume that a dead cell will reside in the water-column for only six days. This short residence time is based on a sinking speed of 1 to 2 m/day, which is at the fast end of the scale for algal detritus (Burns and Rosa 1980). During the six days' residence, the dead cell will undergo temperature dependent leaching and lose between 20 and 60% of its cellular organic carbon depending on the water temperature (Cole 1982b). If 100% of the particulate algal detritus (that is the 16.5 g C/m^2-yr in *Figure V.D–1*) consists of cells that die in the water-column, then between 4.3 and 9.7 g C/m^2 would dissolve in the water-column and between 1.8 and 7.3 g C/m^2 would dissolve or be decomposed on the sediments depending on the temperature (*Figure V.D–5*). If the fraction of cells that die in the water-column were lower, decomposition on the sediments would increase correspondingly (*Figure V.D–5*). If the residence time of the dead cell in the water-column were longer than six days (which is

D. Decomposition

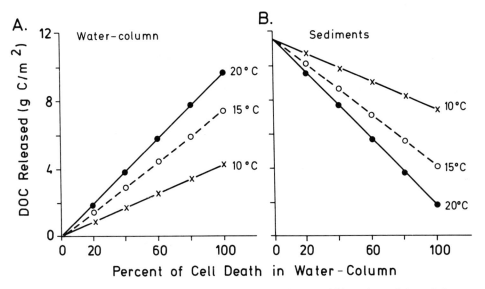

Figure V.D–5. A model to predict the amount of algal DOC that would be released (gross) from particulate algal detritus in **(A)** the water-column and **(B)** on the sediments, depending on the location of cell death and the water temperature. The production of particulate algal detritus is taken as 16.5 g C/m^2-yr (*Figure V.D–1*) and it is assumed that 70% of this detritus will ultimately decompose (see text). The residence time of the dead cell in the water-column is assumed to be only six days (see text).

likely to be the case) this model would tend to underestimate the amount of decomposition in the water-column.

The surface sediments do not appear to contain recognizable phytoplankton. Benthic desmids and diatoms are common in the sediments (*Chapter VII.B*) and some fast-sinking taxa such as the planktonic diatoms from the early spring bloom may reach the sediments intact. It is likely, therefore, that cell death occurs largely in the water-column or soon after contact with the sediments. Zooplankton may also affect the location of cell death. Many cells are damaged during zooplankton grazing. If the zooplankton do ingest most of the algal production, it is likely that the egested material contains a large proportion of dead cells. Some zooplankters, especially copepods, produce compact, fast-sinking fecal pellets. Thus, while grazing by zooplankton would tend to increase mortality in the water-column it also may decrease the residence time of algal detritus within the water-column. In Mirror Lake, rotifers plus cladocerans, neither of which produce compact feces, account for about 80% of zooplankton production (*Chapter V.B.5;* Makarewicz and Likens 1979). It is likely, therefore, that in Mirror Lake, grazing would increase the release of DOC to the water-column.

By using measurements of the age-specific rate of the decomposition of algal detritus in combination with a model of the age structure of particulate detritus in the water-column, it was calculated that about 12 g C/m^2-yr of algal DOC would be released from particulate algal detritus in the water-column, excluding the input from PDOC (Cole 1982b). This result implies that more than 80% of the death of algal cells would occur in the water-column (*Figure V.D–4*). The assumption was made in the model, however, that all of the sestonic POC was derived from phytoplankton. In a recent whole-lake experiment, Hesslein et al. (1980) added 1 Ci of ^{14}C-bicarbonate to a small lake in Ontario, Canada. By following the specific activity (^{12}C/^{14}C) of the sestonic POC they concluded that only 40% of the POC in the water could be derived from recent (within 100 days) primary production. Hesslein et al. (1980) suggested that the excess POC may have come from the watershed or the sediments. In Mirror Lake the allochthonous input of POC is small (5.5 g C/m^2-yr *Chapter V.C*) and much of it enters as large particles such as leaves or twigs that sink quickly once they become waterlogged. If we assume, then, that 60% of the detrital POC in the lake was derived from resuspended sediments and that this material resists decomposition we would calculate that about 4.7 g C/m^2-yr dissolved from particulate algal detritus. This result implies that 65% of algal cell death occurs in the water-column, given an average temperature of 15°C for the ice-free season (*Figure V.D–5*). Because we have underestimated the residence time of dead cells in the water-column (above), 4.7 g C/m^2-yr is a minimum estimate for the dissolution of algal detritus in the water-column (*Figure V.D–1*).

Allochthonous Inputs

The annual inputs of allochthonous DOC and POC have been measured (*Chapter V.D*). We wil assume here that all of the allochthonously supplied POC reaches the sediments and that all of the DOC is injected into the water-column (*Figure V.D–1*). The allochthonous input of DOC actually supplied to the lake, however, is probably smaller than the estimate presented in *Chapter V.D* because most of the fluvial input occurs during snowmelt while the lake is covered with ice. Because the inflowing water (at about 0°C) is less dense than the bulk of the water in the lake (at about 4°C) it is likely that much of this water flows into and out of the lake just under the ice and mixing only little with the remainder of the water in the lake (*Chapter IV.B*).

Turnover of DOC in Mirror Lake

About 12 g C/m^2 of DOC of allochthonous origin and at least 19.5 g C/m^2 (probably more) of DOC of algal origin are added to Mirror Lake each year (*Figure V.D–1*). An export of about 11 g C/m^2-yr as DOC occurs in outflowing water. If the amount of DOC stored in the lake does not change from year to year (which seems to be the case, at least since 1973) then 22 g C/m^2 of DOC must be removed by metabolism or physical-chemical reactions each year.

Some of the DOC of algal origin is metabolized rapidly (20 to 40% per day) and would not be expected to accumulate in lake water (*Figure V.D–4*). The ambient DOC In lake water, however, is quite resistant to oxidation and turns over very slowly, if at all (*Figure V.D–6*). If all of the autochthonous but none of the allochthonous DOC were metabolized by microorganisms and metabolized at a growth yield of 45% (Cole 1982b), bacterial production in the plankton would be about 8.2 g C/m^2-yr. This rate compares with the estimate of bacterial production of 6.25 g C/m^2-yr based on the uptake of ^{35}S-SO_4 (Jordan and Likens 1980). The estimate of the production of bacterioplankton, which is based on the DOC budget (*Chapter V.C*), is an underestimate because that budget did not include the large internal input of DOC from algal growth and decomposition referred to above. The comparison between the measurement of total production by bacterioplankton with the amount of algal organic carbon metabolized by bacterioplankton suggests that most or all of bacterial production in the pelagic region of the lake is supported by algal organic C.

Sedimentation and Benthic Metabolism

We do not have direct measurements of bacterial production on the sediments, and estimates are based upon oxygen consumption and on the difference between the gross and net rates of sedimentation of organic carbon (*Chapter V.C.*). By adding the losses experienced by phytoplankton in the water-column (*Figure V.D.–1*), we estimate here that only 12 g C/m^2-yr of algal organic C reaches the sediment surface. Other potential inputs to the sediments include benthic production (3.6 g C/m^2-yr), allochthonous POC (5.5 g C/m^2-yr), the remains of zooplankton (which are common in the surficial sediments; D. Strayer, personal communication) (up to 2.5 g C/m^2-yr) and the remains of bacterioplankton (up to 8.2 g C/m^2-yr). Clearly bacteria would not reach the sediments if they sank as individual cells, but bacteria could be incorporated into larger particles such as feces or detrital aggregates, which would sink. Finally, the resuspension of previous sediment should be viewed as both one of the inputs to gross sedimentation and as a potential source of detritus to the water-column (*Figure V.D–1*).

D. Decomposition

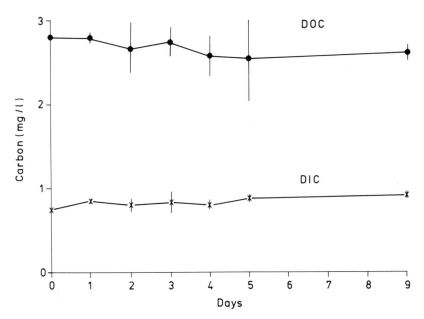

Figure V.D-6. Turnover of ambient lake-water DOC during long incubations in bottles. Lake water from 3 m was incubated in 21 opaque BOD bottles at the *in situ* temperature during August 1980. Dissolved organic C (solid circles) and dissolved inorganic C (X) were measured on each day for triplicate incubations. Mean and 90% confidence limits are shown.

The magnitude of resuspension in Mirror Lake, however, is not known (*Chapter VII.A*). If zooplankton consume some bacterial or detrital C in place of algal C, the amount contributed to the sediments in each group would change correspondingly but the total input would probably not change greatly.

By adding the above terms, gross sedimentation (plus the export of POC in outflowing water) could be as large as 31.8 g C/m²-yr (*Figure V.D–1*). The export of POC in outflowing water is small, about 1.3 g C/m²-yr. Gross sedimentation is, thus, on the order of 30.5 g C/m²-yr, according to this estimate.

We have used wide-mouth jars suspended in the lake to measure the rates of gross sedimentation directly (*Chapter VII.A*). By this technique the estimate of gross sedimentation is at least 48 g C/m²-yr, which is larger than the calculated rate above (30.5 g C/m²-yr). There are several possible reasons for the discrepancy between these estimates. The jars certainly would catch any resuspended material in the lake. Because the sediments represent a huge reservoir of organic C (1,000 g C/m² to a depth of 10 cm; Likens and Bormann 1979) a small amount of resuspension may inject 10 (or more) g C/m² into the water-column. The fact that "out of season" pollen does not become resuspended does not necessarily imply that organic fine particles also do not become resuspended (see *Chapter VII.A*). It is likely that some of the sestonic C in Mirror Lake is derived from the sediments; on the average there is about four times as much detrital as algal C in the water-column (Jordan and Likens 1975). Further, the sedimentation jars may "overtrap" because of hydrodynamic properties (*Chapter VII.A*). If the extent of overtrapping were about 30% it, also, would account for the discrepancy.

From our experiments on algal decomposition we find that about 30% of the particulate algal detritus (again, the 16.5 g C/m²-yr from *Figure V.D–1*) will resist dissolution during one year. The net sedimentation of algal detritus, then, would be about 5 g C/m²-yr. We do not have any data on the fate of the other inputs to the sediments, but the long-term accumulation rate of organic C in the sediments has been about 12.8 g C/m²-yr (mean rate from 1840 to the present; *Chapter VII.A*). If the current rate of net sedimentation were similar to the long-term average rate, algal detritus would contribute about 40% of the accumulated organic C.

The Fate of Algal C in Mirror Lake

While we have a relatively good understanding of the fate of algal C in Mirror Lake it is still not possible to assign firm, quantitative rates to several of the important transfers. We do not know, for example, the causes of algal mortality in nature, whether cell death occurs in the water or on the sediments, and whether zooplankton selectively consume living or detrital material (*Figure V.D*–1). Further, the relative distribution of decomposition in the water-column and on the sediments is not yet clear; new methods are needed to obtain direct, accurate measurements of microbial production in the sediments.

Do our sediment traps overestimate sedimentation by catching resuspended material or have we overestimated the decomposition of phytoplankton in the water-column? What is the fate of the fluvial and pluvial DOC? Does it simply flow through the lake as indicated in *Figure V.D*–1, or does it enter microbial food webs?

With these reservations in mind we can discuss the fate of algal organic carbon in the Mirror Lake ecosystem. About 32% of net primary production escapes the algal cells as gross PDOC and 10.5% and 16.5% of net primary production dissolve from particulate algal detritus in the water-column and on the sediments, respectively. Thus, about 60% of net primary production fluxes through a soluble stage; this soluble carbon is metabolized by microorganisms in the water-column and on the sediments (*Figure V.D*–1). Zooplankton may consume (net) as much as 29% of net primary production but would consume a smaller fraction of algal production if they also ingested bacteria or nonalgal detritus (*Figure V.D*–1). Finally, about 12% of net primary production (or up to 30% of the gross sedimentation of algal organic carbon) will remain undecomposed in the sediments at the end of one year.

Summary

Algal growth and decomposition is a major source of DOC for the sediments and water-column of Mirror Lake. This internal input of DOC (about 20 g C/m^2-yr, possibly more) is larger than the allochthonous input of DOC (12 g C/m^2-yr). Because most of the DOC of algal origin is metabolized rapidly, it never accumulates in the water and, thus, has been overlooked in previous studies. In contrast, the DOC actually present in the water is remarkably stable and is metabolized very slowly if at all.

Bacterioplankton respire about as much algal carbon as do zooplankton (10 g C/m^2-yr). Superficially the grazing food web (based on living particles) and the microbial food web (based on DOC and the dissolution of detrital POC) are separate and function independently. There are, however, at least two critical points of intersection: (1) zooplankton probably speed the release of DOC from algal cells by mechanical disruption, thus, making it available to microbes; and (2) bacteria that have grown at the expense of algal DOC may be eaten by zooplankton. The distinction, then, between the grazing and detrital food webs is not sharp.

Chapter VI

Mirror Lake—Ecologic Interactions

A. The Littoral Region

Robert E. Moeller, Rhoda A. Walter, David L. Strayer and Bruce J. Peterson

Much of our information from Mirror Lake is descriptive. Many hours have been devoted to sampling planktonic and benthic organisms; collecting and chemically analyzing water, plants and animals; monitoring precipitation, lake temperatures and light attenuation; compiling annual production for selected organisms and whole trophic levels. The parameters and processes we have described are part of the biogeochemical interaction that the lake represents. Some of these interactions have been examined analytically in Mirror Lake, others can be inferred—often rather speculatively—from limnological experience elsewhere. Few cause-and-effect relationships can be laid out in detail, but our survey of organisms and abiotic factors is broad enough to attempt a comprehensive overview of the lake.

The overview begins at the lake's margin, proceeds along the bottom from shallow water (the *littoral* region, this chapter) into deep water (the *profundal* region, *Chapter VI.B*), then expands vertically into the water-column (the *pelagic* region, *Chapter VI.C*) and concludes with a synthesis for the ecosystem (*Chapter VI.D*). This subdivision of the lake into three regions (*Figure VI.A–1*) is an analytical convenience consistent with the limnological tradition of separating the sampling and study of free-swimming planktonic populations from that of benthic populations. Limnological tradition is not consistent on criteria for distinguishing between the littoral and profundal benthic regions: The lake bottom intersects continuous gradients of a multitude of semi-independent physico-chemical factors to which organisms respond in patterns that are modulated by interactions among the organisms themselves.

By littoral zone, freshwater ecologists usually mean a shallow, nearshore area colonized by large benthic plants. Often there is an obvious zonation of aquatic plants developed around small ponds or sheltered embayments of larger lakes—large emergent reeds, rushes or other grasslike plants rooted at the shoreline are replaced at water depths of 1 to 2 m by floating-leaved species, and these in turn give way to completely submersed plants that extend, in clear lakes, to depths of 10 m or more. The littoral region of Mirror Lake is superficially simple, since emergent plants are absent and floating-leaved species poorly developed. The submersed vegetation, however, is extensive, zoned according to water depth, and relatively rich in species if not in biomass (*Chapter V.A.4*).

The littoral region in Mirror Lake is really a benthic zone underlying the nearshore portion of the pelagic water-column. Except for a few waterlilies, the large aquatic plants do not grow more than 60 cm above the lake bottom; most are no more than 10 cm tall. In contrast, the submersed vegetation of hardwater lakes often extends to the surface from depths of 2 to 5 m (Rickett 1924; Lang 1967), adding a biologic structure to the water-column that increases the species richness and productivity of free-swimming communities found there (Lim and Fernando 1978).

In the following discussion, we emphasize

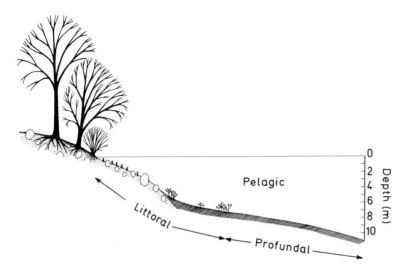

Figure VI.A–1. The pelagic, littoral benthic and profundal benthic zones as they have been recognized in Mirror Lake. The diagram depicts the characteristically abrupt transition from forest to lake.

the relation of littoral interactions to the major environmental gradients that lie across the lake bottom.

Light

Solar irradiance is attenuated rapidly in Mirror Lake (*Chapter IV.B*), although the lake is relatively clear in comparison with many small lakes that have become eutrophic or that are heavily loaded with suspended matter or dissolved humic compounds. At the limit of the macrophytes (7.2 m), light has been reduced to about 6% of the surface irradiance in summer. Short-lived benthic algae, capable of growing and completing their life cycle more rapidly than the macrophytes, take advantage of periods of greater-than-normal water clarity to extend their depth-distribution. Mats of blue-green algae have been noted even in the deepest part of the lake for extended periods of the year, though these may have been growing heterotrophically. We treat the entire extent of macrophyte colonization (0 to 7.2 m) as *littoral,* and deeper substrates simply as *profundal.* In very deep lakes, a *littoriprofundal* zone can be recognized between the limit of vascular plants and the deepest penetration of benthic algae (Hutchinson 1975).

Light is an obvious requirement for autotrophic plant growth, and therefore a factor that may limit plant distributions. Light cues, by way of daylength or the spectral quality of irradiance, may control the temporal pattern of an organism's activity and ultimately its spatial distribution. For example, the changing spectral distribution of light with increasing depth has been suggested to limit reproduction of the benthic macroalga, *Nitella flexilis,* and thereby set its depth limit (Stross 1979). Animals that depend on visual acuity to find food may be restricted to the littoral zone, and to diurnal activity. In summer, the most common fish in Mirror Lake (yellow perch and smallmouth bass) are active during the day (*Chapter V.A.9*), but colorful little piles of crushed crayfish appendages scattered across the littoral region attest to some temporal overlap with the mainly nocturnal crayfish.

Temperature

Seasonal temperature changes in the littoral region of a temperate lake parallel those in the air above the lake. Because of the thermal stratification that develops within the lake (*Chapter IV.B*), the profundal sediments become isolated from convective heating in late spring and summer, damping the seasonality. The thermal stratification of the water-column during summer creates a sharp environmental gradient across the bottom between depths of 4 and 7 m. In Mirror Lake, any benthic organism that requires a warm season (above 15°C) to complete its life cycle is necessarily restricted to the littoral zone, even if it has no requirement for light, or for plants as food or habitat.

In clear lakes, thermal stratification may play

A. The Littoral Region

a role along with light in defining the littoral zone. Vascular plants rarely grow beneath a thermocline, but may grow at depths that only enter the epilimnion during the course of the summer, as described for *Utricularia purpurea* (Moeller 1980; *Chapter V.A.4*). Many of the rosette plants in Mirror Lake are evergreen, but their actual growing season is restricted to late spring through early autumn. A seasonal succession of shorter-lived algal populations probably occurs within the epiphytic and benthic algal communities (cf. Stockner and Armstrong 1971).

Long-lived animals, such as fish and the larger insect larvae, can be relatively inactive during the winter, when plant production in the water-column and sediment surface is low. Little feeding is needed to survive the winter, if respiration rates become very low at temperatures of 1 to 5°C. The physical environment of the littoral region in winter is less harsh than that on land, where freezing and desiccation threaten survival. Only at depths of less than 0.75 m can ice ever contact the sediment surface, and this tendency is offset by heat conducted out of the "warm rim" of substrates during winter (Likens and Johnson 1969). The seasonal variations in populations of the common benthic invertebrates (*Chapters V.A.6 and V.A.7*) seem to be less pronounced than those of the zooplankton (*Chapter V.A.5*). Probably this relative stability of numbers reflects a continued supply of food, even in winter, by way of bacteria and fungi of the detrital food chain (Rich and Wetzel 1978).

Currents and Waves

Aside from small currents created by the inlet streams at high discharge, water movements in the lake are controlled by winds blowing across the lake. Direct wave action on shallow sediments is limited to depths of <3 m on the southern and eastern shores of the lake. Wind-driven currents, however, are important characteristics of the littoral zone. They distribute dissolved nutrients and suspended particles, which may be actively or passively removed by sessile, filter-feeding invertebrates. Water movements across the lake bottom may cause suspended seston to become impacted within the firmly attached community of epibenthic algae. There the seston, including living phytoplankton, would provide a larger food source for the invertebrate grazers than benthic primary production (*Chapter V.B.4*) alone supplies. Currents ultimately sweep fine-grained particulate matter away from the upper littoral zone. This transported particulate matter is added to the profundal sediment (*Chapter VII.A*).

Currents also enhance an exchange of gases, notably oxygen, between the water-column and lake bottom. The littoral fauna slightly resembles that of nearby streams, e.g. in its caddisflies and mayflies, owing to the well-oxygenated water moving past the lake bottom. Wind-driven currents determine the depth of the thermocline (4 to 7 m), below which the exchange of heat and gases with the atmosphere is sharply reduced.

Sediments

The benthic habitat is biologically distinct from the pelagic; benthic organisms utilize the lake bottom in various ways. Some organisms use the lake bottom as a platform from which to filter the suspended plankton. These filter-feeders include sponges, bryozoans, hydras, sessile rotifers and protozoa, and the large clam, *Elliptio complanata,* in Mirror Lake. Other organisms are functionally dependent on food sources found on the bottom, such as the benthic plants and the microflora colonizing organic detritus.

The littoral portion of the lake bottom in Mirror Lake includes a variety of sediment types, compared with the ubiquitous fine-grained organic mud, or gyttja, of the profundal zone. Sand mixed with gravel and cobbles predominates, with frequent boulders up to 2 m in diameter and scattered patches of terrestrial leaf debris. The deeper part of the littoral zone (5 to 7.5 m) contains the marginal reaches of the gyttja, which is locally colonized by the macrophytes *Potamogeton berchtoldii* and *Nitella flexilis.* We portray in *Figure III.A–6* major features of substrate patterning—not the local variations and exceptions that are important to organisms and that are encountered by ecologists trying to sample benthic communities.

Many organisms actually live in the sediment. For these the physical and chemical nature of the sediment is a distribution-determining variable. The motile algae (e.g. diatoms), protozoa and other tiny invertebrates (*Chapter V.A.7*) living within the interstitial water or attached to sediment particles are likely to be sensitive to whether their subsurface environment is oxygenated (as in shallow-water sands and gravels) or not (the shallow organic deposits and deeper gyttja). Some large invertebrates, such as the mayflies, *Ephemera* and *Hexagenia*, and the caddisfly, *Phylocentropus*, excavate permanent U-shaped tubes within firm sand. Crayfish spend the day in burrows alongside rocks or beneath fallen tree trunks. For these invertebrates the lake bottom offers concealment from free-swimming predators, such as fish and salamanders.

Other organisms live on, or rooted in the sediment. These include many plants—probably more than 100 species of algae as well as the 23 species of macrophytes (*Chapter V.A.4*). For attached plants and sessile, filter-feeding animals, the range of available substrates, including the macrophytes themselves, represents a gradient of community organization that probably accounts in part for the greater biologic diversity of littoral communities compared with the profundal communities (e.g. the benthic macrofauna, *Chapter V.A.6*; and microfauna, *Chapter V.A.7*).

Algal communities on rocks or macrophytes support a full complement of grazers and detritus-feeders, notably amphipods and orthoclad chironomids. And these in turn are food for predators like tanypod chironomids and odonates. A carnivorous macrophyte, *Utricularia purpurea*, is both host to an attached epiphytic community and consumer of some of the smaller invertebrate grazers and predators that patrol its photosynthetic stems. Because of the diversity of biologic interactions within the benthic communities, it is not clear which factors most influence the spatial distribution and the abundance of individual species.

Some organisms are not strictly tied to the lake bottom, but are appropriately viewed as part of the littoral community. Fish in Mirror Lake are concentrated near shore. These are warm-water species, which in summer avoid the cold waters of the profundal. They feed on benthic invertebrates, as well as on larger zooplankton and terrestrial insects blown or washed into the lake (*Chapter V.A.9*). Mirror Lake today has no resident muskrat or beaver colony, but these amphibious mammals likely were present before humans settled along the shore. The zooplankton in nearshore waters is enriched with species that avoid the pelagic. These include the large predatory cladoceran, *Polyphemus pediculus* (*Chapter V.A.5*), as well as occasional chydorid Cladocera that spend much of their time in the benthic region. Whereas some benthic organisms disperse as briefly free-swimming larvae, many planktonic algae and zooplankton outlast unfavorable times of year as benthic resting stages. The basically terrestrial adults of the American toad (*Bufo americanus*) breed in the lake in the late spring, and their tadpoles develop as a component of the littoral community. In contrast, the red-spotted newt, whose larvae and mature adults are both aquatic, spends two or more years before maturity as the terrestrial eft (*Chapter V.A.8*). Most benthic insects emerge as adults to mate above the lake's surface, where they are fed upon by birds and bats. Thus, the littoral communities are loosely tied biologically to both the surrounding landscape and the overlying water-column through the feeding activities and the life histories of some of their populations.

Nutrient Availability

Nutrients entering the lake by way of inlet streams or shallow seepage pass through the littoral zone. Benthic plants have a chance to incorporate these elements before they reach the pelagic zone. The density of large aquatic plants is greater in the immediate vicinity of the western inlets than elsewhere, and it is mainly on these plants that appreciable amounts of epiphytic algae occur. Green tufts of filamentous algae occur where inlet water passes across the nearshore substrate, graphic testimony to the fertilizing effect of inflowing water.

It is well appreciated that the metabolism of lakes depends on continued input of nutrients. The actual level of biologic activity, notably primary productivity, depends on the rate of recycling within the lake, which is some multiple of

A. The Littoral Region

the primary rate of nutrient input, or loading. The littoral sediment is a site of intensive nutrient cycling, and this recycling process is probably more significant to the benthic organisms in Mirror Lake than their proximity to nutrients entering the lake for the first time. Some of the benthic blue-green algae in Mirror Lake fix nitrogen into biologically utilizable phases, but this process can only be estimated for the lake as a whole (Moeller and Roskoski 1978; *Chapter IV.E*).

Fee (1979) has found that the proportion of a lake's surface area that lies in contact with the summer epilimnion is an important parameter controlling the volumetric rate of primary productivity (g C/m^3-yr) in the epilimnetic plankton. Using data for several small lakes of the Canadian Shield, which are similar to Mirror Lake, Fee (1979) found that a regression of annual mean volumetric phytoplanktonic productivity (g C/m^3-yr) on the ratio A_e/V_e, where A_e was the area of lake bottom in contact with the summer epilimnion and V_e the volume of the epilimnion, accounted for 90% of the lake-to-lake variability in annual production. This simple relationship accounted for more of the variability than did regressions of primary production on annual loading of phosphorus, even though the lakes were known to be phosphorus limited during the summer. Fee's explanation was that the ratio A_e/V_e can be viewed as an approximation to the probability that a suspended epilimnetic particle (e.g. an alga) will settle onto the littoral sediment within the epilimnion, as opposed to sinking through the thermocline. Only a small portion of nutrients lost to the hypolimnion are likely to circulate back into the epilimnion, which includes most of the euphotic zone, during the summer growing season. Such evidence points to the importance of littoral benthic metabolism in nutrient cycling even in lakes like Mirror Lake, where the benthic vegetation itself is only a minor part of the total primary productivity.

During their annual cycle of growth and decay, rooted macrophytes may move nutrients like nitrogen and phosphorus from the sediment to attached communities and to the plankton (Barko and Smart 1981; Carpenter 1980). Such movement is a component of nutrient cycling within the lake; its significance should be evaluated in comparison with the sum total of cycling, rather than with the primary input of nutrients to the lake. When nutrient release is temporally displaced from sedimentation within the littoral zone (e.g. seasonally or at different stages of geologic infilling), the large rooted vegetation may become a net source of nutrients for other organisms. This nutrient pathway cannot be very large today in Mirror Lake, because the rooted macrophytes are not abundant.

The sparsity of the macrovegetation may reflect the absence of an accumulated sedimentary nutrient reserve from which to draw potentially limiting elements like nitrogen and phosphorus. Erosion of the drainage basin—that is, of the granitic bedrock and the till and soil derived from it—produces dissolved salts and some coarse sand, but little silt and clay. Consequently there is little chance for fine organic detritus to be buried within the littoral zone, or for regenerated ionic nutrients to be retained chemically in the substrate. The organic residue of biologic production is soon swept to the profundal gyttja, beyond the range of most benthic vegetation. Eventually during the lake's development, infilling will have removed the deep-water *sediment trap* that the profundal region represents; then sedimentation will accelerate near shore, and macrophytes may become more abundant.

Biological Structure

When the distribution of benthic organisms is plotted with respect to depth (*Figure V.A.4–2; Figure V.A.6–2* and *3; Figures V.A.7–2* and *3*; etc.), a gradient emerges that we assume reflects biologic interactions within the broad physico-chemical gradients discussed above. The littoral region contains a large part of this community gradient, including the transition from the communities of sandy bottom to those of the gyttja.

A conspicuous feature of the gradient within the littoral zone is the impoverishment of intermediate depths (3 to 5 m) in both macrophytes and macroinvertebrates. In part, the impoverishment in invertebrates reflects the habitat that large aquatic plants create for benthic invertebrates at shallower depths, although the unstudied distribution of benthic algae might contribute equally to the pattern. In part, the impoverishment may represent parallel responses of plants and animals to the physicochemical gradients that we have discussed. The depths of 3 to 5 m represent a zone that is suboptimal for the plants and animals characteristic of the 0- to 3-m zone and suboptimal for those characteristic of the gyttja. It is a transition zone lacking specially-adapted benthic community.

An understanding of the biologic structure that we glimpse through surveys of plants and animals can emerge only as precise answers are obtained to the following sort of questions: (1) To what extent do biologic interactions (e.g. grazing of plants by invertebrates, predation among invertebrates, predation of fish on invertebrates) contribute to the observed depth-distributions? (2) Which interactions control the relative abundances of the individual benthic species? (3) To what extent do nutrient limitations of benthic primary productivity determine the relatively low abundance of all trophic groups represented in the littoral region?

Strayer et al. (1981) point out that the low density of yellow perch in Mirror Lake may limit the abundance of the unionid clam, *Elliptio complanata*, which lives in sandy substrates of the littoral region. They suggest that the relative sparsity of adult clams (*Chapter V.A.6*) might reflect a limiting number of hosts for the larvae, which are parasitic on the gills of perch. Evidence for low adult mortality, and against the idea that adult growth might be stunted (e.g. by a deficiency in calcium or a shortage of food), suggested a population control at the level of reproduction or larval survival.

The benthic vegetation of Mirror Lake is sparse compared with that of many other lakes. Algal growth on rocks is low in terms of both biomass and productivity (*Chapters V.A.3* and *V.B.4*). The fact that respiration of the benthic community on littoral cobbles tends to exceed, over a 24-h period, the daily primary production, suggests that allochthonous and planktonic organic matter is a significant food source within the littoral region (*Chapter V.B. 4*). The macrophyte vegetation is less abundant than in some other oligotrophic lakes (*Chapter V.A.4*). Although there is some experimental evidence that at least one species (*Lobelia dortmanna*) is nutrient-limited in mid-summer, other factors may well contribute to the low overall density of littoral macrovegetation (*Chapter V.B.3*). Epiphytes on the macrophytes are apparently like those in Lake Kalgaard, Denmark—low in biomass and much less productive, per unit bottom area, than the plants they colonize (Søndergaard and Sand-Jensen 1978). It is easy to suppose that these benthic plants, like the pelagic phytoplankton (*Chapter V.B.2*), are strongly limited by the supply of phosphorus and nitrogen during the summer.

The abundance of benthic plants also is influenced by the intensity of grazing. Benthic algae are readily consumed by invertebrates. The abundance of benthic algae sometimes has been shown to be correlated inversely to the intensity of consumption by amphipods (Hargrave 1970) or crayfish (Flint and Goldman 1975). Submersed macrophytes likewise may be virtually absent when crayfish are abundant (Abrahamsson 1966; Dean 1969; Flint and Goldman 1975). In Mirror Lake neither macrophytes nor crayfish are abundant. Except for an aquatic caterpillar that causes appreciable mortality of one of the rosette species (*Chapter V.B.3;* Fiance and Moeller 1977), we have no clear impression of the significance of herbivores in controlling the density of large aquatic plants in Mirror Lake. Small invertebrates are relatively abundant in the littoral benthic region (*Chapter V.A.6 and 7*) and presumably have a significant impact on the abundance and dynamics of benthic algae. Moderate grazing may actually stimulate productivity of the benthic algal community, owing to nutrient release from the remains of ingested algae. Since the abundance of invertebrate grazers in turn must be affected by the intensity of predation upon them (especially by fish or salamanders), the abundance and productivity of benthic plants must be viewed as the outcome of an interaction with the entire littoral food web as well as the supply of nutrients.

Summary

The littoral region of Mirror Lake is a benthic region that extends from the rocky lake shore to a water depth of 7.2 m. It is sparsely vegetated with submersed macrophytes and benthic algae. In summer the littoral region is well illuminated (6 to 100% of the irradiance penetrating the lake surface), warm (seasonal maxima of 17 to 26°C) and well oxygenated (100 to 120% of saturation). Wind-driven currents prevent fine organic sediment from accumulating over the sand, gravel, cobbles and boulders of the shallower portion of the region. A fine-detritus gyttja predominates below 5 m. Nutrient regeneration within the littoral region helps to sustain planktonic productivity; planktonic production is partially respired along with benthic production and allochthonous organic matter in the littoral region.

The littoral region is biologically rich in species. It contains many species of (1) plants, only 23 of which are macrophytes, (2) invertebrates that feed directly on the benthic vegetation, (3) sedentary invertebrates that filter organic particles from the water-column, (4) detritivorous invertebrates feeding on or in the lake bottom, and (5) predatory vertebrates as well as invertebrates. The taxonomic diversity of both macroinvertebrates and microinvertebrates is greater in the littoral region than in the pelagic region. In agreement with Mirror Lake's classification as an unproductive, oligotrophic lake, littoral populations of large aquatic plants, benthic algae, macro- and microinvertebrates and predatory vertebrates (fish and salamanders) are relatively small, compared with those of eutrophic lakes. Heterotrophic metabolism within the littoral region is of a similar magnitude, on a unit area basis, to that within the profundal zone, despite the low productivity of benthic plants.

B. The Profundal Region

RHODA A. WALTER, ROBERT E. MOELLER, DAVID L. STRAYER and JONATHAN J. COLE

The profundal region lies below the littoral region, its upper boundary corresponding to the lower limit of macrophyte vegetation. This region is characterized by bare unvegetated mud that harbors a community of heterotrophic organisms.

In Mirror Lake, the profundal region extends from 7.2 to 11 m. The upper limit corresponds fairly well with the average upper limit of the hypolimnion. The sediment of the profundal region of Mirror Lake is much more homogeneous than that of the littoral region, and is composed almost entirely of gyttja—a very fine, highly organic mud (*Figure V.B.6–1*). To a SCUBA diver, the surface of the gyttja appears as a uniform flocculent layer, with no boulders or other objects rising through it.

Surface sediment temperatures are slightly (usually <0.5°C) cooler in summer and warmer in winter than the overlying water. Sediment temperatures are generally less than 10°C although they may reach 13°C or higher in late summer or just after autumnal overturn (*Chapters IV.B* and *C*).

During summer and winter stratification, the profundal region of Mirror Lake is overlain by relatively quiet water. Because of this stratification, the lack of appreciable photosynthesis, and the relatively large biochemical oxygen demand of the hypolimnion and the sediments, concentrations of oxygen gradually decrease throughout the stratification period (*Chapter IV.B*). By late in the stratification period (late summer and late winter), the surface sediments become anoxic, especially in the deepest part of the lake (deeper than 9 m). Oxygen concentrations 0.5 m above the sediment surface are often much higher than at the sediment-water interface, resulting in a steep diffusion gradient of dissolved oxygen just above the sediment surface (*Figure VI.B–1*).

Fish are absent in the profundal region of Mirror Lake during the summer, but may be present during autumnal overturn and during the early winter when the water is well oxygenated and often slightly warmer than water in the littoral regions. Although no macrophytes occur in the profundal region, benthic algae—particularly desmids and diatoms—are common in the upper profundal region. Mats of blue-green algae (*Oscillatoria* and related genera) also occur, especially in late summer. In fact, these mats have been observed as deep as 10.5 m (D. Strayer, unpublished data), where light is very low (<1% of radiation entering the surface of the lake) and oxygen is extremely low or ab-

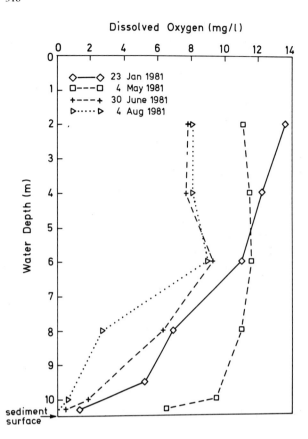

Figure VI.B-1. Dissolved oxygen profiles in a 10.5-m water-column in Mirror Lake, showing the steep diffusion gradient near the sediment surface at various seasons.

sent. Such mats must photosynthesize anaerobically, as Rich (1979) has postulated for algal mats growing under similar conditions in a softwater lake in Connecticut.

Macrofauna of the Profundal Region

Compared with the littoral region, species diversity of the benthic macrofauna (as well as microfauna; see *Chapter V.A.7*) is greatly reduced in the profundal region of Mirror Lake, probably owing to the lack of habitat diversity resulting from the homogeneous, unvegetated substrate, to the low year-round temperatures, and particularly, to the low oxygen concentrations that develop in the profundal region in the summer. The dominant benthic macroinvertebrates in the the profundal sediments are phantom midge larvae (*Chaoborus*), oligochaetes and two taxa of chironomid midge larvae—*Chironomus anthracinus* and Tanytarsini (*Figures V.A.6–2* and *3*). The Tanytarsini are restricted to the uppermost reaches of the profundal region. In the summer there is a marked decrease in abundance and biomass of the benthic macroinvertebrates with increasing depth within the profundal region, probably owing to the increasingly low oxygen concentrations. In winter, this distribution is reversed, with maximum numbers occurring in the deepest part of the lake. At all times of the year, the biomass in the profundal region is dominated by *C. anthracinus* (*Figures V.A.6*–2D and 4B). Except for the predatory midge larvae of the Tanypodinae and Chaoboridae, the benthic macrofauna in the profundal region are detritus-feeders. *Chaoborus*, *C. anthracinus* and oligochaetes are associated with environments deficient in oxygen (Mundie 1957) and are characteristic fauna of the profundal region of eutrophic lakes (Stahl 1959). Thus, because environmental conditions in the profundal region of Mirror Lake approach those of a eutrophic lake, the profundal fauna resembles a "eutrophic" fauna, in both composition and density.

B. The Profundal Region

Roles of Benthic Macrofauna

Since oligotrophic lakes are, by definition, unproductive, the abundance, biomass and productivity of their biota (fish, macrophytes, plankton and benthic invertebrates) are usually low in both the littoral and profundal regions when compared with eutrophic lakes. Although Mirror Lake exhibits low abundance, biomass and production for most biotic components—consistent with its classification as an oligotrophic lake—the benthic invertebrates are disproportionately abundant throughout the lake (see *Table VI.B*–1). This apparent anomaly must result at least in part from the eutrophic nature of the profundal region in Mirror Lake.

Table VI.B–1. Secondary Production of the Zoobenthos ("P"), Expressed as a Fraction of Net Organic Inputs (Autochthonous Primary Production + Allochthonous Inputs − Fluvial Outputs), for Several Lakes[a,b]

Lake	Mean Depth (m)	Net Organic Inputs (g C/m^2-yr)	"P"
Mirror, USA	5.8	49	12[i]
Tundra Pond, USA	~0.1	26[c]	7[g]
Myvatn, Iceland	2.3	330[c]	6[g]
Kiev Reservoir, USSR	4.0	280[e,f]	6[f,h]
Findley, USA	7.8	12	6[d]
Esrom, Denmark	12.3	160[c]	6[g]
Marion, Canada	2.3	110	3[d]
Paajarvi, Finland	14.4	60[e]	3[h]
Mikolajskie, Poland	11.0	260[e]	2[g]
Red, USSR	6.6	140[e,f]	1.4[g]
Dalnee, USSR	—	260[e,f]	1[f]
Naroch, USSR	9.0	160[c,e,f]	0.8[f,g]
Wingra, USA	2.4	610	0.4[d]
Rybinsk Reservoir, USSR	4.0	93[e,f]	0.3[f]

[a] Because of substantial methodological differences among the studies, the data are only approximately comparable.
[b] From Strayer 1984.
[c] Not including allochthonous inputs.
[d] "Aquatic insects" only.
[e] Converted from kcal m^2/yr by dividing by 10.
[f] Vegetative season only (May–October).
[g] Macrobenthos.
[h] Meio- + macrobenthos.
[i] All benthic metazoans.

Since the standing crop and possibly production of benthic macroinvertebrates in Mirror Lake is higher than would be expected, these organisms may play a more important role in the Mirror Lake system than in other more typical oligotrophic or even eutrophic lakes. Although the respiration of the benthic invertebrates is probably of minor importance in Mirror Lake (*Chapter V.B.6*), there are a number of other processes in which benthic macroinvertebrates play key roles. These roles are common to all lakes, although the magnitude depends on the particular lake in question. These roles are not restricted to the profundal region, and the following discussions generally apply to the entire benthic region of a lake.

Benthic macrofauna may serve a function in aeration and mixing of the sediment similar to that of terrestrial earthworms. Movements causing aeration and mixing may be more important than the actual respiration of benthic macrofauna in determining the rate of oxygen uptake in lake sediments (Edwards 1958; Neame 1975). Such movements are of three types (Kleckner 1967 as cited in Stockner and Lund 1970): (1) burrowing and tube-building (McCall et al. 1979; Rogaar 1980); (2) ingestion and egestion at different levels in the sediment. For example, oligochaetes feed beneath the sediment surface and deposit feces on the surface (Brinkhurst et al. 1969), while chironomids feed on the surface and deposit feces beneath the sediment surface (Jonasson 1972); (3) respiratory and feeding currents. Such currents bring oxygenated water into the sediment, thus increasing the depth of the oxidized layer (Edwards 1958; Davis 1974b; Hargrave 1975b).

In the profundal region of Mirror Lake, chironomids and oligochaetes generally inhabit the top 4 cm, while *Chaoborus* often penetrates to 12 cm. Experiments at several water depths in Mirror Lake show that mixing of sediment by benthic macroinvertebrates occurs to 2 to 4 cm over the course of one year (*Chapter VII.A.2*). Mixing appears to be deepest in the deepest regions of the lake (≥8-m water depth), where the sediment is the least consolidated. Close examination of the sediment in the profundal region of Mirror Lake reveals that it is composed almost entirely of feces, indicating that the entire sediment surface is being continuously reworked by benthic organisms. These feces are colonized inside and out by vast numbers of bacteria—10^9 bacteria per gram dry weight of mud (J. Cole, unpublished data).

Benthic macrofauna may play a large role in cycling of nutrients and energy within a lake. Activities of the benthic macrofauna within the sediments may cause the uptake of nutrients from the water-column or the release of nutrients to the water-column directly by way of feeding and respiratory currents (Wood 1975) or indirectly by exposing reduced sediments to the sediment-water interface (Rossolimo 1939). Tubificids can accelerate uptake of phosphorus by the sediments (Davis et al. 1975a; Gardner et al. 1981), while both tubificids and chironomids have been shown to release phosphorus from the sediments to the water-column (Gallepp et al. 1978; Gardner et al. 1981). Such nutrient release could be important, especially during spring and autumnal overturn when these nutrients are circulated throughout the water-column. Use of benthic invertebrates as food by benthic feeding fish—such as yellow perch, smallmouth bass and suckers—and by newts is another important mechanism for nutrient cycling, especially in the littoral regions.

In addition to contributing to energy flow and nutrient cycling, benthic macrofauna may act in a regulatory capacity for the benthic community. The benthic macrofauna probably function like the terrestrial soil macrofauna that break up detritus and make it more suitable for colonization by fungi and bacteria, and thus increase the rate of energy flow and nutrient release throughout the system. Mechanical disruption during the respiration and feeding activities of the benthic macrofauna must hasten release of dissolved organic carbon (DOC) from particulate algal detritus that has settled onto the sediment. Because this DOC is a major carbon source for the benthic bacteria (*Chapter V.D*), microbial metabolism may be partially regulated by activities of the benthic macrofauna. In addition, secretions of the salivary glands of benthic macroinvertebrates may stimulate bacterial growth (Rybak 1969). As would be expected, then, bacterial numbers are usually high in the surface sediments and begin to decline only below the macrofaunal burrowing layer (Hayes 1964). In addition to affecting bacterial numbers, the presence of macrofauna (at least amphipods) can cause an increase in algal production (Hargrave 1970). The importance of such interactions in Mirror Lake is not known.

Insect migration and emergence may be important mechanisms for "uphill movement" (*sensu* Leopold 1941; see *Chapter I*) of nutrients and energy from the profundal region to the epilimnion, especially during lake stratification, as well as to the surrounding watershed. Migration of *Chaoborus* and chironomid larvae throughout the water-column provides opportunity for cycling of nutrients by means of defecation, predation by fish or newts, or natural death. Migration of *Chaoborus* and chironomid larvae and pupae involves exchanges of energy as well as nutrients through respiration and feeding activities. In addition, migration causes nightly competitive interaction between the insect larvae and the zooplankton community. Thus, migration acts as a biologic link between the communities of the benthic environment and those of the water-column.

Cycling of nutrients to the water-column and to the surrounding watershed by means of insect emergence appears to be an important role of the benthic macrofauna in Mirror Lake. The amount of N and P lost annually from Mirror Lake by emerging insects (males only) is substantial (*Table V.B.6*–3), especially when compared with annual inputs of these nutrients to Mirror Lake (*Table IV.E*–12). For P, the amount lost as a result of emergence is approximately 18% of the total annual natural input (ignoring sewage) for the lake. In addition, an amount of N and P twice as large as that lost from the lake by way of emergence is recycled within the lake through dead pupae, dead female adults and pupal skins. Some of this matter is blown to the sides of the lake where it is eaten or readily decomposed by fungi and bacteria. The rest sinks through the water-column, where it is attacked by bacteria and fungi, and thus enters the life of the plankton once again. Since N and P are limiting nutrients for the phytoplankton in Mirror Lake (Gerhart 1973; *Chapter V.B.2*), these nutrients must have a rapid rate of cycling within the lake. It appears that the benthic macrofauna play an important role in this cycling.

B. The Profundal Region

Interactions Among the Benthic Macrofauna

At any given time, each major group of benthic macrofauna appears to reach its peak in abundance at a different depth in Mirror Lake (*Table VI.B*–2), except for the Sphaeriidae, which have a bimodal distribution. Relative positions of the different taxa in January remain the same as in July while the peaks shift downward. This spatial and temporal separation in the benthic environment resembles the vertical and temporal stratification in the plankton (see *Chapter VI.C*), and could be a result of evolution, physiology and competition. Such habitat or niche separation must reduce competition between benthic macroinvertebrates, resulting in greater diversity in the benthic region as a whole.

Predator-prey interactions may also affect species distributions. *Chaoborus* larvae apparently can feed on oligochaetes (Swuste et al. 1973) and may do so in the winter when zooplankton abundance is low (Jonasson 1972). The fact that oligochaetes have never been found at >10 m in Mirror Lake may be due to predation by *Chaoborus* larvae in winter, when *Chaoborus* is extremely abundant in this region, as well as to year-round predation by chironomids (Tanypodinae and Chironomini; Loden 1974) and *Chaoborus*. Growth of *Chaoborus* larvae between January and March supports this hypothesis (see *Chapter V.A.6*). Larvae of the Tanypodinae are important predators throughout the lake.

Population regulation may also occur from within a species—for example, *Chironomus anthracinus* eggs are eaten by larvae of that species in Lake Esrom, eliminating overlapping generations (Jonasson 1972).

Interactions Between Benthic Macrofauna and Vertebrate Predators

Fish are usually the most important predators on benthic macroinvertebrates in a lake. Yellow perch is the main predator in Mirror Lake owing to its benthic feeding habits and its large numbers in the lake. Amphipods are an important source of food for all of the benthic-feeding fish in the upper littoral region, as well as for the newt, an important predator in <2 m of water. The most intense predation on benthic insects occurs when larvae and pupae migrate through the pelagic region or emerge from it; at these times not only the benthic-feeding fish, but also pelagic fish such as smallmouth bass feed extensively on benthic invertebrates. Some organisms such as *Chironomus anthracinus* and the mayfly *Hexagenia* are utilized extensively as food *only* when they are emerging (Mazsa 1973).

Despite the fairly small populations of fish in Mirror Lake, their annual consumption—20 to 40% of production of zooplankton and benthic macroinvertebrates (*Chapter V.B.7*)—is high enough that they must exert substantial predation pressure on the benthic macroinvertebrates. During the summer, fish predation in Mirror Lake is restricted to the littoral region above the thermocline, and predation on the profundal benthic macrofauna is negligible except during migration and emergence. During autumnal overturn, however, predation in the profundal region is probably high, as has been shown for Lake Esrom (Jonasson 1972), because temperature and oxygen conditions no longer prevent fish from utilizing this region.

Table VI.B–2. Water-depth Interval Where Each Taxon Reaches its Peak Abundance in Mirror Lake in Summer and Winter

Water Depth (m)	July	January
0–1	Sphaeriidae	—[a]
1–2	Amphipoda	—
2–3	Tanypodinae, *Elliptio*	—
3–4	None	—
4–5	None	—
5–6	Sphaeriidae	—
6–7	Tanytarsini	Sphaeriidae
7–8	Oligochaeta	None
8–9	*Chironomus anthracinus*	Oligochaeta
9–10	*Chaoborus*	*Chironomus anthracinus*
>10	None	*Chaoborus*

[a] — no information available.

Autumn and winter mortality of benthic macroinvertebrates as a result of fish predation in Mirror Lake has not been documented, but it may be substantial; in other lakes yellow perch are known to feed on benthic organisms of the profundal region throughout the winter (Pearse and Achtenberg 1918). Fish predation may be responsible for the large mortality of *C. anthracinus* between January and March in Mirror Lake (see *Chapter V.A.6*).

Summary

The profundal region of Mirror Lake occupies the deeper portion of the lake bottom (7.2 to 11 m). It is characterized by the absence of macrophytes, a sediment composed of gyttja, which is often anoxic during the latter part of the summer stratification period, a benthic microflora some of which may be photosynthesizing anaerobically and a benthic fauna of low diversity. The dominant macrofauna in this region—*Chaoborus, Chironomus anthracinus* and oligochaetes—are characteristic fauna of eutrophic lakes.

Most of the interactions occurring in the profundal region of a lake also occur in the littoral regions. The benthic macrofauna of all lakes play roles in aeration and mixing of the sediment, exchange of nutrients across the sediment-water interface, cycling of nutrients and energy within the lake as a result of migration, predation by fish and insect emergence, removal of nutrients and energy from the lake through insect emergence, and regulation of other components of the benthic community (such as bacterial and algal production). Macroinvertebrates may play a more important role in the functioning of the Mirror Lake ecosystem than they do in other more typical oligotrophic or eutrophic lakes. The benthic macrofauna appear to be important in the cycling of nitrogen and phosphorus within the lake and to its surrounding watershed.

Throughout the benthic regions of Mirror Lake, the niches of the dominant benthic invertebrates are separated by depth and time. Competition and predation occur between the various taxa of benthic macrofauna. Fish are important predators on the benthic invertebrates, especially in the littoral regions and during insect migration and emergence from the profundal region.

C. The Pelagic Region

JOSEPH C. MAKAREWICZ, GENE E. LIKENS and MARILYN J. JORDAN

The pelagic or open water portion of a lake, historically, has received the most attention from limnologists. The pelagic region is the environmental integrator of the lake, as the open water is often the first to respond to changes in the external environment, particularly in a small lake, and as a fluid, solvent and catalyst, water physically links all other components of the aquatic ecosystem. Organisms within the water-column are inseparable from the water of their physical environment. Odum (1962) pointed out this obvious, but very important structural difference between a pelagic ecosystem and a terrestrial (or benthic) ecosystem. That is, the structure and ecologic classification of a terrestrial ecosystem are determined by the biota (usually vegetation), whereas the pelagic ecosystem is dominated by water. If all of the trees were removed from a forest, it no longer would be (or look like) a forest, but if all the plankton were removed from a lake it would still *look* like a lake from the shoreline. Conversely, if all the water were removed from a lake, it would no longer be a lake. Thus, the ecologic interactions occurring in the pelagic region are of utmost importance to the metabolism, biogeochemistry and development of a lake-ecosystem.

During most of the year the pelagic region consists essentially of two "lakes" of open water, the epilimnion and the hypolimnion as separated by the thermocline. For a short period in the autumn (usually late October and November, *Figure IV.C–1*) and for an even shorter and

C. The Pelagic Region

inconsistent period in the spring, the entire pelagic region exists as one well-mixed water body. In winter the pelagic region is covered by ice and snow some 40 to 75 cm thick, which reduces light penetration and the effects of wind. An inverse thermal stratification is typical of the water-column during winter.

The epilimnion of Mirror Lake is well lighted, warm in summer and cold in winter, low in nutrients, well oxygenated and well mixed. Large variations in these parameters are common throughout the year. For example, the temperature at the surface of the lake ranges from <0°C when there is an ice-cover to about 28°C on exceptionally hot, calm days. The epilimnion reaches a maximum thickness of 6 to 7 m during the summer (*Figure IV.C–1*).

In contrast, the hypolimnion is relatively lower in light intensity, colder and richer in nutrients, particularly during late summer and early autumn before overturn; it has lower concentrations of dissolved oxygen (often anaerobic in the deepest parts) and weak mixing, with horizontal velocities exceeding the vertical ones. There commonly is a much smaller amplitude of change in some environmental conditions, such as temperature or light, in the hypolimnion. For example, the temperature of water in the hypolimnion varies only from about 4°C to 13°C during the course of a year. In contrast, dissolved oxygen varies from zero to 100% of saturation during the year and thus is much more variable than in the epilimnion.

A major problem for organisms in the pelagic region is to remain suspended. Most aquatic organisms have a density that is slightly greater than water, and so tend to sink. The reduction in size and development of spines, bristles, etc. increase the surface area to volume ratio and retard sinking rates in plankton. Inclusion of lighter substances, such as fats and oils, within the protoplasm also slows sinking. However, as stressed by Hutchinson (1967, 282) convective motion is a primary means for a weakly swimming, constantly sinking organism to maintain a favorable position within the upper trophogenic zone of a lake. At surfaces, such as the mud-water interface, organisms are provided with a relatively stable site for attachment and an increased source of nutrients. Even at the air-water interface, the relatively high surface tension of water provides a favorable habitat for attachment, suspension or support, as well as the first opportunity to utilize nutrients or dissolved organic matter entering the lake in direct precipitation. Indeed, the pleustonic, including neuston and benthic communities, are adapted to utilize these special interfaces of the pelagic region.

The ecologic interactions between the bacterioplankton, phytoplankton, zooplankton, fish and salamanders are related to or governed by these environmental variables. Community structure and dynamics contribute to the structure and function of the entire ecosystem.

Interactions Between Bacteria and Phytoplankton

Interactions between bacterioplankton and phytoplankton are potentially of major trophic importance but have received little study in either Mirror Lake or other fresh waters (Cole 1982a). Phytoplankton may lose DOC from photosynthetically-fixed carbon, as measured in short-term incubations with ^{14}C (Wiebe and Smith 1977; Nalewajko and Schindler 1976). This exudated DOC is rapidly metabolized by bacteria (Wiebe and Smith 1977; Derenbach and Williams 1974; Wright 1975; Cole and Likens 1979). Thus, algal losses of DOC could provide a major part of the organic C substrate available to planktonic bacteria (Cole et al. 1982). In Mirror Lake algal losses of DOC averaged 18.5% of measured ^{14}C uptake, or 8.7 g C/m^2-yr, which could have provided 50 to 100% of the C metabolized by the planktonic bacteria (Jordan and Likens 1980).

The surfaces of viable, growing algae seem to be free of attached bacteria (Rheinheimer 1974; Droop and Elson 1966). However, moribund and dead phytoplankton are presumably quickly colonized and decomposed by bacteria. In incubated Mirror Lake water, 25 to 30% of ^{14}C-labeled phytoplankton detritus was mineralized to CO_2 by bacteria within 500 h (Cole and Likens 1979). In Lake Bysjon, Sweden, peak bacterial numbers occurred immediately following the collapse of algal blooms, presumably because of the sudden input of decomposable substrate (Coveney et al. 1977).

Bacteria could potentially have strong negative effects on algae because of competition for nutrients, particularly N and P. Rhee (1972) found that the growth of the alga, *Scenedesmus* sp., in laboratory cultures was severely limited in the presence of the bacterium *Pseudomonas* sp., whereas the growth of the bacteria was unaffected by the algae. The inhibition of algal growth was due to the more rapid uptake of P by the bacteria and was not due to bacterial metabolites or autolysis products. Under most natural conditions aquatic bacteria are probably limited by organic C, not by N or P, and this fact plus the ability of algae to store P enhances the ability of algae to compete with bacteria for P (Rhee 1972). In water bodies receiving large amounts of allochthonous organic C, the resultant high bacterial growth could deplete the amount of N and P available to the algae. This interaction has been demonstrated for Mirror Lake by B. Peterson (unpublished data). When bottles of Mirror Lake water enriched with 140 μg/liter N and 30 μg/liter P (plus $NaHCO_3$ to prevent CO_2 depletion) were incubated *in situ* for four days, concentrations of chlorophyll and particulate organic C increased two- to threefold. In bottles to which 4 mg/liter organic C (glucose, peptone and yeast extract) was also added, the chlorophyll concentration increased only 36%. When 20 mg/liter of organic C were added, the chlorophyll concentration actually decreased by 19%. The concentration of particulate organic C in the organic-enriched bottles increased two- to sevenfold because of the abundant growth of heterotrophs, including fungi.

Other more subtle interactions as yet unknown may also occur between algae and bacteria. When various species of axenic laboratory-cultured algae were grown with and without added bacteria, some were stimulated and others were inhibited or unaffected (Delucca and McCracken 1977). Wiebe and Smith (1977) postulated the existence of "interspecific regulatory control loops" among plankton since the rates of DOC production and utilization were much greater for isolated size fractions of sea water than for unfractionated sea water. Since the response they observed could have been an artifact of the size-fractionation process, the existence of feedback loops remains an interesting, but hypothetical possibility.

Interactions Between Phytoplankton

Phytoplankton represent 2.5% (*Table VI.C–1*) of the total biomass (as organic carbon) in the pelagic region of Mirror Lake. Species diversity is high (>230 species) with the following groups represented: Cyanophyceae (23 species), Chlorophyceae (82), Euglenophyceae (10), Chrysophyceae (57), Diatomeae (30), Cryptophyceae (18) and the Peridineae (18) (see *Table V.A.2–2*). By contrast, tropical Lake Lanao in the Philippines, a much larger lake (357 km^2), has only 70 pelagic species (Lewis 1979). The Chrysophyceae and Cryptophyceae are poorly represented in Lake Lanao.

A distinct periodicity in phytoplankton biomass is observed throughout the year (*Figures V.A.2–1* and *V.B.5–3*). Biomass is small in the winter because of reduced algal growth during this period of low temperature and light, then rises dramatically with increased light conditions and builds to a spring maximum, determined by the amount of nitrogen and phosphorus (*Chapter V.B.2*) in Mirror Lake. Generally, in oligotrophic lakes the algal biomass in summer is often low, followed by a second maximum in autumn (Wetzel 1975). This pattern does not occur in oligotrophic Mirror Lake. After the exponential increase in spring biomass, the algal biomass continues at a high level and may even rise in some years until August. A progressive decline in biomass occurs through autumn and into winter with a minimum in March and April. The seasonal pattern of pelagic phytoplankton in Mirror Lake resem-

Table VI.C–1. Percent of Total Biomass (as Organic Carbon) in the Pelagic Region of Mirror Lake, New Hampshire[a]

	% of Total Pelagic Biomass
Phytoplankton	2.3
Zooplankton	1.2
Bacteria	0.5
Fish	1.6
Salamanders	<0.1
Particulate Detritus	8.1
Dissolved Organic Carbon	86.2

[a] From *Chapter V.C*.

bles that of high altitude lakes, which have a conspicuous single summer bloom of algae (Wetzel 1975).

Within the seasonal pattern, the species composition changes in a regular manner each year (*Figure VI.C–1*). That is, there is a replacement or succession of species and groups of species with time. In spring, the Cryptophyceae are dominant and are succeeded by the Chrysophyceae and Diatomeae. By July the blue-green alga (Cyanophyceae) are dominant, with the Peridineae dominant in August. This dominance of the Peridineae in 1969 is caused by a bloom of a large species of *Gymnodinium* spp. In other summers, this species was not as prevalent. With cooling of the water in the autumn, the Cryptophyceae reappear and are followed, as in the spring, by the Chrysophyceae and Diatomeae.

This succession of phytoplankton species is controlled by physical (e.g. light, temperature, turbulence), chemical (e.g. availability of nutrients, growth promoting substances, vitamins, antibiotics) and biologic factors (e.g. uptake kinetics, algal species, parasitism, predation and competition) (Golterman 1975).

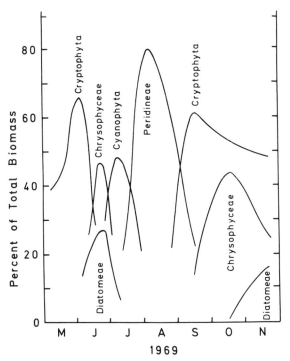

Figure VI.C–1. Succession of phytoplankton in Mirror Lake.

Some research on seasonal periodicity and succession of *Dinobryon* has been done in Mirror Lake (Lehman 1976). At certain times, *D. divergens* dominates the net phytoplankton of Mirror Lake, but not during the vernal excess of nutrients. Lehman (1976) postulates that *Dinobryon* simply cannot keep pace with other more rapidly growing species during this period. When the bloom organisms reduce nutrient concentrations to a low level, *Dinobryon*, because of its more effective mechanism of phosphate uptake at low ambient levels, has a competitive advantage over other algae with a resulting increase in abundance of *Dinobryon*. The length of time that *Dinobryon* is abundant will be influenced by environmental factors such as occurrence of high temperatures, high concentrations of potassium (which may be toxic), continued depletion of nutrients (Lehman 1976) and grazing pressure from zooplankton (Tappa 1965).

Other examples of the subtle interactions that can occur between phytoplankton are provided by the blue-green algae. Some species of blue-green algae release metabolites, e.g. allelopathic substances, that inhibit growth of diatoms (Keating 1977, 1978). Once a critical, effective concentration is reached, allelopathy provides a mechanism by which blue-green algae may eventually predominate in a lake.

Schindler (1977) and Fee (1979) have suggested that low N:P ratios are conducive to growth of nitrogen-fixing, blue-green algae. Often by late summer, epilimnetic concentrations of inorganic nitrogen in oligotrophic lakes have decreased because of uptake by various algal populations. Blue-green algae capable of fixing atmospheric nitrogen then have a competitive advantage over nonfixers and may increase in abundance.

A blue-green algal requirement for sodium, first shown by Emerson and Lewis (1942), is well established in the literature with an optimum growth rate at 5 mg/liter (Fogg et al. 1973). Evidence is accumulating that growth of blue-green algae can be stimulated by increased levels of sodium at levels (2 to 10 mg Na/liter) commonly observed in lakes and may provide another competitive advantage (Baybutt and Makarewicz 1981). Sodium, as well as K^+, Ca^{2+}, and Mg^{2+}, may act as catalysts that stimulate phosphate uptake and polyphosphate body formation in some species of Chlorophyceae and Cyanophyceae (Shieh and Barber 1971; Jensen and Sicko-Goad 1976; Mohleji and Verhoff 1980; Rhee 1972; Peverly and Adamec 1978; Lawry and Jensen 1979). Although blue-green algae are not currently abundant in Mirror Lake, sodium loading is increasing from the northeast tributary because of salt usage during winter on a nearby interstate highway (*Chapters IV.E and IX*). Concentrations of sodium in the northeast tributary are currently ~12 mg/liter, whereas only nine years ago they were ~2 mg/liter (*Figure IX–3*).

Interactions Between Zooplankton, Phytoplankton and Bacteria

Zooplankton represent 1.2% of the total pelagic biomass in Mirror Lake (*Table VI.C–1*). Zooplankton production peaks in the spring and continues to be high through autumn followed by a decrease to a winter minimum (*Figure V.B.5–3*). This seasonal pattern of production of zooplankton closely follows the seasonal pattern of phytoplankton biomass, their predominant source of food. Abundance of zooplankton does not always correlate with abundance of phytoplankton, and sometimes zooplankton are abundant when phytoplankton are nearly absent (Moss 1980). Because zooplankton can assimilate organic detritus and bacteria, they are partly buffered from low availability of algae as food. These other linkages in the food web of lake-ecosystems are important, as algal productivity is often inadequate to support the herbivorous zooplankton community (Nauwerck 1963).

The maximum size particle that can be ingested by zooplankton is directly related to body size, but the minimum size particle that can be efficiently filtered has not been determined (Hall et al. 1976). Therefore, the significance of natural bacteria as a food source for zooplankton is largely unknown. *Daphnia* and *Ceriodaphnia* (Cladocera) can efficiently filter cultured bacteria (McMahon and Rigler 1965; Gophen et al. 1974; Lampert 1974; Haney 1973). In contrast, *Diaptomus gracilis* and *D. graciloides* (Copepoda) could not filter dispersed bacteria (Malovitskaya and Sorokin 1961). Natural bacteria (0.05 to 0.10 μm^3) are much smaller than cultured bacteria (1 μm^3) and, thus, may be less efficiently filtered. Using direct count techniques, Peterson et al. (1978) found that several species of *Daphnia* filtered natural bacteria at about 30% of the rate for yeast (3.5 μm). Fenchel (1975) used direct counting to measure the rate of disappearance of bacteria from feeding vacuoles in small benthic protozoans and found that zooflagellates consumed natural bacteria almost exclusively. The few other studies demonstrating zooplankton grazing on natural bacteria used ^{14}C-labeled plankton (Saunders 1969, 1972; T. Colton, unpublished data) and were inconclusive because algae also may have been labeled, and some filtered bacteria may have been attached to particles.

Attempts to measure grazing of bacteria by zooplankton have just begun in Mirror Lake. Cladocera and rotifers account for 81% of the zooplankton production in Mirror Lake and were more likely to graze bacteria than were copepods, which account for the remaining 19%. The average total flux of organic C through the Cladocera and rotifers, including respiration, was about 11 g C/m²-yr (0.81 × [2 + 12 g C/m²-yr]). If the Cladocera and rotifers grazed the seston indiscriminately, bacteria would comprise 5% of this C flux, or 0.55 g C/m²-yr, because bacteria accounted for 5% of the potentially available sestonic C (*Table V.C–1*). The Cladocera may, in fact, graze free-living bacteria less efficiently than larger particles, as suggested by the work of Peterson et al. (1978). The selectivity of the rotifers is unknown. Although the rate of grazing of bacteria by the zooplankton community in Mirror Lake is not known, it is probably significant. The average production by planktonic bacteria in Mirror Lake was probably 3 to 8 g C/m²-yr (Jordan and Likens 1980) of which only 2.4 g C/m²-yr could be accounted for by sinking or losses in drainage waters. The remaining loss of C was likely due to grazing by zooplankton and protozoa.

Ciliate protozoans may also serve as an important link between zooplankton and bacteria in the freshwater food chain. In nature, ciliate protozoans ingest bacteria (Fenchel 1968) and themselves are subject to zooplankton predation (Porter et al. 1979). This link by which dissolved organic matter, bacteria and detritus enter the planktonic food web is most likely to be important during the declining phase of algal blooms, near the metalimnion, at chemoclines and above the sediments (Sorokin and Paveljeva 1972), when bacterial populations are abundant enough to sustain ciliate populations. Unfortunately, this aspect of the pelagic food web is poorly known.

Generally, rotifers are thought to feed on small particles (Edmondson 1965) with a general affinity for flagellates (Lewis 1979). Small planktonic rotifers, though, are known to capture relatively large flagellates (Pourriot 1965). The planktonic rotifers feed largely by sedimenting seston particles into their mouth by the action of their coronal cilia (Pourriot 1965). However, research by Gilbert and Starkweather (1977) and Starkweather and Gilbert (1977) indicates that rotifers can also use different feeding mechanisms (e.g. seizing, rupturing and ingestion of particulate parts) to consume food particles of various sizes and shapes.

Bacteria alone can be used as food by some planktonic rotifers, but the demographic indices for laboratory populations fed with bacteria are lower than those of populations fed with algae (Pourriot 1977). Although the abundance and duration of occurrence of several detritus-eating rotifers increase with the abundance of detritus and bacteria (Hillbricht 1961; Hillbricht-Ilkowska 1964), there is not much information about the dietary importance of detritus in rotifers in relation to particle size or origin in freshwater habitats (Pourriot 1977).

Zooplankton can change the species composition of the phytoplankton community and influence seasonal periodicity by differential grazing. Porter (1973) enclosed and sealed lake water in large polyethylene bags and resuspended the bags in Fuller Pond, Kent, Connecticut for several days. In some bags the larger crustaceans were experimentally removed, while in other bags the zooplankton populations were increased severalfold by addition of plankton previously caught with a net. The major effect of grazers, which included filter-feeding and raptorial crustacean herbivores, was to decrease populations of cryptomonads, certain diatoms and nannoplankton, increase gelatinous green algae populations and not have an effect on the blue-green alga, *Anabaena*.

Large-sized species of phytoplankton are not completely immune from grazing. A selective predation of large prey by smaller predators, such as protozoa, is possible. Protozoa can move into colonies of large green algae and feed on individual cells with pseudopodia or suction organelles (Canter and Lund 1968). In tropical Lake George, blue-green algae, which are generally considered to be infrequently grazed, are fed upon and digested by a raptorial copepod (Moriarty et al. 1973).

Interactions Between Fish and Zooplankton

Only five species of fish are found in Mirror Lake (*Chapter V.A.9*). The effect of fish predation on zooplankton species composition has been demonstrated in the U.S. by Brooks and Dodson (1965) and others. With introduction of the obligate, planktivorous alewife into a Connecticut lake, the zooplankton community was changed from one of large zooplankton (*Epischura, Daphnia* and *Mesocyclops*) generally greater than 1 mm in size to ones (*Ceriodaphnia, Tropocyclops* and *Bosmina*) 1 mm or smaller (Brooks and Dodson 1965).

Yellow perch in Mirror Lake are not obligate planktivores. They eat large zooplankters but will shift to alternate food sources. However, in both the obligate and facultative planktivores, size selection of prey is characteristic (Werner and Hall 1974).

If planktivorous fish are present, size selection of zooplankton may predominate. In this situation, large zooplankton dominate and smaller zooplankter decrease in number. Predominance of larger zooplankton is related to size-selective predation of the smaller plankton by carnivorous zooplankton (Dodson 1974). For example, larger copepods and the midge larvae, *Chaoborus,* can inflict significant mortalities on smaller zooplankton (e.g. McQueen 1969; Confer 1971).

Vertical Movements: Sedimentation and Vertical Migration

Because the density of most freshwater plankton is greater than that of water, the organism will sink in undisturbed water unless the organism actively maintains itself by swimming.

There is a continuous "rain" of living, dying and dead matter (zooplankton, phytoplankton, bacteria, fecal matter, etc.) from the pelagic that is deposited at the profundal. In Mirror Lake this average sedimentation rate is 7.6 g C/m^2-yr, which is 7.2% of the annual organic carbon flux. As the dead material is sinking, bacteria invade it, forming detritus. Of the particulate organic matter produced in the trophogenic zone, 75 to 95% is decomposed by bacteria by the time it reaches the sediment surface (Wetzel 1975). Remineralization of nutrients allows uptake by bacteria and phytoplankton, recycling nutrients and energy into the food web. Deposition of undecayed blue-green algae and diatoms can be fed upon by *Chironomus anthracinus,* a species common in the benthic regions of Mirror Lake, and aid in growth (Jonasson 1972).

With the crustacean zooplankton, large upward vertical movements in the water-column occur with darkness. Light is generally agreed to be the stimulus for this behavior. Because of the massive movement of a large component of the zooplankton community, a marked increase in grazing pressure on the algal and bacterial community of the epilimnion should occur. Similarly, *Chaoborus,* which also migrates from the sediments toward the surface with darkness, grazes on zooplankton, perhaps preferentially on cyclopoid copepods (Smyly 1972). *Chaoborus* themselves become victims to fish on this nightly movement. From an ecosystem point of view, vertical migration and sedimentation provide a link between the epilimnion and hypolimnion and between the pelagic and benthic habitat by which nutrients and energy are cycled.

Nutrient Cycling

Nutrient regeneration is the release of soluble organic or inorganic nutrients by or from organisms or their remains (Johannes 1968). At least three mechanisms operate in the pelagic region to return nutrients to the available nutrient pool: (1) direct release by algae; (2) excretion by zooplankton; and (3) enzymatic hydrolysis or organic compounds excreted or produced by autolysis of dead plankton (Rigler 1973). For example, the return of phosphorus to solution may be largely due to phytoplankton excretion (Lean and Nalewajko 1976; Lean and Charlton 1976).

We have estimated the regeneration of phosphorus by zooplankton in Mirror Lake using the extensive data on zooplankton (*Chapter V.A.5*) and Peters and Rigler's (1973) model, which predicts P regeneration rates of zooplankton. Peters and Rigler's model considers water temperature, food concentration, phosphorus content of the phytoplankton and biomass of zooplankton. All parameters were entered as monthly means for appropriate water depths.

Excretion of P by zooplankton in Mirror Lake ranged from 102 in March to 471 mg P/m^2-month in June. In oligotrophic Lake George, New York, LaRow and McNaught (1978), using excretion values obtained by LaRow et al. (1975) observed an excretion range of soluble reactive P of 0.33 to 1.29 mg P/m^3-day for the crustacean community between June and November. If it were assumed that 90% of the P released in the Peters and Rigler model is orthophosphate (Peters 1972), then the excretion range for the crustacean community of oligotrophic Mirror Lake for the same period is similar to that in Lake George (0.58 to 1.66 mg P/m^3-day). During August in the eutrophic Bay of Quinte, Lake Ontario (Johnson and Owen 1971). Peters (1975) calculated a P-excretion range of 30 to 69 mg P/m^2-day by the multiple regression model and experimentally determined the rate to be 77 mg P/m^2-day for the crustacean community. In oligotrophic Mirror Lake in August, excretion of the crustacean community was considerably lower (8.5 mg P/m^2-day). In eutrophic Lake 227 and oligotrophic Lake 302 in the ELA of Canada, crustacean community excretion (7.0 and 2.8 mg P/m^2-day, respectively; Peters 1975) was also similar to excretion by natural assemblages of zooplankton in Mirror Lake.

Lehman (1980b) suggested that P excretion, as predicted by the Peters and Rigler model for natural assemblages (Peters 1975), is high compared with his values for Lake Washington. Indeed in Mirror Lake, excretion of natural assemblages in September (9.2 mg P/m^2-day) is higher than that determined by Lehman (1980a) for the same month (4 mg P/m^2-day). Yet considerable seasonal and yearly variation in excretion occurs in Mirror Lake, with some values lower than those observed by Lehman. For

C. The Pelagic Region

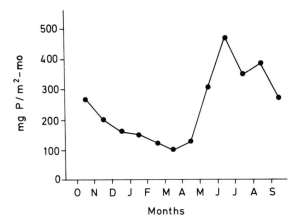

Figure VI.C–2. Zooplankton excretion of phosphorus in Mirror Lake.

example, the mean monthly excretion rate for March is 3.2 mg P/m²-day with values as low as 1.7 (March) and 2.3 (April) mg P/m²-day for individual years. Furthermore, Lehman (1980a,b) did not consider rotifers, although he believed they made important contributions to the total excretion rate of the zooplankton community. This omission strongly suggests that his excretion values may be low. See Peters and Rigler (1973), Bishop and Barlow (1975) and Lehman (1980b) for reviews of this subject.

A seasonal cycle in excretion of P from the Mirror Lake zooplankton community is evident with a peak in June and a minimum in late winter during March (*Figure VI.C–2*). Lehman (1980a) suggests a similar cycle for Lake Washington but provides no seasonal data to corroborate his suggestion. Seasonally, an inverse relationship in excretion rates exists between the rotifers and the copepods (*Figure VI.C–3*). Rotifers predominate during the winter, with values reaching as high as 61.2% of the P excreted by the total zooplankton community in January. A partial explanation for the decrease in P regeneration by copepods during the winter may be provided by a life history phenomenon (i.e. *Mesocyclops edax* overwinters in the sediments in Mirror Lake; Makarewicz and Likens 1979). During the spring and summer, the copepods become predominant reaching a high of 57% of the P excreted by the zooplankton community in April and May. The cladocerans never accounted for more than 16% of the P regenerated by the zooplankton community.

On average, copepods accounted for the largest percentage (48%) of the P recycled (*Table VI.C–2*), whereas the cladocerans contributed the smallest (12%). This observation was surprising since the cladocerans were the dominant group within Mirror Lake in regard to production (Makarewicz and Likens 1979). However, LaRow and McNaught (1978) observed the same relationship for copepods and cladocerans in Lake George. Functionally within Mirror Lake, copepods account for only 17% of the zooplankton production (Makarewicz and Likens 1979) but contribute 48% of the P ex-

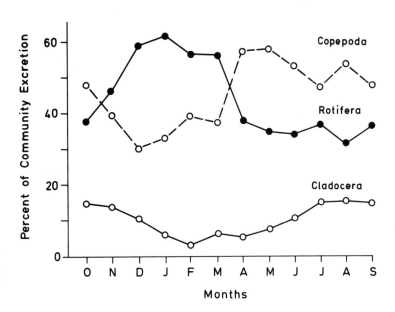

Figure VI.C–3. Seasonal contribution of Copepoda, Cladocera and Rotifera to the zooplankton community excretion of phosphorus.

creted by the zooplankton community. Cladocerans, on the other hand, have a high productivity (41% of the total zooplankton community production; Makarewicz and Likens 1979) and thus are potentially important as a food source to higher trophic levels. However, they apparently are not major contributors to nutrient cycling within the lake.

Perhaps most surprising is the observation that the rotifers account for over 40% of the P excreted by zooplankton in Mirror Lake (*Table VI.C-2*). This observation should be viewed with some caution, however, since Peters and Rigler (1973) did not consider rotifers in their experiments or in their construction of the multiple regression model. However, the results of Taylor and Lean (1981) on microzooplankton indicate that regeneration of P is not appreciably different for the microzooplankton than for the larger crustacean zooplankton.

The question arises as to just how important zooplankton cycling of P is, compared with allochthonous inputs to the lake-ecosystem. Lehman (1980a,b) observed that nutrient release by crustaceans was an order of magnitude higher than summer allochthonous inputs to Lake Washington. Similarly in Mirror Lake, zooplankton excretion of P was more than an order of magnitude higher than allochthonous inputs for each season of the year (*Table VI.C-3*). An interesting aspect of nutrient cycling by zooplankton is that allochthonous inputs are high during the winter and cycling of P by zooplankton is low; during the spring, allochthonous inputs are low and cycling is high. Thus during the spring, the availability of a limiting nutrient in Mirror Lake (Gerhart 1975) may be maintained to phytoplankton by zooplankton regeneration of P. The annual P requirement of Mirror Lake phytoplankton lies in the 0.4 (based on a 90:1 ratio for Mirror Lake seston; see *Chapter IV.B*) to 0.9 g P/m^2-yr range [based on 40:1 ratio (Vallentyne 1974)],

Table VI.C-2. Mean Annual Excretion of P by Copepoda, Cladocera and Rotifera in Mirror Lake

Zooplankton Excretion	mg P/m^2-yr	% of Total
Copepoda	1,394	48
Cladocera	353	12
Rotifera	1,174	40

Table VI.C-3. Seasonal Allochthonous Inputs and Zooplankton Excretion of P in Mirror Lake[a]

	Allochthonous Inputs	Zooplankton Excretion
	mg P/m^2	mg P/m^2
Spring	25	898
Summer	47	1,008
Autumn	20	628
Winter	41	383
Total	133	2,921

[a] Allochthonous inputs are from Likens and Loucks (1978).

which is higher than allochthonous inputs of P. Regeneration of P by some other source, such as the zooplankton community or the sediments, is required to maintain the primary productivity of the lake-ecosystem.

Dead limnetic phytoplankton can release 35 to 60% of their P in a single day (Golterman 1964, 1973). Zooplankton also are thought to release P rapidly to the water after death (Marshall and Orr 1961). Annually in Mirror Lake 72 mg P/m^2-yr (Makarewicz and Likens 1979), a value in the same order of magnitude as natural allochthonous inputs (*Table IV.E-16*), is potentially available to phytoplankton through remineralization of decomposing zooplankton tissue.

Zooplankton Community Dynamics (Niche Structure)

An organism exists in a matrix influenced by the surrounding chemical, physical and biologic variables that we define as the environment. Yet each species has its own position in a community in relation to other species—its niche. In Mirror Lake a vertical tow with a zooplankton net during summer stratification could yield up to 23 species of zooplankton. According to the competitive exclusion principle (Hardin 1960), no two species can coexist in the same niche; that is, one species would be successful and prosper, the other would become extinct. In a forest or a stream, the heterogeneity of the

C. The Pelagic Region

environment (e.g. shrubs vs. trees, riffles vs. pool, etc.) is obvious. Often, niche differentiation of species in forest or stream habitats is related to adaptations of organisms to this heterogeneous environment. In a lake-ecosystem, however, the environment appears homogeneous to the casual observer. Why then does one species not outcompete and eliminate the others in Mirror Lake by competitive exclusion? If we assume the competitive exclusion principle is a viable theory, the obvious answer is that each species must have a different niche.

A species' niche may be defined by a number of biologic (e.g. food size), chemical (e.g. oxygen concentration) and physical (e.g. temperature) variables. A multidimensional space often called the niche hyperspace (Hutchinson 1958) is delineated when these variables are graphed as axes. Along each axis, values range from low to high and can be thought of as environmental gradients. Each species is adapted to a portion of this gradient. The upper and lower boundaries of all the environmental variables for a given species identify the portion or the volume of the niche space a species may utilize, be affected by or occupy. This multidimensional volume is termed the hypervolume (Hutchinson 1958) (*Figure VI.C*–4a).

The niche hypervolume of a species defined by three environmental variables would appear as a cube or rectangle. This model of the niche assumes that over the entire environmental gradient, each species' response to environmental variables is uniform, as if there were only one genotype. Because it is the response of the range of genotypes in the population to the complex of environmental variables that is of concern, a better characterization of the species' niche may be obtained by the addition of a population response (*Figure VI.C*–4b). Thus within the hypervolume, the species' response to niche variables is expressed by population measurements resulting in a niche response surface (Whittaker 1975; Whittaker et al. 1973).

In Mirror Lake, biologic, physical and chemical components interact to produce a multitude of environmental variables, integrated to form a functioning whole. Within the boundaries of this ecosystem, each species population of the zooplankton community is subjected to and interacts within a range of these environmental variables. Changing patterns of the possibly im-

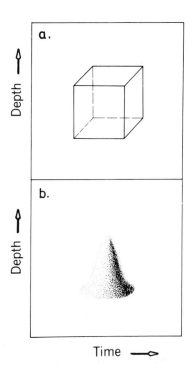

Figure VI.C–4. Niche hypervolume (**a**) and niche response surface (**b**) of a hypothetical zooplankter. In (**a**) the cube describes a volume where the zooplankter actually exists or lives—the realized niche of Hutchinson (1958). The model provides no information on how well the species' population is doing at different depths or seasons. By plotting productivity and connecting points of equal productivity, the niche response surface is delineated (**b**). The response surface can be visualized as a three-dimensional, bell-shaped curve. Because a measure of success (productivity) relative to environmental conditions at each depth and season is included, a more realistic characterization of a species population niche is provided than with a sharply bound boxlike hypervolume (see Whittaker 1975).

portant variables such as temperature, light intensity, water chemistry, predation pressure and trophic interactions in lakes correspond with depth and time. The depth and time axes may be consequently thought of as complex gradients, each comprising numerous factor gradients that may change separately or together. Particular components of the complex gradients (such as temperature and abundance of phytoplankton) may ascend and descend with time over a yearly cycle.

As a measure of the species' response to environmental variables in Mirror Lake, we chose productivity because it seemed most appropriate as a measure of the species' role and fitness in the community. Although other measures of species' success or well being (e.g. frequency of occurrence or density) could be used, productivity expressed the species' use of resources for individual and population growth and permitted a comparison of species of widely different size and metabolism on a single scale. As such, productivity of each species served as a measure of the species role (niche) in the ecosystem.

When mean monthly production (*Chapter V.B.5*) of each species is plotted on axes of depth and time and isolines are drawn connecting values of equal productivity, major features of the species' niche are identified in the niche response surfaces (*Figure VI.C–5*).

The first step in analyzing the niche response surfaces was to divide the zooplankton community into carnivores and herbivores. Species considered to be predators are grouped in *Figure VI.C–5A*: *Mesocyclops edax* (Confer 1971), *Cyclops scutifer* (Alimov et al. 1972; Vardapetyan 1972) and *Asplanchna priodonta* (Monakov 1972). *M. edax* and *C. scutifer* possess one generation per year (i.e. univoltine) and are present in the lake throughout the year but possess a niche center restricted to a specific season. Adults of *C. scutifer* occur during the spring, while adults of *M. edax* occur in the summer. Similarly, when the nauplii of *C. scutifer* are present in the water-column, the nauplii of *M. edax* are in diapause in the sediments. Since the copepodids and adults are carnivorous and nauplii are probably filter-feeders, the alternation of life history (i.e. the adults of each species co-occur with the immature of the other species) reduces competition between the two ecologically-similar species. Each of the predaceous species, *M. edax* and *C. scutifer*, possesses its own distinctive niche response, clearly separated in time and space from that of the other species and from *A. priodonta* (*Figure VI.C–5*).

A further division of feeding behavior can be based on the size of the food utilized by the herbivorous zooplankton. The herbivorous zooplankton feed on detritus and bacteria as well as on algae. Rotifers, copepod nauplii and *Bosmina* feeding on particles in the 1- to 5-μm range are classified as microconsumers, while

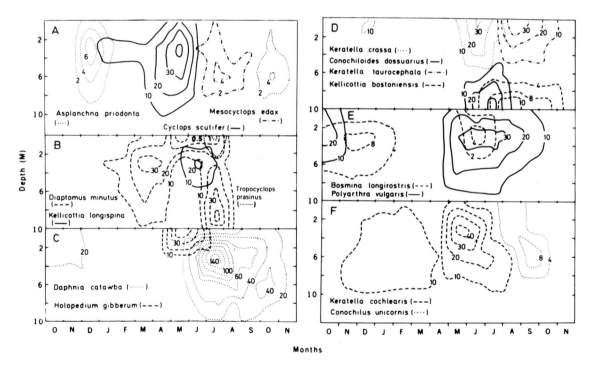

Figure VI.C–5. Niche response surfaces of Mirror Lake zooplankton (From Makarewicz and Likens 1975, *Science* 190:1000–1003. Copyright © 1975 by The American Association for the Advancement of Science. Used with permission.)

the calanoid copepods, *Daphnia* and *Holopedium* are considered to be macroconsumers feeding on material in the 3- to 20-μm size range (Hillbricht-Ilkowska et al. 1972). A difference in niche as a result of feeding behavior can separate the individual species of the macro- and microconsumer groups.

Within the macroconsumers, the niche responses of *Holopedium gibberum*, *Daphnia catawba* and *Diaptomus minutus* are separated in time and space (*Figure VI.C-5B* and C). In relation to depth (at a given time) or in relation to time (at a given depth), the distributions of these species form bell-shaped curves, except for one case. *D. minutus* has a bimodal curve of distribution resulting from the growth of adults in May and of nauplii in the hypolimnion during July. The different feeding behaviors of the life stages (adults as macroconsumers and nauplii as microconsumers) of this species result in two different niche-response surfaces. As was demonstrated with alternation of life history of *M. edax* and *C. scutifer*, it is critical to know the details of the life history of a species to carefully interpret the niche response surfaces.

The niches of the rotatorian microconsumers *Keratella taurocephala*, *Kellicottia bostoniensis*, *Keratella crassa* and *Conochiloides dossuarius* are distinctly separated into the top 2 m of the lake and into the hypolimnion (*Figure VI.C-5D*). Further examination of *Figure VI.C-5D* suggests that the response surfaces of any remaining summer-occurring rotifers should lie in the strata of water between 2 m and the hypolimnion if no food selectivity was occurring. In fact, the niche centers of the remaining microconsumers occur during the summer. The niche center of *Conochilus unicornis* (*Figure VI.C-5F*) is generally located in the metalimnion in September, whereas *Kellicottia longispina* (*Figure VI.C-5B*), *Keratella cochlearis* (*Figure VI.C-5F*) and *Polyarthra vulgaris* (*Figure VI.C-5E*) are at 3 to 6 m in the spring and early summer.

The niche centers of *P. vulgaris* and *K. cochlearis* do overlap. However, their niches may be separated by virtue of differences in feeding behavior. Edmondson (1965) has demonstrated that the abundance of the alga *Chrysochromulina* is reflected in the reproductive rate of *Keratella* but not of *Polyarthra*. *Polyarthra*'s reproductive rate increased with the abundance of the larger alga *Cryptomonas*. Sachse (1911) also concluded that *Polyarthra* sp. feed mainly on *Cryptomonas*. Edmondson's (1965) study strongly suggests that *P. vulgaris* and *K. cochlearis* are feeding on organisms of different sizes. Even though the niche hypervolumes do overlap in Mirror Lake, they are apparently separated by a division of food (size) resources. Differences of diet between populations allow rotifers to avoid, at least in part, competition for food (Pourriot 1977).

Edmondson (1965) did not find any difference in feeding behavior between *K. longispina* and *K. cochlearis*. Any separation of niche would have to arise by some other mechanism. In Mirror Lake these two species are separated by time and space (*Figure VI.C-5B* and F).

Bosmina (*Figure VI.C-5E*) ingests particles in the same size range as the other Cladocera but feeds most intensively on particles in the size range 1 to 3 μm, which is also the optimum food range of rotifers (Gliwicz 1969). Thus, *Bosmina* does not fit into the arbitrary distinction (between macro- and microconsumers) but is adapted to a niche intermediate with (and overlapping broadly with) these groups. The degree of overlap of niches and the weak population (production) response of *B. longirostris* are probably reflections of intense competition between *B. longirostris* and the other herbivores in the lake. During the winter when *Daphnia* and *Holopedium* are not present in the plankton, *B. longirostris* has a higher standing crop than in the summer. Moreover, a higher standing crop of *B. longirostris* is particularly evident in winters when the food supply is high.

Species populations of zooplankton are thus scattered in the niche hyperspace. The broad overlap of the niche response surfaces of some species indicates that there is still some competition among species. Part of the overlap of niche response surfaces could be due to niche differences other than separation in time and space and feeding behavior, such as vertical migration.

However, our analysis of the vertical migration of rotifers in Mirror Lake indicates that this mechanism does not reduce competition or separate niches. For example, the niche response surfaces and the niche centers of *K. taurocephala* and *K. bostoniensis* are located in August on the time axis but are separated into the epilimnion and hypolimnion, respectively, on the depth axis. There is little overlap of the niche response surfaces of these two species. These species have a very limited vertical migration (*Figure VI.C-6A*); thus, there is little overlap over a 24-h period. Examination of *P. vulgaris* and *K. longispina,* two species whose niche centers are not located in the month of August but whose niche response surfaces do overlap during August, indicates a similar pattern of vertical migration (*Figure VI.C-6B*). Competition between these species does not appear to be reduced by vertical migration in Mirror Lake. Instead, niches are separated and competition is reduced by the scattering of the niche response surfaces and their centers along the depth and time axes. Possible mechanisms of vertical separation of niche response surfaces include the limitation of vertical migration by temperature-dependent photoresponses (Smith and Baylor 1953) and avoidance of temperature extremes (Calaban and Makarewicz 1982; Kikuchi 1937).

Even so, overlap is not a contradiction of the exclusion principle; some overlap is to be expected, as formulated by May and MacArthur (1972). Niche differentiation is revealed not in sharp boundaries between competing species but in relative differences in the location of species' centers in the niche hyperspace.

The overlapping curves together form a community continuum in time (*Figure VI.C-7*) and depth. Certain species seem to have the modes of their distributions grouped together along the temporal complex gradient. This grouping is apparent and not real. If another dimension, depth, were considered, a series of population surfaces overlapping with one another, but with no two distributions alike, would be evident. This finding parallels Ramensky's (1924) and Gleason's (1926) concepts of species individuality and community continuity.

The principle of species individuality may be stated as follows (as quoted by Whittaker 1975): "Each species is distributed in its own way, according to its own genetic, physiological and life-cycle characteristics and its way of relating to both physical environments and interactions with other species; hence, no two species are alike in distribution." Lane (1978) comments, "Much of the support for the individualistic concept has come from studies of terrestrial plant communities" and argues that the concept is not relevant to zooplankton communities. It is true that the individualistic concept was originally advanced in studies of plant communities, but much subsequent work has suggested that it is applicable to a wide range of plant and animal taxa. Whittaker's (1952) early support for the individualistic concept was based on studies of forest trees and insects. McIntosh (1963) reviewed population distributions on a variety of gradients for copepods, insects, protozoans and herbaceous plants illustrating individualistic responses of species. Numerous studies during the 1950s and 1960s by Curtis and Whittaker

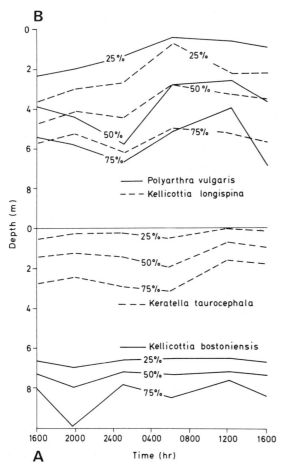

Figure VI.C-6. Vertical migration patterns of selected zooplankton species in Mirror Lake during August. Migration patterns were determined by the quartile method of Pennak (1943). (From Makarewicz and Likens 1978, *Science* 200:458–463. Copyright © 1978 by the American Association for the Advancement of Science. Used with permission.)

C. The Pelagic Region

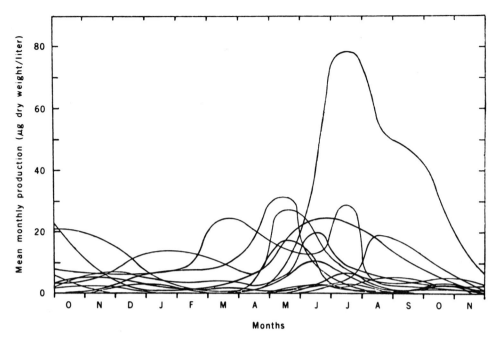

Figure VI.C–7. Distribution of zooplankton populations in Mirror Lake along the complex gradient of season. (From Makarewicz and Likens 1975, *Science* 190:1000–1003. Copyright © 1975 by The American Association for the Advancement of Science. Used with permission.)

and their students (see McIntosh 1958, 1967 and Whittaker 1967 for a review) based on forest herbs, as well as trees, prairie grasses and herbs and soil fungi, supported the individualistic concept and introduced community continuum or gradient approaches. Also, the individualistic concept was supported by studies of benthic and littoral marine organisms (Stephens 1933; MacGinitie 1939; Sanders 1960) and of phytoplankton (Karentz and McIntire 1977). Numerous studies of birds (e.g. Bond 1957; Terbough 1970) also have supported the individualistic concept in addition to the simple example of zooplankton discussed in this section.

Lane (1978) repeats a common criticism of studies supporting the individualistic concept—that the results are an artifact of the sampling approach or the measurements used. The studies supporting the individualistic concept have used very different sampling methods, measurements of species populations and methods of data analysis. They range from direct gradient analyses, in which the species' population responses are measured along one or more measured spatial or environmental gradients, to indirect gradient analyses in which the community samples themselves provide the gradient for ordering the species (Whittaker 1967, 1975). The number and diversity of studies supporting the individualistic concept suggest that the artifact argument is unwarranted.

Lane (1978) asserts that sne finds "macroscopic properties" or "consistent indices" that "are characteristic of a type of community" and "vary among different types of communities." To accomplish this, Levins' (1968) equation, which measures only "niche overlap" or mutual occurrence, is utilized to imply species integration. As Wiens (1977) and Gilyarov and Matveev (1977) point out, this use is untenable, and it is not clear how Lane recognizes a community type and distinguishes between types. However, the major point is that Lane's position rests on the concept of characteristic properties of an integrated community type. Her method of demonstrating the community type or its integrity is of doubtful merit and subtracts from its credibility. Further discussion on this controversy may be found in Makarewicz and Likens (1978).

The division of niche hyperspace now apparent in zooplankton of Mirror Lake is a result of predation (Hall et al. 1976), competitive displacements (Snell 1979) and possible extinctions of species in the past. Competition among species certainly exists. However, competition is reduced by the fact that the species select food of various sizes and by the partial separation in space and time of each species' niche response. In general, the limnetic zooplankton community of Mirror Lake can be conceived of as a system of interacting populations whose niche centers have evolved toward dispersion in relation to the complex gradients of food size, depth and time.

Summary

The pelagic region of Mirror Lake is characterized by >230 species of phytoplankton, >23 species of zooplankton and 5 species of fish. A distinct periodicity of phytoplankton biomass and zooplankton productivity with maxima in the summer and minima in the winter is evident. Numerous interactions occur between the biotic components that determine abundance, species composition, species diversity and succession.

Phytoplankton lose DOC and nutrients, which may be rapidly taken up by bacteria. Phytoplankton also require nutrients and actively compete with bacteria for nutrients. The succession of phytoplankton species is controlled by physical, chemical and biologic factors. For example, *Dinobryon* is not abundant until other species have decreased the available phosphate. Because of *Dinobryon*'s more effective mechanism of phosphate uptake than species dominating earlier in a bloom condition, it has a competitive advantage and can become abundant in lakes. Allelopathy, ability to fix atmospheric nitrogen and sodium stimulation of phosphate uptake and polyphosphate formation provide a competitive advantage to blue-green algae under certain environmental conditions.

Zooplankton feed on bacteria, protozoans, phytoplankton and other zooplankton. The significance of zooplankton feeding on bacteria and organic detritus in an aquatic ecosystem is not known, but is probably significant as algal productivity often cannot support herbivorous zooplankton. Ciliate protozoans may serve as an important link between zooplankton and bacteria in the food web. Generally, rotifers are thought to feed on small particles including bacteria, organic detritus and phytoplankton. Feeding by zooplankton can affect phytoplankton species' composition and influence seasonal periodicity by differential grazing. Similarly, size selection and predation by fish and carnivorous zooplankton can alter composition of the zooplankton community.

Although physical circulation of water is minimal between the hypolimnion and epilimnion during summer stratification, a movement of nutrients and energy between these strata does occur. Sedimentation provides a continuous "rain" of material and nutrients to the hypolimnion and eventually to the sediments. As living, dying and dead particulate organic matter sinks, it is a source of energy and nutrients for bacteria, phytoplankton and zooplankton in deeper waters. Vertical migration of zooplankton into the surficial waters should increase grazing pressure on algae in the evening and possibly cycle nutrients from the hypolimnion into the epilimnion by excretion.

A seasonal cycle in regeneration of P from the zooplankton of Mirror Lake was evident with a peak in June and a minimum in late winter. Excretion of P by zooplankton is an order of magnitude higher than allochthonous inputs for each season of the year. Functionally within Mirror Lake, copepods accounted for only 17% of the zooplankton production but contributed 48% of the P excreted by the zooplankton community. Cladocerans, on the other hand, are important in Mirror Lake considering their high productivity but are not major contributors to P cycling within the lake. Rotifers accounted for over 40% of the P excreted by zooplankton in Mirror Lake. Regeneration of P by zooplankton and remineralization of P by decomposition of zooplankton tissue undoubtedly play a major role in maintaining the metabolism of lake-ecosystems by making P available to primary producers whose productivity could not be maintained by allochthonous inputs alone.

Our study of how the zooplankton species in the pelagic community of Mirror Lake relate to one another has led us to conclude that each species has its own position or niche, with a central location that differs from those of other species. This division of the niche hyperspace now apparent in the zooplankton of Mirror Lake is a result of evolutionary processes including reduction of competition, predation, competitive displacement and possible extinction of species in the past. The zooplankton community of Mirror Lake is considered to be structured as a system of interacting populations, whose niche centers have evolved toward dispersion in relation to the complex gradients of depth, time and food within the ecosystem. Studies of the density, behavior and metabolism of all zooplankton species interacting within the multitude of environmental variables in an aquatic ecosystem are required to unravel the complex relationships between organisms in a community. Community research should, when it can, go beyond expressing only the degree of structure and should investigate how and why communities are structured.

D. The Lake-Ecosystem

Gene E. Likens

The Mirror Lake ecosystem is a beautiful component of the landscape in the Hubbard Brook Valley. However, its peaceful setting and appearance belie the intricate structural, metabolic and biogeochemical complexities and interconnections, both internal and external, that constitute its existence. Clearly, an axiom of long-term ecologic studies is that the greater the understanding about the system, the clearer are the questions that invite detailed study. Long-term descriptive data and experience are absolutely essential to formulate *meaningful* questions (hypotheses), particularly at the ecosystem level where complexity and variability are the rule. Long, relative to ecosystem research, usually means several years at numerous levels of complexity and organization. Much of the appreciation of the real beauty of Mirror Lake comes from an understanding of how the various parts contribute to a functioning ecosystem, day by day and over long periods.

We began studies of Mirror Lake in 1963 as a part of the Hubbard Brook Ecosystem Study. Ideally all components of the ecosystem (e.g. metabolism, nutrient flux and cycling, population dynamics) should have been studied simultaneously over the intervening years. Such a program was not possible because of limitations in human resources and funds. In an attempt to partially overcome this deficiency, in 1969 we began a comprehensive and detailed study of the sediments of the lake, to extend our time frame of inquiry to the beginnings of the lake, some 14,000 yr ago. Our belief was that study of this much longer period would provide a better understanding and integration of the complex ecologic interactions affecting the lake and its environment. The results of those studies are presented in *Chapter VII*.

Thus, our current ecosystem study is flawed by gaps in discrete information (e.g. fungi and protozoa) or in coincident timing of component studies. As a result, various cause-and-effect relations in the Mirror Lake ecosystem are poorly known or only speculative. Nevertheless, the ecosystem approach taken in these studies provides a conceptual overview whereby critical and significant questions amenable to experimental testing can be readily identified. Testing of hypotheses is basic to science and major advances in our understanding occur as a result. Experimentation at the ecosystem level of organization is very exciting and informative. If hypotheses are carefully drawn and tests are critically done, then experiments at the ecosystem level are particularly meaningful because the unit is a realistic one, i.e. a functional component of the landscape.

Early manipulations of entire lake ecosystems by C. Juday (Juday and Schloemer 1938) and by A. D. Hasler and colleagues in Wisconsin (e.g. Johnson and Hasler 1954), more elaborate studies by D. W. Schindler and colleagues in the Experimental Lakes Area of Ontario (see Schindler 1980a) and experimental studies of watershed-ecosystems at Hubbard Brook (e.g. Bormann and Likens 1979; Likens et al. 1970; Hornbeck et al. 1975) have provided major new insights and thrusts for ecosystem research. The opportunity to experimentally alter an entire natural ecosystem (e.g. by nutrient enrichment or destruction of vegetation) and compare the results with an undisturbed reference (or *control*) system, as done so successfully in laboratory studies, provides an extremely powerful and sound scientific tool. Because whole-ecosystem studies are expensive in time and money, and because the opportunity to do long-term ecosystem manipulation is rare, aquatic scientists have frequently chosen to isolate by plastic tubes or "corrals" columns of water or sediment for experimental study within a lake (e.g. Lund 1972). We had to adopt this approach in our experimental studies of Mirror Lake (e.g. Gerhart 1973) because we do not have ownership of the lake and cannot disturb it, and also because it is the only lake in the Valley and there is no appropriate reference system.

Nevertheless, Mirror Lake is ostensibly typical of many lakes in northern New England and the Precambrian Shield of Canada. Typical can be a very ambiguous term to apply to an ecosystem, but the meaning here is in reference to a more casual observation, rather than to a scientific one.

In this chapter, I summarize some of the major features of metabolism and nutrient flux and cycling in the Mirror Lake ecosystem.

Ecosystem Structure, Metabolism and the Food Web

Traditionally, ecologists have referred to food chains when discussing who eats whom in nature. However, following Lindeman's (1942) classic studies of Cedar Bog Lake and from an ecosystem point of view, it makes much more sense to refer to food webs. This change follows for several reasons: (1) the food habits of many organisms are unknown, diverse or changing, and thus it is difficult to categorize animals as strict herbivores, carnivores, etc.; (2) food eaten by larger consumers may have cycled through several other consumers previously by way of detritivore loops and, thus, it is exceedingly difficult to identify levels of consumption such as primary, secondary and so forth; and (3) allochthonous input of organic matter of unknown trophic level may be assimilated by bacteria and fungi, which in turn may be consumed by larger consumers, or it may be assimilated directly by animals, thus the "chain" of who eats whom becomes meaningless relative to trophic levels. The concept of a food web is clearly much more appropriate in dealing with the complexity of an ecosystem.

For these reasons and others, it also seems appropriate to categorize the principal components of a food web for an ecosystem simply as *producers* (autotrophic and chemosynthetic) and as heterotrophic *consumers* (subdivided into micro and macro, if desirable or helpful). Allochthonous inputs of organic matter also may contribute significantly to the maintenance of food webs, particularly in aquatic ecosystems (e.g. Likens 1972b).

The total metabolism of a lake-ecosystem depends upon the availability and utilization of potential energy stored in organic matter. Carbon fixed by photosynthetic (phytoplankton, periphyton and macrophytes) and chemosynthetic organisms constitutes the autochthonous input. In addition, various allochthonous sources (e.g. precipitation, terrestrial litter) may contribute significant amounts of organic matter to lakes (see *Chapter V.C*). Potentially, all of this organic matter is available to consumers including bacteria and fungi, and thus may be important in expanding detritivore food webs and in maintaining the overall metabolic status of lake-ecosystems. From an ecosystem point of view, the productivity of a lake is more than just the photosynthesis of phytoplankton (see Likens 1972b). Therefore, we have adopted a new approach (and some new terminology) for the ecosystem studies of metabolism in Mirror Lake. All reduced carbon compounds, which can provide energy for consumers are referred to as *Ecosystem Source Carbon* (ESC). The overall productivity of the ecosystem is the rate of input from internal and external sources per unit time of all reduced carbon compounds that can provide energy for consumers (Likens 1972b).

A food web for organic carbon in the Mirror Lake ecosystem is shown in *Figure VI.D–1*. Twenty-one percent of the annual ESC is provided from allochthonous sources. The remaining 79% is divided between autotrophic producers (phytoplankton 87%, periphyton 6%, macrophytes 4%) and chemosynthetic bacteria (3%).

Approximately 0.34% of the average incident solar radiation (11.1×10^9 kcal/ha) is fixed by photosynthetic plants in Mirror Lake. Phyto-

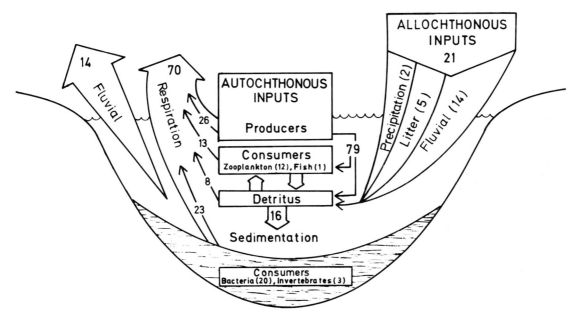

Figure VI.D–1. Food web for the Mirror Lake ecosystem. Sedimentation represents net sedimentation.

plankton is at least 20-fold more efficient at fixing carbon (gross productivity) per unit biomass than are the other producers in the lake (*Table VI.D–1*).

Makarewicz and Likens (1979) have shown a direct relationship between zooplankton production and net primary production (phytoplankton, periphyton plus macrophytes) in eight lakes, including Mirror Lake (*Figure VI.D–2*). Allochthonous inputs are known for only three of the lakes; however, it seems apparent from these data and from the fact that the linear relation was strengthened when net periphyton and macrophyte production were added (Makarewicz and Likens 1979), that ecologic efficiency (energy transfer between trophic levels) increased with increases in autochthonous inputs, and that inputs of organic carbon other than from phytoplankton may be utilized directly or indirectly by zooplankton. Ecologic efficiency is high in productive and very unproductive lakes, but low in mesotrophic lakes. Peterson et al. (1978) and others have shown that various zooplankton use bacteria as food under natural conditions, so that ESC entering detrital food webs and ultimately being consumed by zooplankton is one way that the system becomes more productive. In aquatic ecosystems, more interactions must be considered than only those between zooplankton and phytoplankton.

Conventional ecologic efficiency, i.e. zooplankton production/net phytoplankton production would be 4.4%, whereas on an ecosystem basis (zooplankton production/ESC), the efficiency would be 3.1% for Mirror Lake. Actual efficiencies under field conditions, however, are extremely difficult to determine because of the *reuse* of organic carbon in detritivore food webs

Table VI.D–1. Production/Biomass (P/B) Ratios for Various Communities of Organisms in Mirror Lake[a]

Organisms	P/B
Phytoplankton	150
Epilithic Algae	6.6
Macrophytes	1.8
Zooplankton	13
Benthic Macroinvertebrates	4.9
Salamanders	2.5
Fish	0.5–1.0[b]

[a] Gross production is used for plants.
[b] Based on Waters 1977.

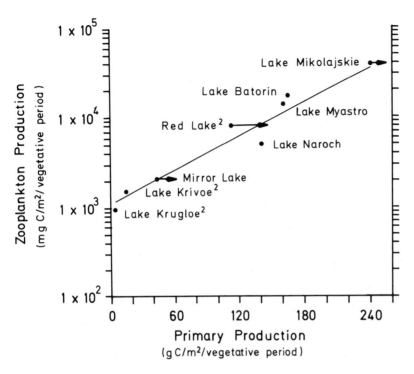

Figure VI.D–2. Zooplankton production vs. net primary production of phytoplankton, periphyton and macrophytes in various lakes. Y = antilog 3.032 + 0.0063 X, r = 0.88, n = 8. [2] Represents net phytoplankton plus macrophyte production only. Arrows indicate the amount of allochthonous inputs of reduced carbon. (Modified from "Structure and function of the zooplankton community of Mirror Lake, New Hampshire," by J. C. Makarewicz and G. E. Likens, *Ecological Monographs*, 1979, 49(1): 109–127. Copyright © 1979 by The Ecological Society of America. Used with permission.)

as referred to above, and because benthic invertebrates also consume the products of net phytoplankton production.

The benthic region of Mirror Lake is generally heterotrophic, largely surviving on the "rain" of organic matter from the water-column above. For example, respiration in the cobbled areas of the littoral zone exceeds net epilithic periphyton production by more than two times (*Chapter V.B.4*). In the profundal zone all of the macroinvertebrates are detritus-feeders, except for two predators—*Chaoborus* and a tanypodinian chironomid. The detritus-feeding chironomid, *Chironomus anthracinus*, dominates the macroinvertebrate biomass during all seasons (*Chapter V.A.6*). Some microinvertebrates in the sediments of Mirror Lake (e.g. the oligochaete, *Chaetogaster diastrophus*) eat significant amounts of living periphyton. In fact, based on analyses of gut contents, Strayer (1984) has estimated that net production of the periphyton may need to be as large as 10 g C/m²-yr to support this community of consumers.

The benthic invertebrates are relatively abundant for an oligotrophic lake and the profundal zone and overlying hypolimnion are more characteristic of a eutrophic lake. The relatively small volume of the hypolimnion is "fed" and enriched by particulate organic matter (e.g. feces, dead organisms) falling from the much larger volume of water above. Thus, the profundal zone can metabolically support a large heterotrophic community and the deepest regions of Mirror Lake are more eutrophic because of this volume-concentrating effect.

Benthic insects are preyed upon most intensively as they pass through or emerge from the pelagic zone. Fish annually consume 20 to 40% of the production by zooplankton and benthic invertebrates in the lake (*Chapter V.B.7*).

Because detailed studies of organic carbon flux and metabolism have been done in various ecosystems of the Hubbard Brook Valley, an unusual opportunity exists to compare the organic carbon balances of terrestrial and stream-ecosystems with the Mirror Lake ecosystem (Likens and Bormann 1979). The relative contribution of various organic components to lake-, stream- and forest-ecosystems in the Hubbard Brook Valley are given in *Table VI.D–2*.

Clearly, autochthonous inputs dominate in the forest- and lake-ecosystems, whereas allochthonous inputs dominate in the heterotrophic stream-ecosystem. Ecosystem respira-

D. The Lake-Ecosystem

tion is large and roughly similar in the forest- and lake-ecosystems, whereas fluvial throughput is the major loss in the stream-ecosystem. Respiration losses of organic carbon in the stream-ecosystem are approximately one-half of the values for the forest- or lake-ecosystems (*Table VI.D–2*). From an ecosystem point of view, it is interesting and significant that such a small percentage of the organic matter, which is input annually, accumulates in these various ecosystems.

The concentration of organic matter varies appreciably throughout the volume of these ecosystems. For example, the concentration is low in the water-column and high in the sediments of Mirror Lake. Patches or layers of increased concentration are common.

Dissolved organic carbon (DOC) and organic detritus often play central roles in ecosystem food webs (see Wetzel 1975). DOC may be excreted by living plants and animals or may be produced in the decomposition of dead organic matter. Assimilation of DOC by bacteria and fungi (and possibly higher forms) then provides an important mechanism for cycling or reusing energy that ostensibly would be lost from a food chain. Calculation of ecologic efficiencies with and without DOC obviously would be very different. Apparently, bacterioplankton could be supported solely by primary production of phytoplankton, but it is obvious that the only way that such a large biomass of bacteria in the entire lake-ecosystem could be maintained is through utilization of the DOC and detritus reservoirs (*Table VI.D–3*). Some 69% of the allochthonous inputs and 79% of the total standing stock of organic carbon in the water-column of Mirror Lake were in the form of DOC. Dead organic matter and DOC clearly dominate the total biomass in the lake-ecosystem on both an areal and volumetric basis (*Table VI.D–3*). Organic detritus is mostly accumulated in the sediments. The high proportion of DOC in the lake-ecosystem is one of the striking differences between lake-, stream- and forest-ecosystems within the Hubbard Brook Valley (Likens and Bormann 1979).

Table VI.D–3. Average Distribution of Living and Dead Biomass in the Mirror Lake Ecosystem[a]

Component	(g C/m^2)[b,c]	(g C/m^3)[c,d]
Biota	7.4	1.29
Bacteria	(4.1)[g]	(0.71)
Plant	(2.1)	(0.37)
Animal	(1.2)[f]	(0.21)
Detritus	1,000	174
DOC[e]	13.8	2.4

[a] Modified from Likens and Bormann 1979.
[b] See *Chapter V.C*.
[c] To a depth of 10 cm in the sediments.
[d] Average depth of ecosystem = 5.75 m.
[e] Water-column only.
[f] Does not include benthic microinvertebrates.
[g] 0.08 in water-column; 4.0 in sediments.

Table VI.D–2. Relative Contribution of Annual Organic Carbon Flux for Different Hubbard Brook Ecosystems on an Areal Basis[a]

	Forest	Stream	Lake
Inputs			
% Autochthonous	>99	< 1	79
% Allochthonous	< 1	>99	21
% Terrestrial Litter	0	47	5
% Precipitation	< 1	< 1	2
% Dissolved Organic Carbon	< 1	43	15
Outputs			
% Respiration	86	37	70
% Fluvial (including ground water)	< 1	63	14
% Dissolved Organic Carbon	< 1	42	13

[a] Modified from Likens and Bormann 1979.

Nutrient Standing Stocks and Cycling

An analysis of the standing stocks of organic carbon, nitrogen and phosphorus in the Mirror Lake ecosystem shows the dominance of the lake water (dissolved) and sediment reservoirs (*Table VI.D–4*). Surprisingly, bacteria represent a large reservoir of carbon, nitrogen and phosphorus in the lake-ecosystem. Fish and salamanders are enriched in nitrogen and phosphorus and energy (kcal) per unit carbon relative to the other components of the lake-ecosystem, but their total biomass is rela-

Table VI.D-4. Average Standing Stock (kg and kcal) and Ratios (Weight Basis) of Carbon, Nitrogen and Phosphorus in the Mirror Lake Ecosystem[a]

Component	kcal × 10^{3b}	C (kg)	N (kg)	P (kg)	C:N:P (by wt)
Dissolved[c]	33,600	3,360	124	2.5	1,344 : 50 : 1
Seston	2,420	247[d]	32.3	2.48	100 : 13 : 1
Bacterioplankton	120	12	2.1[e]	0.3[e]	40 : 7 : 1
Phytoplankton	411	56	9.8[e]	1.4[e]	40 : 7 : 1
Epilithic Algae	570	57	3.0	0.075	760 : 40 : 1
Macrophytes	2,020	202	14	0.8	252 : 17 : 1
Zooplankton	313	30	4.9	0.84	36 : 5.8 : 1
Fish	408	36	8.2	2.7	13.1 : 3.0 : 1
Salamanders	10	0.77	0.13	0.055	14.1 : 2.4 : 1
Benthic Macroinvertebrates	1,050	105	25.5	3.0	35 : 8.5 : 1
Benthic Bacteria	6,000	600	105[e]	15[e]	40 : 7 : 1
Sediment (top 10 cm)	733,000	150,000	10,570	1,280	117 : 8.3 : 1

[a] Divide by 15×10^{-4} to convert to kcal/m² or by 0.15 to convert to mg/m².
[b] Computed by converting C to dry weight (C ÷ 0.45) and multiplying by mean caloric values from Cummins and Wuycheck (1971).
[c] Organic plus inorganic (DIC = 1,290 kg; DOC = 2,070 kg; NH_4-N = 21.6 kg; NO_3-N = 16.4 kg; organic N = 86 kg; PO_4-P = 1.7 kg.
[d] Excluding zooplankton, *Chapters IV.B* and *V.C*.
[e] Assuming a weight ratio of 40:7:1 for C:N:P.

tively small. In contrast, the water, and to a lesser extent the epilithic algae, are deficient in nitrogen and particularly phosphorus (*Table VI.D*–4).

The productivity of phytoplankton is simultaneously limited by the availability of inorganic nitrogen and phosphorus in Mirror Lake (Gerhart 1973; *Chapter V.B.2*). Excretion by zooplankton (*Chapters V.B.5* and *VI.C*) and mineralization of organic matter by bacteria (*Chapter V.D*) are important sources of inorganic nitrogen and phosphorus for photosynthetic producers. The average turnover rate for seston in Mirror Lake probably varies from seconds and minutes for phosphorus, to minutes and hours for nitrogen and hours and weeks for carbon. In contrast, the average turnover rate for total dissolved substances in the Mirror Lake *ecosystem* (standing stock/input) varies from 1.32 yr for carbon to 0.67 yr for nitrogen and to 0.63 yr for phosphorus. Theoretical turnover rates would be shorter if all inputs to the ecosystem (e.g. *Table IV.E*–16) were considered, however, some inputs (e.g. terrestrial litter) may have a long lag time before inorganic nutrients become available for cycling by way of mineralization. If the system were in steady state relative to mineralization of inputs, this delay would not be a concern. However, many inputs are highly episodic, e.g. leaf fall from deciduous trees or flooding, and the system is probably not in steady state during most time periods of biologic interest.

The importance of climatic, edaphic and morphometric factors in determining the structure, metabolism and biogeochemistry of lake-ecosystems has been stressed by various authors (e.g. Rawson 1939; Deevey 1940; Mortimer 1942; Carpenter 1983; Likens and Bormann 1979). Carpenter (1983) argues that nutrient cycling may be strongly influenced by basin morphometry as expressed in two ratios: (1) mean depth, $\bar{z}$, to maximum depth, z_{max}, and (2) sediment surface area in contact with the epilimnion to epilimnetic volume. The surface sediment below the epilimnion to epilimnetic volume ratio declines with the depth ratio in lakes like Mirror Lake. As depth ratio decreases, Carpenter (1983) would predict that nutrient cycling from the sediment surface, lake productivity and sediment accretion would all increase. Thus, a depth ratio of 0.523 in Mirror Lake would suggest that all of these ecosystem functions are relatively active in Mirror Lake.

Fee (1979) has also stressed the importance of littoral sediments for nutrient cycling in similar oligotrophic lakes of the ELA. This zone in Mirror Lake is characterized by sand, gravel and cobbles. Upon casual observation, it appears to be a rather barren, uninteresting area. However, such an impression is far from the actual situation. The littoral sediments of Mirror Lake are sites of intense activity. Nutrient exchange by cation exchange, diffusion and bioturbation occurs at the sediment-water interface, and seepage of water and chemicals passes in both directions through these sediments (C. Asbury, personal communication). Active resuspension of recent inorganic sediments and organic particulate matter occurs here during storms or during autumn circulation (*Chapter VII*). Mineralization of dead organic matter may be very rapid in this relatively high-energy, oxygenated zone. Numerous benthic invertebrates and periphyton colonize this area and compete for organic and inorganic nutrients. Nutrients may be "pumped" from the littoral sediments by macrophytes, but the magnitude of nutrient pumping by macrophytes is probably not very large in Mirror Lake given the size and extent of the macrophyte community (*Chapter V.B.3*). Relative to the functioning of the entire ecosystem, this shallow zone should be the focus of intensive and detailed process-level studies.

Nutrient Flux

There are differences in chemistry between the various tributaries to Mirror Lake (*Chapters IV.B and IV.E*). Some of the differences may be explained by differences in historical land use, biogeochemical variations and human disturbance within the watersheds. In addition, morphometric and hydrologic differences between the watersheds may underlie other less apparent differences in chemistry, or compound those referred to above. For example, Watershed W is elongate, whereas Watershed NW is round (*Table III.A*–1). As a result, stream channel length in W is longer than NW, peak discharges in streamflow may occur faster in NW and potential for stream channel drainage rather than direct drainage to the lake is greater in NW than in W. Because transport of particulate matter is exponentially related to streamflow and P is exported from the terrestrial watersheds primarily as particulate matter (Meyer and Likens 1979), export of P by Watershed NW may be much more variable and ultimately greater in amount than from Watershed W, particularly under certain hydrologic regimes. Unfortunately, our sampling design was not adequate to evaluate these relationships in the Mirror Lake drainage basin.

Sewage effluents from septic tanks apparently provide a small but significant input of phosphorus, chloride, sodium and possibly nitrogen (NH_4^+) to Mirror Lake (*Table IV.E*–16). Moreover, inputs of road salt through the Northeast watershed is an obvious anthropogenic input of sodium and chloride (*Chapters II, IV.E and IX*). Currently, mats of blue-green algae (e.g. *Oscillatoria*) are widespread on the deep (>9 m) profundal sediments (D. Strayer, personal communication) and blue-green filamentous green algae (e.g. *Spirogyra*) are common near mouths of inlet streams and on littoral sediments of Mirror Lake (*Chapter V.A.2*).

In 1958, Provasoli suggested that monovalent ions, e.g. Na^+ and K^+, might be important, in addition to nitrogen and phosphorus in stimulating blooms of blue-green algae. Although there is some correlational support for this hypothesis, almost no direct evidence is available. Baybutt and Makarewicz (1981) have shown, using multivariate and regression analysis, that only sodium concentrations were correlated with blue-green algal abundance in nearshore waters of Lake Michigan. Provasoli (1969) and Ward and Wetzel (1975) cite some additional examples where increased concentrations of Na^+ and K^+ were correlated with blooms of blue-green algae in various lakes (see also *Chapter VI.C*). Clearly this hypothesis should be tested in relation to the simultaneous limitation by N and P (*Chapter V.B.2*) in Mirror Lake. The various anthropogenically induced changes now occurring in Mirror Lake would provide an appropriate and interesting system for such tests.

The timing of input is very important for utilization of nutrients by the aquatic biota. Relatively large inputs occur during the winter and early spring (*Chapter IV.E*) when Mirror Lake is cold and poorly lighted. Also, much of these inputs pass quickly through the lake to the outlet and presumably have little effect on the metabolism of organisms in the lake. Thus, judging the potential productivity from annual inputs or budgets of nutrients may result in a misleading view of lake metabolism. Short-term fluxes such as those immediately following a large summer rainstorm may be much more important for the metabolism of natural oligotrophic lakes.

The balance for inputs and outputs of dissolved Ca^{2+}, Mg^{2+}, Na^+, K^+ and sulfur shows no significant accumulation overall (*Table IV.E*–14), and yet the fact that these elements are being accumulated in the sediments (*Table IV.E*–16; *Chapter VII*) would suggest that the majority of the sedimented material for these elements originated *ultimately* from allochthonous inputs of particulate matter or from dry deposition (*Table IV.E*–16). This useful and interesting insight into the significant sources of nutrients and functioning of the lake-ecosystem is provided from an ecosystem analysis using the budget approach and presents important hypotheses for subsequent testing.

The large net retention of dissolved inorganic N, Si and P (*Table IV.E*–15) demonstrates the important biologic role these nutrients play in the metabolism and biogeochemistry of Mirror Lake. In addition, sediments may act as a major sink for such limiting nutrients, which are being transported "downhill" in the landscape with flowing water. Insect emergence (*Chapter V.B.6*) and salamander migration (*Chapter V.B.7*) represent two ways in which nutrients may be moved "uphill" (sensu Leopold 1949) from aquatic to terrestrial ecosystems.

Summary

An ecosystem represents an intricate balance of the interactions between the living and nonliving components. Energy (reduced carbon) and nutrients are added from autochthonous (photosynthetic and chemosynthetic) and allochthonous sources. The energy is oxidized (respired) within the ecosystem by consumers and nutrients are cycled by assimilation, excretion, decomposition and other biogeochemical reactions.

A food web, rather than food chain, more accurately describes the biotic and structural complexity of energy transformations occurring in aquatic ecosystems.

The metabolism of an aquatic ecosystem depends ultimately upon all the sources that can provide energy for consumers. Thus, the concept, *Ecosystem Source Carbon* (ESC), is adopted. Some 21% of the ESC for Mirror Lake comes from allochthonous sources and 79% comes from autochthonous sources. Phytoplankton are much more efficient at fixing carbon than other plants in Mirror Lake and, overall, approximately 0.34% of the incoming solar radiation is fixed annually by photosynthetic plants. Some 70% of the annual ESC is respired by consumers.

The sediments are the major reservoir of organic matter and nutrients in Mirror Lake. This habitat is the site of intense activity by invertebrate consumers.

Primary productivity in Mirror Lake is limited by the availability of both inorganic nitrogen and phosphorus. Excretion by zooplankton and mineralization reactions provide important means of cycling these nutrients within the ecosystem. Nutrient turnover rates are generally very short for organisms and much longer for the ecosystem as a whole.

Chapter VII

Paleolimnology

Margaret B. Davis, M. S. (Jesse) Ford, and Robert E. Moeller

A. Sedimentation

The process of sediment deposition couples lake metabolism and inputs from the watershed to the sediment accumulating on the lake bottom, which in turn provides a record of ecosystem events through time. The sediment deposit grows slowly, perhaps somewhat irregularly, as particles are washed into the lake and as dissolved compounds in the lake water are converted by organisms into particulate forms.

Three intergrading types of sediment are being deposited in Mirror Lake today. In the center of the basin is (1) a flocculent organic mud, rich in diatom frustules but lacking mineral particles larger than silt-sized grains. The chemistry of this "gyttja" (Lundquist 1927) and the biologic remains preserved in it are the principal sources of paleolimnological information. The substrates exposed in shallower water are (2) coarse, inorganic deposits dominated by sands and gravels. These deposits include boulders, cobbles, gravel and sand originating within the original basin, as well as coarse inwashed inorganic debris, analogous to that trapped behind weirs of the Experimental Watersheds at Hubbard Brook (Likens et al. 1977; Bormann et al. 1974). Inwashed particulates have been deposited over these original sediments near the lake's inlets, but elsewhere have generally been mixed with material eroded and redeposited within the basin. We have not tried to establish the relative proportions of inwashed and redeposited littoral deposits. The littoral deposits are more difficult to sample than the gyttja because of the difficulty in penetrating them, are difficult to date because of the low C content and are probably of locally irregular stratigraphy.

A third sediment type is (3) an intermittent layer of decaying leaves and sticks from terrestrial plants. These deposits are highly localized and usually thin (1 to 10 cm). They occur along all parts of the lake shore, but are especially abundant along the northern, western and northeastern shores. Usually this debris overlies coarse sand and gravel, but in some places the decomposing leaves form a thicker coarse detrital gyttja that intergrades with the midlake gyttja.

During the early history of the lake a fourth type of sediment, (4) inorganic silt and sand of a characteristic gray color, was deposited widely across the lake bottom. Near shore these sediments are sometimes coarse and sandy; even near the center of the basin, layers of sand are interbedded with sediment that is composed largely of silt and clay-sized particles (80 to 90% by weight). These gray sediments contain little organic matter (*Chapter VII.E*). They were carried into the lake by streams, slopewash or solifluction, at a time when erosion rates were high and water velocities were sufficient to carry relatively dense but fine-grained materials into the center of the lake basin. These deposits may include an unknown quantity of eolian silt originating from the glacial outwash deposits in the nearby Pemigewasset River Valley.

1. Late-Glacial and Holocene Sedimentation

MARGARET B. DAVIS and M. S. (JESSE) FORD

Basin Morphometry

Sedimentation began 14,000 yr B.P. when a buried ice block melted forming a lake basin (*Chapter III.A*). The shape of the basin at this time was inferred from cores and probes through the sediment (Davis and Ford 1982). Nine cores bottomed in gray silt and sand older than 11,000 yr; other cores and probes gave information on minimum depth at that time. From these data we have made a generalized map of the lake basin found in glacial drift, prior to infilling with lake sediment. The shape of the basin (*Figure VII.A.1*–1) was influenced by the thickness of glacial drift and the shape of the ice block; it bears little relationship to the underlying bedrock surface (*Chapter III.A*). The early lake basin was much the same shape as the modern lake, except that the steep slopes that now characterize only the shallower parts of the lake extended to the deepest part of the lake basin, 23 m below the level of the modern lake surface. Although the level of the lake surface at that time is not known, we assume it was similar to the modern, relatively high lake level.

The cumulative thickness of sediment is summarized in *Figure VII.A.1*–2 as a function of water depth. The variable u is the percentage of the lake area deeper than x; it varies between 100 (where x is zero) and zero (where x is the maximum depth). The difference between the plot for the original basin and that for the modern basin represents infilling. Values of u for the shallower areas of the basin have changed very little because gray silt is the only kind of sediment that accumulated there.

Total volume of sediment was calculated from a map of sediment thickness. Areas of lake bottom underlain by 0, 1, 2 . . . 12 m of sediment were traced from the map and square meters of area were calculated. The total volume of sediment is approximately one-half million cubic meters, which would make a uniform blanket of sediment 3.5 m deep if spread evenly over the basin. Instead, almost half the basin has 1 m of sediment or less, and other parts of the basin have up to 12 m of sediment.

Sediments deposited prior to 11,000 yr B.P.

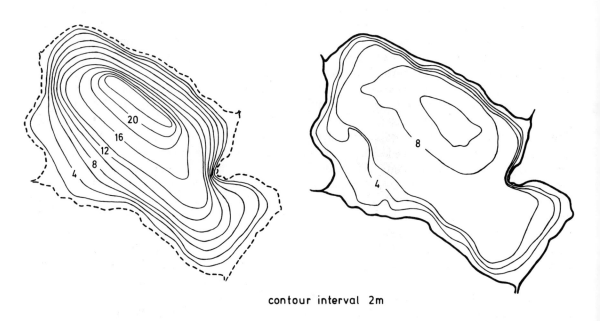

A. 14,000 yr BP   B. present

Figure VII.A.1–1. Bathymetry of Mirror Lake. **A.** 14,000 yr ago; and **B.** at the present time.

A. Sedimentation

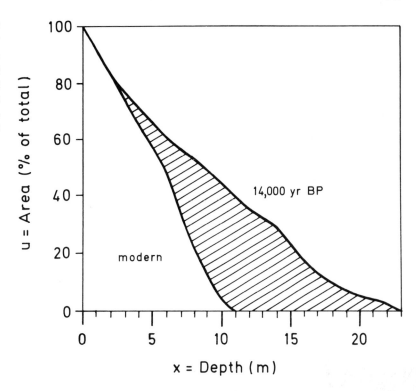

Figure VII.A.1–2. Plot of u vs. x for Mirror Lake 14,000 yr B.P. and at the present time. In this area vs. depth plot, u is the percentage of lake bottom deeper than x, where x is the depth of overlying water. The shaded region represents accumulated gyttja.

are primarily gray silt and sand. Cores and probes show that these sediments are distributed widely over the original lake basin, covering about 85% of the basin area (*Figure VII.A.1*–3A). Probes by a SCUBA diver revealed gray sand between large boulders in shallow water; often the gray silt or sand is covered with a thin layer of coarse sand or very recent organic sediment (*Figure VII.A.1*–4). By contrast, during most of the Holocene, organic sediments were deposited over only about 50% of the basin, or in water that is now deeper than 6 m (*Figure VII.A.1*–3B). Damming of the lake in the nineteenth century raised the level 1 to 2 m; postsettlement organic sediments now accumulate over about 68% of the lake basin in water deeper than 5 m (*Figure VII.A.1*–3C). These relationships suggest that the depth to the midsummer thermocline, which is now at approximately 4 to 6 m, is an important parameter controlling the pattern of gyttja accumulation.

Pattern and Rates of Sediment Accumulation

Accumulation rates of sediment (mg accumulated/cm^2-yr) were calculated at five sites in Mirror Lake (numbered sites on *Figure VII.A.1*–3B). Nine radiocarbon dates were obtained from the Central core, and a polynomial curve was fitted by least squares regression to a plot of age vs. depth. The equation of this curve was used to estimate the age of all points in the core, and the first derivative was used to estimate deposition time, i.e. the years represented by 1-cm thickness of sediment (cf. Likens and Davis 1975). Pollen was counted in many levels in the cores collected at all five sites, and the pollen diagrams were cross-correlated with the pollen diagram from the radiocarbon-dated Central core. In this way it was possible to estimate with only a few hundred years' error the age of 10 to 20 levels in each of the other cores. A few levels in the other cores (indicated by circled points in *Figure VII.A.1*–5) were also dated by radiocarbon.

The deepest sediments in core #782 are older than the deepest levels recovered in the Central core. These older sediments were radiocarbon-dated and the pollen was analyzed. The deepest levels in the other cores were then cross-correlated with the dated levels in core #782, using relative abundance of pollen taxa, loss-on-ignition and total pollen concentration. Once a number of levels had been dated, polynomial equations were fitted by regression analysis to the plots of age vs. depth in each of these four

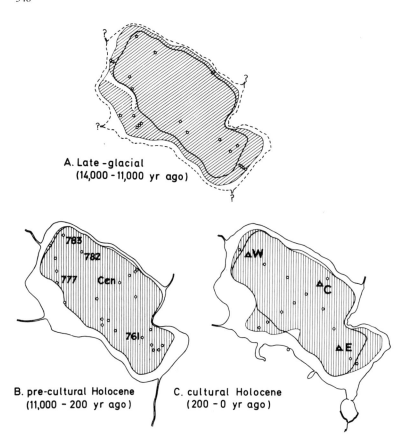

Figure VII.A.1–3. Maps of Mirror Lake. Hachured area indicates the zone of sediment deposition at three stages of lake history. **A.** Late-glacial gray silts and fine sand, showing locations of cores or probes penetrating this deposit. Water level at this stage is not known; dashed line shows the present level. **B.** The Holocene gyttja, up to the cultural period. The cores used to map the deposit are indicated, and the five cores discussed in detail are identified. Water level was about 1.5 m below its present-day level as the cultural period began. **C.** The postsettlement ("cultural") gyttja, showing locations of surface cores and the three sediment-trap sites. The modern lake shore is drawn. Note that the present 6-m water depth contour is included in all maps as a reference.

cores. The first derivatives of the equations were used to estimate deposition time. The dry bulk density (mg/cm^3), percentage organic matter (loss-on-ignition at 500 ± 25°C), and percentage inorganic matter (weight remaining after ignition) were determined at various levels in all five cores. The dry bulk density divided by deposition time (yr/cm) is the rate of sediment accumulation (as mg/cm^2-yr). Detailed curves for the deposition of organic and inorganic matter at each site through time have been presented elsewhere (Davis and Ford 1982; *Chap-*

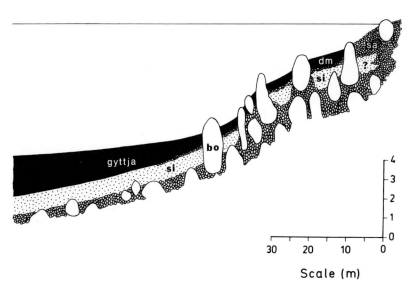

Figure VII.A.1–4. A generalized reconstruction of shallow-water sediment stratigraphy at the eastern end of the transect of surface cores. Gyttja intergrades with a coarse organic debris (dm) in shallow water. The late-glacial silt/sand (si) extends into very shallow water, where it lies buried under sands and gravel (sa) originating within the basin. The limit of the late-glacial deposit and its maximal thickness (≥0.7 m) are unknown. Note the late-glacial deposit between boulders (bo) in shallow water.

A. Sedimentation

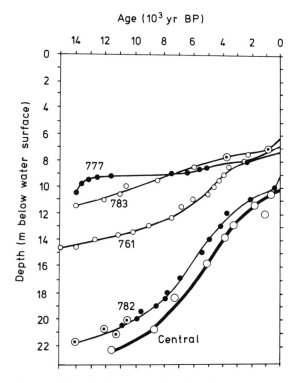

Figure VII.A.1–5. Age vs. depth below water surface for five cores collected from the five stations indicated in *Figure VII.A.1*–3B. Large circles (Central) and circled points (782, 783) indicate radiocarbon-dated levels. Age estimates for other levels were obtained by cross-correlation with radiocarbon-dated cores.

ter VII.E) and are summarized in the following discussion.

The accumulation rate for inorganic sediment was relatively high, averaging 53 mg/cm^2-yr, at about 13,000 yr B.P. Accumulation was fairly uniform from site to site in the deeper parts of the lake but was slightly greater in shallow water. At site #777, the extremely high rates prior to 13,000 yr B.P. apparently reflect at least one episode of sand deposition. (The high rate could also be an artifact caused by miscorrelation of pollen-poor sediment with #782). Between 12,000 and 11,000 yr B.P. the accumulation rate for inorganic sediment fell sharply and remained low throughout the Holocene.

Inputs of inorganic particulate matter to the lake were calculated by extrapolating the average accumulation rate 13,000 years ago in the five cores to the entire area receiving sediment at that time (about 85% of the lake area). Annual inputs of inorganic sediment to the entire lake were 67,800 kg/yr. Assuming that the size of the drainage area was similar to that at the time of settlement (107 ha), 650 kg of particulate matter per ha of drainage area would have been eroded and transported to the lake each year. We assume the estimate includes only inorganic particulates such as clay, silt and sand. Diatoms are absent from sediments of this age. Presumably dissolved inorganics, including all the dissolved silica, left the system through the outlet or subsurface seepage.

The export rate thus calculated for the late-glacial period can be contrasted with modern rates for forested and for disturbed watersheds at Hubbard Brook. The late-glacial rate of 650 kg/ha-yr is more than 30 times the modern particulate inorganic exports from forested watersheds, which average about 20 kg/ha-yr (Likens et al. 1977). Three years after clear cutting and suppression of regrowth on Watershed 2 of the HBEF, particulate exports had risen to 327 kg/ha-yr (Bormann et al. 1974), one-half the rate during late-glacial time in the landscape surrounding Mirror Lake. The export rate from Watershed 2 was affected by flushing of material accumulated in the streambed and may represent a maximal rate, possible only for a short time period (Bilby and Likens 1980).

The deposition of organic matter averaged 1.2 mg/cm^2-yr during the late-glacial period of silt deposition at sites where silt accumulated, or 1.0 mg/cm^2-yr for the lake as a whole. This is almost two-thirds of the Holocene deposition rate: 3.2 mg/cm^2-yr in the gyttja zone, or 1.6 mg/cm^2-yr for the lake as a whole. Our estimates of organic matter are based on loss-on-ignition, which may overestimate organic content of highly inorganic sediment containing clays (Mackereth 1966). Parallel loss-on-ignition and organic C determinations of the late-glacial silt from Mirror Lake (*Chapter VII.E*) suggest that an overestimate on the order of twofold has occurred. It is evident, nevertheless, that organic deposition during the late-glacial, especially in the period after 12,000 yr B.P., was appreciably greater than the highly inorganic nature of the sediment suggests.

It has often been supposed that productivity of lakes was low under cold, late-glacial climatic conditions (Hutchinson and Wollack 1940; Deevey 1953; Livingstone and Boykin 1962). Indeed, the presence of subarctic and arctic plant species in the late-glacial terrestrial flora, and geomorphic evidence for formerly massive frost action at high elevations, indicate that the past regional climate of New England was colder than today. Charophyte oospores and pollen from aquatic macrophytes occur in the late-glacial sediment, however, indicating that the lake was not sterile during this period, even though diatoms are not present in the sediment (*Chapter VII.B*). Productivity was apparently higher than implied by the low organic content; diatoms could have dissolved in sediment pore water undersaturated with respect to silica. The surprisingly high deposition rate of organic matter may also reflect unusually effective preservation resulting from rapid burial in dense, silty sediments, or from anoxic conditions in and above the sediment.

Between 12,000 and 11,000 yr B.P., the nature of the sediment changed greatly (see *Chapter VII.B,C,D* and *E* for details of chemistry and biota), as did the pattern of sediment deposition within the lake. Inputs of inorganic particulate matter decreased rapidly, and by 10,000 yr B.P. accumulation rates for inorganics were similar to modern rates. This striking change in sediment export from the watershed is correlated with the local arrival of spruce trees 11,300 yr B.P. (Davis et al. 1980; Davis and Ford 1982; see also *Chapter VII.E*). Meanwhile inputs of sedimentary organic matter increased, and the percentage loss-on-ignition of the sediment began to rise, increasing steadily throughout the Holocene from about 10% 10,000 yr B.P. to 34 to 40% in presettlement time (*Figure VII.A.1–8*). Diatoms first appeared in the sediment about 11,000 yr B.P. (*Chapter VII.B*). During the Holocene most of the inorganic sedimentary material, at least in the deeper part of the lake basin, was diatomaceous silica, and thus represents biologic precipitation of dissolved lake-water silica.

The sediments of the post-glacial period are much lower in density than the late-glacial silts. Wet bulk density is close to 1, whereas wet bulk density for late-glacial silt is 1.5 to 2.2 g/cm^3. Because of its low density, post-glacial sediment was apparently much more easily resuspended from the shallower portions of the lake bottom and focused into the deeper parts of the lake. In deep water (site #782 and Central, *Figure VII.A.1*–3B), net accumulation rates of sediment were as much as six times greater than at intermediate depths (#782), while no sediment at all accumulated for several thousand years at site #777. Sediment accumulation ceased in all parts of the lake basin less than 6 m below the modern water surface.

Several cores taken at water depths of 6 to 8 m contained layers of sand as much as 5 cm thick interbedded with organic sediment. An apparently similar layer of sand is exposed today on the lake bottom at water depths of 5 to 7 m near the southwest shore of the lake. The sand deposits may have accumulated as the result of unusual storms that produced waves or currents of sufficient velocity to move sand out from shore (*Chapter III.B.2*). Alternatively, the lake level may have fallen, changing formerly deep-water sites into sites sufficiently shallow for deposition of sandy littoral sediments. A third possibility is that the sand layers represent a lag deposit, with organic fines winnowed away. This last possibility seems unlikely, since the sand content of the organic muds in most parts of the lake is quite low. It seems more likely that the sand resulted from unusual events that carried beach deposits out into the lake basin.

Sediment focusing appears to have been a continuous process throughout the Holocene history of the lake. We visualize at least two different mechanisms, one dependent on unusual weather and the other continuous. We believe that both processes have been important at Mirror Lake.

The first mechanism of sediment focusing is the episodic erosion and resuspension of sediment several centimeters thick during storms. The frequency of storms may have varied through time, and as the basin filled the locations of sites subject to erosion may have changed. Sediments, once resuspended by this means, may have been moved laterally by deep wave action, especially in the autumn when the lake is not density-stratified. Resuspended sediments also may flow as density currents into the deeper portions of the basin, where layers of sediment of uniform age several centimeters thick may be deposited over a small area (Ludlam 1974). Sediment deposition appears continuous at most sites in Mirror Lake, but the homogeneous color and texture would make it difficult to identify layers that had originated in

A. Sedimentation

this way. Von Damm et al. (1979) have identified a 6-cm layer of sediment with uniform ^{210}Pb content at 24- to 30-cm depth in the core they studied from near the deepest part of Mirror Lake. The pollen content of the sediment in this layer is uniform and is consistent with the idea that the sediment was deposited within a short time interval. Gorham and Sanger (1972) have described a local surficial sediment layer from Stewart's Dark Lake, Wisconsin, which is characterized by unusual pigment concentrations, suggesting transport as a unit from a shallower part of the lake. A different line of evidence is provided by the stratigraphy at site #777, where there is an unconformity separating sediments deposited 11,500 and 7,600 yr B.P. During at least the latter part of this time, site #777 may have been a locality that was especially prone to erosion. The sharp contact suggests that sediment accumulated there and was subsequently eroded away.

A second mechanism for focusing is the continuous annual erosion and resuspension of flocculent organic sediment from the littoral zone of the lake. Resuspended sediment is mixed in the water-column during seasonal overturn and transported throughout the lake. In the late autumn and after the lake freezes in winter, the sediment is redeposited. This process has been demonstrated to occur in Mirror Lake by sediment trap studies (see *Chapter VII.A.2*). The resuspended sediment comes from the littoral zone (Davis 1968), but it is eventually redeposited in a thinner, although uneven, layer over most of the basin. Sediment redeposited in the littoral zone will be resuspended again the following year, whereas reworked sediment is seldom if ever resuspended from the profundal parts of the basin. The net effect of the process of annual resuspension from the littoral zone, combined with redeposition over a wider area, is lesser net accumulation in the littoral zone than in the profundal zone. In the extreme case (such as Mirror Lake) the littoral zone may have zero net accumulation of sediment. The increment of sediment accumulated in the profundal zone each year is the sum of sediment produced that year plus a smaller quantity of nearly contemporaneous sediment transported from the littoral zone.

The deposition of resuspended sediment is influenced by water currents, and therefore is not uniform in time or space. Although the net outcome is continuous accumulation in the profundal zone, the rates of accumulation differ from site to site and vary through time at individual sites. Holocene rates of accumulation of organic sediment were lower, for example, at the two sites of intermediate depth, #777 and #783, than at the three deep-water sites (*Figure VII.A.1*-6). We know that even when net accumulation of sediment occurred at #777, it was a site where water movement winnowed away organic sediment; the sediments there have consistently lower percentage loss-on-ignition than contemporaneous sediment at other sites. The other site at intermediate depth, #783, accumulated organic sediment at rates similar to those at deep-water sites until 8,000 yr B.P. After 8,000 yr B.P., however, average accumulation rates at this site dropped to less than those found in deeper water. This change seems to represent a permanent change in the velocity of water currents in this part of the lake.

During the Holocene, the pattern of sediment accumulation has been changing continuously through time, probably because the pattern of water movement within the lake is sensitive to the constantly changing morphometry of the infilling lake basin. The velocity of water movement may also have been affected by the climatic regime and by the vegetation cover around the lake. The three deep-water sites have accumulated more sediment during the Holocene than the two sites at intermediate depths. All three show peak rates of accumulation between 7,000 and 4,000 yr B.P. Before 5,500 yr B.P., large amounts of organic sediment accumulated at site #782, but more recently rates were consistently higher at the Central site, resulting in a greater total accumulation there by the time of settlement 200 yr B.P. This shift in relative accumulation rates, illustrated diagrammatically for four sites along a west-east transect in *Figure VII.A.1-6*, occurred even though accumulation rates at all sites declined after 4,000 yr B.P. Clearly, the time of maximum accumulation in any one core depends on processes of deposition and is not a clear indication of lakewide changes in sediment inputs. Frequently paleolimnologists have inferred lake productivity, or changes in inputs from the watershed, from changes in organic content or accumulation rates in a single core.

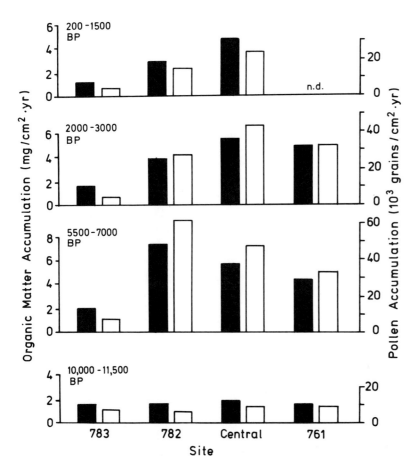

Figure VII.A.1–6. Rates of accumulation of organic matter (solid histograms) and pollen (open histograms) at four sites along a west-east transect in Mirror Lake. Accumulation rates are averaged within four time intervals. Rates were not determined at site #761 for the interval 1,500 to 200 yr B.P.

Our comparisons of several cores show that greater knowledge of sedimentary patterns within the basin *must* be established before inferences are drawn about events in the lake basin as a whole.

The total input of inorganic sediment during the Holocene, calculated by averaging the accumulation in all five cores, was approximately 12 mg/cm²-yr. When extrapolated over the area of the lake basin accumulating sediments (50%), total inputs were 9,200 kg/yr, equivalent to an export rate of 86 kg/ha-yr from the watershed. This estimate, unlike the late-glacial estimate, includes much of the dissolved silica from the watershed because diatoms, which convert dissolved silica into siliceous cell walls, are present in large numbers in the sediment (*Chapter VII.B*). Currently, dissolved silica enters Mirror Lake through inlet streams, but less than half of this leaves by way of the outflow (*Chapter IV.E*). The Mirror Lake watershed now exports dissolved silica at the rate of approximately 60 kg/ha-yr and particulate inorganic matter at the rate of about 20 kg/ha-yr. These exports are close to the 86 kg/ha-yr we estimate as the average Holocene export rate based on sedimentary accumulation of inorganic matter.

A late-Holocene decline in organic matter accumulation was originally reported for the Central core in 1975 (Likens and Davis 1975). Lehman (1975) analyzed this result by making certain assumptions about sediment accumulation patterns in Mirror Lake and applying a mathematical model to generate estimates for post-glacial inputs to the lake. The significant prediction from Lehman's model was that even though the accumulation rate of organic matter decreased at the Central site after about 5,000 yr B.P., lakewide inputs were still increasing to a maximum at about 4,000 yr B.P., after which they remained constant (*Figure VII.A.1–7*).

The model of sediment accumulation Lehman used to arrive at this conclusion was one in which sediment accumulated in parallel

A. Sedimentation

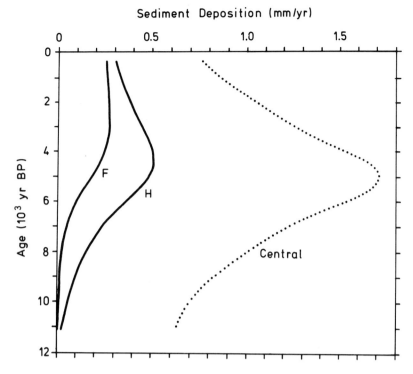

Figure VII.A.1–7. Hypothetical rates of sediment accumulation in Mirror Lake as extrapolated from the Central core (Central) to the entire lake basin, using Lehman's frustum (F) and hyperboloid (H) models of deposition. Neither model produces an accurate lakewide rate of accumulation owing to depositional irregularities not written into simple geometrical models.

horizontal layers over an ever-increasing area of the basin (frustum model). Our subsequent work, however, has demonstrated that this model is inappropriate for Mirror Lake, since even the oldest sediments arc found to occur near the margins of the lake. We propose two alternative models of sediment accumulation. One is the hyperboloid model described by Lehman, which would allow for slow accumulation even at the edge of the lake. Using the same data base, that is, the sedimentation rates at the Central site, we have calculated lakewide inputs (*Figure VII.A.1–7*). Significantly, we find that this model makes a very different prediction from the frustum model, that lakewide inputs began a continuous decline about 5,500 yr B.P. A third model alternative to both the frustum and hyperboloid models makes no assumptions about the symmetry of sediment accumulation, i.e. allows for accumulation in irregular layers, and therefore does not permit lakewide deposition to be computed from a single core.

The three models can now be tested with data from the four additional cores. All of the cores show an irregular decline in accumulation rates for both organic and inorganic sediment during the last 4,000 yr. We have attempted to learn whether this decline is (1) the result of sediment focusing at a time of constant inputs, i.e. decreased accumulation at our sites and increased accumulation elsewhere in the basin, or whether it represents (2) a lakewide event of significance to the ecosystem as a whole, such as a decline in productivity or a decline in allochthonous inputs to the lake. The first alternative is suggested by the frustum model, the second by the hyperboloid. Both were discussed by Likens and Davis (1975) as possible explanations for the pattern observed at the Central site.

To distinguish sediment distribution phenomena from changes in lakewide sediment inputs, we have used pollen accumulation rates. Pollen is a conservative sedimentary constituent of density similar to the organic sediment matrix. Pollen is generally well preserved in the sediments of Mirror Lake; differential rates of accumulation cannot be attributed to differential decomposition. Furthermore, inputs of pollen to the lake can be assumed to be more or less constant for the last 9,000 yr, when mixed deciduous forests grew on the surrounding landscape. Inputs of pollen prior to that time were lower, because spruce forests produce fewer pollen grains (Davis et al. 1973). We show in *Figure*

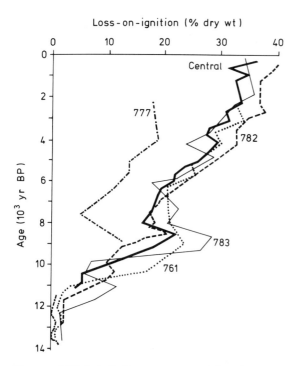

Figure VII.A.1-8. Percentage weight loss-on-ignition vs. age for five cores from Mirror Lake.

VII.A.1-6 the pollen accumulation rates at four sites, averaged for the same 1,500-yr intervals for which organic matter accumulation is shown. Pollen accumulation was low and fairly uniform during the spruce period 10,000 to 11,500 yr B.P. Rates of accumulation rose during the Holocene, and the difference in rates between intermediate and deep-water sites demonstrates focusing. During the interval 7,000 to 5,500 yr B.P., pollen accumulation was highest at site #782; but during the intervals 3,500 to 2,000 and 1,500 to 200 yr B.P., pollen accumulation was highest at the Central site. This result confirms our conclusion that sediment deposition patterns have shifted from one site to another within the basin through time, resulting in deposition in uneven layers within the basin. Redeposition has brought more sediment and more pollen each year to one site than to another for periods lasting several thousands of years. Then, as the pattern of water circulation has shifted, redeposited sediment has been focused at different sites within the basin for long periods. Several shifts of this kind have occurred during the history of the lake. The pattern of pollen accumulation leads us to reject both the hyperboloid and frustum models as accurate predictors of sedimentation within the lake. Both models assume deposition in even layers, whereas our data show that sediment tends to be deposited unevenly, with a pileup of sediment first at one site and then at another within the lake.

Pollen accumulation rates decline at the three sites shown in *Figure VII.A.1-6* beginning 3,500 yr B.P., paralleling the decline in accumulation of organic sediment. It seems unlikely that production of pollen by the vegetation would have declined during this long interval; we assume that pollen inputs to the lake remained more or less constant, since the vegetation was much the same. A similar conclusion was reached by R. B. Davis et al. (1975b) for a lake in Maine. The decline in accumulation rates for both pollen and organic sediment appears, therefore, to result from the process of focusing, which distributed more sediment throughout this interval to other sites within the basin. Of course this argument is weakened by the fact that we have not detected an increase in sediment accumulation at this time at any of the sites we studied in detail; nor have we found sites where sedimentation was initiated between 3,500 to 200 yr B.P. However, not all areas of the lake basin have been studied intensively. An increase in sedimentation at intermediate depths (6 to 8 m) would have to be quite large, on the order of 20%, to account for the decline of sediment accumulation throughout the deeper parts of the basin. We believe that sediment focusing is the likely explanation for the decrease in accumulation at the Central site, although the pattern is not adequately described by a simple geometrical model.

During the mid- and late Holocene, the character of the sediment changes everywhere in the lake basin. All of the sites studied in detail show similar trends in percentage loss-on-ignition. At the four sites shown in *Figures VII.A.1-6* and 8 the loss-on-ignition changes gradually from about 20% 6,000 yr B.P. to 35% in presettlement time (*Figure VII.A.1-9*). As all of the cores show a decrease after 3,500 yr B.P. in the accumulation rates for both organic matter and inorganic matter, the change in loss-on-ignition indicates that there is a relatively greater

A. Sedimentation

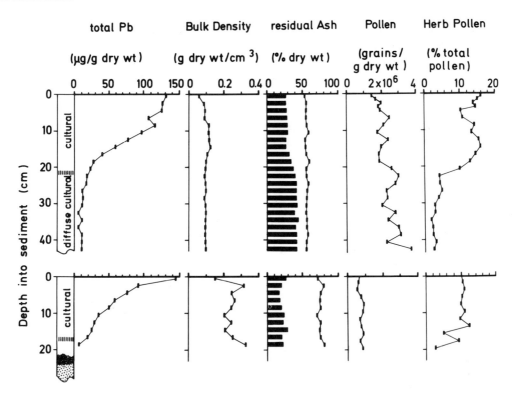

Figure VII.A.1-9. Stratigraphy of two surface cores. The upper core (REM no. 4) is from the Central sediment-trap site in 10.1 m of water. The lower core (REM no. 25) is from the eastern margin of the gyttja zone, in 5.4 m of water. Residual ash is the residue left behind after HNO_3 digestion and HCl extraction. The residual ash is subdivided into amorphous silica (black) and mineral particles (clear). Note that the cultural gyttja in the lower core overlies a thin layer of sand and, below that, the late-glacial silt/sand.

decline in deposition of inorganic matter than organic matter. The change in ratio appears to represent a significant lakewide change of some kind, either in inputs to the lake, or in the way those inputs were processed by biotic or chemical processes within the lake. A community change that affected the frequency of diatoms relative to other algae might produce this result, or a change in conditions affecting preservation of diatomaceous silica. A number of possible interpretations are discussed in *Chapter VII.F.*

2. Contemporary Sedimentation

Robert E. Moeller

The Surficial Sediments: Stratigraphy, Focusing and Rate of Accumulation

The most recent sediments can be related to the lake as we have known it during the period 1965 to 1982, though information regarding the spatial pattern, rate and composition of the sediment accumulating today also aids the interpretation of older deposits sampled as long cores. In Mirror Lake we have pursued three convergent approaches to modern sedimentation: (1) stratigraphic analysis of 16 short cores (<50 cm long) taken with a piston corer across the mapped zone of gyttja deposition; (2) analysis of a 6-yr sequence of sediment collected periodically as it settled into wide-mouth jars (*sediment traps*) exposed 1 m above the lake bottom; and (3) computation of the amounts of several chemical elements deposited annually in the lake as the difference between inputs to the lake and measured outputs. The major objectives have been to assess the impact of human disturbance on sedimentation, and to test whether these three approaches yield mutually consistent results (Moeller, Davis and Likens, in preparation).

From the surface cores we have calculated the amount of sediment, and of some of its constituent elements, that has been deposited since 1840 A.D. The sediment traps record the seasonality of deposition, which cannot be deduced from the sediment stratigraphy owing to substantial mixing of surface mud by benthic invertebrates. Material collected by the sediment traps is partially, but not completely, decomposed. After studying the rate of solubilization of trapped particulates, we have made rough estimates for both the actual and the net (long-term) flux of matter to the sediments. The nutrient budgets for the lake as a whole (*Chapter IV.E*) provide an estimate of long-term sedimentation that is independent of depositional patterns within the lake, and thus provide a check on rates of elemental deposition calculated from cores and from sediment traps.

The rate at which sediment has accumulated within the mapped gyttja zone has been determined on the basis of a single datable pollen horizon, which can be found in all cores. At a depth into the sediment that varies from 12 to 25 cm, a substantial increase in the deposition of herb pollen relative to tree pollen occurred that suggests a major change within the local or regional vegetation. Below this level herbs make up roughly 3% of the total terrestrial pollen. Above, herbs average 10 to 15%. Europeans did not settle in the Pemigewasset Valley near Mirror Lake until about 1770 A.D. Local forest clearing and agriculture appear to have left only a sparse record in the pollen until the mid-nineteenth century, when the local population rose to a maximum that probably represented an overpopulation for the generally infertile, quickly exhausted soils (Likens 1972c; see also *Chapter III.B.3*). This population maximum can be associated with the major rise in herb pollen grains within the sediment; it was soon followed by a massive depopulation (Goldthwait 1927).

The date assigned to the rise in herb pollen, 1840 A.D. ± 30 yr, is imprecise owing to a major inconsistency in dating. Based on local population trends, the major increase in herbs should have occurred before 1820, as settlers cleared farms within the Pemigewasset Valley and established a sawmill at the outlet of Mirror Lake (Likens 1972c; see also *Chapter III.B.3*). The observed increase in herbs occurred just as the concentration of lead (Pb) in the sediments rose above its background level, and shortly before a rapid rise began (*Figure VII.A.1–9*). (The coincidence of the herb rise with the first increase of Pb is consistent [±2 cm] from core to core—we have used the Pb rise in place of pollen counts to locate the 1840 level in many of the surface cores.) The problem, however, is that the Pb rise is attributable to regional or distant industrial activity and should not have occurred before 1860 (Edgington and Robbins 1976; Appleby and Oldfield 1979). Geochemical dating (^{210}Pb) of a single deep-water core by Von Damm et al. (1979) indicated a date of 1884 for the herb rise. Different assumptions applied to the same data suggest a slightly earlier date of 1870 (F. Oldfield, personal communication). Thus, the chemical evidence suggests that the herb rise dates to 1860–1880 A.D. Is it possible that the peak of activity in the towns of Thornton and Woodstock, dating from 1820 to 1860 (Likens 1972c; see also *Chapter III.B.3*), left almost no impression in the pollen record? We think not, but the geochemical evidence for a later date is equally compelling. Our date of 1840 is therefore a compromise. At two other sites in more heavily farmed parts of New Hampshire, the herb rise lies several centimeters below the first increase in Pb (R. Moeller, unpublished data). At those sites there is no reason to suppose that the herb rise does not reflect local clearing early in the eighteenth century.

Aside from the changes in the pollen spectrum at the 1840 horizon, there is little evidence in cores from deep water to indicate that any dramatic disturbance occurred within the drainage basin (*Figure VII.A.1–9*). The pollen concentration decreased above the 1840 horizon, suggesting dilution in a more rapidly accumulating matrix. The dry bulk density and the residual-ash (the portion of the sediment that resists extraction in hydrochloric acid following digestion in concentrated nitric acid) do not change. The constancy of residual-ash, at about 55% of the mass of dried mud, does hide a significant

A. Sedimentation

increase in the proportion contributed by mineral particles compared with diatomaceous silica. Settlement around Mirror Lake has not, however, brought about the dramatic erosion of silts and clays from plowed land that occurred where agriculture was more intense (e.g. Davis 1976a; Brugam 1978). The deposition of dry matter since 1840 has averaged 140 g/m²-yr at sites deeper than 9 m. By comparison, the rate during the preceding 200 yr was about 120 g/m²-yr in a long core from the same region of the lake (Likens and Davis 1975).

One aspect of sedimentation that has changed dramatically is the spatial pattern of deposition. Early in the settlement period, shortly before the major herb rise, something happened within the lake that caused gyttja to accumulate where it had not before. Presumably this event was the construction of a small dam at the outlet to provide water power for a sawmill (Likens 1972c; see also *Chapter III.B.3*). Maintained till the present, probably irregularly, such dams have raised the lake surface by 1 to 2 m. The sedimentation limit has likewise risen, allowing gyttja to accumulate over about 68% of the present lake area, compared with about 50% before the dam. Cores from the southern and eastern margins of the gyttja zone contain a post-1840 layer comparable in thickness to that of the mid-lake sites. The stratigraphy of the marginal deposits, revealed down to a depth of 1 m below the substrate surface by a SCUBA diver using a soil auger, shows contemporary gyttja accumulating almost directly on top of late-glacial silt and sand (*Figure VII.A.1–4*). Of course, the gyttja from shallow water is sandier than mid-lake gyttja; the dry bulk density is much higher than in mid-lake, and the pollen concentration is lower (*Figure VII.A.1–9*).

A more useful representation of sedimentary patterns considers the mass of sediment deposited locally, in addition to its thickness. Although the thickness of the post-1840 layer is relatively uniform along a transect from the western to the eastern shore (*Figure VII.A.2–1*), the mass of sediment deposited is greatest near shore. Although organic carbon and amorphous silica are fairly uniform, mineral matter accumulated much more rapidly near shore than in mid-lake (*Figure VII.A.2–1*). Amorphous silica is defined here as the SiO_2 that dissolves from the residual-ash upon heating with

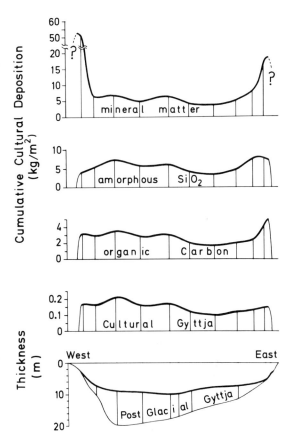

Figure VII.A.2–1. West-to-East cross-section of the cultural gyttja (thickness and mass of major constituents deposited since 1840 A.D., corrected for mixing) compared with the thickness of all Holocene gyttja. The thickness of the Holocene gyttja is based on cores and probes and sometimes includes some of the late-glacial silt/sand.

0.1N KOH for 1.5 h. Amorphous silica is composed largely of diatom frustules, so it is a biogenic product of lake metabolism.

Chemical analysis of the gyttja cannot provide accurate values for the total deposition in the lake of elements that are associated with coarse particulate inputs, especially inorganic particulates. Coarse particulate matter that washed into the lake is deposited in part within the shallow zone above the sedimentation limit defined by the accumulated gyttja. The coarse particulate matter contained in the gyttja, especially the inorganic matter near the margin of the gyttja zone, probably includes material (e.g. sand) that has been transported from basin deposits at the shoreline. Shoreline erosion has increased since human alteration of the lake's water level and continues conspicuously along the southern shore today.

The contemporary deposition of gyttja follows a pattern that differs in two respects from the longer established pattern preceding settlement: (1) the area over which *permanent* deposition occurs has increased; and (2) the intensity of differential focusing into deeper portions of the gyttja zone has decreased markedly. The layer of gyttja deposited since 1840 tends to be only slightly thinner near the margin of the accumulation zone than in the center of the lake, whereas the whole thickness of post-glacial gyttja decreases dramatically toward the margin of the gyttja zone (*Figure VII.A.2*–1). Any attempt to deduce changes in landscape erosion since settlement must recognize that *an unchanging rate of gyttja formation in this lake would show up as a decline in accumulation at a midlake site.*

At any site, the mass of sediment accumulated above the 1840 pollen horizon provides a mean annual rate of deposition for the 138 yr between 1840 and 1978, when the surface cores were taken. Probably the deposition rate has varied within this period, a problem that has been addressed at one site by geochemical dating techniques (Von Damm et al. 1979). One complicating problem that we have recognized in Mirror Lake is the mixing of surface sediments by benthic invertebrates. Exotic pollen grains deposited experimentally at the surface of gyttja *in situ* at depths of 6, 7, 8, and 10 m were moved downward 2 cm (at 6 and 7 m) or 4 cm (at 8 and 10 m) during the course of one year (Moeller et al. 1984). The *Chaoborus* that are common in mid-lake gyttja (*Chapter V.A.6*) are known to swim through watery surface sediments, even to depths of 5 to 10 cm (LaRow 1976). They may be responsible for the observed mixing at the 10-m site. Other organisms are abundant in gyttja deposited at depths less than 8 m, including oligochaetes and chironomids. The large tubificids, so renowned for their sediment reworking ability (R. B. Davis 1974a; Robbins et al. 1977, 1979), occur at intermediate depths. The net effect of sediment mixing is that a significant portion of the sediment situated immediately above the apparent onset of the major herb rise in a core actually predates the horizon (R. B. Davis 1974a).

From our 16 surface cores we have calculated the mean rate of deposition in the lake of dry matter and selected chemical elements (*Table VII.A.2*–1). These are separated into two phases: detrital minerals (which successively resist solubilization in concentrated HNO_3 and dilute KOH) and the collected sum of all other phases. The latter phases are considered "reactive"—they may have participated in the biogeochemistry of lake metabolism. These values for sedimentation can be compared with numbers presented in the nutrient budgets (*Table IV.E*–13). For the discussion of nutrient budgets, permanent sedimentation was extrapolated from deposition rates at the deepest part of the lake, taking into account the area of sediment accumulation. This procedure underestimates the deposition of elements like Si, Na, K, Mg and Ca, which occur in detrital minerals. Sand and silt-sized mineral particles resist focusing, and thus accumulate more rapidly near the margin of the gyttja zone.

Values for permanent sedimentation calculated as the difference between measured inputs to the lake and measured outputs (see data in *Table IV.E*–13) agree somewhat better with the extrapolated values ("permanent sedimentation" of *Table IV.E*–13) for elements like Si, Ca, Mg, and especially Na and K, rather than with the actually measured values (*Table VII.A.2*–1). Three factors contribute to the discrepancy: (1) particulate inputs to the lake were extrapolated from exports of the undisturbed watershed at Hubbard Brook and must be underestimates for the more disturbed lake watersheds; (2) the cultural layer of gyttja accumulated over the last 140 yr, including periods of more intense disturbance than today; and (3) detrital minerals deposited in the gyttja may also originate within the lake during erosion of the shoreline.

Origin of the Sediment

Algae, zooplankton and bacteria function as producers, ingesters and decomposers of particulate matter as they gradually sink through the water-column. Most bacteria are so small that they could remain suspended virtually indefinitely. Some algae are neutrally buoyant while alive. Zooplankton, as well as many flagellated algae, swim upward to stay in the water-

A. Sedimentation

Table VII.A.2–1. Sedimentary Deposition in Mirror Lake Calculated from the Mass of Gyttja Accumulated Since Approximately 1840 A.D.[a]

	Dry Mass	Si	C	N	P	Ca	Mg	Na	K
	kg/yr								
Reactive Phases[b]	8,900	1,800	1,920	139	15	44	47	1.6	25
Detrital Minerals[c]	7,800	2,200[d]	0	0	1.0	73	21	219	180
Total	16,700	4,000	1,920	139	16	117	68	221	205

[a] Gyttja accumulation has been corrected for mixing.
[b] Includes organic matter and associated metals, acid-soluble authigenic minerals and exchangeable ions, dilute-alkali-soluble silica and detrital minerals dissolved by hot conc. HNO_3/HCl.
[c] Estimated by mass as (acid-insoluble ash) − (amorphous SiO_2).
[d] Estimated as 46.8% of the mass of detrital minerals volatilized on heating with HF.

column. But lysed algal cells, zooplankton feces and molted carapaces, organic matter from the lake's watershed—in short, dead organic matter—all sink or are washed out of the lake. Dead diatoms sink more rapidly than living cells and their inorganic siliceous frustules reach the lake bottom along with fine mineral particles eroded by water or wind from surrounding lands. Microscopic examination of the gyttja reveals mineral particles, numerous identifiable diatoms and terrestrial pollen grains, a few zooplankton remains, and scattered cell walls, stomates and tracheids from terrestrial plants. At low magnification (e.g. 100 times), the bulk of the gyttja appears as an amorphous, humus-colored matrix that represents, in undetermined proportions, a residue from allochthonous organic inputs and autochthonous organic production.

The amount of any visually or chemically determined component (e.g. pollen grains or total organic carbon) that reaches the lake bottom depends both on the supply of that component to the water-column and on its susceptibility to degradation while sinking. Components that do not change in amount while sinking or, afterwards, in the sediments can be called "conservative." Terrestrial pollen grains, mineral particles, and diatomaceous silica are presumably nearly conservative in Mirror Lake. Individual diatom taxa may not be conservative, if grazing zooplankton and benthic deposit-feeders fragment some frustules beyond recognition. Particulate organic carbon is readily changed to dissolved organic compounds (by cell lysis) and to dissolved carbon dioxide (by heterotrophic respiration), as discussed in *Chapter V.D.*

Few components of sediment are truly conservative after they have reached the lake bottom. Organic matter is decomposed to progressively more resistant residues, over a period of many years. Diatom frustules tend to dissolve, though slowly. In many lakes—including Mirror Lake—diatoms are preserved (*Chapter VII.B*), apparently because the interstitial water of the sediment matrix becomes saturated with dissolved silica before appreciable diatom dissolution has occurred (cf. Tessenow 1966). In some other lakes, such as Lake Michigan, diatoms dissolve nearly completely (Parker and Edgington 1976). Chemical and biochemical processes occurring within the surficial sediment alter the compounds that bind elements in particulate form. And if light reaches the sediment surface, benthic diatoms grow and die, leaving behind frustules that never sank through the water-column. The net effect of these diagenetic processes is to produce the "permanent" sediment that accumulates below the biogeochemically active zone (the upper 10 to 20 cm). This sediment is chemically different from the sinking particulate matter that first reached the lake bottom. It has been depleted of soluble compounds and consequently appears enriched in substances that resist resolubilization, like mineral particles, diatoms and pollen.

One approach to the study of sedimentation is to catch the sinking matter in the water-column, before it reaches the sediment surface. Decomposition can be partially arrested by frequently changing the traps and drying the collected sediment. The seasonality of deposition is evident from such data, and analysis of the sediment may demonstrate that at some times of year the trapped sediment includes material that was recently resuspended from the surficial sediment (Davis 1968, 1973; Pennington 1974; Bonny 1976).

At three sites in Mirror Lake (*Figure VII.A.1*–3) we suspended wide-mouth jars (3 liters in volume) from submersed floats, so that they hung open 1 m above the sediment. These traps were usually changed at 4- to 6-week intervals. In one experiment, parallel sets of traps were changed at frequencies that varied from 2 to 80 days, in order to estimate the rate at which material reaching the traps was resolubilized. One objective of the variable-exposure experiment was to determine by how much the regular exposure series underestimated the actual flux of nonconservative components, such as organic carbon, to the lake bottom.

Particulate matter sinking toward the bottom of a lake might be visualized as a vertical *rain* of descending particles. However, even in such quiet waters as the hypolimnion of a stratified lake during calm weather, the movement of particles is governed also by horizontal currents and by the turbulent eddies inherent to a natural water body. The net downward component of motion that sinking represents is likely to be much smaller than horizontal components and turbulent motions. Accurate measurements of sedimentation require a trap designed to collect matter at a rate corresponding to the net sinking rate, regardless of currents flowing across or into the trap.

The type of trap used in Mirror Lake is not ideal, for reasons pointed out by Bloesch and Burns (1980). First, the diameter of the body exceeds that of the mouth (140 vs. 90 mm). Such traps tend to overestimate sedimentation relative to the nominal area of the mouth, though the error may be relatively small in quiet water. A test in Mirror Lake during fall overturn suggested that this effect is negligible. Second, the height of the trap (h) is only 2.6 times the diameter of the mouth (d). Such a trap is vulnerable to periods of unusual turbulence (e.g. during wind storms), when settled matter may be resuspended and removed from the trap. In Mirror Lake the trapping rate of our standard trap ($h/d = 2.6$) was the same as that for shorter ($h/d = 1.1$) and taller traps ($h/d = 3.8$) during autumnal circulation, when currents should have been relatively strong (Moeller and Cole, in preparation). This finding testifies, in effect, to the weakness of water movements in the deeper regions of Mirror Lake. In Lake Erie, with a vastly larger fetch exposed to the wind, resuspension occasionally occurred from traps of $h/d = 5$ (Bloesch and Burns 1980). Finally, our traps were routinely protected with a coarse fish net, stretched across the mouth. This net complicates the geometry of trapping in an unknown manner. Measurements were impossible during autumn, however, unless fish were excluded from the traps (Moeller and Likens 1978).

Seasonal Pattern and Resuspension

Seasonality is evident in data from sediment traps at the Central site (*Figure VII.A.2*–2). Sedimentation was low during the winter, but varied substantially from year to year. When the lake is frozen, both allochthonous and autochthonous sources of particulate sediment are sharply curtailed: runoff from the watersheds is reduced, plankton production is reduced, and the lake bottom is protected from wind-driven currents that might resuspend shallow-water sediments. Sedimentation was higher and relatively consistent during six summers of measurements (see Moeller and Likens 1978 for the first three years of data). Sedimentation remained high or increased during the autumn, when the lake was destratified. Terrestrial pollen extracted from the collected sediment demonstrated the presence of resuspended material in the variable autumnal peak.

Tree pollen was abundant in sediment trapped during the spring, when it was "in season" as defined by flowering, as well as during the fall, when it was clearly "out of season" (*Figure VII.A.2*–2). Little deposition of tree pollen occurred between mid-July and mid-September. Some tree pollen presumably was washed or blown into the lake during the autumn (cf. Bonny 1976), but most of the trapped pollen represents sediment resuspended from the lake bottom. A different pattern characterized the deposition of *Ambrosia* (ragweed) pollen. This herb flowers in later summer or early

A. Sedimentation

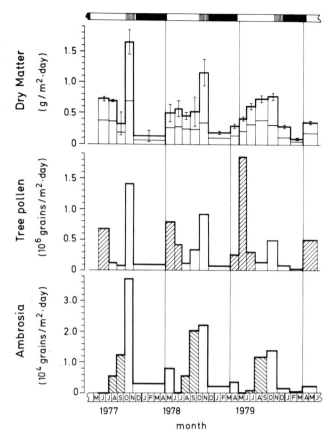

Figure VII.A.2–2. Seasonal sedimentation of dry matter and pollen at the Central sediment-trap station. Rates for total dry matter are sums of rates of organic matter (lower part of histogram) and ash (upper part of histogram). A 95% confidence interval is drawn about the mean (n = 3 to 4 traps). The annual depositional sequence for tree pollen and *Ambrosia* (ragweed) pollen is separated into *in season* collections corresponding to the flowering period (shaded histograms) and *out-of-season* collections that include resuspension and stream-transported pollen. The bar at the top of the figure indicates thermal stratification (clear), autumn overturn (vertical lines) and ice-cover (black).

autumn. Its in-season peak cannot be separated unambiguously from the subsequent period of resuspension, but very little *Ambrosia* was deposited between mid-May and mid-July (*Figure VII.A.2*–2). A small peak during the spring indicates resuspension of material that had previously settled in shallow water, perhaps at the end of the previous autumnal circulation. Spring circulation in Mirror Lake is typically short, and may not involve the entire water-column (*Chapter IV.B*).

Pollen tracers introduced naturally into the sedimenting seston therefore show that little resuspension of older sediment occurs between mid-May and mid-September, when the lake is thermally stratified. Unlike Mirror Lake, shallow unstratified lakes are subjected to intermittent sediment resuspension throughout the summer (Davis 1973; Gasith 1976). If resuspension takes place at the actual site of trap collections, the deposition measured in traps would, over an annual period, greatly exceed that observed as long-term accumulation in sediment cores from the collection site (Davis 1973). Because our sediment traps indicate deposition rates consistent with the mean rate since 1840 A.D. at all three sites, we hypothesize that the modest amount of autumnal resuspension detected in Mirror Lake takes place at the margin of the gyttja zone and over the lake bottom above the limit of permanent sediment accumulation. Autumnal resuspension probably scrubs fine organic, pollen-containing sediment from the inorganic substrates exposed beneath shallow water, and thereby defines the sedimentation limit. It is possible that rare windstorms of unusual severity contribute to defining the sedimentation limit. Resuspension from midlake gyttja would be limited to pollen and other particles not yet firmly entangled in the soft, but cohesive, surficial mud.

Decomposition During Sediment Formation

The interpretation of data from sediment traps hinges on the different behavior of conservative vs. nonconservative constituents. The mean annual deposition of residual-ash, a mixture of

Table VII.A.2–2. Sedimentation at Three Sites in Mirror Lake[a]

	Organic Carbon			Residual-Ash		
	West	Central	East	West	Central	East
Seston Deposition (1977 to 1980)			g/m^2-yr			
Sum of Trap Collections	40 ± 19	36 ± 8	30	95 ± 75	71 ± 28	55
Accumulated Gyttja (1840 to 1978)						
Apparent Rate since 1840	24	24	17	98	86	85
Corrected for Mixing[b]	21	19	15	87	69	73

[a] Annual mean values from sediment-trap collections (with 95% confidence interval; n = 3 at West and Central, n = 2 at East) are compared with values from surface cores at the collecting sites.
[b] Assuming a 2-cm mixing depth at West and East, 4 cm at Central.

diatoms and mineral particles that had been assumed to be conservative (Moeller and Likens 1978), does in fact agree with its long-term accumulation rate since 1840 A.D. (*Table VII.A.2–2*). In contrast, the deposition of organic carbon in traps is two times the long-term accumulation at all sites. A recent increase in carbon sedimentation is conceivable, but organic carbon is by no means conservative. Decomposition occurs as the organic matter sinks and continues after it reaches the lake bottom or, analogously, enters a sediment trap. A value for organic carbon deposition measured by sediment trap is some arbitrary intermediate—dependent on the length of time traps were exposed—that underestimates the actual flux of organic carbon to the sediments, yet overestimates the *permanent* deposition below the biologically-active zone.

Evidence of the contrasting behavior of organic carbon and residual-ash was provided by the series of parallel sediment trap exposures (*Figure VII.A.2–3*). Residual-ash proved to be nearly, not completely, conservative. A small but statistically significant loss of diatomaceous silica did occur between the 80-day collection and its residue after one year of further exposure in permeable plastic containers (Moeller and Cole, in preparation). Organic matter decomposed continuously.

The decomposition curve from *Figure VII.A.2–3* provides a means of estimating the actual flux of organic carbon to the lake bottom. If the annual trap collection (*Table VII.A.2–2*) is assumed to arise from a series of uniform 40-day exposures (which is approximately true during the summer, when most sedimentation takes place), then the total flux of organic carbon to the lake bottom would exceed the measured amount by almost twofold. The actual flux would be roughly 67 g C/m^2-yr in the gyttja zone, or 46 g C/m^2-yr for the lake as a whole. This value is too high to be reconciled with the budgetary terms for organic carbon supply and for planktonic respiration by zooplankton and bacteria (*Chapters V.C and V.D*). Errors could lie anywhere in our reconstruction of the decomposition curve, in assuming a 100% collecting efficiency for the sediment traps, and within the rest of the carbon budget. It is clear, however, that a major portion of the lake's organic carbon supply reaches the bottom. There it supports the abundant benthic fauna (*Chapters V.A.6 and V.A.7*) and microflora and thus indirectly causes the progressive deoxygenation of the lower hypolimnion observed during the summer (*Chapter IV.B*). Mirror Lake stands in contrast to deep oligotrophic lakes, in which most decomposition occurs before sinking organic matter reaches the lake bottom (Hargrave 1975a).

At the Central site in Mirror Lake, estimates of sedimentation obtained with sediment traps are consistent with the longer-term rate of accumulation as gyttja (*Table VII.A.2–3*). This agreement emerges when reductions are made for the decomposition of organic matter after reaching the lake bottom (a major 39% correction applied to organic C), and for the downward mixing of the dated pollen horizon (a smaller 18% correction applied to all sediment constituents). Because the rate at which gyttja has accumulated may have changed within the

A. Sedimentation

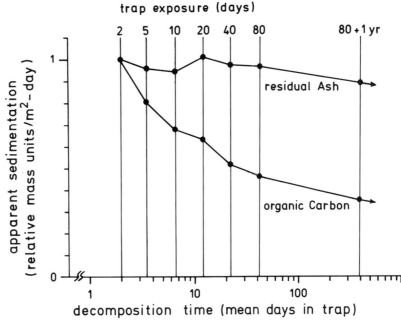

Figure VII.A.2–3. The effect of length of sediment-trap exposure on the apparent sedimentation rate. Residual-ash is nearly conservative—that is, independent of length of exposure. Organic carbon is partially resolubilized during exposure. Decomposition time is equal to one-half the exposure plus one or two days for settling.

138-yr period since the pollen horizon, the agreement does not prove unequivocally that the two methods of measuring sedimentation are equivalent. And if the major herb rise dates to 1870 instead of 1840, the long-term accumulation rate would exceed that measured by sediment traps. Von Damm et al. (1979) concluded that their ^{210}Pb profile indicated a great acceleration of gyttja accumulation in the past 25 yr, in part attributable to episodic redeposition of sediment eroded elsewhere in the basin. Such accelerated deposition is certainly not consistent with the sediment trap collections over the years 1977 to 1980.

Sedimentation varies from year to year, owing especially to the intensity of resuspension during the autumn. As with many measurements at the ecosystem level, three years of sediment trap data are too few to provide precise annual means. The degree of imprecision varies among sedimentary constituents. The large variability of residual-ash deposition does not even include the additional variability caused by rare events such as the major flood of

Table VII.A.2–3. Sedimentation at the Central Sediment-trap Station[a]

	Organic C	Residual-Ash	Pollen[b]
	(g/m²-yr)		(10⁶ grains/m²-yr)
Seston Deposition (1977 to 1980)			
Sum of Trap Collections	36 ± 8	71 ± 28	170 ± 64
Corrected to 2-Day-Old Sediment	72	71	170
Corrected to 1-Yr Sediment	22	66	170
Accumulated Gyttja (1840 to 1978)			
Apparent Rate since 1840	24	86	270
Corrected for Mixing[c]	19	69	220

[a] Seston sedimentation into traps is compared with the rate of sediment accumulation at the collection site during the postsettlement period (*Table VII.A.2-4*). A 95% confidence interval is calculated for seston deposition (n = 3 years).
[b] Pollen is assumed to be conservative.
[c] Assuming a 4-cm mixing depth.

December 1973, which deposited an additional 1–2 years' worth of residual-ash (at the mean rate for 1977 to 1980) in the Central sediment traps during a single exposure period (Moeller and Likens 1978). Deposition of organic matter is fairly regular from year to year, but pollen has a variable influx because of weather and the biologic conditions that control the intensity of flowering by individual taxa (Bonny 1980).

Changes in Deposition Caused by Human Disturbance

The settlement of people of European origin within the watersheds of thousands of small North American lakes signaled the onset of environmental changes that, in some lakes, have exceeded changes caused by any natural process since the climatic amelioration at the beginning of the Holocene. Sediments deposited early during the settlement period commonly contain more clastic mineral matter than precultural sediment (Sasseville and Norton 1975; Davis 1976a; Adams and Duthie 1976; Brugam 1978; Manny et al. 1978; Birch et al. 1980). These particles were eroded as land was cleared of its natural vegetation and—at some sites—plowed. Erosion of forest soils leads to the fluvial transport of much organic matter as well as mineral particles (Likens et al. 1970). Dissolved nutrients carried along with the particulate matter may cause organic production to increase within lakes downstream. Occasionally nutrient enrichments so stimulate lacustrine productivity that organic deposition accelerates more rapidly than mineral deposition. Sewage draining into a small English lake, for instance, caused a dramatic eutrophication; the deposition of organic sediment increased enough to reverse the trend to less organic sediment that had earlier been caused by agriculture (Pennington 1978a). In other cases of severe cultural eutrophication, the deposition of eroded mineral matter has more than kept up with higher organic deposition (e.g. Manny et al. 1978). Sometimes the settlement horizon is masked by parallel increases of both mineral and organic matter (Sasseville and Norton 1975). In all cases, sedimentation rates must be determined before and after settlement in order to assess the nature and severity of human disturbance of the drainage basin.

In Mirror Lake, the major herb rise coincides with an increase in the concentration of mineral particles (*Figure VII.A.1*–9). Concentrations of other chemical elements (*Chapter VII.E*) and of biologic remains such as diatoms (*Chapter VII.B*), Cladocera (*Chapter VII.C*) and terrestrial pollen, tend to change across this same horizon. The surface cores are separable into: (1) the "cultural" layer, which includes the herb rise and all sediments overlying it; (2) a "diffuse cultural" zone, which contains some herb pollens and dates to the earliest decades of settlement; and (3) the precultural zone, which intergrades imperceptibly with the diffuse cultural zone and is perhaps not sampled in some of the cores.

Sedimentation rates calculated above and below the herb rise (*Table VII.A.2*–4) show that the changes at the herb rise are due mainly to increased deposition of mineral particles. The unchanged deposition rates for some constituents (organic C, N, P), and the decreased deposition of diatoms and amorphous silica, would be puzzling if we had not established the change in depositional pattern within the lake at that time. The spatial extension of the zone where gyttja accumulates (from 50 to 68% of the present lake area) and the observed relaxation of focusing within that zone mean that the average rate of gyttja accumulation in the lake as a whole has risen by perhaps 50% since 1840. We cannot be more precise, because the pattern of deposition just prior to the settlement period has not been determined in detail.

The changed pattern of deposition above the settlement horizon is only one of two related historical developments that impair our ability to reconstruct the intensity of disturbance within the drainage basin. To augment the flow of water over the outlet dam, which had raised the water level by 1 to 2 m and thereby modified the depositional pattern, early settlers diverted a portion of Hubbard Brook into the lake. This diversion continued from about 1850 to 1920 (Likens 1972c). The dimensions of the surviving canal are too small to pass peak discharges of Hubbard Brook, but most of the lower summer flow (about six times 10^5 m^3/mo; G. E. Likens, personal communication) could have entered the lake. For much of the cultural period, then, the lake's watershed was increased in an ill-defined manner. The influx of eroded soil from

A. Sedimentation

Table VII.A.2–4. Changes in the Composition and Deposition Rate of Sediment at Deep-water Sites (10–10.5 m) Following Nineteenth-Century Settlement[a]

	Concentration		Deposition Rate		
	Precultural	Cultural	Precultural	Cultural	
Major Sediment Fractions[b]					
Total Dry Matter	100.0	100.0 % dry wt	118	124	g/m²-yr
Organic Matter	35.8	33.4	42.2	41.4	
Acid-Soluble Ash	7.1	11.3	8.4	14.0	
Amorphous Silica	41.2	27.5	48.6	34.1	
Detrital Minerals	15.9	27.8	18.8	34.5	
Individual Chemical Elements[c]					
C	15.5	15.2	18.3	18.8	
N	1.38	1.23	1.63	1.53	
S	0.44	0.45	0.52	0.56	
P	0.20	0.18	0.24	0.22	
Si	27.1	23.1	32.0	28.6	
Al	2.78	4.76	3.28	5.91	
Ca	0.47	0.55	0.55	0.68	
Mg	0.18	0.32	0.21	0.40	
Na	0.36	0.61	0.42	0.76	
K	0.44	0.88	0.52	1.09	
Biologic Remains					
Diatoms[d]	2.1	0.8 10⁹/g	250	99	10⁹/m²-yr
Chrysophyte Cysts[d]	0.47	0.20 10⁹/g	55	25	10⁹/m²-yr
Pollen Grains[b]	2.6	1.8 10⁶/g	310	220	10⁶/m²-yr
Cladocera[e]	0.86	0.56 10⁵/g	100	69	10⁵/m²-yr

[a] Because of the relaxation of focusing during the cultural period, the increase in lakewide rates of deposition following settlement is obscured at mid-lake sites.
[b] Core 4 data; cultural = 0–10 cm, precultural = 22–28 cm.
[c] Precultural = mean of Central long core (25 cm) and core 4 (22–28 cm); cultural = mean of core 4 (1–10 cm) and Moeller and Likens 1978 (0–10 cm).
[d] See *Chapter VII.B*.
[e] See *Chapter VII.C*.

this expanded watershed included mineral particles, soil organic matter, and pollen grains—but evidently few diatoms.

Mirror Lake has so far been spared major consequences of landscape disturbance. There has never been extensive tillage of its stony, infertile, often steep watershed. The forest was cut selectively throughout the nineteenth century, but only locally cleared at low elevations near the lake. The forest was cut extensively early in the twentieth century, but subsequent revegetation was not interrupted. Intact forest soils with medium-aged successional forest cover most of the watershed today (*Chapter II*). Road construction during 1969–70 probably increased particulate inputs to the lake for at least several years, but the cultural sediments as a whole give no indication of dramatic, sustained changes from precultural conditions. Further analysis of the cultural stratigraphy may reveal changes within the cultural period. The increased sedimentation since the herb rise is similar to increases observed following Neolithic forest clearing in Britain (Pennington 1978a, b) but smaller than increases associated with permanent forest removal for agriculture or local urbanization (Davis 1976a; Brugam 1978; Pennington 1978a, b).

Summary

Cores and probings at numerous stations in Mirror Lake show that the original basin extended 23 m below the present-day lake surface. Half of the original lake volume has been filled with sediments deposited asymmetrically in the deepest part of the basin. Four types of sediment occur in Mirror Lake: (1) highly inorganic silts and fine sand deposited over much of the lake during late-glacial times, (2) an organic gyttja deposited over the central half or two-thirds of the lake during Holocene and postsettlement time, (3) sands and gravels deposited in shallow water and intermixed with basin-forming till and outwash, and (4) small patches of terrestrial organic debris deposited in shallow water.

Sedimentation rates were calculated for several cores from the relationship of absolute age to depth in the core. Age was determined by cross-correlation of pollen profiles to a composite ^{14}C-dated pollen profile. The results show that both sedimentation rates and spatial patterns of deposition have varied during the lake's history. During late-glacial time (14,000 to 11,000 yr B.P.), silts and fine sand were deposited over 85% of the lake basin at an average rate within the deposition zone of 530 g of inorganic matter/m^2-yr and 6 g organic matter/m^2-yr. Inorganic matter was composed overwhelmingly of mineral particles. During the precultural Holocene (11,000 to 200 yr B.P.), gyttja was deposited over 50% of the present-day basin at the mean rate of 120 g inorganic matter/m^2-yr and 32 g organic matter/m^2-yr. Inorganic matter was dominated by diatomaceous silica. After settlement occurred in the nineteenth century, the zone of gyttja deposition increased to 68% of the present lake area. Deposition within that zone has averaged 123 g inorganic matter/m^2-yr and 41 g organic matter/m^2-yr. Mineral particles and diatomaceous silica contribute equally to the inorganic matter.

Wind-driven currents established during autumnal turnover and during severe storms apparently prevent the long-term accumulation of gyttja at depths shallower than the thermocline in mid-summer (4 to 6 m). The increase in lake level during the cultural period (1 to 2 m) has caused a shoreward expansion of the zone where gyttja accumulates. Within the zone of accumulation, sediment is differentially focused to deeper parts of the basin, but in an irregular pattern that has changed several times during the Holocene. Because sediment accumulates in uneven layers, simple mathematical models of sediment accumulation (e.g. Lehman 1975) are only conceptually useful; they cannot quantitatively *correct* accumulation rates measured only at a mid-lake site to rates integrated over the entire lake.

Contemporary sedimentation rates for the lake as a whole calculated from sediment traps and surface cores are similar to one another. Sediment traps collect an incompletely decomposed sediment at a rate that is intermediate between the gross flux of matter to the sediment surface and the long-term accumulation as gyttja. Sedimentation of both inorganic and organic matter calculated for the lake as a whole has increased significantly, but not dramatically, since farming and logging began in the vicinity of Mirror Lake.

B. Diatoms

JOHN W. SHERMAN

Investigations of biologic remains in post-glacial lake sediments provide insight into the natural evolution of lake-ecosystems and their drainage basins. Sediments deposited during historical times contain information about the kinds and degrees of change brought about by settlement, urbanization and industrialization.

Dominant among the primary producers of lakes are the single-celled diatoms (Bacillariophyceae), algae whose silica walls are often preserved in sediments. The small size and large numbers of diatoms, coupled with their species diversity and specific ecologic requirements make them excellent bioindicators.

Early studies of deposits from the U.S. containing diatom assemblages involved the identification of freshwater-bearing strata in New Jersey (Woolman 1891, 1892) and oil-bearing formations in Florida (Hanna 1933). Patrick (1943) associated fluctuations in the diatom content of Linsley Pond sediments with climatic change in Connecticut. Since these early papers, many authors have used diatoms to unravel the paleoecology of lake basins both in the U.S. and worldwide. However, the multidisciplinary approach, such as that employed by Hutchinson and his associates in their study of Lago di Monterosi (Hutchinson et al. 1970), has the advantage of bringing together many lines of evidence to more completely characterize the history of specific lake basins. A similar multidisciplinary approach has been taken in the studies of Mirror Lake, as this chapter examines the post-glacial diatom assemblages occurring in cores from the central basin of Mirror Lake for the purpose of further defining the paleoecology of the Hubbard Brook Valley.

Additional information can be found in a more comprehensive study of New England lake sediments (Sherman 1976).

Methods

Samples for diatom analysis were taken from the 12.4-m Swedish foil core and from the 68-cm Davis-Doyle core obtained by M. B. Davis for pollen analysis (*Chapter VII.E*). The details of sampling and analysis are given in Sherman (1976). Permanent slides made from sediments taken from 124 segments of the long core (every 10 cm) were scanned at medium magnification to identify those intervals to be used for detailed diatom counts. Intervals chosen were based upon the apparent rate of assemblage change between successive horizons.

Taxon identifications and tabulations were made at a magnification of 1,200 diameters. Traverses of the slides were made until 1,000 individuals were recorded. Specimens were scarce below 1,200 cm—shorter counts were used for assemblage analysis. Species' percentages, numbers of cells per gram dry weight inorganic sediment and the Shannon-Wiener diversity index (Shannon and Weaver 1963) were calculated for each completed diatom count.

Preliminary examination of Mirror Lake sediments revealed the presence of large numbers of Chrysophycean cysts, that is, siliceous remains of resting stages of Chrysophycean algae. Total numbers of cysts were obtained during cell counts of diatoms, and their numbers per gram dry weight of inorganic sediment were recorded. No attempt was made to separate these forms taxonomically.

Results and Discussion

A total of 232 freshwater diatom taxa were identified during the analysis of 30 core intervals (*Table VII.B–1*). Detailed enumeration of these taxa can be found in Sherman (1976). Relative abundances (as percentages) of taxa representing 5% or more of the assemblage in at least one interval, along with selected forms present in amounts greater than 1% in at least one interval are diagrammed in *Figure VII.B–1A, B and C*. The number of taxa per 1,000 cells (an indicator or the *richness* of respective assemblages) and Shannon-Wiener diversity indices (H'—an indicator of the *evenness* of species' abundances) are graphed in *Figure VII.B–1C* along with the numbers of diatom cells per gram dry weight inorganic sediment. Zone designations indicated on *Figure VII.B–1* are those of Likens and Davis (1975).

Zone I—Late Glacial

The deepest sediments sampled by the long core are devoid of diatoms. The coarse size and angularity of these sediments suggest that they were derived from glacial tills of the lake basin. Rapid inorganic sedimentation would be expected to occur immediately after glacial retreat, as suggested by Likens and Davis (1975). Lack of diatoms below the 1,230-cm sample may be explained by minimal diatom production (resulting from low nutrient concentrations and low temperature), by dilution of cells by inorganics, or by a combination of these factors. Light-limiting high turbidity may also have depressed diatom production. A few fragments of diatom frustules were found at 1,240 cm; these valves may have been destroyed by mechanical agitation of coarse sediment prior to burial or may have been introduced to the deeper sediments from above by mechanical disturbance. Similar fragments were found at 1,230 cm among intact valves.

Table VII.B-1. Taxonomic Listing of Mirror Lake Diatoms

Division—Bacillariophyta
  Class—Bacillariophyceae
    Order—Eupodiscales
    Family—Coscinodiscaceae

Cyclotella Kütz.
Cyclotella bodanica Eulenst.
Cyclotella bodanica v. *lemanensis* O. Müll.
Cyclotella comta (Ehr.) Kütz.
Cyclotella operculata (Ag.) Kütz.
Cyclotella stelligera Cl. u. Grun.

Melosira Ag.
Melosira ambigua (Grun.) O. Müll.
Melosira distans v. *alpigena* Grun.
Melosira distans v. *lirata* (Ehr.) Bethge (stat. 1)
Melosira distans v. *lirata* (Ehr.) Bethge (stat. 2)
Melosira distans v. *lirata* (Ehr.) Bethge (stat. 3)
Melosira fennoscandica Cl. Eul. (stat. 1)
Melosira fennoscandica Cl. Eul. (stat. 2)
Melosira italica v. *subarctica* O. Müll. (stat. 1)
Melosira italica v. *subarctica* O. Müll. (stat. 2)
Melosira italica v. *tenuissima* (Grun.) O. Müll.
Melosira italica v. *valida* Grun.
Melosira tenella Nygaard
Melosira varians C. A. Ag.
Melosira sp. 1
Melosira sp. 2
Melosira sp. 3
Melosira sp. 4
Melosira sp. 5

    Order—Rhizosoleniales
    Family—Rhizosoleniaceae

Rhizosolenia Ehr.
Rhizosolenia sp. 1

    Order—Fragilariales
    Family—Fragilariaceae

Tabellaria Ehr.
Tabellaria fenestrata (Lyngb.) Kütz.
Tabellaria flocculosa (Roth) Kütz.

Diatoma Bory
Diatoma anceps (Ehr.) Kirchn.
Diatoma hiemale v. *mesodon* (Ehr.) Grun.

Meridion Ag.
Meridion circulare (Greg.) Ag.
Meridion circulare v. *constrictum* (Ralfs) V. H.

Fragilaria Lyngb.
Fragilaria bicapitata A. Mayer
Fragilaria brevistriata Grun.
Fragilaria brevistriata v. *inflata* (Pant.) Hust.
Fragilaria construens (Ehr.) Grun.
Fragilaria construens v. *binodis* (Ehr.) Grun.
Fragilaria construens v. *venter* (Ehr.) Grun.
Fragilaria crotonensis Kitton
Fragilaria inflata (Heid.) Hust.
Fragilaria leptostauron (Ehr.) Hust.
Fragilaria leptostauron v. *dubia* (Grun.) Hust.
Fragilaria pinnata Ehr.
Fragilaria vaucheriae (Kütz.) Peters.
Fragilaria vaucheriae v. *capitellata* (Grun.) Patr.
Fragilaria virescens Ralfs

Synedra Ehr.
Synedra delicatissima W. Sm.
Synedra delicatissima v. *angustissima* Grun.
Synedra rumpens v. *familiaris* (Kütz.) Hust.
Synedra ulna (Nitz.) Ehr.
Synedra ulna v. *danica* (Kütz.) V. H.
Synedra sp. 1

Asterionella Mass.
Asterionella formosa Hass.
Asterionella formosa v. *gracillima* (Hantz.) Grun.

    Order—Eunotiales
    Family—Eunotiaceae

Eunotia Ehr.
Eunotia bidentula W. Sm.
Eunotia curvata (Kütz.) Lagerst.
Eunotia diodon Ehr.
Eunotia elegans Ostr.
Eunotia flexuosa Breb. Ex Kütz.
Eunotia hexaglyphis Ehr.
Eunotia incisa W. Sm. Ex Greg.
Eunotia maior (W. Sm.) Rabh.
Eunotia meisteri Hust.
Eunotia pectinalis (O. F. Müll.?) Rabh.
Eunotia perpusilla Grun.
Eunotia praerupta Ehr.
Eunotia serra v. *diadema* (Ehr.) Patr.
Eunotia sudetica O. Müll.
Eunotia trinacria v. *undulata* Hust.
Eunotia vanheurckii Patr.
Eunotia vanheurckii v. *intermedia* (Krasske Ex Hust.) Patr.
Eunotia sp. 1

    Order—Achnanthales
    Family—Achnanthales

Achnanthes Bory
Achnanthes calcar Cl.
Achnanthes clevei Grun.
Achnanthes clevei v. *rostrata* Hust.
Achnanthes didyma Hust.
Achnanthes exigua Grun.
Achnanthes exigua v. *heterovalva* Krasske

Table VII.B-1 Continued

Achnanthes flexella (Kütz.) Brun
Achnanthes lanceolata (Breb.) Grun.
Achnanthes lanceolata v. *abbreviata* Reim.
Achnanthes lanceolata v. *apiculata* Patr.
Achnanthes lanceolata v. *elliptica* Cleve
Achnanthes lapponica Hust.
Achnanthes laterostrata Hust.
Achnanthes microcephala (Kütz.) Grun.
Achnanthes minutissima Kütz.
Achnanthes peragalli Brun & Herib.
Achnanthes peragalli v. *fossilis* Temp. & Perag.
Achnanthes pinnata Hust.
Achnanthes stewartii Patr.
Achnanthes suchlandti Hust.
Achnanthes sp. 1
Achnanthes sp. 2
Cocconeis Ehr.
Cocconeis diminuta Pant.
Cocconeis placentula v. *lineata* (Ehr.) V. H.

Order—Naviculales
Family—Naviculaceae

Anomoeoneis Pfitz.
Anomoeoneis serians v. *brachysira* (Breb. Ex Kütz.) Hust.
Anomoeoneis vitrea (Grun.) Ross
Caloneis Cl.
Caloneis ventricosa v. *minuta* (Grun.) Patr.
Caloneis ventricosa v. *subundulata* (Grun.) Patr.
Caloneis sp. 1
Diploneis Ehr.
Diploneis finnica (Ehr.) Cl.
Diploneis marginestriata Hust.
Diploneis oblongella (Naeg. Ex Kütz.) Ross
Diploneis oculata (Breb.) Cl.
Diploneis puella (Schum.) Cl.
Frustulia Rabh.
Frustulia rhomboides v. *amphipleuroides* (Grun.) Cl.
Frustulia rhomboides v. *saxonica* (Rabh.) Det.
Navicula Bory
Navicula acceptata Hust.
Navicula begeri Krasske
Navicula cocconeiformis Greg. Ex Grev.
Navicula cryptocephala Kütz.
Navicula explanata Hust.
Navicula festiva Krasske
Navicula globulifera Hust.
Navicula gottlandica Grun.
Navicula gysingensis Foged
Navicula hambergii Hust.
Navicula hassiaca Krasske
Navicula helensis Schulz
Navicula helmandensis Foged
Navicula hustedtii Krasske
Navicula ignota v. *anglica* Lund
Navicula jarnefeltii Hust.
Navicula lacustris Greg.
Navicula latens Krasske
Navicula mediocris Krasske
Navicula menisculus v. *upsaliensis* (Grun.) Grun.
Navicula minima Grun.
Navicula muralis Grun.
Navicula mutica Kütz.
Navicula paucivisitata Patr.
Navicula peregrina (Ehr.) Kütz.
Navicula placenta Ehr.
Navicula pseudoexilissima Hust.
Navicula pseudoscutiformis Hust.
Navicula pupula Kütz.
Navicula pupula v. *capitata* Skv. & Meyer
Navicula pupula v. *mutata* (Krasske) Hust.
Navicula pupula v. *rectangularis* (Greg.) Grun.
Navicula pusio Cl.
Navicula radiosa Kütz.
Navicula radiosa v. *parva* Wallace
Navicula regularis Hust.
Navicula rotaeana (Rabh.) Grun.
Navicula salinarum v. *intermedia* (Grun.) Cl.
Navicula schmassmannii Hust.
Navicula seminulum Grun.
Navicula simula Patr.
Navicula subatomoides Hust.
Navicula viridula v. *linearis* Hust.
Navicula vulpina Kütz.
Navicula sp. 1
Navicula sp. 3
Navicula sp. 4
Navicula sp. 5
Navicula sp. 6
Navicula sp. 7
Navicula sp. 8
Neidium Pfitz
Neidium affine (Ehr.) Pfitz.
Neidium affine v. *longiceps* (Greg.) Cl.
Neidium hitchcockii (Ehr.) Cl.
Neidium iridis (Ehr.) Cl.
Neidium sp. 1
Pinnularia Ehr.
Pinnularia abaujensis v. *rostrata* (Patr.) Patr.
Pinnularia acrosphaeria W. Sm.
Pinnularia biceps Greg.
Pinnularia biceps v. *minor* (Boye Pet.) A. Cl.
Pinnularia braunii (Grun.) Cl.
Pinnularia intermedia (Lagerst.) Cl.
Pinnularia legumen (Ehr.) Ehr.
Pinnularia maior v. *transversa* (A. S.) Cl.

Table VII.B-1 *Continued*

Pinnularia microstauron (Ehr.) Cl.
Pinnularia polyonca (Breb.) O. Müll.
Pinnularia subcapitata Greg.
Pinnularia viridis (Nitz.) Ehr.
Stauroneis Ehr.
Stauroneis anceps Ehr.
Stauroneis anceps f. *gracilis* Rabh.
Stauroneis anceps f. *linearis* (Ehr.) Hust.
Stauroneis livingstonii Reim.
Stauroneis nobilis v. *baconiana* (Stodd.) Reim.
Stauroneis phoenicenteron (Nitz.) Ehr.
Stauroneis sp. 1

Family—Cymbellaceae

Amphora Ehr.
Amphora ovalis v. *libyca* (Ehr.) Cl.
Amphora ovalis v. *pediculus* Kütz.
Amphora perpusilla Grun.

Cymbella Ag.
Cymbella affinis Kütz.
Cymbella cesati (Rabh.) Grun.
Cymbella gaeumanni Meis.
Cymbella heteropleura (Ehr.) Kütz.
Cymbella laevis Naeg. Ex Kütz.
Cymbella latens Krasske
Cymbella lunata W. Sm.
Cymbella microcephala Grun.
Cymbella minuta v. *silesiaca* (Bleisch Ex Rabh.) Reim.
Cymbella naviculiformis Auerswald
Cymbella pusilla Grun.
Cymbella sinuata Greg.
Cymbella turgida (Greg.) Cl.
Cymbella ventricosa Kütz.

Family—Gomphonemaceae

Gomphonema Ag.
Gomphonema acuminatum v. *coronata* (Ehr.) W. Sm.
Gomphonema angustatum (Kütz.) Rabh.
Gomphonema angustatum v. *producta* Grun.
Gomphonema clevei Fricke

Gomphonema gracile v. *naviculoides* (W. Sm.) Grun.
Gomphonema intricatum Kütz.
Gomphonema intricatum v. *dichotoma* (Kütz.) Grun.
Gomphonema intricatum v. *pumila* Grun.
Gomphonema olivaceoides Hust.
Gomphonema parvulum v. *exilis* Grun.
Gomphonema truncatum v. *turgidum* (Ehr.) Patr.
Gomphonema sp. 1

Order—Bacillariales
Family—Epithemiaceae

Rhopalodia O. Müll.
Rhopalodia gibba (Ehr.) O. Müll.

Family—Nitzschiaceae

Bacillaria Gmelin
Bacillaria paradoxa Gmelin
Nitzschia Hass.
Nitzschia acicularis W. Sm.
Nitzschia dissipata (Kütz.) Grun.
Nitzschia frustulum v. *perminuta* f. *constricta* Hust.
Nitzschia frustulum v. *perpusilla* Grun.
Nitzschia gracilis Hantzsch
Nitzschia mirabilis Cl. Eul.
Nitzschia romana Grun.
Nitzschia subcommunis Hust.
Nitzschia tropica Hust.
Nitzschia vermicularis Grun.
Nitzschia sp. 1
Nitzschia sp. 2

Order—Surirellales
Family—Surirellaceae

Campylodiscus Ehr.
Campylodiscus hibernicus v. *hungaricus* Cl. Eul.
Stenopterobia Breb.
Stenopterobia intermedia Lewis
Surirella Turpin
Surirella angustiformis Hust.
Surirella delicatissima Lewis
Surirella robusta Ehr.

The first appearance of a diatom assemblage is at 1,230 cm (~11,000 yr B.P.), consisting almost exclusively of the periphytic diatoms *Fragilaria pinnata* and *Cocconeis diminuta* (*Figure VII.B*–1B and C, respectively) with relatively few specimens of *Melosira, Navicula* and other *Fragilaria* species. The ecologies of the dominant forms during this period suggest that the pH of the water was alkaline (7.5 to 8.0) and that the dominant growth was in the littoral region of the early lake. A conspicuous lack of true planktonic forms indicates that diatom production in open water was minimal; in addition, lack of planktonic forms suggests that the water

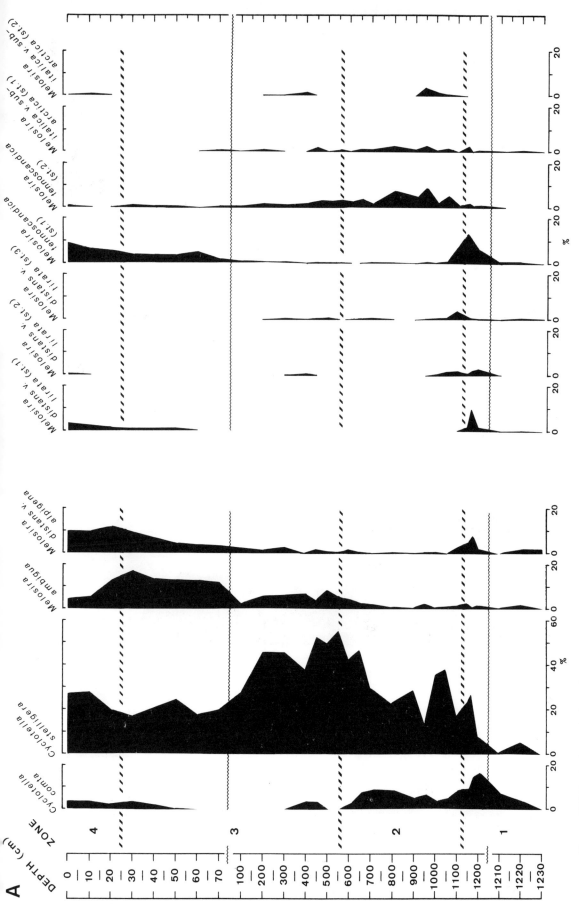

Figure VII.B–1. A. Relative abundances (percentages) of commonly occurring diatom taxa in Mirror Lake cores. **B.** Relative abundances (percentages) of commonly occurring diatom taxa in Mirror Lake cores. **C.** Relative abundances (percentages) of commonly occurring diatom taxa, diatom diversity, and the numbers of diatom cells and Chrysophycean cysts in Mirror Lake cores.

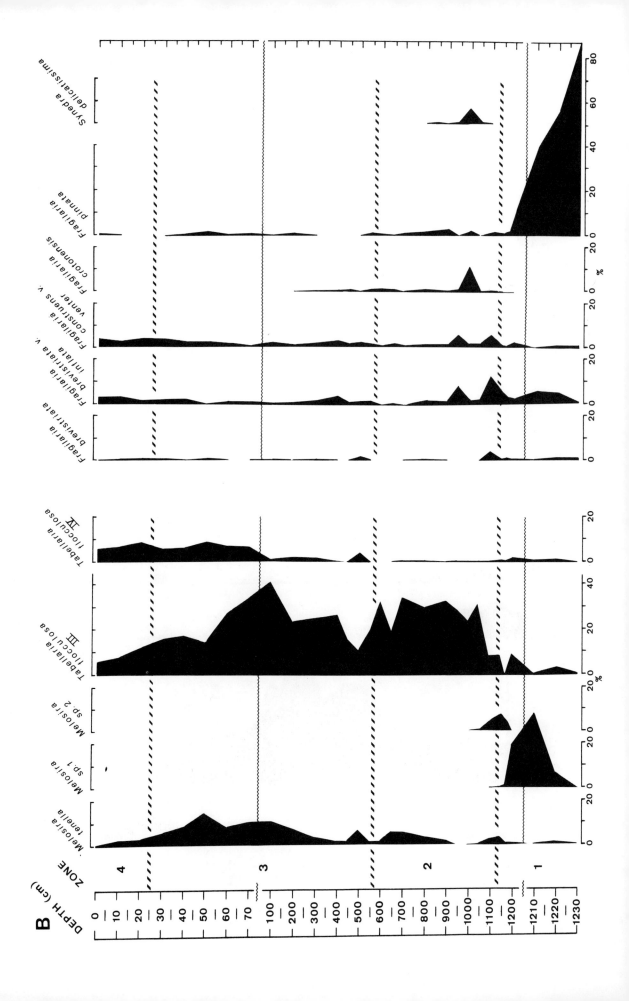

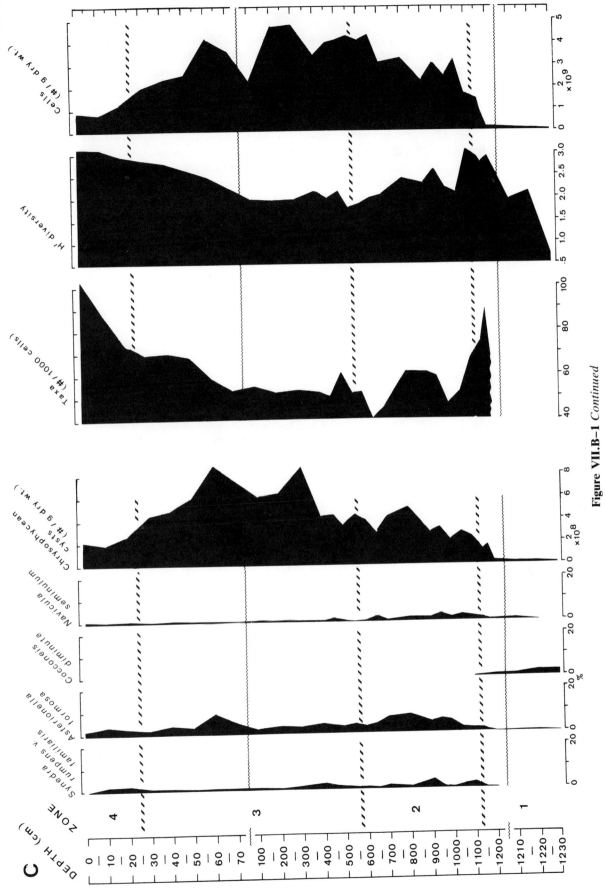

Figure VII.B-1 *Continued*

may have been quite shallow. *Fragilaria pinnata,* accounting for well over one-half of the total number of cells in the earliest assemblages, has been associated with oxygen-rich water (Lowe 1974) and oligotrophic conditions (Bradbury and Megard 1972), although *F. pinnata* is generally considered to be widely distributed in fresh water (Patrick and Reimer 1966).

The lowest H' diversity values of the core are found in the deepest samples of Zone I (less than 1.0 to about 2.0). Total diatom cell production was minimal as evidenced by very low numbers of cells per gram dry weight of inorganic sediment ($<0.1 \times 10^9$ cells from 1,230 to 1,210 cm). Influx of inorganic sediment to the lake during this early stage would have had the effect of *diluting* the concentrations of cells deposited at the location of the core. Nevertheless, the simplicity of the diatom assemblage suggests that autochthonous production by diatoms within the lake was minimal.

Several changes in the chemistry and physics of Mirror Lake water are indicated by diatom assemblages above 1,230 cm. The sparse, alkaline-indicating, littoral assemblage is rapidly replaced by a more diverse group of littoral and planktonic species that are characteristic of more neutral to slightly acid water. Abundance patterns for *Cyclotella comta* and *Tabellaria flocculosa* (*Figure VII.B*–1A and B, respectively), both planktonic forms, testify to the increasing productivity in open waters of Mirror Lake. Numerous species of *Fragilaria* and *Navicula* (all not included on *Figure VII.B*–1) suggest that the littoral habitats were becoming increasingly productive at this time. An unusual association of *Melosira* taxa (especially those of the *M. distans* group, *Figure VII.B*–1A) suggests diversification in both open water and littoral environments.

The pH requirements of the common taxa found above 1,200 cm (~10,400 yr B.P.) suggest that the pH during this time may have been very near neutral, as several of the commonly occurring taxa are known to favor slightly alkaline conditions (*Cyclotella comta, C. stelligera, Fragilaria brevistriata*), while others favor slightly acidic conditions (*Melosira distans* v. *lirata, Fragilaria virescens, Tabellaria flocculosa*).

Shannon-Wiener diversity values rise from 1,230 to 1,220 cm, fall slightly at 1,210 cm and rise sharply near the upper limit of Zone I (*Figure VII.B*–1C). The second highest number of taxa per 1,000 cells for the entire profile is reached at 1,170 cm (90 taxa), approximately 10,200 yr B.P. Total numbers of diatom cells increase above 1,200 cm, apparently indicating increased primary production within the lake, an observation supported by increasing levels of fossil pigments throughout Zone I (*Chapter VII.D*).

The pattern of changes within the diatom assemblages of Zone I suggests that prior to the establishment of the surrounding forest (1,125 cm in the pollen profile) Mirror Lake was becoming less alkaline, more productive and more supportive of planktonic flora. These changes were continuous across the upper limits of Zone I.

Zone II—Early Post-Glacial

Diatom assemblages of Zone II indicate that a variety of planktonic and periphytic populations of diatoms were well established in the lake. Numbers of taxa per 1,000 cells ranged from 40 at 650 cm (~5,500 yr B.P.) to 67 at 1,100 cm (~9,400 yr B.P.). Shannon-Wiener diversities exceeded 2.0 from 1,100 to 700 cm. During the period of deposition of Zone II sediments, the euplanktonic (normally suspended in open water) forms *Cyclotella comta* (*Figure VII.B*–1A) and *Tabellaria flocculosa* (*Figure VII.B*–1B) and the tychoplanktonic (normally associated with periphytic habitats but often found suspended in open water) form *Cyclotella stelligera* (*Figure VII.B*–1A), along with members of the genera *Melosira, Fragilaria, Synedra, Asterionella* and *Navicula* were the most common diatoms in the lake. The coexistence of *Cyclotella comta* (pH optimum 7.0 and above), *C. stelligera* (pH optimum 7.5 to 8.0) and *Tabellaria flocculosa* (pH optimum 5.0 to 5.3) suggests that Mirror Lake remained in a near neutral state during this time period.

Total numbers of cells generally increased throughout Zone II, reaching 4.3×10^9 cells at 650 cm (~5,500 yr B.P.). The diatom evidence points to a time of continually increasing primary production in waters of moderate nutrient levels and near-neutral pH. These conclusions are supported by pigment analyses (*Chapter*

B. Diatoms

VII.D), which suggest significant allochthonous organic input early in Zone II shifted towards autochthonous contributions later in the period.

The cause for the abrupt increases of *Fragilaria crotonensis,* indicative of moderate nutrient enrichment, and *Synedra delicatissima* (*Figure VII.B*–1B) at 1,000 cm remains unknown. It should be noted that these taxa peak in abundance simultaneously with high sulfur and high chlorophyll content in the long core (*Chapters VII.D and E*), evidence which, in total, suggests that a temporary condition of high nutrient input and algal productivity was in effect at approximately 1,000 cm (~8,000 yr B.P.).

Zone III—Late Post-Glacial

In discussing the character of diatom assemblages in Zone III, it is useful to distinguish two sections of the profile: that between 550 cm (base of Zone III) and 200 cm, and that between 200 cm and 25 cm (top of Zone III). Below 200 cm, the assemblages indicate very little change in the kinds of species common in the lake from those of Zone II. *Cyclotella comta* decreased in importance relative to *Melosira ambigua* while *Melosira distans* v. *alpigena* became slightly more common (*Figure VII.B*–1A). With the exception of an approximate increase of 20% in species numbers at 500 cm, numbers of taxa determined remained quite constant from 550 to 200 cm. Based on diatom cell content of the sediments, apparent peak diatom productivity in the lake may have been reached by the 650-cm (~5,500 yr B.P.) level of Zone II and maintained through the 200-cm level. Equally significant, however, is the inverse abundance relationship displayed by the two most dominant taxa *Cyclotella stelligera* and *Tabellaria flocculosa* (*Figure VII.B*–1A and B, respectively).

A recent diatom survey of several Minnesota lakes (Koppen 1973) indicates that the genus *Tabellaria* is common in northeastern lakes in Minnesota and rare in the more productive, alkali-rich southwestern lakes. Several species and strains of *Tabellaria* are replaced in the alkali-rich lakes by species of *Cyclotella*. In Zone II of the Mirror Lake core, in relative terms, *Tabellaria flocculosa* was more prevalent than *Cyclotella stelligera* below 650 cm. From 650 to 200 cm, *C. stelligera* increases in importance relative to *T. flocculosa*. Results of Koppen's work would suggest that the alkalinity of Mirror Lake water may have increased during the time sediments were deposited from 650 to 200 cm. Increasing alkalinity of the water was most likely independent of significant changes in pH, as the diatom flora as a whole indicates that near neutral pH values were maintained throughout Zone II and to the 200-cm level of Zone III. Higher alkalinity coupled with increased densities of diatom cells in the sediment and other evidence for increasing productivity within the lake (*Chapters VII.D and E*) lead to a conclusion that nutrient levels within the lake and associated algal productivity of the lake were highest from approximately 5,500 to 2,200 yr B.P. (650 to 200 cm).

In the upper portion of Zone III, above 200 cm, a number of changes within the diatom assemblages may be seen. Floristically, *Cyclotella stelligera* declines in importance while *Tabellaria flocculosa* and several *Melosira* species including *M. ambigua, M. distans* v. *alpigena, M. fennoscandica* and *M. tenella* become more important. Near the upper limits of Zone III, *Cyclotella comta* and *M. distans* v. *lirata* appear in significant numbers. An important result of Koppen's investigations has been the definition of strains of *Tabellaria flocculosa* based on optical and electron microscopy (Koppen 1975). A form of *Tabellaria flocculosa* defined by Koppen as strain III is an elongate, planktonic diatom found both in the low conductivity lakes in the extreme northeastern part of Minnesota as well as in lakes toward the southwestern part of the state, indicating that this form is somewhat tolerant of increasing conductivity. Strain IV is defined as a shorter form of *T. flocculosa* confined to the northeastern, low conductivity lakes (Koppen 1978). Koppen's taxonomic distinction between these two strains has been followed during this investigation as is indicated by the separate stratigraphies of *Figure VII.B*–1B. *Tabellaria flocculosa* strain IV is of minor importance in the core below 200 cm. Above 100 cm, however, strain IV becomes much more common. The ecologic requirements of these two *Tabellaria* strains suggest that Mirror Lake water became lower in conductivity over the past 1,500 years. Unfortunately, pH ranges and trophic requirements of many of the diatoms common during this interesting period in the history of Mirror Lake are

Plate 1. 1. Chrysophycean cysts: Mirror Lake core MLlc, 750-cm depth, 2,200 ×. **2.** Chrysophycean cyst: Mirror Lake core MLlc, 750-cm depth, 2,400 ×. **3.** *Cyclotella comta* (Ehr.) Kütz. (Cell is resting on valve surface, view is of upper valve surface and girdle; above right, note inner surface of valve of same species.): Mirror Lake core MLlc, 1,150-cm depth, 2,400 ×. **4.** *Cyclotella comta* (Ehr.) Kütz. (inner surface of valve): Mirror Lake core MLlc, 950-cm depth, 2,100 ×. **5.** *Cyclotella stelligera* Cl. u. Grun. (inner surface of valve): Mirror Lake core MLlc, 550-cm depth, 2,400 ×.

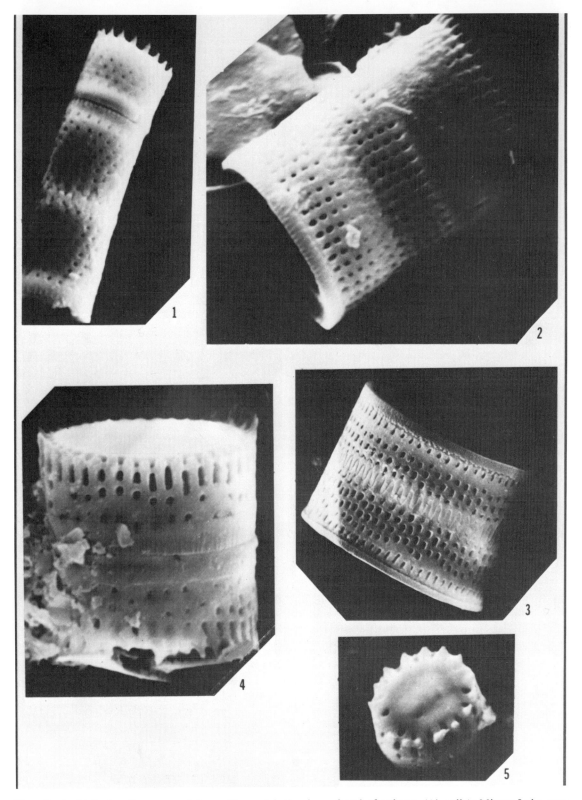

Plate 2. 1. *Melosira distans* v. *alpigena* Grun. (girdle view of end of colony, 1½ cells): Mirror Lake core MLsc, 10-cm depth, 5,000 ×. **2.** *Melosira distans* v. *lirata* (Ehr.) Bethge (Stat. 1). (girdle view, 1½ cells): Mirror Lake core MLlc, 1,000-cm depth, 5,000 ×. **3.** *Melosira distans* v. *lirata* (Ehr.) Bethge (Stat. 1). (girdle view of junction of 2 one-half cells): Mirror Lake core MLsc, 10-cm depth, 5,000 ×. **4.** *Melosira distans* v. *lirata* (Ehr.) Bethge (Stat. 2). (girdle view, complete cell): Mirror Lake core MLlc, 1,200-cm depth, 5,000 ×. **5.** *Melosira distans* v. *lirata* (Ehr.) Bethge (Stat. 3). (oblique view of valve surface and portion of girdle): Mirror Lake core MLlc, 1,100-cm depth, 5,100 ×.

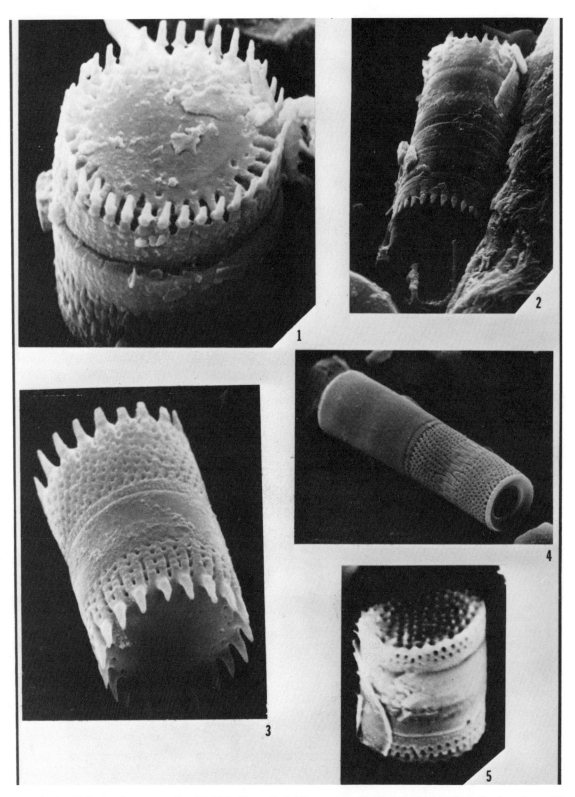

Plate 3. 1. *Melosira fennoscandica* Cl. Eul. (Stat. 1). (oblique view of valve surface and girdle, junction with second cell visible lower left): Mirror Lake core MLlc, 1,150-cm depth, 6,000 ×. **2.** *Melosira fennoscandica* Cl. Eul. (Stat. 1). (oblique view of valve surface of one cell in colony of 3 cells, upper surface in view are girdles and junctions between cells): Mirror Lake core MLlc, 1,150-cm depth, 2,400 ×. **3.** *Melosira italica* v. *subarctica* O. Müll. (Stat. 1). (oblique view of valve surface and girdle of single cell): Mirror Lake core MLlc, 950-cm depth, 5,000 ×. **4.** *Melosira italica* v. *subarctica* O. Müll. (Stat. 1). (oblique view of one complete cell with one-half cells attached at each end; conspicuous is the internal extension of the sulcus of the one-half cell in foreground): Mirror Lake core MLlc, 950-cm depth, 2,400 ×. **5.** *Melosira tenella* Nygaard (oblique view of 2 cells): Mirror Lake core MLlc, 550-cm depth, 5,000 ×.

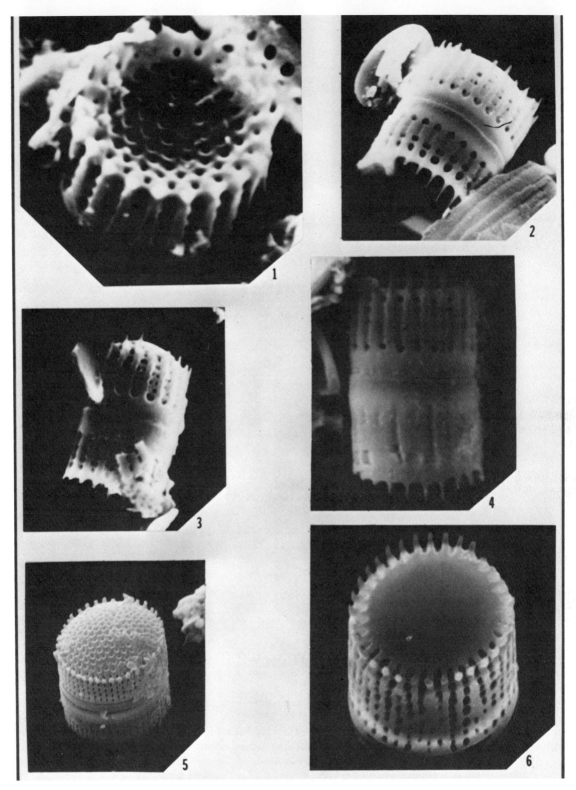

Plate 4. 1. *Melosira* sp. 1 (oblique view of one-half cell, concave valve surface seen with regularly placed punctae): Mirror Lake core MLlc, 1,200-cm depth, 9,000 ×. **2.** *Melosira* sp. 1 (oblique view of single cell, girdle view conspicuous with convex valve surface seen in upper right, concave valve surface in lower left): Mirror Lake core MLlc, 1,200-cm depth, 4,800 ×. **3.** *Melosira* sp. 1 (girdle view of single cell, bulge of convex valve surface seen in upper right): Mirror Lake core MLlc, 1,200-cm depth, 4,600 ×. **4.** *Melosira* sp. 2 (girdle view of single cell, similarity of girdle structures with *Melosira* sp. 1 is demonstrated): Mirror Lake core MLlc, 1,150-cm depth, 5,200 ×. **5.** *Melosira* sp. 4 (oblique view of single cell): Mirror Lake core MLlc, 1,150-cm depth, 5,000 ×. **6.** *Melosira* sp. 5 (oblique view of one-half cell): Mirror Lake core MLsc, 2-cm depth, 4,900 ×.

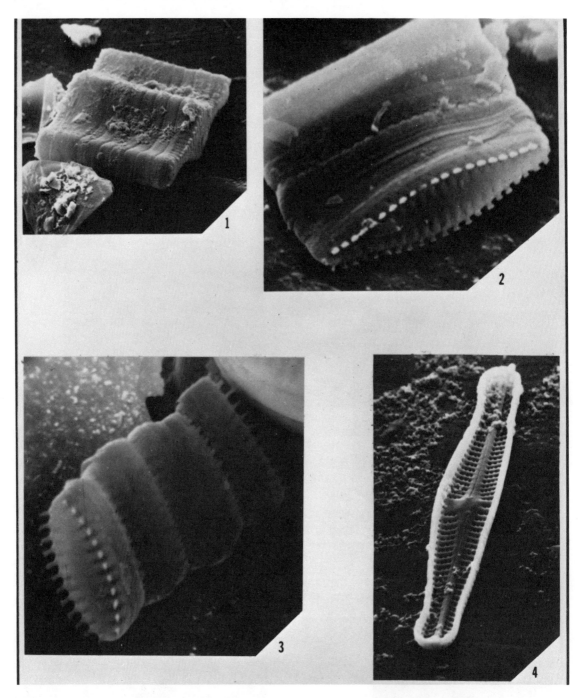

Plate 5. 1. *Tabellaria flocculosa* (Roth) Kütz. (oblique view of single cell, valve surface with striae seen at right, upper surface composed of edges of numerous intercalary bands): Mirror Lake core MLlc, 10-cm depth, 2,400 ×. **2.** *Fragilaria brevistriata* v. *inflata* (Pant.) Hust. (oblique view of colony, valve surface of cell at lower right seen with marginal extensions [teeth or spines] effective in colony formation): Mirror Lake core MLlc, 950-cm depth, 5,000 ×. **3.** *Fragilaria construens* v. *venter* (Ehr.) Grun. (oblique view of 4-cell colony, long spines or teeth extending from valve margins are not evident in optical views): Mirror Lake core MLlc, 1,100-cm depth, 4,900 ×. **4.** *Achnanthes microcephala* (Kütz.) Grun. (valve view, raphe valve): Mirror Lake core MLlc, 950-cm depth, 5,900 ×.

either too broad or too poorly understood to provide supplementary information.

Concurrent with floristic changes in the upper levels of Zone III, numbers of species increase. Progressively larger numbers of periphytic and littoral taxa contribute to the *richness* of the diatom remains. This apparent increase in diversity may well be a result of declining diatom productivity as cells produced by open-water blooms became less abundant. Relative contributions of diatom cells from the more diverse littoral habitats would then increase in importance in the preserved sediments. Total cell contributions to the sediments are seen to decrease near the upper limits of Zone III concurrent with increasing *apparent* diversity (*Figure VII.B*–1C).

Zone IV—Recent Cultural

The trend discussed above, increasing "apparent" diversity and lower total diatom production, is seen to continue throughout Zone IV. Otherwise, there is little change in the diatom assemblages of the recent cultural period from those of Zone III above 50 cm. In particular, there is no evidence of progressive lake acidification in recent times based on the ecologic preferences of the surface and near-surface diatom assemblages. In surface samples of Mirror Lake sediment there are very diverse assemblages of periphytic and littoral diatoms. In the uppermost sample (0 to 2 cm) of the Davis-Doyle core there were 103 taxa represented in 1,000 cells. This abrupt increase in species richness is especially surprising in light of the tendency for species diversity profiles to decline in less compacted surface sediments (Smol 1981). The widely varying habitat preferences of members of this surface assemblage suggest that diatoms from communities developed in a wide variety of habitats are contributing to the basin sediments at the present time. One possible explanation could be an increase in the supply of allochthonous diatom frustules to the lake over the past few years.

Chrysophycean Cysts

Densities of Chrysophycean cysts in the Mirror Lake cores closely parallel diatom cell densities (*Figure VII.B*–1C). The presence of these cysts in lake sediments is believed to be of paleoecologic importance because the group as a whole is thought to be best developed in nutrient-poor water (Leventhal 1970). In Lago di Monterosi, Levanthal related reductions in cyst numbers to an increase in the eutrophic nature of the lake.

As evidenced by Chrysophycean remains, the Chrysophyceae were most abundant from approximately 3,000 to 1,000 yr B.P. (300 to 60 cm). Although evidence suggests that Mirror Lake became more nutrient rich from 5,500 to 2,200 yr B.P., it is likely that the trophic status of the lake never advanced beyond an oligomesotrophic state.

"Apparent" Productivity of Mirror Lake

Increases in densities of diatom cells in the sediment and the floristic patterns suggest that nutrient enrichment of Mirror Lake took place from 5,500 to 2,200 yr B.P. Increased densities of diatoms and Chrysophycean cysts during this period could be explained by assuming that larger populations of these algae did exist in the lake, or that allochthonous inorganic input to the lake merely diluted the remains of essentially stable populations of these algae both early and late in the history of the lake. It is reasonable to postulate that both factors have contributed to the observed density patterns of both diatom cells and Chrysophycean cysts in the light of the chemical and physical descriptions of the long core presented elsewhere (*Chapter VII.E*).

Summary

Analyses of diatom assemblages in 12.3 m of post-glacial sediments cored from the central basin of Mirror Lake support the general history of the lake basin proposed by Likens and Davis (1975; *Chapter VII.F*). Earliest diatom remains in sediment dated 11,000 yr B.P. indicate the existence of a slightly alkaline, oligotrophic lake supporting a moderate littoral diatom flora but impoverished in planktonic forms. Concurrent with forestation of the drainage basin, the lake supported increasing numbers of planktonic diatoms with assemblages indicating the lake was becoming more nutrient rich and more neutral. From 5,500 to 2,200 yr B.P., subtle changes in the relative abundances of diatoms along with increases in the total numbers of preserved frustules suggest that lake productivity peaked as conductivity of the water increased. Diatom remains indicate that the pH of the lake has remained quite stable (near neutral to slightly acid) throughout Zones II, III and IV although alkalinity and conductivity have declined over the past 1,500 years. Apparent increases in diatom diversity in surface and near-surface samples are most probably related to declining open-water diatom productivity as cells produced from plankton blooms have become less abundant relative to those supplied from littoral habitats. Basin disturbances occurring in the recent cultural period (Zone IV) have not apparently influenced the natural diatom communities of Mirror Lake.

C. Animal Microfossils

CLYDE GOULDEN and GAY VOSTREYS

Cladocera, a small group of branchiopod Crustacea, form an important component of the populations of the plankton and littoral habitats of freshwater lakes. The exoskeleton of many species is preserved in the sediments of lakes, and can be easily identified. Species of the Bosminidae, a diminutive planktonic form, and of the Chydoridae, substrate species that inhabit aquatic macrophytes, filamentous algae or burrow in the sediment of the littoral zone, are well preserved in lake sediments. Although these two groups are only a part of a complex of organisms present, their study can provide some insight into the development of the lake-ecosystem with time.

Unfortunately, other groups, which can be preserved in sediments, including nondiatom algae, Protozoa, flatworms, ostracods, midges and Cladocera, other than the bosminids and chydorids (Frey 1964), are poorly represented in the Mirror Lake sediments.

Methods

Subsamples from the core were oven-dried at 110°C for 24 h, cooled in a desiccator for 24 h and then weighed. Most of the samples were at least 1 g dry wt sediment. The sediment was dispersed in a 10% solution of potassium hydroxide and, after the heated sample cooled to room temperature, it was sieved through a 37-μm mesh, brass-nickel screen to eliminate most of the silt and clay-sized particles.

After sieving, the sample was concentrated to either 5 or 10 ml. A series of 0.05-ml subsamples was withdrawn, partially evaporated on microscope slides, and mounted in 0.1 ml of glycerin jelly stained with gentian violet. Microfossils on the slides were counted at 125× magnification. Identification of all remains was based on descriptions by Frey (1958, 1959) or from a slide collection of cladoceran exoskeletal remains maintained in our laboratory.

The abundance of microfossils was normalized to a sample weight of 1 g dry wt. Replicate counts on subsamples from the same sediment samples were used to calculate the standard error of the mean.

Results

Likens and Davis (1975; *Chapter VII.F*) divided the history of Mirror Lake into four periods as follows:

Zone I: Late glacial (9,000 to 14,000 yr B.P.)
Zone II: Early post-glacial (5,000 to 9,000 yr B.P.)
Zone III: Late post-glacial
Zone IV: Recent cultural horizon.

In addition, they have identified peaks in autogenic organic production at 1,050 cm and between 750 to 550 cm. Changes in the estimated absolute abundance and in the relative abundance of cladoceran microfossils follow this pattern and, therefore, will be discussed in this context.

Only two species of planktonic Cladocera have an abundant fossil record in Mirror Lake, *Bosmina longirostris* and *Eubosmina tubicen*. Mandibles and ephippia of *Daphnia* are scattered throughout the sediment core but are never sufficiently common to identify trends in

C. Animal Microfossils

occurrence. No other planktonic animals are represented. This absence obviously is unfortunate because their distributions might have yielded some interesting comparisons with the modern plankton community and provided a partial test of ideas presented by Makarewicz and Likens (1975, 1979). The littoral Cladocera, particularly the chydorids, have a good fossil record throughout the core. Unfortunately, the modern littoral habitat has not been as well studied as the plankton.

Table VII.C-1. The Number of Zooplankton and Littoral Species in Each Zone of a Sediment Core from Mirror Lake

	Number of Species	
	Plankton	Littoral Chydorids
Zone I	5	20
Zone II	6	28
Zone III	6	27
Zone IV	3	21

Zone I—Late Glacial

In the late-glacial sediments, plankton and littoral cladoceran fossils are rare until 1,150 cm when both begin to increase (*Figure VII.C-1*). This increase continues into Zone II. *Bosmina longirostris* and *Eubosmina tubicen* are both present, the latter most common.

Zone I is characterized by low lake productivity, presumably determined by low temperatures and high rate of inorganic sedimentation (Likens and Davis 1975), suggesting high lake turbidity. But the cold climate did not inhibit the immigration of organisms and the formation of an aquatic ecosystem with numerous species. Twenty species of chydorids and four planktonic species are present in the sediments of Zone I (*Table VII.C-1*).

At the top of Zone I, the bosminids and chydorids both increase with a peak at 1,150 cm followed by a decrease to 1,125 cm. *Bosmina longirostris* increased sharply, though still not the dominant bosminid. *Alonella nana,* a diminutive animal inhabiting underwater plants, was the most abundant chydorid. The other common chydorids, *Acroperus harpae, Alona af-*

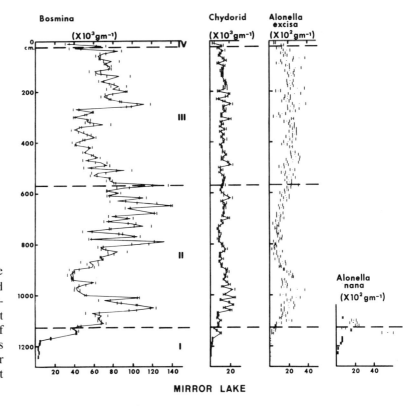

Figure VII.C-1. Abundance of *Bosmina*, chydorids, and of *Alonella excisa* and *Alonella nana* in a sediment core from the deepest part of Mirror Lake. Abundance as number of microfossils per gram dry weight of sediment (bars represent ±S.E.).

finis and *Rhynchotalona falcata* either live on aquatic macrophytes or in the sediment. This assemblage suggests a well-developed littoral zone.

Zone II—Early Post-Glacial

Zone II, which dates from 9,000 to 5,000 yr B.P., begins with a sharp increase in abundance of cladoceran microfossils, reaching a peak at 1,050 cm (*Figure VII.C*–1). The number of fossils of bosminids declines above 1,010 cm and remains low until 890 cm. Above this point, numbers increase again to a continuously high abundance from 800 to 560 cm, but there is a great deal of variability between successive sediment levels. Chydorids are more variable in abundance between 1,050 and 1,000 cm, though they never have as sharp a peak in abundance as do the bosminids (*Figure VII.C*–2).

The relative abundance of littoral animals (chydorids) increases from 960 to 900 cm (*Figure VII.C*–3), the only period of persistent abundance, but this is largely due to the low absolute abundance of bosminids rather than an increase in chydorids. Thus, there is no indication that the water level may have been lower. Instead these changes primarily suggest a lower lake productivity.

There are continuous changes in the relative abundance of the two bosminids throughout Zone II (*Figure VII.C*–3). In general, *Bosmina* is most common between 1,010 and 780 cm, while the two species are almost equally abundant from 780 to 580 cm.

The most striking changes in species abundance occurred among the littoral animals. *Chydorus brevilabris* (Frey 1980), a ubiquitous species, sharply increased in number from 1,070 to 1,040 cm and was abundant until 910 cm. It had a peak at 820 cm that may not be significant as it represents only one sediment level, but between 720 to 610 cm, *C. brevilabris* was again very abundant. *Alonella excisa* was most common from 1,050 to 920 cm. It did not increase at 720 cm. *Rhynchotalona falcata* was the

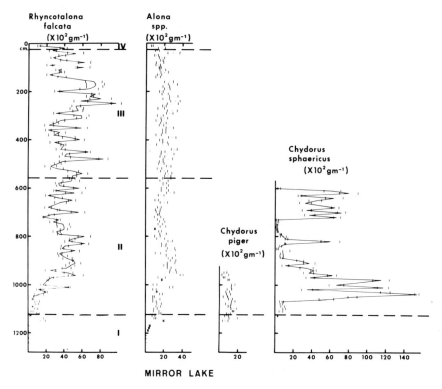

Figure VII.C–2. Abundance of *Rhynchotalona*, *Alona* spp., *Chydorus piger* and *Chydorus brevilabris* in a sediment core from the deepest part of Mirror Lake. Abundance as number of microfossils per gram dry weight of sediment (bars represent ±S.E.).

C. Animal Microfossils

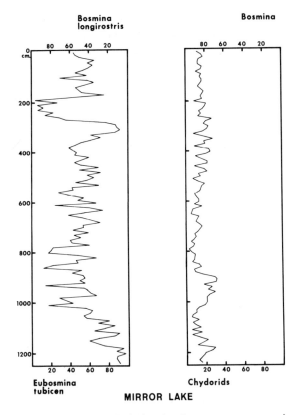

Figure VII.C–3. Variation in the percent composition of *Bosmina longirostris* vs. *Eubosmina tubicen* and of total bosmina vs. chydorids in a sediment core from the deepest part of Mirror Lake.

most abundant chydorid inhabiting sediment throughout Zone II. This species is almost totally restricted to sandy littoral shelves that have an organic matter content below 5% (Goulden 1971).

The increase in *Chydorus* at approximately 5,600 yr B.P. 720 cm was accompanied by a decline in *R. falcata*. The abundance of *C. brevilabris* could have been due to filamentous algal mats growing over the sandy shelves that would have served as excellent substrate for *Chydorus*. These algal mats could have increased the organic matter of the sand and made the area less suitable for *R. falcata*. *C. brevilabris* may also be abundant in the plankton during blooms of blue-green algae; however, we found no fossil blue-green algae.

Zone III—Late Post-Glacial

The southern shore of Mirror Lake is now sandy (*Chapter VII.A*; Moeller 1975) and has *R. falcata* in abundance. we suggest that the physical character of the littoral was established in Zone II, at 960 cm or by 8,000 yr B.P. But the increase in *C. brevilabris* about 5,600 yr B.P. indicates that the lake did not fully attain its present character until later.

Bosmina and *Eubosmina* declined in abundance in the lower half of Zone III, from 560 to 270 cm. A sharp increase occurred at 250 cm. The number of fossils continued high up to Zone IV. Chydorid counts indicated almost no qualitative change in the littoral zone throughout Zone III that could cause changes in species relative abundance, though numbers of fossils in successive layers were most variable between 340 and 190 cm. *R. falcata* was most abundant between 250 and 160 cm.

There were sharp changes in the relative abundance of the bosminids from 360 to 160 cm (*Figure VII.C–3*). The relative abundance of *Eubosmina* increased from approximately 40 to 90% between 380 and 320 cm. Because the total abundance of bosminids declined at this time, this actually means that *Bosmina* declined much more than *Eubosmina*. From 280 to 190 cm, a reversal in abundance, with *Bosmina* the most abundant form, occurred. This reversal was associated with a sharp increase in total bosminid fossils in the sediment; the succession then was first due to a decrease and then an increase in *Bosmina longirostris* while *Eubosmina* numbers changed little.

Zone IV—Recent Cultural Horizon

In general, all Cladocera decline in abundance in the recent sediments. This zone includes only two sediment levels examined for Cladocera, which is insufficient to identify trends. *Eubosmina tubicen* appears to be more common than *Bosmina longirostris*. Among the littoral chydorids the sand-dwelling species *Rhynchotalona falcata* was the most common.

Discussion

Our data on the distribution and abundance of cladoceran microfossils in the Mirror Lake core support the division of the lake's history into four zones (Likens and Davis 1975; *Chapter III.B.1*). We have identified three periods of increased lake productivity. Peak lake productivity at 1,050 cm and between 750 and 560 cm was identified by Likens and Davis (1975). In addition, the Cladocera suggest a third period of high lake productivity at 250 cm. This corresponds with the zone of greatest abundance of diatoms (*Chapter VII.B*) and, therefore, must be real.

Microfossils of planktonic Cladocera, specifically of the bosminids, were more abundant than the littoral chydorid Cladocera. This suggests that the lake has never had a well-developed littoral zone of aquatic macrophytes and has been low in primary productivity. Two species of bosminids were abundant, *Bosmina longirostris* and *Eubosmina tubicen*. Very few data are available on the ecology of *E. tubicen* because it has only recently been discovered to occur in the U.S. (Goulden and Frey 1963; Deevey and Deevey 1971). *Bosmina longirostris* is broadly distributed geographically, but the forms assigned to this species may represent several cryptic species (Manning et al. 1978). We do not know which forms of this species were present in the lake. As a result, we can say very little about ecologic factors that may have induced species' successions as occurred in Zone III of the Mirror Lake core. The poor preservation of daphnids precludes our drawing conclusions about the effect of size-selective predation on the zooplankton.

The most striking changes among the species of Chydoridae occur for *Chydorus brevilabris* and *Rhynchotalona falcata*. *Chydorus brevilabris* was abundant throughout most of Zone II, probably as a result of changes in the extent of aquatic macrophyte productivity or owing to an extensive growth of mats of algae. We found no evidence that blue-green algae were ever abundant in the lake, which could otherwise explain the pattern of abundance of *Chydorus*.

The littoral chydorid Cladocera fauna of Mirror Lake has been dominated primarily by *Rhynchotalona falcata* for most of the lake's history (*Figure VII.C–2*). We have a reasonable amount of information about this species (Goulden 1971). It is restricted to sandy sediments of the littoral zone that have a very low organic content. In Lake Lacawac, Pennsylvania, this species is very abundant on waveswept shores where organic matter seldom accumulates. *Rhynchotalona* was rarely found inhabiting sediments with an organic content greater than 10% of dry wt. At present, the littoral of Mirror Lake is similar to this, its littoral shore is composed largely of sand and *Rhynchotalona* is very abundant. Mirror Lake probably attained this character early in its history and, with the exception of periods when *Chydorus* was abundant in Zone III, has not been significantly altered.

Summary

Microfossils of Cladocera are distributed throughout the sediment core from Mirror Lake, first appearing after 14,000 yr B.P. A total of 34 species were encountered, six planktonic forms and 28 littoral chydorid species. *Bosmina longirostris* and *Eubosmina tubicen* were the dominant microfossils of zooplankton species; whereas, *Rhynchotalona falcata* and *Chydorus brevilabris* were the most common littoral Cladocera.

Zooplankton species were much more abundant than were the littoral species, generally planktonic forms composed more than 80% of the cladoceran assemblages. No definitive succession of the two zooplankton species was evident.

Among the littoral species, species that inhabit aquatic macrophytes were the dominant forms during the very early history of the lake, while sediment forms were dominant during the mid- and late history of the lake. The macrophyte inhabiting species included a succession from *Alonella nana*, to *Alonella excisa*, and then *Chydorus brevilabris*. *Rhynchotalona falcata* was the most abundant littoral cladoceran during most of the history of Mirror Lake. This species is restricted in its habitat to sandy sediments with very low organic content. This species is common in the lake at the present time. Its distribution in the sediment core suggests that the present character of the lake developed at least 9,000 yr B.P. and the lake has remained relatively unchanged since that time.

D. Fossil Pigments

Gene E. Likens and Robert E. Moeller

Remains of plant pigments in the sediments of lakes have been used to deduce the origin of those sediments (e.g. Gorham and Sanger 1967; Sanger and Gorham 1972; Likens and Davis 1975). It has been proposed that such sedimentary, or "fossil," pigments and their derivatives could be used to differentiate between autochthonous and allochthonous origins as well as to give information about the trophic status of the lake at the time the sediments were deposited (e.g. Gorham 1960; Gorham et al. 1974; Sanger and Gorham 1972; Gorham and Sanger 1975, 1976). Thus, this approach was applied to the study of sediments in Mirror Lake with the hope of elucidating more about the post-glacial history of the lake and the interactions between the lake and its watershed.

Methods

Fossil pigments were extracted in the dark with appropriate solvents (see Gorham and Sanger 1967; Sanger and Gorham 1972; Likens and Davis 1975) from samples obtained from the Central core taken in the deepest part of Mirror Lake (see *Chapters VII.A* and *VII.E*). Pigment concentration was expressed as units/g of organic matter, where one unit was equivalent to an absorbance of 1.0 in a 10-cm cell when dissolved in 100 ml of appropriate solvent (see Sanger and Gorham 1972). Values are presented here as concentration (units/g organic matter) within the Central core and as net deposition or influx to the sediments (units/m²-yr). Focusing of sediments within the central portion of the basin has not been quantitatively considered in these analyses, but would affect the overall deposition rate for the lake basin, particularly in the last 4,000 years (see Davis and Ford 1982; *Chapter VII.A* and *F*).

Results and Discussion

Although the rate was increasing, accumulation of organic matter in the sediments prior to 9,500 yr B.P. was very slow. Likewise, the sediments contained a low concentration of pigments (*Figure VII.D*-1) and net deposition was small (*Figure VII.D*-2). Based on analyses and interpretations by Gorham et al. (1974), these data suggest that biologic productivity in Mirror Lake and its watershed was low, but increasing during this late-glacial period.

Between 9,000 and 7,000 yr B.P. there was a marked increase in the influx of organic matter to the sediments (see *Chapter VII.E*) and a progressive increase in concentration and net deposition of fossil pigments. The increase in pigment concentration (*Figure VII.D*-1) also suggests increasing autochthonous productivity. The concentration of chlorophyll degradation products peaked about 8,000 yr B.P. and, although variable, remained high until about 5,500 yr B.P. In contrast, net deposition of chlorophyll did not reach a maximum until about 5,500 yr B.P. (*Figure VII.D*-2). The pigment ratio (chlorophyll/carotenoids) during this period may indicate that the proportion of allochthonous organic matter was high during the first part of the period (9,000 to 7,000 yr B.P.). The ratio declined after about 6,500 yr B.P. (whereas Ca and P increased; *Chapter VII.E*), suggesting a decline in allochthonous inputs and increased autochthonous inputs.

The forest surrounding Mirror Lake was disturbed 4,800 yr ago by a sudden decline in hemlock, followed by a succession of birch, then beech and sugar maple (*Chapter III.B.1*). After 1,000 yr (~4,000 yr B.P.) hemlock recovered and beech, birch, and sugar maple decreased in abundance again. The sudden decline in hemlock about 5,000 yr ago can be seen in pollen diagrams throughout eastern and midwestern U.S. (Davis 1965; Brubaker 1973), suggesting a widespread disaster such as the spread of a pathogen. This major change from coniferous to deciduous vegetation, involving the precipitous death of numerous trees, might have affected the lake-ecosystem. However, few changes were found in the sediment that are clearly correlated with it. There was no significant change in the pigment concentrations (or in

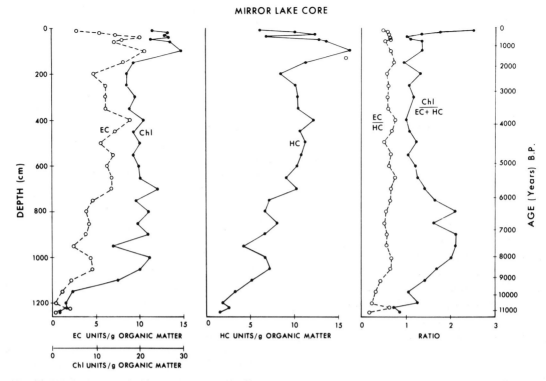

Figure VII.D–1. Concentration of fossil pigments and pigment ratios in the sediment profile of Mirror Lake. (Modified from Likens and Davis 1975.) Chl = chlorophyll derivatives; EC = epiphasic carotenoids; and HC = hypophasic carotenoids.

their ratio). Net deposition of hypophasic carotenoids peaked at about 4,800 yr B.P. (*Figure VII.D–2*), but the peak deposition of pigments around this time may be a feature of the Central core alone, since the peak deposition of organic matter as a whole varies over a period of millennia among the three analyzed deep-water cores (*Figure VII.E–4*).

The ratio of chlorophyll derivatives to carotenoids is lower from about 6,000 to almost 200 yr B.P. than either earlier or later, possibly indicating an interval of increased autochthonous

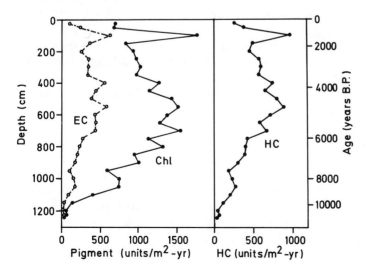

Figure VII.D–2. Deposition of fossil pigments in the sediment profile in Mirror Lake. Chl = chlorophyll derivatives; EC = epiphasic carotenoids; and HC = hypophasic carotenoids.

D. Fossil Pigments

Table VII.D-1. Average Pigment Concentrations (Units/g Organic Matter) and Pigment Ratios in Organic Debris Collected from Streams and Small Pools[a]

Material	Chlorophyll (Chl)	Epiphasic Carotenoids (EC)	Hypophasic Carotenoids (HC)	$\dfrac{\text{Chl}}{\text{EC + HC}}$	$\dfrac{\text{HC}}{\text{EC}}$
Debris for Northern Hardwoods	1.426	0.190	0.940	1.26	0.20
Debris from White Pine	0.542	0.311	0.628	0.597	0.46
Debris from Hemlock	0.630	0.251	0.524	0.830	0.49

[a] Modified from Likens and Davis (1975).

inputs. This ratio might also have been lowered by a decrease in allochthonous inputs or a change from coniferous to deciduous sources of the allochthonous component (Gorham and Sanger 1967). This decrease in the pigment ratio after 6,500 yr B.P. closely follows the end of the pine period of forest development (*Chapter III.B.1*). The pigment ratios in present-day litter do not support such an interpretation (*Table VII.D-1*), although this coarse particulate organic matter is presumably not the predominant form in which allochthonous pigments are deposited in the profundal sediment.

In contrast to concentrations, the net deposition of pigments generally declined from about 5,000 to 2,000 yr B.P. (*Figure VII.D-2*), which probably was caused by changes in the focusing of sediments within the central portion of the basin (Likens and Davis 1975; Davis and Ford 1982; *Chapter VII.A*). The explanation for the large increase in concentrations between 2,000 and 1,000 yr B.P. (*Figure VII.D-1*), which involves a peak in deposition shortly before 1,000 yr B.P. (*Figure VII.D-2*), is unknown.

Organic matter deposited during the climatic change of 11,000 to 10,000 yr B.P. contains low concentrations of chlorophyll derivatives and a ratio of chlorophyll to carotenoids approaching 1.0. These characteristics are both consistent with modern sediments deposited in relatively unproductive lakes in England (Gorham et al. 1974) and Minnesota (Gorham and Sanger 1976), where allochthonous organic matter may predominate. Since 9,000 yr B.P. the pigment concentrations have been much higher, suggesting a predominance of well-preserved autochthonous organic matter. However, since 9,000 yr B.P. the ratio of chlorophyll derivatives to the sum of epiphasic plus hypophasic carotenoids, chlorophyll/(EC + HC), has ranged from 1 to 2, levels characteristic of well-decomposed, perhaps allochthonous, organic matter. This paradoxical occurrence of high pigment ratios with high chlorophyll concentrations precludes confident interpretation of the Mirror Lake stratigraphy based on pigment characteristics from other lakes that have been studied.

Where pigment ratios greater than or equal to 2 have been found in lake sediments, they were associated with low chlorophyll concentrations, suggesting a long period of decomposition under lighted, aerobic conditions. These sediments include profundal mud from an unproductive, dystrophic Finnish lake (Huttunen et al. 1978; point H in *Figure VII.D-3*), periodic levels within the stratigraphy of Windermere, England (Belcher and Fogg 1964) and sedge peat in Kirchner Marsh, Minnesota (Sanger and Gorham 1972). The Holocene concentrations of chlorophyll derivatives in Mirror Lake (see *Figure VII.D-3*) exceed those even in productive English and Minnesota lakes and equal that in a eutrophied, meromictic Finnish Lake, where pigment preservation should be optimal (Huttunen and Tolonen 1975; point L in *Figure VII.D-3*).

According to arguments presented by Sanger and Gorham (1972), Gorham et al. (1974) and Gorham and Sanger (1975, 1976), the concentration of chlorophyll derivatives in the organic fraction of lake sediment ranges downward from that in living tissue (approximately 20 to 40 units/g organic matter; Gorham and Sanger 1975) as a cumulative response to bio- and photochemical degradation that takes place

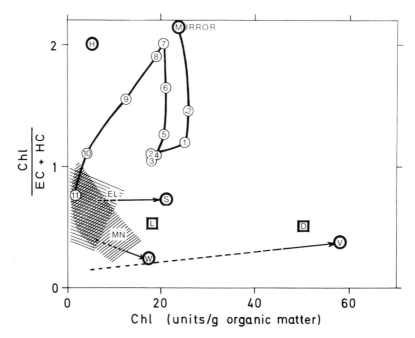

Figure VII.D-3. Pigment characteristics of Mirror Lake core (smoothed trajectory plotted at 1,000-yr intervals) compared with surficial sedimentary pigments of other North Temperate lakes: sixteen English lakes (EL—Gorham et al. 1974), 42 Minnesota lakes (MN—Gorham and Sanger 1976), Stewart's Dark Lake, Wisconsin (D—Gorham and Sanger 1972), Lake Hakojarvi, Finland (H—Huttunen et al. 1978), Lake Lovojarvi, Finland (L—Huttunen and Tolonen 1975), Shagawa Lake, Minnesota (S—Gorham and Sanger 1976), Lake Varese, Italy (V—Adams et al. 1978) and Wintergreen Lake, Michigan (W—Manny et al. 1978). Rectangular symbols indicate meromictic lakes; arrows portray hypereutrophication over the last one to two centuries. Carotenoid data have been slightly altered to increase comparability in some cases. Chl = chlorophyll derivatives; EC = epiphasic carotenoids; and HC = hypophasic carotenoids.

before and during burial. Carotenoids seem to be somewhat more susceptible to degradation than chlorophyll derivatives, at least after the early phases of death and decay, when high proportions of all pigments are lost (Gorham and Sanger 1975). Thus the ratio chlorophyll/(EC + HC) is higher in soil humus than leaf litter (Gorham and Sanger 1975), in profundal sediment of less productive compared with more productive lakes (Gorham et al. 1974; Gorham and Sanger 1976), and in littoral compared with profundal sediments of particular lakes (Gorham and Sanger 1972). In contrast to soil humus (Gorham and Sanger 1975) and leaf litter collected from streams (*Table VII.E*-1), the profundal sediment from Mirror Lake has levels of chlorophyll derivatives demonstrating little, if any, greater degradation of chlorophyll relative to that of organic matter as a whole.

It is plausible, but unlikely, that the high chlorophyll levels in organic matter deposited after 9,500 yr B.P. can be attributed to physical aggregation of the pigment-bearing fraction of otherwise pigment-poor allochthonous organic matter. More likely, unknown characteristics of autochthonous pigment production [exceptionally high chlorophyll/(EC + HC) ratios in living phytoplankton?] or sediment formation (especially good chlorophyll preservation, especially poor carotenoid preservation?) are responsible for the discordant content of the sediment in chlorophyll derivatives compared with carotenoids.

The increase in pigment concentration between 10,000 and 7,000 yr B.P., along with its increased deposition, probably reflects an increasing productivity within Mirror Lake that exceeded any increase in allochthonous inputs from the developing forest. Cultural eutrophication during the past one to two centuries has been recorded in the surficial sediments of some lakes as strong increases in chlorophyll concentration without major shifts in the pigment ratio (points associated with arrows in *Figure VII.D–3*). In Mirror Lake the cultural period coincides with a strong increase in the ratio, with a decrease in chlorophyll concentration in the surface sediments.

Deducing historical productivity, erosion and transport and other features of lake- and watershed-ecosystems from analyses of lacustrine sediments is an important, but difficult, problem in ecosystem analysis. We have looked to the sediments of Mirror Lake for relationships between lake and watershed throughout the past 14,000 years. Many workers have used only one or two criteria to interpret the post-glacial history (e.g. productivity) of a lake. Evaluating several parameters simultaneously has made us quite cautious, yet hopeful relative to the interpretations. For example, chlorophyll degradation products, pigment ratios, P, N and C influx, and percentages of organic matter are not consistently correlated with one another (*Chapter VII.E*). Taken alone, none of these criteria seems to be a sufficient indicator of lake productivity. Taken together and combined with other chemical data they do provide information on the history of the lake, but the interpretation is often very difficult (see *Chapter VII.F*).

The interpretation of fossil pigment concentration, deposition rates or ratios is complicated and depends not only on the variable inputs from allochthonous and autochthonous sources, but also on preservation and redeposition of sediment. We found that concentrations and ratios for pigments in organic debris from various aquatic habitats within the drainage basin were appreciably different from values obtained in the sediments of Mirror Lake (*Table VII.D–1*). Moreover, coniferous trees, such as hemlock, along stream channels might dominate the allochthonous inputs of organic debris from a drainage basin otherwise characterized by deciduous vegetation. This influence would be most obvious in stream-water inputs of particulate matter, particularly if there were no overland runoff from the drainage basin. Interpreting the proportion of dissolved vs. particulate matter inputs that accumulate as sediments in a lake is a particularly perplexing problem. Thus, it is not yet possible to indicate quantitatively how pigments get to the bottom of a lake or to indicate, definitively, what was their origin. Once deposited, organic matter may continue to decompose at variable rates depending upon the organic substrate and on environmental conditions, including resuspension by water movements and mixing with older sediments by biologic activity.

Summary

Concentrations and net deposition of pigment derivatives were low in sediment prior to 10,000 yr B.P. A marked rise in pigment concentrations occurred from about 10,000 to 8,000 yr B.P., followed by a maximal net deposition of pigments between 4,000 and 5,500 yr B.P. The ratio of chlorophyll derivatives to epiphasic plus hypophasic carotenoids reached a maximum of 2 between about 8,500 and 6,000 yr B.P., stabilized at a lower level of about 1 during the period 5,000 to 500 yr B.P., then returned to a value exceeding 2 in the surficial sediment.

The pigment ratio, which is unusually high for profundal sediments of a post-glacial lake, might indicate the predominance of allochthonous organic matter throughout the sedimentary record. High concentrations of pigments are, however, more consistent with predominance of better-preserved autochthonous organic matter, particularly during the interval 5,000 to 500 yr B.P. The unusual co-occurrence of high pigment concentrations with a high ratio of chlorophyll to carotenoids suggests that: (1) allochthonous organic matter with a high pigment ratio is physically sorted during sedimentation to produce pigment-rich sediment in the profundal zone, or more likely, (2) phytoplankton in Mirror Lake produce less carotenoid pigment than algae in other lakes studied, or most likely, (3) carotenoids are less well preserved in Mirror Lake sediments than the excellently preserved chlorophyll derivatives—in marked contrast to the proportionately better preservation of carotenoids in most eutrophic lakes. Until the processes of pigment production, transport, degradation and burial in lakes are better understood, the clear trends in the Mirror Lake core cannot be interpreted confidently.

E. Chemistry

GENE E. LIKENS and ROBERT E. MOELLER

Inferring the history of aquatic ecosystems from analyses of lacustrine sediments is one of the more difficult, yet important, problems in paleolimnology (Likens and Davis 1975). We have looked to the sediments of Mirror Lake for relationships between lake and watershed, both of which apparently have been nutrient-poor and relatively unproductive throughout the lake's history.

Few studies have attempted to relate sediment chemistry to changes in a lake like Mirror Lake, which extensive modern limnological investigation (*Chapters IV.B* and *V.B*) shows to be nutrient-poor and unproductive at most trophic levels. The sediment of many lakes records changes associated with cultural eutrophication (e.g. Hutchinson et al. 1970). These changes have often been much greater than other, more natural, changes occurring during a lake's history. Cultural eutrophication has not been a major perturbation in Mirror Lake, and our interest in the sedimentary record lies in the possibility of recognizing more subtle changes that might ultimately reflect changes in the vegetation and soils of the watershed.

Permanent sedimentation in a lake results from the difference between the total inputs (from the atmosphere as well as the immediate drainage basin) and the total outputs (fluvial and atmospheric) from the lake. The chemical form of elements in the sediment is largely dictated by the biogeochemical transformations that occur within the water-column and at the sediment-water interface of the lake. Thus, the sediments contain information about the conditions that existed in the lake as well as its watershed and airshed throughout the history of the lake. The challenge is to decipher and interpret this information because frequently the clues are subtle or indirect. Here we attempt to decipher the clues to lake and watershed processes that are provided by the chemical stratigraphy in the sediments of Mirror Lake.

Methods

In July 1969 a continuous core of sediments (12.42 m long and approximately 6.6 cm in diameter) was obtained near the center of Mirror Lake with a Swedish foil corer, yielding a maximum age (^{14}C) of 11,300 ± 200 yr (see Likens and Davis 1975). Most of the chemical analyses reported here are from this core (Central).

Some chemical analyses have been made on additional cores of the sediments obtained since 1969. In the autumn of 1974, an 80-cm *freeze core* was obtained by J. Sherman from a water depth of 10.9 m in Mirror Lake. Sediment and interstitial waters were frozen *in situ* on a hollow pipe filled with a freezing solution of methyl alcohol and dry ice. The position of the core relative to the sediment-water interface was adjusted so that a few centimeters of clear ice formed on the pipe while it was embedded in the sediments (see Sherman 1976 for additional details). Other cores were obtained by M. Davis and colleagues with a Livingstone-type piston-corer (Wright et al. 1965) at various locations within the basin (see *Chapter VII.A*; Davis and Ford 1982). Of these, core #782 was obtained from the northwest part of the basin at a water depth of 8.9 m, and is referred to most extensively here.

Samples were analyzed for chemical content on a per unit weight or per unit volume basis. Volumetric samples were obtained (1) by cutting a cross-section from the core with a very fine wire and trimming the edges with a thin plastic ruler to form a small block, or (2) by packing sediment into a porcelain spoon of 1-ml capacity. The various chemical analyses were done on subsamples within (±5-cm depth) the Central core.

Samples from the freeze-core were dried at 80°C and ashed at 600°C for 2 h (Sherman 1976). All other samples were dried at 100°C and ashed at 550°C. Matter lost on ignition was taken to be organic and the remaining ash as inorganic. Organic carbon was determined on dried sediment by the method of Menzel and Vaccaro (1964) and nitrogen by a modified Kjeldahl procedure (core #782) or by gas chromatography following high-temperature combustion in oxygen with a Carlo-Erba model 1104 Elemental Analyzer (Central core). Total sulfur and phosphorus were determined on samples from the Central and Livingstone-cores by the sodium

E. Chemistry

carbonate fusion method (Kanehiro and Sherman 1964). Phosphorus was extracted from the freeze-core by a wet ashing procedure (Sherman 1976). The ash component of the sediment was successively fractionated into (1) that portion soluble in hot 6 N HCl, (2) the amorphous silica, determined as 2.14 times the Si dissolved by 0.1 N KOH upon heating for 1.5 h at 95°C (modified from Tessenow 1975), and (3) the detrital mineral portion, which resisted the preceding treatments.

Total Ca, Mg, Na, K, Cu, Zn, Mn, Al, Fe, Cd, Cr and Pb were determined by atomic absorption spectrophotometry on aliquots from an HF digestion procedure (Pawluck 1967). In samples from the freeze-core, HNO_3 was used to dissolve any pyrite prior to the HF digestion (Sherman 1976). Extractable Ca, Mg, Na, K, Cu, Zn, Mn and Fe were determined on paired samples (each 1 g dry wt/25-ml solution) using 1 N ammonium acetate (pH 7) or 1 N acetic acid (pH 2.3). Samples were shaken for 4 h and allowed to stand overnight to facilitate the extraction.

The Sediment Profile

Likens and Davis (1975) divided the Central core from Mirror Lake into four zones, based on pollen and chemical data. Zone I, the spruce-dominated late glacial, extended upward from the bottom of the core (1,242 cm) to about 1,125 cm and spanned about 2,000 yr (11,300 to 9,300 yr B.P.). Zone II, the early part of the Holocene, extended from 1,125 cm to about 560 cm and spanned about 4,500 yr (9,300 to 4,800 yr B.P.). Zone III, the late part of the Holocene, extended from 560 cm to about 25 cm and spanned about 4,600 yr (4,800 to 200 yr B.P.). Zone IV, the recent cultural horizon, extended from 25 cm to the mud-water interface and spanned about 200 yr (200 yr B.P. to present). Recent data (*Chapter VII.A*) suggest that the cultural horizon began about 1840 A.D.

Sediments nearly 14,000 yr old have now been found at many locations in the lake (*Chapter VII.A*). This older sediment apparently fell out of the Central core during its recovery. Thus, for purposes of discussion, a fifth zone can be added: the *tundra-dominated late-glacial zone,* which extended from about 14,000 yr B.P. to initial spruce arrival on the watershed (11,500 yr B.P.). The late-glacial spruce period extended from 11,500 to 10,000 yr B.P. in the form of an open spruce forest or woodland (*Chapter III.B.1*). In this book, the term "late-glacial" is restricted to sediment older than 10,000 yr B.P.

The ^{14}C age of sediments at various depths in the Central core and in core #782 are shown in *Figure VII.E*–1. The B.P. notation applied to precultural sediment refers to uncorrected radiocarbon years "before present (1950 A.D.)," where the half-life of ^{14}C is taken as 5,570 yr (see Krishnaswami and Lal 1978 for general information on radiometric dating procedures in lacustrine sediments).

Our analyses and interpretations are based on physical and chemical parameters within the lacustrine sediment profile. Because of the detailed chemical analysis of surficial sediments obtained by the freeze core, the cultural period is discussed separately and more fully at the end of this chapter.

Bulk Density and Major Sediment Fractions

Bulk density of the sediments decreased sharply during the late-glacial period, and then only gradually throughout the remainder of the profile (*Figure VII.E*–2). The decrease in bulk density during the late glacial reflects the diminishing input of particulate inorganic matter and the addition of progressively greater amounts of organic matter and amorphous silica (*Figure VII.E*–3) of lower specific density. The slow decrease in density after 9,000 yr B.P. parallels a continuing gradual decrease in detrital mineral matter and increase in organic matter, but also could reflect progressive consolidation of the wet sediment as newer sediment accumulated above it.

In the center of Mirror Lake, the Holocene sediment had a high content of amorphous (meaning "noncrystalline") silica. Only 30 to 40% of the mass of ash is composed of inwashed mineral particles. The amorphous silica is largely thought to be diatom frustules, but the analytical procedure employed also includes SiO_2 that had been complexed with organic matter as well as some SiO_2 from clay minerals (cf. Tessenow 1966; Follett et al. 1965). The small amount of amorphous SiO_2 in silt more than 12,000 yr old (*Figure VII.E*–3) proba-

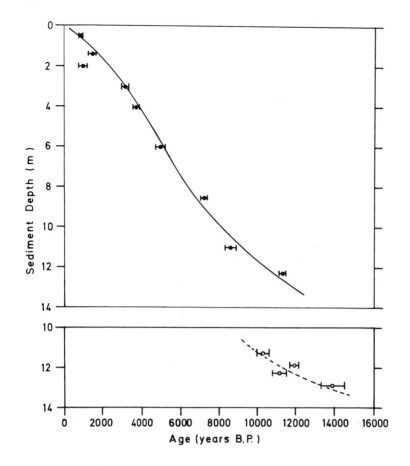

Figure VII.E–1. Relationship between depth and radiocarbon age for the Central core (——) and for core #782 (---) from Mirror Lake. Polynomials were fitted to the points for each core by the least-squares method. The regression of age (yr) on depth (cm) for the Central core yielded the following relationships: yr = 56.54 + 13.64 (cm) − 1.341·10^{-2} (cm)^2 + 7.734·10^{-6} (cm)^3; Deposition time (yr/cm) = 13.64 − 2.681·10^{-2} (cm) + 2.320·10^{-5} (cm)^2. Data for the 200-cm depth were excluded and the point 0.0 was assumed.

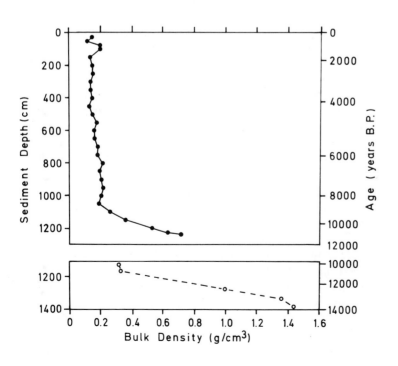

Figure VII.E–2. Bulk density as a function of depth in the Central core (●——●) and core #782 (○---○) from Mirror Lake. Bulk density is the dry weight of matter contained in 1 cm^3 of wet sediment. Note separate time scale for core #782.

E. Chemistry

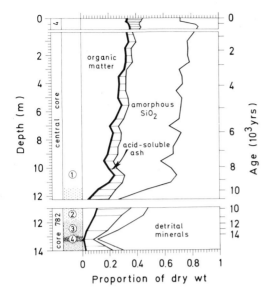

Figure VII.E-3. Fractionation of dried sediment into organic matter, acid-soluble ash, amorphous silica and detrital minerals. The composite core for the deep-water sediment of Mirror Lake is made up of the Central core, minus its top- and bottommost levels, a surficial core (#4; see also *Figure VII.A–9*) spanning the cultural horizon, and the bottom of core #782. Sediment types: ① diatomaceous fine-detritus gyttja, ② gyttja rich in mineral silt, ③ mineral silt, and ④ fine sand.

bly reflects dissolution of some clay-sized silicate minerals.

These data on the sediment stratigraphy suggest that the earliest lake may have been quite different from now, with the major change occurring within the late-glacial and earliest Holocene. After 9,000 yr B.P., the sediment changed little, though a trend to more organic, less mineral-rich sediment has proceeded until European settlement. The cultural horizon is marked by a decrease in amorphous SiO_2 and an increase in mineral particles, but only a slight decrease in organic matter.

Bulk Chemistry

Detailed chemical analyses reveal patterns that reflect and amplify the major trend of decreasing mineral matter and increasing organic matter. The chemical data provide a basis for more subtle reconstruction of lake and watershed conditions. Chemical data are presented in two forms: (1) as total concentration of an element per unit weight of dried sediment at a particular level (e.g. g/g), and as (2) local rate of deposition (g/m²-yr), calculated as the product of an element's concentration at a particular level, the dry bulk density of the sediment (g/cm³), and the rate of sediment accumulation (cm/yr) at that level. Accumulation rate is derived from the smoothed curve of ^{14}C ages (*Figure VII.E-1*).

Both forms of data need to be interpreted with caution. Bulk chemical analysis ignores the elemental form; major differences in interpretation might result from knowing more about changes in the sedimentary form, or phase, of particular elements. Some insights are provided by the separation of extractable phases in the following section. Rates of chemical deposition may be misleading, because our single analyzed core (Central) cannot provide an unbiased record of elemental deposition for the lake as a whole, which is what we want to know, rather than the local deposition at one midlake site. There are some major similarities in the rates of deposition of total organic and inorganic matter at three widely separated sites within the central part of the basin (*Figure VII.E-4*), but particular peaks or valleys in these curves (and by extension in the rates of deposition of particular organic or mineral-related elements in the Central core) may not be reliable evidence of changes in lakewide deposition, especially on a scale of less than several thousand years. This problem of changing pattern of sediment deposition, termed "focusing" in Mirror Lake, is discussed elsewhere (*Chapter VII.A*; Likens and Davis 1975; Davis and Ford 1982; Davis et al. 1984). Changes in elemental concentration are probably less sensitive than deposition rates to changes in focusing, because the gyttja is spatially fairly uniform (cf. *Figure VII.A-5*; Davis et al. 1984).

Late-glacial sediment is very different from the Holocene gyttja. The inorganic silt and sand at the bottom of core #782 have relatively high concentrations of Ca, Mg, K, Na, Al, Fe and Mn (*Figure VII.E-5*). These elements are constituents of the most common minerals making up the bedrock and till (quartz, plagioclase feldspar, biotite mica and muscovite mica) in the Hubbard Brook Valley (see *Chapter II*). Slightly lower metal concentrations in the layer of fine sand near the bottom of core #782 must

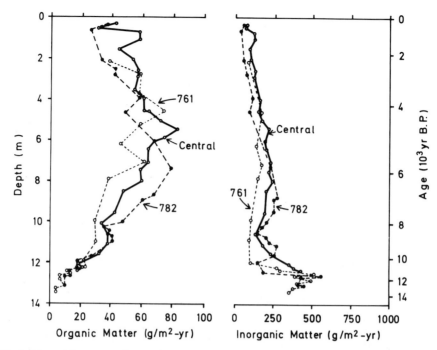

Figure VII.E–4. Deposition rate of organic matter and inorganic matter (ash) at three deep-water sites in Mirror Lake. (Redrawn from Davis and Ford 1982.)

reflect a greater proportion of quartz, which contains only SiO_2. This fluctuation indicates the effect of grain-size distribution on the mineral composition, and thus on the elemental composition of the sediment. Organic C and Kjeldahl N are very low in the earliest sediment. The low C:N ratio of the early inorganic sediment (see *Figure VII.E–7*) suggests that inorganic NH_4^+, adsorbed to detrital minerals, is a significant component of the Kjeldahl N (Mackereth 1966). Sulfur and phosphorus, though potentially associated with organic matter, occur at relatively high levels in the late-glacial sediment. Primary detrital sulfide and P-containing minerals are presumably present. The P content of the lowermost silt (0.11%) is the same as that in unweathered bedrock of the Littleton Formation (0.10%; Johnson et al. 1968).

The elemental concentrations of Ca, Mg, Na, K, Al, Fe and Mn in the earliest inorganic sediment are generally intermediate between that found in till groundmass and the unweathered Littleton Formation schists that make up the local bedrock (Johnson et al. 1968). This observation supports the assumption that the initial substrate in which soils developed in the Hubbard Brook Valley was a mixture of exposed bedrock and till, the till being a pulverized and slightly weathered derivative of the local bedrock.

Deposition of inorganic matter at 13,000 yr B.P. is estimated at 67,800 kg/yr for the lake as a whole (*Chapter VII.A*). This value is much higher than that implied earlier (Likens and Davis 1975), when inorganic matter was thought to have been focused strongly in the center of the basin. Instead, the inorganic sediments of the late-glacial period were deposited over an area even larger than that of the succeeding gyttja. Instead of exaggerating the lake-wide decrease of inorganic sedimentation that took place between 12,000 and 10,000 yr B.P., calculations of inorganic deposition in the Central core (*Figure VII.E–4*) underestimate the change.

If the inorganic matter deposited 13,000 yr B.P. originated within the 107-ha watershed existing prior to 1969, it would represent an output of 650 kg/ha-yr (Davis and Ford 1982)—more than 30 times the particulate inorganic output from a forested watershed (W6) at Hubbard Brook, and twice the maximal output from the experimentally-deforested watershed (W2: Bormann et al. 1974). The high output of fine sand, silt and clay-sized particles reflects the great

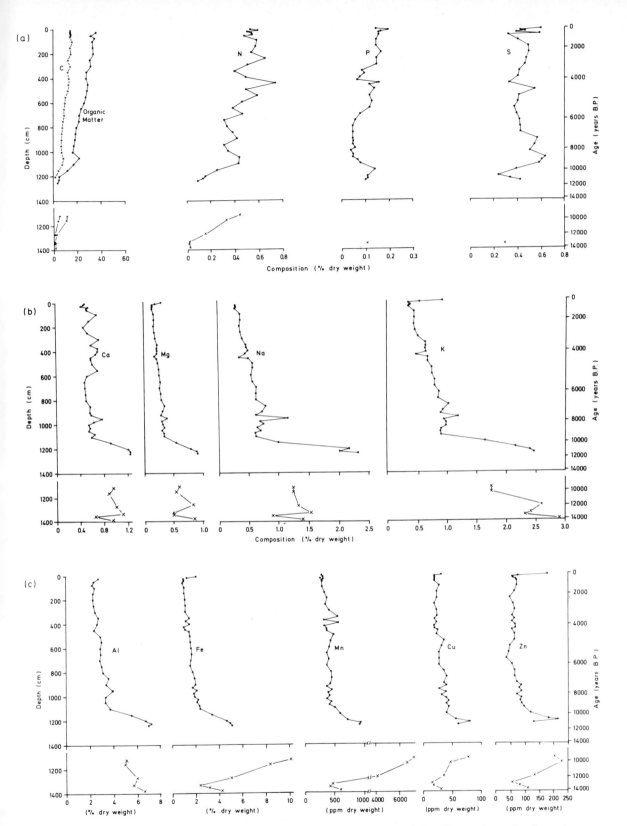

Figure VII.E–5. Concentrations (percentage or parts per million) in dried sediment from the Central core (●——●) and core #782 (×——×) of Mirror Lake: (**a**) Carbon, total organic matter (from loss-on-ignition), nitrogen, phosphorus and sulfur; (**b**) Calcium, magnesium, sodium and potassium; (**c**) Aluminum, iron, manganese, copper and zinc. The difference between 100 and percent of organic matter equals percent of ash at each depth. Data are plotted according to depth below the sediment-water interface, so the time scales for the two cores are different.

availability of glacially-pulverized bedrock during the three or more millennia between melting of the continental glacier and stabilization of local soils. Stabilization apparently did not occur until about 10,000 yr B.P., when presumably climate became similar to that today. By that time, frost-heaving of incipient soils must have ceased, and a mixed conifer-aspen-birch forest had coalesced across what had been tundra and open spruce woodland (*Chapters III.B.1* and *VII.F*).

It may be that part of the late-glacial inorganic deposit arrived as windblown sand and dust. Quartz grains deposited within a relatively mineral-rich zone of the early Holocene gyttja (7,900 yr B.P.) have a surface indicating transport by wind (U. M. Cowgill, personal communication). These grains may represent redeposition of the late-glacial sediment. In the littoral zone, fine sand and silt of late-glacial age are protected from erosion only by a thin veneer of Holocene sand (*Figure VII.A–4*).

The change from highly inorganic late-glacial sediment to moderately organic gyttja began about 11,500 yr B.P. as the density of vegetation, now including spruce trees, increased in the watershed. Elements associated with inorganic minerals (Al, Na, K, Ca, Mg, Fe, Mn) decreased in both concentration and deposition (*Figures VII.E–5* and 6). Nitrogen and C increased sharply in concentration (*Figure VII.E–5*). It already has been pointed out (*Chapter VII.A*) that the low organic content in the late-glacial sediment is somewhat misleading. The high sedimentation rate of this material, combined with its wide distribution in the lake, mean that the deposition of organic matter in the lake as a whole increased less during the silt-gyttja transition than the change in the Central core suggests. Deposition of phosphorus was very high early in the late glacial, but if the P was associated principally as detrital minerals, there may have been little P available to aquatic metabolism (Livingstone and Boykin 1962; Mackereth 1966). More significantly, P concentration (*Figure VII.E–5*) and deposition (*Figure VII.E–6*) did not decrease until 10,000 yr B.P., suggesting that dissolved P leached

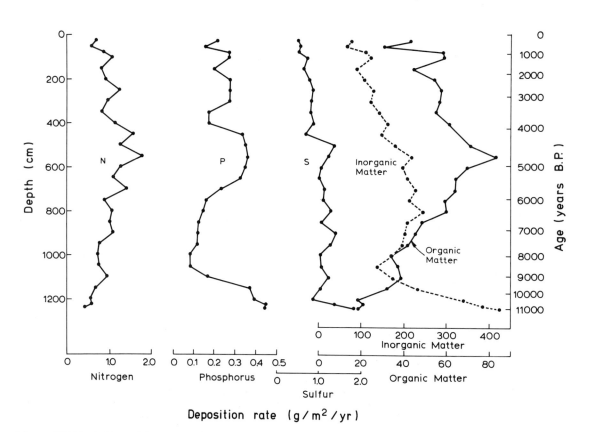

Figure VII.E–6. Deposition rates of nitrogen, phosphorus and sulfur compared with rates of organic matter and inorganic matter in the Central core from Mirror Lake. (From Likens and Davis 1975.)

E. Chemistry

from developing forest soils may have been increasingly available to aquatic metabolism during the period 12,000 to 10,000 yr B.P. After 10,000 yr B.P., however, P concentration and deposition decreased precipitously, suggesting that the terrestrial vegetation and soils had become major sinks for P released by weathering.

The late-glacial period (14,000 to 11,000 yr B.P.) was a time of major change within the lake and its watershed, ultimately related to the progressive climatic warming that occurred during this period. There is no biologic imprint on the earliest silt, but by 12,000 yr B.P. the chemical stratigraphy reveals a biologic ecosystem at work. Levels of Fe and Mn in the silty gyttja at the bottom of the Central core (*Figure VII.E–5*) are similar to levels in composite till (Johnson et al. 1968). In core #782, concentrations of these elements for the same time period (10,500 to 11,500 yr B.P.) are much enriched. The enrichment developed progressively, however. The oldest silt does resemble the till in its contents of Fe and Mn.

Our interpretation of the spatial difference in Fe and Mn concentrations is that increasingly anoxic conditions developed in the deepest waters of the early lake, probably during a long ice-covered period. Iron and Mn settling to the bottom in forms other than crystalline minerals did not accumulate at the very bottom of the lake, but migrated by way of the deoxygenated water-column to slightly shallower sediments (e.g. the site of core #782), where they were precipitated and retained (cf. Tessenow 1975). At 11,000 yr B.P., the water-column above the site of the Central core would have been 2 to 3 m deeper than that at the #782 site (23 vs. 21 m below present lake level). Apparently only a small volume of the lake became sufficiently deoxygenated that Mn and Fe escaped from the surficial sediment. At 11,000 yr B.P., Cu and Zn were enriched at both sites by twofold above levels in the oldest silt (*Figure VII.E–5*). Evidently their secondary particulate phases (complexed with organic matter, adsorbed onto clays, precipitated as sulfides) were much less sensitive to redox conditions of the overlying water.

The Holocene trend of increasingly organic sediment was subtly interrupted during the interval 6,500 to 8,500 yr B.P. by slightly lower organic content (*Figures VII.A–8* and *VII.E–5*).

Although deposition of both organic matter and inorganic matter was gradually accelerating at all three deep-water sites (*Figure VII.E–4*), at the Central site this interval was characterized by exceedingly low P concentration, and hence low P deposition. The following millennia (6,500 to 4,000 yr B.P.) brought maximal rates of organic influx at all of the deep-water sites. These sediments had higher P concentrations and—probably—represented a higher lakewide influx of P than had occurred earlier in the Holocene. If the rate of P deposition was positively correlated with lacustrine productivity, Mirror Lake must have been more productive in the mid-Holocene than earlier.

After 4,000 yr B.P., a general decline of sediment deposition (both organic and inorganic) is evident and may represent, in part, a decrease in focusing (Davis et al. 1984), with presumably greater accumulation at shallower sites (see *Chapter VII.F*). Unlike the concentrations of Na, K and Mg, which decline in parallel with the decline in detrital minerals, concentrations of Ca are virtually constant throughout the Holocene. Apparently the sediments contain a complex of Ca with organic matter, as suggested for some other lakes (Mackereth 1966; Koljonen and Carlson 1975). Phosphorus concentration increased during the later Holocene at a rate greater than that of organic matter. In fact, the influx of P did not decrease during the late Holocene.

Elemental Ratios

Four elemental ratios are plotted in *Figure VII.E–7* to suggest the absence of major chemical changes in Mirror Lake during the Holocene. The elements selected are associated with noncrystalline phases, at least during the Holocene. Carbon, N and S are important constituents of organic matter. Carbon and N must occur principally as organic matter. There is no evidence for inorganic carbonate in Mirror Lake sediments, even at the bottom of the cores. In any case, the C determinations are specific to organic C. The C:N ratio is remarkably constant throughout the Holocene, suggesting the absence of major fluctuations in the source of the organic matter. Lower C:N ratios in late-glacial silts are probably caused by relatively large amounts of ammonium-N in the generally low concentrations of total N.

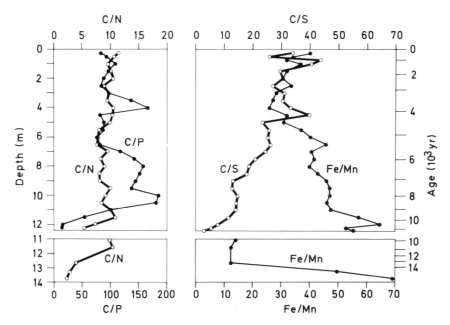

Figure VII.E–7. Chemical ratios (mass/mass) in the sediment profile of the Central core from Mirror Lake. Phosphorus and sulfur values are not available for core #782.

The low C:P ratios at the bottom of the Central core must reflect the presence of P in detrital minerals. The period of phosphorus-poor gyttja during the early Holocene (6,000 to 9,000 yr B.P.) shows up as the interval of high C:P (120–180). Throughout most of the late Holocene, the C:P ratio has been 80–100. The stoichiometry of C:N:P during the later Holocene (by mass approximately 100:10:1) is remarkably similar to that of suspended seston in Mirror Lake during the summer (100:13:1—B. Peterson, unpublished data, *Chapters V.A.3* and *V.B.4*), indicating that the loss of organic matter during sediment formation (*Figure VII.A–12*) parallels similar losses of N and P.

Sulfur is relatively abundant in the earliest inorganic silts (0.3%); apparently primary sulfide minerals were deposited along with the more abundant silicate minerals. In the early Holocene, the high concentration of S combined with the relatively low concentration of C (*Figure VII.E–5*) gives low C:S ratios of 10 to 15. Probably secondary sulfides of iron are present, precipitated biochemically in surficial sediment or in the anaerobic hypolimnion. The rate of S deposition in the Central core (*Figure VII.E–6*) is remarkably constant during the Holocene, suggesting that the decrease in S concentration after 7,000 yr B.P. might merely reflect dilution in the increasing deposition of sulfur-poor organic matter that characterized the mid-Holocene.

Like S, Fe and Mn are constituents of both primary crystalline minerals (in this case silicates as well as sulfides and oxides) and secondary phases produced in the lake or watershed (organic complexes, oxyhydroxides and sulfides). Unlike S, Fe and Mn concentrations more closely parallel that of inorganic matter throughout the Holocene. The Fe:Mn ratio in the Central core decreased during the Holocene as the sediments became increasingly organic, but not enough to suggest major changes in the redox conditions at the sediment surface (see Mackereth 1966). We have concluded, from the differences in Fe and especially Mn concentrations between sediment deposited 11,000 yr B.P. in the Central core (Fe:Mn = 54) and that deposited in core #782 (Fe:Mn = 13) that the deepest waters of the lake became sufficiently anaerobic to inhibit the accumulation of secondary oxyhydroxides. This condition apparently prevailed throughout the Holocene.

Another comparison of elemental ratios is presented in *Table VII.E–1*. In this case, the six selected elements (Al, Fe, Ca, Na, K, Mg) all are major structural components of silicate minerals. Magnesium, the least abundant, was arbitrarily selected as the denominator. We have already pointed out the similarities of chemical

E. Chemistry

Table VII.E-1. Mass Ratios of Percent Composition of Elements in the Hubbard Brook Valley[a]

	Al	Fe	Ca	Na	K	Mg
Mirror Lake						
Sediment Depth (cm)						
1,330–1,160[b]	9.3 ± 1.3	8.8 ± 3.3	1.70 ± 0.31	2.32 ± 0.44	3.76 ± 0.57	1
1,240–1,200[c]	8.0 ± 0.23	5.6 ± 0.9	1.41 ± 0.04	2.50 ± 0.14	2.69 ± 0.01	1
1,100– 600[c]	10.8 ± 0.14	6.2 ± 0.12	1.82 ± 0.04	2.20 ± 0.08	2.94 ± 0.04	1
550– 100[c]	13.4 ± 0.34	6.6 ± 0.19	3.16 ± 0.13	2.24 ± 0.03	2.88 ± 0.04	1
75– 25[c]	17.6 ± 0.15	7.5 ± 0.09	3.55 ± 0.41	2.18 ± 0.03	2.79 ± 0.01	1
Till Groundmass[d]	12.5	6.0	1.17	2.17	4.50	1
Composite Bedrock[d]	7.6	4.0	1.27	1.45	2.64	1
Weathered Residue[e]	51	6	4	10	24	1
Current Terrestrial Export of Dissolved and Particulate Materials[f]	1.01	0.19	4.17	2.24	0.72	1
Dissolved Export Only[f]	0.63	—	4.35	2.30	0.60	1
Particulate Export Only[g]	7.3	3.4	1.1	1.3	2.7	1

[a] Values as means ± standard deviation of the mean. (Modified from Likens and Davis 1975.)
[b] Core #782.
[c] Central core.
[d] 20% Littleton Formation, 20% till groundmass and 60% Kinsman Formation (Johnson et al. 1968).
[e] A_2 soil horizon (Johnson et al. 1968).
[f] Likens et al. 1977.
[g] Likens et al. 1977; the inorganic particulate export is assumed to have the chemical composition of composite bedrock.

concentrations between the earliest silt in core #782 and the composite bedrock, the assumed original parent material for soil development (Johnson et al. 1968). The mass ratios are consequently also similar. Iron is significantly enriched in the silt (approximately twofold); otherwise the deviations can be interpreted as a 30 to 50% depletion of Mg in the silt as compared with the other elements.

During the course of soil development, the parent material has been depleted of elements to differing degrees (Johnson et al. 1968), leaving a weathered horizon (A_2) below the surficial organic layers (A_1) and, below that, a zone where some of the leached or physically elutriated materials have been redeposited. Losses of Ca, Mg, Na and K from the soil—and to the lake—principally involve dissolved forms (Likens et al. 1977). Aluminum and Fe may travel complexed with dissolved and particulate organic matter (cf. Johnson et al. 1981). Only a small part of the dissolved input of these metals is likely to be converted to an insoluble particulate phase and end up in the sediment of Mirror Lake. Thus, it is not surprising that the elemental composition of the Holocene gyttja bears little resemblance to the total inputs to the lake, which are predominantly in dissolved form (Table VII.E-1). In contrast, a large part of the particulate losses from the watershed accumulate as sediment, though only particles smaller than fine sand (silt and clay) regularly reach the profundal gyttja.

The Na:Mg and K:Mg ratios of the Holocene gyttja did not change with time. These elements may form secondary clay minerals in the soil, but probably not in the sediment of northern lakes (Jones and Bowser 1978; Dean and Gorham 1976). They occur primarily as silicate minerals in lake sediment (Mackereth 1966; Engstrom and Wright 1984). The constancy of the Na and K ratios near the values characteristic of unweathered bedrock or till (Table VII.E-1) suggests that physical erosion within the stream channel—not overland flow across the soil surface—has been the source of the detrital mineral component of the gyttja throughout the Holocene.

Relative to Mg, Al and Ca became progressively enriched during the Holocene, possibly because they were deposited as a stable complex with organic matter. The ratio of Fe:Mg has changed only slightly during the Holocene, inconclusive evidence that Fe, which does form secondary phases in lake sediment, mainly occurs in detrital minerals (at least in the Central core). Because of the potential impact of phases other than detrital minerals on the bulk chemical composition of the gyttja, the data in *Table VII.E*–1 are of limited value in deciphering changes in the type of mineral matter transported to the lake during the Holocene.

Extractable Cations

The treatment of dried sediment with 1 N ammonium acetate and 1 N acetic acid was intended to extract metals held as cations to negatively charged sites on organic matter and mineral particles. However, the treatment also dissolves some secondary particulate phases, but fails to completely separate detrital mineral phases from secondary phases. Nevertheless, its application to the gyttja from Mirror Lake has helped to establish lower limits for the concentrations of secondary phases.

For most elements (Ca, Mg, Na, Mn, Cu) the two extractions gave essentially the same results (*Figure VII.E*–8). Ammonium ion dislodged more K than hydrogen ion did, whereas hydrogen ion released substantially more Fe and Zn. Tessenow (1975) found ammonium acetate to dissolve significant Fe only from sulfide-rich sediment, although the soluble phase was not identified. Even the acetic acid treatment is ineffective at dissolving secondary Fe and Mn oxyhydroxides (Chester and Hughes 1967). The small amount of extractable Fe in the Mirror Lake sediment might represent FeS. Extractable Fe did not exceed 12% of the total Fe, however (*Figure VII.E*–9). Approximately half of the total Mn is extractable, in agreement with findings elsewhere (Delfino and Lee 1968; Tessenow 1975). The high extractability of Mn demonstrates secondary deposition, perhaps adsorption. The decrease in the Fe:Mn ratio during the Holocene is possibly a result of proportionally greater accumulation of Mn in secondary phases than of Fe. The Mn enrichment is much less extreme than that observed at the late-glacial Holocene transition in core #782, where both total Mn and total Fe were enriched above levels in basal silt, and the Fe:Mn ratio was reduced to 12 (*Figure VII.E*–5).

In the gyttja of Mirror Lake, which is noncalcareous, the two extractions release equal, and substantial, amounts of Ca. Calcium carbonate is soluble in the acetic acid, but only sparingly in the ammonium acetate (Berglund and Malmer 1971). The extractable fraction of total Ca increased rapidly from 10 to 40% during the early Holocene, as the concentration of detrital minerals decreased. A more gradual increase continued throughout the Holocene, reaching 70 to 90% by the cultural horizon. These data confirm that only part, perhaps a small part, of the total Ca is present as detrital minerals, in contrast to Na, K and Mg (*Figure VII.E*–9). The relative extractability of major cations in Mirror Lake (Ca $\gg$ Mg $>$ Na $\cong$ K) is the same as that in a small Norwegian lake (Kjensmo 1968). Mackereth (1966) noted a stratigraphic correlation of Ca with organic matter in English lakes within relatively neutral, calcite-containing drainage basins—e.g. Esthwaite and Windermere. In contrast, Ca was correlated with other detrital mineral elements such as Na, K and Mg in Ennerdale, which is surrounded by crystalline bedrock and soils that became acidic early in its history. The nature of the assumed complex between Ca and organic matter needs to be studied further, especially to determine if it is a product of soil processes or of lacustrine metabolism.

The changes in concentration of extractable metals during the Holocene (*Figure VII.E*–8) represent small portions of the total elemental concentration. Nevertheless, two points stand out: (1) The late Holocene gyttja contains extractable Ca, Mg, Na and K equal to or greater than concentrations found at any earlier time in the Holocene. Extractable Ca is correlated with increasing organic matter but the abrupt increase in extractable Na about 2,500 yr B.P. suggests that some more cryptic process is involved. (2) The early Holocene gyttja contains extractable Fe, Mn, Cu and Zn at levels equal to or greater than those found later during the Holocene, although the pattern of decline has differed among elements. Extractable Fe concentrations of 0.15 to 0.2% of dry weight (7 to 12% of total Fe) during the early Holocene, compared with 0.04 to 0.1% (3 to 7% of total Fe) after 6,000 yr B.P., might represent a greater abundance of FeS during that part of the early Holocene (6,000 to 9,000 yr B.P.) when total P concentration was especially low.

E. Chemistry

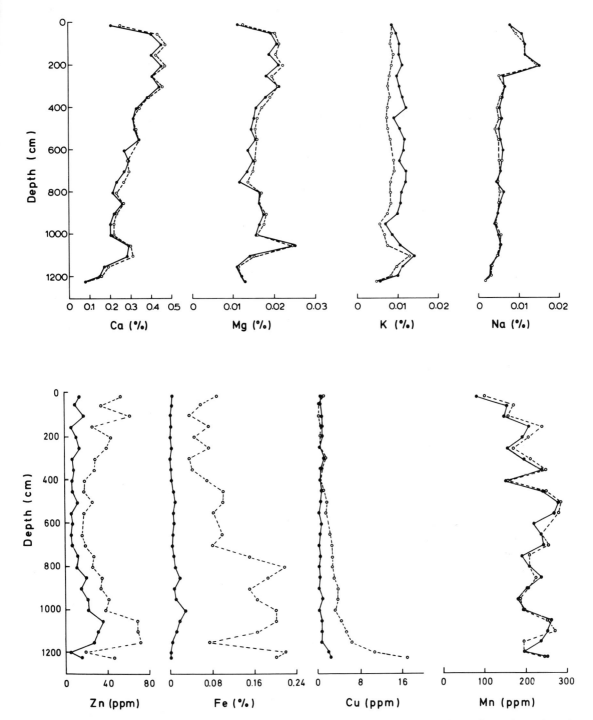

Figure VII.E–8. Concentrations of metals extractable by ammonium acetate (●——●) and acetic acid (○---○) from sediments in the Central core from Mirror Lake. These extractable amounts are varying portions (*Figure VII.E–9*) of the total elemental concentrations (*Figure VII.E–5*).

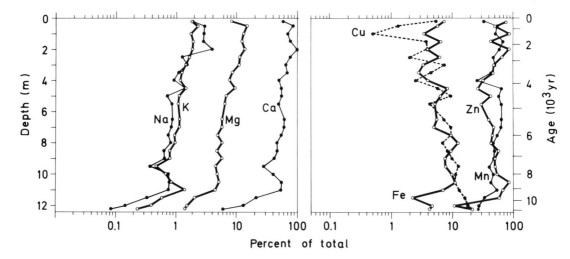

Figure VII.E–9. Extractability of metals from the sediments of Mirror Lake. Data represent the percentage of each element that was extracted into 1 N acetic acid at each level in the Central core.

The Cultural Sediment

Following settlement within the lake's watershed in the early nineteenth century, there has been a significant increase in the deposition of lacustrine sediment. The increase is obscured at individual sites in the middle of the lake, because it also involved a relaxation of focusing and an expansion of the zone where gyttja has accumulated (*Chapter VII.A*). Although the organic content of cultural sediment is only slightly lower than that of precultural sediment, there has been a significant increase in the content of detrital minerals and a compensatory decrease in amorphous SiO_2 (*Figure VII.E–3; Table VII.A–1*). The recent increase in detrital mineral matter is also reflected in bulk chemistry. Moreover, concentrations of some heavy metals increase toward the surface of the sediments, presumably owing to human activities (*Figure VII.E–10*).

Elements associated primarily with silicate minerals (Na, K, Mg) increase above 20 cm. The cultural pollen horizon (approximately 1840 A.D.) was not located in this particular core, but lies at 20 to 25 cm in other cores from the deepest part of the basin. Calcium increases proportionately less, being enriched as a secondary organic complex both above and below the cultural horizon. Within the cultural horizon, the concentrations of extractable Ca and Mg are abruptly lower than precultural levels (*Figure VII.E–8*). It seems, therefore, that the extractable phase of these elements, unlike the silicate phase, does not originate in stream channels of the watershed. Perhaps it forms in the lake, or in watershed soils as a complex with dissolved humic and fulvic acids. In either case, the extractable phase is diluted by the increased deposition of silicate minerals and, presumably, particulate organic matter from the stream channels.

The increases in concentrations of the trace elements Cd, Pb and Zn (but not Cu and Cr) are proportionately much larger than the increases of the lighter metals. Increases of heavy metals such as Cd, Pb and Zn in recent sediments are a general phenomenon in North Temperate lakes, even in remote regions, and have been attributed largely to long-distance transport through the atmosphere (Galloway and Likens 1979; Rippey et al. 1982). There are no local sources of such pollution for the lake, although the watershed is lightly populated (*Chapter III.B.3*).

There is no chemical evidence to suggest a change in lake productivity associated with European settlement. The P concentration does not change throughout the cultural horizon. In these relatively organic sediments, most P is likely bound as organic matter or adsorbed to Fe oxyhydroxides (Williams et al. 1971; Jones and Bowser 1978). The analytical method used on the freeze core probably did not include P of detrital minerals (see Sherman 1976). Increases in sediment P have been observed in lakes

E. Chemistry

Figure VII.E-10. Chemical stratigraphy of surficial cores from the deepest part of Mirror Lake, showing changes associated with settlement (see also *Figure VII.A-9* and *Table VII.A-4*). Organic matter and phosphorus were determined on a piston core (collected in 1974) (○---○); other elements were determined on a freeze core (●——●). (Modified from Sherman 1976.)

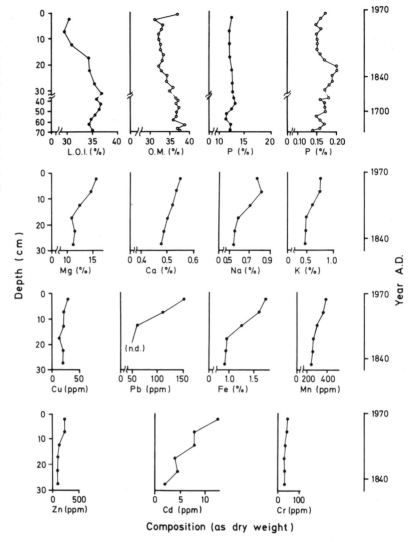

strongly enriched by sewage (e.g. Shapiro et al. 1971; Rippey et al. 1982), but any change in P deposition in Mirror Lake involves mainly an increase in the area of deposition and does not show up as an increase in P concentration in cores from the middle of the lake. The decrease in amorphous SiO_2 and diatom frustules in the cultural sediment (*Table VII.A-1*) reflects dilution. If there were any increase in diatom productivity, it would be obscured in the wider distribution of cultural sediment and the proportionately greater deposition of detrital minerals and organic matter.

The marked increase in concentration of Fe and Mn in the surficial sediment probably represents transitory pooling of Fe and Mn migrating upward, perhaps as oxyhydroxides, below the aggrading surface of the sediment. In part, detrital Fe and Mn containing minerals must be responsible for the increase. Because of the great reactivity of these elements at redox boundaries such as the sediment surface, their peaks in the surficial gyttja cannot be accorded stratigraphic significance.

Discussion

The interpretation of chemical profiles within the organic sediments of lakes is hindered by very incomplete understanding of the processes that form the sediment. An initial analytical step is to separate components into autochthonous phases and allochthonous phases. Allochthonous phases (e.g. detrital minerals, soil organic matter) may have influenced lake metabolism during passage to the sediment, as substrates for ion exchange or chelation; their presence in sediment might also be indicative of changing climate or levels of nutrient-loading. Autochthonous phases (organic matter in general, diatom frustules, secondary minerals and adsorbed ions) directly reflect aquatic metabolism according to rules that are poorly understood.

We have analytically separated diatomaceous silica within the total inorganic component of the sediment. This separation provides, by difference, a more accurate estimate of the inwashed component of detrital, mostly silicate, minerals. The stratigraphic behavior and inextractability of bulk Na, K and Mg suggest predominant phases as silicate minerals. Calcium and P are believed to shift from detrital mineral phases in the late glacial to more autochthonous, secondary phases during the early Holocene. The chemical forms of Fe, Mn and Al during the Holocene are not clear. Obviously there is need for more detailed chemical and mineralogical analysis of the Mirror Lake stratigraphy. Our present information is used, however, to address four important limnological questions.

Origin of Organic Matter in the Sediments

Allochthonous is mixed with autochthonous organic matter in bulk analyses of C and N. Is the deposition of sedimentary organic matter more nearly reflective of output from the vegetated watershed, or does it represent the residue of aquatic metabolism? A quick calculation employing data from the organic carbon budget (*Chapter V.C*; Jordan and Likens 1975) demonstrates that allochthonous inputs (2,602 kg C/yr) could contribute all of the organic matter deposited in the cultural gyttja (1,920 kg C/yr lakewide: *Table VII.A*–1), but only under the extreme assumption that all of the particulate allochthonous organic matter (1,027 kg C/yr) and the majority of the *dissolved* allochthonous organic matter (1,575 kg C/yr) are deposited as sediment. This assumption implies that the substantial outflow of dissolved organic carbon from the lake (1,630 kg C/yr) is entirely autochthonous. Hesslein et al. (1980) have shown that only 6% of the epilimnetic dissolved organic carbon in a Canadian oligotrophic lake is equilibrated with phytoplanktonic productivity over a 100-day period; by implication much of the dissolved organic matter was allochthonous, on its way through the lake and out. Considering that some of the allochthonous input to the lake must be respired and a larger part lost to outflows, autochthonous organic matter must be well represented in the gyttja, a conclusion consistent with the large deposition of chlorophyll derivatives (*Chapter VII.D*). A similar calculation and conclusion apply to the peak deposition of organic matter in the mid-Holocene (7,000 to 4,000 yr B.P.), when the average local deposition at five coring sites was 44 g organic matter/m^2-yr (Davis and Ford 1982; *Figure VII.E*–4 for three deep-water sites). This average local deposition represents about 20 g C/m^2-yr, or 1,500 kg C/yr for the lake as a whole if the area of deposition was 7.5 ha (*Chapter VII.A*).

The C:N ratio of 9 to 10 in Mirror Lake gyttja (*Figure VII.E*–7) also suggests that autochthonous organic matter is well represented, perhaps predominant, in the Holocene. Ratios greater than 10 are thought to demonstrate significant contribution of humus from acidic soils (Hanson 1959, 1960), with ratios reaching 15 to 20 only in humic-stained lakes of apparently high allochthonous inputs of dissolved and colloidal organic matter (Hanson 1960; Alhonen 1970; Huttunen et al. 1978). Ratios in the profundal sediment of temperate lakes are typically 8 to 15 (e.g. Hanson 1960; Mackereth 1966; Alhonen 1970; Brunskill et al. 1971; Dean and Gorham 1976; Flannery et al. 1982), with the lower ratios often found in lakes of higher autochthonous productivity (e.g. Flannery et al. 1982) where allochthonous inputs are diluted (but see Gorham et al. 1974). A significant limitation of C:N ratios as evidence for the origin of sedimentary organic matter lies in the generally lower C:N ratio of humus in many soils (10 to 15: cf. Kononova 1966; Jenny 1980) and the uncertainty that C:N ratios of the organic matter transported to and deposited in a lake are as high as those of bulk soil humus. The C:N ratio of the humus fraction of present-day soils at Hubbard Brook is 24 (Gosz et al. 1976), typical of acid mor humus.

E. Chemistry

Utility of Deposition and the Problem of Focusing

The value of computing rates of local deposition (influx) for individual chemical components of the sediment lies (1) in removing spurious negative correlations inherent in concentration, and (2) in producing a basis for quantitative comparisons among stratigraphic levels and between different lakes, as well as for situating permanent sedimentation within biogeochemical cycles. An adequate spatial sampling across the lake basin is required to obtain reliable values for lakewide deposition, however. For Mirror Lake, chemical deposition can be calculated only for the Central core. These values must consistently underrepresent the deposition within the zone of gyttja accumulation of detrital minerals (higher rate of deposition at the margin of the gyttja: *Figure VII.A–10*) and would inaccurately portray deposition of such elements as Fe, Mn and possibly P if these were either locally depleted or enriched owing to their migration within redox gradients (cf. Tessenow 1975). Nevertheless, some bulk chemical analyses from Mirror Lake (Moeller and Likens 1978) indicate that elemental concentrations of C, N, P, S, Ca, Mg, Na, K and Al near the margin of the gyttja are within $\pm 30\%$ of levels at a water depth of 10 m, illustrating the relative compositional uniformity of the surficial gyttja across its zone of accumulation.

Davis and Ford (1982) demonstrated that the deposition of organic and inorganic matter has varied with space and time in Mirror Lake. Within the Holocene, spatial variation principally involved greater deposition of gyttja in three cores from deep water compared with two cores near the margin of the gyttja. This spatial *focusing* was not constant during the lake's history (*Chapter VII.A*). Temporal variation involved irregular changes in deposition at the two shallower sites—related to type of sediment and probably to water depth—and a more regular shift in the pattern of focusing among the deep-water sites. The patterns of deposition at these five sites (too few, especially at the margin of the gyttja zone, to precisely calculate lakewide deposition) are helpful in relating the deceptively detailed curves from the Central site to conditions in the lake as a whole.

For example, the decreasing deposition of all metals during the transition to Holocene climate (11,000 to 9,000 yr B.P.) is underestimated by local deposition in the Central core, because the zone of deposition in the late-glacial extended into what is now a nonsedimentary littoral zone. The increasing concentration and deposition of organic matter in the Central core during this period exaggerates the increase in organic deposition, because organic deposition during the late glacial between 11,000 and 13,000 yr B.P. was perhaps as high as one-third of the mean Holocene rate (*Chapter VII.A*). Furthermore, the steadily increasing rate of organic deposition at mid-lake sites (*Figure VII.E–4*) may hide a period of lower lakewide deposition, approximately 8,500 to 9,500 yr B.P., which is suggested by the unusually high concentration and rate of deposition of pollen at the deep-water sites, and by the depositional hiatus at one of the marginal sites (Davis and Ford 1982).

The higher deposition of organic C, N and P in the mid-Holocene (4,000 to 7,000 yr B.P.) was not an artifact of especially strong focusing. Deposition of organic matter was high at all three deep-water sites as well as one marginal site. Pollen concentrations were less than, or equal to, levels both earlier and later. The gradually declining rate of deposition of most elements in the Central core during the late Holocene parallels that of pollen (Davis and Ford 1982; *Figure VII.A–6*) and might be caused by a relaxation of focusing (Davis et al. 1984). If so, the deposition of C, N, P, S and Ca in the lake as a whole may have been essentially constant since 7,000 yr B.P.

Evidence of Trophic Development

Clues to trophic conditions in the lake include biologic as well as chemical evidence (discussed in *Chapter VII.F*). The chemical evidence involves two sorts of interrelated data: analyses of elements or compounds that are important constituents of organic matter in the sediment and for which the relative proportions and actual deposition rates might reflect primary productivity in the lake; and analyses of other elements whose presence and rate of deposition indirectly suggest conditions of nutrient supply or benthic metabolism ultimately correlated with productivity.

Examples of direct evidence are the deposition rates of C, N, pigments (*Chapter VII.F*), amorphous silica, and possibly P, Ca and S. These substances can be considered a fairly direct residue of autotrophic production. However, interpretation is hindered by three major uncertainties: the unknown magnitude of allochthonous sources, possible changes in the efficiency of decomposition before permanent burial (and any longer-term diagenesis in the sediment), and variations in the pattern of focusing that make actual deposition difficult to calculate. These parameters suggest higher productivity in the mid-Holocene than the early Holocene, when local deposition of organic matter was lower at all coring sites and pigments made up a smaller part of the total organic matter. Since this mid-Holocene peak, deposition of organic components, except P, has decreased at midlake sites. The decrease is thought to result in part from a change in focusing, but the consistently higher organic matter and P content, suggest that chemical changes also are involved.

Indirect evidence regarding trophic conditions comes from the deposition of detrital minerals (represented by the elements Na, K, Mg and, in the late glacial, Ca, P, S, etc.). These minerals did not contribute to lake metabolism, but their rate of deposition reflects physical erosion. Certainly in the cultural period, there has been an input of nutrients reflecting human settlement (*Chapter IV.E*) along with the increased input of detrital minerals. Mackereth (1966) argued, however, that periods of greater physical erosion corresponded to periods of lesser chemical weathering (in an absolute as well as relative sense). It seems reasonable to accept an inverse relation between inputs of detrital minerals and inputs of dissolved nutrients during the late-glacial period, when the colder climate would have retarded chemical weathering and the inorganic soils must have provided fewer organic chelators for trace elements.

Oxygen concentrations in the water above the profundal sediments may also influence productivity by controlling the efficiency with which P is cycled back to the water-column from the surficial sediment. Near the deepest part of the basin (Central core) throughout the Holocene, the sediment surface became deoxygenated long enough during the course of each year that only small amounts of secondary Fe and Mn accumulated. Under oxygenated conditions, phosphate released during microbial metabolism can become adsorbed to contemporaneously formed Fe oxyhydroxide. Under deoxygenated conditions, phosphate as well as organic phosphorus may escape, especially if enough sulfide is present to cause Fe to precipitate as FeS instead of an oxyhydroxide.

During much of the early Holocene (6,000 to 9,000 yr B.P.), the P concentration in gyttja in the center of the basin (i.e. Central core) was exceptionally low (0.05% of dry weight) and the C:P ratio was high (approximately 160 by weight). During that time period, both the concentration of extractable Fe and its proportion of total Fe were higher than later in the Holocene (*Figures VII.E–8* and *9*). Although total S was only slightly higher, and only during the first part of this period (*Figure VII.E–5*), precipitation of a small portion of the total Fe (which must include much Fe in detrital minerals) as FeS may have allowed a greater proportion of the sedimented P to be returned to lake metabolism. Ultimately such cycled P, which represents only one of many cycling pathways in the ecosystem, would have been deposited elsewhere (e.g. in shallower water, or possibly outside the lake). The paradoxical problem with P deposition is that times of lower deposition, especially when they create a high C:P ratio, could reflect higher efficiency of cycling from the profundal sediment and, conceivably, higher productivity. In Mirror Lake, the increased rate of deposition of organic matter in the mid-Holocene is associated with rising, not decreasing, P concentration—and a C:P ratio of approximately 85. The time of low organic deposition in the earliest Holocene (approximately 10,000 yr B.P.) coincides with P concentrations that represent a substantial enrichment over the decreasing background from detrital minerals, suggesting that P was being less efficiently used in primary productivity.

Sediment as an Inwashed Soil?

The Holocene sediment of British lakes has been interpreted as essentially an inwashed soil, in which the bulk chemistry much more closely reflects changes in the drainage basin than changes in lacustrine metabolism and sedimentation. Gorham et al. (1974) concluded that the organic sediment deposited in the more productive, in fact somewhat eutrophied, lakes was predominantly autochthonous, but allochthonous organic matter may predominate in the less productive lakes (Mackereth 1966; Cranwell 1973). Paleoecological studies from many British lakes seem consistent with Mackereth's contention: major vegetational changes caused by climate or human disturbance appear in profundal lake sediments as changes in bulk chemistry (concentrations and deposition of organic C, N, K, Na, Mg, etc.), interpretable as intensity of soil erosion and transport to the lake (Pennington 1981a, b). The generally subtle changes in Mirror Lake during the Holocene may represent watershed processes, but the sediment is certainly not an inwashed soil.

First, the organic deposition in Mirror Lake is thought to be largely autochthonous, at least during the mid- and late-Holocene. Evidence for its autochthonous content includes its relatively low C:N ratio (~9 to 10), high content of preserved pigments (*Chapter VII.D*), and magnitude of deposition compared with allochthonous particulate inputs to the modern lake. Second, much of the ash component of the profundal sediment is amorphous silica (40 to 60% of the ash since 9,000 yr B.P.). Not all of the amorphous silica represents diatom frustules, but amorphous silica is only a minor component of the podzolic soil of the area (e.g. 5% amorphous Si in the A_1 humus layer, 0.5% in the A_2 leached horizon; Von Damm et al. 1979). Finally, the detrital mineral fraction of the profundal gyttja seems to have originated during erosion of the stream channels into the till groundmass, not by lateral (overland) transport of surficial layers of the soil. The relative proportions of Na, K and Mg do not change significantly during the Holocene; the gyttja resembles till groundmass more closely than surficial soil (A_2 horizon) throughout the Holocene.

Severe disturbance of watershed soils would presumably cause substantial transport of bulk soil to the lake (e.g. after ploughing or excavation), but even the cultural sediment of Mirror Lake does not seem to represent such a deposit. Inputs of soil organic matter have been recognized in lake profiles as discrete zones of high C:N, often with anomalously old ^{14}C dates and pollen indicative of soils or of vegetational disturbance (Pennington 1975, 1981a; Tolonen 1980). A substantial portion of the sediment might represent essentially allochthonous matter, but this matter is likely to be enriched in particles and originally dissolved compounds (e.g., organic matter) that washed *out* of the soil.

Summary

The chemical content of various cores from the sediment of Mirror Lake have been used to deduce the past history of the lake, as well as its airshed and watershed.

Although lacustrine sediments have different origins, they ultimately result from the difference between total inputs and total outputs for the ecosystem.

The gyttja sediment is spatially rather uniform in Mirror Lake although focusing of sediments may seriously complicate interpretation of lake history.

The sediment record suggests that the earliest lake (late glacial, 14,000 to 10,000 yr B.P.) was quite different from the present lake. For example, (1) the bulk density of the sediments decreased sharply during the late-glacial period, suggesting more control of erosion by the terrestrial watershed; (2) the earliest sediments contained high concentrations of Ca, Mg, K, Na and Al, and low concentrations of organic C and N; and (3) some of the sediments may have been windblown.

With warming of climate, development of vegetation, stabilization of soils and formation of organic debris dams in stream channels, input of eroded particulate matter was reduced during the Holocene. Only 30 to 40% of the ash during the Holocene was inwashed mineral particles. The terrestrial vegetation must have been a sink for P, as the input also markedly decreased during this period. Mirror Lake apparently was more biologically productive at 6,500 to 4,000 yr B.P. with maximal deposition of organic matter. Increasing anoxia of the hypolimnion affected the accumulation of Fe and Mn in the sediments. Overall, the Holocene was characterized by an absence of major chemical changes in the sediments.

Elemental composition of the sediments during the Holocene bears little resemblance to total inputs to the lake because most inputs were in dissolved form. The intepretation of sediment chemistry as reflecting inputs of dissolved substances vs. particulate matter is one of the more difficult challenges in paleolimnology. Erosion of particulate matter in the Hubbard Brook Valley occurs primarily from stream channels rather than from overland flow. Thus, the formation and maintenance of organic debris dams play a major role in this process.

Cultural sediments showed (1) a significant increase in deposition, (2) increased concentration of heavy metals (e.g. Pb) from human pollution of the airshed and watershed, and (3) no obvious change in lake productivity.

In addition, the following general questions were addressed: (1) What is the origin of the organic matter in the sediments? (2) How does focusing affect the utility of deposition values? (3) What trophic changes occurred? and (4) Is the sediment an inwashed soil?

F. Paleoecology of Mirror Lake and its Watershed

MARGARET B. DAVIS, ROBERT E. MOELLER, GENE E. LIKENS, M. S. (JESSE) FORD, JOHN SHERMAN, and CLYDE GOULDEN

Few studies have attempted to integrate sedimentary parameters in a lake with the history of events on the lake's watershed. Such attempts are ambitious and difficult because the interpretations of changes in individual parameters are often ambiguous, and correlation with watershed processes is usually uncalibrated and not rigorous. We attempted such a study in Mirror Lake, largely because of an unusually long and detailed analysis of outputs from forested and manipulated watersheds in the Hubbard Brook Valley (e.g. Likens et al. 1977; Bormann and Likens 1979; *Chapter II*). Various parameters were studied in the sediments (*Chapters VII.A, B, C, D and E*); the following discussion is a summary and commentary on these results.

The discussion is organized around four categories of information that can be derived from lake sediments. The first concerns the sediments themselves—whether they were transported into the lake from the watershed or precipitated within the lake from dissolved substances, whether they were produced near the coring site or moved from shallow to deep water. The second category concerns the lake—its biology and its limnological characteristics such as depth, temperature, transparency, and nutrient concentration. Emphasis is placed on changes in productivity within the lake, as evidenced by various biologic and chemical indicators in the sediments. A third category of information concerns the watershed of the lake, its vegetation cover and factors that have affected it, the development of soil profiles, the loss rates of particulate matter, and the export of dissolved nutrients. Finally we have considered the implications of the Mirror Lake record regarding the terrestrial and aquatic ecosystems, whether they are stable, or whether they have changed through time. If nutrient outputs from the watershed changed over time, was the change caused by the biologic portion of the ecosystem or by changes in the quantities of nutrients available to be leached from the parent material? The nearby Experimental Watersheds in the HBEF provide comparative data on exports from forested and disturbed landscapes; these comparisons have been useful for interpreting the sedimentary record.

We have divided the historical record into six time periods for convenience in discussion. These designations differ slightly from the zones used in some other chapters in this book and from a previous publication by Likens and Davis (1975). Likens and Davis based their zonation on stratigraphic changes in the Central core: Zone I—11,000 to 9,700 yr B.P.; II—9,700 to 4,800; III—4,800 to 200 and IV—200 to 0 yr B.P. The new zones reflect major differences in the vegetation on the watershed: Tundra period 14,000 to 11,500 yr, Spruce period 11,500 to 10,000, Early Holocene 10,000 to 5,000, Late Holocene I 5,000 to 2,000, Late Holocene II 2,000 to 140, and Settlement period

F. Paleoecology of Mirror Lake and its Watershed

140 to 0 yr B.P. An additional limnological transition is recognized at 8,500 yr B.P. These zones correspond to Central core depths as follows: Tundra period (not recovered), Spruce period 12.4–11.7 m, Early Holocene 11.7–5.8 m, Late Holocene 5.8–0.25 m, Settlement period 0.25–0 m. Although most of the chemical and biologic analyses were carried out on the Central core, we have incorporated recently obtained knowledge of sedimentation patterns over the entire lake basin into this summary of all paleoecologic work at Mirror Lake.

The Sediments: Constituents and Pattern of Deposition

Fourteen thousand years ago the Mirror Lake basin sloped steeply to an off-centered deepest point about 23 m below the present lake surface. By 10,000 yr B.P., the beginning of the Holocene, about 1 m of gray lacustrine silt and fine sand had been deposited across approximately 85% of the present lake area. With the climatic change that marked the transition to the Holocene, deposition of gyttja began. This flocculent, more organic sediment was deposited over a more restricted, deeper portion of the basin than was the silt. The pattern of differential deposition—termed focusing—has varied during the Holocene. The inorganic fraction of the gyttja includes diatom frustules as well as fine mineral particles washed in from the watershed. Organic matter and noncrystalline silica (mostly diatom frustules) currently are deposited fairly uniformly across the gyttja zone, but detrital minerals are enriched at the periphery. The present more-or-less uniform pattern of organic deposition within the zone of gyttja contrasts with the pattern earlier in the Holocene, when deposition was greatest at mid-lake sites.

Focusing results in part from wind-driven currents, which resuspend sediment, preventing long-term accumulation above the depth of the summer thermocline. Currents are most effective in autumn, when no thermocline is present to shield the gyttja surface (see *Chapters IV.B* and *IV.C*). A modest amount of sediment is resuspended each autumn, almost entirely from the shallower parts of the lake. Focusing is also accomplished by rare events that move sediment in density currents into the lake center. Focusing creates a sampling problem for paleoecologists. Deposition rates (net accumulation per cm^2 per yr) of sediment constituents can be determined in single cores, but these rates are primarily of local significance; inputs to the entire basin can be estimated only when enough cores are studied to reveal how focusing patterns have changed through time to produce the entire sediment body.

Pollen and ^{14}C dating of five widely spaced cores through late-glacial silt and Holocene gyttja of Mirror Lake have demonstrated past focusing patterns (Davis and Ford 1982; *Chapter VII.A*). These data are augmented by 15 cores and as many probes from other parts of the basin. The data from the five cores do not describe the focusing process precisely, but they are sufficient to outline its major features:

1. The late-glacial deposit was not focused—it underlies more than 85% (12 ha) of the present-day lake area in roughly uniform thickness. Mean deposition rates in the three mid-lake cores are representative of much of the basin.
2. The early Holocene gyttja was strongly focused. Only four of the five cores include early Holocene sediment, which was deposited over an unusually small area within the basin. About 40% of the basin was receiving sediment, which accumulated most rapidly in the deepest part of the basin.
3. Focusing relaxed somewhat in the mid-Holocene. At this time 50% of the basin (7.5 ha) was receiving sediment. Even so, deposition was most rapid near the center, as recorded in the three mid-lake cores. Peak rates of deposition occurred at different times in each of these three cores. Apparently the pattern of bottom currents changed as the lake filled in and the locus of most rapid deposition shifted from west to east across the basin through time.
4. Focusing gradually relaxed further during the late Holocene. Deposition rates for pollen and for gyttja declined at all three mid-lake sites. The decline in deposition rates at these three sites is attributed to increasing deposition at the periphery of the 7.5-ha zone of accumulation.
5. With an increase in lake level of 1 to 2 m in the settlement period beginning 140 yr ago, deposition became nearly uniform within a larger, 10.2-ha zone of accumulation.

Sediment constituents have been studied primarily in the Central core (*Chapters VII.A, B, C, D* and *E*). However, between-core comparisons of pollen concentrations and percentages, as well as data on loss-on-ignition, indicate that the sediments are well mixed and nearly uniform in composition, even though rates of deposition have varied from one part of the lake to another (Davis et al. 1984). This means that changes in *concentrations* and *proportions* of different constituents of the sediment observed in the Central core should be typical of sediment throughout the lake. *Rates of deposition* in the Central core, however, are primarily of local significance because of sediment focusing. To interpret these rates, we have attempted to incorporate in this chapter all that we know about patterns of focusing in the lake as a whole.

Understanding of the history of focusing in Mirror Lake has been augmented by studies of modern sedimentary processes (with traps and short cores), and comparative studies of pollen deposition rates at six additional lakes in the region (*Chapter VII.A.2*; Davis et al. 1984). The pollen data were particularly helpful because they imply that inputs of pollen from the regional terrestrial vegetation have been more or less constant for the last 9,000 yr. This pattern of constant pollen input to the lake as a whole means that variations in rates of pollen deposition in the Central core at Mirror Lake 9,000 to 150 yr B.P. are probably due to focusing rather than changes in inputs to the lake. Changes in pollen deposition (P_t) around its long-term mean at the Central site for this period ($P_{\bar{x}}$ = 40,000 grains/cm^2-yr) serve, in effect, as a rough index to the intensity of focusing: $f_t = P_t/P_{\bar{x}}$. The effect of changes in the focusing pattern then can be removed from deposition rates of organic matter and other constituents having depositional patterns closely correlated with that of pollen: $C_t = A_t/f_t$, where C_t and A_t are corrected and actual deposition rates, respectively, at time t (see *Figure VII.F*–1 and 2). This calculation does not remove the two- to three-fold overestimation of lakewide deposition rates inherent in the Holocene record at the

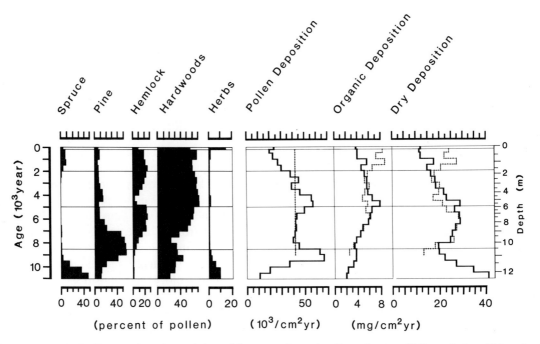

Figure VII.F–1. Pollen stratigraphy and deposition rates from the Central core of Mirror Lake. Abbreviated pollen diagram and deposition rates for pollen, organic matter and total dry matter are presented for 500-yr intervals. The dotted lines represent estimated deposition rates corrected for focusing. The correction is based on pollen deposition rates, under the assumption that without the effects of focusing pollen deposition rates would have remained constant, because pollen inputs to the lake were constant during the Holocene.

F. Paleoecology of Mirror Lake and its Watershed

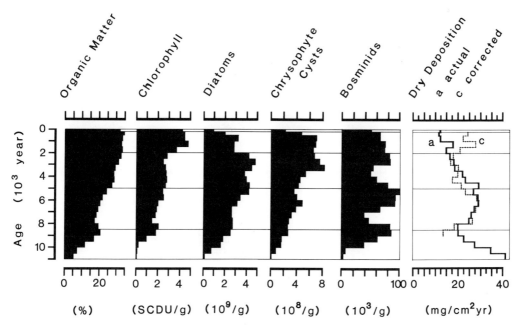

Figure VII.F–2. Sediment constituents reflective of lake productivity from the Central core of Mirror Lake. Values are averaged for 500-yr intervals. Deposition rates for dry matter are the observed rate of local dry matter deposition (**a**), and an estimated rate corrected for focusing (**c**).

Central site (see Davis and Ford 1982; *Chapter VII.A*). However, it does correct the Central core for the conspicuous relaxation in focusing during the late Holocene (feature 4 above), facilitating comparisons of deposition rates during early and late Holocene time at the Central core site.

The increase in deposition rates at the margin of the gyttja zone that we assume occurred in the late Holocene cannot be verified in data from the five analyzed cores, only three of which are adequately dated within the 2,500- to 0-yr B.P. range (see *Figure VII.A–5*). Extension of a focusing correction based on pollen to other sediment constituents is reasonable but speculative. Thus, a focusing correction has not been applied in *Chapters VII.B, C, D,* and *E* but is taken into account here whenever deposition rates are utilized in reconstructing the history of the lake.

Clues to the lake's history come from the stratigraphic sequences of biologic remains and chemical analyses, studied mainly in the Central core (*Chapters VII.B, C, D* and *E*). The plankton left an abundant record biased toward fossilizable forms such as diatom frustules, chrysophyte cysts and cladoceran exoskeletons (*Figure VII.F–2* and *3*). Remains of nonpelagic littoral and benthic organisms are numerically much less abundant than those of planktonic species, but represent the majority of all species of both diatoms and cladocera occurring in the core. Low productivity in the littoral zone compared with the pelagic zone is a probable explanation for the very low frequency of the littoral species, since autumnal resuspension focuses virtually all undecomposed littoral detritus onto the profundal gyttja (cf. Mueller 1964). Remains of larger benthic invertebrates, such as head capsules of chironomids and *Chaoborus,* have not yet been studied. These are not abundant in the Central core—evidence, perhaps, that a late summer anoxia, characteristic of the deepest part of the water-column (*Chapter IV.B*), limited the distribution of these invertebrates throughout the Holocene just as it does today (*Chapter V.A.6*). Seeds and spores of aquatic macrophytes from the margin of the gyttja zone (core #783) provide evidence for the floristic development of the macrophyte community (*Figure VII.F–4*; also *Chapter V.A.4*).

The biases of the fossil record are evident when fossils are compared with the modern planktonic community (*Chapters V.A.2* and *5; V.B.2* and *5*). Diatoms are significant components of the plankton in spring and autumn, and

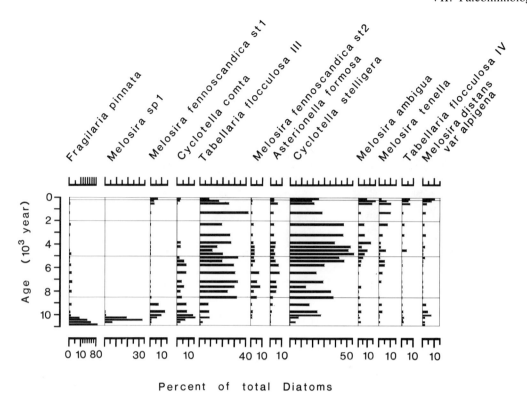

Figure VII.F–3. Diatom stratigraphy from the Central core of Mirror Lake. Twelve of the more than 200 taxa found in the sediment are replotted from *Chapter VII.B*.

collectively contribute as much as 20 to 40% of the annual phytoplankton production. Chrysophytes collectively are at least as productive as the diatoms in the modern lake, but are apparently underrepresented in the sediments. However, not all cells produce cysts. The bias within the zooplankton record is severe; neither of the dominant Cladocera (*Daphnia*, *Holopedium*), nor any of the rotifers or copepods, preserve well enough to be readily retrieved and counted (Deevey 1964). The two closely related bosminid Cladocera that are well preserved (*Bosmina longirostris* and *Eubosmina tubicen*) collectively contribute only 3% of the modern zooplankton production. Because these organisms are not dominant in the modern lake, changes in the deposition rates of diatoms and bosminids do not necessarily have any correlation with past changes in planktonic productivity.

The chemical data provide numerous possibilities, and problems for interpretation. The sedimentary organic matter is believed to be largely autochthonous (*Chapter VII.E*). This conclusion runs counter to a common supposition that the organic matter preserved as sediment in relatively unproductive lakes of temperate forested regions is predominantly allochthonous (Mackereth 1966; Pennington 1981a,b). Low C:N ratios (8 to 11 by mass) and the modern organic carbon budget (*Chapter V.C*) both support an autochthonous source; but it is the high content of chlorophyll degradation products (*Chapter VII.D*) that most strongly suggests an autochthonous origin.

The lakewide input of autochthonous organic matter, if it can be estimated from deposition rates or inferred from pigments (cf. Wetzel 1970; Adams and Duthie 1976), serves as an index of paleoproduction. The reliability of organic matter and chlorophyll in the Central core from Mirror Lake is enhanced by evidence from Fe and Mn distributions (*Figure VII.F–5* and *Chapter VII.E*) that anoxic conditions that favor preservation have developed during part of the year above the deepest sediments since lateglacial times. This suggests that changes in preservation have been relatively minor.

F. Paleoecology of Mirror Lake and its Watershed

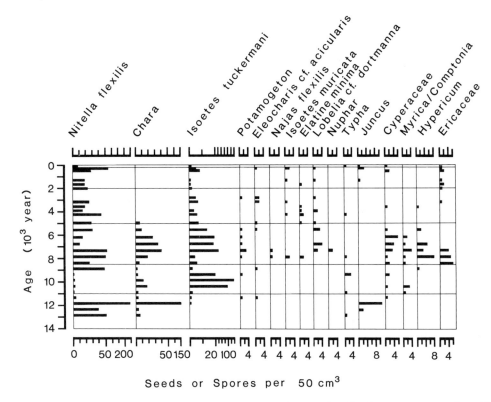

Figure VII.F–4. Stratigraphy of macrophyte remains from Mirror Lake. Seeds, spores, or anthers (*Myrica/Comptonia* only) were extracted from sediment of core #783, from a water depth of 6.2 m. Taxa to the right of *Nuphar* are emergent, shoreline, or possibly terrestrial species. Definitely terrestrial taxa are not included. Note changes in scale for the sometimes abundant cryptogams at left of diagram.

Allochthonous contributions are more difficult to evaluate. If they did not change greatly during the Holocene, the focusing-corrected deposition rates of organic matter and the chlorophyll in the Central core will be useable indices to lakewide productivity. Diatoms and bosminids are potentially subordinate components of their respective communities, whereas chlorophyll degradation products do have the advantage of reflecting the entire autotrophic assemblage.

The ash component was separated chemically into detrital minerals, amorphous silica, and acid-soluble salts (*Figure VII.E–3*). Only the detrital mineral fraction directly reflects physical erosion and transport of particulate matter from the watershed. This fraction is correlated closely with total Na, K, and Mg (e.g. Ford 1984; Engstrom and Wright 1984; Mackereth 1966; Pennington 1981a,b). Because the concentration of amorphous silica in the Central core does not parallel the Holocene rise and fall of the sum of diatoms plus chrysophyte cysts in the same core, this fraction may include some amount of nondiatomaceous silica (e.g. amorphous complexes or crystalline clays). In contrast to Na, K, and Mg, the elements Ca and P occur both as detrital minerals and as secondary phases, perhaps organic complexes. The deposition rates of Ca and P parallel the deposition of detrital minerals in the late-glacial sediment, while they parallel the deposition of organic matter in the Holocene (*Figures VII.E–5 and 6*).

Outline of Lake Development

The Late-Glacial Lake (14,000 to 11,000 yr B.P.)

During the first three millennia, the lake was part of a cold, tundralike environment. Remnants of the original ice block may or may not

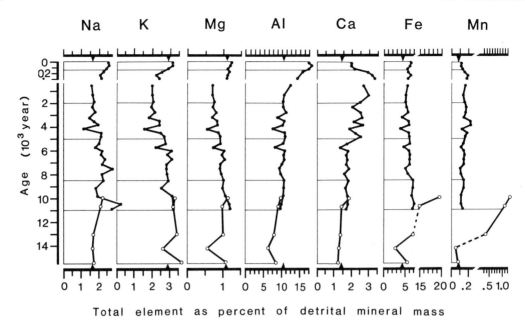

Figure VII.F-5. Stratigraphic trends of major rock-forming metals from Mirror Lake. Total elemental concentrations in dry matter from the Central core and core 4 (●——●) or core 782 (○——○) are recalculated as if the elements occurred exclusively within the detrital mineral component. The metal concentrations in composite till/bedrock are indicated on each concentration scale as a reference.

have persisted in the sands underlying the earliest lacustrine sediments. The widespread distribution throughout the basin of gray silt and fine sand suggests transport by rapidly moving water flowing through the incipient lake basin. Part of this material could have been transported by wind from the thinly vegetated landscape, especially the floodplain of the nearby Pemigewasset River.

By 12,000 yr B.P., Mirror Lake was probably a deep, clear body of slightly alkaline water. Deposition of inorganic silts continued. Silt and fine sand were transported by streams selectively washing fine material from the incipient, frost-heaved soils developing on the mantle of glacial till and drift. The concentration of organic matter in the silt was low but the total amount deposited in the lake was surprisingly high; lakewide inputs of organics were probably about one-third of the mean Holocene rate (*Chapter VII.A*). Benthic macroalgae, such as *Nitella flexilis* and *Chara,* were present and perhaps abundant as early as 12,000 to 13,000 yr B.P. *Nitella* can grow in cold, clear water at great depths (more than 10 m).

The apparent absence of diatoms in the silt from this time is puzzling. The outer, siliceous walls are missing from *Isoetes* megaspores deposited before 11,000 yr B.P. Perhaps these and the siliceous diatom frustules dissolved subsequent to burial. Conceivably the lake water had low concentrations of Si because of low rates of chemical weathering in the periglacial landscape.

By 11,000 yr B.P. the distribution of Fe and Mn (*Figure VII.F-5; Chapter VII.E*) suggests that anoxic conditions developed periodically above the deepest sediments (Central core) but not above slightly shallower sediment (core 782). Such evidence of heterotrophic metabolism is consistent with the moderate rate of organic deposition and the evidence for benthic vegetation.

The high influx of inorganic silts does not imply a similarly high availability of dissolved phases of nutrients like N, P, and trace metals. These elements are transported to the lake today by means of atmospheric deposition, and in stream water as dissolved ions, organic compounds and organic chelates. Even by 11,000 yr B.P., when the diatom flora suggests that lake water was more alkaline than today, productivity still may have been limited strongly by nutrient shortages during the cool summers. On one hand, leaching of P and trace metals from weathering sites within the soil may have been inhibited until accumulating soil organic pools initiated a sustained output of dissolved organic P compounds and chelated metals. And N, on the other hand, was not available within the mineral soil to be released by weathering. It could only have come from the watershed through leaching of organic soils, as soils progressively accumulated N, both from precipitation and N-fixation on the watershed. Herbs and shrubs known to fix atmospheric N were present in the late-glacial tundra, but rates of N-fixation may have been low until 10,500 yr B.P. when alder became established.

The supply of nutrients probably controlled the productivity of the late-glacial lake, judging from studies on modern arctic lakes. Char Lake is an alkaline, hardwater lake of the Canadian high arctic (Schindler et al. 1974b). Weathering of the local sedimentary bedrock yields high concentrations of dissolved Ca, Na, etc., but inputs of both P and N are extremely low, and aquatic productivity is severely limited (Kalff et al. 1975). Although temperature only barely exceeds 4°C during the short summer period of adequate illumination, enrichment with N and P can cause large increases in productivity (Schindler et al. 1974a). The important point is that low temperature limits productivity only indirectly, by inhibiting soil development and the chemical mobilization of P, N, and possibly trace elements from the watershed soils. Low temperature may secondarily inhibit nutrient regeneration within the lake.

Climatic and Limnological Transition (12,000 to 8,500 yr B.P.)

By 11,000 yr B.P. the early, still silty gyttja in the Central core contained a sparse diatom assemblage dominated by the periphytic species *Fragilaria pinnata*. The planktonic bosminids also were present, however, and in the same ratio to the littoral chydorids as has prevailed throughout the Holocene (~8.5:1). By 10,000 yr B.P., the diatom assemblage was quantitatively dominated by a few planktonic species— a change that reflects the great increase in abundance of planktonic forms rather than a substantial decrease in the concentration of the littoral forms. The taxonomic richness of Cladocera and diatoms has been essentially unchanged since 10,000 yr B.P. The immigration phases of community development (Goulden 1969) must have occurred during the tundra period, and were completed before 11,400 yr B.P., the age of the oldest sediment that was recovered in the Central core.

During this period of climatic and sedimentologic change, the increasing concentrations of organic matter, chlorophyll derivatives, diatoms, chrysophyte cysts, and bosminids (*Figure VII.F–2*) all suggest increasing productivity. The increase in focusing at this time, however, caused the increases in deposition rates to be exaggerated in the mid-lake cores. Productivity probably increased less than two- or threefold, judging from the whole-lake deposition of organic matter; the lake must have still been oligotrophic. The earliest diatoms (especially *Fragilaria pinnata*) indicate more alkaline conditions than prevailed later in the Holocene, but the difference is slight. The macrophytes *Nitella flexilis* and *Isoetes tuckermani* suggest that conditions even then were circumneutral. Diatoms were a more important fraction of the phytoplankton than they are today, judging from the ratios of diatoms:chrysophyte cysts and diatoms:chlorophyll units.

The transition period was marked by a number of chydorid Cladocera and diatoms that disappeared or became comparatively rare by the mid-Holocene. A still larger number of taxa have been important components of their assemblages throughout the Holocene. These include Cladocera (*Bosmina longirostris, Eubosmina tubicen, Rhynchotalona falcata, Alonella excisa*), diatoms (*Cyclotella stelligera, Tabellaria flocculosa* strain III, *Asterionella formosa, Melosira* spp.) and macrophytes (*Nitella flexilis, Isoetes tuckermani*).

The Early Holocene (8,500 to 5,000 yr B.P.)

The trend of slowly increasing concentrations of organic matter, chlorophyll derivatives, diatoms, chrysophytes and bosminids continued. The rate of dry matter deposition increased at mid-lake sites during this period; the evidence, once focusing is taken into account, suggests increasing productivity. The lake at 5,000 to 6,000 yr B.P. may have been similar to the modern lake: a moderately productive, still oligotrophic body of soft water that had nonetheless become more productive than the lake in late-glacial or transition periods.

The relaxation of sediment focusing after 8,000 yr B.P. could have resulted from a rise in lake level related to climatic changes that occurred at this time. In particular, a rise in summer rainfall or a decrease in evaporation rates during the summer months could have resulted in higher lake levels in the autumn when mixing and sediment resuspension occurs. The resulting rise in the levels affected by wind-driven currents could have permitted the gyttja zone to expand within the lake basin. A decline in charcoal and other dry climate indicators suggest a more moist and/or summer-cool climate starting by 7,000 yr B.P.; certainly these changes were underway by 5,000 yr B.P. (*Chapter III.B.1*).

The increase in productivity during the early Holocene from oligotrophic to somewhat more productive conditions might have been stimulated by warmer temperatures and/or increased nutrient supply. Early work on natural eutrophication (e.g. Strøm 1928; Hutchinson and Wollack 1940; Deevey 1942) had suggested that the amount of available nutrients was slowly augmented through more efficient internal cycling. Alternatively, the nutrient capital may have increased owing to accelerated rates of nutrient leaching from the watershed. The mass ratios of C:N:P in the sediment changed during the early Holocene from about 170:20:1 to 85:10:1 (*Figure VII.E–7*), principally from increasing concentration of P. As productivity increased, the deposition of P increased even more rapidly. The opposite pattern would have been expected if more efficient P cycling alone had been responsible for the increased production (cf. Adams and Duthie 1976). Since most of the N and P currently entering Mirror Lake is ultimately retained in the sediment (*Chapter IV.E*), deposition rates for these elements (*Figures VII.E–6* and *VII.F–1*) may be rough indices of their past inflow. Significantly, the implied increase in nutrient inflow is correlated with the shift on the watershed from birch, pine, oak and softwood species to forests of hardwoods and hemlock. The mechanism by which these forest changes might be related to changed nutrient output is discussed in a later section.

During this 3,500-yr period of increasing productivity, the phytoplankton community seems to have been remarkably stable. Diatoms were more important relative to chrysophytes and other algae than they are today. Chrysophyte cysts and "anonymous" algae decreased only slightly during this period. The diatom assemblage, indicative of circumneutral conditions, was dominated by *Cyclotella stelligera* and *Tabellaria flocculosa* strain III, with subordinate amounts of *Asterionella formosa, Cyclotella comta*, and *Melosira fennoscandica*. This assemblage is very similar to today's.

Seeds and megaspores of several macrophytes characteristic of nutrient-poor, acidic-to-circumneutral waters first appeared during this period (e.g. *Lobelia dortmanna, Elatine minima* and *Isoetes muricata*). Changing conditions within the littoral zone caused the nearly synchronous disappearance, about 5,200 yr B.P., of two species prominent in the early Holocene: the macroalga *Chara* and the cladoceran *Chydorus brevilabris. Chydorus* species occur in many benthic habitats. They often live on macrophytes, including *Chara* (Flössner 1964; Williams and Whiteside 1978). Although most *Chara* are restricted to alkaline waters, independent evidence of circumneutral water in the mid-Holocene in Mirror Lake convinces us that this *Chara* must have been one of those that are not so restricted (e.g. *Chara furcata*: *Chapter V.A.4*). Its disappearance at 5,200 yr B.P., following the appearance of additional neutral or acid-water species, does suggest that some threshold related to water quality had finally been crossed.

The Late Holocene. I. (5,000 to 2,000 yr B.P.)

The mid-Holocene, 5,000 yr B.P., is a convenient sedimentologic dividing point. During the earlier Holocene, concentrations of trophic indicators increased in parallel with increasing deposition rates of these same indicators at the mid-lake sites. At the mid-lake sites, the average rate of dry matter deposition (also of organic matter and chlorophyll derivatives) declined slightly after 6,000 yr B.P., but this decline is apparently a local effect. During the late Holocene, however, concentrations of organic matter and chlorophyll derivatives continued to increase, but their deposition rates at mid-lake sites declined. We believe that focusing relaxed at this time, and that the apparent decrease in deposition is entirely due to this change in sedimentation pattern. The constancy of the pollen concentration, while pollen deposition rates decline in the Central core, suggests that this is the case. If lakewide deposition of organics and chlorophyll remained constant, productivity also may have been essentially constant throughout this interval. Chemically, the late Holocene was characterized by gyttja distinctly enriched in P, slightly and irregularly richer in Ca, but depleted in S. At the Central core site the deposition of elements associated with detrital minerals (Na, K, Mg) decreased steadily, owing to relaxed focusing and decreased erosion and transport from the watershed.

During the period 5,000 to 2,000 yr B.P., the proportions of major diatoms fluctuated, but the assemblage as a whole was only slightly changed—still a community of circumneutral waters. Chrysophyte cysts increased significantly relative to diatom abundance.

A conspicuous oscillation in the relative abundances of the two bosminids (*Figure VII.C–3*) occurred between 4,000 and 2,000 yr B.P., including a shift from 80% *Eubosmina tubicen* to 80% *Bosmina longirostris* at 2,900 yr B.P. This change is not reflected in the sediment chemistry nor in the diatoms. Some biologic change within the plankton community must have occurred, perhaps a shift in nondiatomaceous phytoplankton or a change in the species of predators present in the lake.

The Late Holocene. II. The Presettlement Lake (2,000 to 140 yr B.P.)

Shortly after 2,000 yr B.P., the chlorophyll concentration increased abruptly, without a concomitant increase in total organic matter. The change could represent improved preservation, although we believe preservation was relatively good throughout the Holocene (*Chapter VII.D*). Alternatively, the increased chlorophyll represents a further increase in productivity. Chlorophyll concentrations increased more than deposition of diatoms and chrysophytes, suggesting a relative increase in other kinds of algae that leave no morphological fossil record.

The diatom assemblage changed slightly during this period, perhaps indicating slight acidification (e.g. the increase in *Tabellaria flocculosa* strain IV). Concurrently the diversity and species richness increased substantially (*Figure VII.B–1C*): two of the dominant species declined in abundance, and as a result uncommon benthic species were better represented. A gradual decline in the concentration of both diatoms and chrysophyte cysts and a slight change within the diatom assemblage occurred in the centuries prior to European settlement around Mirror Lake. These changes do not consistently parallel changes in the Cladocera nor in the sediment chemistry. As far as we can tell, the lake was only slightly different chemically from the mid-Holocene lake. The diatoms had hinted at greater alkalinity/pH of the lake water in the late-glacial than in the early Holocene. Since a number of species that are thought to be sensitive to water chemistry persist throughout the Holocene, changes in alkalinity/pH were small, and the lake was circumneutral throughout the Holocene.

The Settlement Period (140 yr B.P. to Present)

Clearly, settlement of nineteenth century farmers in the watershed caused an increase in the concentration and deposition of detrital minerals. Disturbance of soils and stream channels occurred, and it continues today. The declining concentrations of diatoms and chrysophyte cysts within the postsettlement sediment reflect dilution by inputs of detrital minerals. Gyttja accumulated over an increased area of the lake bottom after settlement. This change demonstrates a relaxation of focusing, and resulted in an apparent decline of deposition rates in individual cores. When the increased area of gyttja and the decreased intensity of focusing are taken into account, the rate of deposition of phytoplanktonic remains is essentially constant before and after settlement 140 yr B.P.

Since settlement, there have been changes in the relative importance of some diatoms, particularly *Melosira* species, but these changes do not suggest major shifts in environmental conditions. In Mirror Lake there has not been a shift to dominance by species that were previously rare or absent, as has occurred where lakes or their watersheds have been more severely disturbed (e.g. Bradbury and Winter 1976; Manny et al. 1978; Elner and Happey-Wood 1980; Smol and Dickman 1981). The unusually great diversity of diatom species in the most recent sediment (*Figure VII.B*–1C) is provocative, however. In Mirror Lake this increasing frequency of rare species might reflect changing chemical conditions within the inlet streams, which have been somewhat enriched as well as acidified within recent decades. During the 1970s, the pH of lake water ranged seasonally from about 6.5 to 7.0 (G. E. Likens, unpublished data), still consistent with the diatom evidence for circumneutral, or at most only slightly acidic conditions. The diatoms recorded in contemporary plankton collections (*Chapter V.A.2*) apparently belong to taxa recorded from the surficial sediment, but differences in nomenclature (*Tabellaria* species) and identification (*Cyclotella* species) confuse this comparison.

The present oligotrophic character of the lake, as well as its Holocene record, indicate that no more than a modest cultural increase in productivity has occurred. The change is small relative to the natural eutrophication (oligo-mesotrophic range), which took place during the first half of the Holocene.

The Watershed

The Mirror Lake watershed has undergone a series of major changes since the continental ice sheet melted 14,000 yr B.P. A number of these changes were caused directly by climate. Arctic to subarctic conditions prevailed in the region until 10,000 yr B.P. (Davis et al. 1980). Between 9,000 and 5,000 yr B.P. the mean annual temperature was about 2°C higher than now and summers may have been hot and dry (Davis et al. 1980). Only within the last 1,000 yr have regional temperatures fallen to levels similar to, or even slightly cooler than, today's.

The initiation of tree growth on the late-glacial tundra, the gradual development and deepening of soil profiles, and—many thousands of years later—the disturbance of these soils by farming and logging, changed export rates from the watershed to the lake and affected the aquatic ecosystem. These changes correlate with changes in the lake. In addition, several large changes in the abundances of species within the forest community occurred during the history of Mirror Lake. These might have been important in the functioning of the forest ecosystem; we have endeavored to test the idea that they affected nutrient exports and thus had an effect on the aquatic portion of the ecosystem as well.

Tundra Period (14,000 to 11,500 yr B.P.)

Particulate exports from the watershed have been assessed from estimates of whole lake inputs of inorganic debris. During late-glacial time biologic precipitation of dissolved inorganic material was ineffective; the inorganic sediment represents particulate inputs alone. During the late-glacial tundra phase, 13,000 yr B.P., exports per unit area of watershed were more than 30 times higher than exports of particulate inorganic matter from present-day forested watersheds in the HBEF (Davis and Ford 1982; cf. Likens et al. 1977). The high loss rate during late-glacial time implies mass-wasting processes, including frost action, that were effective in removing fine material from the glacial deposits that mantle the watershed. These materials were exported from the watershed in streams that probably contained no significant organic debris dams (Likens and Bilby 1982; Bilby 1981).

Frost-heaving on bare soils is effective in New England even under the present climatic regime. We presume that during late-glacial time solifluction lobes and other frost features developed on the slopes surrounding Mirror Lake, disrupting incipient soil profiles and bringing fresh glacial till composed of unsorted coarse and finely divided rock to the surface. Silt-sized particles would have been selectively removed from this material and carried to the lake. Wind also may have transported silt to the lake.

We picture a tundra landscape with discontinuous herbaceous plant cover. Some of the tundra plants (such as *Dryas*) were mat-forming and nitrogen-fixing plants capable of withstanding considerable frost-heaving and disturbance (Crocker and Major 1955). Cations and other nutrients might have been leached rapidly from the finely divided rock flour, within which the individual particles were still without weathering rind. The amounts of dissolved nutrient from this source that reached the lake are unknown, as we have no direct measure in the sediment of dissolved inputs to the lake. Even indirect measures such as siliceous fossils are not present. However, the indications from diatoms in slightly younger sediment for moderate alkalinity of the lake water, and a pH of 7.5 to 8, suggest significant inputs from this source. Any dissolved cations that entered the lake left through the outlet; only those elements that came into the lake as particulate mineral matter are preserved as sediment.

By 12,000 yr B.P. shrubs were growing on the watershed together with the tundra herbs. Presumably a mosaic of habitats had developed by this time, with patches of acid humus in protected areas, and patches of bare, newly exposed glacial drift on steep slopes and on other sites subject to intense frost action. A mosaic of habitats is indicated by the terrestrial flora; late-glacial macrofossil assemblages from other sites in New England contain a mixture of calcicolous species like *Dryas*, and acidophilous species like *Vaccinium uliginosum* (Argus and Davis 1962; Miller and Thompson 1979). Shrubs included willow, juniper, heaths and the nitrogen-fixing shrub *Shepherdia canadensis*. Although soil development was occurring, including weathering and the addition of nitrogen compounds and humus to the soil, export rates of particulate matter remained high.

Spruce Period (11,500 to 10,000 yr B.P.)

Spruce began to grow on the watershed 11,500 yr B.P. Spruce needles were deposited in Mirror Lake, and the rate of accumulation of spruce pollen rose steeply, indicating a regional increase in abundance. However, the pollen assemblages and the low pollen deposition rates indicate that spruce grew in open stands or in groves rather than as continuous forest (Davis 1967; Davis et al. 1980).

The appearance of trees on the watershed is coincident with a dramatic decline in exports of particulate matter (Davis and Ford 1982). We speculate that spruce logs provided debris for dams in the streams, increasing the storage capacity within stream channels and slowing loss rates from the watershed (Likens and Bilby 1982; Keller et al. 1982). Berglund and Malmer (1971) interpret a similar decrease in silt inputs to a Swedish lake as evidence that a continuous humus cover had become established on the landscape, and Tutin (1969) interprets late-glacial organic sediment in Blelham Tarn as evidence for incipient soil profiles, later disrupted when clay and moss layers were deposited in the tarn during the Younger Dryas cold stadial. However, in Mirror Lake there is no simple or direct connection between the decline in export rates of inorganic particulates and soil stabilization. The correlation is instead with the first occurrence of trees in the watershed vegetation. The trees apparently played a direct role in the reduction of exports, through the mechanism of debris dams, and/or by increasing evapotranspiration and thus reducing runoff. Tree roots could also have played a role in stabilizing slopes, although it can also be argued that trees could not have become established on sites where solifluction was still very intense. The development of a complete humus cover and stable soil profile in the vicinity of Mirror Lake was apparently a gradual process. The process must have been well underway before trees arrived. We are not sure when the process was completed. By 10,000 yr B.P. the climate was as warm as today (Spear 1985; Davis 1983), and mixed conifer-deciduous forests developed near the lake, although the forests may still have been rather open. Green alder, which is capable of N-fixation (Crocker and Major 1955), was present as well as aspen, birch and other trees requiring open sites for establishment. The elevated deposition rates for charcoal in sediment older than 7,000 yr (*Figure III.B.1–3*) suggest that fire occasionally occurred, although charcoal deposition rates were at least an order of magnitude lower than charcoal deposition rates in recent centuries in the Upper Great Lakes region, suggesting lower fire frequency (Heinselman 1973; Swain 1973).

Organic as well as inorganic particulate matter was exported from the watershed during late-glacial time. The deposition rate of organic matter to the lake as measured by the organic content of the sediment cores was surprisingly high—perhaps one-third the rate of input during the Holocene (Davis and Ford 1982). If these inputs were entirely allochthonous, exports would have had to equal 12 kg/ha-yr, a rate similar to modern particulate organic exports from forested watersheds at HBEF (Likens et al. 1977). The export rate was undoubtedly lower than this, because some production was occurring in the lake. Thus, although export rates of inorganics were many times higher than from modern forests, export rates for organics were lower than modern rates at HBEF. Low export rates for organics relative to inorganics are compatible with the view that the watershed lacked continuous vegetation cover and well-developed soils.

Terrestrial plants that require neutral or alkaline soils grew on the watershed during the tundra period and spruce period, but not during the Holocene. These include *Dryas integrifolia* (identified from macrofossils) and *Picea glauca*, whose presence was inferred from the occurrence of large-sized spruce pollen. Cupressineae pollen could represent either *Thuja* (arbor vitae), which grows on nutrient-rich soils, or *Juniperus*, which can tolerate acid, nutrient-poor soils.

The Holocene (10,000 to 140 yr B.P.)

Developmental changes in the terrestrial ecosystem occurred as a humus cover was formed and soil profiles matured. Although trees began to grow on the watershed as early as 11,500 yr B.P., forests did not form until the Holocene. The early forests were probably quite open in character, judging from the abundance of shade-intolerant tree species, such as aspen. The increase in abundance of trees after 10,000 yr B.P. undoubtedly increased transpiration and reduced stream outflow. A drainage network developed and the probable presence of organic debris dams in the streams could have increased their storage capacity for eroded particulate matter and reduced the severity of erosion during floods.

Export rates of inorganic particulates began to fall 12,000 to 11,000 yr B.P., reaching low and relatively stable levels by 10,000 yr B.P. Meanwhile the deposition of dissolved inorganics increased, primarily silica in the form of diatom frustules. Throughout the Holocene, the summed input to the lake of inorganic particulates plus amorphous Si was closely similar to the modern export rate of particulates plus dissolved Si from the watershed (Davis and Ford 1982); exports throughout the Holocene were therefore at a reasonable rate for a forested watershed in this region (*Chapters II* and *VII.A*).

The Hemlock Decline (4,800 yr B.P.)

Sediments deposited 4,800 yr B.P. throughout eastern North America record a sudden decline in pollen of hemlock. The rapid die-off of hemlock trees was probably caused by a forest pathogen that spread rapidly throughout what was then the geographical range of hemlock (Davis 1981a). Hemlock pollen percentages at this time were higher than today. Hemlock often forms dense groves around lakes and we assume that hemlock grew abundantly in the forests surrounding Mirror Lake. It therefore seems surprising that the replacement of hemlock by hardwoods had little apparent effect on the sediment record. The forests must have been disrupted, at least temporarily, by the death of a dominant species and its replacement through processes of secondary succession.

Disturbed and early successional communities often have higher rates of nutrient loss (Likens et al. 1970; Vitousek and Reiners 1975; Gorham et al. 1970; Bormann and Likens 1979). At Berry Pond, Massachusetts, Whitehead and Crisman (1978) and Rochester (1978) detected changes in the sediments that they interpret as evidence for ephemeral limnological changes caused by the hemlock decline. Similar changes correlated with the hemlock decline have not been found at Mirror Lake, but the sampling interval for chemistry and diatoms was not intensive enough to reveal changes persisting less than two to four centuries.

The decline in populations of hemlock on the watershed occurred within a decade, judging from the distribution of pollen in annually laminated sediments at another site in New Hampshire (Allison et al. 1984). The increased light on the forest floor probably permitted germination of birch seedlings, and presumably other successional species, such as cherry, that are less well represented by pollen. Rapid replacement of hemlock by birch is indicated, because the concentration of birch pollen increased in the sediment at the very same level where the concentration of hemlock pollen declined. Apparently the forest growing at Hubbard Brook during the mid-Holocene was able to replace the function of a single forest species rapidly, without large nutrient loss.

Soil Development During the Holocene

Exports of organic matter to the lake were apparently relatively low throughout the Holocene, while autochthonous inputs to the sediment rose to a maximum about 6,000 yr B.P. remaining near this level until the time of settlement. The changing rate of production in the lake as well as changes in sediment chemistry point to quantitative changes in the amounts of nutrients exported from the watershed 10,000 to 6,000 yr B.P. These changes might be related to maturation of soil profiles, which was apparently occurring at the same time.

The initial increase in productivity of the lake 10,000 yr B.P. might have been influenced directly or indirectly by climatic warming, but the continuing rising trend in production between 10,000 and 6,000 yr B.P. suggests increasing inputs of nutrients, as discussed in the preceding section. The development of forest vegetation, followed by the replacement of tree species characteristic of early successional forests and xeric communities by northern hardwood forest thus correlated with increased exports of nutrients to the lake. A larger supply of organic N and ammonium in the more mature soils may be indicated; increasing exports of P are also suggested by the sediment chemistry. Both nutrients may have entered the lake partly in association with dissolved and particulate organics from the forest soils (Meyer and Likens 1979; Meyer et al. 1981). The implications of this finding to our understanding of ecosystem functioning are discussed in the next section.

The chemical composition of the sediment suggests that erosion within the channels of inlet streams has been the main source of detrital minerals for the lake throughout the Holocene, including the settlement period. The relative proportions of total Na, K and Mg in the Central core changed very little during the lake's history. If these elemental concentrations are expressed as a percentage of the detrital mineral component of the sediment (which is the form in which they largely occur), then the strong chemical resemblance between the late-glacial silt, the Holocene gyttja and the composite soil material is evident (*Figure VII.F*–5). The slight, parallel decrease in content of all three metals during the Holocene reflects a slight decrease in the readily weathered minerals, and an enrichment in quartz or possibly a secondary Al mineral, a change that may reflect progressive weathering of soils.

Another line of evidence for stream channels as the source of the material is the more or less constant ratios of Mg:Na:K. If erosion of an increasingly mature A-2 horizon had occurred, we would have expected depletion of Mg relative to both Na and K (*Table VII.E*–1). Or if deposition of materials from the B horizon had occurred, we would have expected depletion of Na relative to K (cf. Johnson et al. 1968). Neither of these trends occur. The accelerated erosion resulting from human activities after 140 yr B.P. was also largely restricted to the stream channels, since the mass ratio of metals in post-settlement gyttja (Na:K:Mg = 2.1:2.6:1) remains similar to that in composite parent material (Na:K:Mg = 1.5:2.6:1).

Exports of Ca, however, had a different history. This element became enriched in the Holocene gyttja, especially after 5,000 yr B.P. Calcium is not exclusively a component of the detrital minerals, because much Ca in Mirror Lake sediment is extractable by dilute acid (*Figure VII.E*–9). Currently, most Ca is exported from forested watersheds in a dissolved phase (*Chapter II*). During the Holocene it was deposited as sediment in an unknown form, perhaps in dead biota or as an organic complex formed in the lake. Increased deposition of Ca in the late Holocene implies increased formation of this hypothetical organic material; it says very little about export from the watershed. Today, the Ca preserved in sediment is only about 4% of the total amount entering the lake (*Table IV.E*–16).

Aluminum accumulates in forest soils, often complexing with organic matter (A-2 horizon) or forming secondary aluminosilicates or oxides. Its concentration in the sediment increases in the late Holocene relative to other cations (except Ca). The increase suggests (1) that progressive soil development increased the availability of particulate soil Al for transport to the lake; or, more probably, (2) that leaching of Al from the soil accelerated in the late Holocene, with a correlated increase in Al deposition within the lake. Because Al tends to become complexed with organic matter and other ligands in stream water or within the lake, Al leached from soils is likely to be deposited within the lake (Johnson et al. 1981). A change from mull soils to mor soils, with increased production of organic acids, might have accelerated leaching of Al and also Ca from the soils. The present soils are mor soils, but we do not know if or when a shift from mull to mor occurred. Above the settlement horizon, Ca concentrations are abruptly lower (*Figure VII.F–5*), but Al increased slightly. Whereas the Al-containing complexes apparently come largely from the watershed—and thus increased during watershed erosion, the Ca-containing organic complex in the sediment seems to be an autochthonous lake product, which was diluted within the increasingly allochthonous material after settlement.

It is conceivable that the increased deposition of Al in the late Holocene is related to the expansion of spruce on the watershed, which occurred within the last 2,000 yr. Although the increase in Al relative to other metals appears to begin at least 5,000 to 6,000 yr B.P., there has been a larger increase since 2,000 yr B.P. Climatic changes in the last half of the Holocene could have accelerated soil changes. These climatic changes included a gradual decline in temperature, especially a decrease in summer insolation (Kutzbach 1981), which might have increased soil moisture and encouraged the development of a thick humus layer. The climatic trend in temperature seems to have started about 5,000 yr ago and apparently crossed a critical threshold for spruce 2,000 yr ago.

Still more suggestive of changes in exports from the watershed, possibly related to progressive soil development, is the change in the composition of the lake biota. The diatom record suggests a very slight change in water chemistry that included a small drop in alkalinity/pH in the late-glacial and early Holocene, and a possible decline in the late Holocene. The changes are slight, as the lake appears to have been circumneutral throughout, and many diatom species persisted throughout this interval. The trend parallels a cumulative depletion in the soil of readily weathered cations; possibly it is associated with increasing inflow of dissolved organic acids to the lake from watershed soils.

During the Holocene, at the same time the lake was becoming slightly less alkaline or more acidic, diatoms apparently decreased in importance within the phytoplankton community as a whole. Today the phytoplankton also includes abundant chrysophytes (which are underrepresented in sediment by their cysts), as well as abundant forms represented in our sediment analyses only as chlorophyll derivatives and total organic matter (e.g. cryptomonods, dinoflagellates, chloromonads, nonheterocystous blue-greens, and green algae—*Chapter V.A.2*). During the last 11,000 yr, the phytoplankton has shifted away from diatoms, evidently towards chrysophytes and increasingly to these other algal groups.

Today, planktonic diatoms are common during the spring and autumn in Mirror Lake, but some species occur year-round (*Chapter V.A.2*). Their subordinate role in the plankton as a whole is not caused by a Si-limitation; the epilimnetic concentration of dissolved Si remains greater than 1 mg/liter, well above the level at which diatom growth is inhibited (0.1 to 0.2 mg/liter: Lund 1964; Schelske and Stoermer 1971). In lakes that are rich in P and N, spring diatoms may exhaust the supply of dissolved Si, giving way to summer populations of green and blue-green algae (Lung 1964; Schelske and Stoermer 1971; Tilman et al. 1982). But in Mirror Lake it may be competition for P or N that leads to the summer decline in diatoms (cf. Lehman et al. 1975; *Chapter V.B.2*). Since factors controlling the seasonal succession as well as geographic distribution of phytoplankton are still poorly known, we will not hazard a more precise interpretation of the Holocene trend in Mirror Lake. Although most of the dissolved Si entering the lake is retained in the sediment (*Chapter IV.E*), largely as diatom frustules, diatoms do not exhaust the pelagic pool of dissolved Si today, and there is no reason to suppose that their abundance in the past has been limited by Si.

The diatom stratigraphy in Mirror Lake exhibits a pattern characteristic of small lakes developed upon crystalline bedrock, or nutrient-poor drift, following glacial retreat (Round 1964a, Alhonen 1970; Pennington et al. 1972; Huttunen et al. 1978; Walker 1978). Sparse, floristically impoverished assemblages of mainly littoral forms in the late-glacial shift rapidly, following climatic warming, to more diverse, abundant, and often planktonic assemblages indicating circumneutral or slightly acidic conditions. These persist with minor change unless basin infilling, dystrophication, or human disturbance radically alters the lake environment. Remains of aquatic macrophytes tend to confirm the general trend from more to less alkaline conditions (Vasari and Vasari 1968; *Figure VII.F–4*).

Most workers accept sparse late-glacial assemblages as evidence for suppressed productivity (e.g. Round 1964a; Crabtree 1969). Lakes that remain alkaline into the Holocene (depending on soil chemistry, such as the presence or absence of carbonates) may later become relatively productive (but cf. Wetzel 1970; Manny et al. 1978). In Mirror Lake, however, the low deposition rates of organic matter and chlorophyll derivatives suggest that the early Holocene was a time of low productivity. Productivity increased gradually, a trend we attribute to increasing nutrient inputs from the developing forest soils. We believe that the slight late Holocene shift to lower alkalinity/pH, and lower deposition rates of most productivity indices at mid-lake sites (*Figure VII.F–2*) should not be used as evidence for decreased productivity in the lake. The decreasing intensity of focusing makes the trend at mid-lake sites misleading; lake productivity was sustained in the late Holocene.

During the first few millennia of the Holocene, the lake appears to have been unproductive because of low supplies of N and P from the watershed. It seems possible that the growth of forest vegetation and the development of soils during the Holocene led to accelerated weathering of P from the parent soils and fixation and accumulation of N in the soils; ultimately an increasing supply of N and P reached the lake. Because there is no strong evidence for a decline in lake productivity, we conclude that watershed output of N and P probably remained near current levels since 6000 yr B.P.

Settlement Period (140 to 0 yr B.P.)

The minimal disturbance of soils within the Mirror Lake watershed contrasts with the paleolimnological record at many European and North American lakes where agricultural activity was more intense, or where the watershed was more erodible. Even during Neolithic times, forest clearance at European sites sometimes resulted in massive exports of particulate inorganics and soil humus (Pennington 1978a). Inorganic exports from a watershed of meadows and plowed fields in southern Michigan increased tenfold after deforestation (Davis 1976a). Construction of roads and homes 25 years ago around Linsley Pond, Connecticut caused massive exports of subsurface soil (Brugam 1978). Although the increased concentration of metals, including Al, in postsettlement sediment at Mirror Lake (*Figure VII.F–5*) indicates some soil input, perhaps from the A-2 horizon, little of the watershed has been plowed or disturbed by construction. The soil profile has been more or less preserved. At Mirror Lake, soil weathering has not been changed by disturbance, and organic nutrient reserves in vegetation and soil detritus are largely intact. Although some human sewage is believed to seep into the lake (*Chapter IV.E*), the currently unproductive lake has not yet been significantly eutrophied.

Ecosystem-Level Effects

An important question is whether ecosystems can persist, unchanged, over long periods. The question is difficult to answer, because even a slight imbalance between inputs and losses, too small to detect with instantaneous measurements, could be sufficient to change the system dramatically if continued for a long time.

The sedimentary record permits long-term observations of the Mirror Lake ecosystem including the lake and its watershed. The sediments provide a time series of measurements extending back over the last 14,000 yr. Our methods are analogous to those used for the Experimental Watersheds at Hubbard Brook, with the lake functioning as a giant settling basin for measuring materials coming off the watershed. However, the record of the watershed in limnic sediments is complex because sediments are composed of input materials that have been processed, one way or another, by the lake's biota, and mixed with autochthonous materials produced within the lake. Most dissolved components escape measurement by leaving the lake through the outlet or deep seepage. Sorting out the various components and making an estimate of dissolved and particulate exports from the watershed is a major challenge for paleolimnologists.

The first stages of the lake's history, the late-glacial tundra and spruce periods, were times of rapid change. Both lake and watershed were colonized by a succession of species that probably changed the functioning of the system. Severe climate may have limited the productivity of the biota, and physical processes such as frost action disturbed the watershed. Large quantities of particulates were exported to the lake.

The early years of the Holocene also witnessed changes as closed forest became established and soils developed on the watershed. The productivity of the lake increased until about 6,000 yr B.P. and has remained more or less constant since. The quantities of materials exported from the watershed have remained low and fairly constant, but there appears to have been a qualitative change in exports, which may have been related to soil development. The change appears to have caused the increase in lake productivity 10,000 to 6,000 yr B.P. Additional impacts of changes in the terrestrial system on the lake chemistry and biota are very slight, however, possibly involving a lowering in the pH or decrease in alkalinity in what remained a circumneutral lake, while levels of lake productivity remained more or less the same.

Input-output budgets for chemical elements in the Hubbard Brook Experimental Watersheds are presented in *Chapter II* (see also Likens et al. 1977). Because outputs exceed inputs for most elements (except N and P), the supply of those elements in minerals in the surficial parent soil would eventually be exhausted if these processes continued indefinitely. This outcome is delayed periodically by tree-throw, which mixes the soil profile and renews the supply of unweathered material. Obviously the losses of readily weathered minerals in the last 10,000 yr have been small relative to the supply, and the quantitative change in exports from the watershed has affected the lake only slightly.

Only a few changes in the vegetation on the watershed are correlated with changes in the lake, suggesting a complex relationship between the nature of the vegetation and the kinds and amounts of nutrients lost from the system. The presence of trees is important in reducing particulate loss, probably because of biotic control of erosion. It is less clear why the change from open forests of aspen, fir, birch, pine and oak to closed northern hardwood forests should coincide with increased supplies of N and P in the lake. Increased export of N and P as the forest changed from pioneer species to hardwood forest runs counter to the idea, articulated by Odum (1969), that early successional communities leak nutrients while more "mature" ecosystems conserve nutrients effectively in nearly closed nutrient cycles. The change from pioneering vegetation to mature forest must have involved a change in the size of the available nutrient pool within the terrestrial system. The small proportional losses from the forest nutrient cycles represent substantial inputs to the lake. Changes in these proportions, or in the soil organic pools from which elements like N are leached, may be the cause of changes observed in the lake sediment.

Analogy with modern successional systems is not really appropriate. The successional communities that were studied by Vitousek and Reiners (1975) were disturbed sites from which old-growth forest had been removed. Soil profiles were only partially disrupted, and the chemistry of the soil was the product of many years of weathering. Humus was present with its store of nutrients. Early Holocene forests, in contrast, were an aggrading system. Soils had thin or incomplete humus cover, developed in the arctic or subarctic conditions of the preceding 4,000-yr-long late-glacial period. Frequent disturbance had disrupted incipient soil profiles resulting in incomplete weathering of the underlying mineral soil.

Likens et al. (1977), Gorham et al. (1979), Vitousek (1977) and Bormann and Likens (1979) agree that an aggrading forest system accumulates biomass and soil organic detritus, thereby storing large quantities of nutrients. The high rate of uptake implies high rates of weathering, and/or release from organic detritus or inputs from other sources. It is unclear whether high weathering rates would continue in the steady-state system; if so, the loss rates would be higher than from the aggrading system. Current data from the White Mountains show that old-age forests can have smaller losses than disturbed or aggrading forests (Martin 1979). However, succession occurs on the scale of hundreds of years, whereas we are considering ecosystem development over thousands of years.

Whatever the situation for modern forests, which may or may not be analogous, the Mirror Lake data suggest that productivity increased, implying an *increase* in the nutrient supply to the lake as the vegetation on the watershed changed from aggrading, open, pioneer woodland to steady-state northern hardwood and hemlock forests. During this same period soil profiles developed and humus accumulated. This conclusion stems from our belief that organic matter in the lake sediment is principally autochthonous, and that the apparent decrease in organic deposition since 6,000 yr B.P. at mid-lake sites is an artifact of focusing.

Mirror Lake can be considered a stable or changing system depending upon the criterion used. Productivity was low in the late-glacial, rising in the early Holocene to a plateau 6,000 yr B.P. Thereafter there was little change in total productivity, and if this parameter is the most important, the lake displays a pattern similar to the "trophic equilibrium" described by Deevey 40 yr ago (Deevey 1942). Conversely, the alkalinity/pH of the lake may have decreased slightly during its lifetime. The terrestrial system has a similar history of rising productivity and biomass; once closed forest became established in the early Holocene, the productivity of the terrestrial vegetation probably remained more or less constant. Although a major perturbation occurred about 4,800 yr B.P., when hemlock declined in abundance, the forest-ecosystem quickly recovered and forest was reestablished without major nutrient loss.

Mirror Lake is a relatively pristine system, with little human influence in its 14,000-yr history. Even the postsettlement changes in the last two centuries have been small, involving changes in lake level and at most a slight increase in productivity. Postsettlement erosion from the watershed consisted mainly of material washed in from the stream beds, rather than by overland flow that eroded and transported soil profiles. Many lakes with detailed paleolimnological records have had a history strongly influenced by human disturbance. The impacts are usually so large that they dominate the record and produce much more dramatic changes than occurred at Mirror Lake. Extensive changes caused by human activity make it difficult to assess natural processes.

Mirror Lake and its watershed have had a relatively short lifetime. Glacial ice retreated from this area 14,000 yr B.P. Given the normal length of interglacial intervals (15,000 to 20,000 yr), climatic changes are to be expected within the next few millennia which will have major impact, changing the forest in ways that will doubtless affect the lake. Most paleolimnological studies have been done in similar temperate regions of youthful landscape. Quite a different perspective is gained on older landscapes, where soil processes can proceed to the point where nutrients become limiting to terrestrial vegetation. In these cases terrestrial productivity declines with time in a process termed "retrogressive succession." A chronosequence of dunes in Australia, ranging in age from late Holocene to over 400,000 yr (Walker et al. 1981) provides a spectacular illustration, but the phenomenon also has been described in Denmark (Iversen 1964, 1969). Similar changes in forest type owing to nutrient depletion and increasing acidity have not been observed in North America. The rates of nutrient loss from the watershed of Mirror Lake are so small that little effect on the terrestrial system would be expected during a typical interglacial of 10,000 to 20,000 yr. When glacial ice eventually overrides the site, new unweathered material will be exposed and the cycle will begin again.

Summary

During late-glacial time (14,000 to 10,000 yr B.P.) Mirror Lake formed as a kettle-hole lake in a tundra landscape dominated by cold climate. It was a relatively deep (23 m), slightly alkaline, probably clear body of water in which productivity was strongly limited. Cold climate and immature and unstable soils limited plant growth on the watershed. In the lake, productivity was probably limited by a shortage of fixed nitrogen, perhaps compounded by unavailability of other nutrients. Benthic macroalgae grew in Mirror Lake, but its early ecology is otherwise unknown prior to 11,000 yr B.P., when the earliest analyses of diatoms and Cladocera reveal a community that already may have been well developed. Particulate inorganic matter (detrital minerals) were exported from the watershed by water and possibly wind. At 13,000 yr B.P. the export rate was 30 times higher than from modern forested watersheds. After spruce trees became established 11,500 yr B.P., export of particulate inorganic matter decreased sharply. Stream channels were stabilized through the cessation of frost-heaving and solifluction, and through the incorporation of spruce logs into large debris dams.

By 10,000 yr B.P. the climate had warmed to modern temperatures. Numerous light-demanding trees and shrubs formed an open forest on the watershed, and a continuous soil humus layer presumably developed. The profundal sediment had shifted from an inorganic gray silt to an organic gyttja as particulate inorganic inputs decreased and autochthonous productivity increased. The lake was still unproductive, with a predominantly planktonic diatom flora indicating circumneutral conditions. Compared with later conditions, strong focusing of the gyttja to the deepest part of the basin suggests periods of lower water level, at least seasonally, for the period 10,000 to 8,000 yr B.P. Climate was warm and/or dry—the mean annual temperature may have been 2°C higher than today. Exports of particulate inorganic matter fell to rates typical of modern forests at HBEF. Diatomaceous silica represents the principal mechanism by which some of the dissolved export from the watershed was retained as sediment.

Productivity of the lake continued to increase until 6,000 yr B.P., judging from the deposition rates for organic matter and chlorophyll derivatives. The subsequent decline in deposition rates at mid-lake sites may be explained as a progressive relaxation of focusing as sediment was deposited in a more uniform layer over the surface of the gyttja. Productivity seems to have remained more or less constant from 6,000 to 140 yr B.P. Nevertheless, there is evidence from diatoms for a slight decrease in alkalinity and/or pH over the course of the Holocene. Diatoms decreased in relative importance within the phytoplankton community, but remained a flora of circumneutral pH.

The progressive decrease in alkalinity/pH presumably reflects a parallel decrease in the surface area of readily weatherable silicate minerals within the soil profile. Significantly, the export of production-limiting nutrients like N and P to the lake does not seem to have decreased. Export of N and P apparently increased 8,000 to 6,000 yr B.P. during the gradual replacement of pine-birch forest by northern hardwood-hemlock forest. Climatic cooling began about 5,000 yr B.P. and apparently crossed a threshold for spruce 2,000 yr B.P., at which time spruce began to increase on the watershed.

In the lake there was a further decrease in alkalinity/pH 2,000 yr B.P., along with increased abundance of chrysophytes and other nondiatomaceous phytoplankton. This change could be related to changes in the organic humus layer of the soil profile, but we have no direct record of the soil humus and do not know when the present acid soil humus and podsolic soils began to develop.

Settlement of the surrounding watershed in the nineteenth century (nominally dated to 140 yr B.P.) led to increased erosion of particulate inorganic matter from disturbed stream channels and banks. Any small increase in productivity of the lake was obscured by a relaxation of focusing, and increase in the area of lake bottom receiving sediment, attributable to dams at the outlet that raised the water level 1 to 2 m. Human impact has been slight: soils were largely left intact, and the forest has regrown following cutting. The aquatic flora of diatoms and macrophytes remains similar to that prevailing through most of the Holocene. An unusual species-richness of diatoms in the surface sediment may reflect the onset of changes in the watershed, possibly associated with chemical changes of the inlet streams. The second-growth forest is floristically similar to the pre-settlement forest, but spruce, beech and hemlock are much less abundant, while birch is much more abundant.

Although there have been fluctuations in the relative abundances of fossil diatoms and Cladocera during the Holocene, the lake seems to have changed remarkably little during the past

6,000 yr. Progressive weathering and cumulative leaching of the forest soil, probably associated with the buildup of large detrital nutrient pools in acidic mor humus, is reflected in the slight decrease in alkalinity/pH recorded in the diatoms. There is no other evidence for depletion of soils in nutrients to the extent that production of the watershed or lake decreased. At this site, such a "retrogression" apparently requires more than 10,000 yr to take effect, perhaps more than the expected 15,000-yr duration of the current interglacial period.

Chapter VIII

The Aquatic Ecosystem and Air-Land-Water Interactions

GENE E. LIKENS

Lakes are open ecosystems, and as such their metabolism and biogeochemistry strongly reflect the inputs of energy, water and chemicals from their surroundings. Outputs from lakes further connect these aquatic ecosystems with other ecosystems in the biosphere (*Figure I.A.1*–3). Climate (particularly solar radiation, temperature and amount of precipitation), geologic substrate, terrestrial vegetation, land use and air pollution are dominant factors that regulate the quantity and quality of the major inputs, and hence the metabolic and biogeochemical status of a lake.

Most lakes in northern North America are relatively young, dating from the last glacial advance of the Pleistocene era some 8,000 to 15,000 years ago. Following the formation of a basin, lakes immediately begin to accumulate sediments. This process has relentlessly decreased the volume of lakes in the intervening years. Thus, with time, all lakes become filled with sediments and, therefore, ultimately age to extinction. Rates of filling are highly variable and depend on basin morphometry and volume, external inputs, and rates of sedimentation and decomposition. Eutrophication, particularly cultural eutrophication, has been confused with this natural aging function, however it is a separate process. Both processes are induced and regulated by external inputs, particularly those associated with disturbance of the watershed or airshed for the lake. However, trends in eutrophication are not linear and are reversible. Patterns of eutrophication in many (if not most) lakes have been highly variable. Paleolimnological records (e.g. Edmondson 1972; Hutchinson 1969; Hutchinson et al. 1970; Horie 1969; Stockner 1972; Likens and Davis 1975; *Chapter VII*) indicate that some lakes have become more productive with time, others less productive, and in many lakes the pattern has fluctuated with time. Eutrophication can accelerate filling, and likewise changes in basin morphometry can affect the overall biogeochemistry and metabolic status of a lake. Thus, ecosystem volume becomes an important consideration in analyses of lake systems (Likens and Bormann 1972), but the process of filling and eutrophication should be clearly differentiated (Likens and Bormann 1979).

We and others have proposed that the drainage area or landscape surrounding a lake-ecosystem is the functional ecosystem unit. To this should be added the adjacent atmosphere, or airshed (cf. Likens and Bormann 1974a, 1979). A lake cannot be separated from its watershed and airshed in attempts to understand its structure, metabolism, biogeochemistry and particularly its management. Accelerated eutrophication is usually caused by human activities within the watershed or airshed that overwhelm natural regulatory processes (see *Chapter IX*).

There are a few critical biogeochemical linkages (e.g. precipitation and streamflow and seepage inputs) between a lake and its watershed and its airshed (see *Chapters I* and *II*). Thus, quantitative studies of the terrestrial watershed and its outputs must be done in conjunction with studies of lake metabolism and nutrient cycling to allow researchers to understand or to predict trends in eutrophication. Too few models of so-called lake dynamics include an understanding of the vital inputs from the terrestrial watershed and from the atmosphere. A few examples are given here:

A. Direct Atmospheric Input to Lakes

The quantitative importance of meteorologic inputs of chemicals to aquatic and terrestrial ecosystems has only recently been documented (Gorham 1958, 1961; Likens 1975a; Likens et al. 1977; Matheson and Elder 1976; Braekke 1976; Dochinger and Seliga 1976; Hutchinson and Havas 1980). These inputs are of particular ecologic importance for nitrogen, phosphorus, organic carbon and strong acids. For Mirror Lake, about 13% of the phosphorus and 56% of the nitrogen inputs come from direct precipitation plus airborne terrestrial litter (ignoring input from sewage; *Table IV.E*–16). Of major ecologic significance is the recent finding that concentration and/or deposition of NO_3 in precipitation have increased up to tenfold in the eastern U.S. and northwestern Europe during the past three decades or so (Likens 1976; Söderlund 1977; Brimblecombe and Stedman 1982; Likens et al. 1977). Increases have diminished since 1972 to 1974 at Hubbard Brook (*Figures II.B*–9 and 10). The increase in concentration is thought to be due to increased generation and widespread dispersal in the atmosphere of gases produced in the combustion of fossil fuels (Galloway and Likens 1981). Because nitrogen is a limiting element in Mirror Lake, along with phosphorus, the effects of these increased meteorologic inputs of N may be to increase productivity, especially if phosphorus inputs to the lake are increased correspondingly (e.g. through sewage inputs).

At the same time, wet and dry deposition may contribute ecologically significant quantities of pesticides, heavy metals and strong acids to the surface of lakes (Likens 1976; Braekke 1976; Dochinger and Seliga 1976; *Ambio* 1976). The toxic effects of these inputs on the metabolism of a lake are poorly known, but the potential and real impacts of strong acids in precipitation, and acidifying substances in dry deposition—"acid rain"—on lakes and streams in the eastern U.S., southeastern Canada and northwestern Europe are currently of serious concern (Likens et al. 1979). Impairment or extinction of fish populations in thousands of lakes in these areas have been ascribed to direct and indirect effects of acid deposition (see Braekke 1976; Schofield 1976; Drabløs and Tollan 1980).

In addition, there have been reports of deleterious effects, in recently acidified lakes, on various other aquatic organisms (e.g. salamanders; Pough 1976), including zooplankton (Almer et al. 1974; Hendrey and Wright 1975; Raddum et al. 1980; Sprules 1975; Roff and Kwiatkowski 1977; Yan and Strus 1980; Crisman et al. 1980). We examined ten small headwater lakes in the White Mountains of New Hampshire (including Mirror Lake) and ten in the Adirondack Mountains of New York for possible effects of acidification on zooplankton (Confer et al. 1983). We found that diversity and biomass of zooplankton were reduced in the more acid lakes, with a loss of 2.4 species and about 23 mg dry wt/m^2 per unit decrease in pH. Various Cladocera (with the exception of *Holopedium* and *Polyphemus*), and the copepods, *Epischura lacustris*, *Mesocyclops edax* and *Cyclops scutifer*, were abundant at higher pH values but rare or absent in lakes with a pH below 5. It appears that the effects of acidification are widespread at various trophic levels in lakes, and that direct inputs of hydrogen ion by way of precipitation can be the dominant source of acid in some lakes.

Some 94% of the total hydrogen-ion input to Mirror Lake is added in precipitation that falls directly on the surface of the lake (*Table IV.E*–16). The proportion of the total input of hydrogen ion that lakes may receive from direct precipitation depends on the ratio of lake surface area to watershed area and on the chemistry of the runoff. The acidity of runoff waters is determined by the difference between all sources and sinks for hydrogen ion within the terrestrial ecosystem. At Hubbard Brook, 52% of the total hydrogen-ion sources for the terrestrial ecosystem can be attributed to atmospheric inputs (Driscoll and Likens 1982).

The increased nitrate concentration in precipitation referred to above also plays a role in the acid precipitation problem. Not only is NO_3^- a nutrient in aquatic and terrestrial ecosystems, but it is also potentially an acid-forming anion.

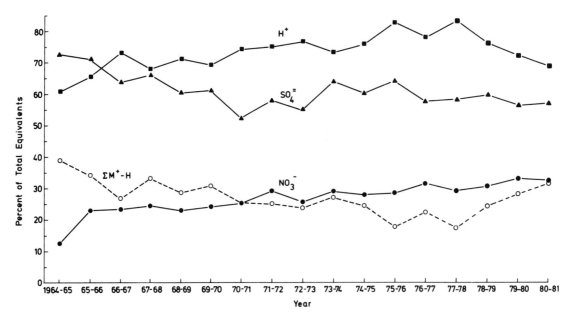

Figure VIII–1. Percent of ionic composition of precipitation at Hubbard Brook during 1964 to 1981. ΣM^+ is the sum of all cations in precipitation. (Modified from Likens et al. 1977. Used with the permission of the publisher and the authors.)

Based on stoichiometry, the nitrate contribution to the total measurable acidity in precipitation in the Hubbard Brook Valley has increased from about 15% in 1964–65 to more than 30% in 1980–81 (*Figure VIII–1*). A concurrent drop in the contribution of sulfuric acid also has occurred during this period. The relative annual change in the proportions of these two acids in precipitation has decreased since about 1970–71. The contribution of hydrogen ion to total cations has increased from 1964 to 1981, but this change is largely due to the decrease in total cation equivalents during the period, of which hydrogen ion is the major constituent (*Figure VIII–1*).

B. Fluvial Input to Lakes

Drainage streams are the obvious link between terrestrial and aquatic ecosystems in humid regions, although direct runoff (overload flow and seepage) is also important (*Chapter IV.D*). Streams, even more than lakes, reflect the characteristics of their drainage basins. As the landscape changes (naturally or through disturbance by humans), conditions in drainage streams reflect these changes.

Terrestrial organic matter, such as leaves and twigs, is the dominant input, other than water, to small headwater streams in the Hubbard Brook Valley. Much of this organic matter entering streams is mechanically or biologically degraded to smaller particles, dissolved substances or gases during its sojourn in the channel. For example, in Bear Brook, a tributary to Hubbard Brook (*Figure II.A–1*), 69% of coarse (>1 mm) particulate organic C, 78% of coarse particulate organic N and 73% of coarse particulate organic P inputs to the stream were exported in the fine particulate (0.45 μm to 1 mm), dissolved (<0.45 μm) or gaseous fraction. Organic C was exported primarily in the gaseous or dissolved fraction, P in the fine particulate fraction and N in the dissolved fraction (Meyer et al. 1981). Material balance studies of Bear Brook showed that 36% of the total annual organic carbon inputs are respired to CO_2 and 64% are transported downstream (Fisher and Likens 1973).

Stream-ecosystems in the Hubbard Brook Valley also store organic matter (Bilby and Likens 1980) and nutrients like phosphorus (e.g. Meyer and Likens 1979). However, the storage depends on the loading rate, and storage capacity in shallow streams without major impoundments is limited and temporary. Accumulations of terrestrial organic debris in stream channels (organic debris dams) form small impoundments, which reduce the kinetic energy of the moving water and enhance the storage capacity of the stream. Organic debris dams are common in headwater streams of the Hubbard Brook Valley (Bilby and Likens 1980; *Table II*–9). Nutrients may accumulate or be transformed in the sediments of the impoundments, as well as in the organic dams themselves.

Transport of N, P and organic C in streams is dependent on concentration and discharge. During a year at Hubbard Brook, discharge can vary over four or five orders of magnitude, while concentration of dissolved substances remains relatively constant except for marked seasonal changes in a few ions, e.g. nitrate concentration (see *Chapter II*). Thus, transport of dissolved N, P and C is directly proportional to stream discharge (and season, for N). In contrast, concentration of particulate matter increases exponentially with increasing discharge (see *Chapter II*); thus, transport of N, P and C in particulate matter is exponentially related to stream discharge.

Based on mass balances for Bear Brook (Meyer et al. 1981), phosphorus losses from streams at Hubbard Brook occur essentially as fluvial export, and 82% are in the particulate fraction. Hence, P export from streams is most strongly affected by the hydrology (Meyer and Likens 1979). In contrast, fluvial export of organic C is 64% of total loss from the stream, and only 32% of this fluvial export is as particulate matter. Therefore, organic C export is much less susceptible to hydrologic fluctuations than is P. Nitrogen occupies an intermediate position, since export is essentially all fluvial but exports of particulate N represent only 4% of total loss. Thus, solely on the basis of discharge relationships, export/input ratios for streams probably would increase in the order: organic $C < N < P$, during an extremely wet year or for a period after a disturbance (e.g. clear cutting) that increased discharge (Meyer et al. 1981).

The quality and quantity of outputs from undisturbed stream-ecosystems depends on density and activity of aquatic organisms, the length of the stream channel and/or retention time within the channel, type and condition of banks, successional state and type of vegetation cover, shading, presence of impoundments, type of substrate, slope, season, hydrologic events, etc. (cf. Likens and Bormann 1974a, 1979; Likens et al. 1977; Manny and Wetzel 1973).

Stream-ecosystems receive, alter and respond to the inputs from terrestrial systems. The processing and storage of these inputs within the stream-ecosystem then determines overall what the quality and quantity of stream-water outputs from the entire drainage system will be. Therefore, terrestrial effluents that drain directly into a lake could differ significantly, both qualitatively and quantitatively, from those that are transported to the lake by streams. The effects of these factors on lake metabolism and biogeochemistry are poorly known.

C. Effects of Disturbance on the Land-Water Linkage

Disturbances in the airshed or watershed may increase or alter inputs to the aquatic ecosystem or may disrupt the capacity of the terrestrial ecosystem to regulate outputs of water, dissolved substances and particulate matter (see *Chapters II; III.B.2* and *IX*). The effects and management implications of local disturbances such as clear cutting, agriculture, sewage effluents and road salt are discussed in *Chapter IX*.

Various land-use patterns in terrestrial ecosystems result in significantly different concentrations of N and P in drainage waters (*Figure VIII*–2). As a result, the potential impact on the receiving lake-ecosystems will reflect these changes in complex ways, depending on the quantity and quality of inputs, flushing time of the lake, timing of the inputs, etc. About 20 years ago, Vollenweider (1968) developed a simple, empirical model to relate such inputs (loading rates) to the trophic status of lakes. This model served as a landmark in providing some order in the complexity of different inputs to a diversity of natural aquatic ecosystems.

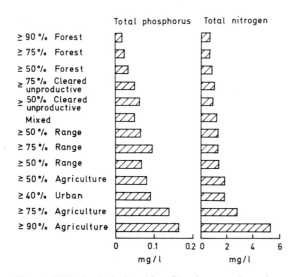

Figure VIII–2. Relationship of land use and nutrient concentrations in streams. (From Omernik 1977.)

The numerous refinements that have been made to Vollenweider's original model (e.g. Dillon and Rigler 1975; Vollenweider 1976; Oglesby and Schaffner 1978) allow a much clearer understanding of these linkages between outputs from terrestrial ecosystems and inputs to aquatic ecosystems.

We have adopted an experimental approach to clarify certain aspects of these land-water linkages in the Hubbard Brook Valley. To test the role of organic debris dams in regulating outputs from forested and disturbed watersheds, we experimentally removed all organic debris dams from a 175-m stretch of a second-order stream in W5 (Bilby and Likens 1980). In the absence of these dams, output of dissolved organic carbon (<0.50 μm) increased by 18%, fine particulate organic carbon (0.50 μm to 1 mm) export increased by 632% and coarse particulate organic carbon (>1 mm) export increased by 138%, for a total export of 1,940 kg C/yr without dams vs. 791 kg C/yr with organic debris dams (Bilby and Likens 1980).

Experimental enrichment of small streams with dissolved organic carbon (McDowell 1982), inorganic phosphorus (Meyer 1978) and inorganic and organic nitrogen (Sloane 1979) has demonstrated the important role of stream-ecosystems in modifying export from a forested watershed. Empirical and simulation models have been developed to further describe the role of terrestrial and stream-ecosystems in reg-ulating outputs of water (Federer and Lash 1978), phosphorus (Meyer and Likens 1979) and dissolved organic matter (McDowell 1982).

D. Atmospheric Inputs to Watersheds Affecting Inputs to Lakes

One of the major questions relative to acid precipitation is: What is its effect on fluvial outputs from terrestrial watersheds? Continuing in the experimental mode in our analysis of natural ecosystems, we have acidified sections of a small stream (Norris Brook) in the Hubbard Brook Valley (Hall et al. 1980). The stream was acidified to a pH of 4 or so in various experiments to simulate the effects of acidification from snowmelt during the spring or to approximate the hydrogen-ion concentration in ambient precipitation. Numerous and complex biogeochemical changes occurred as a result of this manipulation of the stream.

In its undisturbed state, the experimental stretch of Norris Brook had a pH consistently greater than 5.4, a relatively high population and diversity of macroinvertebrates, and contained brook trout. Following acidification of the stream with H_2SO_4 to a pH of 4, stream-water concentrations of Al, Ca, Mg, K and probably Mn, Fe and Cd were increased, downstream drift of immature insects was increased, emergence of adult mayflies, some stoneflies and some true flies was decreased, periphyton biomass was increased, hypomycete fungal densities were decreased, a basiomycete fungus increased in abundance and most trout moved downstream to areas of higher pH (Hall et al. 1980). The increased concentrations of ionic Al is particularly significant ecologically because of its toxicity to fish (Schofield and Trojnar 1980; Driscoll et al. 1980) and to invertebrates (M. Havas, personal communication). Obviously, these changes could have important effects on the inputs to lakes.

Driscoll and Likens (1982) have attempted to predict how changes in atmospheric loading of hydrogen ion could affect the annual weighted mean pH of headwater streams at Hubbard Brook (*Figure VIII*–3). Such changes, if they

D. Atmospheric Inputs to Watersheds Affecting Inputs to Lakes

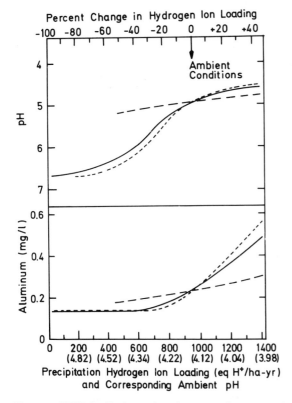

Figure VIII–3. Estimated changes in annual weighted mean stream-water pH and Al for different loadings of hydrogen ion in precipitation (131 cm of precipitation, annual average). Projections are shown for loading of nitric acid (–––), sulfuric acid (-----) and a combination of nitric (0.28) and sulfuric (0.72) acids (——) corresponding to ambient precipitation at Hubbard Brook. (Modified from Driscoll and Likens 1982.)

occurred, undoubtedly would bring about some of the increased outputs observed in the Norris Brook acidification experiment, plus additional watershed effects, for example, increased leaching of ionic Al.

A final point—as has been stressed, stream-ecosystems can modify outputs. Thus, if disturbance occurs at appreciable distance from the receiving water body, the effects (e.g. outputs) may be much different, both qualitatively and quantitatively, than if the disturbance occurs in the watershed nearby the lake. Significant longitudinal variation in the chemistry of stream water has been observed in the Hubbard Brook Valley (e.g. Johnson et al. 1981). This result can even occur when air pollutants are applied uniformly to various-sized drainage areas, because water retention times can be different, resulting in stream-water outputs that are correspondingly different along the course of the stream (Driscoll and Likens 1982).

Summary

Lake-ecosystems are connected to other ecosystems in the biosphere by a variety of inputs and outputs.

Lake basins fill with time. Eutrophication may contribute to filling, but is a separate, reversible process, resulting from large or unregulated inputs. A lake cannot be separated from its drainage basin (watershed) or airshed with regard to its functional biogeochemistry, metabolism or management. A lake reflects the developmental state and condition (disturbance) of its watershed, drainage stream and airshed.

Direct meteorologic inputs can add significant amounts of nutrients (e.g. N and P) or toxic substances (e.g. acids) to lakes. Increased nitrate concentrations in precipitation over the past few decades may have contributed to greater productivity of lakes like Mirror Lake. Conversely, some 94% of the total hydrogen-ion input to Mirror Lake is added in direct precipitation. As such, acid precipitation is directly affecting Mirror Lake, but no trends in acidity of the lake water have been discernible over the period 1963 to 1981.

Stream discharge is one of the major fluvial inputs to lakes. Outputs of water, dissolved substances and particulate matter from terrestrial ecosystems can be altered or regulated by stream-ecosystems, but closely reflect the conditions of the watershed. Disturbance within the terrestrial ecosystem or stream-ecosystem can increase the inputs to lakes. Increased levels of nutrients and toxic substances in wet and dry deposition can indirectly affect these fluvial inputs through direct effects on the terrestrial ecosystem. Thus, lake-ecosystems can be affected by human activities at great distances (e.g. air pollution) or within the watershed.

Experimental manipulations of streams have been done at Hubbard Brook to elucidate some of these relationships. These studies have led to various models for the role of stream-ecosystems in regulating fluvial outputs.

Chapter IX

Air and Watershed Management and the Aquatic Ecosystem

F. Herbert Bormann and Gene E. Likens

The structure, function and development of a lake-ecosystem is not only a reflection of internal processes, but is strongly influenced by biotic and abiotic inputs into the system (Bormann and Likens 1967; Likens and Bormann 1972). For a lake like Mirror Lake, inputs might be thought of as direct additions, such as water and wet or dry deposition added directly from the atmosphere, or sewage or fertilizers added directly to the lake, or as outputs from the lake's watershed added by stream and groundwater flow into the lake. The relative significance of direct atmospheric inputs vs. watershed outputs is heavily influenced by the lake surface/watershed surface ratio, with a large ratio generally indicating greater importance of atmospheric inputs. This important relation is frequently overlooked in efforts to predict or model inputs to lakes. Even today, many workers assume that direct atmospheric inputs to lakes are relatively unimportant to metabolic and biogeochemical functions. Such is not the case, as they frequently dominate these processes (see Likens 1975a; *Chapter IV.E*).

All of these inputs can be strongly influenced by humans. For example, in developing a federal policy for air pollution where Mirror Lake is but a tiny consideration in a vast regional decision (Bormann 1982), atmospheric inputs to the lake would be changed. Also local management practice, direct or by default, done on the Mirror Lake watershed can have direct, immediate and strong consequences on the lake, as well as regionally through deterioration of overall recreational quality (i.e. diminished resource). Here we briefly consider potential effects of both air pollution and watershed management on inputs entering the lake-ecosystem.

A. Developmental State of the Forested Watershed-Ecosystem

Regulation of the outputs (water, nutrients and particulate matter) from the northern hardwood ecosystem is highly influenced by the developmental status of the forest. This status can be altered by natural developmental processes or by managerial decisions which, in turn, alter outputs and hence the watershed's influence on lake dynamics. Consequently, watershed managers, to understand the impact of watershed management on lake dynamics, must have not only an understanding of the biogeochemistry of the terrestrial ecosystem currently occupying the watershed but also of its developmental relationships.

The developmental behavior of the northern hardwood ecosystem surrounding Mirror Lake and its relationship to various biogeochemical output parameters has been explored in great detail in the Hubbard Brook Valley (Likens et al. 1977; Bormann and Likens 1979).

We (Bormann and Likens 1979) have developed a biomass accumulation model for the northern hardwood ecosystem which, after an initial disturbance like clear cutting, projects a continuum of four developmental phases: Reorganization, Aggradation, Transition and the "Shifting-Mosaic Steady State." Hydrologic and biogeochemical output is strongly influenced by the developmental phase of the north-

A. Developmental State of the Forested Watershed-Ecosystem

ern hardwood ecosystem (*Figure IX*–1). Clear cutting or a windstorm that blew down all trees in a forest would trigger the Reorganization Phase. This phase, which is characterized by a drop in total biomass (living and dead) that continues for a decade or two, starts with a dramatic loss of regulation over hydrology and biogeochemistry (*Figure IX*–1). We have examined this response through experimental studies and studies of commercial clear cuts (Bormann and Likens 1979) and have shown that shortly after disturbance, nutrient concentrations in drainage water can increase by an order of magnitude (*Figure IX*–2) and that liq-

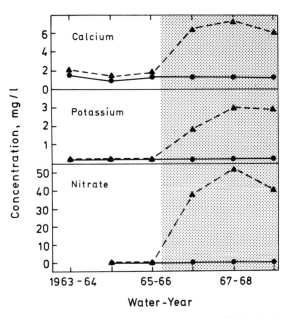

Figure IX–2. Annual weighted average of dissolved ions in stream water: (●——●) forested Watershed 6; (▲---▲) Watershed 2. Shaded area represents period when Watershed 2 was devegetated. (From Bormann and Likens 1979. Used with the permission of the publisher and the authors.)

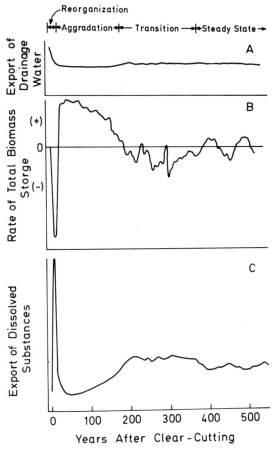

Figure IX–1. A schematic view of changes during 500 years after clear cutting of a 60-yr-old northern hardwood forest: **A.** export of drainage water; **B.** rate of total biomass storage, with the ecosystem (+) accumulating biomass and (−) losing biomass; **C.** the export of dissolved substances in drainage water. (From Bormann and Likens 1979. Used with the permission of the publisher and the authors.)

uid water output can increase about 25%. This combination of increased yield of water at elevated nutrient concentration produces a very large output of chemicals. Regulation over outputs is fairly rapidly regained and in a decade or two the Aggradation Phase, characterized by increasing total (living and dead) biomass begins. We have discussed the output relations of this phase in detail in *Chapter II*. Of all phases, this one exercises maximum regulation over ecosystem outputs of water, dissolved substances and particulate matter. Outputs are smaller and more regulated during this phase than during any other.

Most of the Mirror Lake watershed is currently in the Aggradation Phase and, as a consequence, inputs to the lake from the forested parts of the watershed may be judged to be near the minimum achievable from the northern hardwood ecosystem with the least hydrologic delivery to the lake per unit of precipitation falling on the watershed, the lowest total dissolved substance concentrations and the least additions of organic and inorganic particulate matter per storm peak. Inputs from the watershed are nutrient-poor, averaging 25–35 mg/liter total dissolved substances (*Chapter IV.B*), thus, contributing to Mirror Lake's oligotrophy.

A watershed manager, by doing nothing, may choose to let ecosystem development proceed, in which case outputs (inputs to the lake) will change. Given enough time, two or three centuries after disturbance, and no further large-scale disturbance or major change in climate, the aggrading ecosystem will go through a Transition Phase and develop into the "Shifting-Mosaic Steady State." During the Transition Phase, we propose a substantial drop in living biomass as the more or less even-aged aggrading forest becomes an all-aged forest characterized by a mosaic of differently-aged patches (*Figure IX*–1).

Our developmental model (Bormann and Likens 1979) proposes a decrease in living biomass of about 100 mt/ha during the Transition Phase. Such a drop in biomass should be accompanied by increased export of nutrients in stream water as the biomass decays. In addition, net storage no longer takes place in the biomass of the Transition Phase and, hence, nutrients normally stored in the biomass of the Aggradation Phase now might be expected to pass through the system and be exported in stream water. Taking both of these sources of nutrients into account, our calculations suggest that, for example, average annual stream-water concentrations for N and Ca during the Transition Phase might increase 1 or 2 mg/liter. In the Steady-State Phase, where system biomass is thought to steady off at a lower level than that achieved toward the end of the Aggradation Phase and where no net biomass accumulation occurs (*Figure IX*–1), average annual increases in weighted average stream-water concentrations for both N and Ca are predicted to be about 1 mg/liter over those of the Aggradation Phase. In both cases, no significant change in particulate matter export is expected, hence P output, which occurs primarily in particulate form, would be expected to remain very small and unchanged.

In summary, the Hubbard Brook Biomass Accumulation Model predicts that forest-ecosystem development beyond the Aggradation Phase would lead to increased nutrient concentrations in stream water draining into Mirror Lake. These increases, however, are thought to be small, generally less than 1 mg/liter for most nutrients (phosphorus probably would be an order of magnitude smaller). Thus, it seems unlikely that development beyond the Aggradation Phase will alter significantly the generally oligotrophic inputs to the lake from the forest.

B. Forest Harvesting

The effects of forest harvesting on lake dynamics are difficult to predict because they involve at least three effects that may complement each other. One is the effect of cutting the trees, which removes leaf surfaces and thereby changes both active (e.g. photosynthesis and transpiration) and passive (e.g. impaction and reflection) regulating processes (Bormann and Likens 1979). The second effect is the disturbance of soil surfaces resulting from moving logs from the site of cutting to a landing or loading area. Third is the effect of building temporary logging roads. At Hubbard Brook, we have studied the first of these effects in great detail (Bormann and Likens 1979; Pierce et al. 1970).

Disruption of Biogeochemical Regulation by Clear Cutting

The experimental deforestation of W2 at Hubbard Brook for a 3-yr period indicated that clear cutting can temporarily disrupt the high degree of regulation over hydrology and biogeochemistry exercised by the aggrading forest-ecosystem (i.e. it triggers the Reorganization Phase, which is at first characterized by large increases in dissolved substances and hydrologic output). During the 3-yr period, streamflow was increased by more than 30% and snowmelt was advanced by several days in the spring, average stream-water concentrations of nitrate, potassium, calcium and aluminum rose 40 times, 11 times, 5 times and 5 times, respectively (Likens et al. 1970; Bormann and Likens 1979). Annual export of phosphorus was increased tenfold, mostly from increases in particulate phosphorus (Hobbie and Likens 1973); hydrogen-ion concentration increased fivefold following deforestation. Stream-water temperature was elevated by several degrees Celsius over streams

B. Forest Harvesting

Table IX–1. Ratio of Stream-water Concentrations of Commercially Clear-cut Forests to that of Reference Streams Draining Nearby Uncut Forests in the White Mountains of New Hampshire[a]

Years after Cutting	No. of Sites Sampled	Clear Cut/Uncut Ratio		
		Nitrate	Calcium	Potassium
1	4	14.0 ± 5.0	2.4 ± 0.1	3.0 ± 0.7
2	6	9.7 ± 3.8	2.1 ± 0.4	2.6 ± 0.3
3	6	3.8 ± 1.6	1.6 ± 0.3	2.1 ± 0.3
4	2	1.1 ± 1.0	1.4 ± 0.4	2.5 ± 0.1

[a] Values expressed as means ± S.E. Ratios are based on unweighted averages (Pierce et al. 1972; R. S. Pierce, C. W. Martin and C. A. Federer, unpublished data; Bormann and Likens 1979).

draining forested watersheds, and heavy algal blooms of *Ulothrix zonata* appeared, which had never before been seen in Hubbard Brook streams (Likens et al. 1970).

In forested watersheds, such as those at Hubbard Brook, dissolved Al occurs in high concentrations in the headwaters of first-order streams, but it is complexed and nontoxic by the time it reaches lower stream segments (Johnson et al. 1981). Clear cutting increases hydrogen ions in the soil and in the stream (Likens et al. 1970; Bormann and Likens 1979) and mobilizes more Al^{3+} and, thus, increases the possibility that ionic Al, which is toxic to some freshwater invertebrates and fish (Baker and Schofield 1980; Havas and Hutchinson 1982) will reach interconnected lakes.

Our study of commercial clear cuts in northern New Hampshire (Martin et al., in preparation) indicated a response similar to our deforested watershed, but stream-water concentrations were less extreme and recovery to precutting concentrations occurred in about four to five years (*Table IX*–1). Thus, nutrient enrichment from commercial clear cuts that are done in a careful way is not very long lasting. This response is because clear cutting not only results in large increases in the availability of light, water and nutrients at the level of the forest floor but these latter changes stimulate an array of higher plant growth-responses that result in rapid revegetation (Bormann and Likens 1979).

Clear cutting, or similar disturbance, of the northern hardwood ecosystem, in effect, results in a spike of water and nutrients added to an interconnected lake (*Figure IX*–1). However, the effects of such a spike on lake dynamics depend on many factors including the proportion of watershed cut. Watershed managers can diminish this effect by restricting cutting to a small proportion of the watershed, thus, promoting dilution of the nutrient output from the cut area by outputs from uncut portions.

The effects of clear cutting on stream-water chemistry and lake inputs can be moderated still more by simple procedures. One is by clear cutting in strips, where the whole watershed is cut in several cuttings during several years. The theory is that strips that have been cut and are regrowing will remove dissolved nutrients from subsurface water flow that originate in newly-cut strips. At Hubbard Brook, one watershed (W4) was cut in three cuttings over a 4-yr period (e.g. Hornbeck et al. 1975; Likens and Bormann 1974a). During an 11-yr period following the first cut, the export of water and losses of nitrogen, calcium and potassium to streamflow were somewhat greater than a reference uncut watershed, but 25 to 50% less than from a commercial, block clear cut (Hornbeck et al. 1982).

A very simple procedure in clear cutting is to leave a buffer strip of about 10 m of uncut forest beside drainage streams and along lake shores. Hornbeck and Kropelin (1982) noted only modest increases in dissolved substances in a stream draining a whole-tree harvested area with a buffer strip. Tube lysimeters in the soil of the cut area indicated very large increases in nitrate concentrations in soil water of the B-horizon, but these large increases were not measured in stream water. One possible explanation is that roots of the uncut vegetation in the buffer strip removed substantial amounts of nutrients from subsurface drainage water moving toward the stream.

Skid Trails and Temporary Roads

As we discussed in *Chapter II*, the intact aggrading ecosystem exercises considerable regulation over erodibility. Much of this regulation is centered in mechanisms operating in the forest soil. Road-building or log skidding can seriously disrupt these regulating mechanisms, resulting in increased surface runoff and erosion. These activities have the potential to change the pattern of water delivery to interconnected aquatic systems by increasing surface flow and storm peaks and decreasing subsurface flow, and by increasing the delivery of organic and inorganic sediments. For elements such as phosphorus and lead that are tightly held on and in particulate matter of the ecosystem and transported as particulate matter (Meyer and Likens 1979; Meyer et al. 1981; Siccama and Smith, in preparation), logging and road-building can greatly accelerate the movement of these elements relative to elements such as N, S, Mg, Na and Ca, which are transported primarily in the dissolved phase (see also *Chapter VIII*).

The effects of logging and temporary logging roads on interconnected lakes within the logged ecosystem depend on the meteorology, topography and erodibility of the watershed, but can be greatly influenced by the skill, technology and execution of the logging managers. In hilly country, particularly with erodible soil, poorly designed and constructed roads can be a major and long-lasting source of sediment to streams and lakes. This consideration is extremely important as tens of thousands of permanent and temporary roads are constructed annually on public and private forested lands in the United States (Bormann and Likens 1979). In northern hardwood forests of the White Mountains about 2 to 10% of the harvested area is utilized in temporary roads (J. Hornbeck, personal communication). The effects of road-building on lake-ecosystems depends not only on slope, engineering design and execution but also on nearness of the cut area to the lake shore and the ratio of new construction and roaded area to area of the whole watershed draining into the lake. Low ratios result in considerable dilution of erosion effects. In northern New England, sufficient design capability and road-building skill exist such that erosion need not be a serious problem during most logging operations, and rapid recovery of erosion control after logging can be expected (Patric 1976).

C. Highway Construction and Maintenance

The construction and maintenance of permanent roads presents somewhat different problems from temporary logging roads. An example was seen at Hubbard Brook when Interstate 93 was constructed through the Mirror Lake watershed in 1969 (see *Chapter III.B.3*). As a result of adversary debate, a special effort was made to divert drainage from the highway away from the Mirror Lake watershed, but as a result the watershed area of Mirror Lake was reduced by 17% (see *Chapter III.A* and *IV.E*). This has resulted in a reduction of water and total dissolved substances entering the lake. More subtly it has caused an increase in the proportion of inputs resulting from wet and dry fallout entering the lake directly. In effect, the loss of watershed area not only meant a loss in water entering the lake, but also a relatively higher percentage of unbuffered fallout as opposed to buffered output from the surrounding watershed. We elaborate on this point in the next section.

D. Air and Water Pollution

The capacity of the aggrading ecosystem to exercise considerable regulation over biogeochemistry has important implications for the potential effects of air pollutants on the lake. Buffering capacity of the terrestrial ecosystem is clearly seen in the case of lead and hydrogen ion. A budget analysis indicates that 98% of the lead falling on Hubbard Brook watersheds, where precipitation concentrations sometimes exceed U.S. Public Health Standards for drinking water, is retained and stored (at the rate of 27 mg/m²-yr) within the watershed (Siccama and Smith 1978; Smith and Siccama 1981; *Chapter II*). Thus, even though hydrologic input from the watershed exceeds direct precipitation input to the lake by ~3.6 times, lead in direct input exceeds watershed input by 7.7 times. The loss of watershed area owing to the construction of I-93 has probably resulted in higher concentrations of lead entering the lake (Pb directly entering the lake in bulk precipitation + Pb entering in drainage water ÷ total water entering the lake) but less total lead because of a reduction in total water. However, our calculations do not take into account possible increases in dry fallout resulting from lead emissions from vehicles traveling on I-93, which are now closer to the lake.

The terrestrial ecosystem has a similar effect on hydrogen ion. Precipitation at Hubbard Brook averages 4.1 pH, while pH of runoff or ground water entering the lake averages between 5.4 and 6.1 (*Table IV.B–2*). The decrease in acidity is due to exchange reactions within the terrestrial and interconnected stream-ecosystems (Johnson et al. 1981).

The durability of the terrestrial ecosystem's capacity to buffer air pollutants is unknown, but surely there is not an unlimited capacity to store lead or to buffer acid rain.

The use of sodium chloride for deicing purposes presents two interesting management problems at Hubbard Brook. In the negotiations regarding the planning of Interstate 93, special effort was made to divert water carrying road pollutants away from Mirror Lake. Specially constructed ditches carried road drainage to other watersheds. Yet monitoring of a small stream draining a subwatershed adjacent to I-93 has shown that Na⁺ concentrations have increased more than 10 times (*Figure IX–3*). The cause of this increase is unclear, but aerosol impaction of windblown road salt on evergreen vegetation in the adjacent subwatershed during winter months or ground-water drainage from the ditches could be a contributing factor.

One of our studies on the effect of white-tailed deer as a factor in nutrient cycling and energy in northern hardwood ecosystems has revealed a subtle way in which highway maintenance might influence aquatic ecosystems (Pletscher 1982). Terrestrial plants at Hubbard Brook contain very small amounts of sodium, about 0.002% of dry weight (Likens and Bormann 1970), and nutrient budgets for deer reveal that deer obtain only a small proportion, approximately 6 to 12%, of their annual sodium requirement from terrestrial vegetation (Pletscher 1982). Other potential sources of sodium are natural salt licks, apparently absent around Hubbard Brook and aquatic macrophytes, which may have sodium concentrations 500 to 1,000 times greater than those of terrestrial plants (Pletscher 1982; Jordan et al. 1973; Likens and Bormann 1970). A survey for wet sites supporting aquatic macrophytes indicated that such sites were not abundant in the uplands around Hubbard Brook, while an exclosure study of one such site indicated only modest use by deer. On the other hand, heavy use by deer of wet sites beside highways was observed. These sites were found to have high concentrations of sodium resulting from road drainage and substantial use of deicing salt during the winter. Approximately 50 mt/km of NaCl were applied to I-93 near the HBEF during the winter of 1978–79. These observations suggest that salting of roads by humans may be relieving grazing pressure on aquatic macrophytes and/or facilitating penetration of deer into areas where sodium was acting as a limiting factor.

E. Eutrophication Trends in Mirror Lake

Cultural eutrophication is one of the more perplexing problems posed to those responsible for managing or protecting inland bodies of water. Accelerated eutrophication often means greater aquatic productivity (including desirable or undesirable fish spp.), but increased productivity may lead to a succession of undesirable algal species (blue-green algae) or anoxic conditions. An understanding of this delicate interplay between physical, chemical and biologic conditions and the underlying causes is an important management priority. While it is possible to identify the trophic level of a lake, the actual process of eutrophication is rarely studied because the time frame of the process generally

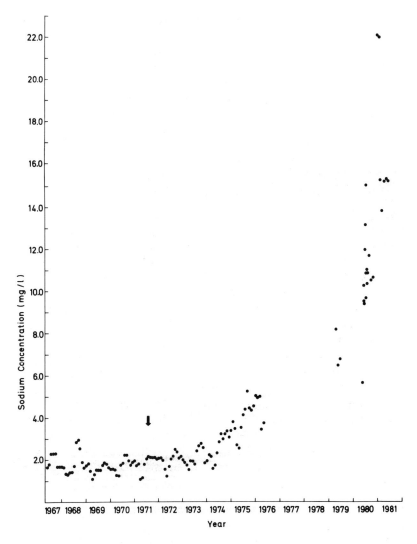

Figure IX–3. Sodium concentration (mg/liter) in the Northeast tributary to Mirror Lake. The arrow denotes the construction of Interstate 93 through the watershed.

exceeds the period of time covered by research. However, our paleolimnological analysis of the sediments in Mirror Lake (*Chapter VII*) has shown that trophic changes have been relatively minor, not unidirectional, and have occurred gradually over the 14,000 years of the lake's history.

In recent years, the level of human activity in the watershed for Mirror Lake has increased markedly. Not only was I-93 extended through the watershed, but at the same time an access road was cut through the forest, bordering the northern shore. This access road cuts directly across the remaining two drainage streams. The land along this road has been subdivided, and to date three new houses have been constructed. Sewage disposal in the area is by septic tank. The subsoil is composed of very porous sand and gravel and inputs of P, NH_4^+, Na^+, Cl^- and K^+ from septic drainage may be appreciable (*Chapter IV.E*).

Use of the lake for recreation has increased dramatically for at least three reasons: (1) the new highway has made the area more easily accessible to people living in the megalopolis (Boston-New York) to the south; (2) a small boys' camp located on the eastern shore of the lake has been converted to a relatively large and active family camp; and (3) the public beach on the southern shore has been improved and is now popular. Because N and P limit the primary production of Mirror Lake (Gerhart 1973; *Chapter V.B.2*), human activity may change the character of the lake in the coming years. Currently, the total input of N and P from all sources is small and the productivity of the lake

E. Eutrophication Trends in Mirror Lake

is very low (Likens and Loucks 1978; Jordan and Likens 1980; *Chapter V.C*); currently, some 54% of the inorganic N and 3% of the inorganic P directly enter the lake with rain and snow. However, NO_3^- concentrations have increased in precipitation, presumably from air pollution during the past 40 years or so (*Chapters II* and *VIII*) and P inputs from domestic sewage currently represent about 39% of the annual P inputs to the lake (*Table IV.E–16*). What will be the effects of these local and regional changes on lake eutrophication?

It is reasonable to expect that species composition of the algae and productivity will change (*Chapter V.B.2*). With increased human influence (e.g. sewage, fertilizers, etc.), drainage inputs may become the main source of both N and P, rather than those in precipitation. Such changes are well known elsewhere but a quantitative understanding of these air-land-water interactions, the process rates underlying them, and historical considerations in oligotrophic lakes are not (Anonymous 1969; Likens 1972c, 1975a; Likens and Bormann 1979).

Specific, recent eutrophication trends have not yet been verified, but there are indications of changes in the lake's metabolism and structure. For example, there appears to have been a significant increase in mid-summer phytoplankton production. A noticeable increase in the abundance of *Spirogyra* sp. has developed in the inlets and on the littoral sediments since 1969 and most dramatically during the past few years. A green alga, *Spaerocystis schroteri*, and two rotifers, *Keratella quadrata* and *Kellicottia bostoniensis*, have become more prevalent in the lake. Finally, the deep-water peak in algal biomass, characteristic of Class-B oligotrophic lakes, appears to have changed in species composition and magnitude (*Chapter V.A.2*). To determine whether or not these observations are part of a general eutrophication trend requires further observation and comparison with historical trends.

Summary

Human activities affecting a lake's airshed and watershed have the potential to cause significant changes in the structure and function of the lake-ecosystem. The relative importance of a lake's airshed vs. its watershed in affecting the lake is determined, in part, by the lake's surface/watershed surface ratio. Understanding the effect of a watershed on the dynamics of a lake is dependent, not only on understanding the biogeochemistry of the terrestrial ecosystem currently occupying the watershed but also on the developmental relationships of the terrestrial ecosystem.

The Hubbard Brook Biomass Accumulation Model projects four developmental phases following severe disturbance of the forest-ecosystem: Reorganization, Aggradation, Transition and the "Shifting-Mosaic Steady State." The Aggradation Phase is characterized by very close regulation over hydrologic and biogeochemical output (input to the lake). The watershed of Mirror Lake is currently occupied mostly by aggrading forest and the very low levels of dissolved substances in forest runoff contribute to the lake's oligotrophy.

Clear cutting or severe windstorms can trigger off the Reorganization Phase, which is characterized by substantial increases in water and nutrient output. These can act as a nutrient spike to the lake. However, in well-done commercial clear cuts where attention is paid to harvesting and road-building, revegetation following cutting limits the increased output to a period of about three to five years, after which outputs of very low dissolved substances dominate. Clear-cutting methods have been developed that have relatively little impact on aquatic systems like Mirror Lake.

Watershed managers may let the watershed regrow naturally to the "Shifting-Mosaic Steady State." The Hubbard Brook Model projects slight and not very significant increases in dissolved substances output with development beyond the Aggradation Phase.

The aggrading terrestrial ecosystem has a strong "filtration" effect on certain air pollutants like hydrogen ion and lead. The terrestrial system acts as a buffer for the aquatic system, however, buffering capacity may be limited.

Construction of a major interstate highway in the Mirror Lake watershed provided an opportunity to study a variety of potential effects. Some were: a change in lake surface/watershed surface ratio, which diminishes the buffering effect of the watershed; an unexpected increase in the movement of sodium into the lake, which might have resulted from unanticipated impaction of sodium in the portion of the Mirror Lake watershed adjacent to the highway or an increase in sodium in ground water moving to a stream draining into the lake.

Accelerated human use of the Mirror Lake watershed (including increased housing) may result in runoff inputs of N and P becoming the main source rather than N and P in precipitation as is presently the case. This could lead to eutrophication of Mirror Lake, but such trends have not yet been verified. The future of Mirror Lake depends not only on local decisions concerning the use of the lake and its watershed, but also on national decisions on pollutant emissions occurring in Mirror Lake's airshed.

References

Compiled by PHYLLIS C. LIKENS

	CHAPTER
Abdirov, C. A., L. G. Konstantinova and N. S. Sagidullaev. 1968. Microbiology of Lake Karateren. *Microbiology* 37:297–300.	V.B.1
Åberg, B., and W. Rodhe. 1942. Uber die Milieufaktoren in einigen sudschwedishen Seen. *Symbolae Botan. Upsalienses* 5:1–256.	IV.B; IV.C
Abrahamsson, S. A. A. 1966. Dynamics of an isolated population of the crayfish *Astacus astacus* Linné. *Oikos* 17:98–107.	VI.A
Adams, M. S., and M. D. McCracken. 1974. Seasonal production of the *Myriophyllum* component of the littoral of Lake Wingra, Wisconsin. *J. Ecol.* 62:457–467.	V.A.4; V.B.3
Adams, M. S., P. Guilizzoni and S. Adams. 1978. Sedimentary pigments and recent primary productivity in northern Italian lakes. *Mem. Ist. Ital. Idrobiol.* 38:267–285.	VII.D
Adams, M. S., J. Titus and M. D. McCracken. 1974. Depth distribution of photosynthetic activity in a *Myriophyllum spicatum* community in Lake Wingra. *Limnol. Oceanogr.* 19:377–389.	V.A.4; V.B.3
Adams, R. W., and H. C. Duthie. 1976. Relationships between sediment chemistry and postglacial production rates in a small Canadian lake. *Int. Rev. Ges. Hydrobiol.* 61:21–36.	VII.A.2; VII.F
Adams, W. P., and J. A. A. Jones. 1971. Observations in the Crystallographic relations of white ice and black ice. *Geophysica* 11:151–163.	IV.C
Ahlstrom, E. H. 1943. A revision of the rotatorian genus *Keratella* with descriptions of three new species of five new varieties. *Bull. Amer. Museum Nat. Hist.* 80:411–457.	V.A.5
Alhonen, P. 1970. The paleolimnology of four lakes in southwestern Finland. *Ann. Acad. Sci. Fenn. Ser. A, III. Geologica-Geographica* 105:1–39.	VII.E; VII.F
Alimov, A. F., V. V. Boullion, N. P. Finogenova, M. B. Ivanova, N. K. Kuzmitskaya, V. N. Nikulina, N. G. Ozerotskovskaya and T. V. Zharova. 1972. Biological productivity of Lakes Krivoe and Krugloe. In *Productivity Problems of Freshwaters,* ed. Z. Kajak and A. Hillbricht-Iłkowska, 39–56. Proc. of IBP-UNESCO Symp., May 1970, Kazimierz Dolny, Poland. Warsaw, Poland: Polish Scientific Publisers.	V.B.5; VI.C
Allee, W. C., A. E. Emerson, O. Park, T. Park and K. P. Schmidt. 1949. *Principles of Animal Ecology*. Philadelphia: W. B. Saunders Co. 232 pp.	V.A.5
Allen, H. R. 1972. Phytoplankton photosynthesis, micronutrient interactions, and inorganic carbon availability in a soft-water Vermont lake. In *Nutrients and Eutrophication,* ed. G. E. Likens, 63–83. American Society of Limnology and Oceanography Special Symposium 1.	V.B.2
Allen, T. F. H. 1977. Scale in microscopic algal ecology: a neglected dimension. *Phycologia* 16:253–257.	IV.A

	CHAPTER
Almer, D., W. Dickson, C. Edstrom, E. Hornstrom and U. Miller. 1974. Effects of acidification of Swedish lakes. *Ambio* 3:330–336.	VIII
Ambio. 1976. Report from the International Conference on the effects of acid precipitation, Telemark, Norway, June 14–19, 1976. *Ambio* 5(5–6):200–264.	VIII
American Public Health Association. 1965. *Standard Methods for the Examination of Water and Wastewater.* 12th ed. New York: American Public Health Association. 269 pp.	V.B.4
———. 1971. *Standard Methods for the Examination of Water and Wastewater.* 13th ed. Washington, DC: American Public Health Association. 874 pp.	V.A.6
Andersen, S. T. 1970. The relative pollen productivity and pollen representation of north European trees and correction factors for tree pollen spectra. *Geological Survey* (Denmark) II Ser. (96):1–96.	III.B.1
———. 1973. The differential pollen productivity of trees and its significance for the enterpretation of a pollen diagram from a forested region. In *Quaternary Plant Ecology,* ed. H. J. B. Birks and R. G. West, 109–116. London: Blackwell Scientific Publications.	III.B.1
Anderson, R. O. 1959. A modified flotation technique for sorting bottom fauna samples. *Limnol. Oceanogr.* 4(2):223–225.	V.A.6
Anderson, R. S., and A. de Henau. 1980. An assessment of the meiobenthos from nine mountain lakes in western Canada. *Hydrobiologia* 70:257–264.	V.A.7
Andronikova, I. N., V. G. Drabkova, K. N. Kuzmenko, N. F. Michaïlova and E. A. Stravinskaya. 1972. Biological productivity of the main communities of the Red Lake. In *Productivity Problems of Freshwaters,* ed. Z. Kajak and A. Hillbricht-Iłkowska, 57–71. Proc. of IBP-UNESCO Symp., May 1970, Kazimierz Dolny, Poland. Warsaw, Poland: Polish Scientific Publishers.	V.A.4
Anonymous. 1969. *Eutrophication: Causes, Consequences, and Correctives.* Proceedings of International Symposium on Eutrophication. Washington, DC: National Academy of Sciences. 661 pp.	IV.E; IX
Appleby, P. G., and F. Oldfield. 1979. Letter to the editors. *Environ. Sci. Technol.* 13:478–480.	VII.A.2
Argus, G. W., and M. B. Davis. 1962. Macrofossils from a late-glacial deposit at Cambridge, Massachusetts. *Amer. Midl. Natur.* 67:106–117.	III.B.1; VII.F
Armitage, P. D. 1968. Some notes on the food of the chironomid larvae of a shallow woodland lake in South Finland. *Ann. Zool. Fenn.* 5:6–13.	V.A.6
Armstrong, R. A. J., and D. W. Schindler. 1971. A preliminary chemical characterization of waters in the Experimental Lakes Area, northwestern Ontario. *J. Fish. Res. Bd. Canada* 28:171–188.	V.A.2
Atema, J. 1971. Structures and functions of the sense of taste in catfish (*Ictalurus natalis*). *Brain, Behav. Evol.* 4:273–294.	V.A.9
Bailey, R. M. 1938. The fishes of the Merrimack Watershed. In *Biological Survey of the Merrimack Watershed,* ed. E. E. Hoover, 149–185. Concord, NH: New Hampshire Fish and Game Department.	V.A.9
Baker, A. L., and A. J. Brook. 1971. Optical density profiles as an aid to the study of microstratified phytoplankton populations in lakes. *Arch. Hydrobiol.* 69:214–233.	V.A.2
Baker, J. P., and C. L. Schofield. 1980. Aluminum toxicity to fish as related to acid precipitation and Adirondack surface water quality. In *Ecological Impact of Acid Precipitation,* ed. D. Drabløs and A. Tollan, 292–293. Proceedings of the First International Conference, Sandefjord, Norway.	IX
Barbour, C. D., and J. H. Brown. 1974. Fish species diversity in lakes. *Amer. Natur.* 108:473–489.	V.A.9
Barko, J. W., and R. M. Smart. 1981. Sediment-based nutrition of submersed macrophytes. *Aquatic Botany* 10:339–352.	VI.A

References

CHAPTER

Barlow, J. P., W. R. Schaffner, F. deNoyelles, Jr. and B. J. Peterson. 1973. Experimental phytoplankton successions as nutrient bioassays. In *Bioassay Techniques and Environmental Chemistry,* ed. G. E. Glass, 299–319. Ann Arbor, MI: Ann Arbor Publishing Co. — V.B.2

Barnes, R. D. 1963. *Invertebrate Zoology.* Philadelphia: W. B. Saunders Co. 632 pp. — V.A.5

Bartos, E. 1951. The Czechoslovak Rotatoria of the order Bdelloidea. *Vestnik Ceskosl. Zoologicke Spolecnost* 15:241–244, 345–500. — V.A.7

Baybutt, R. J., and J. C. Makarewicz. 1981. Multivariate analysis of the Lake Michigan phytoplankton community at Chicago. *Bull. Torrey Bot. Club* 108(2):255–267. — VI.C; VI.D

Bayly, I. E. A. 1963. Reversed diurnal vertical migration of planktonic Crustacea in inland waters of low hydrogen ion concentration. *Nature* 200:704–705. — V.A.5

Beamish, R. J., and H. H. Harvey. 1969. Age determination in the white sucker. *J. Fish. Res. Bd. Canada* 26:633–638. — V.A.9

Belcher, C. F. 1961. The logging railroads of the White Mountains. Part IV, The New Zealand Railroad (1884–1892). *Appalachia* 33:353–374. — III.B.2

Belcher, J. H., and G. E. Fogg. 1964. Chlorophyll derivatives and carotenoids in the sediments of two English lakes. In *Basic Researches in the Fields of Hydrosphere, Atmosphere and Nuclear Chemistry,* ed. Y. Miyake and T. Koyama, 39–48. Tokyo: Maruzen. — VII.D

Bennett, L. 1979. Experimental analysis of the tropic ecology of *Lepidodermella squammata* (Gastrotricha: Chaetonotida) in mixed culture. *Trans. Amer. Microscopal Soc.* 98:254–260. — V.A.7

Benson, A. B., ed. 1937. *Peter Kalm's Travels in North America.* Vol. 1. New York: Dover. 797 pp. — III.B.2

Berg, K. 1937. Contributions to the biology of *Corethra* Meigen (*Chaoborus* Lichtenstein). *Kgl. Dan. Vidensk. Selsk., Biol. Meddel.* 13(11):1–101. — V.A.6

———. 1938. *Studies on the Bottom Animals of Esrom Lake.* Copenhagen: Ejnar Munksgaard. 255 pp. — V.A.6

Berg, K., and I. C. Petersen. 1956. Studies on the humic acid Lake Gribso. *Folia Limnol. Scand.* 8:1–273. — V.A.6

Berglund, B. E. 1969. Vegetation and human influence in south Scandinavia during prehistoric time. *Oikos Suppl.* 12:9–28. — III.B.1

———. 1973. Pollen dispersal and deposition in an area of southeastern Sweden—some preliminary results. In *Quaternary Plant Ecology,* ed. H. J. B. Birks and R. G. West, 117–129. Oxford: Blackwell Scientific Publications. — III.B.1

Berglund, B. E., and N. Malmer. 1971. Soil conditions and late-glacial stratigraphy. *Geol. Fören. Stockholm Förhandl.* 93:575–586. — VII.E; VII.F

Betson, A. R., Jr., W. P. James and J. W. Rouse, Jr. 1978. Evapotranspiration from water hyacinth in Texas reservoirs. *Water Resour. Bull.* 14(4):919–930. — IV.D

Betson, R. P., et al. 1978. Physical basin characteristics for hydrologic analysis. Chapter 7 in the *National Handbook of Recommended Methods for Water Data Acquisition.* Washington, DC: U.S. Govt. Printing Office. 19 pp. — III.A

Bilby, R. E. 1981. Role of organic debris dams in regulating the export of dissolved and particulate matter from a forested watershed. *Ecology* 62:1234–1243. — VII.F

Bilby, R. E., and G. E. Likens. 1979. Effect of hydrologic fluctuations on the transport of fine particulate organic carbon in a small stream. *Limnol. Oceanogr.* 24(1):69–75. — II.B

———. 1980. Importance of organic debris dams in the structure and function of stream ecosystems. *Ecology* 61(5):1107–1113. — II.B; III.B.2; VII.A.1; VIII

Billings, M. P. 1956. *The Geology of New Hampshire.* Part II, *Bedrock Geology.* Concord, NH: New Hampshire State Planning and Development Commission. 200 pp. — II.A; III.A

	CHAPTER
Birch, P. B., R. S. Barnes and D. E. Spyridakis. 1980. Recent sedimentation and its relationship with primary productivity in four western Washington lakes. *Limnol. Oceanogr.* 25:240–247.	VII.A.2
Birkett, L. 1958. A basis for comparing grabs. *J. Cons. Int. Explor. Mer.* 23:202–207.	V.A.6
Birks, H. J. B., and S. M. Peglar. 1980. Identification of Picea pollen of late Quaternary age in eastern North America: numerical approach. *Can. J. Bot.* 58:2043–2058.	III.B.1
Bishop, J. W., and J. P. Barlow. 1975. Phosphorus release by zooplankton. *Limnol. Oceanogr.* 20(1):148–149.	VI.C
Bishop, S. C. 1941. The salamanders of New York. *New York State Museum Bull.* 324:1–365.	V.A.8
Bloesch, J., and N. M. Burns. 1980. A critical review of sedimentation trap technique. *Schweiz. Zeitsch. Hydrol.* 42:1–88.	VII.A.2
Bloesch, J., P. Stadelmann and H. Bührer. 1977. Primary production, mineralization, and sedimentation in the euphotic zone of two Swiss lakes. *Limnol. Oceanogr.* 22(3):511–526.	V.B.6
Blomgren, N., and E. Naumann. 1925. Untersuchungen über die höhere Vegetation des Sees Stråken bei Aneboda. *Lunds Univ. Arsskr., N.F., Avd. 2,* 21(6):1–51.	V.A.4
Bloomberg, W. G. 1950. Fire and spruce. *For. Chron.* 26:157–161.	III.B.2
Bloomfield, J. A., ed. 1978. *Lakes of New York State.* Vol. I, *Ecology of the Finger Lakes* and Vol. II, *Ecology of the Lakes of Western New York.* New York: Academic Press. 499 pp. (I) and 473 pp. (II).	III.B.2
Boardman, N. K. 1977. Comparative photosynthesis of sun and shade plants. *Ann. Rev. Plant Physiol.* 28:355–377.	V.B.2
Bodin, K., and A. Nauwerck. 1968. Produktionsciologische Studien über die Moosvegetation eines klaren Gebirgssees. *Schweiz. Zeitsch. Hydrol.* 30(2):318–351.	V.A.4
Boling, R. H., E. D. Goodman, J. A. Van Sickle, J. O. Zimmer, K. W. Cummins, R. C. Petersen and S. R. Reice. 1975. Toward a model of detritus processing in a woodland stream. *Ecology* 56:141–151.	II.B
Bond, R. R. 1957. Ecological distribution of breeding birds in the upland forests of southern Wisconsin. *Ecol. Monogr.* 27:351–384.	VI.C
Bonny, A. P. 1976. Recruitment of pollen to the seston and sediment of some Lake District lakes. *J. Ecol.* 64:859–887.	VII.A.2
———. 1980. Seasonal and annual variation over 5 years in contemporary airborne pollen trapped at a Cumbrian lake. *J. Ecol.* 68:421–441.	VII.A.2
Bormann, F. H. 1982. Air pollution stress and energy policy. In *New England Prospects: Critical Choices in a Time of Change,* ed. C. H. Reidel, 85–140. Hanover, NH: Univ. Press of New England.	IX
Bormann, F. H., and G. E. Likens. 1967. Nutrient cycling. *Science* 155(3761):424–429.	I.A; I.A.1; II.A; II.B; III.A; IX
———. 1979. *Pattern and Process in a Forested Ecosystem.* Springer-Verlag New York Inc. 253 pp.	I.A; I.A.1; II.A; II.B; III.B.1–3; IV.E; VI.D; VII.F; IX
Bormann, F. H., G. E. Likens and J. S. Eaton. 1969. Biotic regulation of particulate and solution losses from a forested ecosystem. *BioScience* 19(7):600–610.	II.B; V.C
Bormann, F. H., G. E. Likens and J. M. Melillo. 1977. Nitrogen budget for an aggrading northern hardwood forest ecosystem. *Science* 196(4293):981–983.	II.B
Bormann, F. H., G. E. Likens, T. G. Siccama, R. S. Pierce and J. S. Eaton. 1974. The export of nutrients and recovery of stable conditions following deforestation at Hubbard Brook. *Ecol. Monogr.* 44:255–277.	II.B; V.C; VII.A; VII.A.1; VII.E

References

	CHAPTER
Bormann, F. H., T. G. Siccama, G. E. Likens and R. H. Whittaker. 1970. The Hubbard Brook Ecosystem Study: composition and dynamics of the tree stratum. *Ecol. Monogr.* 40(4):373–388.	II.A; III.B.1–3
Borutskii, E. V. 1950. Materialy po dinamike biomassy makrofivov ozer. I. *Trudy Vses. Gidrobiol. Obshch.* 2:43–55.	V.B.3
Borutzky, E. V. 1939. Dynamics of the total benthic biomass in the profundal of Lake Beloie (in Russian with English summary). *Trudy Limnol. Sta. Kosine* 22:196–218.	V.A.6
Botkin, D. B., J. F. Janak and J. R. Wallis. 1972. Some ecological consequences of a computer model of forest growth. *J. Ecol.* 60:849–872.	III.B.1
Bouton, N. 1856. *The History of Concord.* Concord, N.H.: Benning W. Sanborn, Publ. 786 pp.	III.B.3
Boyd, C. E. 1970. Production, mineral accumulation and pigment concentrations in *Typha latifolia* and *Scirpus americanus*. *Ecology* 51:285–290.	V.B.3
———. 1971. The limnological role of aquatic macrophytes and their relationship to reservoir management. In *Reservoir Fisheries and Limnology*, 153–166. Amer. Fish. Soc. Spec. Publ. 8.	V.B.3
Bozniak, E. G., and L. L. Kennedy. 1968. Periodicity and ecology of the phytoplankton in an oligotrophic and eutrophic lake. *Can. J. Botany* 46:1259–1271.	V.A.2
Bradbury, J. P., and R. O. Megard. 1972. Stratigraphic record of pollution in Shagawa Lake, northeastern Minnesota. *Geol. Soc. Amer. Bull.* 83:2639–2648.	VII.B
Bradbury, J. P., and T. C. Winter. 1976. Areal distribution and stratigraphy of diatoms in the sediments of Lake Sallie, Minnesota. *Ecology* 57:1005–1014.	VII.F
Bradley, E., and R. V. Cushman. 1956. *Memorandum report on geologic and ground-water conditions in the Hubbard Brook wastershed, New Hampshire.* On file at Northeastern Forest Experiment Station, Durham, NH. 15 pp.	III.A
Braekke, F. W., ed. 1976. *Impact of Acid Precipitation on Forest and Freshwater Ecosystems in Norway.* SNSF Project Research Report FR6/76. Oslo, Norway. 111 pp.	VIII
Bretschko, G. 1973. Benthos production of a high-mountain lake: Nematoda. *Verh. Internat. Verein. Limnol.* 18:1421–1428.	V.A.7
Brettum, P. 1971. Fordeling og biomasse av *Isoetes lacustris* og *Scorpidium scorpiodes* i Øvre Heimdalsvatn, et høyfjellsvann i Sør-Norge. *Blyttia* 29:1–11.	V.A.4
Brimblecombe, P., and D. H. Stedman. 1982. Historical evidence for a dramatic increase in the nitrate component of acid rain. *Nature* 298:460–462.	VIII
Brinkhurst, R. O. 1964. Observations on the biology of lake dwelling Tubificidae. *Arch. Hydrobiol.* 60:385–418.	V.A.6
———. 1970. Distribution and abundance of tubificid (Oligochaeta) species in Toronto Harbour, Lake Ontario. *J. Fish. Res. Board Canada* 27:1916–1969.	V.A.6
Brinkhurst, R. O., K. E. Chua and E. Batoosingh. 1969. Modifications in the sampling procedures as applied to studies on the bacteria and tubificid oligochaetes inhabiting aquatic sediments. *J. Fish. Res. Board Canada* 26:2581–2593.	VI.B
Bristow, J. M., and M. Whitcombe. 1971. The role of roots in the nutrition of aquatic vascular plants. *Amer. J. Bot.* 58:8–13.	V.A.4
Brodie, E. D., Jr. 1968. Investigations on the skin toxin of the red spotted newt, *Notophthalmus viridescens viridescens*. *Amer. Midl. Natur.* 80(1):276–280.	V.A.8
Bromley, S. W. 1935. The original forest types of southern New England. *Ecol. Monogr.* 5:61–89.	III.B.2

	CHAPTER

Brooks, J. L. 1957. The systematics of North American Daphnia. *Mem. Conn. Acad. Arts Sci.* 13:5–180. — V.A.5

———. 1969. Eutrophication and changes in the composition of the zooplankton. In *Eutrophication: Causes, Consequences and Correctives*, 236–255. Washington, DC: National Academy of Sciences. — V.B.5

Brooks, J. L., and S. I. Dodson. 1965. Predation, body size and composition of plankton. *Science* 150:28–35. — VI.C

Brown, J. W. 1958. Forest history of Mount Moosilauke. Part I, Primeval conditions, settlements and the farm-forest economy. Part II, Big logging days and their aftermath (1890–1940). *Appalachia* 24(7):23–32, 221–233. — III.B.2

Brown, W. R. 1958. *Our Forest Heritage*. Concord, NH: New Hampshire Historical Society. — III.B.3

Browne, G. W. 1906. The Merrimack River—The Stone Age. *Granite State Magazine* 1, 193. — III.B.3

Brubaker, L. B. 1973. Modern and ancient patterns of forest deforestation on outwash and till soils in upper Michigan. Ph.D. diss., University of Michigan, Ann Arbor. — VII.D

Brugam, R. B. 1978. Human disturbance and the historical development of Linsley Pond. *Ecology* 59:19–36. — VII.A.2; VII.F

Brundin, L. 1951. The relation of O_2-microstratification at the mud surface to the ecology of the profundal bottom fauna. *Rep. Inst. Freshwater Res. Drottningholm* 32:32–43. — V.A.6

Brunskill, G. J., and S. D. Ludlam. 1969. Fayetteville Green Lake, New York. I. Physical and chemical limnology. *Limnol. Oceanogr.* 14(6):817–829. — V.C

Brunskill, G. J., D. Povoledo, B. W. Graham and M. P. Stainton. 1971. Chemistry of surface sediments of sixteen lakes in the Experimental Lakes Area, northwestern Ontario. *J. Fish. Res. Bd. Canada* 28:277–294. — VII.E

Brunson, R. B. 1959. Gastrotricha. Chapter 17 in *Freshwater Biology*, ed. W. T. Edmondson. 2d ed. New York: John Wiley & Sons. — V.A.7

Bryce, D., and A. Hobart. 1972. The biology and identification of the larvae of the Chironomidae (Diptera). *Entomol. Gaz.* 23:175–217. — V.A.6

Buchanan, R. E., and E. I. Fulner. 1928. *Physiology and Biochemistry of Bacteria*, Vol. 1. Baltimore: Williams and Wilkens. — V.B.1

Bullard, W. E., Jr. 1965. Role of watershed management in the maintenance of suitable environments for aquatic life. In *Biological Problems in Water Pollution*, 265–269. U.S. Dept. of Health, Education and Welfare. Public Health Service Publ. No. 999-WP-25. — III.B.2

Burch, J. B. 1975. *Freshwater Unionacean Clams (Mollusca: Pelecypoda) of North America*. Revised ed. Hamburg, MI: Malacological Publications. 204 pp. — V.A.6

Burnham, J. C., T. Stetak and G. Locher. 1976. Extracellular lysis of the blue-green alga *Phormidium luridum* by *Bdellovibrio bacteriovorus*. *J. Phycol.* 12:306–313. — V.D

Burns, N. M., and F. Rosa. 1980. *In situ* measurement of the settling velocity of organic carbon particles and 10 species of phytoplankton. *Limnol. Oceanogr.* 25:855–864. — V.D

Burton, T. M. 1973. The role of salamanders in ecosystem structure and function in the Hubbard Brook Experimental Forest in New Hampshire. Ph.D. diss., Cornell University. 211 pp. — V.C

———. 1976. An analysis of the feeding ecology of the salamanders (Amphibia, Urodela) of the Hubbard Brook Experimental Forest, New Hampshire. *J. Herp.* 10(3):187–204. — V.B.7

———. 1977. Population estimates, feeding habits and nutrient and energy relationships of *Notophthalmus v. viridescens*, in Mirror Lake, New Hampshire. *Copeia* 1977 (1):139–143. — V.A.8; V.A.9; V.B.7

References

	CHAPTER
Burton, T. M., and G. E. Likens. 1973. The effect of strip-cutting on stream temperatures in the Hubbard Brook Experimental Forest, New Hampshire. *BioScience* 23(7):433–435.	IV.B
———. 1975. Energy flow and nutrient cycling in salamander populations in the Hubbard Brook Experimental Forest, New Hampshire. *Ecology* 56(5):1068–1080.	V.B.7
Buscemi, P. A. 1961. Ecology of the bottom fauna of Parvin Lake, Colorado. *Trans. Amer. Microscopal Soc.* 80:266–307.	V.A.6
Butler, M., M. C. Miller and S. Mozley. 1980. Macrobenthos. Chapter 7 in *Limnology of Tundra Ponds, Barrow, Alaska,* ed. J. E. Hobbie. U.S. IBP Synthesis Series 13. New York: Academic Press. 514 pp.	V.A.6; V.B.6
Calaban, M. J., and J. C. Makarewicz. 1982. The effect of temperature and density on the amplitude of vertical migration of *Daphnia magna*. *Limnol. Oceanogr.* 27(2):262–271.	VI.C
Calvert, J., chairman. 1983. *Acid deposition atmospheric processes in eastern North America*. Washington, DC: National Academy Press.	II.B
Campbell, P. G., and J. H. Baker. 1978. Estimation of bacterial production in fresh waters by the simultaneous measurement of [^{35}S] sulphate and D-[^{3}H] glucose uptake in the dark. *Can. J. Microbiol.* 24:939–946.	V.B.1
Canter, H. M., and J. W. G. Lund. 1968. The importance of protozoa in controlling the abundance of planktonic algae in lakes. *Proc. Linn. Soc. Lond.* 179:203–219.	VI.C
Capblancq, J. 1973. Phytobenthos et productivité primaire d'un lac de haute montagne dans les Pyrénéés centrales. *Ann. Limnol.* 9:193–230.	V.A.4
Carignan, R., and J. Kalff. 1979. Quantification of the sediment phosphorus available to aquatic macrophytes. *J. Fish. Res. Bd. Canada* 36:1002–1005.	V.A.4
Carlander, K. D. 1969. *Handbook of Freshwater Fishery Biology,* Vol. I. Ames: Iowa State Univ. Press. 720 pp.	V.A.9
———. 1977. *Handbook of Freshwater Fishery Biology,* Vol. II. Ames: Iowa State Univ. Press. 409 pp.	V.A.9
Carlson, R. E. 1977. A trophic state index for lakes. *Limnol. Oceanogr.* 22:361–369.	V.A.2
Carpenter, S. R. 1980. Enrichment of Lake Wingra, Wisconsin, by submersed macrophyte decay. *Ecology* 61:1145–1155.	V.B.3; VI.A
———. 1983. Lake geometry: implications for production and sediment accretion rates. *J. Theor. Biol.* 105:273–286.	IV.B; VI.D
Case, T. J., and R. K. Washino. 1979. Flatworm control of mosquito larvae in rice fields. *Science* 206:1412–1414.	V.A.7
Chang, M. 1977. An evaluation of precipitation gage density in a mountainous terrain. *Water Resour. Bull.* 13(1):39–46.	IV.D
Chapman, H. H. 1947. Natural areas. *Ecology* 28:193–194.	III.B.2
Charles, W. N., K. East and T. D. Murray. 1976. Production of larval Tanypodinae (Insecta: Chironomidae) in the mud at Loch Leven, Kinross. *Proc. Roy. Soc. Edinburgh (B)* 75(10):157–169.	V.A.6; V.B.6
Charles, W. N., K. East, D. Brown, M. C. Gray and T. D. Murray. 1974. The production of larval Chironomidae in the mud at Loch Leven, Kinross. *Proc. Roy. Soc. Edinburgh (B)* 74:241–258.	V.A.6; V.B.6
Charlton, E. A. 1856. *New Hampshire As It Is*. Tracy and Co., Claremont, NH. 592 pp.	III.B.3
Chaston, I. 1969. The light threshold controlling the vertical migration of *Chaoborus punctipennis* in a Georgia impoundment. *Ecology* 50:916–920.	V.A.6
Chen, C. T., and F. J. Millero. 1977. The use and misuse of pure water PVT properties for lake waters. *Nature* 266:707.	IV.C

	CHAPTER

Chester, R. and M. J. Hughes. 1967. A chemical technique for the separation of ferro-manganese minerals, carbonate minerals and adsorbed trace elements from pelagic sediments. *Chem. Geol.* 2:249–262. — VII.E

Child, H. 1886. *Gazetteer of Grafton County, New Hampshire 1709–1886.* Syracuse, NH: Syracuse Journal Co., Printers and Binders. 380 pp. — III.B.3

Chittendon, A. K. 1905. Forest conditions of northern New Hampshire. *USDA Bur. For. Bull.* No. 55. 131 pp. — III.B.2

Chodorowski, A. 1959. Ecological differentiation of turbellarians in Harsz-Lake. *Polskie Arch. Hydrobiol.* 6:33–73. — V.A.7

Chua, K. E., and R. O. Brinkhurst. 1973. Evidence of interspecific interactions in the respiration of tubificid oligochaetes. *J. Fish. Res. Bd. Canada* 30:617–622. — V.A.6

Clarke, G. L. 1939. The utilization of solar energy by aquatic organisms. In *Problems of Lake Biology,* ed. F. R. Moulton, 27–38. Publ. Amer. Assoc. Adv. Sci. No. 10. Washington, DC: The Science Press. — IV.B

Clements, F. E. 1916. *Plant Succession.* Carnegie Institute Washington Publication #242. 512 pp. — I.A

Cline, A. C., and S. H. Spurr. 1942. The virgin upland forest of central New England. *Harv. For. Bull.* No. 21. 51 pp. — III.B.2

Coble, D. W. 1975. Smallmouth bass. In *Black Bass Biology and Management,* ed. H. Clepper, 21–33. Washington, DC: Sport Fish. Inst. — V.A.9

Cogbill, C. V., and G. E. Likens. 1974. Acid precipitation in the northeastern United States. *Water Resour. Res.* 10(6):1133–1137. — II.B

Colby, P. J., ed. 1977. Percid International Symposium (PERCIS). *J. Fish. Res. Bd. Canada* 34:1447–1993. — V.A.9

Cole, G. A. 1953. Notes on the vertical distribution of organisms in the profundal sediments of Douglas Lake, Michigan. *Amer. Midl. Natur.* 49:252–256. — V.A.6

———. 1955. An ecological study of the microbenthic fauna of two Minnesota lakes. *Amer. Midl. Natur.* 53:213–230. — V.A.7

Cole, G. A., and J. C. Underhill. 1965. The summer standing crop of sublittoral and profundal benthos in Lake Itasca, Minnesota. *Limnol. Oceanogr.* 10(4):591–597. — V.A.6

Cole, J. J. 1982a. Interactions between bacteria and algae in aquatic ecosystems. *Ann. Rev. Ecol. Syst.* 13:291–314. — V.B.1; VI.C

———. 1982b. Microbial decomposition of algal organic matter in an oligotrophic lake. Ph.D. diss., Cornell University, Ithaca, NY. 279 pp. — V.B.1; V.D

Cole, J. J., and G. E. Likens. 1979. Measurements of mineralization of phytoplankton detritus in an oligotrophic lake. *Limnol. Oceanogr.* 24:541–547. — V.D; VI.C

Cole, J. J., G. E. Likens and D. L. Strayer. 1982. Photosynthetically-produced dissolved organic carbon: an important carbon source for planktonic bacteria. *Limnol. Oceanogr.* 27(6):1080–1090. — V.D; VI.C

Cole, J. J., W. H. McDowell and G. E. Likens. 1984. Sources and molecular weight of "dissolved" organic carbon in an oligotrophic lake. *Oikos* 42:1–9. — V.D

Collins, L. W., and G. E. Likens. 1969. The effect of altitude on the distribution of aquatic plants in some lakes of New Hampshire, U.S.A. *Verh. Internat. Verein. Limnol.* 17:154–172. — V.A.4

Confer, J. L. 1971. Intrazooplankton predation by *Mesocyclops edax* at natural prey densities. *Limnol. Oceanogr.* 16:663–666. — VI.C

Confer, J. L., T. Kaaret and G. E. Likens. 1983. Zooplankton diversity and biomass in recently acidified lakes. *Can. J. Fish. Aquatic Sci.* 40(1):36–42. — VIII

Costerton, J. W., G. G. Geesey and K. J. Cheng. 1978. How bacteria stick. *Sci. Amer.* 238:86–95. — V.A.1

Coveney, M. J., G. Cronberg, M. Enell, K. Larsson and L. Olofsson. 1977. Phytoplankton, zooplankton and bacteria—standing crop and production relationships in a eutrophic lake. *Oikos* 29:5–21. — VI.C

References

	CHAPTER
Covington, W. W. 1976. Secondary succession in northern hardwoods: forest floor organic matter and nutrients and leaf fall. Ph.D. diss., Yale University, New Haven, CT. 117 pp.	II.A
Crabtree, K. 1969. Post-glacial diatom zonation of limnic deposits in North Wales. *Mitt. Internat. Verein. Limnol.* 17:165–171.	VII.F
Cranwell, P. A. 1973. Chain-length distribution of n-alkanes from lake sediments in relation to post-glacial environmental change. *Freshwater Biol.* 3:259–265.	VII.E
Craven, R. E. 1969. Benthic macroinvertebrates of Boomer Lake, Payne County, Oklahoma. *Southwest. Nat.* 14(2):221–230.	V.A.6
Crawford, C. C., J. E. Hobbie and K. L. Webb. 1974. Utilization of dissolved free amino acids by estuarine microorganisms. *Ecology* 55:551–563.	V.B.1
Crisman, T. L., R. L. Schulze, P. L. Brezonik and S. A. Bloom. 1980. Acid precipitation: the biotic response in Florida lakes. In *Proc. Internat. Conf. on Ecological Impact of Acid Precipitation,* ed. D. Drabløs and A. Tollan, 296–297. Oslo, Norway.	VIII
Crisp, D. J. 1971. Energy flow measurements. In *Methods for the Study of Marine Benthos.* ed. N. A. Holme and A. D. McIntyre, 197–279. IBP Handbook No. 16. Oxford: Blackwell Scientific Publications. 334 pp.	V.A.6; V.B.6
Crocker, R. L., and J. Major. 1955. Soil development in relation to vegetation and surface age at Glacier Bay, Alaska. *J. Ecol.* 43:427–448.	III.B.1; VII.F
Crossman, E. J., and G. E. Lewis. 1973. An annotated bibliography of the chain pickerel, *Esox niger* (Osteichthyes: Salmoniformes). *Misc. Publ. Roy. Ontario Museum.* 81 pp.	V.A.9
Cummins, K. W., and J. C. Wuycheck. 1971. Caloric equivalents for investigations in ecological energetics. *Mitt. Int. Ver. Theore. Angew. Limnol.* 18. 158 pp.	V.B.6; VI.D
Cunningham, L. 1972. Vertical migration of *Daphnia* and copepods under the ice. *Limnol. Oceanogr.* 17(2):301–303.	V.A.5
Curry, L. L. 1952. An ecological study of the Family Tendipedidae of two fresh-water lakes in Isabella County, Michigan. Ph.D. diss., Michigan State College of Agriculture and Applied Sciences, East Lansing, MI.	V.A.6
Curtis, J. T. 1959. *The Vegetation of Wisconsin.* Madison: Univ. Wisconsin Press. 657 pp.	V.A.4
Daft, M. J., and W. D. P. Stewart. 1973. Light and electron microscope observations on algal lysis by bacterium CP-1. *New Phytol.* 72:799–808.	V.D
Daley, R. J., and J. E. Hobbie. 1975. Direct counts of aquatic bacteria by a modified epifluorescence technique. *Limnol. Oceanogr.* 20:875–882.	V.B.1
Davies, B. R. 1974. The planktonic activity of larval Chironomidae in Loch Leven, Kinross. *Proc. Roy. Soc. Edinburgh (B)* 74:275–283.	V.A.6
Davies, R. W., and V. J. E. McCauley. 1970. The effect of preservatives on the regurgitation of gut contents by Chironomidae (Diptera) larvae. *Can. J. Zool.* 48:519–522.	V.A.6
Davis, M. B. 1958. Three pollen diagrams from central Massachusetts. *Amer. J. Sci.* 256:540–570.	III.B.1
———. 1965. Phytogeography and palynology of northeastern United States. In *The Quaternary of the United States,* ed. H. E. Wright and D. G. Frey, 377–401. Princeton, NJ: Princeton Univ. Press.	VII.D
———. 1967. Late-glacial climate in northern United States: a comparison of New England and the Great Lakes Region. In *Quaternary Paleoecology,* ed. E. J. Cushing and H. E. Wright, Jr., 11–43. New Haven, CT: Yale Univ. Press.	VII.F

	CHAPTER
———. 1968. Pollen grains in lake sediments: redeposition caused by seasonal water circulation. *Science* 162:796–799.	VII.A.1; VII.A.2
———. 1969. Climatic changes in southern Connecticut recorded by pollen deposition at Rogers Lake. *Ecology* 50:409–422.	III.B.1
———. 1973. Redeposition of pollen grains in lake sediment. *Limnol. Oceanogr.* 18:44–52.	VII.A.2
———. 1976a. Erosion rates and land-use history in southern Michigan. *Environ. Conserv.* 3:139–148.	VII.A.2; VII.F
———. 1976b. Pleistocene plant geography. *Geoscience and Man* 13:13–26.	III.B.1
———. 1981a. Outbreaks of forest pathogens in Quaternary history. *Proc. of IV Internat. Palynology Conference* 3:216–227. Lucknow, India.	III.B.1; VII.F
———. 1981b. Quaternary history and the stability of forest communities. In *Forest Succession—Concepts and Application,* ed. D. G. West, H. H. Shugart and D. B. Botkin, 132–153. Springer-Verlag New York Inc.	III.B.1
———. 1983. Holocene vegetational history of the eastern United States. In *Late-Quaternary Environments of the United States.* Vol. 2. *The Holocene,* ed. H. E. Wright, Jr., 116–181. Minneapolis: University of Minnesota Press.	III.B.1; VII.F
Davis, M. B., and M. S. (J.) Ford. 1982. Sediment focusing in Mirror Lake, New Hampshire. *Limnol. Oceanogr.* 27(1):137–150.	III.A; III.B.1; VII.A; VII.A.1; VII.D; VII.E; VII.F
Davis, M. B., L. B. Brubaker and T. Webb III. 1973. Calibration of absolute pollen influx. In *Quaternary Plant Ecology,* ed. H. J. B. Birks and R. G. West, 9–25. London: Blackwell Scientific Publications.	III.B.1; VII.A.1
Davis, M. B., R. E. Moeller and J. Ford. 1984. Sediment focusing and pollen influx. In *Lake Sediments and Environmental History,* ed. E. Y. Haworth and J. W. G. Lund, 261–293. Leicester, England: University of Leicester Press.	VII.E; VII.F
Davis, M. B., R. W. Spear and L. C. K. Shane. 1980. Holocene climate of New England. *Quat. Res.* 14:240–250.	III.B.1; VII.A.1; VII.F
Davis, R. B. 1967. Pollen studies of near-surface sediments in Maine lakes. In *Quaternary Paleoecology,* ed. E. J. Cushing and H. E. Wright, 143–173. New Haven, CT: Yale Univ. Press.	III.B.1
———. 1974a. Stratigraphic effects of tubificids in profundal lake sediments. *Limnol. Oceanogr.* 19:466–488.	VII.A.2
———. 1974b. Tubificids alter profiles of redox potential and pH in profundal lake sediment. *Limnol. Oceanogr.* 19(2):342–345.	VI.B
Davis, R. B., and T. Webb III. 1975. The contemporaneous distribution of pollen in eastern North America: a comparison with the vegetation. *Quat. Res.* 5:395–434.	III.B.1
Davis, R. B., D. L. Thurlow and F. E. Brewster. 1975a. Effects of burrowing tubificid worms on the exchange of phosphorus between lake sediment and overlying water. *Verh. Internat. Verein. Limnol.* 19(1):382–394.	VI.B
Davis, R. B., T. E. Bradstreet, R. Stuckenrath, Jr. and H. W. Borns, Jr. 1975b. Vegetation and associated environments during the past 14,000 years near Moulton Pond, Maine. *Quat. Res.* 5:435–466.	VII.A.1
Dawson, F. H. 1976. The annual production of the aquatic macrophyte *Ranunculus penicillatus* var. *calcareus* (R. W. Butcher) C. D. K. Cook. *Aquat. Bot.* 2:51–73.	V.B.3
Day, G. 1953. The Indian as an ecological factor in the northeastern forest. *Ecology* 34:329–346.	III.B.2
Dean, J. L. 1969. *Biology of the crayfish* Orconectes causeyi *and its use for control of aquatic weeds in trout lakes.* U.S. Dept. Interior, Fish and Wildlife Service, Bureau of Sport Fisheries and Wildlife, Tech. Paper 24. 15 pp.	VI.A
Dean, W. E., and E. Gorham. 1976. Major chemical and mineral components	VII.E

References

	CHAPTER
of profundal surface sediments in Minnesota lakes. *Limnol. Oceanogr.* 21:259–284.	
Deevey, E. S., Jr. 1940. Limnological studies in Connecticut, 5. A contribution to Regional Limnology. *Amer. J. Sci.* 238(10):717–741.	VI.D
———. 1941. The quantity and composition of the bottom fauna of thirty-six Connecticut and New York lakes. *Ecol. Monogr.* 11:413–455.	V.A.6
———. 1942. Studies on Connecticut Lake sediments. III. The biostratonomy of Linsley Pond. *Amer. J. Sci.* 240:233–264, 313–324.	VII.F
———. 1953. Paleolimnology and climate. In *Climatic Change,* ed. H. Shapley, 273–318. Cambridge, MA: Harvard Univ. Press.	VII.A.1
———. 1964. Preliminary account of fossilization of zooplankton in Rogers Lake. *Verh. Internat. Verein. Limnol.* 15:981–992.	VII.F
Deevey, E. S., Jr., and G. B. Deevey. 1971. The American species of *Eubosmina* Seligo (Crustacea, Cladocera). *Limnol. Oceanogr.* 16:201–218.	V.A.5; VII.C
Deevey, E. S., Jr., and R. F. Flint. 1957. Post-glacial hypsithermal interval. *Science* 125:182–184.	III.B.1
Delfino, J. J., and G. F. Lee. 1976. Chemistry of manganese in Lake Mendota, Wisconsin. *Environ. Sci. Technol.* 2:1094–1100.	VII.E
Delucca, R., and M. D. McCracken. 1977. Observations on interactions between naturally-collected bacteria and several species of algae. *Hydrobiologia* 55:71–75.	VI.C
de March, L. 1975. Nutrient budgets for a high arctic lake (Char Lake, N.W.T.). *Verh. Internat. Verein. Limnol.* 19:496–503.	V.C
———. 1978. Permanent sedimentation of nitrogen, phosphorus, and organic carbon in a high Arctic lake. *J. Fish. Res. Bd. Canada* 35:1089–1094.	V.C
Den Hartog, C., and S. Segal. 1964. A new classification of the water-plant communities. *Acta Bot. Neerl.* 13:367–393.	V.A.4
deNoyelles, F., Jr., and W. J. O'Brien. 1978. Phytoplankton succession in nutrient enriched experimental ponds as related to changing carbon, nitrogen and phosphorus conditions. *Arch. Hydrobiol.* 84:137–165.	V.B.2
deNoyelles, F., Jr., R. Knoechel, D. Reinke, D. Treanor and C. Altenhofen. 1980. Continuous culturing of natural phytoplankton communities in the Experimental Lakes Area: effects of enclosure, *in situ* incubation, light, phosphorus, and cadmium. *Can. J. Fish. Aquat. Sci.* 37:424–433.	V.A.2; V.B.2
Derenbach, J. B., and P. J. Williams. 1974. Autotrophic and bacterial production: fractionation of plankton populations by differential filtration of samples from the English Channel. *Mar. Biol.* 25:263–269.	V.B.1; VI.C
Dermott, R. M., J. Kalff, W. C. Leggett and J. Spence. 1977. Production of *Chironomus, Procladius* and *Chaoborus* at different levels of phytoplankton biomass in Lake Memphremagog, Quebec–Vermont. *J. Fish. Res. Bd. Canada* 34:2001–2007.	V.A.6; V.B.6
Devol, A. H., and R. C. Wissmar. 1978. Analysis of five North American lake ecosystems. V. Primary production and community structure. *Verh. Internat. Verein. Limnol.* 20:581–586.	V.C
Dillon, P. J., and F. H. Rigler. 1975. A simple method for predicting the capacity of a lake for development based on lake trophic status. *J. Fish. Res. Bd. Canada* 32(9):1519–1531.	VIII
Dochinger, L. S., and T. A. Seliga, eds. 1976. *Proceedings of The First International Symposium on Acid Precipitation and the Forest Ecosystem.* USDA Forest Service General Tech. Report NE-23. 1,074 pp.	VIII
Dodson, S. I. 1974. Zooplankton competition and predation: an experimental test of the size-efficiency hypothesis. *Ecology* 55:605–613.	VI.C
Doetsch, R. N., and T. M. Cook. 1973. *Introduction to Bacteria and Their Ecobiology.* Baltimore: Univ. Park Press. 371 pp.	V.A.1

CHAPTER

Donaldson, J. R. 1967. Phosphorus budget of Iliamna Lake, Alaska, as related to the cyclic abundance of sockeye salmon. Ph.D. diss., University of Washington, Seattle.   I.A.2

Downing, J. A. 1979. Aggregation, transformation and the design of benthos sampling programs. *J. Fish. Res. Bd. Canada* 36:1454–1463.   V.A.6

Drabkova, V. G. 1965. Dynamics of the bacterial number generation time and production of bacteria in the water of Red Lake (Punnus-Jarvi). *Microbiology* 34:933–938.   V.B.1

Drabløs, D., and A. Tollan, eds. 1980. *Ecological Impact of Acid Precipitation*. Proc. International Conference, Sandefjord, Norway, 11–14 March 1980. 383 pp.   VIII

Driscoll, C. T., and G. E. Likens. 1982. Hydrogen ion budget of an aggrading forested ecosystem. *Tellus* 34:283–292.   II.B; IV.E; VIII

Driscoll, C. T., J. P. Baker, J. J. Bisogni and C. L. Schofield. 1980. Effect of aluminum speciation on fish in dilute acidified waters. *Nature* 284:161–164.   VIII

Driver, H. E., and W. C. Massey. 1957. Comparative studies of North American Indians. *Trans. Amer. Phil. Soc.* 47:165–456.   III.B.3

Droop, M. R., and K. G. Elson. 1966. Are pelagic diatoms free from bacteria? *Nature* 211:1096–1097.   VI.C

Dugdale, R. C. 1956. Studies in the ecology of the benthic Diptera of Lake Mendota. Ph.D. diss., University of Wisconsin, Madison. 99 pp.   I.A.2; V.A.6

Dumont, H. J., I. V. Velde and S. Dumont. 1975. The dry weight estimate of biomass in a selection of Cladocera, Copepoda, and Rotifera from the plankton, periphyton, and benthos of continental waters. *Oecologia* 19:75–97.   V.A.5

Dunne, T., and R. D. Black. 1970. Partial area contributions to storm runoff in a small New England watershed. *Water Resour. Res.* 6(5):1296–1311.   IV.D

Eaton, J. S., G. E. Likens and F. H. Bormann. 1973. Throughfall and stemflow chemistry in a northern hardwood forest. *J. Ecol.* 61(2):495–508.   II.B

———. 1978. The input of gaseous and particulate sulfur to a forest ecosystem. *Tellus* 30:546–551.   II.B; IV.E

———. 1980. Wet and dry deposition of sulfur at Hubbard Brook. In *Effect of Acid Precipitation on Terrestrial Ecosystems,* ed. T. C. Hutchinson and M. Havas, 69–75. NATO Conf. Series 1: Ecology 4. New York: Plenum Publishing Corp.   II.B

Edgington, D. N., and J. A. Robbins. 1976. Records of lead deposition in Lake Michigan sediments since 1800. *Environ. Sci. Technol.* 10:266–273.   VII.A.2

Edmondson, W. T. 1946. Factors in the dynamics of rotifer populations. *Ecol. Monogr.* 16(9):357–372.   V.B.5

———. 1965. Reproductive rates of planktonic rotifers as related to food and temperature in nature. *Ecol. Monogr.* 35:61–111.   VI.C

———. 1972. Nutrients and phytoplankton in Lake Washington. In *Nutrients and Eutrophication: The Limiting-Nutrient Controversy,* ed. G. E. Likens, 172–193. Special Symposium, American Society of Limnology and Oceanography 1.   VIII

———. 1974. Secondary production. *Mitt. Internat. Verein. Limnol.* 20:229–272.   V.B.5

———. 1980. Secchi disk and chlorophyll. *Limnol. Oceanogr.* 25(2):378–379.   IV.B

———, ed. 1959. *Freshwater Biology.* 2d ed. New York: John Wiley & Sons. 1,248 pp.   V.A.5

Edmondson, W. T., and G. G. Winberg. 1971. *Methods for the Assessment of Secondary Productivity in Fresh Waters.* I.B.P. Handbook No. 17. Oxford: Blackwell Scientific Publications. 358 pp.   V.B.5

	CHAPTER

Edwards, R. W. 1958. The effect of larvae of *Chironomus riparius* Meigen on the redox potentials of settled activated sludge. *Ann. Appl. Biol.* 46(3):457–464.   VI.B

Eggleton, F. E. 1931a. Limnetic distribution and migration of Corethra larvae in two Michigan lakes. *Pap. Mich. Acad. Sci. Arts Lett.* 15(1931):361–388.   V.A.6

———. 1931b. A limnological study of the profundal bottom fauna in certain fresh-water lakes. *Ecol. Monogr.* 1(3):231–331.   V.A.6

———. 1934. A comparative study of the benthic fauna of four northern Michigan lakes. *Pap. Mich. Acad. Sci. Arts Lett.* 20:609–644.   V.A.6

———. 1936. Productivity of the profundal benthic zone in Lake Michigan. *Pap. Mich. Acad. Sci. Arts Lett.* 22:593–611.   V.A.6

Egler, F. E. 1940. Berkshire plateau vegetation, Massachusetts. *Ecol. Monogr.* 10:145–192.   III.B.2

Elliott, J. M. 1971. *Some Methods for the Statistical Analysis of Samples of Benthic Invertebrates.* Freshwater Biological Association Scientific Publication No. 25. 144 pp.   V.A.5; V.A.6

Elner, J. K., and C. M. Happey-Wood. 1980. The history of two linked but contrasting lakes in North Wales from a study of pollen, diatoms and chemistry in sediment cores. *J. Ecol.* 68:95–121.   VII.F

Emerson, R., and C. M. Lewis. 1942. The photosynthetic efficiency of phycocyanin in Chrococcus, and the carotenoid participation in photosynthesis. *J. Ger. Physiol.* 25:579–595.   VI.C

Emig, J. W. 1966. Brown bullhead. In *Freshwater Fisheries Management,* ed. A. Calhoun, 463–475. Sacramento: California Dept. Fish Game.   V.A.9

Engstrom, D. R., and H. E. Wright, Jr. 1984. Chemical stratigraphy of lake sediments as a record of environmental change. In *Lake Sediments and Environmental History,* ed. E. Y. Haworth and J. W. G. Lund. Univ. Leicester Press. pp. 11–67.   VII.E; VII.F

Eriksson, F. 1974. Makrofyter. In *Ekosystemets Struktur i Sjön Vitalampa,* ed. F. Eriksson, J.-Å. Johansson, P. Mossberg, P. Nyberg, H. Olofsson and L. Ramberg, 9–25. Klotenprojektet Rap. 4, Scripta Limnol. Upsaliensia 370.   V.A.4

Evans, F. C. 1956. Ecosystem as a basic unit in ecology. *Science* 123:1127–1128.   I.A

Fajen, O. 1975. The standing crop of smallmouth bass and associated species in Courtois Creek. In *Black Bass Biology and Management,* ed. H. Clepper, 231–239. Washington, DC: Sport Fish. Inst.   V.A.9

Fallon, R. D., and T. D. Brock. 1980. Planktonic blue-green algae: production, sedimentation and decomposition in Lake Mendota, Wisconsin. *Limnol. Oceanogr.* 25:72–88.   V.D

Fassett, N. C. 1930. The plants of some northeastern Wisconsin lakes. *Trans. Wisconsin Acad. Sci. Arts Lett.* 25:157–168.   V.A.4

Federer, C. A. 1969. New landmark in the White Mountains. *Appalachia* (December), 586–594.   III.B.3

———. 1973. *Annual cycles of soil and water temperatures at Hubbard Brook.* U.S. Forest Service, Northeastern Forest Experiment Station, USDA Forest Service Res. Note NE-167. 7 pp.   II.A, B

Federer, C. A., and D. Lash. 1978. *BROOK: A hydrologic simulation model for eastern forests.* Research Report No. 19. Water Resour. Res. Center, Univ. New Hampshire, Durham. 84 pp.   II.B; IV.E; VIII

Fedorenko, A. Y., and M. C. Swift. 1972. Comparative biology of *Chaoborus americanus* and *Chaoborus trivittatus* in Eunice Lake, British Columbia. *Limnol. Oceanogr.* 17(5):721–730.   V.A.6

	CHAPTER
Fee, E. J. 1976. The vertical and seasonal distribution of chlorophyll in lakes of the Experimental Lakes Area, northwestern Ontario: implications for primary production. *Limnol. Oceanogr.* 21:767–783.	V.A.2; V.B.2
———. 1978a. A procedure for improving estimates of *in situ* primary production at low irradiances with an incubator technique. *Verh. Internat. Verein. Limnol.* 20:59–67.	V.B.2
———. 1978b. Studies of hypolimnion chlorophyll peaks in the Experimental Lakes Area, northwestern Ontario. *Fish. Mar. Serv. Tech. Report* 754:1–25.	V.A.2; V.B.2
———. 1979. A relation between lake morphometry and primary productivity and its use in interpreting whole-lake eutrophication experiments. *Limnol. Oceanogr.* 24(3):401–416.	V.A.2; V.B.2; VI.A; VI.C; VI.D
———. 1980. Important factors for estimating annual phytoplankton production in the Experimental Lakes Area. *Can. J. Fish. Aquatic Sci.* 37:513–522.	V.A.2; V.B.2
Fee, E. J., J. A. Shearer and D. R. DeClercq. 1977. In vivo *chlorophyll profiles from lakes in the Experimental Lakes Area, northwestern Ontario*. Can. Fish. Mar. Serv. Tech. Report 703; 136 pp.	V.A.2; V.B.2
Fenchel, T. 1968. The ecology of marine microbenthos. II. The food of marine benthic ciliates. *Ophelia* 5:73–121.	VI.C
———. 1975. The quantitative importance of the benthic microfauna of an arctic tundra pond. *Hydrobiologia* 46:445–464.	VI.C
———. 1978. The ecology of micro- and meiobenthos. *Ann. Rev. Ecol. Syst.* 9:99–121.	V.A.7
Fernald, M. L. 1950. *Gray's Manual of Botany*. 8th ed. New York: American Book Company. 1,632 pp.	V.A.4
Fiance, S. B., and R. E. Moeller. 1977. Immature stages and ecological observations of *Eoparargyractis plevie* (Pyralidae: Nymphulinae). *J. Lepidop. Soc.* 31:81–88.	V.A.4; V.A.6; V.B.3; VI.A
Ficke, J. F. 1972. *Comparison of evaporation computation methods, Pretty Lake, La Grange County, northeastern Indiana*. U.S. Geol. Surv. Prof. Paper 686-A. 49 pp.	IV.D
Ficke, J. F., D. B. Adams and T. W. Danielson. 1977. *Evaporation from seven reservoirs in the Denver water-supply system, central Colorado*. U.S. Geol. Surv. Water-Resour. Investigations 76-114. 170 pp.	IV.D
Ficken, J. H. 1967. *Winter loss and spring recovery of dissolved solids in two prairie pothole ponds in North Dakota*. U.S. Geol. Surv. Prof. Paper 575-C:228–231.	IV.C
Findenegg, I. 1964. Types of planktonic primary production in the lakes of the Eastern Alps as found by the radioactive carbon method. *Verh. Internat. Verein. Limnol.* 15:352–359.	V.B.2
———. 1966. Factors controlling primary productivity, especially with regard to water replenishment, stratification, and mixing. In *Primary Productivity in Aquatic Environments,* ed. C. R. Goldman, 105–119. Mem. Ist. Ital. Idrobiol., 18 Suppl. Berkeley: Univ. California Press.	V.A.2; V.B.2
Findley, D. L. 1978. *Seasonal successions of phytoplankton in seven lake basins in the Experimental Lakes Area, northwestern Ontario, following artificial eutrophication: data from 1974 to 1976*. Fish. Mar. Serv. Tech. Report 1466. 41 pp.	V.A.2
Findley, D. L., and H. J. Kling. 1979. *A species list and pictoral reference to the phytoplankton of central and northern Canada*. Parts I and II. Can. Fish. Mar. Serv. Tech. Report 1503. 619 pp.	V.A.2
Fisher, D. W., A. W. Gambell, G. E. Likens and F. H. Bormann. 1968. Atmospheric contributions to water quality of streams in the Hubbard Brook Experimental Forest, New Hampshire. *Water Resour. Res.* 4(5):1115–1126.	II.B
Fisher, S. G. 1970. Annual energy budget of a small forest stream ecosystem:	II.B

Bear Brook, West Thornton, New Hampshire. Ph.D. diss., Dartmouth College. 97 pp.

Fisher, S. G., and G. E. Likens. 1972. Stream ecosystem: organic energy budget. *BioScience* 22(1):33–35.   I.A.2

———. 1973. Energy flow in Bear Brook, New Hampshire: an integrative approach to stream ecosystem metabolism. *Ecol. Monogr.* 43(4):421–439.   II.B; V.C; VIII

Flannery, M. S., R. D. Snodgrass and T. J. Whitmore. 1982. Deepwater sediments and trophic conditions in Florida lakes. *Hydrobiologia* 92:597–602.   VII.E

Flett, R. J., D. W. Schindler, R. D. Hamilton and N. E. R. Campbell. 1980. Nitrogen fixation in Canadian Precambrian Shield lakes. *Can. J. Fish. Aquatic Sci.* 37(3): 494–505.   IV.E

Flint, R. W., and C. R. Goldman. 1975. The effects of a benthic grazer on the primary productivity of the littoral zone of Lake Tahoe. *Limnol. Oceanogr.* 20:935–944.   VI.A

Flössner, D. 1964. Zur Cladocerenfauna des Stechlinsee-Gebietes II. Okologische Untersuchungen über die litoralen Arten. *Limnologica* (Berlin) 2:35–103.   VII.F

Fogg, A. J. 1874. *The Statistics of Gazetteer of New Hampshire.* Concord, NH: D. L. Guernsey. 668 pp.   III.B.3

Fogg, G. E., W. D. P. Stewart, P. Fay and A. E. Walsby. 1973. *The Blue-Green Algae.* New York: Academic Press. 495 pp.   VI.C

Follett, E. A. C., W. J. McHardy, B. D. Mitchell and B. F. L. Smith. 1965. Chemical dissolution techniques in the study of soil clays: Part I. Clay Minerals 6:23–34.   VII.E

Forbes, S. A. 1887. The lake as a microcosm. Reprinted 1925. *Bull. Illinois Nat. Hist. Surv.* 15:537–550.   I.A

Forcier, L. K. 1973. Seedling pattern and population dynamics and the reproductive strategies of sugar maple, beech and yellow birch at Hubbard Brook. Ph.D. diss., Yale University, New Haven, CT.   II.A

———. 1975. Reproductive strategies and the co-occurrence of climax tree species. *Science* 189:808–810.   II.A

Ford, M. S. (J.) 1984. The influence of lithology on ecosystem development in New England: A comparative paleoecological study, Ph.D. diss., Univ. of Minnesota, Minneapolis.   VII.F

Forrest, W. W. 1969. Energetic aspects of microbial growth. In *Microbial Growth,* 65–86. 19th Symp. Soc. Gen. Microbiol.   V.B.1

Forsyth, D. J. 1978. Benthic macroinvertebrates in seven New Zealand lakes. *New Zealand J. Marine and Freshwater Res.* 12(1):41–50.   V.A.6

Fowler-Billings, K., and L. R. Page. 1942. *The Geology of the Cardigan and Rumney Quadrangles, New Hampshire.* Concord, NH: New Hampshire Planning and Development Commission.   III.A

Frank, E. C., and R. Lee. 1966. *Potential solar beam irradiation on slopes.* U.S. Forest Service Res. Paper RM-18. 116 pp.   II.A

Freeze, R. A. 1972. Role of subsurface flow in generating surface runoff. 2. Upstream source areas. *Water Resour. Res.* 8(5):1272–1283.   IV.D

Frey, D. G. 1958. The late-glacial cladoceran fauna of a small lake. *Arch. Hydrobiol.* 54:209–275.   VII.C

———. 1959. The taxonomic and phylogenetic significance of the head pores of the Chydoridae (Cladocera). *Int. Rev. Ges. Hydrobiol.* 44:27–50.   VII.C

———. 1964. Remains of animals in quaternary lake and bog sediments and their interpretation. *Arch. Hydrobiol. Beih., Ergebn. Limnol.* 2:1–114.   VII.C

———. 1980. On the plurality of *Chydorus sphaericus* (O. F. Müller) (Cladocera, Chydoridae), and designation of a neotype from Sjaelsø, Denmark. *Hydrobiologia* 69:83–123.   VII.F

CHAPTER

Frissell, S. S., Jr. 1973. The importance of fire as a natural ecological factor in Itasca State Park, Minnesota. *Quat. Res.* 3(3):397–407.   III.B.2

Fry, F. E. J. 1957. The aquatic respiration of fish. In *The Physiology of Fishes.* Vol. 1, *Metabolism,* ed. M. E. Brown, 1–63. New York: Academic Press.   V.C

Fujita, T. T. 1977. Wind storm map based on aerial survey and mapping by T. T. Fujita on 13–14 July 1977. A separate map distributed by T. T. Fujita. Department of Geophysical Science, University of Chicago.   III.B.2

Galkovskaya, G. A. 1965. Planktonnye kolovratki i ikh rol v productivosti vodoemov. Dissertation thesis, Biel. Gos. Univ., Lenina, Minsk. 19 pp.   V.B.5

Gallepp, G. W., J. F. Kitchell and S. M. Bartell. 1978. Phosphorus release from lake sediments as affected by chironomids. *Verh. Internat. Verein. Limnol.* 20(1):458–465.   VI.B

Galloway, J. N., and G. E. Likens. 1979. Atmospheric enhancement of metal deposition in Adirondack lake sediments. *Limnol. Oceanogr.* 24:427–433.   VII.E

———. 1981. Acid precipitation: the importance of nitric acid. *Atmos. Environ.* 15(6):1081–1085.   VIII

Gangopadhyaya, M., G. E. Harbeck, Jr., T. J. Nordenson, M. H. Omar and V. A. Uryvaev. 1966. *Measurement and estimation of evaporation and evapotranspiration.* World Meteorolog. Organiz. Tech. Note 83. 121 pp.   IV.D

Gardner, W. S., T. F. Nalepa, M. Quigley and J. M. Malczyk. 1981. Release of phosphorus by certain benthic invertebrates. *Can. J. Fish. Aquatic Sci.* 38:978–981.   VI.B

Gasith, A. 1975. Allochthonous organic matter and organic matter dynamics in Lake Wingra, Wisconsin. Ph.D. diss., University of Wisconsin, Madison.   V.C

———. 1976. Seston dynamics and tripton sedimentation in the pelagic zone of a shallow eutrophic lake. *Hydrobiologia* 51:225–231.   VII.A.2

Gasith, A., and A. D. Hasler. 1976. Airborne litterfall as a source of organic matter in lakes. *Limnol. Oceanogr.* 21:253–258.   V.C

Geen, G. H., T. G. Northcote, G. F. Hartman and C. C. Lindsey. 1966. Life histories of two species of catostomid fishes in Sixteenmile Lake, British Columbia, with particular reference to inlet stream spawning. *J. Fish. Res. Bd. Canada* 23:1761–1788.   V.A.9

George, M. G., and C. H. Fernando. 1970. Diurnal migration in three species of rotifers in Sunfish Lake, Ontario. *Limnol. Oceanogr.* 15:218–223.   V.A.5

Gerhart, D. Z. 1973. Nutrient limitation and production of phytoplankton in Mirror Lake, West Thornton, New Hampshire. Ph.D. diss., Cornell University, Ithaca, NY. 160 pp.   V.B.2; V.B.5; VI.B; VI.D; IX

———. 1975. Nutrient limitation in a small, oligotrophic lake in New Hampshire. *Verh. Internat. Verein. Limnol.* 19:1013–1022.   V.B.3; VI.C

Gerhart, D. Z., and G. E. Likens. 1975. Enrichment experiments for determining nutrient limitation: four methods compared. *Limnol. Oceanogr.* 20:649–653.   V.B.2

Gerlaczyńska, B. 1973. Distribution and biomass of macrophytes in Lake Dgał Maly. *Ekol. Polska* 21:743–752.   V.A.4

Gessner, F. 1948. The vertical distribution of phytoplankton and the thermocline. *Ecology* 29:386–389.   V.A.2

———. 1959. *Hydrobotanik.* II. *VEB Deutsch.* Berlin: Verlag Wiss. 701 pp.   V.A.4

Gibbs, R. J. 1967. Amazon River: environmental factors that control its dissolved and suspended load. *Science* 156(3783):1734–1737.   II.B

Gilbert, J. J. 1968. Dietary control of sexuality in rotifer *Asplanchna brightwell,* Gosse. *Physiol. Zool.* 41:14–43.   V.A.5

Gilbert, J. J., and P. L. Starkweather. 1977. Feeding on the rotifer *Brachionus calyciflours.* I. Regulatory mechanisms. *Oecologia* 48:125–131.   VI.C

References

CHAPTER

Gilbert, J. J., and G. A. Thompson. 1968. Alpha tocopherol control of sexuality and polymorphism in the rotifer *Asplanchna*. *Science* 159:734–736.   V.A.5

Gilyarov, A. M., and V. F. Matveev. 1977. Spatial overlap in the zooplankton community of Lake Glubokoe. *The Soviet J. Ecol.* 8(4):291–386.   VI.C

Gleason, H. A. 1926. The individualistic concept of the plant association. *Bull. Torrey Bot. Club* 53:7–26.   VI.C

Gliwicz, M. Z. 1969. Studies on the feeding of pelagic zooplankton in lakes of varying trophy. *Ekol. Polska Ser. A* 17:663–708.   VI.C

Godshalk, G. L., and R. G. Wetzel. 1978. Decomposition of aquatic angiosperms. III. *Zostera marina* L. and a conceptual model of decomposition. *Aquatic Botany* 5:329–354.   V.B.3

Goldman, J. C., J. J. McCarthy and D. G. Peavey. 1979. Growth rate influence on the chemical composition of phytoplankton in oceanic waters. *Nature* 279:210–215.   IV.A

Goldthwait, J. W. 1927. A town that has gone downhill. *Geographical Rev.* 17(4):527–552.   III.B.3; VII.A.2

Goldthwait, J. W., L. Goldthwait and R. P. Goldthwait. 1951. *The geology of New Hampshire*. Part I, *Surficial geology*. Concord, NH: New Hampshire Planning and Development Commission. 83 pp.   III.A

Goldthwait, R. P. 1976. Past climates on "The Hill." Part I, When glaciers were here. Part II, Permafrost fluctuations. *Mt. Washington Observatory Bull.* (March) 12–16, (June) 329–382.   III.B.1

Golterman, H. L. 1964. Mineralization of algae under sterile conditions by bacterial breakdown. *Verh. Internat. Verein. Limnol.* 15:544–548.   V.D; VI.C

———. 1973. Vertical movement of phosphate in freshwater. In *Environmental Phosphorus Handbook,* ed. E. J. Griffith et al., 509–538. New York: John Wiley & Sons.   VI.C

———. 1975. *Physiological Limnology: An Approach to the Physiology of Lake Ecosystems*. Amsterdam: Elsevier Scientific Publishing Corporation. 489 pp.   VI.C

Goodlett, J. C. 1954. Vegetation adjacent to the border of the Wisconsin Drift in Potter County, Pennsylvania. *Harv. For. Bull.* No. 25. 91 pp.   III.B.2

Goodyear, C. P., and C. E. Boyd. 1972. Elemental composition of the largemouth bass (*Micropterus salmoides*). *Trans. Amer. Fish. Soc.* 101(3):545–547.   V.B.7

Gophen, M., B. Z. Cavari and T. Berman. 1974. Zooplankton feeding on differentially labeled algae and bacteria. *Nature* 247:393–394.   VI.C

Gorham, E. 1958. Observations on the formation and breakdown of the oxidized microzone at the mud surface in lakes. *Limnol. Oceanogr.* 3:291–298.   VIII

———. 1960. Chlorophyll derivatives in surface muds from the English lakes. *Limnol. Oceanogr.* 5:29–33.   VII.D

———. 1961. Factors influencing supply of major ions to inland waters, with special reference to the atmosphere. *Geol. Soc. Amer. Bull.* 72:795–840.   VIII

Gorham, E., and J. Sanger. 1967. Plant pigments in woodland soils. *Ecology* 48(2):306–308.   VII.D

———. 1972. Fossil pigments in the surface sediments of a meromictic lake. *Limnol. Oceanogr.* 17:618–621.   VII.A.1; VII.D

———. 1975. Fossil pigments in Minnesota lake sediments and their bearing upon the balance between terrestrial and aquatic inputs to sedimentary organic matter. *Verh. Internat. Verein. Limnol.* 19:2267–2273.   VII.D

———. 1976. Fossilized pigments as stratigraphic indicators of cultural eutrophication in Shagawa Lake, northeastern Minnesota. *Geol. Soc. Amer. Bull.* 87:1638–1642.   VII.D

Gorham, E., P. M. Vitousek and W. A. Reiners. 1979. The regulation of   VII.F chemical budgets over the course of terrestrial ecosystem succession. *Ann. Rev. Ecol. Syst.* 10:53–84.

Gorham, E., J. W. G. Lund, J. E. Sanger and W. E. Dean, Jr. 1974. Some relationships between algal standing crop, water chemistry, and sediment chemistry in the English lakes. *Limnol. Oceanogr.* 19:601–617.   VII.D; VII.E

Gosz, J. R., G. E. Likens and F. H. Bormann. 1972. Nutrient content of litter fall on the Hubbard Brook Experimental Forest, New Hampshire. *Ecology* 53(5):769–784.   V.C

———. 1976. Organic matter and nutrient dynamics of the forest floor in the Hubbard Brook forest. *Oecologia* (Berl.) 22:305–320. Springer-Verlag New York Inc. Also in *The Belowground Ecosystem: A Synthesis of Plant Associated Processes,* ed. J. K. Marshall, 311–319. Science Series No. 26, Ft. Collins, Colorado (1977).   IV.E; VII.E

Gosz, J. R., R. T. Holmes, G. E. Likens and F. H. Bormann. 1978. The flow of energy in a forest ecosystem. *Sci. Amer.* 238(3):92–102.   II.A

Goulden, C. E. 1969. Developmental phases of the biocoenosis. *Proc. Nat. Acad. Sci. USA* 62:1066–1073.   VII.F

———. 1971. Environmental control of the abundance and distribution of the Chydorid Cladocera. *Limnol. Oceanogr.* 16:320–331.   VII.C

Goulden, C., and D. Frey. 1963. The occurrence and significance of lateral head pores in the genus *Bosmina* (Cladocera). *Int. Rev. der Gesamten Hydrobiol.* 48:513–522.   VII.C

Graham, S. A. 1941. Climax forests of the upper peninsula of Michigan. *Ecology* 22:355–362.   III.B.2

Grahn, O., H. Hultberg and L. Landner. 1974. Oligotrophication—a self-accelerating process in lakes subjected to excessive supply of acid substances. *Ambio* 3:93–94.   V.A.4

The Granite Monthly. July 1897. Vol. XXIII, No. 1.   III.B.3

Green, R. H. 1980. Role of a unionid clam population in the calcium budget of a small Arctic lake. *Can. J. Fish. Aquatic Sci.* 37:219–224.   V.A.6

Griffin, J. B. 1964. The Northeast woodlands area. In *Prehistoric Man in the New World,* ed. J. D. Jennings and E. Norbeck, 223. Chicago: University of Chicago Press.   III.B.3

Grill, E. V., and F. A. Richard. 1961. Nutrient regeneration from phytoplankton decomposition in sea water. *J. Mar. Res.* 22:51–69.   V.D

Grout, A. J. 1931. *Moss Flora of North America.* III, Part 2. New York: A. J. Grout. Published privately.   V.A.4

Gunn, J. M., S. U. Qadri and D. C. Mortimer. 1977. Filamentous algae as a food source for the brown bullhead (*Ictalurus nebulosus*). *J. Fish. Res. Bd. Canada* 34:396–401.   V.A.9

Haines, D. A., W. J. Johnson and W. A. Main. 1975. *Wildlife atlas of the northeastern and north central states.* USDA Forest Service General Tech. Report NC-16. 25 pp.   III.B.2

Hale, S. W., et al. 1885. *Report of the Forestry Commission of New Hampshire,* 91–100. Concord, New Hampshire: P. B. Cogsbill.   III.B.2

Hall, C. A. S. 1972. Migration and metabolism in a temperate stream ecosystem. *Ecology* 53(4):585–604.   I.A.2

Hall, D. J., S. T. Threlkeld, C. W. Burns and P. H. Crowley. 1976. The size-efficiency hypothesis and the size structure of zooplankton communities. *Ann. Rev. Ecol. Syst.* 7:177–208.   VI.C

Hall, K. J., and K. D. Hyatt. 1974. Marion Lake International Biological Program from bacteria to fish. *J. Fish. Res. Bd. Canada* 31:893–911.   V.C

References

CHAPTER

Hall, R. J., and G. E. Likens. 1981. Chemical flux in an acid-stressed stream. *Nature* 292(5821):329–331.   IV.E; V.A.6

Hall, R. J., G. E. Likens, S. B. Fiance and G. R. Hendrey. 1980. Experimental acidification of a stream in the Hubbard Brook Experimental Forest, New Hampshire. *Ecology* 61(4):976–989.   II.B; VIII

Hamilton, A. L. 1965. An analysis of a freshwater benthic community with special reference to the Chironomidae. Ph.D. diss., University of British Columbia, Vancouver. 93 pp.   V.A.6

Hamilton, A. L., W. Burton and J. F. Flannagan. 1970. A multiple corer for sampling profundal benthos. *J. Fish. Res. Bd. Canada* 27(1):1867–1869.   V.A.6

Haney, J. F. 1973. An *in situ* examination of the grazing activities of natural zooplankton communities. *Arch. Hydrobiol.* 72:87–132.   VI.C

Hanna, G. D. 1933. Diatoms of Florida peat deposits. *Ann. Report Florida State Geol. Surv.* 23–24:68–119.   VII.B

Hanson, H. C. 1972. The accuracy of groundwater contour maps. *Water Resour. Res.* 8(1):201–204.   IV.D

Hanson, K. 1959. The terms Gyttja and Dy. *Hydrobiologia* 13:309–315.   VII.E

———. 1960. Lake types and lake sediments. *Verh. Internat. Verein. Limnol.* 14:285–290.   VII.E

Harbeck, G. E., Jr. 1962. *A practical field technique for measuring reservoir evaporation utilizing mass-transfer theory*, 101–105. U.S. Geol. Surv. Prof. Paper 272-E.   IV.D

Harbeck, G. E., Jr., M. A. Kohler, G. E. Koberg, H. B. Phillips, J. E. Allen, B. J. Huebert, T. J. Nordenson, W. E. Fox and W. U. Garstka. 1958. *Water-loss investigations—Lake Mead studies*. U.S. Geol. Surv. Prof. Paper 298. 100 pp.   IV.D

Hardin, G. 1960. The competitive exclusion principle. *Science* 131:1292–1297.   VI.C

Hargrave, B. T. 1969. Epibenthic algal production and community respiration in the sediments of Marion Lake. *J. Fish. Res. Bd. Canada* 26(8):2003–2026.   V.A.6; V.B.4; V.B.6; V.C

———. 1970. The effect of a deposit-feeding amphipod on the metabolism of benthic microflora. *Limnol. Oceanogr.* 15(1):21–30.   VI.A; VI.B

———. 1973. Coupling carbon flow through some pelagic and benthic communities. *J. Fish. Res. Bd. Canada* 30:1317–1326.   V.C

———. 1975a. The importance of total and mixed-layer depth in the supply of organic material to bottom communities. *Symp. Biol. Hung.* 15:157–165.   VII.A.2

———. 1975b. Stability in structure and function of the mud-water interface. *Verh. Internat. Verein. Limnol.* 19(2):1073–1079.   VI.B

Harring, H. K., and F. J. Myers. 1928. The rotifer fauna of Wisconsin. IV. The Dicranophorinae. *Trans. Wisconsin Acad. Sci. Arts Letters* 23:667–808.   V.A.7

Harris, G. P. 1980. Temporal and spatial scales in phytoplankton ecology. Mechanisms methods, models and management. *Can. J. Fish. Aquatic Sci.* 37:877–900.   IV.A

Hart, G. E., Jr., R. E. Leonard and R. S. Pierce. 1962. *Leaf fall, humus depth, and soil frost in a northern hardwood forest*. U.S. Forest Service, Northeastern Forest Experiment Station, Forest Research Notes, No. 131. 3 pp.   II.B

Hasler, A. D. 1947. Eutrophication of lakes by domestic drainage. *Ecology* 28:383–395.   IV.E; V.B.2

Hasler, A. D., and B. Ingersoll. 1968. Dwindling lakes. *Nat. Hist.* 77(9):1–6.   IV.E

Hasler, A. D., and G. E. Likens. 1963. Biological and physical transport of radionuclides in stratified lakes. In *Radioecology*, ed. V. Schultz and A. W. Klement, Jr., 463–470. New York: American Institute of Biological Sciences and Reinhold Publ. Corp.   I.A.2

Havas, M., and T. C. Hutchinson. 1982. Aquatic invertebrates from the Smoking Hills, N.W.T.: Effect of pH and metals on mortality. *Can. J. Fish. Aquatic Sci.* 39:890–903.   IX

	CHAPTER

Hayes, F. R. 1964. The mud-water interface. *Oceanogr. Mar. Biol. Ann. Rev.* 2:121–145. — VI.B

Healey, F. P., and L. L. Hendzel. 1980. Physiological indicators of nutrient deficiency in lake phytoplankton. *Can. J. Fish. Aquatic Sci.* 37:442–453. — V.B.2

Healey, M. C. 1977. *Experimental cropping of lakes. 4. Benthic communities.* Fisheries and Marine Service Tech. Report No. 711. 48 pp. — V.A.6

Healy, W. R. 1970. Reduction of neoteny in Massachusetts populations of *Notopthalmus viridescens*. *Copeia* 1970(3):578–581. — V.A.8

Heede, B. H. 1972a. Flow and channel characteristics of two high mountain streams. USDA Forest Service Research Paper RM-96, Rocky Mountain Forest and Range Experiment Station, Ft. Collins, Colorado. — II.B

———. 1972b. Influences of a forest on the hydraulic geometry of two mountain streams. *Water Resour. Bull.* 8:523–530. — II.B

Heinselman, M. L. 1973. Fires in the virgin forests of the Boundary Waters Canoe Area, Minnesota. *Quat. Res.* 3(3):329–382. — III.B.1; III.B.2; VII.F

Helfman, G. S. 1979. Twilight activities of yellow perch, *Perca flavescens*. *J. Fish. Res. Bd. Canada* 36:173–179. — V.A.9

———. 1981. Twilight activities and temporal structure in a freshwater fish community. *Can. J. Fish. Aquatic Sci.* 38(11):1405–1420. — V.A.9

Hellebust, J. A. 1974. Extracellular products. In *Algal Physiology and Biochemistry,* ed. W. D. P. Stewart, 838. Botanical Monographs 10. Blackwell Scientific Publications. — V.D

Hellquist, C. B. 1980. Correlation of alkalinity and the distribution of *Potamogeton* in New England. *Rhodora* 82:331–344. — V.A.4

Hendrey, G. R., and R. F. Wright. 1975. Acid precipitation in Norway: effects on aquatic fauna. *J. Great Lakes Res.* 2 (Suppl.):192–207. — VIII

Henry, J. D., and J. M. A. Swan. 1974. Reconstructing forest history from live and dead plant material: an approach to the study of forest succession in southwest New Hampshire. *Ecology* 55:772–783. — III.B.2

Henson, E. B. 1954. The profundal bottom fauna of Cayuga Lake. Ph.D. diss., Cornell University, Ithaca, NY. 108 pp. — V.A.6

Hesslein, R. H., W. S. Broecker, P. D. Quay and D. W. Schindler. 1980. Whole-lake radiocarbon experiment in an oligotrophic lake at the Experimental Lakes Area, northwestern Ontario. *J. Can. Fish. Aquatic Sci.* 37:454–463. — V.D; VII.E

Hewlett, J. D., and C. A. Troendle. 1975. *Non-point and diffused water sources—a variable source area problem.* Amer. Soc. Civil Eng., Irrigation and Drainage Division, Proc. Symp. on Watershed Management, Logan, Utah, August 1975. — IV.D

Hill, H. 1967. A note on temperature and water conditions beneath ice in spring. *Limnol. Oceanogr.* 12:550–552. — IV.C

Hillbricht, A. 1961. The character of free swimming rotatoria bred in aquaria. *Ekol. Polska, Ser. A* 9(3):39–58. — VI.C

Hillbricht-Iłkowska, A. 1964. The influence of the fish population in the biocoenosis of a pond, using Rotifera fauna as an illustration. *Ekol. Polska Ser. A* 12(28):453–503. — VI.C

Hillbricht-Iłkowska, A., I. Spondiewska, T. Weglenska and A. Karabin. 1972. The seasonal variation of some ecological efficiencies and production rates in the plankton community of several Polish lakes of different trophy. In *Productivity Problems of Freshwaters,* Z. Kajak and A. Hillbricht-Iłkowska, 111–128. Proc. of IBP-UNESCO, Symp., Kazimierz Dolny, Poland. May 1970. — VI.C

Hilsenhoff, W. L. 1966. The biology of *Chironomus plumosus* (Diptera: Chironomidae) in Lake Winnebago, Wisconsin. *Ann. Entomol. Soc. Amer.* 59(3):465–473. — V.A.6

References

	CHAPTER
Hilsenhoff, W. L., and R. P. Narf. 1968. Ecology of Chironomidae, Chaoboridae, and other benthos in fourteen Wisconsin lakes. *Ann. Entomol. Soc. Amer.* 61(5):1173–1181.	V.A.6
Hobbie, J. E., and C. Crawford. 1969. Respiration corrections for bacterial uptake of dissolved organic compounds in natural waters. *Limnol. Oceanogr.* 14:528–532.	V.B.1
Hobbie, J. E., and G. E. Likens. 1973. Output of phosphorus, dissolved organic carbon, and fine particulate carbon from Hubbard Brook watersheds. *Limnol. Oceanogr.* 18(5):734–742.	II.B; IX
Hobbie, J. E., and R. F. Wright. 1966. Competition between planktonic bacteria and algae for organic solutes. In *Primary Productivity in Aquatic Environments*, ed. C. R. Goldman, 175–185. Univ. California Press.	V.B.1
Hobbie, J. E., R. J. Daley and S. Jasper. 1977. Use of Nuclepore filters for counting bacteria by fluorescence microscopy. *Appl. Environ. Microbiol.* 33:1225–1228.	V.A.1
Holmes, C. H. 1915. The Indians of New Hampshire. *The Granite Monthly* 47:85–88.	III.B.3
Holopainen, I. J., and L. Paasivirta. 1977. Abundance and biomass of the meiozoobenthos in the oligotrophic and mesohumic lake Pääjärvi, southern Finland. *Ann. Zool. Fenn.* 14:124–134.	V.A.7
Horie, S. 1969. Asian lakes. In *Eutrophication: Causes, Consequences, Correctives*, 98–123. Washington, DC: National Academy of Sciences, National Academy of Sciences/National Research Council. Publ. 1700.	VIII
Hornbeck, J. W., and W. Kropelin. 1982. Nutrient removal and leaching from a whole-tree harvest of northern hardwoods. *J. Environ. Qual.* 11(2):309–316.	IX
Hornbeck, J. W., G. E. Likens and J. S. Eaton. 1976. Seasonal patterns in acidity of precipitation and the implications for forest-stream ecosystems. In *Proc. The First Internat. Symp. on Acid Precipitation and the Forest Ecosystem*, ed. L. S. Dochinger and T. A. Seliga, 597–609. USDA Forest Service General Tech. Report NE-23; Also in *Water, Air, and Soil Pollut.* 7:355–365 (1977).	II.B
Hornbeck, J. W., G. E. Likens, R. S. Pierce and F. H. Bormann. 1975. Strip cutting as a means of protecting site and streamflow quality when clearcutting northern hardwoods. In *Proc. 4th North American Forest Soils Conf. on Forest Soils and Forest Land Management*, ed. B. Bernier and C. H. Winget, 208–229. August 1973. Quebec.	VI.D; IX
Hornbeck, J. W., C. W. Martin, R. S. Pierce, G. E. Likens, J. S. Eaton and F. H. Bormann. 1982. Impacts of even-age management on nutrient and hydrologic cycles of northern hardwood forests. *Bull. Ecol. Soc. Amer.* 63(2):158.	IX
Hough, A. F. 1936. A climax forest community on East Tionesta Creek in northwestern Pennsylvania. *Ecology* 17:9–28.	III.B.2
Hough, A. F., and R. D. Forbes. 1943. The ecology and silvics of forests in the high plateaus of Pennsylvania. *Ecol. Monogr.* 13:299–320.	III.B.2
Hounam, C. E. 1973. *Comparison between pan and lake evaporation*. World Meteorological Organization, Tech. Note 126. 52 pp.	IV.D
Hubbs, C. L., and R. M. Bailey. 1938. The small-mouthed bass. *Cranbrook Inst. Sci. Bull.* 10:1–89.	V.A.9
Hulings, N. C., and J. S. Gray. 1971. A manual for the study of meiofauna. *Smithsonian Contrib. Zool.* 78:1–83.	V.A.7
Hultberg, H. 1976. Thermally stratified acid water in late winter—a key factor inducing self-acceleration processes which increase acidification. *Water, Air, and Soil Pollut.* 7:279–294.	II.B
Hunter, J. G. 1970. *Production of arctic char* (Salvelinus alpinus Linnaeus) *in a small arctic lake*. Fish. Res. Bd. Canada Tech. Report No. 231. 190 pp.	V.A.6

	CHAPTER
Hurd, D. H. 1892. *Town and city atlas of the State of New Hampshire*. Boston: D. H. Hurd and Co.	III.B.3
Hurlbert, S. H. 1970. Predator responses to the vermillion-spotted newt (*Notophthalmus viridescens*). *J. Herp.* 4(1–2):47–55.	V.A.8
Hutchinson, G. E. 1957. *A Treatise on Limnology, Vol. I, Geography, Physics and Chemistry*. New York: John Wiley & Sons. 1,115 pp.	IV.B; IV.C
———. 1958. Concluding remarks. *Cold Spring Harbor Symp. Quant. Biol.* 22:415–427.	VI.C
Hutchinson, G. E. 1967. *A Treatise on Limnology*, Vol. II. *Introduction to Lake Biology and the Limnoplankton*, John Wiley & Sons, New York. 1,115 pp.	V.A.2; V.A.5; V.A.6; V.A.7; V.B.2; V.B.5; VI.C
———. 1969. Eutrophication, past and present. In *Eutrophication: Causes, Consequences and Correctives*, 17–26. Washington, DC: National Academy of Sciences/National Research Council. Publ. 1700.	VIII
———. 1975. *A Treatise on Limnology*. Vol. III, *Limnological Botany*. New York: Wiley-Interscience. 660 pp.	V.A.4; V.B.3; VI.A
Hutchinson, G. E., and A. Wollack. 1940. Studies on Connecticut lake sediments. II. Chemical analysis of a core from Linsley Pond, North Branford. *Amer. J. Sci.* 238:493–517.	VII.A.1; VII.F
Hutchinson, G. E., E. Bonatti, U. M. Cowgill, C. E. Goulden, E. A. Leventhal, M. E. Mallett, F. Margaritora, R. Patrick, A. Racek, W. A. Robak, E. Stella, J. B. Wart-Perkins and T. R. Wellman. 1970. Ianula: an account of the history and development of the Lago di Monterosi, Latium, Italy. *Trans. Amer. Phil. Soc.* 60(4):1–178.	VII.B; VII.E; VIII
Hutchinson, T. C., and M. Havas, ed. 1980. *Effects of Acid Precipitation on Terrestrial Ecosystems*. NATO Conf. Series 1: Ecology 4. New York: Plenum Publishing Corp. 654 pp.	VIII
Huttunen, P., and K. Tolonen. 1975. Human influence in the history of Lake Lovojärvi, S. Finland. *Finskt Museum* 1975:68–117.	VII.D
Huttunen, P., J. Meriläinen and K. Tolonen. 1978. The history of a small dystrophied forest lake, southern Finland. *Pol. Arch. Hydrobiol.* 25(1/2):189–202.	VII.D; VII.E; VII.F
Hyman, L. H. 1949. *The Invertebrates: Platyhelminthes and Rhynchocoela*. New York: McGraw-Hill. 550 pp.	V.A.7
———. 1951. *The Invertebrates: Acanthocephala, Aschelminthes, and Entoprocta*. New York: McGraw-Hill. 572 pp.	V.A.7
Hynes, H. B. N., and M. J. Coleman. 1968. A simple method of assessing the annual production of stream benthos. *Limnol. Oceanogr.* 13(4):569–573.	V.B.6
Idso, S. B. 1973. The concept of lake stability. *Limnol. Oceanogr.* 18:681–683.	IV.C
Ikusima, I. 1966. Ecological studies on the productivity of aquatic plant communities. II. Seasonal changes in standing crop and productivity of a natural, submerged community of *Vallisneria denseserrulata*. *Bot. Mag.* (Tokyo) 79:7–19.	V.B.3
Il'ina, L. K. 1973. The behavior of perch (*Perca fluviatilis* L.) underyearlings of different ecological groups in the progeny of a single pair of spawners. *J. Ichthyol.* 13:294–303.	V.A.9
Iversen, J. 1929. Studien über die pH-Verhältnisse dänischer Gewässer und ihren Einfluß auf die Hydrophytenvegetation. *Bot. Tidskr.* 40:277–326.	V.A.4
———. 1964. Retrogressive vegetational succession in the postglacial. *J. Ecol.* 52:59–70.	VII.F
———. 1969. Retrogressive development of a forest ecosystem demonstrated in pollen diagrams from fossil mor. *Oikos Suppl.* 12:35–49.	VII.F

References

	CHAPTER
———. 1973. The development of Denmark's nature since the last glacial. *Danish Geol. Surv. Ser.* 5(7-C):1–126.	III.B.1
Izvekova, E. I. 1971. On the feeding habits of chironomid larvae. *Limnologica* (Berlin) 8:201–202.	V.A.6
Jackson, G. A. 1980. Phytoplankton growth and zooplankton grazing in oligotrophic oceans. *Nature* 284:439–441.	IV.A
Jacobson, G. L., Jr., and R. H. Bradshaw. 1981. The selection of sites for paleovegetational studies. *Quat. Res.* 16:80–96.	III.B.1
Järnefelt, H. 1952. Plankton als Indikator der Trophiegrappen der Seen. *Ann. Acad. Sci. Fenn. Ser. A* 18:1–29.	V.B.2
———. 1956. Zur limnologie einiger Gewässer Finnlands. XVI. Mit besodnderer Berucksichtigung des Planktens. *Ann. Zool. Soc. Zool. Bot. Fenn.* 17:1–201.	V.A.2
Jassby, A. D. 1975. Dark sulfate uptake and bacterial productivity in a subalpine lake. *Ecology* 56:627–636.	V.B.1
Jenny, H. 1980. *The Soil Resource*. Springer-Verlag New York Inc. 377 pp.	VII.E
Jensen, T., and L. Sicko-Goad. 1976. *Aspects of phosphate utilization by blue-green algae*. Springfield, VA: Corvallis Env. Res. Lab. EPA-600/3-76-103. NTIS. 103 pp.	VI.C
Jensen, V. S. 1939. *Edging white pine lumber in New England*. USDA Forest Service. Northeastern Forest Experiment Station Tech. Note No. 26. 2 pp.	III.B.2
JIBP-PF Research Group of Lake Biwa. 1975. Productivity of freshwater communities in Lake Biwa. In *Productivity of Communities in Japanese Inland Waters*, ed. S. Mori and G. Yamamoto, 1–45. JIBP Synthesis Vol. 10. Contribution 151, Otsu Hydrobiol. Station, University of Kyoto.	V.B.1
Johannes, R. E. 1968. Nutrient regeneration in lakes and oceans. In *Advances in Microbiology of the Sea*, Vol. 1, ed. M. R. Droop and E. J. Wood, 203–213. New York: Academic Press.	V.B.1; VI.C
Johannessen, M., T. Dale, E. T. Gjessing, A. Henriksen and R. F. Wright. 1976. *Proceedings of the International Symposium on Isotopes and Impurities in Snow and Ice*. Internat. Assoc. of Hydrological Science, Grenoble, France. August 1975. Internat. Assoc. Hydrol. Sci. Publ. 118.	II.B
Johnson, M. G., and R. O. Brinkhurst. 1971. Production of benthic macroinvertebrates of Bay of Quinte and Lake Ontario. *J. Fish. Res. Bd. Canada* 28:1699–1714.	V.B.6
Johnson, M. G., and G. E. Owen. 1971. The role of nutrients and their budgets in the Bay of Quinte, Lake Ontario. *J. Water Pollut. Contr. Fed.* 43:836–853.	VI.C
Johnson, M. G., M. F. P. Michalski and A. E. Christie. 1970. Effects of acid mine wastes on phytoplankton communities of two northern Ontario lakes. *J. Fish. Res. Bd. Canada* 27:425–444.	V.A.2
Johnson, N. M., and G. E. Likens. 1967. Steady-state thermal gradient in the sediments of a meromictic lake. *J. Geophys. Res.* 72:3049–3052.	IV.C
Johnson, N. M., and D. H. Merritt. 1979. Convective and advective circulation of Lake Powell, Utah–Arizona, during 1972–75. *Water Resour. Res.* 15:873–884.	IV.C
Johnson, N. M., J. S. Eaton and J. E. Richey. 1978. Analysis of five North American lake ecosystems. II. Thermal energy and mechanical stability. *Verh. Internat. Verein. Limnol.* 20:562–567.	IV.C
Johnson, N. M., C. T. Driscoll, J. S. Eaton, G. E. Likens and W. H. McDowell. 1981. "Acid rain," dissolved aluminum and chemical weathering at the Hubbard Brook Experimental Forest, New Hampshire. *Geochim. Cosmochim. Acta* 45(9):1421–1437.	II.B; VII.E; VII.F; VIII; IX

	CHAPTER
Johnson, N. M., G. E. Likens, F. H. Bormann, D. W. Fisher and R. S. Pierce. 1969. A working model for the variation in stream water chemistry at the Hubbard Brook Experimental Forest, New Hampshire. *Water Resour. Res.* 5(6):1353–1363.	II.B
Johnson, N. M., G. E. Likens, F. H. Bormann and R. S. Pierce. 1968. Rate of chemical weathering of silicate minerals in New Hampshire. *Geochim. Cosmochim. Acta* 32:531–545.	VII.E; VII.F
Johnson, W. E., and A. D. Hasler, 1954. Rainbow trout production and dystrophic lakes. *J. Wildl. Management* 18(1):113–134.	VI.D
Jonasson, P. M. 1955. The efficiency of sieving technique for sampling freshwater bottom fauna. *Oikos* 6:183–207.	V.A.6
———. 1958. The mesh factor in sieving techniques. *Verh. Internat. Verein. Limnol.* 13:860–866.	V.A.6; V.A.7
———. 1972. Ecology and production of the profundal benthos in relation to phytoplankton in Lake Esrom. *Oikos Suppl.* 14. Copenhagen: Munksgaard. 139 pp.	V.A.6; V.B.6; VI.B; VI.C
———. 1979. The Lake Myvatn ecosystem, Iceland. *Oikos* 32:289–305.	V.B.6
Jones, B. F., and C. J. Bowser. 1978. The mineralogy and related chemistry of lake sediments. In *Lakes: Chemistry, Geology, Physics,* ed. A. Lerman, Springer-Verlag New York Inc.	VII.E
Jordan, M. J., and G. E. Likens. 1975. An organic carbon budget for an oligotrophic lake in New Hampshire, U.S.A. *Verh. Internat. Verein. Limnol.* 19(2):994–1003.	IV.E; V.B.1; V.B.2; V.C; V.D; VII.E
———. 1980. Measurement of planktonic bacterial production in an oligotrophic lake. *Limnol. Oceanogr.* 25(4):721–732.	V.B.1; V.B.2; V.C; V.D; VI.C; IX
Jordan, M. J., and B. J. Peterson. 1978. Sulfate uptake as a measure of bacterial production. *Limnol. Oceanogr.* 23:146–150.	V.B.1
Jordan, P. A., D. B. Botkin, A. S. Dominski, H. S. Lowendorf and G. E. Belovsky. 1973. Sodium as a critical nutrient for the moose of Isle Royale. In *Proc. 9th North American Moose Conference and Workshop,* 13–42.	IX
Juday, C. 1908. Some aquatic invertebrates that live under anaerobic conditions. *Trans. Wisconsin Acad. Sci. Arts Lett.* 16:10–16.	V.A.6
———. 1921. Observations on the larvae of *Corethra punctipennis* Say. *Biol. Bull.* (Woods Hole) 40(5):271–286.	V.A.6
———. 1934. The depth distribution of some aquatic plants. *Ecology* 15:325.	V.A.2
———. 1940. The annual energy budget of an inland lake. *Ecology* 21:438–450.	V.B.6
———. 1942. The summer standing crop of plants and animals in four Wisconsin lakes. *Trans. Wisconsin Acad. Sci. Arts Lett.* 34:103–135.	V.A.4; V.A.6
Juday, C. and C. L. Schloemer. 1938. Effect of fertilizers on plankton production, and on fish growth in a Wisconsin lake. *Progressive Fish Culturist* 40:24–27.	VI.D
Judd, W. W. 1953. A study of the population of insects emerging as adults from the Dundas Marsh, Hamilton, Ontario, during 1948. *Amer. Midl. Natur.* 49:801–824.	V.A.6
———. 1960. A study of the population of insects emerging as adults from South Walker Pond at London, Ontario. *Amer. Midl. Natur.* 63:194–210.	V.A.6
———. 1961. Studies of the Byron Bog in Southwestern Ontario. XII. A study of the population of insects emerging as adults from Redmond's Pond in 1957. *Amer. Midl. Natur.* 65:89–100.	V.A.6
———. 1964. A study of the population on insects emerging as adults from Saunders Pond at London, Ontario. *Amer. Midl. Natur.* 71:402–414.	V.A.6
Kajak, Z., and K. Dusoge. 1970. Production efficiency of *Procladius choreus* Mg. (Chironomidae, Diptera) and its dependence on the trophic conditions. *Pol. Arch. Hydrobiol.* 17:217–224.	V.A.6
Kajak, Z., and B. Ranke-Rybicka. 1970. Feeding and production efficiency of	V.A.6

References

CHAPTER

Chaoborus flavicans Meigen (Diptera, Culicidae) larvae in eutrophic and dystrophic lake. *Pol. Arch. Hydrobiol.* 17(3):225–232.

Kajak, Z., A. Hillbricht-Iłkowska and E. Pieczynska. 1972. The production processes in several Polish lakes. In *Productivity Problems of Freshwaters,* ed. Z. Kajak and A. Hillbricht-Iłkowska, 129–147. Proc. of IBP-UNESCO Symp., Kazimierz Dolny, Poland. 6–12 May 1970.   V.B.6

Kalff, J., and R. Knoechel. 1978. Phytoplankton and their dynamics in oligotrophic and eutrophic lakes. *Ann. Rev. Ecol. Syst.* 9:475–495.   V.A.2; V.B.2

Kalff, J. and H. E. Welch. 1974. Phytoplankton production in Char Lake, a natural polar lake, and in Meretta Lake, polluted polar lake, Cornwallis Island, Northwest Territories. *J. Fish. Res. Bd. Canada* 31(5):621–636.   V.C

Kalff, J., H. J. Kling, S. H. Holmgren and H. E. Welch. 1975. Phytoplankton, phytoplankton growth and biomass cycles in an unpolluted and in a polluted polar lake. *Verh. Internat. Verein. Limnol.* 17:487–495.   VII.F

Kanehiro, Y., and G. D. Sherman. 1965. Fusion with sodium carbonate for total elemental analysis. In *Methods of Soil Analysis,* ed. C. A. Black et al., 953–958. Madison, WI: Amer. Soc. Agronomy Inc.   VII.E

Kann, E. 1940. Ökologische Untersuchungen an Litoralalgen ostholsteinischer Seen. *Arch. Hydrobiol.* 37:177–269.   V.A.3

Kansanen, A., and R. Niemi. 1974. On the production ecology of isoetids, especially *Isoëtes lacustris* and *Lobelia dortmanna* in Lake Pääjärvi, southern Finland. *Ann. Bot. Fenn.* 11:178–187.   V.B.3

Kansanen, A., R. Niemi and K. Overlund. 1974. Pääjärven makrofyytit. *Luonnon Tutkija* 78(4/5):437–574.   V.A.4

Karentz, D., and C. D. McIntire. 1977. Distribution of diatoms in the plankton of Yoquina estuary, Oregon. *J. Phycol.* 13:379–388.   VI.C

Keast, A. 1979. Patterns of predation in generalist feeders. In *Predator-Prey Systems in Fisheries Management,* ed. H. Clepper, 243–255. Washington, DC: Sport Fish. Inst.   V.A.9

Keating, K. I. 1977. Allelopathic influence on blue-green blown sequence in a eutrophic lake. *Science* 196:885–886.   VI.C

———. 1978. Blue-green algal inhibition of diatom growth: transition from mesotrophic to eutrophic community structure. *Science* 199:971–973.   VI.C

Keller, E. A., A. MacDonald and T. Tally. 1982. Streams in the coastal redwood environment: the role of large organic debris. In *Symposium on Watershed Rehabilitation in Redwood National Park and other coastal areas,* ed. R. Coats. John Muir Inst. Center for Natural Resource Studies and National Park Service.   VII.F

Kerekes, J. J. 1974. Limnological conditions in five small oligotrophic lakes in Terra Nova National Park, Newfoundland. *J. Fish. Res. Bd. Canada* 31:555–583.   V.A.2

Kiefer, D. A., O. Holm-Hansen, C. R. Goldman, R. Richards and T. Berman. 1972. Phytoplankton in Lake Tahoe: deep-living populations. *Limnol. Oceanogr.* 17:418–422.   V.A.2; V.B.2

Kikuchi, K. 1937. Studies in the vertical distribution of the plankton crustacea. I. A comparison of the vertical distribution of the plankton crustacea in six lakes in middle Japan in relation to the underwater illumination and the water temperature. *Rec. Oceanogr. Wrks. Japan* 10:17–42.   VI.C

Kilbourne, F. W. 1916. *Chronicles of the White Mountains.* Boston: Houghton Mifflin Co. 434 pp.   III.B.2; III.B.3

Kilham, P., and G. E. Likens. 1968. *Penetration of light in the lakes of Grafton County, New Hampshire.* Concord, NH: New Hampshire Department of Resources and Economic Development. 93 pp.   IV.B

Kimerle, R. A., and N. H. Anderson. 1971. Production and bioenergetic role of the midge *Glyptotendipes barbipes* (Staeger) in a waste stabilizaiton lagoon. *Limnol. Oceanogr.* 16(4):646–659.   V.A.6; V.B.6

	CHAPTER
Kitchell, J. F., J. F. Koonce and P. S. Tennis. 1975. Phosphorus flux through fishes. *Verh. Internat. Verein. Limnol.* 19:2478–2484.	V.B.7
Kjensmo, J. 1968. Late and post-glacial sediments in a small meromictic lake Svinsjøen. *Arch. Hydrobiol.* 65:125–141.	VII.E
Kleckner, J. F. 1967. The role of the bottom fauna in mixing lake sediments. M.S. thesis, University of Washington, Seattle. 61 pp.	VI.B
Klemer, A. R. 1976. The vertical distribution of *Oscillatoria aghardii* var. *isothrix*. *Arch. Hydrobiol.* 78:343–362.	V.A.2
Kling, H. J., and S. K. Holmgren. 1972. *Species composition and seasonal distribution of phytoplankton in the Experimental Lakes Area, northwestern Ontario*. Fish. Mar. Serv. Tech. Report 337. 55 pp.	V.A.2; V.B.2
Knoechel, R., and F. J. deNoyelles, Jr. 1980. Analysis of the response of hypolimnetic phytoplankton in continuous culture to increased light or phosphorus using track autoradiography. *Can. J. Fish. Aquatic Sci.* 37:434–441.	V.A.2; V.B.2
Koehl, M. A. R., and J. R. Strickler. 1981. Copepod feeding currents: food capture at low Reynolds number. *Limnol. Oceanogr.* 21:674–683.	V.A.5
Kohler, M. A., T. J. Nordenson and D. R. Baker. 1959. *Evaporation maps for the United States*. U.S. Weather Bureau Tech. Paper 37. 13 pp.	IV.E
Kohler, M. A., T. J. Nordenson and W. E. Fox. 1955. *Evaporation from pans and lakes*. U.S. Weather Bureau, Res. Paper 38. 20 pp.	IV.D
Kolasa, J. 1977. III. Turbellaria and Nemertini. In *Bottom Fauna of the Heated Konin Lakes,* ed. A. Wroblewski, 29–48. Monogr. Fauny Polski 7.	V.A.7
———. 1979. Ecological and faunistical characteristics of Turbellaria in the eutrophic Lake Zbechy. *Acta Hydrobiol.* 21:435–459.	V.A.7
Koljonen, T., and L. Carlson. 1975. *Behavior of the major elements and minerals in sediments of four humic lakes in south-western Finland*. Fennia 137. 47 pp.	VII.E
Kononova, M. M. 1966. *Soil Organic Matter*. 2d English ed. Oxford: Pergamon Press. 544 pp.	VII.E
Konopka, A. 1981. Influence of temperature oxygen, and pH on a metalimnetic population of *Oscillatoria rubescens*. *Appl. Environ. Microbiol.* 42:102–108.	V.A.2
Konstantinov, A. S. 1971. The group effect in the respiration of larvae of *Chironomus dorsalis*. Hydrobiol. J. (English Translation *Gidbrobiol. Zh.*) 7(5):84–86.	V.A.6
Koppen, J. D. 1973. Distribution of the species of the diatom genus *Tabellaria* in a portion of the northcentral United States. Ph.D. diss., Iowa State University, Ames. 270 pp.	VII.B
———. 1975. A morphological and taxonomic consideration of *Tabellaria* (Bacillariophyceae) from the northcentral United States. *J. Phycol.* 11(2):236–244.	VII.B
———. 1978. Distribution and aspects of the ecology of the genus *Tabellaria* Ehr. (Bacillariophyceae) in the northcentral United States. *Amer. Midl. Natur.* 99(2):383–397.	VII.B
Kormondy, E. J. 1969. *Concepts of Ecology*. Englewood Cliffs, NJ: Prentice-Hall Inc.	I.A.2
Krishnaswami, S., and D. Lal. 1978. Radionuclide limnochronology. Chapter 6 in *Lakes: Chemistry, Geology, Physics,* ed. A Lerman, Springer-Verlag New York Inc.	VII.E
Krokhin, E. M. 1967. Effect of size of escapement of sockeye salmon spawners on the phosphate content of a nursery lake. *Fish. Res. Bd. Canada Trans.* Ser. No. 1186:31–54.	I.A.2
Kutzbach, J. E. 1981. Monsoon climate of the early Holocene: climate experiment with the earth's orbital parameters for 9,000 years ago. *Science* 214:59–61.	VII.F

References

	CHAPTER
Kuznetsov, S. I. 1968. Recent studies on the role of microorganisms in the cycling of substances in lakes. *Limnol. Oceanogr.* 13:211–224.	V.B.1
Lamarck, J. B. 1802. *Hydrogéologie, ou Recherches sur l'influence qu'ont les eaux sur la surface du globe terrestre; sur les causes de l'existence du bassin des mers, de son déplacement et de son transport successif sur les différénts points de la surface de ce globe; surfin sur les changements que les corps vivants exercent sur la nature et l'état de cette surface.* Paris, An X.	I.A
Lampert, W. 1974. A method for determining food selection by zooplankton. *Limnol. Oceanogr.* 19:995–998.	VI.C
Lane, P. A. 1978. Zooplankton niches and the community structure controversy. *Science* 200:458–460.	VI.C
Lang, G. 1967. Die Ufervegetation des westlichen Bodensees. *Arch. Hydrobiol. Suppl.* 32(4):437–574.	V.A.4; VI.A
Langbein, W. B., C. H. Hains and R. C. Culler. 1951. *Hydrology of stockwater reservoirs in Arizona.* U.S. Geol. Surv. Circular 110. 18 pp.	IV.D
Larkin, P. A. 1979. Predator-prey relations in fishes: an overview of the theory. In *Predator-Prey Systems in Fisheries Management,* ed. H. Clepper, 13–22. Washington, DC: Sport Fish. Inst.	V.A.9
LaRow, E. J. 1968. A persistent diurnal rhythm in *Chaoborus* larvae. I. The nature of the rhythmicity. *Limnol. Oceanogr.* 13(2):250–256.	V.A.6
———. 1969. A persistent diurnal rhythm in *Chaoborus* larvae. II. Ecological significance. *Limnol. Oceanogr.* 14(2):213–218.	V.A.6
———. 1970. The effect of oxygen tension on the vertical migration of *Chaoborus* larvae. *Limnol. Oceanogr.* 15(3):357–362.	V.A.6
———. 1976. Biorhythms and the vertical migration of limnoplankton. In *Biological Rhythms in the Marine Environment,* ed. P. J. De Coursey, 225–238. Columbia: Univ. S. Carolina Press.	VII.A.2
LaRow, E. J., and D. C. McNaught. 1978. Systems and organismal aspects of phosphorus remineralization. *Hydrobiologia* 59(3):151–154.	VI.C
LaRow, E. J., J. W. Wilkinson and K. Kumar. 1975. The effect of food concentration and temperature on respiration and excretion in herbivorous zooplankton. *Verh. Internat. Verein. Limnol.* 19:966–973.	VI.C
Larson, D. W. 1972. Temperature, transparency and phytoplankton productivity in Crater Lake, Oregon. *Limnol. Oceanogr.* 17:410–417.	V.B.2
Larsson, U., and A. Hagstrom. 1979. Phytoplankton exudate release as an energy source for the growth of pelagic bacteria. *Marine Biology* 52:199–206.	V.D
Lawacz, W. 1969. The characteristics of sinking materials and the formation of bottom deposits in a eutrophic lake. *Mitt. Int. Ver. Theore. Angew. Limnol.* 17:319–331.	V.B.6
Lawry, N., and T. Jensen. 1979. Deposition of condensed phosphate as an effect of varying sulfur deficiency in the *Cyanobacterium Synechococcus* sp. (*Anacystis nidulans*). *Arch. Microbiol.* 120:1–7.	VI.C
Lean, D. R. S., and N. M. Charlton. 1976. A study of phosphorus and its kinetics in a lake ecosystem. In *Environmental Biochemistry,* Vol. 1, ed. J. O. Nriagu, 283–294. Ann Arbor, MI: Ann Arbor Science.	VI.C
Lean, D. R. S., and C. Nalewajko. 1976. Phosphate exchange and organic phosphorus excretion by freshwater algae. *J. Fish. Res. Bd. Canada* 33:2805–2811.	VI.C
Learner, M. A., and D. W. B. Potter. 1974. The seasonal periodicity of emergence of insects from two ponds in Hertfordshire, England, with special reference to the Chironomidae (Diptera: Nematocera). *Hydrobiologia* 44:495–510.	V.A.6

	CHAPTER
Lechevalier, H. A., and D. Pramer. 1971. *The Microbes*. Philadelphia: J. B. Lippincott. 507 pp.	V.A.1
LeCren, E. D. 1955. Year to year variation in the year class strength of *Perca fluviatilis*. *Verh. Internat. Verein. Limnol.* 12:187–192.	V.A.9
Lee, D. R. 1977. A device for measuring seepage flux in lakes and estuaries. *Limnol. Oceanogr.* 22(1):140–147.	IV.D
Lehman, J. T. 1975. Reconstructing the rate of accumulation of lake sediments: the effect of sediment focusing. *Quat. Res.* 5:541–550.	IV.B; VII.A.1; VII.A.2
———. 1976. Ecological and nutritional studies on Dinobryon Ehrenb.: seasonal periodicity and the phosphate toxicity problem. *Limnol. Oceanogr.* 21(5):646–658.	VI.C
———. 1980a. Nutrient recycling as an interface between algae and grazers in freshwater communities. In *Evolution and Ecology of Zooplankton Communities*, ed. W. C. Kerfoot, 251–263. Spec. Symp. Vol. 3, Limnol. Oceanogr.	VI.C
———. 1980b. Release and cycling of nutrients between planktonic algae and herbivores. *Limnol. Oceanogr.* 25(4):620–632.	VI.C
Lehman, J. T., and D. Scavia. 1982. Microscale patchiness of nutrients in plankton communities. *Science* 216:729–730.	IV.A
Lehman, J. T., D. B. Botkin and G. E. Likens. 1975. The assumptions and rationales of a computer model of phytoplankton population dynamics. *Limnol. Oceanogr.* 20:343–364.	V.B.2; VII.F
Leopold, A. 1941. Lakes in relation to terrestrial life patterns. In *A Symposium on Hydrobiology*, 17–22. Univ. Wisconsin Symp., Vol. on Hydrology. Madison.	I.A.2; IV.E; VI.B
———. 1949. *A Sand County Almanac*. Oxford: Oxford University Press. 226 pp.	I.A.2; VI.D
Leopold, L. B., M. G. Wolman and J. P. Miller. 1964. *Fluvial Processes in Geomorphology*. San Francisco: W. H. Freeman. 522 pp.	II.B
Leventhal, E. A. 1970. The Chrysomonadina. In *Ianula: An Account of the History and Development of the Lago di Monterosi, Latium, Italy*, ed. G. E. Hutchinson et al. *Trans. Amer. Philos. Soc.* (Philadelphia) 60(4):123–142.	VII.B
Levins, R. 1968. *Evolution in Changing Environments*. Princeton, NJ: Princeton Univ. Press.	VI.C
Lewis, W. M. 1979. *Zooplankton Community Analysis: Studies on a Tropical System*. Springer-Verlag New York Inc. 143 pp.	V.B.5; VI.C
Liddicoat, J. C. 1970. Steady-state thermal gradients in New England lake sediments: their applicability for determining geothermal heat flow. M.S. thesis, Dartmouth College, Hanover, NH.	IV.C
Likens, G. E. 1972a. *The chemistry of precipitation in the central Finger Lakes Region*. Water Resources and Marine Sciences Center Tech. Report 50. Cornell University, Ithaca, NY. 61 pp.	IV.E
———. 1972b. Eutrophication and aquatic ecosystems. In *Nutrients and Eutrophication*, ed. G. E. Likens, 3–13. American Society of Limnology and Oceanography. Lawrence, KS. Special Symposia, Vol. I.	V.B.6; V.C; VI.D
———. 1972c. Mirror Lake: its past, present and future? *Appalachia* 39(2):23–41.	III.A; III.B.1–3; V.A.4; VII.A.2; IX
———, ed. 1972d. *Nutrients and Eutrophication*. American Society of Limnology and Oceanography. Lawrence, KS. Special Symposia, Vol. I. 328 pp.	IV.E; IX
———. 1973. Primary production: freshwater ecosystems. *Human Ecology* 1(4):347–356.	II.A
———. 1975a. Nutrient flux and cycling in freshwater ecosystems. In *Mineral Cycling in Southeastern Ecosystems*, ed. F. G. Howell, J. B. Gentry and M. H. Smith, 314–348. Augusta, GA: ERDA Symp. Series CONF-740513. May 1974.	I.A.; IV.E; VIII; IX

References

	CHAPTER
———. 1975b. Primary production of inland aquatic ecosystems. In *Primary Productivity of the Biosphere,* ed. H. Lieth and R. H. Whittaker, 314–348. Springer-Verlag New York Inc.	IV.A
———. 1976. Acid precipitation. *Chemical and Engineering News* 54:29–44.	VIII
———. 1983. A priority for ecological research. *Bull. Ecol. Soc. Amer.* 64(4):234–243.	II.B; IV.E
———. 1984. Beyond the shoreline: a watershed-ecosystem approach. *Verh. Internat. Verein. Limnol.* 22:1–22.	I.A.2; II.B
Likens, G. E., and R. E. Bilby. 1982. Development, maintenance and role of organic debris dams in streams. In *Sediment Budgets and Routing in Forested Drainage Basins,* ed. F. J. Swanson, R. J. Janda, T. Dunne and D. N. Swanston, 122–128. USDA Forest Service General Tech. Report PNW-141.	II.B; III.B.2; VII.F
Likens, G. E., and F. H. Bormann. 1970. Chemical analyses of plant tissues from the Hubbard Brook ecosystem in New Hampshire. *Yale Univ. School For. Bull.* 79. 25 pp.	IX
———. 1972. Nutrient cycling in ecosystems. In *Ecosystem Structure and Function,* ed. J. Wiens, 25–67. Corvallis: Oregon State Univ. Press.	I.A; I.A.1; IV.E; VIII; IX
———. 1974a. Effects of forest clearing on the northern hardwood forest ecosystem and its biogeochemistry. In *Proc. First Internat. Congress Ecology,* 330–335. September 1974. The Hague: Centre Agric. Publ. Doc. Wageningen.	II.B; III.B.3; VIII; IX
———. 1974b. Linkages between terrestrial and aquatic ecosystems. *BioScience* 24(8):447–456.	I.A; I.A.2; III.B.2; V.A.4
———. 1975. An experimental approach in New England landscapes. In *Proc. INTECOL Symp. on Coupling of Land Water Systems, 1971,* ed. A. D. Hasler, 7–29. Leningrad. Springer-Verlag New York Inc.	I.A.1
———. 1979. The role of watershed and airshed in lake metabolism. In *Lake Metabolism and Management,* ed. W. Rodhe, G. E. Likens and C. Serruya, 195–211. Papers Emanating from the Jubilee Symp. of Uppsala Univ., Sweden. Arch. Hydrobiol. Beih. Ergebn. Limnol. 13.	I.A; II.A; III.B.2; IV.E; V.D; VI.D; VIII; IX
Likens, G. E., and M. B. Davis. 1975. Post-glacial history of Mirror Lake and its watershed in New Hampshire, U.S.A.: an initial report. *Verh. Internat. Verein. Limnol.* 19(2):982–993.	III.A; III.B.1; V.A.9; V.C; VII.A.1; VII.A.2; VII.B; VII.C; VII.D; VII.E; VII.F; VIII
Likens, G. E., and J. J. Gilbert. 1970. Notes on quantitative sampling of natural populations of planktonic rotifers. *Limnol. Oceanogr.* 15:816–820.	V.A.5; V.B.5
Likens, G. E., and A. D. Hasler. 1962. Movements of radiosodium (Na^{24}) within an ice-covered lake. *Limnol. Oceanogr.* 7:48–56.	IV.C
Likens, G. E., and N. M. Johnson. 1969. Measurement and analysis of the annual heat budget for the sediments in two Wisconsin lakes. *Limnol. Oceanogr.* 14:115–135.	IV.B; IV.C; VI.A
Likens, G. E., and O. L. Loucks. 1978. Analysis of five North American lake ecosystems. III. Sources, loading and fate of nitrogen and phosphorus. *Verh. Internat. Verein. Limnol.* 20:568–573.	IV.E; V.B.2; V.B.6; V.C; VI.C; IX
Likens, G. E., and R. A. Ragotzkie. 1965. Vertical water motions in a small ice-covered lake. *J. Geophys. Res.* 70:2333–2344.	IV.C
———. 1966. Rotary circulation of water in an ice-covered lake. *Verh. Internat. Verein. Limnol.* 16:126–133.	IV.C
Likens, G. E., F. H. Bormann and N. M. Johnson. 1969. Nitrification: importance to nutrient losses from a cutover forested ecosystem. *Science* 163(3872):1205–1206.	IV.E
———. 1972. Acid rain. *Environment* 14(2):33–40.	II.B
Likens, G. E., E. S. Edgerton and J. N. Galloway. 1983. The composition and deposition of organic carbon in precipitation. *Tellus* 35B:16–24.	II.B; V.C

Reference	Chapter
Likens, G. E., F. H. Bormann, N. M. Johnson, D. W. Fisher and R. S. Pierce. 1970. Effects of forest cutting and herbicide treatment on nutrient budgets in the Hubbard Brook watershed-ecosystem. *Ecol. Monogr.* 40(1):23–47.	IV.B; VI.D; VII.A.2; VII.F; IX
Likens, G. E., F. H. Bormann, N. M. Johnson and R. S. Pierce. 1967. The calcium, magnesium, potassium and sodium budgets for a small forested ecosystem. *Ecology* 48(5):772–785.	II.B
Likens, G. E., F. H. Bormann, R. S. Pierce, J. S. Eaton and N. M. Johnson. 1977. *Biogeochemistry of a Forested Ecosystem*. Springer-Verlag New York Inc. 146 pp.	I.A.1; II.A; II.B; III.A; IV.B; IV.C; IV.E; VII.A.1; VII.E; VII.F; VIII; IX
Likens, G. E., F. H. Bormann, R. S. Pierce, J. S. Eaton and R. E. Munn. 1984. Long-term trends in precipitation chemistry at Hubbard Brook, New Hampshire. *Atmos. Environ.* 18(12):2641–2647.	II.B
Likens, G. E., R. F. Wright, J. N. Galloway and T. J. Butler. 1979. Acid rain. *Sci. Amer.* 241(4):43–51.	II.B; VIII
Lim, R. P., and C. H. Fernando. 1978. Production of Cladocera inhabiting the vegetated littoral of Pinehurst Lake, Ontario, Canada. *Verh. Internat. Verein. Limnol.* 20:225–231.	VI.A
Lindegaard, C., and P. M. Jonasson. 1979. Abundance, population dynamics and production of zoobenthos in Lake Myvatn, Iceland. *Oikos* 32:202–227.	V.A.6
Lindeman, R. L. 1942. Seasonal distribution of midge larvae in a senescent lake. *Amer. Midl. Natur.* 27:428–444.	V.A.6; VI.D
Lindequist, A. W., and C. C. Deonier. 1942. Emergence habits of the Clear Lake gnat. *J. Kansas Entomol. Soc.* 15:109–120.	V.A.6
———. 1943. Seasonal abundance and distribution of larvae of the Clear Lake gnat. *J. Kansas Entomol. Soc.* 16:143–149.	V.A.6
Linsley, R. K., M. A. Kohler and J. L. H. Paulhus. 1958. *Hydrology for Engineers*. New York: McGraw-Hill. 340 pp.	IV.D
Livingstone, D. A., and J. C. Boykin. 1962. Vertical distribution of phosphorus in Linsley Pond mud. *Limnol. Oceanogr.* 7:57–62.	VII.A.1; VII.E
Loden, M. S. 1974. Predation by chironomic (Diptera) larvae on oligochaetes. *Limnol. Oceanogr.* 19:156–159.	VI.B
Lohammer, G. 1938. Wasserchemie und hohere Vegetation schwedischer Seen. *Symb. Bot. Upsal.* 3(1). 253 pp.	V.A.4
Lorenzen, C. J. 1967. The determination of chlorophyll and pheo-pigments: spectrophotometric equations. *Limnol. Oceanogr.* 12:342–345.	V.B.4
Lorimer, C. G. 1977. The presettlement forest and natural disturbance cycle of northeastern Maine. *Ecology* 58:139–148.	III.B.2
Lowe, R. L. 1974. *Environmental requirements and pollution tolerance of freshwater diatoms*. National Environmental Research Center Office of Research and Development, U.S. EPA. EPA-670/4-74-005.	VII.B
Ludlam, S. D. 1974. Fayetteville Green Lake, New York. 6. The role of turbidity currents in lake sedimentation. *Limnol. Oceanogr.* 19:656–664.	VII.A.1
Ludlum, D. M. 1963. *Early American hurricanes, 1492–1870*. Boston: American Meteorological Society. 198 pp.	III.B.2
Lund, J. W. G. 1964. Primary production and periodicity of phytoplankton. *Verh. Internat. Verein. Limnol.* 15:37–57.	VII.F
———. 1965. The ecology of the freshwater phytoplankton. *Biol. Rev.* 40:231–293.	V.B.2
———. 1969. Phytoplankton. In *Eutrophication: Causes, Consequences and Correctives*, 306–330. Washington, DC: National Academy of Sciences.	V.B.2
———. 1972. Preliminary observations on the use of large experimental tubes in lakes. *Verh. Internat. Verein. Limnol.* 18:71–77.	VI.D
Lundbeck, J. 1926. Die Bodentierwelt Norddeutschen Seen. *Arch. Hydrobiol. Suppl.* 7:1–473.	V.A.6

References

CHAPTER

Lundquist, G. 1927. Bodenablagerungen und Entwicklungstypen der Seen. In *Die Binnengewasser*. 2, ed. A. Thienemann, Stuttgart: E. Schweizerbart'sche Verlag. 122 pp. — VII.A

Luther, A. 1960. Die Turbellarien Ostfennoskandiens. I. Acoela, Catenulida, Macrostomida, Lecithoepitheliata, Prolecithophora, und Proseriata. *Fauna Fenn.* 7:1–155. — V.A.7

Lutz, H. J. 1930. The vegetation of Heart's Content, a virgin forest in northwestern Pennsylvania. *Ecology* 11:1–29. — III.B.2

———. 1940. Disturbance of forest soil resulting from uprooting of trees. *Yale Univ. School For. Bull.* No. 45. — III.B.2

Macan, T. T. 1958. The temperature of a small stony stream. *Hydrobiologia* 12:89–106. — IV.B

———. 1970. *Biological Studies of the English Lakes*. London: Longman Group Ltd. 620 pp. — V.A.2

MacDonald, W. W. 1956. Observations on the biology of chaoborids and chironomids in Lake Victoria and on the feeding habits of the "Elephant-Snout Fish" (*Mormyrus kannume* Forsk.). *J. Anim. Ecol.* 25:36–53. — V.A.6

MacGinitie, C. D. 1939. Littoral marine communities. *Amer. Midl. Natur.* 21:28–55. — VI.C

Mackereth, F. J. H. 1966. Some chemical observations on post-glacial lake sediments. *Phil. Trans. Roy. Soc. B* 250:165–213. — III.B.1; VII.A.1; VII.E; VIII.F

MacLean, D. W. 1960. *Some aspects of the aspen-birch-spruce-fir type of Ontario*. Department of Forestry, Forest Research Division, Tech. Note No. 94. — III.B.2

Maeda, O., and S. Schimura. 1973. On the high density of a phytoplankton population found in a lake under ice. *Int. Rev. Ges. Hydrobiol.* 58:673–685. — V.B.2

Magnin, E., and A. Stanczykowska. 1971. Quelques donneés sur la croissance, la biomasse, et la production annuelle de trois mollusques Unionidac de la région de Montreal. *Can. J. Zool.* 49:491–497. — V.A.6

Maier, R. 1973. Produktions und Pigmentanalysen an *Utricularia vulgaris* L. In *Ökosystemforschung*, ed. H. Ellenberg, 87–101. Berlin: Springer-Verlag. — V.B.3

Maissurow, D. K. 1941. The role of fire in the perpetuation of virgin forests of northern Wisconsin. *J. Forestry* 39:201–207. — III.B.2

Maitland, P. S., and P. M. G. Hudspith. 1974. The zoobenthos of Loch Leven, Kinross and estimates of its production in the sandy littoral area during 1970 and 1971. *Proc. Roy. Soc. Edinburgh (B)* 74:219–239. — V.A.6; V.B.6

Makarewicz, J. C. 1974. The community of zooplankton and its production in Mirror Lake, New Hampshire. Ph.D. diss., Cornell University, Ithaca, NY. 206 pp. — IV.B; V.A.5; V.A.6; V.C

———. 1976. *Generation times of rotifers in lakes of varying trophic status*. Deposited at the Edmund Niles Huyck Preserve, Rensselaerville, Albany County, NY. — V.B.5

Makarewicz, J. C., and G. E. Likens. 1975. Niche analysis of a zooplankton community. *Science* 190:1000–1003. — VI.C; VII.C

———. 1978. Zooplankton niches and the community structure controversy. *Science* 200:461–463. — VI.C

———. 1979. Structure and function of the zooplankton community of Mirror Lake, New Hampshire. *Ecol. Monogr.* 49(1):109–127. — V.A.5; V.A.9; V.B.5; V.C; V.D; VI.C; VI.D; VII.C

Malovitskaya, L. M., and Y. I. Sorokin. 1961. Experimental investigation of the feeding of *Diaptomus* (crustacea, Copepoda) with the use of C^{14} (in Russian). *Trud. Inst. Biol. Vodokhr.* 4:262–272. — VI.C

Malueg, K. W. 1966. An ecological study of *Chaoborus*. Ph.D. diss., University of Wisconsin, Madison. 231 pp. — V.A.6

	CHAPTER
Maly, E. J., S. Schoenholtz and M. T. Arts. 1980. The influence of flatworm predation on zooplankton inhabiting small ponds. *Hydrobiologia* 76:233–240.	V.A.7
Manning, B. J., W. C. Kerfoot and E. M. Berger. 1978. Phenotypes and genotypes in cladoceran populations. *Evolution* 32:365–374.	VII.C
Manny, B. A., and R. G. Wetzel. 1973. Diurnal changes in dissolved organic and inorganic carbon and nitrogen in a hardwater stream. *Freshwater Biol.* 3:31–43.	VIII
Manny, B. A., R. G. Wetzel and R. E. Bailey. 1978. Paleolimnological sedimentation of organic carbon, nitrogen, phosphorus, fossil pigments, pollen, and diatoms in a hypereutrophic, hardwater lake: a case history of eutrophication. *Pols. Arch. Hydrobiol.* 25(1/2):243–267.	VII.A.2; VII.D; VII.F
Maristo, L. 1941. Die Seetypen Finnlands auf floristischer und vegetationsphysiognomischer Grundlage. *Ann. Bot. Soc. Zool. Bot. Fenn. Vanamo* 15(5):1–314.	V.A.4
Marshall, S. M., and A. P. Orr. 1961. On the biology of *Calanus finmarchichus,* 12. The phosphorus cycle: excretion, egg production, autolysis. With an addendum, "The turnover of phosphorus by *Calanus finmarchus*" by R. J. Conoven. *J. Mar. Biol. Assoc. U.K.* 41:463–488.	VI.C
Martin, C. W. 1979. Precipitation and streamwater chemistry in an undisturbed forested watershed in New Hampshire. *Ecology* 60:36–42.	VII.F
Matheson, D. H., and F. C. Elder, ed. 1976. *Atmospheric contribution to the chemistry of lake waters.* First Specialty Symposium of the International Association for Great Lakes Research. 28 September–1 October 1975, Geneva Park Conf. Centre, Ontario, Canada. J. Great Lakes Res. Suppl. 1 to Vol. 2. 266 pp.	VIII
Matteson, M. R. 1948. Life history of *Elliptio companatus* (Dillwyn, 1817). *Amer. Midl. Natur.* 40:690–723.	V.A.6
May, R. M., and R. H. MacArthur. 1972. Niche overlap as a function of environmental variability. *Proc. National Acad. Sci. U.S.A.* 69:1109–1113.	VI.C
Mazsa, D. J. 1973. Studies on the fish populations of Mirror Lake, New Hampshire. M.S. thesis, Cornell University, Ithaca, NY. 172 pp.	V.A.8; V.A.9; V.B.7; VI.B
McCall, P. L., M. J. S. Tevesz and S. F. Schwelgien. 1979. Sediment mixing by *Lampsilis radiata siliquoidea* (Mollusca) from western Lake Erie. *J. Great Lakes Res.* 5:105–111.	V.A.6; VI.B
McDowell, W. H. 1982. Mechanisms controlling the organic chemistry of Bear Brook, New Hampshire. Ph.D. diss., Cornell University, Ithaca, NY. 152 pp.	II.B; V.C; VIII
McIntosh, R. P. 1958. Plant communities. *Science* 128:115–120.	VI.C
———. 1963. Ecosystems, evolution and relational patterns in the Great Smoky Mountains. *Ecol. Monogr.* 22:1–44.	VI.C
———. 1967. The continuum concept of vegetation. *Biol. Rev.* 33:130–187.	VI.C
McIntyre, A. D. 1969. Ecology of marine meiobenthos. *Biol. Rev.* 44:245–290.	V.A.7
McLusky, D. S., and A. McFarlane. 1974. The energy requirements of certain larval chironomid populations in Loch Leven, Kinross. *Proc. Roy. Soc. Edinburgh (B)* 74:259–264.	V.B.6
McMahon, J. W., and F. H. Rigler. 1965. Feeding rate of *Daphnia magna* Straus if different foods labeled with radioactive phosphorus. *Limnol. Oceanogr.* 10:105–113.	VI.C
McNaught, D. G. 1971. Appendix to Clarke-Bumpus plankton sampler. In *Secondary Productivity in Fresh Waters,* ed. W. T. Edmondson and G. G. Winberg, 11–12. Oxford and Edinburgh: Blackwell Scientific Publications.	V.A.5
McQueen, D. J. 1969. Reduction of zooplankton standing stocks by predaceous *Cyclops bicuspidatus thomasi* in Marion Lake, British Columbia. *J. Fish. Res. Bd. Canada* 26:1605–1618.	VI.C

References

	CHAPTER

Menzel, D. W., and N. Corwin. 1965. The measurement of total phosphorus in seawater based on the liberation of organically bound fractions by persulphate oxidation. *Limnol. Oceanogr.* 10:280–281. — IV.B

Menzel, D. W., and R. F. Vaccaro. 1964. The measurement of dissolved and particulate carbon in seawater. *Limnol. Oceanogr.* 9(1):138–142. — V.C; VII.E

Merritt, R. W., and K. W. Cummins. 1978. *An Introduction to the Aquatic Insects of North America.* Dubuque, IA: Kendall/Hunt Publishing Co. 441 pp. — V.A.6

Meyer, J. L. 1978. Transport and transformation of phosphorus in a stream ecosystem. Ph.D. diss., Cornell University, Ithaca, NY. 226 pp. — II.B; VIII

Meyer, J. L., and G. E. Likens. 1979. Transport and transformation of phosphorus in a forest stream ecosystem. *Ecology* 60(6):1255–1269. — I.A.2; II.B; VI.D; VII.F; VIII; IX

Meyer, J. L., G. E. Likens and J. Sloane-Richey. 1981. Phosphorus, nitrogen and organic carbon flux in a headwater stream. *Arch. Hydrobiol.* 91(1):28–44. — I.A.2; II.B; VII.F; VIII; IX

Miller, N. G., and G. G. Thompson. 1979. Boreal and western North American plants in the late Pleistocene of Vermont. *J. Arnold Arboretum* 60:167–218. — III.B.1; VII.F

Miller, R. B. 1941. A contribution to the ecology of the Chironomidae of Costell Lake, Algonquin Park, Ontario. University of Toronto Studies, Biol. 49. Publ. *Ontario Fish. Res. Lab.* 60:5–63. — V.A.6

Mills, E. L., J. L. Forney, M. D. Clady and W. R. Schaffner. 1978. Oneida Lake. In *Lakes of New York State.* Vol. II, *Ecology of the lakes of western New York,* 367–451. New York: Academic Press. — III.B.2

Mitchell, J. M., Jr. 1977. The changing climate. In *Studies in Geophysics: Energy and Climate,* 51–58. Wahmal Acad. Sciences. — III.B.1

Moeller, R. E. 1975. Hydrophyte biomass and community structure in a small, oligotrophic New Hampshire lake. *Verh. Internat. Verein. Limnol.* 19:1004–1012. — V.A.4; VII.C

———. 1978a. Carbon-uptake by the submerged hydrophyte, *Utricularia purpurea. Aquatic Botany* 5:209–216. — V.A.4; V.B.3

———. 1978b. The hydrophytes of Mirror Lake: a study of vegetational structure and seasonal biomass dynamics. Ph.D. diss., Cornell University, Ithaca, NY. 212 pp. — III.A; IV.B; V.A.4; V.A.6; V.B.6

———. 1978c. Seasonal changes in biomass, tissue chemistry and net production of the evergreen hydrophyte, *Lobelia dortmanna. Can. J. Bot.* 56:1425–1433. — V.A.4; V.B.3

———. 1980. The temperature-determined growing season of a submerged hydrophyte: tissue chemistry and biomass turnover of *Utricularia purpurea. Freshwater Biol.* 10:391–400. — V.A.4; V.B.3; VI.A

Moeller, R. E., and G. E. Likens. 1978. Seston sedimentation in Mirror Lake, New Hampshire, and its relationship to long-term sediment accumulation. *Verh. Internat. Verein. Limnol.* 20:525–530. — VII.A.2; VII.E

Moeller, R. E., and J. P. Roskoski. 1978. Nitrogen-fixation in the littoral benthos of an oligotrophic lake. *Hydrobiologia* 60(1):13–16. — V.A.3; V.B.4; VI.A

Moeller, R. E., F. Oldfield and P. G. Appleby. 1984. Biological sediment mixing in its stratigraphic implication in Mirror Lake (New Hampshire, U.S.A.). *Verh. Internat. Verein. Limnol.* 22(1):567–572. — VII.A.2

Mohleji, S., and F. Verhoff. 1980. Sodium and potassium ions effects on phosphorus transport in algal cells. *J. Water Pollut. Contr. Fed.* 52(1):111–125. — VI.C

Moke, C. B. 1946. *The geology of the Plymouth Quadrangle, New Hampshire.* Concord, NH: New Hampshire State Planning and Development Commission. 21 pp. — III.A

Molnau, M., W. J. Rawls, D. L. Curtis and C. C. Warnick. 1980. Gage density and location for estimating mean annual precipitation in mountainous areas. *Water Resour. Bull.* 16(3):428–432. — IV.D

	CHAPTER
Monakov, A. V. 1972. Review of studies on feeding of aquatic invertebrates conducted at the Institute of Biology of Inland Waters, Academy of Sciences. *J. Fish. Res. Bd. Canada* 29:363–383.	VI.C
Monard, A. 1920. La Faune profonde du Lac de Neuchatel. *Société Neuchateloise des Sciences Naturelles Bull.* 44:65–235.	V.A.7
Monheimer, R. H. 1972. Heterotrophy by plankton in three lakes of different productivity. *Nature* 236:436–464.	V.B.1
———. 1974. Sulfate uptake as a measure of planktonic microbial production in freshwater ecosystems. *Can. J. Microbiol.* 20:825–831.	V.B.1
Mooney, J. 1928. *The Aboriginal Population of America North of Mexico.* Smithsonian Misc. Coll. 80(4), Publ. 2955. Washington, DC: Smithsonian Institution. 40 pp.	III.B.3
Moore, G. M. 1939. A limnological investigation of the microscopic benthic fauna of Douglas Lake, Michigan. *Ecol. Monogr.* 9:537–582.	V.A.7
Moore, J. W. 1974. Benthic algae of southern Baffin Island. III. Epilithic and epiphytic communities. *J. Phycol.* 10:456–462.	V.A.3
Morgan, A. H. 1930. *Field Book of Ponds and Streams.* New York: G. P. Putnam's Sons. 448 pp.	V.A.6
Morgan, K., and J. Kalff. 1975. The winter dark survival of an algal flagellate—*Cryptomonas erosa* (Skuja). *Verh. Internat. Verein. Limnol.* 19:2734–2740.	V.A.2
Morgan, N. C., and A. B. Waddell. 1961. Insect emergence from a small trout loch and its bearing on the food supply of fish. *Sci. Invest. Freshwater Salmon Fish. Res. Scott. Home Dep.* 25:1–39.	V.A.6
Moriarty, D. J. W., J. P. E. C. Darlington, T. G. Dunn, C. M. Moriarty and M. P. Tarlin. 1973. Feeding and grazing in Lake George, Uganda. *Proc. Roy. Soc. London B* 184:277–346.	VI.C
Morris, I., C. M. Yentsch and C. S. Yentsch. 1971. Relationship between light carbon dioxide fixation and dark carbon fixation by marine algae. *Limnol. Oceanogr.* 16:854–858.	V.B.1; V.C
Mortimer, C. H. 1942. The exchange of dissolved substances between mud and water in lakes. *J. Ecol.* 30:147–201.	VI.D
Moss, B. 1980. *Ecology of Freshwaters.* New York: Halsted Press, John Wiley & Sons. 332 pp.	VI.C
Moyle, J. B. 1945. Some chemical factors influencing the distribution of aquatic plants in Minnesota. *Amer. Midl. Natur.* 34:402–420.	V.A.4
Mueller, W. P. 1964. The distribution of cladoceran remains in surficial sediments from three northern Indiana lakes. *Invest. Indiana Lakes & Streams* 6:1–63.	VIII.F
Müller, P. 1980. Effects of artificial acidification on the growth of periphyton. *Can. J. Fish. Aquatic Sci.* 37:355–363.	V.A.2
Muller, R. N. 1975. The natural history, growth and ecosystem relations of *Erythronium americanum* Ker. in the northern hardwood forest. Ph.D. diss., Yale University, New Haven, CT. 178 pp.	II.A
Mundie, J. H. 1957. The ecology of Chironomidae in storage reservoirs. *Trans. Roy. Entomol. Soc. London* 109(5):149–232.	V.A.6; VI.B
Munther, G. L. 1970. Movement and distribution of smallmouth bass in the Middle Snake River. *Trans. Amer. Fish. Soc.* 99:44–53.	V.A.9
Nalepa, T. F., and M. A. Quigley. 1983. Abundance and biomass of the meiobenthos in nearshore Lake Michigan with comparisons to the macrobenthos. *J. Great Lakes Res.* 9:530–547.	V.A.7
Nalewajko, C., and D. R. S. Lean. 1972. Growth and excretion in planktonic algae and bacteria. *J. Phycol.* 8:361–366.	V.D

CHAPTER

Nalewajko, C., and D. W. Schindler. 1976. Primary production, extracellular release, and heterotrophy in two lakes in the ELA, northwestern Ontario. *J. Fish. Res. Bd. Canada* 33:219–226. — VI.C

Naumann, E. 1932. Grundzuge der regionalen Limnologie. In *Die Binnengewasser 11*, ed. A. Thienemann. 176 pp. — V.A.4

Nauwerck, A. 1963. Die Beziechungen zwischen zooplankton und phytoplankton im See Erken. *Symb. Bot. Upsal.* 17(5):1–163. — VI.C

Neame, P. A. 1975. Oxygen uptake of sediments in Castle Lake, California. *Verh. Internat. Verein. Limnol.* 19:792–799. — V.B.6; VI.B

Neff, E. L. 1977. How much rain does a rain gage gage? *J. Hydrol.* 35(3/4):213–220. — IV.D

Neuwirth, F. 1973. Experiences with evaporation pans at a shallow steppe lake in Austria. In *Hydrology of Lakes,* 290–297. Helsinki Symposium, International Association of Hydrological Sciences, Publ. 109. — IV.D

New Hampshire Water Supply and Pollution Control Commission. 1975. *Nutrient removal effectiveness of a septic tank-leaching field system (1970–1973).* Staff Report No. 65 (to the New England Interstate Water Pollution Control Commission). 145 pp. — IV.E

Newell, I. M. 1947. A systematic and ecological study of the Halacaridae of eastern North America. *Bull. Bingham Oceanogr. Coll.* 10:1–232. — V.A.7

Ney, J. J. 1978. A synoptic review of yellow perch and walleye biology. In *Coolwater Fishes of North America,* ed. R. L. Kendall, 1–12. Washington, DC: Amer. Fish. Soc. — V.A.9

Nichols, D. S., and D. R. Keeney. 1976. Nitrogen nutrition of *Myriophyllum spicatum:* uptake and translocation of ^{15}N by shoots and roots. *Freshwater Biol.* 6:145–154. — V.A.4

Niewolak, S. 1974. Production of bacterial biomass in the water of Ilawa lakes. *Acta Hydrobiol. Krakow* 16:101–112. — V.B.1

Northcote, T. G. 1964. Use of a high-frequency echo sounder to record distribution and migration of *Chaoborus* larvae. *Limnol. Oceanogr.* 9:87–91. — V.A.6

Nygaard, G. 1958. On the productivity of the bottom vegetation in Lake Grane Langsø. *Verh. Internat. Verein. Limnol.* 13:144–155. — V.A.4

O'Brien, W. J., and F. deNoyelles, Jr. 1976. Response of three phytoplankton bioassay techniques in experimental ponds of known limiting nutrient. *Hydrobiologia* 49:65–76. — V.B.2

Oden, B. J. 1979. The freshwater littoral meiofauna in a South Carolina reservoir receiving thermal effluents. *Freshwater Biol.* 9:291–304. — V.A.7

Odum, E. P. 1953. *Fundamentals of Ecology.* 1st ed. Philadelphia: W. B. Saunders Co. 546 pp. — I.A

———. 1959. Fundamentals of Ecology. 2nd edition. Philadelphia and London: W. B. Saunders and Co., 546 pp. — V.B.5

———. 1962. Relationship between structure and function in the ecosystem. *Japanese J. Ecol.* 12(3):108–118. — VI.C

———. 1969. The strategy of ecosystem development. *Science* 164:262–270. — VII.F

———. 1971. *Fundamentals of Ecology.* 3rd ed. Philadelphia: W. B. Saunders Co. 574 pp. — I.A

Oglesby, R. T. 1978. The limnology of Cayuga Lake. 1–120. In *Lakes of New York State.* Vol. I, *Ecology of the Finger Lakes,* New York: Academic Press. — III.B.2

Oglesby, R. T., and W. R. Schaffner. 1978. Phosphorus loadings to lakes and some of their responses. Part 2, Regression models of summer phytoplankton standing crops, winter total P, and transparency of New York lakes with known phosphorus loadings. *Limnol. Oceanogr.* 23:135–145. — V.A.2; VIII

	CHAPTER

Okland, J. 1964. The eutrophic lake Borrevann (Norway)—an ecological study on shore and bottom fauna, with special reference to gastropods, including a hydrographic survey. *Folia Limnol. Scand.* 13:1–337. — V.A.6

Oliver, D. R. 1971. Life history of the Chironomidae. *Ann. Rev. Entomol.* 16:211–230. — V.A.6

Olsen, S. 1950. Aquatic plants and hydrospheric factors. I. Aquatic plants in SW-Jutland, Denmark. *Svensk. Bot. Tidskr.* 44:1–34. — V.A.4

Olsen, S., and D. G. Chapman. 1972. Ecological dynamics of watersheds. *BioScience* 22(3):158–161. — IV.E

Omernik, J. M. 1977. Nonpoint source—stream nutrient level relationships: a nationwide study. 28–29. U.S. Environmental Protection Agency, Environmental Res. Laboratory, Corvallis. Ecological Research Series, EPA-600/3-77-105. — VIII

Otsuki, A., and R. G. Wetzel. 1972. Coprecipitation of phosphate with carbonates in a marl lake. *Limnol. Oceanogr.* 17:763–767. — V.A.4

Overbeck, J. 1979. Dark CO_2 uptake—biochemical background and its relevance to *in situ* bacterial production. *Arch. Hydrobiol. Beih. Ergebn. Limnol.* 12:38–47. — V.B.1

Overbeck, J., and R. J. Daley. 1973. Some precautionary comments on the Romanenko technique for estimating heterotrophic bacterial production. *Bull. Ecol. Res. Comm.* (Stockholm) 17:342–344. — V.B.1

Paasivirta, L. 1975a. Distribution and abundance of Halacaridae (Acari: Trombidiformes) in the oligotrophic Lake Pääjärvi, southern Finland. *Ann. Zool. Fenn.* 12:119–125. — V.A.7

———. 1975b. Insect emergence and output of incorporated energy and nutrients from the oligotrophic Lake Pääjärvi, southern Finland. *Ann. Zool. Fenn.* 12:126–140. — V.B.6

Paasivirta, L., and J. Sarkka. 1978. Effects of pulp mill and municipal effluents and humus loads on the macro- and meiozoobenthos of some Finnish lakes. *Verh. Internat. Verein. Limnol.* 20:1779–1788. — V.A.7

Paerl, H. W. 1978. Microbial organic carbon recovery in aquatic systems. *Limnol. Oceanogr.* 23:927–935. — V.B.1

Palfrey, J. G. 1859. *History of New England.* Boston: Little, Brown & Co. 659 pp. — III.B.3

Parker, J. E., and D. N. Edgington. 1976. Concentration of diatom frustules in Lake Michigan sediment cores. *Limnol. Oceanogr.* 21:887–893. — VII.A.2

Parma, S. 1971. *Chaoborus flavicans* (Meigen) (Diptera, Chaoboridae): an autecological study. Ph.D. diss., Groningen, Netherlands. 128 pp. — V.A.6

Patalas, K. 1970. Primary and secondary production in a lake heated by a thermal power plant. In *Proc. 16th Ann. Tech. Meeting,* 1267–1271. Mt. Prospect, IL: Inst. Environ. Sci. — V.B.5

———. 1971. Crustacean plankton communities in forty-five lakes in the Experimental Lakes Area, northwestern Ontario. *J. Fish. Res. Bd. Canada* 28(2):231–244. — V.A.5

Paterson, C. G., and C. H. Fernando. 1971. Studies on the spatial heterogeneity of shallow water benthos with particular reference to the Chironomidae. *Can. J. Zool.* 49(7):1013–1019. — V.A.6

Paterson, C. G., and K. F. Walker. 1974. Seasonal dynamics and productivity of *Tanytarsus barbitarsis* Freeman (Diptera: Chironomidae) in the benthos of a shallow, saline lake. *Aust. J. Mar. Fresh-water Res.* 25:151–165. — V.A.6; V.B.6

Patric, J. H. 1976. Soil erosion in eastern forests. *J. Forestry* 78:268–272. — IX

Patrick, R. 1943. The diatoms of Linsley Pond, Connecticut. *Proc. Acad. Nat. Sci. Philadelphia* 95:53–110. — VII.B

References

CHAPTER

Patrick, R., and C. W. Reimer. 1966. *Diatoms of the United States*. Vol. I. Philadelphia: Monogr. Acad. Nat. Sci., No. 13. — VII.B

Pawluk, S. 1967. Soil analyses by atomic absorption spectrophotometry. *Atomic Absorption Newsletter* 6(3):53–56. — VII.E

Payne, W. J. 1970. Energy yields and growth of heterotrophs. *Ann. Rev. Microbiol.* 24:17–52. — V.B.1

Pearsall, W. H. 1920. The aquatic vegetation of the English lakes. *J. Ecology* 8:163–199. — V.A.4

———. 1921. The development of vegetation in the English lakes, considered in relation to the general evolution of glacial lakes and rock basins. *Proc. Roy. Soc. London B* 92:259–284. — V.A.4

Pearse, A. S., and H. Achtenberg. 1918. The habits of yellow perch in Wisconsin lakes. *Bull. Bur. Fish.* 36:297–366. — V.A.6; VI.B

Penman, H. L. 1948. Natural evaporation from open water, bare soil and grass. *Proc. Roy. Soc. London Ser. A* 193(1032):120–145. — IV.D

Pennak, R. W. 1943. An effective method of diagramming diurnal movements of zooplankton organisms. *Ecology* 13:216–231. — VI.C

———. 1978. *Fresh-Water Invertebrates of the United States*. New York: John Wiley & Sons, Inc. 803 pp. — V.A.5; V.A.6; V.A.7

Pennington, W. 1974. Seston and sediment formation in five lake district lakes. *J. Ecology* 62:215–251. — VII.A.2

———. 1975. *The effect of Neolithic man on the environment in North-west England: the use of absolute pollen diagrams*. Council for British Archeology, Research Report No. 11:74–86. — VII.E

———. 1978a. The impact of man on some English lakes: rates of change. *Pols. Arch. Hydrobiol.* 25:429–437. — VII.A.2; VII.F

———. 1978b. Responses of some British lakes to past changes in land use on their catchments. *Verh. Internat. Verein. Limnol.* 20:636–641. — VII.A.2

———. 1981a. Records of a lake's life in time: the sediments. *Hydrobiologia* 79:197–219. — III.B.1; VII.E; VII.F

———. 1981b. Sediment composition in relation to the interpretation of pollen. *IVth Internat. Palynol. Conf.*, Lucknow, India (1976–77) 3:188–213. — VII.E; VII.F

Pennington, W., E. Y. Haworth, A. P. Bonny and J. P. Lishman. 1972. Lake sediments in northern Scotland. *Phil. Trans. Roy. Soc. London B* 264:191–294. — VII.F

Peters, R. H. 1972. Phosphorus regeneration by zooplankton. Ph.D. diss., University of Toronto. 205 pp. — VI.C

———. 1975. Phosphorus regeneration by natural populations of limnetic zooplankton. *Verh. Internat. Verein. Limnol.* 19:273–279. — VI.C

Peters, R. H., and F. H. Rigler. 1973. Phosphorus release by *Daphnia*. *Limnol. Oceanogr.* 18(6):821–839. — VI.C

Peterson, B. J. 1978. Radiocarbon uptake: its relation to net particulate carbon production. *Limnol. Oceanogr.* 23:179–184. — V.C

———. 1980. Aquatic primary productivity and the $^{14}CO_2$ method: a history of the productivity problem. *Ann. Rev. Ecol. Syst.* 11:359–385. — V.C

Peterson, B. J., J. E. Hobbie and J. F. Haney. 1978. *Daphnia* grazing on natural bacteria. *Limnol. Oceanogr.* 23:1039–1045. — VI.C; VI.D

Peverly, J., and J. Adamec. 1978. Association of potassium and some other monovalent cations with occurrence of polyphosphate bodies in *Chlorella pyrenoidosa*. *Plant Physiol.* 62:120–126. — VI.C

Pflieger, W. L. 1975. *The Fishes of Missouri*. Jefferson City, MO: Missouri Dept. Conservation. 336 pp. — V.A.9

Pierce, R. S. 1967. Evidence of overland flow on forest watersheds. In *Internat. Symp. on Forest Hydrology Proc.*, ed. W. E. Sopper and H. W. Lull, Oxford: Pergamon Press. — II.A; 11.B

CHAPTER

Pierce, R. S., J. W. Hornbeck, G. E. Likens and F. H. Bormann. 1970. Effects of elimination of vegetation on stream water quantity and quality. In *Internat. Symp. on Results of Research on Representative and Experimental Basins,* 311–328. Wellington, New Zealand: Internat. Assoc. Sci. Hydrol. — IX

Pierce, R. S., C. W. Martin, C. C. Reeves, G. E. Likens and F. H. Bormann. 1972. Nutrient loss from clearcuttings in New Hampshire. In *Symp. on Watersheds in Transition,* 285–295. Ft. Collins, CO. — IX

Pilgrim, S. L. A., and R. D. Harter. 1977. *Spodic horizon characteristics of some forest soils in the White Mountains, New Hampshire.* New Hampshire Agricultural Experiment Station Bulletin No. 507. — II.A

Pletscher, D. H. 1982. White-tailed deer and nutrient cycling in the Hubbard Brook Experimental Forest, New Hampshire. Ph.D. diss., Yale University, New Haven, CT. 145 pp. — IX

Porter, K. G. 1973. Selective grazing and differential digestion of algae and zooplankton. *Nature* 244:179–180. — VI.C

———. 1977. The plant-animal interface in freshwater ecosystems. *Amer. Sci.* 65(2):159–170. — V.B.5

Porter, K. G., J. Gerritsen and J. Orcutt, Jr. 1982. The effect of food concentration on swimming patterns, feeding, behavior, ingestion, assimilation, and respiration by *Daphnia. Limnol. Oceanogr.* 27(5):935–949. — IV.A

Porter, K. G., M. L. Pace and J. F. Battey. 1979. Ciliate protozoans as links in freshwater planktonic food chains. *Nature* 277(5697):563–565. — VI.C

Porter, S. C., and G. H. Denton. 1967. Chronology of neo-glaciation in the North American Cordillera. *Amer. J. Sci.* 265:177–210. — III.B.1

Postgate, J. R. 1973. The viability of very slow-growing populations: a model for the natural ecosystem. *Bull. Ecol. Res. Comm.* (Stockholm) 17:287–292. — V.B.1

Potter, D. W. B., and M. A. Learner. 1974. A study of the benthic macroinvertebrates of a shallow eutrophic reservoir in South Wales with emphasis on the Chironomidae (Diptera); their life-histories and production. *Arch. Hydrobiol.* 74:186–226. — V.A.6; V.B.6

Potzger, J. E., and W. A. Van Engel. 1942. Study of the rooted aquatic vegetation of Weber Lake, Vilas County, Wisconsin. *Trans. Wisconsin Acad. Sci. Arts Lett.* 34:149–166. — V.A.4

Pough, F. H. 1976. Acid precipitation and embryonic mortality of spotted salamanders, *Ambystoma maculatum. Science* 192:68–70. — VIII

Pourriot, R. 1963. Utilisation des algues brunes, uniallulaire pour l'élèvage des rotiferes. *C. R. Hebd. Sianc. Acad. Sci. Paris* 256:1603–1605. — V.D

———. 1965. Recherches sur l'ecologie des rotifers. *Vie Milieu Suppl.* 21:1–224. — VI.C

———. 1977. Food and feeding habits of rotifers. In *Proc. of First Internat. Rotifer Symp. Arch. Hydrobiol. Beih.* 8, ed. C. E. King, 243–260. — VI.C

Prejs, K. 1970. Problems of the ecology of benthic nematodes (Nematoda) of Mikolajskie Lake. *Ekol. Polska* 18:225–242. — V.A.7

Proctor, M. A. 1930. *The Indians of the Winnipesauke and Pemigewasset Valleys.* Franklin, NH: Towne and Robie Publishers. 67 pp. — III.B.2

Prokopowich, J. 1979. *Chemical characterization of epilimnion waters in the Experimental Lakes Area, northwestern Ontario.* Can. Fish. Mar. Serv. Tech. Report 873. 41 pp. — V.A.2

Provasoli, L. 1958. Nutrition and ecology of protozoa and algae. *Ann. Rev. Microbiol.* 12:279–308. — IV.D

Provasoli, L. 1969. Algal nutrition and eutrophication. In *Eutrophication: Causes, Consequences and Correctives,* 574–593. Washington, DC: National Academy of Sciences, Publication 1700. — IV.D

Purcell, E. M. 1977. Life at low Reynolds number. *Amer. J. Physics* 45:3–11. — IV.A

References

	CHAPTER
Raddum, G. G., A. Hobaek, E. R. Lomsland and T. Johnsen. 1980. Phytoplankton and zooplankton in acidified lakes in south Norway. In *Ecological Impact of Acid Precipitation,* ed. D. Drabløs and A. Tollan, 332–333. First International Conference. Oslo, Norway.	VIII
Ragotzkie, R. A., and G. E. Likens. 1964. The heat balance of two Antarctic lakes. *Limnol. Oceanogr.* 9:412–425.	IV.C
Ramensky, L. G. 1924. Die Grundgesetz massigkeitin in Aufbau der Vegetationsdecke (Vêstnik opỹ tnago dêla Stredne-Chernoz Obl., Voronezh), *Bot. Centralbl. Neue Folge* 7:453–455.	VI.C
Ranta, E., and J. Sarvala. 1978. Spatial patterns of littoral meiofauna in an oligotrophic lake. *Verh. Internat. Verein. Limnol.* 20(2):886–890.	V.A.6
Raspopov, I. M. 1972. Zur Methodik der Bestimmung der Jahresproduktion der Makrophyten in den Seen der nordwestlichen UdSSR. *Verh. Internat. Verein. Limnol.* 18:171–175.	V.B.3
Raven, J. A. 1970. Exogenous inorganic carbon sources in plant photosynthesis. *Biol. Rev.* 45:167–221.	V.A.4; V.B.3
Rawson, D. S. 1930. The bottom fauna of Lake Simcoe and its role in the ecology of the lake. University of Toronto Studies Publ. *Ontario Fish. Res. Lab.* 40:1–183.	V.A.6
Rawson, D. S. 1939. Some physical and chemical factors in the metabolism of lakes. *Am. Assoc. Adv. Sci. Publication* 10:9–26.	VI.D
———. 1941. Soils as a primary influence in the biological productivity of lakes. *Trans. Canadian Conservation Assoc.* 1941:78–87.	V.A.6
———. 1953. The bottom fauna of Great Slave Lake. *J. Fish. Res. Bd. Canada* 10(8):486–520.	V.A.6
———. 1955. Morphometry as a dominant factor in the productivity of large lakes. *Verh. Internat. Verein. Limnol.* 12:164–175.	V.A.6
Redfield, A. C., B. H. Ketchum and F. A. Richards. 1963. The influence of organisms on the composition of sea water. In *The Sea.* Vol. 2, ed. M. N. Hill, 26–77. New York: Interscience.	IV.B
Rhee, G-Yull. 1972. Competition between an alga and an aquatic bacterium for phosphate. *Limnol. Oceanogr.* 17(4):505–514.	VI.C
Rheinheimer, G. 1974. *Aquatic Microbiology.* New York: John Wiley & Sons Inc. 184 pp.	VI.C
Rich, P. H. 1979. Differential CO_2 and O_2 benthic community metabolism in a soft-water lake. *J. Fish. Res. Bd. Canada* 36:1377–1389.	VI.B
Rich, P. H., and R. G. Wetzel. 1978. Detritus in the lake ecosystem. *Amer. Natur.* 112:57–71.	V.A.1; VI.A
Rich, P. H., R. G. Wetzel and N. V. Thuy. 1971. Distribution, production and role of aquatic macrophytes in a southern Michigan marl lake. *Freshwater Biol.* 1:3–21.	V.A.4; V.B.3; V.B.4
Richard, P. 1970. Atlas pollinique des arbres et de quelques arbustes indigenes du Quebec. *Naturalist Canadian* 97:1–34.	III.B.1
Richey, J. E., R. C. Wissmar, A. H. Devol, G. E. Likens, J. S. Eaton, R. G. Wetzel, W. E. Odum, P. H. Rich, O. L. Loucks, R. T. Prentki and N. M. Johnson. 1978. Carbon flow in four lake ecosystems: a structural approach. *Science* 202(4373):1183–1186.	V.C
Richman, S. T., J. Smayda and M. K. Melnik. 1980. Direct observations of *Daphnia* mouthpart responses to algal food composition of varying quality. Abstracts of Papers submitted for presentation at 43rd Annual Meeting of the American Society of Limnology and Oceanography.	IV.A
Rickett, H. W. 1922. A quantitative study of the larger aquatic plants of Lake Mendota. *Trans. Wisconsin Acad. Sci. Arts Lett.* 20:501–527.	V.A.4
———. 1924. A quantitative study of the larger aquatic plants of Green Lake, Wisconsin. *Trans. Wisconsin Acad. Sci. Arts Lett.* 21:381–414.	V.A.4; VI.A

	CHAPTER

Riggs, H. C. 1973. *Regional analysis of stream flow characteristics*. U.S. Geol. Surv. Techniques of Water-Resources Investigations, Book 4, Chapter B3. 15 pp. — IV.D

Rigler, F. H. 1973. A dynamic view of the phosphorus cycle in lakes. In *Environmental Phosphorus Handbook,* ed. E. J. Griffith et al., 539–572. New York: John Wiley & Sons Inc. — VI.C

———. 1978. Limnology in the high Arctic: a case study of Char Lake. *Verh. Internat. Verein. Limnol.* 20(1):127–140. — V.A.6; V.B.6

Rippey, B., R. J. Murphy and S. W. Kyle. 1982. Anthropogenically derived changes in the sedimentary flux of Mg, Cr, Ni, Cu, Zn, Hg, Pb and P in Lough Neagh, Northern Ireland. *Environ. Sci. Tech.* 16:23–30. — VII.E

Rittenberg, S. C. 1969. The roles of exogenous organic matter in the physiology of chemolithotrophic bacteria. *Adv. Microbiol. Physiol.* 3:159–196. — V.A.1

Rixen, J.-U. 1968. Beitrag zur Kenntnis der Turbellarien-fauna des Bodenseegebietes. *Arch. Hydrobiol.* 64:335–365. — V.A.7

Robbins, J. A., J. R. Krezoski and S. C. Mozley. 1977. Radioactivity in sediments of the Great Lakes: post-depositional redistribution by deposit-feeding organisms. *Earth Planet. Sci. Lett.* 36:325–333. — VII.A.2

Robbins, J. A., P. L. McCall, J. B. Fisher and J. R. Krezoski. 1979. Effect of deposit feeders on migration of ^{137}Cs in lake sediments. *Earth Planet. Sci. Lett.* 42:277–287. — VII.A.2

Robbins, W. H., and H. R. MacCrimmon. 1974. *The Blackbass in America and Overseas.* Sault St. Marie, ON: Biomanagement and Research Enterprises. 196 pp. — V.A.9

Roberts, R. B., D. B. Cowie, P. H. Abelson, E. T. Bolton and R. J. Britten. 1955. *Studies of biosynthesis in* Escherichia coli. Carnegie Inst. Wash. Publ. 607. 521 pp. — V.B.1

Roberts, W. J., and J. B. Stall. 1967. *Lake evaporation in Illinois.* Illinois State Water Survey Report of Investigation 57. 44 pp. — IV.D

Rochester, H. 1978. Late-glacial and post-glacial diatom assemblages at Berry Pond, Massachusetts in relation to watershed ecosystem development. Ph.D. diss., Indiana University, Bloomington. — VII.F

Rodhe, W. 1958. The primary production in some lakes: some results and restrictions of the ^{14}C method. *Repp. Proc. Verb. Réun. Cons. Perm. Internat. Explor. Mer.* 144:122–128. — V.B.2; V.C

Rodhe, W., J. E. Hobbie and R. T. Wright. 1966. Phototrophy and heterotrophy in high mountain lakes. *Verh. Internat. Verein. Limnol.* 16:302–313. — V.A.2

Roff, J. C., and R. E. Kwiatkowski. 1977. Zooplankton and zoobenthos communities of selected northern Ontario lakes of different acidities. *Can. J. Zool.* 55:899–911. — VIII

Rogaar, H. 1980. The morphology of burrow structures made by tubificids. *Hydrobiologia* 71:107–124. — VI.B

Romanenko, V. I. 1964. Heterotrophic accumulation of CO_2 by aquatic bacteria. *Microbiology* 33:610–614. — V.B.1; V.C

Rosenberg, D. M., A. P. Wiens and B. Bilyj. 1980. Sampling emerging Chironomidae (Diptera) with submerged funnel traps in a new northern Canadian reservoir, Southern Indian Lake, Manitoba. *Can. J. Fish. Aquatic Sci.* 37(6):927–936. — V.A.6

Rossolimo, L. 1939. The role of *Chironomus plumosus* larvae in the exchange of substances between the deposits and the water in a lake. *Arb. Limnol. Sta. Kossine* 22:35–52. — VI.B

Round, F. E. 1957a. Studies on bottom-living algae in some lakes of the English Lake District. Part I, Some chemical features of the sediments related to algal productivities. *J. Ecol.* 45:133–148. — V.A.3

References

	CHAPTER

———. 1957b. Studies on bottom-living algae in some lakes of the English Lake District. Part II, The distribution of Bacillariophyceae on the sediments. *J. Ecol.* 45:343–360. — V.A.3

———. 1957c. Studies on bottom-living algae in some lakes of the English Lake District. Part III, The distribution on the sediments of algal groups other than the Bacillariophyceae. *J. Ecol.* 45:649–664. — V.A.3

———. 1960. Studies on bottom-living algae in some lakes of the English Lake District. Part IV, The seasonal cycles of the Bacillariophyceae. *J. Ecol.* 48:529–547. — V.A.3

———. 1961a. Studies on bottom-living algae in some lakes of the English Lake District. Part V, The seasonal cycles of the Cyanophyceae. *J. Ecol.* 49:31–38. — V.A.3

———. 1961b. Studies on bottom-living algae in some lakes of the English Lake District. Part VI, The effect of depth on the epipelic algal community. *J. Ecol.* 49:245–254. — V.A.3

———. 1964a. The diatom sequence in lake deposits: some problems of interpretation. *Verh. Internat. Verein. Limnol.* 15:1012–1020. — VII.F

———. 1964b. The ecology of benthic algae. In *Algae and Man,* ed. D. F. Jackson, 138–184. New York: Plenum Press. — V.A.3

Rowe, G. T. 1971. Benthic biomass and surface productivity. In *Fertility of the Sea.* Vol. 2, ed. J. D. Costlow, Jr., 441–454. New York: Gordon and Breach. — V.B.6

Rowe, J. S., and G. W. Scotter. 1973. Fire in the boreal forest. *Quat. Res.* 3(3):444–464. — III.B.2

Russell, F. S. 1927. The vertical distribution of plankton in the sea. *Biol. Rev.* 12:213–262. — V.A.5

Rybak, J. I. 1969. Bottom sediments of the lakes of various trophic types. *Ekol. Polska Ser. A* 17:611–662. — VI.B

Sachse, R. 1911. Beiträger zur Biologie litoraler Rädertiere. *Int. Revue Ges. Hydrobiol. Hydrogr.* 3:43–87. — VI.C

Saether, O. A. 1970. A survey of the bottom fauna in lakes of the Okanagan Valley, British Columbia. Unpublished manuscript. *Fish. Res. Bd. Can. Tech. Report.* 196 pp. — V.A.6

Sakamoto, M. 1966. Primary production by phytoplankton community in some Japanese lakes and its dependence on lake depth. *Arch. Hydrobiol.* 62:1–28. — V.A.2; V.B.2

Samuelsson, G. 1934. *Die Verbreitung der höheren Wasserpflanzen in Nordeuropa.* Acta Phytogeogr. Suec. 6. 211 pp. — V.A.4

Sand-Jensen, K. 1975. Biomass, net production and growth dynamics in an eelgrass (*Zostera marina* L.) population in Vellerup Vig, Denmark. *Ophelia* 14:185–201. — V.B.3

Sand-Jensen, K., and M. Søndergaard. 1978. Growth and production of isoetids in oligotrophic Lake Kalgaard, Denmark. *Verh. Internat. Verein. Limnol.* 20:659–666. — V.A.4; V.B.3

———. 1979. Distribution and quantitative development of aquatic macrophytes in relation to sediment characteristics in oligotrophic Lake Kalgaard, Denmark. *Freshwater Biol.* 9:1–11. — V.A.4; V.B.3

Sandberg, G. 1969. A quantitative study of chironomid distribution and emergence in Lake Erken. *Arch. Hydrobiol. Suppl.* 35:119–201. — V.A.6

Sanders, H. L. 1960. Benthic studies in Buzzards Bay. III. The structure of the soft-bottom community. *Limnol. Oceanogr.* 5:138–153. — VI.C

Sanger, J. E., and E. Gorham. 1972. Stratigraphy of fossil pigments as a guide to the post-glacial history of Kirchner Marsh, Minnesota. *Limnol. Oceanogr.* 17(6):840–854. — VII.D

CHAPTER

Sarvala, J. 1979. Benthic resting periods of pelagic cyclopoids in an oligotrophic lake. *Holarctic Ecol.* 2:88–100.   V.A.7

Sasseville, D. R., and S. A. Norton. 1975. Present and historic geochemical relationships in four Maine lakes. *Limnol. Oceanogr.* 20:699–714.   VII.A.2

Saunders, G. W. 1969. Some aspects of feeding in zooplankton. In *Eutrophication: Causes, Consequences and Correctives*, 556–573. Washington, DC: National Academy of Sciences.   VI.C

———. 1972. The transformation of artificial detritus in lake water. *Mem. Ist. Ital. Idrobiol. Suppl.* 29:261–288.   VI.C

———. 1976. Decomposition in freshwaters. In *The Role of Terrestrial and Aquatic Organisms in Decomposition Process*, ed. J. M. Anderson and A. Macfadyen, 341–373. British Ecol. Soc. Symp. 17. Oxford: Blackwell Scientific Publications.   V.B.1

Scheider, W. A., J. J. Moss and P. J. Dillon. 1978. Measurement and uses of hydraulic and nutrient budgets. In *Proc. of the National Conference on Lake Restoration*, 77–83. Minneapolis, MN. August 1978. EPA 440/5-79-001.   IV.D

Schelske, C. L., and E. F. Stoermer. 1971. Eutrophication, silica depletion, and predicted changes in algal quality in Lake Michigan. *Science* 173:423–424.   VII.F

Schindler, D. W. 1971. Carbon, nitrogen, and phosphorus and the eutrophication of freshwater lakes. *J. Phycol.* 7:321–329.   V.A.4; V.B.2

———. 1972. Production of phytoplankton and zooplankton production in Canadian Shield Lakes. In *Proc. IBP-UNESCO Symp. on Productivity Problems in Freshwater*, 311–331. Kazimierz-Dolney, Poland: Pol. Acad. Sci.   V.A.5; V.B.2; V.B.5

———. 1973. Experimental approaches to limnology—an overview. *J. Fish. Res. Bd. Canada* 30(10):1409–1413.   IV.E

———. 1976. Biogeochemical evolution of phosphorus limitation in nutrient-enriched lakes of the Precambrian Shield. In *Environmental Biogeochemistry*. Vol. 2, ed. J. O. Nriagu, 647–664. Ann Arbor, MI: Ann Arbor Science Publ. Inc.   V.B.2

———. 1977. Evolution of phosphorus limitation in lakes. *Science* 195:260–262.   V.B.2; VI.C

———. 1978. Factors regulating phytoplankton production and standing crop in the world's freshwaters. *Limnol. Oceanogr.* 23:478–486.   V.B.2

———. 1980a. Evolution of the Experimental Lakes Project. *Can. J. Fish. Aquatic Sci.* 37(3):313–319.   VI.D

———. 1980b. Experimental acidification of a whole lake: a test of the oligotrophication hypothesis. In *Ecological Impact of Acid Precipitation*, ed. D. Drabløs and A. Tollan, 370–374. SNSF Project, Oslo, Norway.   V.A.2

Schindler, D. W., and F. J. Fee. 1974. Experimental Lakes Area: whole-lake experiments in eutrophication. *J. Fish. Res. Bd. Canada* 31:937–953.   V.A.2; V.B.2

Schindler, D. W., and S. K. Holmgren. 1971. Primary production and phytoplankton in the Experimental Lakes Area, northwestern Ontario, and other low-carbonate waters, and a liquid scintillation method for determining ^{14}C activity in photosynthesis. *J. Fish. Res. Bd. Canada* 28:189–201.   V.A.2; V.B.2

Schindler, D. W., and J. E. Nighswander. 1970. Nutrient supply and primary production in Clear Lake, eastern Ontario. *J. Fish. Res. Bd. Canada* 27:2009–2036.   V.A.2

Schindler, D. W., and B. Noven. 1971. Vertical distribution and seasonal abundance of zooplankton in two shallow lakes of the Experimental Lakes Area, northwestern Ontario. *J. Fish. Res. Bd. Canada* 28(2):245–256.   V.A.5

Schindler, D. W., E. J. Fee and T. Ruszczynski. 1978. Phosphorus input and its consequences for phytoplankton standing crop and production in the Experimental Lakes Area and in similar lakes. *J. Fish. Res. Bd. Canada* 35(2):190–196.   V.B.2

References

CHAPTER

Schindler, D. W., V. E. Frost and R. V. Schmidt. 1973. Production of epilithiphyton in two lakes of the Experimental Lakes Area, northwestern Ontario. *J. Fish. Res. Bd. Canada* 30:1511–1524. — V.A.4; V.B.4

Schindler, D. W., T. Ruszczynski and E. J. Fee. 1980. Hypolimnion injection of nutrient effluents as a method for reducing eutrophication. *Can. J. Fish. Aquatic Sci.* 37:320–327. — V.A.2; V.B.2

Schindler, D. W., R. V. Schmidt and R. A. Reid. 1972. Acidification and bubbling as an alternative to filtration in determining phytoplankton production by the ^{14}C method. *J. Fish. Res. Bd. Canada* 29:1627–1631. — V.C

Schindler, D. W., F. A. J. Armstrong, S. K. Holmgren and G. J. Brunskill. 1971. Eutrophication of lake 227, Experimental Lakes Area, northwestern Ontario, by addition of phosphate and nitrate. *J. Fish. Res. Bd. Canada* 28:1763–1782. — V.B.2

Schindler, D. W., J. Kalff, H. E. Welch, G. J. Brunskill, H. Kling and N. Kritsch. 1974a. Eutrophication in the high arctic—Meretta Lake, Cornwallis Island (75°N Lat.). *J. Fish. Res. Bd. Canada* 31:647–662. — V.B.2; VII.F

Schindler, D. W., H. E. Welch, J. Kalff, G. J. Brunskill and N. Kritsch. 1974b. Physical and chemical limnology of Char Lake, Cornwallis Island (75°N Lat.). *J. Fish. Res. Bd. Canada* 31:585–607. — VII.F

Schindler, D. W., R. W. Newbury, K. G. Beaty and P. Campbell. 1976. Natural water and chemical budgets for a small Precambrian lake basin in central Canada. *J. Fish. Res. Bd. Canada* 33:2526–2543. — IV.E; V.C

Schofield, C. L. 1976. Acid precipitation: effects on fish. *Ambio* 5:228–230. — V.A.2; VIII

Schofield, C. L., and J. R. Trojnar. 1980. Aluminum toxicity to brook trout (*Salvelinus fontinalis*) in acidified waters. In *Polluted Rain*, ed. T. Y. Toribara, M. W. Miller and P. E. Morrow, 341–365. New York: Plenum Press. — VIII

Scott, D. C. 1955. Activity patterns of perch, *Perca flavescens,* in Rondeau Bay of Lake Erie. *Ecology* 36:320–327. — V.A.9

Scott, W. B., and E. J. Crossman. 1973. Freshwater Fishes of Canada. *J. Fish. Res. Bd. Canada* 184:1–920. — V.A.9

Scott, W., and D. F. Opdyke. 1941. The emergence of insects from Winona Lake. *Invest. Indiana Lakes Streams* 2:4–14. — V.A.6

Sculthorpe, C. D. 1967. *The Biology of Aquatic Vascular Plants.* London: Arnold Publ. 610 pp. — V.A.4

Secrest, H. C., A. J. MacAlong and R. C. Lorenz. 1941. Causes of the decadence of hemlock at the Menominee Indian Reservation, Wisconsin. *J. Forestry* 39:3–12. — III.B.2

Seddon, B. 1972. Aquatic macrophytes as limnological indicators. *Freshwater Biol.* 2:107–130. — V.A.4

Sedell, J. R., F. J. Triska, J. D. Hall, N. H. Anderson and J. H. Lyford. 1973. *Sources and fates of organic inputs in coniferous forest streams.* Contribution 66, Coniferous Forest Biome. Internat. Biol. Prog., Oregon State University, Corvallis. 23 pp. — II.B

Seuss, Dr. 1958. *Horton Hears a Who.* New York: Random House. — IV.A

Shannon, C. E., and W. Weaver. 1963. *The Mathematical Theory of Communication.* Urbana: Univ. Illinois Press. — VII.B

Shapiro, J., W. T. Edmondson and D. E. Allison. 1971. Changes in the chemical composition of sediments of Lake Washington, 1958–1970. *Limnol. Oceanogr.* 17:437–452. — VII.E

Shaw, J. B. 1965. Growth and decay of lake ice in the vicinity of Schefferville (Knob Lake, Quebec). *Arctic* 18:123–132. — IV.C

Shearer, J. A., and D. R. DeClercq. 1977. *Light extinction measurements in the Experimental Lakes Area—1976 data.* Fish. Mar. Serv. Data Report 33:IV + 103 pp. — V.A.2

Sheih, C. M., M. L. Wesely and B. B. Hicks. 1979. Estimated dry deposition — IV.E velocities of sulfur over the eastern United States and surrounding regions. *Atmos. Environ.* 13:1361–1368.

Sheldon, R. B., and C. W. Boylen. 1976. Submergent macrophytes: growth under winter ice cover. *Science* 194:841–842. — V.B.3

———. 1977. Maximum depth inhabited by aquatic vascular plants. *Amer. Midl. Natur.* 97:248–254. — V.A.4

Sherman, J. 1976. Post-Pleistocene diatom assemblages in New England lake sediments. Unpublished diss., University of Delaware, Newark. 312 pp. — V.A.3; VII.B; VII.E

Shieh, Y. J., and J. Barber. 1971. Intracellular sodium and potassium concentrations and net cation movements in *Chlorella pyrenoidosa. Biochim. Biophysica. Acta.* (Aust.) 233:594–603. — VI.C

Shih, S. F. 1980. Water budget computation for a shallow lake—Lake Okeechobee, Florida. *Water Resour. Bull.* 16(4):724–728. — IV.D

Siccama, T. G. 1971. Presettlement and present forest vegetation in northern Vermont with special reference to Chittendon County. *Amer. Midl. Natur.* 85:153–172. — III.B.2

Siccama, T. G., and W. H. Smith. 1978. Lead accumulation in a northern hardwood forest. *Environ. Sci. Tech.* 12:593–594. — II.B; IX

Siccama, T. G., F. H. Bormann and G. E. Likens. 1970. The Hubbard Brook Ecosystem Study: productivity, nutrients and phytosociology of the herbaceous layer. *Ecol. Monogr.* 40:389–402. — II.A

Slack, H. D. 1965. The profundal fauna of Loch Lomond, Scotland. *Proc. Roy. Soc. Edinburgh Sec. B Biol.* 69:272–297. — V.A.6

Sloane, J. 1979. Nitrogen flux in small mountain streams in New Hampshire. M.S. thesis, Cornell University, Ithaca, NY. 131 pp. — VIII

Smith, D. M. 1946. Storm damage in New England forests. M.S. thesis, Yale University, New Haven, CT. — III.B.2

Smith, F. E., and E. R. Baylor. 1953. Color responses in the cladocera and their ecological significance. *Amer. Natur.* 37:49–55. — VI.C

Smith, V. H. 1979. Nutrient dependence of primary productivity in lakes. *Limnol. Oceanogr.* 24:1051–1064. — V.A.2; V.B.2

Smith, W. H., and T. G. Siccama. 1981. The Hubbard Brook Ecosystem Study: biogeochemistry of lead in the northern hardwood forest. *J. Environ. Qual.* 10(3):323–333. — II.B; IX

Smol, J. P. 1981. Problems associated with the use of "species diversity" in paleolimnological studies. *Quat. Res.* 15:209–212. — VII.B

Smol, J. P., and M. D. Dickman. 1981. The recent histories of three Canadian Shield lakes: paleolimnological experiment. *Arch. Hydrobiol.* 93:83–108. — VII.F

Smyly, W. J. P. 1968. Some observations on the effect of sampling technique under different conditions on numbers of some freshwater planktonic Entomostroca and Rotifera caught by a water bottle. *Natural History* 2:569–575. — V.A.5

———. 1972. The crustacean zooplankton, past and present, in Estwaite water. *Verh. Internat. Verein. Limnol.* 18:320–326. — VI.C

Snell, T. W. 1979. Intraspecific competition and population structure in rotifers. *Ecology* 60(3):494–502. — VI.C

Söderlund, R. 1977. NO_x pollutants and ammonia emissions—a mass balance for the atmosphere over Northwest Europe. *Ambio* 6(2–3):118–122. — VIII

Solomon, A. M., D. C. West and J. A. Solomon. 1981. Simulating the role of climatic change and species immigration in forest succession. In *Forest Succession—Concepts and Application,* ed. D. G. West, H. H. Shugart and D. B. Botkin, 154–177. Springer-Verlag New York Inc. — III.B.1

Søndergaard, M., and S. Laegaard. 1977. Vesicular-arbuscular mycorrhiza in some aquatic vascular plants. *Nature* 268:232–233. — V.A.4

Søndergaard, M., and K. Sand-Jensen. 1978. Total autotrophic production in — V.A.4; V.B.3; VI.A oligotrophic Lake Kalgaard, Denmark. *Verh. Internat. Verein. Limnol.* 20:667–673.

Sorokin, J. I. 1961. The heterotrophic assimilation of carbonic acid by microorganisms. *J. Gen. Biol.* 22:265–272.   V.B.1; V.C

———. 1966. On the role of chemosynthesis and bacterial biosynthesis in water bodies. In *Primary Productivity in Aquatic Environments,* ed. C. R. Goldman, 187–205. Berkeley/Los Angeles: Univ. California Press. IBP-FP Symp.   V.A.1

Sorokin, Y. I., and N. Donato. 1975. On the carbon and sulfur metabolism in the meromictic lake Faro. *Hydrobiologia* 47:241–252.   V.B.1

Sorokin, Y. I., and H. Kadota. 1972. *Techniques for the Assessment of Microbial Production and Decomposition in Fresh Waters.* IBP Handbook No. 23. Oxford: Blackwell Scientific Publications. 112 pp.   V.A.1

Sorokin, Y. I., and E. B. Paveljeva. 1972. On the quantitative characteristics of the pelagic ecosystems of the Dalnee Lake (Kamchatka). *Hydrobiologia* 40:519–552.   VI.C

Spear, R. W. 1981. Late-Pleistocene and Holocene history of high-elevation plant communities in the White Mountains, New Hampshire. Ph.D. diss., University of Minnesota, Minneapolis. 215 pp.   III.B.1; VII.F

———. In press. Vegetational history of the alpine and subalpine zones of the White Mountains of New Hampshire. *Ecol. Monogr.*   VII.F

Speare, E. A. 1964. *New Hampshire Folk Tales.* Littleton, NH: Courier Printing Co. Inc. 429 pp.   III.B.3

Spence, D. H. N. 1964. The macrophyte vegetation of freshwater lochs, swamps and associated fens. In *The vegetation of Scotland,* ed. J. H. Burnett, 306–425. Edinburgh: Oliver and Boyd.   V.A.4

———. 1967. Factors controlling the distribution of freshwater macrophytes with particular reference to the lochs of Scotland. *J. Ecol.* 55:147–170.   V.A.4

Sprules, W. G. 1975. Midsummer crustacean zooplankton communities in acid-stressed lakes. *J. Fish. Res. Bd. Canada* 32:389–395.   VIII

Spurr, S. H. 1956a. Forest associations in the Harvard Forest. *Ecol. Monogr.* 26:245–262.   III.B.2

———. 1956b. Natural restocking of forests following the 1938 hurricane in central New England. *Ecology* 37:443–451.   III.B.2

Stahl, J. B. 1959. The developmental history of the chironomid and *Chaoborus* faunas of Myers Lake. *Invest. Indiana Lakes Streams* 5:47–102.   VI.B

———. 1966. The ecology of *Chaoborus* in Myers Lake, Indiana. *Limnol. Oceanogr.* 11:177–183.   V.A.6

Stainton, M. P. 1973. A syringe gas-stripping procedure for gas-chromatographic determination of dissolved inorganic organic carbon in fresh water and carbonates in sediments. *J. Fish. Res. Bd. Canada* 30:1441–1445.   V.B.4; V.C

Stainton, M. P., M. J. Capel and F. A. J. Armstrong. 1974. *The chemical analysis of freshwater.* Misc. Spec. Publ. No. 25, Freshwater Institute, Winnipeg, Canada. 119 pp.   IV.B

Stanczykowska, A., and M. Przytocka-Jusiak. 1968. Variations in abundance and biomass of microbenthos in three Mazurian lakes. *Ekol. Polska* 16:539–559.   V.A.7

Starkweather, P. L., and J. J. Gilbert. 1977. Feeding in the rotifer *Brachionus calyciflours*. II. The effect of food density on feeding rates using *Euglena gracilis* and *Rhodotorula glutinus*. *Oecologia* 48:133–139.   VI.C

Stearns, F. S. 1949. Ninety years of change in a northern hardwood forest in Wisconsin. *Ecology* 30:350–358.   III.B.2

Steemann Nielsen, E. 1944. Dependence of freshwater plants on quantity of carbon dioxide and hydrogen ion concentration, illustrated through experimental investigations. *Dansk. Bot. Ark.* 11(8):1–25.   V.A.4

	CHAPTER
———. 1958. Experimental methods for measuring organic production in the sea. *Rapp. P.-V. Reun. Cons. Int. Explor. Mer* 144:38–46.	V.C
Stephens, A. C. 1933. Studies on the Scottish marine fauna; the natural faunistic divisions of the North Sea as shown by qualitative distribution of the molluscs. *Trans. Roy. Soc. Edinburgh* 57:601–616.	VI.C
Stephens, E. P. 1955. Research in the biological aspects of forest production. *J. Forestry* 53:183–186.	III.B.2
———. 1956. The uprooting of trees: a forest process. *Soil Sci. Soc. Amer. Proc.* 20(1):113–116.	III.B.2
Stevenson, L. H. 1978. A case for bacterial dormancy in aquatic systems. *Microbiol. Ecol.* 4:127–133.	V.B.1
Stockner, J. G. 1972. Paleolimnology as a means of assessing eutrophication. *Verh. Internat. Verein. Limnol.* 18:1018–1030.	VIII
Stockner, J. G., and F. A. J. Armstrong. 1971. Periphyton of the Experimental Lakes Area, northwestern Ontario. *J. Fish. Res. Bd. Canada* 28(2):215–229.	V.A.3; VI.A
Stockner, J. G., and J. W. G. Lund. 1970. Live algae in post-glacial lake deposits. *Limnol. Oceanogr.* 15:41–58.	VI.B
Strahler, A. N. 1957. Quantitative analysis of watershed geomorphology. *Trans. Amer. Geophys. Union* 38:913–920.	II.A
Straskraba, M. 1968. Der Anteil der höheren Pflanzen an der Produktion der stehenden Gewässer. *Mitt. Internat. Verein. Limnol.* 14:212–230.	V.B.3
Strayer, D. L. 1984. The benthic micrometazoans of Mirror Lake, New Hampshire. Ph.D. diss., Cornell University, Ithaca, NY. 346 pp.	V.A.6; VI.B; VI.D
Strayer, D. L., J. J. Cole, G. E. Likens and D. Buso. 1981. Biomass and annual production of the freshwater mussel *Elliptio complanata* in an oligotrophic, softwater lake. *Freshwater Biol.* 11:435–440.	V.A.6; VI.A
Strickland, J. D. H. 1960. Measuring the production of marine phytoplankton. *J. Fish. Res. Bd. Canada* 122: 172 pp.	V.A.2; V.A.3
Strøm, K. M. 1928. Recent advances in limnology. *Proc. Linnean Soc. London* 140:96–110.	VII.F
Stross, R. G. 1979. Density and boundary regulations of the Nitella meadow in Lake George, New York. *Aquatic Botany* 6:285–300.	VI.A
Stumm, W. and J. Morgan. 1981. *Aquatic Chemistry. An Introduction Emphasizing Chemical Equilibria in Natural Waters.* 2d ed. New York: John Wiley & Sons Inc. 780 pp.	V.A.1
Suess, E. 1875. *Die Entstehung der Alpen.* Vienna: W. Braumuller.	I.A
Sutton, O. G. 1949. The application of micrometeorology to the theory of turbulent flow over rough surfaces. *Royal Meteorology Soc. Quart. J.* 75(326):336–350.	IV.D
Sverdrup, H. V. 1937. On the evaporation from the oceans. *J. Mar. Res.* 1(1):3–14.	IV.D
Swain, A. M. 1973. A history of fire and vegetation in northeastern Minnesota as recorded in lake sediments. *Quat. Res.* 3:383–396.	III.B.1; III.B.2; VII.F
Swift, M. C., and A. Y. Fedorenko. 1975. Some aspects of prey capture by *Chaoborus* larvae. *Limnol. Oceanogr.* 20:418–426.	IV.A
Swindale, D. N., and J. T. Curtis. 1957. Phytosociology of the larger submerged plants in Wisconsin lakes. *Ecology* 38:397–407.	V.A.4
Swuste, H. F. J., R. Cremer and S. Parma. 1973. Selective predation by larvae of *Chaoborus flavicans* (Diptera, Chaoboridae). *Verh. Internat. Verein. Limnol.* 18:1559–1563.	V.A.6; VI.B
Szczepanski, A. 1965. Deciduous leaves as a source of organic matter in lakes. *Bull. Acad. Pol. Sci. Cl.* II, 13(4):215–217.	V.C
Tansley, A. G. 1935. The use and abuse of vegetational concepts and terms. *Ecology* 16:284–307.	I.A

References

CHAPTER

Tappa, D. W. 1965. The dynamics of the association of six limnetic species of *Daphnia* in Aziscoos Lake, Maine. *Ecol. Monogr.* 35:395–423. — VI.C

Tauber, H. 1967. Investigations of the mode of pollen transfer in forested areas. *Rev. Paleobotany and Palynology* 3:277–286. — III.B.1

Taylor, W. D., and D. R. S. Lean. 1981. Radiotracer experiments on phosphorus uptake and release by limnetic microzooplankton. *Can. J. Fish. Aquatic Sci.* 38:1316–1322. — VI.C

Tebo, L. B. 1955. Bottom fauna of a shallow eutrophic lake, Lizzard Lake, Pocahontas County, Iowa. *Amer. Midl. Natur.* 54:89–103. — V.A.6

Teraguchi, M., and T. G. Northcote. 1966. Vertical distribution and migration of *Chaoborus flavicans* larvae in Corbett Lake, British Columbia. *Limnol. Oceanogr.* 11(2):164–176. — V.A.6

Terbough, J. 1970. Distribution on environmental gradients: theory and a preliminary interpretation of distributional patterns in the avifauna of the *Cordillera vilacabamba*, Peru. *Ecology* 52:24–40. — VI.C

Tessenow, U. 1966. Untersuchungen über den Kieselsäurehaushalt der Binnengewässer. *Arch. Hydrobiol. Suppl.* 32:1–136. — VII.A.2; VII.E

———. 1975. Lösungs-, Diffusions- und Sorptionsprozesse in der Obershicht von Seesedimenten. V. Die Differenzierung der Profundalsedimente eines oligotrophen Bergsees (Feldsee, Hochschwarzwald) durch Sediment-Wasser-Wechselwirkungen. *Arch. Hydrobiol. Suppl.* 47(3):325–412. — VII.E

Tester, A. L. 1932. Food of the small-mouthed black bass (*Mircopterus dolomieui*) in some Ontario waters. *Univ. Toronto Studies Publ. Ontario Fish. Res. Lab.* 46:176–203. — V.A.9

Thienemann, A. 1919. Uber die vertikale Schichtungder Planktons in Ulmener Maar und die Planktonproduction der anderen Eifelmaare. *Verh. Naturh. Ver. Preuss. Rheinl.* 74:103–134. — V.A.5

Thiessen, A. H. 1911. Precipitation for large areas. *Monthly Weather Rev.* 39:1082–1089. — II.B

Thompson, D. Q., and R. H. Smith. 1970. The forest primeval in the Northeast: a great myth. In *Proc. Tall Timbers Fire Ecology Conference No. 10*, 255–256. August 1970, Frederickton, New Brunswick, Canada. Tallahassee, FL: Tall Timbers Research Station. — III.B.2

Thorpe, J. 1977a. Morphology, physiology, behavior, and ecology of *Perca fluviatilis* L. and *Perca flavescens* Mitchill. *J. Fish. Res. Bd. Canada* 34:1504–1514. — V.A.9

———. 1977b. Synopsis of biological data on the perch *Perca fluviatilis* Linnaeus, 1758 and *Perca flavenscens* Mitchill, 1814. *Food Agric. Organ. Fish. Synopsis* 113:1–138. — V.A.9

Thunmark, S. 1931. Der See Fiolen und seine Vegetation. *Acta Phytogeogr. Suec.* 2. 198 pp. — V.A.4

Thut, R. N. 1969. A study of the profundal bottom fauna of Lake Washington. *Ecol. Monogr.* 39:79–100. — V.A.6

Tilman, D., S. S. Kilham and P. Kilham. 1982. Phytoplankton community ecology: the role of limiting nutrients. *Ann. Rev. Ecol. Syst.* 13:349–372. — VII.F

Tilton, L. W., and J. K. Taylor. 1937. Accurate representation of reflectivity and density of distilled water as a function of temperature. *J. Res. U.S. Bur. Stand.* 18:205–214. — IV.C

Tilzer, M. M. 1973. Diurnal periodicity in the phytoplankton assemblage of a high mountain lake. *Limnol. Oceanogr.* 18:15–30. — V.A.2; V.B.2

Tilzer, M. M., H. W. Paerl and C. R. Goldman. 1977. Sustained viability of aphotic phytoplankton in Lake Tahoe (California–Nevada). *Limnol. Oceanogr.* 22:84–91. — V.A.2

Timms, B. V. 1974. Morphology and benthos of three volcanic lakes in the Mt. Gambier district, South Australia. *Aust. J. Mar. Freshwater Res.* 25:287–297. — V.A.6

	CHAPTER
Titman, D., and P. Kilham. 1976. Sinking in freshwater phytoplankton: some ecological implications of cell nutrient status and physical mixing processes. *Limnol. Oceanogr.* 21:409–417.	V.B.2
Tiovonen, H., and T. Lappalainen. 1980. Ecology and production of aquatic macrophytes in the oligotrophic, mesohumic lake Suomunjärvi, eastern Finland. *Ann. Bot. Fenn.* 17:69–85.	V.A.4
Tolonen, K. 1980. Comparison between radiocarbon and varve dating in Lake Lampellonjärvi, South Finland. *Boreas* 9:11–19.	VII.E
Tranter, D. J., and P. E. Smith. 1968. Filtration performance. In *Zooplankton Sampling,* 27–56. Monographs on Oceanographic Methodology 2. Paris, France: UNESCO.	V.A.5
Trautman, M. B. 1957. *The Fishes of Ohio.* Columbus: Ohio State Univ. Press. 638 pp.	V.A.9
Trewartha, G. T. 1954. *Introduction to Climate.* New York: McGraw-Hill Book Co. 402 pp.	II.A
Trimble, G. R., Jr., R. S. Sartz and R. S. Pierce. 1958. How type of soil frost affects infiltration. *J. Soil Water Conserv.* 13(2):81–82.	II.B
Tsukada, M., and E. S. Deevey. 1967. Pollen analysis from four lakes in the southern Maya area of Guatemala and El Salvador. In *Quaternary Paleoecology,* 303–331. New Haven, CT: Yale Press.	III.B.2
Tubb, R. A., and T. C. Dorris. 1965. Herbivorous insect populations in oil refinery effluent holding pond series. *Limnol. Oceanogr.* 10:121–134.	V.A.6
Turner, J. F. 1966. *Evaporation study in a humid region, Lake Michie, North Carolina.* U.S. Geol. Surv. Prof. Paper 272-G. 150 pp.	IV.D
Tutin, W. (Pennington). 1969. The usefulness of pollen analysis in interpretation of stratigraphic horizons, both late-glacial and post-glacial. *Mitt. Internat. Verein. Limnol.* 17:154–164.	III.B.1; VII.F
Utermöhl, H. 1958. Zur Verollkommung der quantitativen phytoplanktonmethodik. *Mitt. Internat. Verein. Limnol.* 9:1–38.	V.B.5
Vallentyne, J. R. 1952. Insect removal of nitrogen and phosphorus compounds from lakes. *Ecology* 33(4):573–577.	I.A.2; IV.E
———. 1974. *The Algal Bowl: Lakes and Man.* Misc. Spec. Publ. 22. Environ. Canada. 185 pp.	VI.C
Vardapetyan, S. M. 1972. Food relations of predatory crustaceans in lake zooplankton. *Ekologiya* 3:38–44.	VI.C
Vasari, Y., and A. Vasari. 1968. Late- and post-glacial macrophytic vegetation in the lochs of northern Scotland. *Acta Bot. Fenn.* 80. 120 pp.	V.A.4; VII.F
Vaughan, A. T. 1965. *New England Frontier—Puritans and Indians 1620–1675.* Boston: Little, Brown and Co. 430 pp.	III.B.3
Vernadsky, W. I. 1929. *La Biosphere.* Libraire Félix Alcan.	I.A
Vinberg, G. C. 1969. Some results of studies on lake productivity in the Soviet Union, conducted as part of the IBP. *Dokl. Akad. Nauk SSSR* 186(1):197–201.	V.B.1
Vitousek, P. M. 1977. The regulation of element concentrations in mountain streams in the northeastern United States. *Ecol. Monogr.* 47:65–87.	VII.F
Vitousek, P. M., and W. A. Reiners. 1975. Ecosystem succession and nutrient retention: a hypothesis. *BioScience* 25:376–381.	II.B; VII.F
Vollenweider, R. A. 1968. *Scientific fundamentals of the eutrophication of lakes and flowing waters, with particular reference to nitrogen and phosphorus as factors in eutrophication,* 182. Water Management Research. Org. Ecol. Coop. Dev. (Paris) Tech. Report DAS/RS1/68.27. 159 pp.	IV.A; IV.E; V.B.2; V.C; VIII

References

	CHAPTER
———. 1969. *Measuring Primary Production in Aquatic Environments.* IBP Handbook 12. Oxford: Blackwell Scientific Publications.	V.B.1; V.C
———. 1976. Advances in defining critical loading levels for phosphorus in lake eutrophication. *Mem. Ist Ital. Idrobiol.* 33:53–83.	VIII
Vollenweider, R. A., and A. Nauwerck. 1961. Some observations on the ^{14}C method for measuring primary production. *Verh. Internat. Verein. Limnol.* 14:134–139.	V.C
Von Damm, K. L., L. K. Benninger and K. K. Turekian. 1979. The ^{210}Pb chronology of a core from Mirror Lake, New Hampshire. *Limnol. Oceanogr.* 24:434–439.	VII.A.1; VII.A.2; VII.E
von Ende, C. N. 1979. Fish predation, interspecific predation, and the distribution of two *Chaoborus* species. *Ecology* 60(1):119–128.	V.A.6
Waksman, S. A., J. L. Stokes and M. R. Butler. 1937. Relation of bacteria to diatoms in sea water. *J. Mar. Biol. Assoc. U.K.* 22:359–373.	V.D
Walker, J. B., G. B. Chandler and J. B. Harrison. 1981. *Report of the Forestry Commission of New Hampshire.* Manchester, NH: J. B. Clarke. 53 pp.	III.B.2
Walker, J., C. H. Thompson, I. F. Fergus and B. R. Tunstall. 1981. Plant succession and soil development in coastal sand dunes of subtropical eastern Austrialia. In *Forest Succession: Concepts and Applications,* ed. D. C. West, H. H. Shugart and D. B. Botkin, 107–131. Springer-Verlag New York Inc. 517 pp.	VII.F
Walker, K. F., and G. E. Likens. 1975. Meromixis and a reconsidered typology of lake circulation patterns. *Verh. Internat. Verein. Limnol.* 19:442–458.	IV.B; IV.C
Walker, R. 1978. Diatom and pollen studies of a sediment profile from Melynllyn, a mountain tarn in Snowdonia, North Wales. *New Phytol.* 81:791–804.	VII.F
Walling, H. F. 1860. *Topographical map of Grafton County, New Hampshire.* Scale 1:70,000. New York.	III.B.3
Walter. R. A. 1976. The role of benthic macrofauna in the structure and function of the Mirror Lake ecosystem. M.S. thesis, Cornell University, Ithaca, NY. 206 pp.	V.A.6; V.A.9; V.B.6; V.C
Ward, A. K., and R. G. Wetzel. 1975. Sodium: some effects on blue-green algal growth. *J. Phycol.* 11:357–363.	VI.D
Warnick, C. C. 1951. Influence of wind on precipitation measurements at high altitudes. *Univ. Idaho Engineering Experiment Station Bull.* 10:6–63.	IV.D
Waters, T. F. 1969. The turnover ratio in production ecology of freshwater invertebrates. *Amer. Natur.* 103:173–185.	V.B.6
———. 1977. Secondary production in inland waters. In *Advances in Ecological Research.* Vol. 10, ed. A. MacFadyen, 91–164. New York: Academic Press.	V.B.7; VI.D
Watts, W. A. 1970. The full-glacial vegetation of northwestern Georgia. *Ecology* 51:17–33.	V.A.4
Wavre, M., and R. O. Brinkhurst. 1971. Interactions between some tubificid oligochaetes and bacteria found in the sediments of Toronto Harbour, Ontario. *J. Fish. Res. Bd. Canada* 28:335–341.	V.A.6
Webb, T., III, S. Rowe, R. H. W. Bradshaw and K. Heide. 1981. Estimating plant abundances from pollen percentages: the use of regression analysis. *Rev. Paleobotany and Palynology* 34:269–300.	III.B.1
Welch, H. E. 1973. Emergence of Chironomidae (Diptera) from Char Lake, Resolute, Northwest Territories. *Can. J. Zool.* 51:1113–1123.	V.A.6; V.B.6
———. 1976. Ecology of Chironomidae (Diptera) in a polar lake. *J. Fish. Res. Bd. Canada* 33:227–247.	V.A.6
Welch, H. E., and J. Kalff. 1974. Benthic photosynthesis and respiration in Char Lake. *J. Fish. Res. Bd. Canada* 31(5):609–620.	V.C

CHAPTER

Werner, E. E., and D. J. Hall. 1974. Optimal foraging and the size selection of prey by the bluegill sunfish (*Leponis macrochirus*). *Ecology* 55:1042–1052.   VI.C

Westlake, D. F. 1965. Some basic data for investigations of the productivity of aquatic macrophytes. In *Primary Productivity in Aquatic Environments,* ed. C. R. Goldman, 229–248. Mem. Ist. Ital. Idrobiol. 18 Suppl. Berkeley: Univ. California Press.   V.B.3

Wetzel, R. G. 1964. A comparative study of the primary productivity of higher aquatic plants, periphyton, and phytoplankton in a large shallow lake. *Int. Rev. Ges. Hydrobiol.* 49:1–61.   V.B.3; V.B.4

——. 1969. The enclosure of macrophyte communities. In *A Manual on Methods for Measuring Primary Production in Aquatic Environments,* ed. R. A. Vollenweider, 81–88. I.B.P. Handbook No. 12, Oxford: Blackwell Scientific Publications.   V.B.3

——. 1970. Recent and postglacial production rates of a marl lake. *Limnol. Oceanogr.* 15:491–503.   VII.F

——. 1975. *Limnology.* Philadelphia: W. B. Saunders Co. 743 pp.   IV.B; V.A.3; V.A.4; V.A.5; V.B.5; V.D; VI.C; VI.D

Wetzel, R. G., and G. E. Likens. 1979. *Limnological Analyses.* Philadelphia: W. B. Saunders Co. 357 pp.   IV.C; V.A.5

Wetzel, R. G., and P. H. Rich. 1973. Carbon in freshwater systems. In *Carbon and the Biosphere,* ed. G. M. Woodwell and E. V. Pecan, 241–263. USAEC Report 720510.   V.C

Wetzel, R. G., P. H. Rich, M. C. Miller and H. L. Allen. 1972. Metabolism of dissolved and particulate detrital carbon in a temperate hard-water lake. *Mem. Ist. Ital. Idrobiol.* 29 Suppl. 185:243.   V.C

Whitehead, D. R., and T. L. Crisman. 1978. Paleolimnological studies of small New England (USA) ponds. I: late-glacial and post-glacial trophic oscillations. *Pol. Arch. Hydrobiol.* 25:471–481.   VII.F

Whittaker, R. H. 1951. A criticism of the plant association and climatic climax concepts. *Northwest Science* 26(1):17–31.   I.A

——. 1952. A study of summer foliage insect communities in the Great Smoky Mountains. *Ecol. Monogr.* 22:1–44.   VI.C

——. 1967. Gradient analysis of vegetation. *Biol. Rev.* 42:207–264.   VI.C

——. 1975. *Communities and Ecosystems.* 2d ed. New York: MacMillan Publishing Co. 387 pp.   V.B.7; VI.C

Whittaker, R. H., S. A. Levin and R. B. Root. 1973. Niche, habitat, and ecotope. *Amer. Natur.* 107:321–338.   VI.C

Whittaker, R. H., F. H. Bormann, G. E. Likens and T. G. Siccama. 1974. The Hubbard Brook Ecosystem Study: forest biomass and production. *Ecol. Monogr.* 44(2):233–254.   II.A

Wich, K. F., and J. W. Mullan. 1958. A compendium of the life history and ecology of the chain pickerel, *Esox niger* (LeSueur). *Mass. Fish. Bull.* 22:1–27.   V.A.9

Wiebe, W. J., and D. F. Smith. 1977. Direct measurement of dissolved organic carbon release by phytoplankton and incorporation by microheterotrophs. *Marine Biology* 42:213–223.   V.D; VI.C

Wiens, J. A. 1977. On competition and variable environments. *Amer. Sci.* 65:590–597.   VI.C

Williams, J. B., and M. C. Whiteside. 1978. Population regulation of the Chydoridae in Lake Itasca, Minnesota. *Verh. Internat. Verein. Limnol.* 20:2434–2487.   VII.F

Williams, J. D. H., J. K. Syers, S. S. Shukla, R. F. Harris and D. E. Armstrong. 1971. Levels of inorganic and total phosphorus in lake sediments as related to other sediment parameters. *Environ. Sci. Tech.* 5:1113–1120.   VII.E

References

CHAPTER

Williams, P. J. 1970. Heterotrophic utilization of dissolved organic compounds in the sea. 1. Size distribution of population and relationship between respiration and incorporation of growth substrates. *J. Mar. Biol. Assoc. U.K.* 50:859–870.   V.B.1

———. 1973. On the question of growth yields of natural heterotrophic populations. *Bull. Ecol. Res. Comm.* (Stockholm) 17:400–401.   V.B.1

Willoughby, L. G. 1974. Decomposition of litter in fresh water. In *Biology of Plant Litter Decomposition*. Vol. 2, ed. C. H. Dickinson and G. J. F. Pugh, 659–681. New York: Academic Press.   V.B.1

Wilson, L. R. 1935. Lake development and plant succession in Vilas County, Wisconsin. I. The medium hard water lakes. *Ecol. Monogr.* 5:207–247.   V.A.4

———. 1939. Rooted aquatic plants and their relation to the limnology of freshwater lakes. In *Problems of Lake Biology*, ed. F. R. Moulton, 107–122. Publ. Amer. Assoc. Advmt. Sci. 10.   V.A.4

———. 1941. The larger aquatic vegetation of Trout Lake, Vilas County, Wisconsin. *Trans. Wisconsin Acad. Sci. Arts Lett.* 33:135–146.   V.A.4

Wilson, M. S., and H. C. Yeatman. 1959. Free-living Copepoda. Chapter 29 in *Freshwater Biology*, ed. W. T. Edmondson. 2d ed. New York: John Wiley & Sons Inc.   V.A.7

Winberg, G. G. 1960. *Rate of metabolism and food requirements of fishes.* Fish. Res. Bd. Canada. Transl. Series 194. 202 pp., 32 tables.   V.B.7

———. 1971a. Methods for calculating productivity. In *Methods for the Assessment of Secondary Productivity in Fresh Waters*, ed. W. T. Edmondson and G. G. Winberg, 296–317. I.B.P. Handbook No. 17. Oxford: Blackwell Scientific Publications.   V.B.5

———. 1971b. *Methods for the Estimation of Production of Aquatic Animals.* New York: Academic Press. 175 pp.   V.B.5; V.B.7

———. 1972. Some interim results of Soviet IBP investigations on lakes. In *Productivity Problems of Freshwaters*, ed. Z. Kajak and A. Hillbricht. Իkowska, 363–382. Proc. of IBP-UNESCO Symp., Kazimierz Dolny, Poland, May 1970.   V.B.5

Winberg, G. G., V. A. Babitsky, S. I. Gavrilov, G. V. Gladky, I. S. Zakharenkov, R. Z. Kovalevskaya, T. M. Mikheeva, P. S. Nevyadomskaya, A. P. Ostapenya, P. G. Petrovich, J. S. Potaenko and O. F. Yakushko. 1972. Biological productivity of different types of lakes. In *Productivity Problems of Freshwaters*, ed. Z. Kajak and A. Hillbricht-Իkowska, 383–404. Warsaw: PWN, Polish Scientific Publishers.   V.A.4; V.B.5

Winer, H. I. 1955. History of the Great Mountain Forest. Ph.D. diss., Yale University, New Haven, CT. 278 pp.   III.B.2

Winter, T. C. 1976. *Numerical simulation analysis of the interaction of lakes and ground water.* U.S. Geol. Surv. Prof. Paper 1001. 45 pp.   IV.D

———. 1978. Numerical simulation of steady state three-dimensional ground water flow near lakes. *Water Resour. Res.* 14(2):245–254.   IV.D

———. 1981a. Effects of water table configuration on seepage through lake beds. *Limnol. Oceanogr.* 26(5):925–934.   IV.D

———. 1981b. Survey of errors for estimating water and chemical balances of lakes and reservoirs. In *Proc. Symp. on Surface-Water Impoundments*, 224–233. American Society of Civil Engineers. June 1980. Minneapolis, MN.   IV.D

———. 1981c. Uncertainties in estimating the water balance of lakes. *Water Resour. Bull.* 17(1):82–115.   IV.D

Wissmar, R. C., and R. G. Wetzel. 1978. Analysis of five North American lake ecosystems. VI. Consumer community structure and production. *Verh. Internat. Verein. Limnol.* 20:587–597.   V.A.6; V.B.6; V.C

Wissmar, R. C., J. E. Richey and D. E. Spyridakis. 1977. The importance of allochthonous particulate carbon pathways in a subalpine lake. *J. Fish. Res. Bd. Canada* 34(9):1410–1418.   V.C

	CHAPTER
Wium-Andersen, W. 1971. Photosynthetic uptake of free CO_2 by the roots of *Lobelia dortmanna*. *Physiol. Plant.* 25:245–248.	V.A.4; V.B.3
Wohlschlag, D. E. 1950. Vegetation and invertebrate life in a marl lake. *Invest. Indiana Lakes Streams* 3(9):321–372.	V.A.6
Wood, K. G. 1956. Ecology of Chaoborus (Diptera: Culicidae) in an Ontario lake. *Ecology* 37(4):639–643.	V.A.6
Wood, L. W. 1975. Role of oligochaetes in the circulation of water and solutes across the mud-water interface. *Verh. Internat. Verein. Limnol.* 19(2):1530–1533.	VI.B
Wood, R. D. 1967. *Charophytes of North America*. West Kingston, RI: Stella Printing. 72 pp.	V.A.4
Wood, T. E. 1980. Biological and chemical control of phosphorus cycling in a northern hardwood forest. Ph.D. diss., Yale University, New Haven, CT. 205 pp.	II.B
Woolman, L. 1891. Geology of artesian wells at Atlantic City, New Jersey. *Proc. Acad. Nat. Sci. Phil.* 42:132–147.	VII.B
———. 1892. A review of artesian well horizons in southern New Jersey. *Annual Report of the State Geologist for the Year, 1891*, 223–232. Geol. Surv. of New Jersey.	VII.B
Worthington, E. G. 1931. Vertical movements of freshwater macroplankton. *Int. Rev. Ges. Hydrobiol. Hydrogr.* 25:394–436.	V.A.5
Wright, H. E., Jr. 1966. Stratigraphy of lake sediments and the precision of the paleo-climatic record. In *Proc. Internat. Symp. on World Climate from 8000 to 0 B.C.*, 157–173.	III.B.1
Wright, H. E., Jr., and M. L. Heinselman. 1973. The ecological role of fire in natural conifer forests of western and northern North America—introduction. *Quat. Res.* 3(3):319–328.	III.B.2
Wright, H. E., D. A. Livingstone and E. J. Cushing. 1965. Coring devices for lake sediments. In *Handbook of Paleontological Techniques*, ed. B. Kummel and D. Raup, 494–520. San Francisco: W. H. Freeman & Co.	VII.E
Wright, R. F. 1974. *Forest fire: impact on the hydrology, chemistry and sediments of small lakes in northeastern Minnesota*. Univ. of Minnesota Limnological Research Center Report No. 10.	III.B.2
———. 1975. Studies on glycolic acid metabolism by freshwater bacteria. *Limnol. Oceanogr.* 20:626–633.	VI.C
Wunderlich, N. O. 1971. The dynamics of density stratified reservoirs. In *Reservoir Fisheries and Limnology*, 219–231. Amer. Fish. Soc. Special Publ. 8.	IV.C
Yamagishi, H., and H. Fukuhara. 1972. Vertical migration of *Spaniotoma akamusi* larvae (Diptera: Chironomidae) through the bottom deposits of Lake Suwa. *Japanese J. Ecol.* 22:226–227.	V.A.6
Yan, N. D., and R. Strus. 1980. Crustacean zooplankton communities of acidic, metal-contaminated lakes near Sudbury, Ontario. *Can. J. Fish. Aquatic Sci.* 37:2282–2293.	VIII
Yeates, G. W. 1979. Soil nematodes in terrestrial ecosystems. *J. Nematology* 11:213–229.	V.A.7
Zaret, T. M., and A. S. Rand. 1971. Competition in tropical stream fishes: support for the competitive exclusion principle. *Ecology* 52:336–342.	V.A.9
Zaret, T. M., and F. S. Suffern. 1976. Vertical migration in zooplankton as a predator avoidance mechanism. *Limnol. Oceanogr.* 21:801–813.	V.A.5; V.A.6
ZoBell, C. E. 1963. Organic geochemistry of sulfur. In *Organic Geochemistry*, ed. I. Breger, 543–578. New York: MacMillan.	V.B.1

Index of Genera and Species

Abies balsamea, 14
Abies sp., 56ff
Abies spiroides, 163
Ablabesmyia mallochi, 209
Ablabesmyia sp., 206
Acanthocyclops vernalis, 208
Acer rubrum, 14, 56
Acer saccharum, 14
Acer sp., 14, 56ff
Achnanthes calcar, 368
Achnanthes clevei v. rostrata, 368
Achnanthes clevei, 368
Achnanthes didyma, 368
Achnanthes exigua, 368
Achnanthes exigua v. heterovalva, 368
Achnanthes flexella, 369
Achnanthes lanceolata, 368
Achnanthes lanceolata v. abbreviata, 369
Achnanthes lanceolata v. apiculata, 369
Achnanthes lanceolata v. elliptica, 368
Achnanthes lapponica, 369
Achnanthes laterostrata, 369
Achnanthes microcephala, 369, 380
Achnanthes minutissima, 369
Achnanthes peragalli v. fossilis, 369
Achnanthes peragalli, 369
Achnanthes pinnata, 369
Achnanthes stewartii, 369
Achnanthes suchlandtii, 369
Achromadora sp., 207
Acroperus harpae, 383
Adineta vaga, 269
Aeolosoma spp., 208
Agrypnia sp., 209
Alaimus sp., 207
Alnus crispa, 57
Alnus rugosa, 178
Alnus sp., 56ff

Alona cf. barbulata, 208
Alona quadrangularis, 208
Alona rustica, 208
Alonella affinis, 208, 383
Alonella excisa, 193, 208, 236, 383, 384, 417
Alonella nana, 208, 383, 386
Ambrosia sp., 360–361
Amoeba sp., 207
Amphichaeta americana, 208
Amphidinium sp., 165
Amphora ovalis v. libyca, 370
Amphora ovalis v. pediculus, 370
Amphora perpusilla, 370
Anabaena circinalis, 163
Anabaena sp., 163, 255, 268, 327
Anabaena spiroides, 163
Anheteromeyenia argyrosperma, 207
Ankistrodesmus braunii, 163
Ankistrodesmus falcatus, 162–163, 168
Ankistrodesmus falcatus v. acicularis, 163
Ankistrodesmus setigerus, 163
Ankistrodesmus spiralis, 163
Anomoeoneis serians v. brachysira, 369
Anomoeoneis vitrea, 369
Anonchus sp., 207
Aphanizomenon flos-aquae, 163
Aphanocapsa delicatissma, 163
Aphanolaimus sp., 207
Aphanotheca clathrata, 163
Argia sp., 209
Arrenurus sp., 208
Arthrodesmus incus, 163
Arthrodesmus phimus, 163
Ascomorpha sp., 193, 207
Aspidiophorus sp., 207
Asplanchna priodonta, 193, 195, 200, 275, 332
Asplanchna sp., 192

Aster accuminatus, 14
Aster novi-belgii, 178
Asterionella formosa 162, 165, 172, 368, 373, 414, 417–418
Asterionella formosa v. gracillima, 368
Asterionella sp., 374
Attheyella obatogamensis, 208
Attheyella sp., 197
Aulodrilus pigueti, 208

Bacillaria paradoxa, 370
Banksiola sp., 209
Basiaeschna sp., 209
Batrachospermum sp., 178, 180, 182, 186, 188
Betula alleghaniensis, 14
Betula sp., 56ff
Bidessus sp., 209
Bitrichia chodatii, 162, 164
Bosmina longirostris, 193, 196–197, 200, 274, 276, 332, 383, 385, 417, 419
Bosmina sp., 327, 332–333, 384
Botryococcus braunii, 162–163
Botryococcus protuberans v. minor, 163
Brillia sp., 209
Bryocamptus minutus, 208
Bryocamptus zschokkei, 208
Bufo americanus, 236, 314

Caenis sp., 208
Caloneis sp., 369
Caloneis ventricosa v. minuta, 369
Caloneis ventricosa v. subundulata, 369
Calothrix sp., 176
Campeloma sp., 208

Campylodiscus hibernicus v. hungaricus, 370
Candona sp., 208
Canthocamptus assimilis, 208
Carteria sp., 163
Castor canadensis, 14
Catostomus commersoni, 241, 289
Cephalodella forficula, 207
Cephalodella spp., 207, 230
Ceratium hirundinella, 162, 165
Ceratophyllum demersum, 181
Ceriodaphnia sp., 326–327
Chaetogaster diastrophus, 208, 340
Chaetonotus spp., 207, 230, 232
Chamaedaphne calyculata, 178
Chaoborus flavicans, 209, 220
Chaoborus punctipennis, 205, 209, 220
Chaoborus sp., 205–206, 214, 217, 219–221, 223, 225–226, 228, 238–239, 285, 287, 288, 318–322, 325, 328, 340, 358
Chara braunii, 180, 183
Chara furcata, 418
Chara sp., 183, 350, 415–416
Chauliodes sp., 209
Chiloscyphus fragilis, 180, 182, 186
Chironomus anthracinus, 209, 211–212, 220, 223–226, 288, 318, 321–322, 328, 340
Chironomus plumosus, 219
Chironomus sp., 209–211, 213
Chlamydomonas sp., 162–163
Chlorella sp., 85
Chlorella vulgaris, 163
Chlorochromonas polymorpha, 164
Chlorohydra sp., 207
Chromadorita sp., 207
Chromulina mikroplankton, 164
Chromulina minima, 164
Chromulina nebulosa, 164
Chromulina ovalis, 164
Chromulina sp. (1), 162, 164, 166
Chromulina woroniniana, 164
Chroococcus dispersus v. minor, 163
Chroococcus limneticus, 163, 166
Chroococcus turgidus, 163
Chroomonas coerulea, 165
Chrysidalis peritaphrena, 164

Chrysidiastrum catenatum, 162–164
Chrysochromulina parva, 162, 164, 166, 168, 255
Chrysochromulina sp., 333
Chrysococcus rufescens, 164
Chrysococcus sp., 162–164, 166, 277
Chrysoikos skujai, 164
Chrysolykos planctonicus, 162, 164
Chrysops sp., 209
Chrysosphaerella longispina, 162, 164, 166, 168–169, 170–172
Chydorus bicornutus, 208
Chydorus nr. brevilabris, 208, 384–386, 418
Chydorus piger, 208, 384
Chydorus sp., 208
Chydorus sphaericus, 384
Cladopelma sp., 209
Cladotanytarsus sp., 209
Clinotanypus sp., 209
Clintonia borealis, 14
Closterium incurvum, 163
Closterium sp., 176, 213
Coccomyxa minor, 163
Coccomyxa sp., 163
Cocconeis diminuta, 369–370, 373
Cocconeis placentula v. lineata, 369
Coelastrum microporum, 163
Colacium sp., 176, 207
Coleps sp., 207
Collotheca sp., 207
Colurella spp., 207
Comptonia sp., 45
Conchapelopia sp., 209
Conochiloides dossuarius, 193, 195, 201, 275–276, 332–333
Conochilus unicornis, 193, 195, 203, 275, 332–333
Cordulia sp., 209
Coscinodiscus sp., 165
Cosmarium depressum, 162–163
Cosmarium depressum v. achondrum, 163
Cosmarium impressulum, 163
Cosmarium nitidulum, 163
Cosmarium sp., 176
Criconemoides sp., 207
Crucigenia tetrapedia, 162–163
Crypotomonas platyuris, 166, 170
Cryptaulax vulgaris, 165
Cryptochironomus sp., 209
Cryptomonas compressa, 165

Cryptomonas erosa, 162, 165, 277
Cryptomonas marssonii, 162, 165–166, 277
Cryptomonas obovata, 165
Cryptomonas ovata, 165
Cryptomonas platyuris, 162, 165
Cryptomonas rostratiformis, 165
Cryptomonas rufescens, 165
Cryptomonas sp., 162, 165–166, 170, 333
Cryptotendipes sp., 209
Cyclops scutifer, 193, 199–201, 204, 271, 275, 332–333, 431
Cyclops strenuus, 201
Cyclops vernalis, 193
Cyclotella bodanica, 165, 365
Cyclotella bodanica v. lemanensis, 368
Cyclotella comensis, 162, 165
Cyclotella comta, 162, 165–166, 171–172, 368, 371, 374–376, 414, 418
Cyclotella glomerata, 165
Cyclotella operculata, 368
Cyclotella sp., 375, 420
Cyclotella stelligera, 371, 374–376, 414, 417–418
Cymatia sp., 209
Cymbella affinis, 370
Cymbella angustat, 165
Cymbella cesati, 370
Cymbella gaeumanni, 370
Cymbella heteropleura, 370
Cymbella laevis, 370
Cymbella lanceolata, 165
Cymbella latens, 370
Cymbella microcephala, 370
Cymbella minuta v. silesiaca, 370
Cymbella naviculiformis, 370
Cymbella pusilla, 370
Cymbella sinuata, 370
Cymbella sp., 165, 370
Cymbella tumidula, 165
Cymbella turgida, 370
Cymbella ventricosa, 370
Cypria turneri, 208

Dactylobiotus grandipes, 207
Daphnia ambigua, 193, 275
Daphnia catawba, 193, 196–198, 204, 269–271, 274–275, 332–333
Daphnia longispina, 201
Daphnia minuta, 275

Index of Genera and Species

Daphnia spp., 85–86, 196, 239, 326–327, 333, 382, 414
Darwinula stevensoni, 208
Dennstaedtia punctilobula, 14
Dero digitata, 208
Dero obtusa, 208
Desmidium sp., 176
Diacyclops nanus, 208
Diaptomus gracilis, 326
Diaptomus graciloides, 326
Diaptomus minutus, 193, 199, 200–204, 332–333
Diatoma anceps, 368
Diatoma hiemale v. mesodon, 368
Diatoma vulgare, 165
Dicranophorus spp., 207
Dicrotendipes spp., 209
Dictyophaerium pulchellum, 162–163
Difflugia sp., 207
Dileptus sp., 207
Dineutus sp., 209
Dinobryon acuminatum, 164
Dinobryon bavaricum, 162, 164
Dinobryon borgei, 162, 164
Dinobryon crenulatum, 164
Dinobryon cylindricum, 162, 164
Dinobryon divergens v. angulatum, 162, 164
Dinobryon divergens v. schauinslandii, 162, 164
Dinobryon divergens, 162, 164, 325
Dinobryon sertularia, 164
Dinobryon sociale, 164
Dinobryon spp., 169, 171–172, 255, 325
Dinobryon suecicum, 162, 164
Dinobryon tabellariae, 162, 164
Diploneis finnica, 369
Diploneis marginestriata, 269
Diploneis oblongella, 369
Diploneis oculata, 369
Diploneis puella, 369
Disparalona acutirostris, 208
Dispora crucigenioides, 163
Dissotrocha aculeata, 207
Dissotrocha macrostyla, 207, 231
Donacia sp., 206, 209, 216
Drepanocladus fluitans, 180, 182, 186, 188
Dromogomphus sp., 209
Drosera rotundifolia, 178
Dryas integrifolia, 54–55, 422
Dryas sp., 55–57, 420–421
Dryopteris spinulosa, 14
Dugesia sp., 207

Elakatothrix gelatinosa, 162–163
Elatine minima, 178–180, 182–184, 186, 190, 415, 418
Eleocharis acicularis, 178–179, 180, 182–183, 186, 190, 418
Eleocharis robbinsii, 180
Elliptio complanata, 206, 208, 215, 222, 228, 313, 316, 321
Elodea canadensis, 181
Enallagma sp., 206, 209
Eoparargyractis plevie, 206, 209, 217
Ephemera sp., 206, 208, 216, 314
Ephemerella sp., 208
Ephydatia mulleri, 207
Epiaeschna sp., 209
Epischura lacustris, 193, 431
Epischura sp., 327
Erignatha clastopsis, 207
Eriocaulon septangulare, 180, 182, 184, 186, 190, 257, 264
Eriocaulon sp., 177, 217
Erkenia subaequiciliata, 277
Erythronium americanum, 14
Esox niger, 240, 289–290
Ethmolaimus sp., 207
Euastrum abruptum v. lagoense, 163
Euastrum binale, 163
Euastrum sp., 176
Eubosmina tubicen, 193, 382–383, 385–386, 417, 419
Eucyclops agilis, 208
Eudorina elegans, 163
Euglena gracilis, 164
Euglena sp., 162, 164, 170
Euglena viridis, 164
Eunapius fragilis, 207
Eunotia curvata, 368
Eunotia diodon, 368
Eunotia elegans, 368
Eunotia flexuosa, 368
Eunotia hexaglyphis, 368
Eunotia incisa, 368
Eunotia maior, 368
Eunotia meisteri, 368
Eunotia pectinalis, 368
Eunotia perpusilla, 368
Eunotia praerupta, 368
Eunotia serra v. diadema, 368
Eunotia sudetica, 368
Eunotia trinacria v. undulata, 368
Eunotia vanheurckii v. intermedia, 368

Eunotia vanheurckii, 368
Eurycercus lamellatus, 193, 236
Eurycercus longirostris, 208

Fagus grandifolia, 14
Fagus sp., 56ff
Fontinalis novae-angliae, 180, 182, 186, 188–189
Forelia sp., 208
Fragilaria bicapitata, 368
Fragilaria brevistriata v. inflata, 368, 372, 380
Fragilaria brevistriata, 368, 372, 374
Fragilaria capucina, 165
Fragilaria construens v. binodis, 368
Fragilaria construens v. venter, 368, 372, 380
Fragilaria construens, 165, 368
Fragilaria crotonensis, 165, 372, 374–375
Fragilaria inflata, 368
Fragilaria leptostauron v. dubia, 368
Fragilaria leptostauron, 368
Fragilaria pinnata, 368, 370, 372, 374, 414, 417
Fragilaria vaucheriae v. capitellata, 368
Fragilaria vaucheriae, 368
Fragilaria virescens, 368, 374
Fragilaria sp., 370, 374
Fraxinus americana, 14
Fraxinus sp., 56ff
Fredericella sultana, 208
Frustulia rhomboides v. saxonica, 369
Frustulia rhomboides v. amphipleuriodes, 368

Galerucella sp., 209, 216
Gastropus minor, 207
Gemellicystis neglecta, 163
Geminella minor, 163
Ghomphonema intricatum, 370
Glenodinium gymnodinium, 165
Glenodinium palustre, 165
Glenodinium pulvisculus, 165
Gleotrichia sp., 176
Gloeococcus schroeteri, 163
Gloeocystis planktonica 162–163, 166, 171–172, 174
Gloeocystis vesiculosa, 163
Glyptotendipes lobiferus, 208–209

Gomphonema acuminatum v.
    coronata, 370
Gomphonema angustatum, 370
Gomphonema angustatum v.
    coronata, 370
Gomphonema angustatum v.
    producta, 370
Gomphonema clevei, 370
Gomphonema gracile v. naviculoides, 370
Gomphonema intricatum v.
    dichotoma, 370
Gomphonema intricatum v.
    pumila, 370
Gomphonema olivaceoides, 370
Gomphonema parvulum v.
    exilis, 370
Gomphonema truncatum v.
    turgidum, 370
Gomphosphaeria lacustris, 163
Gonium pectorale, 162–163
Gonium sp., 176
Gonyostomum semen, 162, 165–166, 170–172
Gratiola aurea, 178–180, 182, 186, 190, 257, 264
Gratiola sp., 217
Gymnodinium fuscum, 165
Gymnodinium helveticum, 165
Gymnodinium lacustre, 165
Gymnodinium mirabile, 162, 165–166, 170–172
Gymnodinium sp., 162, 165, 325
Gymnodinium uberrimum, 165
Gymnozyga moniliformis v.
    maxima, 163
Gyratrix hermaphroditus, 207

Hablotrocha spp., 207
Haliplus sp., 206, 209
Hamamelis virginiana, 178
Helisoma sp., 208
Heptagenia sp., 209
Heterocampa, 236, 238
Heterotanytarsus sp., 209
Heterotrissocladius sp., 209
Hexagenia sp., 209, 216, 239, 241, 314, 321
Holopedium gibberum, 193, 196–197, 200, 273–275, 333, 414, 431
Hyalella azteca, 206, 208, 214, 235
Hyalobryon sp., 255
Hydra americana, 207
Hydra sp., 230
Hydrozetes lacustris, 208
Hygrobates sp., 208
Hypericum ellipticum, 178
Hypericum sp., 183, 415

Ichthydium spp., 207
Ictalurus nebulosus, 240, 289
Ilyocryptus sordidus, 208, 230, 233
Ilyocryptus spinifer, 208
Ilyodrilus templetoni, 208
Ironus sp., 207
Ischnura sp., 209
Isoetes braunii, 180
Isoetes echinospora v. braunii, 180
Isoetes lacustris, 186
Isoetes macrospora, 179–180, 184
Isoetes muricata, 180, 182–183, 186, 190, 257, 415, 418
Isoetes sp., 176, 178–179, 183, 217, 416
Isoetes tuckermani, 179–180, 182–184, 186, 188, 190, 257, 264, 415, 417
Isotomurus palustris, 208

Juncus pelocarpus, 178–180, 182, 184, 186, 190, 264
Juncus sp., 415
Juniperus sp., 422

Katablepharis ovalis, 162, 165–166, 168, 277
Kellicottia bostoniensis, 193, 195–196, 200–201, 275–276, 332–334, 443
Kellicottia longispina, 193, 195, 200–204, 275–276, 332–334
Kellicottia taurocephala, 202–203
Kephyrion boreale, 164
Kephyrion littorale, 164
Kephyrion ovum, 164
Keratella cochlearis, 193, 195, 200, 201–203, 275, 332–333
Keratella crassa, 193, 195, 201, 204, 275, 333
Keratella quadrata, 193, 275, 443
Keratella taurocephala, 193, 195, 201–203, 275, 332–334
Kirchneriella contorta, 163
Kirchneriella elongata, 163

Labrundinia sp., 209
Larix sp., 56ff
Larsia sp., 209
Latona parviremis, 208
Latona setifera, 193, 197, 208

Lauterborniella sp., 209
Lebertia sp., 208
Lecane spp., 207
Lepadella sp., 207
Lepidodermella spp., 207, 232
Lepocinclis sp., 164
Lepus americanus, 14
Limnephilus sp., 206, 209
Limnesia sp., 208
Limnodrilus hoffmeisteri, 208, 214
Littorella unifora, 262
Lobelia dortmanna, 178–180, 182–184, 186, 189, 190–192, 258–265, 316, 415, 418
Lobelia sp., 177, 217
Lobohalacarus sp., 208
Lycopodium lucidulum, 14
Lycopodium selago, 55
Lymnaea sp., 208
Lyngbya nordgaardii, 163

Macrobiotus sp., 207
Macrochaetus spp., 207
Macrocyclops albidus, 193, 208
Macromia sp., 206, 209
Macrostomum sp., 207
Macrothrix laticornis, 208
Macrotrachela sp., 207
Maianthemum canadense, 14
Mallamonas elongata, 164
Mallamonas spp. 172
Mallomonas acaroides, 164
Mallomonas akrokomos, 164
Mallomonas akrokomas v. parvula, 162, 164
Mallomonas caudata, 166, 170–172
Mallomonas globosa, 162, 164
Mallomonas pseudocoronata, 164
Mallomonas reginae, 164
Mallomonas tonsurata, 162, 164, 166
Melosira ambigua, 162, 165, 368, 371, 375, 414
Melosira distans, 374
Melosira distans v. alpigena, 165, 368, 371, 374–375, 377, 414
Melosira distans v. lirata, 368, 371, 374–375, 377
Melosira fennoscandica, 368, 371, 375, 378, 414
Melosira granulata, 165
Melosira italica v. subarctica, 165
Melosira italica v. tenuissima, 368
Melosira italica v. valida, 368

Index of Genera and Species

Melosira sp., 368, 370, 372, 374, 379, 414, 417, 419, 468
Melosira tenella, 368, 372, 375, 378, 414
Melosira varians, 165, 368
Menoidium costatum, 164
Meridion circulare, 368
Meridion circulare v. constrictum, 368
Merismopedia elegans, 163
Merismopedia glauca, 163
Merismopedia sp., 172, 176
Merismopedia tenuissima, 162–163, 166, 168
Mesocyclops edax, 193, 199, 202, 204, 208, 271, 275, 329, 332–333, 431
Mesocyclops sp., 327
Mesostoma spp., 207, 232
Mesovelia sp., 209
Metopus sp., 207
Micrasterias sp., 176, 213
Microcodon clavus, 207
Microcyclops rubellus, 208
Microcyclops varicans, 208
Microcystis aeruginosa, 163
Microdalyellia sp., 207
Micropterus dolomieui, 237, 289–290
Microsectra sp., 209
Microstomum lineare, 207
Microtendipes sp., 209
Monhystera spp., 207
Monhystrella sp., 207
Monochrysis aphanaster, 162, 164, 166, 277
Monochrysis spp., 165
Monodiamesa sp., 209
Monommata astia, 207
Mononchus sp., 207
Monosiga brevicollis, 165
Monospilus dispar, 208
Mougeotia sp., 174–175
Myostenostomum sp., 207
Myrica gale, 178
Myriophyllum spicatum, 181, 189
Myriophyllum tenellum, 180
Mystacides sp., 209

Nais communis, 208
Nais simplex, 208
Najas flexilis, 180, 183, 415
Narpus sp., 206
Navicula acceptata, 369
Navicula begeri, 369
Navicula cocconeiformis, 369
Navicula cryptocephala, 369
Navicula explanata, 369
Navicula festiva, 369
Navicula globulifera, 369
Navicula gottlandica, 369
Navicula gysingensis, 369
Navicula hambergii, 369
Navicula hassiaca, 369
Navicula helensis, 369
Navicula helmandensis, 369
Navicula hustedtii, 369
Navicula ignota v. anglica, 369
Navicula jarnefeltii, 369
Navicula lacustris, 369
Navicula latens, 369
Navicula mediocris, 369
Navicula menisculus v. upsaliensis, 369
Navicula minima, 369
Navicula muralis, 369
Navicula paucivisitata, 369
Navicula peregrina, 369
Navicula placenta, 369
Navicula pseudoexilissima, 369
Navicula pseudoscutiformis, 369
Navicula pupula, 369
Navicula pupula v. capitata, 369
Navicula pupula v. mutata, 369
Navicula pupula v. rectangularis, 369
Navicula pusio, 369
Navicula radiosa, 369
Navicula radiosa v. parva, 369
Navicula regularis, 360
Navicula rotaeana, 369
Navicula salinarum v. intermedia, 369
Navicula schmassmannii, 369
Navicula seminulum, 369, 373
Navicula simula, 369
Navicula sp., 369–370, 374
Navicula subatomoides, 369
Navicula viridula v. linearis, 369
Neidium affine, 369
Neidium affine v. longiceps, 369
Neidium hitchcockii, 369
Neidium iridis, 369
Neidium iridis v. amphigomphus, 165
Neidium sp., 369
Nephrocytium agardhianum, 163
Nephrocytium limneticum, 164
Neurocordulia sp., 209
Nitella flexilis, 180, 182–186, 188–192, 212, 263–264, 312–313, 415–417
Nitella furcata, 180
Nitella gracilis, 180, 182, 186
Nitella sp., 177–178, 416
Nitella tenuissima, 180, 182, 186, 190
Nitzschia acicularis, 165, 370
Nitzschia dissipata, 370
Nitzschia frustulum v. perminuta f. constricta, 370
Nitzschia frustulum v. perpusilla, 370
Nitzschia gracilis, 370
Nitzschia mirabilis, 370
Nitzschia romana, 370
Nitzschia subcommunis, 370
Nitzschia tropica, 370
Nitzschia vermicularis, 370
Nostoc sp., 176, 268
Nothopthalmus viridescens, 197, 234–236, 288–289
Notommata spp., 207
Notonecta sp., 209
Nuphar microphyllum, 180, 190
Nuphar sp., 415
Nuphar variegatum, 180–183, 186, 190, 216, 258–261, 263–265
Nymphaea odorata, 180–182, 186

Ochromonas sp., 162, 165
Odocoileus virginianus, 14
Oecetis sp., 209
Oocystis borgei, 164
Oocystis crassa, 164
Oocystis lacustris, 164
Oocystis parva, 164
Oocystis sp., 166, 172
Oocystis submarina v. variabilis, 164
Ooscystis pusilla, 164
Ophrydium sp., 207
Ophryoxus gracilis, 208
Opisthocystis goetti, 207
Orconectes sp., 206, 208, 217
Oscillatoria agardhii, 163
Oscillatoria agardhii v. isothrix, 163
Oscillatoria limnetica, 163
Oscillatoria limosa, 163
Oscillatoria rubescens, 163
Oscillatoria sp., 176, 317
Oscillatoria subbrevis, 163
Oscillatoria tenuis, 163
Ostrya-Carpinus sp., 56ff
Otomesostoma auditivum, 207
Oxalis montana, 14
Oxyethira sp., 209

Pagastiella sp., 209
Palpomyia sp., 206
Pandorina morum, 164
Parachironomus sp., 209
Paracladopelma sp., 209
Paracyclops affinis, 208
Paracyclops sp., 208
Parakiefferiella sp., 209

Paralauterborniella sp., 209
Paramecium sp., 207
Parastenocaris brevipes, 208
Paratanytarsus sp., 209
Paulschulzia pseudovolvox, 164
Pectinatella magnifica, 206, 208, 217
Pediastrum boryanum, 164
Pediastrum sp., 176
Pediastrum tetras, 164
Perca flavescens, 215, 238, 289–290
Perca fluviatilis, 238–239
Peridinium cinctum, 165
Peridinium inconspicuum, 162, 165
Peridinium palustre, 162, 165
Peridinium plonicum, 162
Peridinium spp., 172
Peridinium willei, 162, 165–166, 170
Peridinium wisconsinense, 165
Phacus longicaudata, 162–163
Phacus sp., 176
Phaenopsectra sp., 209
Phagocata sp., 207
Phylocentropus sp., 209, 314
Picea glauca, 422
Picea rubens, 14
Picea sp., 56ff
Piguetiella blanci, 208
Pinnularia abaujensis v. rostrata, 369
Pinnularia acrosphaeria, 369
Pinnularia biceps, 369
Pinnularia biceps v. minor, 369
Pinnularia braunii, 369
Pinnularia intermedia, 369
Pinnulaira legumen, 369
Pinnularia maior v. transversa, 369
Pinnularia microstauron, 370
Pinnularia polyonca, 370
Pinnularia sp., 213
Pinnularia subcapitata, 370
Pinnularia viridis, 370
Pinus sp., 56ff
Pinus strobus, 14
Piona sp., 208
Pisidium sp., 206, 208
Plethodon cinereus, 288
Pleurotaenium sp., 176
Plumatella fruticosa, 208
Polyarthra spp., 192, 207
Polyarthra vulgaris, 193, 195, 200, 201–204, 275, 332–334
Polycentropus sp., 290
Polymerurus spp., 207
Polypedilum haltere, 209
Polypedilum tritum, 209

Polyphemus pediculus, 193, 197–198, 208, 314
Polyphemus sp., 196, 431
Populus sp., 56, 64
Porohalacarus sp., 208
Porolohmanella sp., 208
Potamogeton berchtoldii, 180, 182, 186, 188–190, 260, 264, 313
Potamogeton epihydrus, 181–182, 186, 190–191
Potamogeton epihydrus v. ramosus, 180
Potamogeton natans, 180
Potamogeton pusillus v. tenuissimus, 180
Potamogeton sp., 177–179, 180, 183, 415
Potamogeton spirillus, 180, 182–184, 186, 190
Potentilla tridentata, 55
Prismatolaimus sp., 207
Pristina aequiseta, 208
Pristina leidyi, 208
Proales sp., 207
Proalinopsis spp., 207
Procladius choreus, 213
Procladius denticulatus, 209
Procladius sp., 209
Prodesmodora sp., 207
Prorhynchella minuta, 207
Prorhynchus stagnalis, 207
Protanypus sp., 209
Psectrocladius spp., 209
Pseudobiotus sp., 207
Pseudochironomus sp., 209
Pseudokephyrion sp., 162, 165
Pseudokephyrion spp., 166, 168
Pseudokephyrion elegans, 162, 165
Pseudokephyrion entzii, 162, 165
Pseudokephyrion spirale, 165
Pseudomonas sp., 85, 324
Pteridium aquilinum, 178
Pyramimonas tetrarhynchus, 164

Quadrigula closterioides, 164
Quadrigula lacustris, 164
Quadrigula pfitzeri, 162–163
Quercus borealis, 14
Quercus sp., 56ff
Quistodrilus multisetosus, 208

Rana p. pipiens, 236
Rhabdolaimus sp., 207
Rhabdomonas incurva, 162–163
Rheocricotopus sp., 209
Rheumatobates sp., 209

Rhizosolenia eriensis, 162, 164
Rhizosolenia sp., 254–255, 368
Rhodomonas lacustris, 165
Rhodomonas minuta v. nannoplanctica, 162, 165
Rhodomonas minuta, 165
Rhodomonas pusilla, 162, 165–166, 168, 172
Rhodomonas tenuis, 165
Rhopalodia gibba, 370
Rhyacodrilus montana, 208
Rhynchomesostoma sp., 207
Rhynchoscolex simplex, 207, 230, 232
Rhynchotalona falcata, 208, 384, 386, 417
Rotaria cf. macrura, 207
Rotaria rotatoria, 207, 269
Rotaria tridens, 207, 230
Rumex sp., 62

Sagittaria graminea, 179, 180, 182, 186, 190, 264
Scapholeberis kingi, 193, 208
Scaridium longicaudatum, 207
Scenedesmus abundans, 162–163
Scenedesmus apiculatus, 164
Scenedesmus bijuga, 162–163
Scenedesmus opoliensis, 164
Scenedesmus quadricauda, 164
Scenedesmus sp., 324
Selenastrum minutum, 164
Sennia parvula, 165
Shepherdia canadensis, 54, 57, 64, 421
Sialis sp., 206, 209
Sida crystallina, 193
Sisyra sp., 206, 209
Slavina appendiculata, 208
Smilicina racemosa, 14
Soldanellonyx sp., 208
Solidago graminifolia, 178
Spaerocystis schroteri, 443
Spaniotoma (Orthocladius) akanusi, 219
Sparganium angustifolium, 180–182, 186, 190
Sparganium fluctuans, 180
Sparganium sp., 177
Specaria josinae, 208
Sphaerium sp., 208
Sphaerozosma excavata v. subquadratum, 164
Sphagnum sp., 178
Spinocosmarium sp., 176
Spirea latifolia, 178
Spirogyra sp., 174–176, 443
Spirosperma nikolskyi, 208
Spondylosium planum, 164
Spongilla lacustris, 207, 216

Index of Genera and Species

Staurastrum arachne, 164
Staurastrum cuspidatum v. divergens, 164
Staurastrum dejectum, 164
Staurastrum lacustre, 164
Staurastrum limneticum, 164
Staurastrum lunatum, 164
Staurastrum megecanthum, 164
Staurastrum muticum, 164
Staurastrum natator, 164
Staurastrum paradoxum, 164
Staurastrum pentacerum, 164
Staurastrum sp., 164, 176
Staurastrum spiculiferm, 164
Staurodesmus sp., 162–163, 176
Staurodesmus, 176
Stauroneis anceps, 370
Stauroneis anceps f. gracilis, 370
Stauroneis anceps f. lilnearis, 370
Stauroneis livingstonii, 370
Stauroneis nobilis v. baconiana, 370
Stauroneis phoenicenteron, 370
Stauroneis sp., 370
Stempellina sp., 209
Stenopterobia intermedia, 270
Stenostomum leucops, 207
Stenostomum spp., 207
Stenostomum unicolor, 207
Stentor spp., 207
Stephanoceras fimbriatus, 207
Stephanodiscus dubius, 165
Stichococcus minor, 164
Stichtochironomus sp., 209
Stylaria lacustris, 208
Subularia aquatica, 180
Surirella angustiformis, 370
Surirella delicatissima, 370
Surirella robusta, 370
Synchaeta sp., 207

Synedra delicatissima, 368, 372, 374–375
Synedra delicatissima v. angustissima, 165, 368
Synedra nan, 165
Synedra radians, 162, 165
Synedra rumpens v. familiaris, 368, 373
Synedra rumpens, 165
Synedra sp., 162, 165
Synedra ulna, 368
Synedra ulna v. danica, 368
Synura adamsii, 162, 165
Synura sp., 169
Synura uvella, 162, 165–166, 169–172

Tabellaria fenestrata, 162, 165–166, 172, 368
Tabellaria flocculosa, 165, 368, 372, 374–375, 380, 414, 417, 419
Tabellaria sp., 375, 420
Tanytarsus spp., 209, 212
Taphrocampa selenura, 207
Taphrocampa sp., 207
Testudinella sp., 207
Tetradesmus smithii, 164
Tetradron asymmretricum, 164
Tetradron minimum, 162–163
Tetradron minimum v. tetralobulatum, 164
Tetrasiphon hydracora, 207
Thalictrum sp., 54
Thuja sp., 422
Tilia americana, 14
Tobrilus sp., 207
Trachelomonas sp., 164
Triaenodes sp., 206, 209
Trichocerca spp., 207
Trichotria sp., 207

Trimalaconothrus novus, 208
Tropocyclops prasinus, 193, 275
Tropocyclops sp., 327
Tsuga canadensis, 14
Tsuga sp., 56ff
Tubifex tubifex, 206, 208, 214
Typha sp., 415

Ulmus americana, 14
Ulmus sp., 56ff
Ulothrix zonata, 439
Ulcinais uncinata, 208
Unionicola sp., 206, 208
Uroglena sp., 169
Uroglena volvox, 165
Uroglenopsis americana, 165
Ursus americanus, 14
Utricularia cornuta, 179–180
Utricularia inflata v. minor, 180
Utricularia minor, 180
Utricularia purpurea, 178, 180, 182, 186, 188–190, 235, 258–265, 313–314
Utricularia sp., 177, 181, 241
Utricularia vulgaris, 260

Vaccinium corymbosum, 178
Vaccinium uliginosum, 56, 421
Vallisneria americana, 181
Vejdovskyella comata, 208
Viburnum alnifolium, 14
Volvox aureus, 164
Vorticella sp., 207
Vulpes fulva, 14

Wierzejskiella sp., 207

Zalutschia spp., 209
Zalutschia zalutschicola, 209

Index of Lakes and Streams

Baikal, Lake, U.S.S.R., 84–85
Batorin, Lake, U.S.S.R., 187, 276, 280
Bay of Quinte, Lake Ontario, 281–282, 328
Bear Brook, New Hampshire, 10, 432–433
Biwa, Lake, Japan, 249
Blelham Tarn, England, 421
Borrevann, Lake, Norway, 223
Boundary Waters Canoe Area, Minnesota, 60, 66, 70–71
Bysjon, Lake, Sweden, 323

Carter Pond, New Hampshire, 54
Castle Lake, California, 286
Cayuga Lake, New York, 104–105, 152, 211–212, 222–223, 297
Cedar Bog Lake, Minnesota, 210
Char Lake, N.W.T., Canada, 211–212, 222, 227, 281–282, 299–301, 417
Clear Lake, California, 227
Clear Lake, Kansas, 210
Clear Lake, Ontario, Canada, 152

Dalnee, Lake, U.S.S.R., 319
Deer Lake Bog, New Hampshire, 54–55
Dgał Maly, Poland, 187

Eagle Lake Bog, New Hampshire, 54
Eglwys Nunydd Lake, South Wales, 211–212, 222–223, 227, 281
English Lake District, 176
Erie, Lake, 88, 360
Erken, Lake, Sweden, 223

Esrom, Lake, Denmark, 212, 217, 222–223, 281–282, 286, 319, 321
Esthwaite Water, England, 402
Experimental Lakes Area (ELA), Ontario, Canada, 154, 161, 166, 168–174, 176, 188, 192, 251–252, 255, 267–268, 315, 328, 337, 343

Findley Lake, Washington, 222, 281–282, 299–301, 318
Fiolen, Lake, Sweden, 191
Flosek, Lake, Poland, 281

George, Lake, New York, 179, 328–329
Great Slave Lake, N.W.T., Canada, 223
Green Lake, Wisconsin, 176, 187

Hakojarvi, Lake, Finland, 390
Horw Bay, Lake of Lucerne, Switzerland, 284

Kalgaard, Lake, Denmark, 187–188, 263, 316
Kempton Park E. Reservoir, England, 227
Keyhole Lake, N.W.T., Canada, 222
Kinsman Pond, New Hampshire, 54
Krivoe, Lake, U.S.S.R., 276, 280
Krugloe, Lake, U.S.S.R., 276, 280
Kiev Reservoir, U.S.S.R., 319

Lanao Lake, Philippines, 324
Lake of the Clouds, Minnesota, 66
Lakes-of-the-Clouds, New Hampshire, 54–55
Lake 227 (ELA), Ontario, Canada, 328
Lake 302 (ELA), Ontario, Canada, 328
Lago di Monterosi, Italy, 367
Latnjajaure, Sweden, 187
Lawrence Lake, Michigan, 187–188, 263, 299–301
Linsley Pond, Connecticut, 210, 223, 367, 425
Little East Pond, New Hampshire, 54
Litle McCauley Lake, Ontario, Canada, 210
Lizard Lake, Iowa, 211–212
Loch Dunmore, Scotland, 227
Loch Leven, Scotland, 211–212, 281–282
Loch Lomond, Scotland, 211–212
Lonesome Lake, New Hampshire, 54–55
Lost Pond, New Hampshire, 54–55, 59
Lower Baker Pond, New Hampshire, 179–180, 184

Marion Lake, B.C., Canada, 222, 281–282, 286, 299–301, 319
Memphremagog, Lake, Quebec/Vermont, 222, 281
Mendota, Lake, Wisconsin, 186–187, 222, 281, 286
Merrimack, Lake, New Hampshire, 49
Merrimack River, New Hampshire, 9, 241
Michie, Lake, North Carolina, 130

Index of Lakes and Streams

Michigan, Lake, 361
Mikolajskie, Lake, Poland, 281, 284, 319
Myastro, Lake, U.S.S.R., 187, 276, 280
Myers Lake, Indiana, 210
Myvatn, Lake, Iceland, 211, 223, 281, 319

Naroch, Lake, U.S.S.R., 187, 276–277, 280, 319
Norris Brook, New Hampshire, 10, 227, 434

Ontario, Lake, 281–282
Opinicon, Lake, Ontario, Canada, 242–243
Øvre Heimdalsvatn, Norway, 187

Pääjärvi, Lake, Finland, 187, 188, 282, 319
Par Pond, South Carolina, 232
Partridge Lake, New Hampshire, 181–182, 189
Parvin, Lake, Colorado, 212, 222
Paul Lake, B.C., Canada, 222
Pemigewasset River, New Hampshire, 9, 10, 40–41, 45, 48–49, 73, 241, 345, 356
Plöner See, Germany, 222
Port-Bielh, France, 187
Punnus-yarvi, U.S.S.R., 187

Rawson Lake, Ontario, Canada, 152, 177, 267–268
Red, Lake, U.S.S.R., 319
Rotsee, Switzerland, 284
Russell Pond, New Hampshire, 180, 189
Rybinsk Reservoir, U.S.S.R., 319

Shagawa Lake, Minnesota, 390
Simcoe Lake, Ontario, Canada 211–212, 222–223
Sniardwy, Lake, Poland, 281
Stewart's Dark Lake, Wisconsin, 351, 390
Sunfish Lake, Ontario, Canada, 201
Suomunjarvi, Finland, 187, 188

Tahoe, Lake, Nevada/California, 159
Taltowisko, Lake, Poland, 281
Third Sister Lake, Michigan, 210
Toronto Harbour, Lake Ontario, 214
Trout Lake, Wisconsin, 187
Tundra Ponds, Alaska, 212, 227, 319

Varese, Lake, Italy, 390
Vechten, Lake, The Netherlands, 210
Vitalampa, Sweden, 187, 188

Warniak, Lake, Poland, 281
Washington, Lake, Washington, 222, 329–330
Waskesiu, Saskatchewan, Canada, 222
Weber Lake, Wisconsin, 187
Werowrap, Lake, Australia, 211–212, 222–223, 281
Windermere, Lake, England, 402
Wingra, Lake, Wisconsin, 187, 222, 281, 299–301, 319
Wintergreen Lake, Michigan, 390

General Index

Achnanthales, 368
Acid (*see also pH*)
  carbonic, 102
  in lakes, 381
  nitric, 18
  in precipitation, 18–19
  in stream water, 19, 147
  sulfuric, 18
Acid Precipitation, 36–38, 431–432, 434, 440–441
  effects on lake, 174, 184
  effects on soil, 434–435
  effects on stream-water chemistry, 434–435, 440–441
Adirondack Mts., New York, 431
Aerosols
  dust, 21
Agricultural Activity (*see also Human Activity, Disturbance*), 425
Alder, 14, 55ff, 421
Algae (*see also Phytoplankton, Periphyton*)
  benthic, 285
  blue-green, 161ff, 176–177, 255ff, 385
  epilithic, 175
  epipelic, 175
  epiphytic, 175, 316
  epipsammic, 175
  organic matter, in, 173
  periphyton, 175–177, 265–268
Alismataceae, 179
Allegheny National Forest, 70
Allochthonous, 415
  defined, 2
  fossil pigments, 387
  organic carbon, 294, 338, 406
Aluminum
  in forest soils, 423
  input-output budget, 24
  in lake sediments, 416, 425
  monthly concentration in precipitation and stream water, 18–22
  in or to sediments, 365, 395ff
  in stream water, 35, 401, 434–435, 438–440
    following forest disturbance, 438–440
Ammonium
  in bulk precipitation, 18ff, 37, 107
  concentration in lake outlet, 100–101
  concentration in lake water, 98–100, 107
  concentration in tributaries to lake, 100–101
  as energy source, 158
  input in bulk precipitation, 22–24, 142–147
  input in runoff to lake, 143–147
  input-output budget, 24, 28–32, 149–155
  in ice, 106–107
  monthly concentration in precipitation, 18–22
  monthly concentration in stream water, 24–26
  output from lake, 147–149
  output in stream water, 26–28
  in sewage, 343, 443
  standing stock in lake water, 98–100
  in stream water, 24ff
  vertical stratification in lake, 98
Amphipoda, 208, 214, 221, 235–236, 321
Anisoptera, 241, 243
Ash, 14, 55ff
Ash Content
  in macrophytes, 264
Aspen, 421–427
Autochthonous, 338
  defined, 2
  fossil pigments, 387
  organic carbon in lake, 294, 406, 409
  sedimentary organic matter, 414

Bacillariophyceae, 366, 368, 370
Bacteria
  assimilation efficiency of, 246
  benthic, 285, 320
  biomass of, 159–160, 293, 324
  chemoautotrophs, 156–158
  doubling times of, 250
  energy content of, 342
  growth rate of, 249–250, 300
  heterotrophs, 156ff
  interactions with bacteria, 323–326
  lake, 156–160
  limiting factors of, 246ff, 324
  numbers of, 159–160
  nutrient content of, 342
  photoautotrophs, 156–158
  production of, 246ff, 294–296, 300–301
  respiration of, 247, 298–299
  washed into lakes, 250
Bass
  Largemouth, 85ff
  Smallmouth, 235–238, 242–245, 290, 312
Basswood, 62ff
Beaver, 314
Bedrock, 40ff, 128, 135, 395, 398, 400
  Kinsman Formation, 14, 40, 400
  Littleton Formation, 14, 40, 44, 396, 400
  saddle, 40–46
  silicate minerals, 14
Beech, 14, 55ff
Benthic Invertebrates (*see Invertebrates*)

General Index

Bicarbonate
  in bulk precipitation, 19
  concentration in lake outlet, 100–101
  concentration in lake water, 98–100, 107
  concentration in tributaries to lake, 100–101
  input in runoff to lake, 143–147
  output from lake, 147–149
  standing stock in lake water, 98–100
Biosphere, 1
Birch, 14, 55ff, 62, 418–427
Birgean Work, 112ff
Bluegill, 85
Bosminids, 382ff, 415
Bottling Works, on Mirror Lake, 80, 83
Bracken Fern, 14, 62ff
Bryophytes, 180
Bullhead, Brown, 237, 240, 243–245, 289

Cadmium
  in surface sediments, 404–405
  in stream water, 434
Calcium
  importance for algae, 325
  in bulk precipitation, 18ff, 37
  concentration in lake outlet, 100–101
  concentration in lake water, 98–100, 107
  concentration in tributaries to lake, 100–101
  in dust, 21
  in fish, 291
  in ice, 106–107
  extracted by acetic acid from lake sediments, 402–404, 423
  extracted by ammonium acetate from lake sediments, 402–404
  input in bulk precipitation, 22–24, 107, 142–147
  input in runoff to lake, 143–147
  input-output budget, 24, 28–32, 149–155, 344
  in litter, 153
  loss in insect emergence, 153
  in macrophytes, 263–264
  monthly concentration in precipitation, 18–22
  monthly concentration in lake water, 98–99
  monthly concentration in stream water, 24–26
  output from lake, 147–149
  output in stream water, 26–28, 423
  precipitation with P, 188
  predicted stream-water concentration, 438
  in or to sediments, 153, 265, 395ff, 416
  standing stock in lake water, 98–100
  in salamanders, 289
  in stream water, 24ff, 35, 401, 423, 434, 437–440
  —following forest disturbance, 438–440
  in surface sediments, 404–405
  vertical stratification in lake, 98
Camps, on Mirror Lake, 80–83
Carbon, Dissolved Inorganic
  in lake water, 102, 104, 293
  uptake of by macrophytes, 181ff
  uptake by periphyton, 265–267
  vertical stratification of, 97–98, 102
Carbon, Dissolved Organic
  in algal excretion, 247, 295, 323
  allochthonous inputs of, to lakes, 296ff, 380
  in ecosystems, 341
  in ground water, 247, 299
  input-output budget, 24, 34
  in lake outflow, 247
  in lakes, 98–100, 102–103, 158, 293, 302ff
  limitation of bacterial growth, 324
  photosynthetically-produced, 302ff
  in precipitation, 19, 34, 247, 294, 297
  regulation of concentration, 33–34, 302ff, 320
  in soil, 33–34
  in stream water, 19, 34, 247, 294, 296
  in throughfall, 34
  turnover in lake, 308–309
Carbon, Organic (see also Dissolved, Dams)
  in bacteria, 293, 342
  in benthic invertebrates, 287–289, 342
  ^{14}C, 54, 87, 246, 254, 269, 293, 295, 302, 304, 323, 393
  cycling of, in lakes, 299–301, 303
  in detritus, 293
  dissolved in lake water, 98–100, 102–103, 302ff, 342
  in fish, 293, 342
  input in bulk precipitation, 142–147, 292ff
  input in runoff to lake, 143–147, 292ff
  input-output budget, 247–249, 292–301, 406
  in lake outlet, 100–101, 299
  in litter, 292ff
  loss in emergence, 282–283, 299
  in macrophytes, 263, 293, 342
  output from lake, 147–149, 300
  particulate in lake water, 103–105, 302ff
  particulate in stream water, 432–433
  in periphyton, 293, 342
  in phytoplankton, 293, 342
  in salamanders, 293, 342
  sedimentation of, 308–310
  in seston, 342
  in or to surface sediments, 342, 357–358, 362–365, 395ff
  in tributaries to lake, 100–101
  turnover in lake, 308–309, 342
  in zooplankton, 293, 342
Carotenoids (see also Pigments, Sediments), 387–391
Caterpillars, 238, 257, 316
Chaoboridae, 209, 226
  abundance of, 205–210
  biology of, 205–210
Charcoal (see Sediments)
Chemical Budget (see also specific chemical)
  annual input-output, 28–32, 149–155, 291ff, 344
Chequamegon National Forest, 70
Cherry, 422
Chironomidae, 210, 217, 219–221, 226, 233–234, 238, 285, 287, 319–320, 358
  abundance of, 210ff
  biology of, 210–214
  biomass of, 210ff
Chironomini, 206, 209, 211, 220, 226
Chestnut decline, 57
Chippewa National Forest, 70
Chloride
  concentration in lake outlet, 100–101
  concentration in lake water, 98–100, 107
  concentration in tributaries to lake, 100–101
  in ice, 106–107
  input in bulk precipitation, 142–147

Chloride (*cont.*)
  input in runoff to lake, 143–147
  input-output budget, 24, 32, 149–155
  in litter, 153
  long-term changes in lake, 99
  loss in insect emergence, 153
  loss from lake, 147–149
  monthly concentration in precipitation, 18–22
  monthly concentration in stream water, 18–22
  in permanent sediments, 153
  in precipitation, 19, 107
  in road salt, 153, 343, 441–442
  in sewage, 153, 343, 443
  in stream water, 19
  standing stock in lake water, 98–100
Chlorophyceae, 163, 324–326
Chlorophyll (*see also Pigments, Sediments*)
  degradation products, 387–391
  in lake water, 103–105, 252ff, 324
  —peaks in, 252ff
  profiles in lake, 169, 172, 273
  in periphyton, 177, 266–267
Chromium
  in surface sediments, 404–405
Chrysophyceae, 164, 324–326, 373
Chydorids, 382ff
Circulation of Water, 89ff
  currents in lake, 119–125, 313, 319
  dimixis, 109–115
  meromixis, 115, 389
  spring meromixis, 91, 109, 123, 126
  vertical, 109ff
Cladocera (*see also Zooplankton*), 192ff, 208, 235–236, 314, 329–330, 339, 364, 382ff
  anatomy of, 194
  biomass of, 196–198, 279–280
  feces of, 307
  grazing by, 326–327
  growth curve of, 197, 270–271
  horizontal distribution of, 198
  production of, 279–280, 303–304
  seasonal distribution of, 198
Climate, 9–14, 55
  air temperature, 11, 58
  change, 58, 399, 411
  solar radiation, 9–11
  wind, 12–14
Coleoptera, 206, 209, 216
Copepoda (*see also Zooplankton*), 192ff, 208, 329–330

  anatomy of, 194
  biomass of, 199, 279–280
  cyclopoid, 230, 233, 270–271
  grazing by, 326–327
  growth of, 270–271
  harpacticoid, 230
  nauplii, 200, 202–203
  production of, 279–280, 303–304
Copper
  extracted by acetic acid from lake sediments, 402–404
  extracted by ammonium acetate from lake sediments, 402–404
  flux through terrestrial ecosystems, 30
  in or to lake sediments, 395ff
  in surface sediments, 404–405
Core (Lake Sediments)
  Swedish foil corer, 367, 392
  freeze-core, 392, 405
  piston, 392, 405
Crayfish, 237, 312, 316
Cryptophyceae, 165, 324–326
Cyanophyceae, 161ff, 176–177, 255ff, 324–326, 385
Cyperaceae, 179, 415

Dams
  on Mirror Lake, 76, 81, 83, 91, 347, 357, 364
  organic debris, 35, 66, 421, 433–434
Dance Halls, on Mirror Lake, 80, 83
Decomposition (*see also Carbon*), 302–310
  of sediment, 361–364
Deep Seepage, 5, 14–15 (*see Seepage*)
  to and from lake, 137–142, 149ff, 343, 349, 432
Deer, 14, 441
Denitrification, 157
  in streams, 35
Density (*see also Temperature, Circulation, Stability*)
  currents, 121–125
  effect of salinity on, 109ff
  effect of temperature on, 109ff
  of organisms, 323
  vertical stratification of, 109ff
Denudation
  cationic, 37
Deposition
  dry, 153–154
Detritus (Organic), 292, 308–309, 315, 324, 339
Diamesinae, 209, 213
Diapause, 332

Diatoms, 161ff, 176, 324, 364, 366ff, 424
  diversity of, 367–382, 420
  of sediments, 366–382, 350–351
Diffusion, 87
Diptera, 209, 232
Dissolved Oxygen
  in hypolimnion, 89, 106, 124, 215, 221, 283, 317ff
  of lake water, 102, 105–106, 246, 265, 323, 374
  near bottom of lake, 286, 313, 319, 408
  profile, 168, 318
  in streams, 35
  vertical stratification of, 97–98, 106, 124, 202, 224
Disturbance (*see also Erosion, Human Activity*), 65ff, 174, 425, 433–435
  fire, 65ff, 60–61
  forest clearing or harvest, 31, 66–67, 74–78, 349, 365, 433, 437–440
  road salt, 99, 433, 441–442
  roads, 80, 147, 365, 440–442
  sewage, 364, 433, 442
  of soil, 425
  wind, 65ff, 109ff, 437
Drift, Glacial, 44ff, 128, 346

Ecosystem
  analysis, 1, 425–427
  approach, 1–8, 52, 337, 430, 434, 436ff
  atmosphere of, 3–7
  available nutrients of, 3–7
  biomass of, 341
  biotic component of, 3–7
  boundaries of, 2ff, 52
  cycling, 2
  development (history) of, 2, 7
  flux, 2
  functional unit, 1–7, 430–431
  intrasystem cycle, 4
  production of, 293ff
  respiration of, 297ff, 340–341
  soil and rock component of, 3–7
  volume of, 2, 430
Ecosystem Source Carbon, 282, 301, 338
Ectoprocta, 208, 217
Elm, 14, 55ff
Elm decline, 57
Emergence
  of insects from lake, 142, 152–153, 218, 225–227, 282–283, 299, 300, 314, 320
  —transport of nutrients from, 282–283, 320, 344

General Index

Energy Content of (Calories)
  bacteria, 342
  benthic invertebrates, 342
  carbon in lake water, 342
  carbon in surface sediments, 342
  fish, 342
  macrophytes, 342
  in organic matter, 282, 287–289
  periphyton, 342
  phytoplankton, 342
  salamanders, 342
  seston, 342
  zooplankton, 342
Ephemeroptera, 208, 216, 243
Ericaceae, 415
Erosion, 18, 34, 401, 424, 440
  Postsettlement, 364–365, 427
  of stream channels, 409, 427
Euglenophyceae, 164, 166, 324–326
Eutrophication (see also Sewage)
  cultural, 174, 251, 299, 364–365, 390, 392
  during Holocene, 420
  of lakes, 88, 418, 430, 441–443
  of profundal region, 221, 284, 287, 318
Evapotranspiration, 15ff, 128–130, 136–142
  annual, 16–18
  energy-budget method for, 129–130
  use of evaporation pan for, 129–130, 138, 140
  from lake, 121, 123, 136–142
  —seasonal, 121, 126
  mass transfer method for, 129–130
Experimental Study of Streams, 432–435
Extinction Coefficient (see Solar Radiation)

Feces, 307, 319, 359
Fermentation, 157–158
Fir, Balsam, 14, 55ff, 426
Fish, 236–245, 314, 317
  biomass of, 289–291, 293, 324
  energy content of, 342
  food of, 242–244
  grazing on zooplankton, 288ff, 340
  growth of, 238–241
  nutrient standing stock of, 291, 342
  P:B ratio, 339
  predation on benthic invertebrates, 226, 321, 340
  relation to food web, 245
  respiration of, 298–300
  role of, 244–245
  species of, 237–241
Focusing (see Sediments), 411ff
Food Chain, 338
Food Web, 338–341
Forest
  buffer strip, 439
  harvesting, 438–440
Forest Ecosystem
  biomass of
    —with time following disturbance, 437–438
  Developmental Phases
    —Aggradation, 31ff, 64, 436–439
    —Reorganization, 31ff, 64, 436–439
    —Steady State, 31ff, 64, 436–439
    —Transition, 31ff, 64, 436–439
Fossil Fuels
  combustion of, 19, 36–38
Franconia Notch, New Hampshire, 54
Frost Heaving (see also Soil), 420
  on bare soils, 420

Gases
  CO_2, 3, 5, 102, 156–158, 181, 183, 188, 191, 246–249, 258, 265–267, 294, 296–297, 302ff, 324
  flux to lake, 153–154
  H_2, 157
  H_2S, 5, 155, 157–158
  in lake water, 105–106
  methane, 5, 157–158
  NH_4, 5
  NO_2, 158
  O_2 (see also Dissolved Oxygen), 156–158
Gastrotricha, 207, 232
Geology (see also Bedrock), 14–15, 40ff
  glacial, 44ff
  topography of lake's drainage basin, 49–52, 346
  watertight (see Hydrology), 14–15
Glacial (Late) Period, 346ff, 367–374, 383–384, 415ff, 420ff
Glacier Bay, Alaska, 57
Glucose, 306
Grass Pollen, 62
Green Mountains, Vermont, 70
Gyttja (see Sediments)

Halacaridae, 208, 233
Heat Content
  advected, 121, 126
  annual budget for lake, 114–117
  budget of ice, 118
  budget for sediments, 117
  geothermal source, 115–117
  of lake sediments, 115ff
  of lake water, 112ff
  sediment flux, 112, 115–117
  seasonal flux in lake, 125–127
  warm rim of lake, 121–122, 313
Heleidae, 209, 216
Hemlock, 14, 55ff, 412ff
  bark, 58, 62, 77–80
  debris, 389
  decline in, 57, 59, 387, 412, 422
Hiawatha National Forest, 70
Holocene Period, 59, 346ff, 422
  soil development, 423–425
  Late Holocene I, 418–419
  Late Holocene II, 419
Hornbeam, 55ff
Hubbard Brook Ecosystem Study, 3–7, 83, 337
Hubbard Brook Experimental Forest, 3, 15, 83
  animals, 14
  climate, 9ff
  geology, 14–15, 40ff
  vegetation, 14, 53ff, 410ff
  watersheds, 420–422, 425, 441
Hubbard Brook Valley, 3, 7, 9ff, 15, 40ff, 337, 432
  animals of, 14, 441
  climate of, 9ff
  chronology of, 83
  geology of, 14–15, 395
  Indians of, 72ff
  paleoecology of, 367
  vegetation of, 14, 53ff, 410ff
Human Activity (see also Disturbance), 62, 65ff, 174, 433–435, 436–443
  agriculture, 62, 74–75, 425, 433, 443
  fertilizer, 443
  fire, 60–61, 65ff
  history, 427, 442
  logging, 31, 59, 63, 66–67, 74–78, 349, 357, 433, 437–440
  recreation, 75, 79, 83, 442
  road salt, 99, 433, 441–442
  settlement in New Hampshire, 71ff, 83, 357, 364–365, 404–405, 425
  tanning leather, 77–80, 83
Huron National Forest, 70

Hurricanes (see also Disturbance), 12, 31, 60, 68–69
Hydrogen Ion (see also Acid)
 in bulk precipitation, 18ff, 107, 431–432
 concentration in lake outlet, 100–101
 concentration in lake water, 98–100, 107
 concentration in tributaries to lake, 100–101
 flux through terrestrial ecosystems, 38
 in ice, 106–107
 input in bulk precipitation, 22–24, 142–147
 input in runoff to lake, 143–147
 input-output budget, 24, 28–32, 149–155
 monthly concentration in precipitation, 18–22
 monthly concentration in stream water, 24–26
 output in stream water, 26–28
 output from lake, 147–149
 standing stock in lake water, 98–100
 in stream water, 24ff, 434–435, 438–440
  —following forest disturbance, 438–440
 vertical stratification in lake, 98
Hydrology, 15ff, 128ff
 annual water budget for watersheds, 16–18
 piezometers, 42, 133, 135
 water budget for lake, 136–142
 watertight substrate, 15, 52–53, 134
Hydrophytes (see Macrophytes)

Ice-Cover, 91–95, 97, 106–107, 110, 113–114, 117–121, 172, 308, 313, 323
 black ice, 106–107, 117
 candle ice, 119
 chemistry of, 106–107
 heat budget of, 118
 period of, 118
 white ice, 106–107, 118
Indians
 in New England, 67, 69ff, 72–73, 83
Inorganic Dissolved Substances (see also specific chemical, Total Dissolved Solids, Stream-water Chemistry)
 in Mirror Lake, 96–103, 142ff
 vs. particulate matter concentration, 31, 433, 440
 regulation of inputs to lakes, 31ff
 relationship of concentration to streamflow, 31
Insecta, 208, 235
Interstate Highway (I-93), 10, 71, 81–83, 93, 99, 147, 440–442
Invertebrates, Aquatic (see also Zooplankton)
 Benthic, 204–234, 228–234
  —abundance of, 213–216, 221–222, 224–225, 229–234
  —biomass of, 213–216, 221–222, 224–225, 229–234, 293
  —cycling of nutrients, 320
  —diversity of, 207–209, 230–231
  —emergence of, 218, 225–227, 282–283, 314, 321
  —energy content of, 342
  —length-dry weight of, 218
  —P : B ratio, 339
  —P : E ratio, 283
  —of profundal region, 318–321
  —production of, 280–285, 319
  —respiration of, 285–286, 298–299
  —role of, 234
  —sampling of, 217–218, 229
  —seasonal distribution of, 223–224
  —spatial distribution of, 218–221, 224–225, 315–316
  —species of, 207–209
  —standing stocks of energy in, 287–289, 342
  —standing stocks of nutrients in, 287–289, 342
  —vertebrate grazing of, 321–322
Iron
 as energy source, 158
 extracted by acetic acid from lake sediments, 402–404
 extracted by ammonium acetate from lake sediments, 402–404
 flux through terrestrial ecosystems, 30
 in or to lake sediments, 395ff, 416
 in stream water, 401, 434
 in surface sediments, 404–405
Isothermal Compression of Water (see also Density), 111, 120

Lakes (see also Circulation and Index of Lakes and Streams)
 Lake Type
  —eutrophic, 88, 174, 177, 187, 222–223, 227, 281–283, 319, 328, 420
  —kettle, 89
  —mesotrophic, 283
  —oligotrophic, 88, 116, 166ff, 174, 177, 179, 187, 191, 222–223, 227, 251, 263, 281–283, 319, 328, 443
Land Use, 82–83, 433–435
Lead
 in bulk precipitation, 32–33, 440–441
 in forest floor, 30–31
 flux through terrestrial ecosystems, 30, 38, 440–441
 input-output budget, 32
 in sediments, 355–356, 363, 404–405
Lepidoptera, 209, 217, 235
Litter (see also specific chemical element)
 input to lake, 143, 152–153, 339, 342, 431
 leaf, 142, 390
 pigment ratios of, 389

Macrophytes, 177–192, 212, 235
 adaptation of, 181ff
 biomass of, 182–183, 185–188, 190, 215, 260–265, 293
 depth distribution of, 186, 312
  —effect of light on, 188ff
  —effect of temperature on, 188ff
 effects of grazing on, 316
 elodeid growth-form of, 180ff
 emergent growth-form of, 187
 energy content of, 342
 isoetid growth-form of, 180ff
 nutrient cycling of, 260–265, 315, 343
 nutrient content of, 342
 nymphaeid growth-form of, 180ff
 P : B ratio, 258–260, 339
 production of, 257–260, 294, 296, 300
 respiration of, 298–301
 rosette growth-form of, 179ff

General Index

seasonal growth pattern of, 259–260
spatial pattern of, 185–188
species of, 178ff
standing stock of nutrients in, 262–263
in tributary inlets, 186
vegetative reproduction of, 190

Magnesium
concentration in lake outlet, 100–101
concentration in lake water, 98–100, 107
concentration in tributaries to lake, 100–101
in dust, 21
extracted by acetic acid from lake sediments, 402–404
extracted by ammonium acetate from lake sediments, 402–404
importance for algae, 325
input in bulk precipitation, 22–24, 107, 142–147
in ice, 106–107
input in runoff to lake, 143–147
input-output budget, 24, 28–32, 149–155, 344
in fish, 291
in litter, 153
in lake sediments, 416
loss in insect emergence, 153
in macrophytes, 264
monthly concentration in precipitation, 18–22
monthly concentration in stream water, 24–26
in precipitation, 18ff, 37
to permanent sediments, 153, 395ff
output in stream water, 26–28
output from lake, 147–149
in salamanders, 298
in stream water, 24ff, 401, 434
standing stock in lake water, 98–100
in or to surface sediments, 358–359, 365, 395ff, 404–405

Manganese
extracted by acetic acid from lake sediments, 402–404
extracted by ammonium acetate from lake sediments, 402–404
flux through terrestrial ecosystems, 30
in or to lake sediments, 395ff, 416
in stream water, 434
in surface sediments, 404–405

Manistee National Forest, 70
Maple, 14, 56ff
Mayflies, 230, 313
Mirror Lake, New Hampshire (see numerous other categories)
depth contours, 90–91
eutrophic hypolimnion of, 221, 284, 287, 318
historical trophic development of, 407–408
littoral region of, 311–317
location of, 9
morphometric characteristics of, 49–52, 89–90, 342, 346, 351
other names of, 72, 83
pelagic region of, 312, 322–337
profundal region of, 312, 317–322
tributaries to, 49, 94, 100, 186, 343, 441–442
water level of, 76, 83, 184, 347, 364
Mineralization, 174, 304–306, 320, 342
effect of temperature on, 305, 307
Mixing (see Circulation)
Model
eutrophication, 433
forest development, 31, 438
nutrient cycling, 3–7, 136, 315
functional linkages, aquatic-terrestrial, 6
hydrology, 16, 137
for lake water residence time, 124–125
for mineralization of particulate organic carbon, 307–308
forest growth, 63
for post-glacial sediment accumulation in lake, 352–353
stream-water chemistry, 31, 37, 435
watershed-ecosystem, 3–7
—aquatic ecosystems, 4, 430–431
—terrestrial ecosystem, 4
Mt. Cushman, New Hampshire, 10
Mt. Kineo, New Hampshire, 9, 10
Mt. Moosilauke, New Hampshire, 54
Mt. Washington, New Hampshire, 48, 54
Muskrat, 257, 314

Nematoda, 207, 232
Niche, 330ff
differentiation, 334–336
hyperspace, 331ff
hypervolume, 331ff
response, 332–336
response surfaces, 332–333
Nickel
flux through terrestrial ecosystem, 30
Nicolet National Forest, 70
Nitrate
in bulk precipitation, 18ff, 107, 431–432, 443
concentration in lake outlet, 100–101
concentration in lake water, 98–100, 107
concentration in stream water, 29, 437–440
—following forest disturbance, 438–440
concentration in tributaries to lake, 100–101
in ice, 106–107
input in bulk precipitation, 22–24, 142–147
input in runoff to lake, 24ff, 143–147
input-output budget, 24, 28–31, 149–155
monthly concentration in precipitation, 18–22
monthly concentration in stream water, 24–26
monthly concentration in lake water, 98–99
output from lake, 147–149
output from stream water, 26–29
standing stock in lake water, 98–100
vertical stratification in lake, 98
Nitrification, 147
Nitrogen (see also Nitrate and Ammonium)
in bacteria, 342
in benthic invertebrates, 287–289, 342
competition for, by algae and bacteria, 324
concentration in stream water related to land use, 434
emissions of, 36
excretion by zooplankton, 86–87
in fish, 291, 342
fixation of, 153–154, 417, 420–421
gaseous uptake and impaction of, 153

Nitrogen (*cont.*)
  input in bulk precipitation, 38, 443
  input-output balance, 344
  in lake water, 342
  in or to lake sediments, 395ff
  turnover rate in lakes, 342
  limitation of phytoplankton, 254–255
  in litter, 153
  loss from lake in insect emergence, 153, 227, 282–283, 320
  macrophyte uptake of, 181, 191
  organic in lake, 153
  particulate in lake water, 103–105
  particulate in stream water, 432–433
  in periphyton, 177, 342
  to permanent sediments, 153
  in phytoplankton, 342
  predicted stream-water concentration, 438
  in salamanders, 289, 342
  in seston, 342
  in soil, 423
  in stream water, 35, 38, 299
  in or to surface sediments, 342, 364–365, 395ff
  turnover rate in lakes, 342
  weathering release of, 417
  in zooplankton, 342
Nutrient (*see specific chemical*)
  flux to lake, 142–149
  standing stock in lake, 341–343

Oak, 14, 55ff, 418–426
Odonata, 209, 216, 235
Oligochaeta, 214, 219–221, 223, 232–234, 285, 287, 318–319, 321, 358
Organic Debris (*see also Dams*)
  dams in streams, 35, 420–422
  pigment ratios, 389–391
Organic Matter (*see also Carbon*)
  in lake sediments, 395ff
  origin in lake sediments, 406
Oribatida, 208, 230, 233
Orthocladiinae, 209, 212, 220, 226
Ostracoda, 208, 230, 233, 236
Ottawa National Forest, 70
Oxidation-Reduction, 156–158
  by bacteria, 156–158
  in sediments, 399
Oxyhydroxides, 400, 402, 404, 408

Paleoecology, 183–184, 410–429, 442
Parasitengona, 208, 216
Particulate Matter (*see also Seston*), 324
  allochthonous inputs of, to lakes, 296ff, 308
  from deforested watersheds, 396
  export in streams, 34, 66, 142, 299, 432–433
  from forested watersheds, 396, 400, 422
  (organic) to lake, 294
  relationship to streamflow, 26–28, 31
  seston, 103–105, 142, 324
  —export from lake, 229
  suspended vs. dissolved, 31, 433, 440
  windblown, 398
Perch, 237–239, 243–245, 289–290, 312, 327
Peridineae, 165, 324–326
Periphyton (*see also Algae*), 175–177
  biomass of, 177, 293
  chlorophyll of, 177
  effects of grazing on, 316
  energy content of, 342
  in inlets, 343
  nitrogen-fixation by, 267–268
  nutrient content of, 342
  nutrient flux and cycling by, 268
  P:B ratio, 339
  production of, 265–268, 294, 296, 300–301
  respiration of, 266–267, 298–301
pH (*see also Hydrogen Ion*)
  in lake outlet, 100–101
  in lake water, 98–100, 202–203, 420, 426
  in lakes, 370, 374–375
  of precipitation, 36–38, 431–432, 434, 440–441
  in tributaries to lake, 100–101
Phosphorus
  in bacteria, 342
  in benthic invertebrates, 287–289, 320, 342
  in bulk precipitation, 18ff, 107
  concentration in lake outlet, 100–101
  concentration in stream water related to land use, 434
  concentration in tributaries to lake, 100–101
  excretion by zooplankton, 86–87, 328–330
  in fish, 291, 342

  in forest floor, 33
  in ice, 106–107
  input in bulk precipitation, 22–24, 142–147, 443
  input to runoff to lake, 143–147, 330, 343
  input-output budget, 24, 149–155, 344
  limitation of phytoplankton by, 254–255, 324
  in litter, 153
  loading of, 315, 330
  loss in insect emergence, 153, 227, 282–283, 320
  macrophyte uptake of, 181, 191
  in macrophytes, 262–264, 342
  particulate in stream water, 432–433
  in periphyton, 342
  in phytoplankton, 342
  in salamanders, 342
  in seston, 342
  in sewage, 343
  in or to surface sediments, 342, 364–365, 395ff, 404–405
  turnover rate in lakes, 342
  in zooplankton, 342
  in lake water, 85, 88, 98–100, 107, 174, 342
Phytoplankton (*also see Algae*), 161–175, 250–257
  biomass of, 161ff, 277, 293, 324
  carbon content of, 173
  cell volume of, 172
  competition of, 424
  energy content of, 342
  interactions with bacteria, 323–326
  migration of, 173
  nutrient content of, 342
  nutrient limitation of, 295–296, 298–301
  occurrence of, 161ff, 424
  P:B ratio, 339
  productivity of, 251–253, 294–296, 300–301, 386
  regulation of, 253–256
  requirement for sodium by, 325–326
  respiration of, 295–296, 298–301
  seasonal distribution of, 167ff
  sinking rate of, 306
  species of, 161ff
  species diversity of, 167–168
  suppression of, 188
  vertical distribution of, 166ff
Pine, 14, 55ff, 412ff
Pickerel, Chain, 237, 240, 243, 245, 289–290

General Index

Pigments (*see also Chlorophyll and Carotenoids*)
  in sediments, 387ff
Pine, 14, 58ff, 412ff
Pinkham Notch, New Hampshire, 59
Pisgah Forest, New Hampshire, 69
Pit and Mound Topography, 18, 69
  tree-throw, 60, 426
Plankton (*see also Bacteria, Phytoplankton, Zooplankton*)
  ecologic interactions of, 323–328, 337ff
  interactions with fish, 327–328
  nutrient cycling of, 328–330
  sedimentation of, 323, 327–328
  vertical migration of, 327–328
Pollen (*see also Sediments*)
  herbaceous, 356ff
  influx to lake, 56–57, 347ff, 356, 363–364, 407
Pollution
  of air, 38, 404, 431–432, 440–442
  of water, 440–441
Polypod fern, 62–63
Popular, 56ff
Potassium
  in bulk precipitation, 18ff, 37, 107
  concentration in lake outlet, 100–101
  concentration in lake water, 98–100, 107
  concentration in tributaries to lake, 100–101
  extracted by acetic acid from lake sediments, 402–404
  extracted by ammonium acetate from lake sediments, 402–404
  in fish, 291
  importance for algae, 325
  input in bulk precipitation, 22–24, 142–147
  input in runoff to lake, 143–147
  input-output budget, 24, 28–32, 149–155, 344
  in ice, 106–107
  in lake sediments, 416
  in litter, 153
  loss in insect emergence, 153
  in macrophytes, 262–264
  macrophyte uptake of, 181
  monthly concentration in precipitation, 18–22
  monthly concentration in stream water, 24–26
  output from lake, 147–149
  output in stream water, 26–28
  to permanent sediments, 153, 395ff
  requirement for blue-green algae, 343
  in salamanders, 289
  in or to sediments, 358–359, 365, 395ff
  in sewage, 443
  standing stock in lake water, 98–100
  in stream water, 24ff, 401, 434, 437–440
  —following forest disturbance, 438–440
  in surface sediments, 404–405
  vertical stratification in lake, 98
Precipitation (*see also Hydrology*)
  annual, 16–18, 128–129, 136–142
  average, 128–129
  monthly, 17–18, 139
  seasonal, 17–18
Precipitation Chemistry (*see also specific element*), 18–24
  concentration, 18–22, 36–38
  effects of elevation on, 18–19
  input of, 22–24
  origin of, 19
  seasonal, 19ff
Precipitation Collection
  bulk, 15ff, 128–129
  dry deposition, 23–24
  —deposition velocity, 23
Productivity (*see also individual groups*)
  algal, 381, 387, 413, 250ff
  bacterial, 246ff
  development of, 407–408
  early Holocene, 418
  invertebrates (benthic), 280ff
  lake, 426
  Late Glacial, 425
  macrophyte, 257ff
  periphyton, 265ff
  primary, 303, 387, 413, 250ff, 257ff, 265ff
  secondary, 268ff, 280ff, 288ff
  vertebrate (aquatic), 288ff
  zooplankton, 268ff
Protein
  content of fish, 289
Protozoa, 207, 382
  benthic, 285

Ragweed, 55ff
Railroads, 83
Ratios
  cation, 401, 423
  C:Chl, 104–105
  C:N, 104–105, 396, 399–400, 406, 409, 414
  C:N:P, 342, 418
  C:P, 104–105, 400, 408
  C:S, 247–248, 400
  elemental, 399–402
  Fe:Mn, 400, 402
  N:P, 104–105, 325
  P:B, 339
  Pigment, 389–391
Residence Time
  for dissolved substances in lake, 149–152, 297, 299, 301, 433
  of lake water, 124ff, 433
Respiration (*see also individual groups*)
  aquatic vertebrates, 288ff, 297ff
  bacterial, 497ff
  invertebrate (benthic), 282, 285ff
  macrophyte, 297ff
  periphyton, 266ff, 297ff
  phytoplankton, 297ff
  zooplankton, 278ff, 297ff
Reynolds Number, 86–88
Rhodomonas pusilla, 277
Rosaceae, 62ff
Rotifers (*see also Zooplankton*), 192ff, 207, 232–233
  anatomy of, 194
  biomass of, 195ff, 279–280
  excretion of P, 303–304
  feces of, 307
  generation time of, 195
  grazing by, 326–327
  growth of, 269–270
  seasonal distribution of, 196
  sequence of generations, 194
  production of, 279–280
Rumex, 62ff
Runoff (*see also Hydrology*)
  infiltration, 18, 34, 52, 131
  overland flow, 52, 131, 432
  regulation of, 31ff
  from watersheds, 16ff, 128, 130–134, 136–143, 432–437

Salamanders, 234–236
  abundance of, 235–236
  biomass of, 289, 293, 324
  distribution of, 235–236
  energy content of, 342
  food of, 235
  in inlets, 235–236
  nutrient content of, 289, 342
  P:B ratio, 339
  respiration of, 298–300
  transport of nutrients, 344

Salt
　from road, 99, 153–154
Sawmills (*see also Human Activity*), 74–77, 83
Schmidt Stability, 110ff, 123
Secchi Disc (*see Solar Radiation*)
Sedge, 62ff
Sediments (*see also Pollen and specific element*)
　age of, 54, 349, 393–395
　bacteria in, 158–160, 342
　bulk density of, 348, 355, 393–395
　chemistry of, 354–355, 357–358, 392–410
　Chrysophycean cysts in, 381, 419, 424
　cores of, 347ff, 356, 392ff
　cultural, 404–405
　decomposition of, 361–364
　extractable cations in lake sediments, 402–404
　focusing of, 283, 349ff, 395, 399, 404, 407, 411ff
　fossil pigments in, 387–391, 408–409
　heat content of, 112ff
　hydrophyte remains in, 183–184
　loss-on-ignition content (ash) of, 354–355, 362–365, 404–405
　organic matter of, 395ff
　origin of, 54, 358–360
　permanent, 358ff
　resuspension of, 308, 343, 351, 360
　　—seasonal pattern of, 360–361
　spruce needles in, 55
　stratigraphy of, 357ff
　　—charcoal, 53, 60–61, 66–67, 418, 421
　　—chemical, 393ff
　　—diatom, 414, 424
　　—macrophyte, remains, 415
　　—pollen, 53ff, 412
　　—rock-forming metals, 416
　surface, 47, 89, 284, 286, 307, 313–314, 319, 342–343, 348, 404–405
　thickness of, 46, 128, 346ff
　volume of, 346
　windblown, 398
Sediment Deposition, 345ff, 407
　of fossil pigments, 388
　during Holocene, 348ff, 364, 402, 406–407
　related to human disturbance, 364–365, 430
　measured by sediment trap, 351, 355ff

　of nutrients, 365
　permanent, 142, 153–154, 298–300, 303, 308–309, 339, 362, 392ff
　rate of, 349ff, 362–363, 394ff, 430
Seepage
　deep seepage from watersheds, 5, 14–15
　to and from lake, 137–142, 149ff, 343, 349, 432
Seston (*see also Particulate Matter*) 313,
　energy content of, 342
　nutrient content of, 342
Settlement Period, 355ff, 364–365, 395, 404–405, 419–420, 425
Sewage Input (*see also Human Activity*), 82, 99, 147, 153–154, 343, 425, 443
Sialidae, 216
Silica, Amorphous
　in or to sediments, 357–358, 365, 393ff, 404–405, 408–409
Silica, Dissolved
　concentration in lake output, 100–101
　concentration in lake water, 98–100, 107
　concentration in tributaries to lake, 100–101
　dissolved in precipitation, 19, 107
　dissolved in stream water, 24ff
　in ice, 106–107
　input in bulk precipitation, 142–147
　input in runoff to lake, 143–147, 352
　input-output budget, 24, 28–30, 149–155, 344
　in litter, 153
　monthly concentration in lake, 98–99
　output from lake, 147–149
　to permanent sediments, 153
　in or to sediments, 358–359, 365
　standing stock in lake water, 98–100
　vertical profile in lake, 98
Silt, Gray, 347ff, 416
Size Relationship, 84ff, 229
Snow, 37
　sublimation of, 130
Snowpack, 17
　meltwater from, 37, 94, 122, 147, 414, 438

Sodium
　in bulk precipitation, 18ff, 107, 431–432
　concentration in lake outlet, 100–101
　concentration in lake water, 98–100, 107
　concentration in tributaries to lake, 100–101
　extracted by acetic acid from lake sediments, 402–404
　extracted by ammonium acetate from lake sediments, 402–404
　in fish, 291
　in ice, 106–107
　input in bulk precipitation, 22–24, 142–147
　input to runoff to lake, 143–147
　input-output budget, 24, 28–32, 149–155, 344
　in lake sediments, 416
　in litter, 153
　long-term changes in lake, 99
　loss in insect emergence, 153
　in macrophytes, 263–264
　monthly concentration in precipitation, 18–22
　monthly concentration in stream water, 24–26
　output in lake, 147–149
　output in stream water, 26–28
　to permanent sediments, 153, 395ff
　requirement by blue-green algae, 325–326, 343
　from road salt, 153, 343, 441–442
　in salamanders, 289
　in or to sediments, 358–359, 365, 395ff
　in sewage, 343, 443
　in surface sediments, 404–405
　standing stock in lake water, 98–100
　in stream water, 24ff, 401
Solar Radiation, 9ff
　albedo, 122, 125
　extinction coefficient in water, 96
　measured by secchi disc, 95
　penetration in lakes, 94–97, 109ff, 168–170, 272, 294, 312, 317, 323–324
Soil, 15, 34, 400
　C:N ratio of, 406
　depth of, 15
　development during Holocene, 423–425
　disturbance of, 438–439
　formation of, 15, 399, 401, 427

General Index

frost, 18, 54, 420
humus, 15, 18, 406, 409, 421, 424, 426
infiltration of water, 18, 34, 131
pH of, 15
stabilization of, 56, 421
washed into lake, 408–409
water storage, 16, 34, 131
Species of (*see also Index of Genera and Species*)
benthic invertebrates, 206–209
diatoms (fossil), 368–370
macrophytes, 180
periphyton, 176
phytoplankton, 162–165
diversity in pelagic zone, 336
zooplankton, 193
Sphaeriidae, 214–215, 220–221, 235–236, 321
Spruce, 14, 55ff, 350, 353–354, 417, 412ff, 421, 424
Spruce Period, 421–422, 426
Stability (*see also Density*)
of lakes, 109ff
Stormflow, 18
Streamflow, 15ff
annual, 16–18
increased following disturbance, 438
measurement of, 18, 130–134
monthly, 17–18
seasonal, 17–18
Stream Water
collection of samples, 18
inputs to lake, 120, 123, 130, 136ff
inputs of particulate matter in, 391
Streams (*see also Erosion, Dams*)
channels, 423
as ecosystems, 7–8, 35–36
experimental study of, 432–435
land-water linkage, 7–8, 35–36
outflow of, 422
overland flow, 427
pH of, 36
watershed export of material by, 420
Stream-water Chemistry
concentration, 19, 24ff, 101, 437
longitudinal variation, 435
output, 26–31, 143ff
relative to streamflow, 25
seasonal, 25–26
Sucker, White, 236, 240, 244–245, 289

Sulfate (*see also Sulfur*)
in bulk precipitation, 18ff, 37, 107
concentration in lake outlet, 100–101
concentration in lake water, 100–101
in ice, 106–107
input in bulk precipitation, 22–24, 142–147
input-output budget, 24, 28–32, 149–155, 344
input in runoff to lake, 143–147
monthly concentration in precipitation, 18–22
monthly concentration in stream water, 24–26
output from lake, 147–149
output from stream water, 26–28
reduction of, 157
in or to sediments, 365, 395ff
standing stock in lake water, 98–100
in stream water, 24ff
vertical stratification in lake, 98
concentration in tributaries to lake, 100–101
Sulfur (*see also Sulfate*)
bacterial uptake of, 247–249
emissions of, 36
in fish, 291
gaseous uptake and impaction of, 153–154
input-output budget for lake, 152–154
in litter, 153
loss from lake in insect emergence, 153
in or to permanent sediments, 153, 395ff
in salamanders, 289
in stream water, 24ff
in throughfall and stemflow, 23
Superior National Forest, 70

Tanneries (*see also Human Activity*), 77–80, 83
Tanytarsini, 209, 211–212, 220, 225–226, 318, 321
Tanypodinae, 209, 213–214, 220, 226, 321
Tardigrada, 207, 230, 233
Temperature
air, 11, 58
of lake, 89–94, 109ff, 272, 312–313, 323–324
of lake sediments, 93–94, 115ff, 317

near bottom of lake, 286
profile, 98, 109ff, 168–169, 202, 224, 252–253, 323 (vertical stratification)
of stream, 35, 94, 438
Thiessen Method, 15
Throughfall, 23, 34
Till, Glacial, 14–15, 395, 400, 409
Topography (*see Geology and Pit*)
Total Dissolved Solids
concentration in lake water, 98–100
effect on water density, 109ff
input in runoff to lake, 143–147, 396, 400
in lake outlet, 100–101
output from lake, 147–149
standing stock in lake water, 98–100
in tributaries to lake, 100–101
Transpiration, 34, 422
Trichoptera, 209, 216, 226, 239
Tripton, 284–285
Tubificids, 214, 320, 358
Tundra Period, 54–55, 415–416, 420–421, 426
Turbellarian, 230, 232

Unionidae, 215
U.S.D.A. Forest Service (*see Hubbard Brook Experimental Forest and Watersheds*), 9
Headquarters site at Hubbard Brook, 10, 11–14, 18, 74, 81, 130, 136, 138–139, 142

Vectors, 3ff
biologic, 3–7
geologic, 3–7, 34, 293, 432–433
meteorologic, 3–7, 22, 34, 293, 431–432
Vegetation
changes in, 53ff, 62–63, 410ff
forest, 423
history of, 53ff, 410ff
type of, 14
Vertebrates (Aquatic)
biomass of, 288–291, 293
grazing of benthic invertebrates, 321–322
nutrient standing stock of, 289–291
production of, 288–291
respiration of, 288–291, 300

Water Budget (*see also Hydrology*)
 for lake, 128ff, 136–142
 for terrestrial watersheds, 16–18
Water Table, 128
 ground water, 128, 134–135, 299–300
Watershed-Ecosystem Approach, 1–8
Watersheds, Experimental, 52, 131
Water-Year, 15
Weathering, 37, 147, 399, 408, 417, 425
 rates of, 426
Weir, 131, 133
Welch Mt., New Hampshire, 57
West Thornton, New Hampshire, 10, 69, 72ff, 356
 population of, 75
White Mountains, New Hampshire, 9, 53–56, 59, 67, 70–71, 62, 431, 440
Wind (*see also Disturbance*), 12ff, 65ff, 109ff
 rose, 13
Woodstock, New Hampshire, 69, 72ff, 356
 population of, 75

Zinc
 extracted by acetic acid from lake sediments, 402–404
 extracted by ammonium acetate from lake sediments, 402–404
 flux through terrestrial ecosystems, 30
 in macrophytes, 263–264
 in salamanders, 289
 in or to sediments, 395ff
 in surface sediments, 404–405
Zones, Sediment (*see also Sediments*), 410–429
Zooplankton (*see also Rotifers, Copepoda, Cladocera*)
 biomass of, 272, 293, 324
 community dynamics of, 330–336
 energy content of, 342
 excretion of N by, 86–87
 excretion of P by, 86–87, 328–330
 filtration rate of, 87
 grazing by, 303–304, 307, 326–327
 growth of, 268ff
 horizontal distribution of, 198, 203
 nutrient content of, 342
 nutrient cycling of, 280
 P:B ratios, 276–278, 339
 P:R ratios, 278–280
 production by, 268ff
 respiration of, 278–280, 298–299, 300
 response to environmental gradients of, 202–203
 sampling of, 200–201
 species of, 193
 vertical migration of, 201–203, 334
  —reverse migration, 201–203

DATE DUE